Daniel W. Beckman

Missouri State University

MARINE ENVIRONMENTAL BIOLOGY AND CONSERVATION

JONES & BARTLETT
LEARNING

World Headquarters
Jones & Bartlett Learning
5 Wall Street
Burlington, MA 01803
978-443-5000
info@jblearning.com
www.jblearning.com

Jones & Bartlett Learning books and products are available through most bookstores and online booksellers. To contact Jones & Bartlett Learning directly, call 800-832-0034, fax 978-443-8000, or visit our website, www.jblearning.com.

Production Credits
Chief Executive Officer: Ty Field
President: James Homer
SVP, Editor-in-Chief: Michael Johnson
SVP, Chief Technology Officer: Dean Fossella
SVP, Chief Marketing Officer: Alison M. Pendergast
Publisher: Cathleen Sether
Senior Acquistions Editor: Erin O'Connor
Senior Associate Editor: Megan R. Turner
Editorial Assistant: Rachel Isaacs
Production Manager: Louis C. Bruno, Jr.
Senior Marketing Manager: Andrea DeFronzo
V.P., Manufacturing and Inventory Control: Therese Connell
Cover Design: Kristin E. Parker
Associate Photo Researcher: Lauren Miller
Permissions and Photo Research Assistant: Amy Rathburn
Composition: Circle Graphics, Inc.
Cover Image: Courtesy of NOAA
Printing and Binding: Courier Kendallville
Cover Printing: Courier Kendallville

Some images in this book feature models. These models do not necessarily endorse, represent, or participate in the activities represented in the images.

ISBN: 978-0-7637-7350-2

Library of Congress Cataloging-in-Publication Data

Beckman, Daniel W.
  Marine environmental biology and conservation / Daniel W. Beckman.—1st ed.
    p. cm.
  ISBN 978-0-7637-7350-2 (alk. paper)
 1. Marine biodiversity conservation. 2. Marine habitat conservation. 3. Endangered ecosystems. I. Title.
  QH91.8.B6B43 2013
  333.95'616—dc23   2011051466

6048

Printed in the United States of America
16 15 14 13 12   10 9 8 7 6 5 4 3 2 1

To my parents, Bill and Anna, for instilling in me an ethic of independence and exploration. To Alex, Chris, Taylor, and Zachary, for diving, fishing, and playing in the ocean with me. And, above all, to Ronda, my loving spouse, moral support, and travel partner.

Brief Contents

Contents

Chapter 5   Tropical Coral Reefs: Environmental Impacts and Recovery 159

Chapter 6   Nearshore Ecosystems: Community Ecology and Habitat Protection 195

Chapter 7   The Open Ocean: Anthropogenic Inputs and Environmental Impacts 225

Chapter 8   Seafloor and Deep-Sea Ecosystems: Resource Harvest and Habitat Protection 255

Chapter 9   Marine Endangered Species: Conservation, Protection, and Recovery 289

Chapter 10   Conservation of Cetaceans . . . 323

Chapter 11   Marine Fisheries: Overharvest and Conservation . . . 349

Chapter 12 Ocean Conservation Laws, Agreements, and Organizations . . 399

Preface

In recent years a new appreciation of environmental and conservation issues for marine ecosystems and resources has emerged. This appreciation has developed, at least in part, in response to a number of issues broadly covered in the news media and widely debated among scientists and the public—including the collapse of many historic fisheries, unprecedented oil spills, mysterious die-offs of corals, unexplained deaths of marine mammals, natural disasters from hurricanes and tsunamis, intense El Niño events, and ongoing climate change and sea-level rise. Our evolving ability to rapidly communicate information concerning such issues has led to vigorous debates over the ethics and the supporting science that drive conservation decisions. The push to establish agreements on the proper human response to these issues has led to a flurry of activity to generate the necessary scientific, cultural, political, and economic knowledge and insight. The persons with arguably the greatest influence in these debates will be current students, who have been experiencing increasing interactions of humans with ocean ecosystems, the explosion of scientific knowledge, and rapidly developing environmental ethic. There is a need to maximize the knowledge base for students, not only those working toward marine-focused careers but also those in other disciplines, to enable informed decisions in their professional and personal lives.

This text was written out of the need for a synthesis of knowledge and unbiased information related to marine environmental and conservation issues; something that is not only useful to students focusing on marine biology or oceanography tracts but also available to those in general biology, environmental science, and other science and environment related fields. Such a text has been previously unavailable. Although this text is approached primarily from a biological perspective, the living portions cannot be divorced from other parts of the ocean environment; thus, fundamental physical, chemical, and geological components are integrated into the coverage. Moreover, marine conservation issues cannot be solved by ignoring the nonscientific aspects of decision-making; therefore, a limited discussion of the political, social, and economic components of conservation is incorporated as well.

The specific goals of this text are to provide an introduction to and understanding of issues concerning marine environmental protection and conservation at the college level. It presumes only a general knowledge of scientific concepts—introductory Chapters 1 and 2 provide adequate background to understand the issues in later chapters.

To better serve upper-level undergraduate and graduate students, or others with a more thorough background in marine or environmental sciences, recent findings from the primary scientific literature are included in the discussion; references to these sources are given at the end of each chapter for those interested in pursuing research findings to their original sources. The goal of this text is not to present the personal opinions of the author or indoctrinate the student with a specific conservation ethic, but to provide the information needed to formulate an educated opinion and, hopefully, to develop solutions to current problems associated with human interactions with the marine environment. Hopefully, students who study this text will become not only future scientists and environmental managers but also political leaders and participants in businesses dependent on marine resources, as well as consumers and users of ocean ecosystems.

Why a text focusing on the marine environment? Although knowledge and conservation methods applied to terrestrial and freshwater systems can often be used to solve marine environmental issues, in many ways conserving marine ecosystems presents unique problems. Many of these problems result from a reliance on the harvest of wild organisms from the seas. Whereas terrestrial food sources are largely agriculture based, marine organisms continue to be harvested primarily from wild populations, presenting unique conservation issues. Another reason that marine ecosystem studies present unique challenges is that the oceans present an environment that is largely foreign to humans. Despite the increased ability to access ocean environments through collection devices, scuba, or submersibles, most of the ocean remains unexplored. This presents a unique conservation dilemma—how to protect an ecosystem we haven't fully defined. A further distinctive influence on ocean protection is that many user groups desire and claim rights of access to the ocean and its resources. Most terrestrial lands are governed by private ownership or have their use tightly regulated through government oversight. However, much of the ocean is still considered, to varying degrees, available for common access, either to citizens of nations bordering the seas or to the global community. Coming to terms with overuse and loss of important marine habitats due to adherence to this philosophy is a phenomenon that conservation policies must address.

Regardless of the arguments for marine-focused text, there is a dilemma in defining the physical boundaries to the marine environment that affects the text coverage. Technically, the marine environment might be considered to end abruptly at the interfaces between ocean waters and the adjacent atmosphere, sediments, and freshwaters. This is unrealistic, however, since many biological, chemical, and physical processes are continuous across these boundaries, and anthropogenic influences that originate beyond the ocean boundaries have a great impact on marine ecosystems. Thus, I have defined the marine environment broadly and included coverage of important anthropogenic effects that might originate beyond the marine environment and impact organisms that may crawl, swim, or fly across ocean boundaries. Coastal processes and influences are a major component of Chapter 3, and inputs from freshwater and terrestrial environments are important to Chapters 4 and 6. Interactions of the atmosphere and its pollutants with the marine environment are important to Chapters 1 and 7, and a section on the influences of climate change is included in each chapter. Many species defined as marine, discussed in Chapter 2, and endangered species, discussed in Chapter 9, spend a portion of their life beyond the oceans.

A further issue in determining coverage for a text focused on the marine environment is the international nature of the subject matter. Although much of the open ocean continues to be governed, at least in theory, as an area commonly accessible to all humans, a large fraction of the scientific research and most of the conservation decisions are made within the constraints of arbitrarily designated political boundaries. Scientifically, we cannot assume that even coastal waters and the matter and creatures they contain can be constrained within political boundaries. Socially and politically, we cannot expect that conservation decisions will be driven by the same ethics and legal systems around the globe. Thus, an attempt has been made to retain a broad international focus for this text. For issues with a national and regional focus, an attempt has been made to include examples from regions across the globe. Any focus on information originating from North American or English-speaking regions is a result of the availability of sources to the author.

Text Organization

Each chapter is organized into sections and subsections that separate topics by physical or biological classifications, habitat and ecosystem types, or categories of environmental and conservation issues. Topics that are pertinent to several chapters may be covered extensively in a single chapter to avoid repetition; in these cases, references are given in the readings to the chapter in which a pertinent

topic is covered. Text boxes are included in each chapter to focus on specific case studies, conservation concerns, recent controversies, or historical lessons. The Study Guide at the end of each chapter includes two types of exercises. Topics for Review contain questions addressing facts and concepts presented in the chapter. Conservation Exercises are designed to require synthesis of information, critical thinking, and development of opinions based on issues presented in the readings. Further Reading provides references used to develop the text or deemed useful for the student to gain further information. When a study author is mentioned by name in the text, a citation is given in the Further Reading section.

Chapters 1 and 2 present an overview of principles of environmental science, oceanography, and marine biology to provide a basic foundation. The material is focused on topics that are important to the understanding of environmental and conservation issues presented in later chapters. Specific conservation concerns and controversies related to the introductory topics are also presented.

Chapters 3 through 8 are organized by habitat and ecosystem type, progressing from the intertidal zone through near-shore ecosystems into the open ocean and deep sea. These chapters present a background description of habitats and ecosystem function, leading up to a discussion of important environmental and conservation issues specific to each ecosystem. Each chapter includes some broader topics that relate to multiple ecosystems. Text boxes present details of specific research studies, stories of conservation successes and failures, and current controversial issues.

Chapters 9 and 10 focus on two groups of organisms that are given special conservation attention. For marine endangered species, covered in Chapter 9, legal protections are broadly discussed and major animal groups are each discussed separately. An overview of the conservation status for each group is given, followed by specific examples of factors leading to endangerment and efforts for recovery and protection. Chapter 10 addresses conservation issues concerning cetaceans, the dolphins and whales. Sections are provided on major controversial issues of whale harvests, other sources of danger to cetaceans, and keeping cetaceans in captivity.

Chapter 11 addresses issues of marine fishery harvest, including the effects of excessive harvest on populations of fishes and other animals, the impacts of fishing on marine habitats and ecosystems, and efforts to conserve populations of harvested marine animals. Text boxes focus on recent scientific efforts to monitor harvested fish populations and give case histories concerning attempts to protect some highly valued and influential fishery species.

Chapter 12 provides an overview of major laws and agreements that address issues of marine conservation and environmental protection. Social and ethical paradigms are addressed as they relate to marine conservation issues. The discussion of laws and agreements is divided into sections on international and national laws for specific regions around the globe. Text boxes include a focus on U.S. fisheries laws and conflicts between international trade laws and conservation efforts.

Instructor Resources

A downloadable **PowerPoint® Image Bank** is available for qualified instructors. These files provide the illustrations, photographs, and tables (to which Jones & Bartlett Learning holds the copyright or has permission to reproduce digitally) inserted into PowerPoint slides. For more information, please contact your Jones & Bartlett sales representative or go to go.jblearning.com/beckman.

Acknowledgments

I thank the scientists, managers, environmental leaders, and colleagues for the studies, reports, and discussions upon which the information in this text is based. Although photographs are acknowledged in the credits section, I thank in particular the scientists and photographers of NOAA, NASA, USGS, and other government agencies for the data, figures, and photographs that serve to enhance the text.

I appreciate the thorough reviews of the early drafts of the chapters that made it a much better text and have enabled it to serve a larger audience:

Kurt Bretsch, Stony Brook University, SUNY
William Evans, University of Notre Dame
Richard Feldman, University of Notre Dame
Romuald N. Lipcius, College of William & Mary
Scott P. Milroy, University of Southern Mississippi
Thomas C. Shirley, Texas A&M University at Corpus Christi, Harte Research Institute
Cory Suski, University of Illinois
Judith Weis, Rutgers University

Special thanks to the exceptional staff at Jones & Bartlett Learning, including Molly Steinbach, Lou Bruno, Megan Turner, Rachel Isaacs, Jessica Acox, Shellie Newell, and Lauren Miller for their patience and assistance in developing this text.

Thanks to Steve Jensen, Alicia Mathis, and others at Missouri State University for encouraging me to develop the courses that led to this text and for allowing me to focus efforts over the past three years to complete this project.

Finally, I am indebted to my teachers and advisors, and the hundreds of students in courses I have taught that led to this text. Their comments, insights, debates and discussion have contributed thoughts and ideas that have been mingling in my brain over the past few decades, finally to become the synthesis that is this book.

Dan Beckman

About the Author

Daniel W. Beckman has been Professor of Biology on the faculty of Missouri State University in Springfield for the past 20 years. There he teaches courses in wildlife management and conservation, ichthyology, fisheries management, fish ecology, and marine conservation. He has taught field and study-away courses on the Alabama and Mississippi Gulf coasts and in Puerto Rico, Belize, and Jamaica. He supervises graduate student research in ecology of streams, reservoirs, and tropical reefs. Dr. Beckman has carried out research and published technical articles in the areas of fish population biology, stream ecology, and marine fisheries and conservation. He has received several university and state awards for his teaching and research accomplishments.

Dr. Beckman earned a bachelor's degree in marine biology and chemistry from the University of North Alabama, completing marine coursework at the Dauphin Island Marine Science Consortium. He received his master's degree in marine sciences from the University of South Carolina, integrating studies as a research student in aquaculture at Kagoshima University, Japan. His Ph.D. in marine sciences is from Louisiana State University.

In conjunction with, and outside of, his academic and scholarly studies, Dr. Beckman travels to islands and coastal regions whenever the opportunity arises. There he enjoys snorkeler, diving, running, and beach combing. He has been active in promoting solutions to marine environmental issues by working with local communities. His work includes promoting the protection of turtle nesting beaches in Japan, working with government agencies to establish sustainable fisheries in the Gulf of Mexico, and, currently, working with community and fisheries leaders in Jamaica to promote sustainable tourism and protection of no-fish marine sanctuaries.

Introduction to Marine Environmental Science and Oceanography

Although a turning point would be difficult to pinpoint, sometime within the past several decades a new environmental awareness was developed with the understanding that conservation of living resources cannot be adequately achieved by focusing only on imperiled species or single components of ecosystems. **Environmental science** was born out of a need to consider all aspects of the physical and living world in solving complex environmental problems. **Environmental biology** is primarily an incorporation of biological processes that realizes the need to embrace environmental processes when studying living resources in the natural world and solving conservation problems. This is especially true in the marine environment where environmental systems processes such as currents, waves, chemical processes, and geologic activity are dynamic and often unpredictable and strong influences on the living components. Although the perspective of this text is biological, rarely can conservation be achieved without considering the non-living environment's effect on life processes.

Marine conservation is concerned with the preservation and management of the ocean environment and the organisms that live within it or are otherwise dependent upon it, that is, the **marine ecosystem** (**Figure 1-1**). The extremely complex, dynamic, and variable ecosystems of the oceans cannot be understood without first considering oceanography. *Oceanography*—the study of the physical, chemical, geologic, and biological processes of the oceans—is the foundation upon which a conservation ethic can be built. This chapter provides a general overview of the non-biological components of oceanography. This includes studies of the water column as well as land areas adjacent to the sea, the continental coastlines, and islands. The physical dynamics of water movements are discussed, including waves, tides and the tidal cycle, and currents occurring at the surface and throughout the water column. Finally, some general aspects of ocean chemistry are covered, introducing basic aspects of the chemistry that comprise the biological components of the sea.

1.1 Geological Oceanography

Geological oceanography or **marine geology** is the study of the sea floor and the lands bordering the sea. Geological oceanographers often study aspects of the formation of the sea floor and coastlines by major geologic processes and sediment formation and accumulation. These processes govern the current mineral makeup of the seafloor and coastline and the physical features of these regions. Marine geology is closely linked with physical and chemical oceanography because physical and chemical processes determine the geology of oceanic regions.

Seafloor Spreading

The sea floor is remarkably complex with a diversity of physical features, some similar to those found in terrestrial

Figure 1-1 A view of the Pacific Ocean and volcanic shoreline of Japan.

regions and others unique to the marine environment (**Figure 1-2**). This region, especially of the deeper ocean basins, was virtually unknown to humans until about the past 50 years. Most of the major physical features are formed through the processes of **seafloor spreading** and **plate tectonics** (**Figure 1-3**). This concept is based on observations that new seafloor is constantly, but slowly, being extruded from the ocean bottom at mid-ocean ridges. Extensive ridges can be seen running through the middle of the Atlantic Ocean bottom and along the east side of the Pacific Ocean up through the North American continent. From a human perspective, these processes are extremely slow. At an average rate of about 1 cm/yr, it takes almost a century for the seafloor to move 1 meter, and it has taken almost 200 million years for the seafloor formed at the mid-Atlantic ridge to move to the U.S. Atlantic coast. As the forces of

seafloor spreading gradually push the seafloor away from the mid-ocean spreading centers, it eventually **subsides**, sinking down below the continental land masses or into deep ocean **trenches**. Some of the deepest of the trenches formed during this process are the western Pacific Ocean, reaching depths as great as 7,000 meters below the average depth of the seafloor. These forces affect the seafloor and the continental land masses worldwide and result in a movement of the large **plates** that interlock to form the ocean basin and support the continents. For example, the North American continent is mostly over one large oceanic plate. Seafloor spreading processes result in the production of some remarkable features, both under the sea and above the sea surface, including mountains, islands, and volcanoes. They also produce some of the most extreme events on the planet: earthquakes and tsunamis.

▪ Seafloor Regions

The features of the sea floor are commonly categorized according to their association with land masses, physical characteristics, and depth below the ocean's surface. Although the divisions between adjacent regions are not always exclusive and discrete, the categories prove a convenient way to make generalizations concerning physical process, ecosystem categories, and conservation needs.

Continental Margins

The Earth's continents might be described as sitting on or riding over the oceanic plates, and the edges of the continents where they meet the sea are called **continental margins** (**Figure 1-4**). This is the location of most of the shallow ocean waters. Because the waters covering the continental margin are typically the most productive and accessible to humans, many of the conservations issues covered in later chapters will be focused in this region. Many of the living

Figure 1-2 Relief map of the Earth. Shades of gray indicate depths and elevations; ocean regions are indicated from light to dark gray, shallow to deep. Note the mid-ocean ridges and other seafloor features. (See color plate 1-2.)

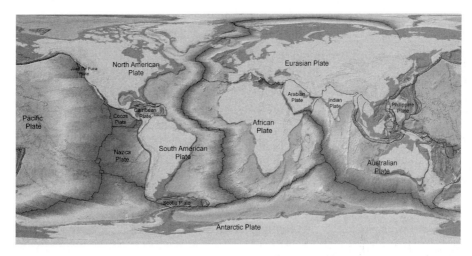

Figure 1-3   Map of the sea floor with evidence of seafloor spreading based on age of ocean sediments. Newer areas of the seafloor near the spreading centers are indicated by dark gray with jagged lines. (See color plate 1-3.)

resources are obtained from water over the continental margin and this is also where about ⅓ of the Earth's oil and gas reserves are located.

Continental margins can be classified according to the plate interactions associated with the margin. **Active margins** are common where the plates are interacting or colliding, such as along the Pacific coast of North America. In this region the intersection of plates continues onto the continent in California, and result in frequent earthquake activities along the U.S. west coast. There is another active margin in the west Pacific; the plate interactions there have resulted in the formation of the many volcanic islands, including the islands that make up Japan. The slope from the coast out to the deep ocean is typically steeper along active margins than along passive margins. Some of the most biologically productive areas of the oceans occur in the deep near-shore

waters along active margins as a result of ocean processes, discussed below.

Passive margins (Figure 1-4) are typically in the middle of one of the ocean plates and, thus, are the source of little geologic activity. The U.S. Atlantic coast is along a passive margin. The slope of the sea floor here is very gradual and the margin is much broader. There is often a buildup of sediments well offshore along passive margins, especially where there is significant river input. The biological productivity can be elevated along passive margins where there are large nutrient inputs from continental sources.

The continental margin can be subdivided into three regions: the shelf, slope, and rise, progressing seaward from the coast. The **continental shelf** is generally the most biologically rich region of the ocean. Several factors contribute to this high productivity. These include the input

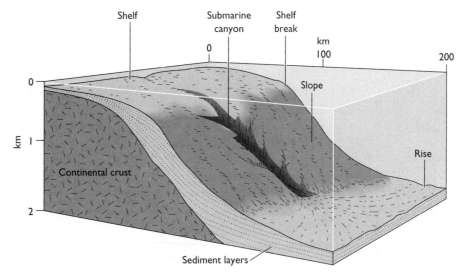

Figure 1-4   Representation of a continental margin, consisting of the continental shelf, slope, and rise. Submarine canyons cut through the continental margin in some regions. (See color plate 1-4.)

of nutrients from rivers flowing onto the shelf or rising up from deeper ocean waters, and the higher sunlight penetration into shallow depths resulting in high rates of photosynthesis. The continental shelf is less than 10% of the ocean bottom. High biological production, biodiversity, and accessibility make this region heavily exploited and impacted by human activities, however. The focus of marine conservation efforts thus is in this region. The downward slope along the shelf is only about 1 degree on average, resulting in only a gradual decrease in depth of less than 2 meter per kilometer moving away from the coast. If the shelf could easily be viewed, it would appear as flat as the flattest plains on the continents. Still, the average depth over the shelf is about 75 meters: deep from a human perspective but very shallow compared to the average depth of the ocean (about 13,000 feet or about 4,000 meters). The width of the continental shelf varies from being almost nonexistent along some active margins, to being hundreds of miles across along some passive margins. This can have an important influence on what types of organisms live above the shelf in different regions of the world. The end of the continental shelf is at the shelf break, where the slope begins to increase onto the continental slope.

The **continental slope** is the region of transition between the flat continental shelf and the deep-sea floor. This slope is somewhat steeper than the shelf, at about 4 degrees. Still, this equates to about 75 meters vertical change per 1 kilometer distance. The slope comprises about 6% of the sea floor. The slope region is richer in features than the shelf, and drop-offs, ridges, and canyons are evident. When first discovered by oceanographers, the submarine canyons were believed to be formed only in association with coastal river input; many are found near rivers and

the canyons may cut into the continental shelf. However, many canyons are present in areas where there is no river input. These, it was determined, are formed by slumping of sediments off the continental slope. This slumping can result in an undersea flow of sediments, and water currents digging into the slope and forming a canyon. These features in the slope are important for fishes and other organisms since they provide shelter and hiding places in an otherwise mostly barren undersea terrain.

The sediments that slump off the continental slope may be transported toward the deep ocean basins. This forms a transitional region dividing the continental margin from the deep sea, called the **continental rise**. The rise is composed of loosely consolidated sediments and ends at the edge of the deep-sea bottom, making up about 5% of the sea floor. The slope is steeper, but similar features occur here as on the continental slope.

Deep Ocean Basins

The **deep ocean basins** (**Figure 1-5**), sometimes called the **abyss**, make up about 40% of the sea floor. Much of this area is extremely flat, flatter even than the continental shelf. This is because most geologic features in the deep ocean basin have been slowly covered with sediments, many of them of biological origin. Although the sediment accumulation is at a very slow rate, many areas of the deep ocean have remained fairly undisturbed for millions of years. This allows sediments accumulations (called **oozes**) up to almost 5 kilometers deep in some regions (though sediment accumulation off river deltas may be as thick as 20 kilometers). The origin of deep-sea sediments varies from region to region, largely as a reflection of ocean chemistry and depth. They are composed of a combination of

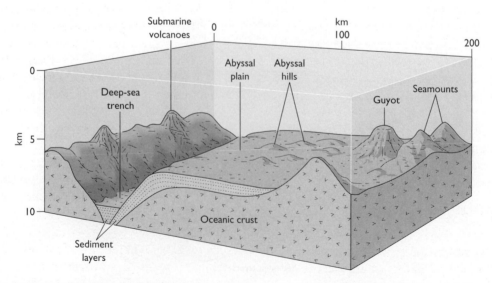

Figure 1-5 Seafloor features of the deep ocean basin. Deep-sea trenches cut into the abyssal plains. Abyssal hills, guyots, seamounts, and submarine volcanoes are scattered over the plains.

fine clays and the remains of organisms, primarily the shells of microscopic organisms that die and sink to the bottom. Abyssal hills break the relief along the ocean basins. These are extinct volcanoes partially exposed above the deep sea sediments. Of course they may have been much higher above the sea floor in past times, but are now mostly buried.

Oceanic Ridges

Oceanic ridges are the mountain ranges of the sea floor. These occur at the spreading centers, often in the center of deep ocean basins. The region associated with the ridges comprises about 30% of the seafloor. The geologic activity at these spreading centers results in mountain chains more extensive than any occurring on the continents, extending about 60,000 kilometers in total length. There is extensive volcanic activity associated with the ridges. Because of the extreme water depths, the ridges are still far below the sea surface, even though the mountains may rise up to 2 kilometers above the seafloor. Because of the relative inaccessibility to humans, little was known of these regions until late in the 20th century, and most are still unexplored. Recent explorations of the ocean ridges have resulted in the discovery of some of the most remarkable geologic features and unique ecosystems on Earth.

Hydrothermal Vents

One type of feature associated with the ocean ridge systems is a geologically active structure called a **hydrothermal vent** (**Figure 1-6**). At these vents, ocean water circulates below the seafloor to eventually spew out in undersea geyser-like formations. The minerals in these waters give them a distinct "smoky" appearance and they are usually named according to the appearance of the water venting from these features, for example, **black smokers** and **white smokers**. Hydrothermal vents support a unique ecosystem that is supported not by **photosynthesis** (the conversion of carbon dioxide into organic compounds using sunlight energy) but by **chemosynthesis** (the synthesis of organic compounds using inorganic chemical reactions as a source of energy). The minerals being released into the water eventually are deposited on the seafloor. These deposits are potentially a valuable recourse for human exploitation, which could lead to conservation issues, but for various reasons are not currently mined commercially. Conservation of these ecosystems will be discussed in Chapter 8.

Undersea Mountains

Away from the ocean ridges there are other, more isolated undersea mountains, called **seamounts**. These are formed by volcanic activity and most are inactive. Many were formed when they were located near a geologically active area like an oceanic **spreading center** at the

Figure 1-6 A black smoker at a mid-Atlantic ridge hydrothermal vent.

mid-ocean ridges, or **hotspot**, a region of upwelling of hot mantle materials (one such hotspot formed the Hawaiian island chain). As the seafloor "conveyor belt" carries these mountains across ocean basins, they gradually get further away from the spreading center or hotspot where they were formed and are carried beneath the ocean's surface; thus, the "relative" sea level gradually changes around the seamount, resulting in the older seamounts being found well beneath the surface of deep ocean waters. Changes in global sea level can also affect whether these mountains are exposed as islands or hidden beneath the sea. With climate change and increasing sea level, low elevation volcanic islands could be flooded and become seamounts if covered by the ocean waters. Those living on such islands are gravely concerned about predicted sea-level rises (see Chapter 3).

Seamounts vary in height, ranging to over 1 km above the sea floor. Some form a relatively sharp peak, but others are flat-topped, called **guyots** (pronounced gē'-ō). The flattened top of guyots is a result of erosion from weathering during the time period they were exposed as islands at the surface. These seamounts and guyots may provide habitats for rich ecosystems relatively near to the surface (but still up to hundreds of meters depth); in some cases they serve as an area for large fish to congregate for feeding or reproduction. Only in recent years have humans been able to access and exploit living resources from these under-sea mountains, leading to concerns about potential over-harvest. Conservation issues in the deep sea are discussed further in Chapter 8.

1.2 Oceanic and Coastal Regions

Sea Coast

The sea coast is the region where the land meets the sea. Although coastal regions are not covered by water at all times, they are so strongly influenced by the ocean that they must be considered in the conservation of marine environments. The access the sea coast gives to the resources of the sea results in high levels of human habitation. The resulting environmental modifications, pollution, and overexploitation of resources results in many conservation issues that will be discussed in late chapters. This section addresses how physical factors governed by oceanic and coastal processes determine the makeup of coastal areas.

Beaches

Beaches are areas where loose particles are deposited along the shore (**Figure 1-7**). Sands accumulate by erosion from terrestrial regions or from near-shore sources such as bottom sediments or coral reefs. Beach materials can range in size from boulders to sand. Typically, sandy beaches develop in areas where the slope of the beach is low; the flatter the beach, the finer the sands. For example, the low-gradient northern Gulf of Mexico coast is dominated by extensive white sandy beaches; the steeper and geologically more

Figure 1-7 Beaches bordered by dunes at Gulf Shores, Alabama in the northern Gulf of Mexico; note the blowout from wave activity associated with 2004 Hurricane Ivan.

recent beaches along the eastern Pacific shoreline are more likely to be composed of gravel or boulders.

On sandy beaches, the energy from the pounding wave action causes continual movement of the sands and a constantly changing beach; in fact, beaches are sometimes referred to as "rivers of sand." Although there may not be an obvious rearrangement of the beach over a short time, if you return to a beach after several years or after a single large storm, the changes can be dramatic. This is one reason that people building homes too close to the ocean often regret it after a few years. Because of the stress from the constant water movement and the lack of nutrients in these areas, beaches are in some ways like deserts. Relatively few macroscopic organisms live their entire lives on or in the beach sands; however, many organisms do depend on beaches in various ways.

Unless there are rocks or cliffs adjacent to the beach, the wave and wind action will accumulate the sands in mounds forming **dunes** above the high tide zone. Sand dunes serve an important job of protecting the regions behind them from the action of the waves. Plants colonize the dunes and help to stabilize them. Behind the dunes, other types of vegetation that are tolerant of the wind and salt water will grow. Loss of this vegetation is a major concern for wildlife living in these areas, and should be a major concern for coastal residents because it protects them from the impact of the sea, especially during storms. Dune and beach ecology and conservation issues are discussed in further detail in Chapter 3.

Deltas

A **delta** is a land structure projecting out onto the continental shelf in a fanlike pattern where a river flows into the sea (**Figure 1-8**). The delta is formed by sediments deposited in a thick layer by the river. They only exist where sediment-laden rivers flow onto a broad continental shelf without much wave or tide action to wash away the sediments. Sediments are usually very fine because the larger, heavier particles settle out from the rivers before they reach the sea. One of the most prominent deltas is the Mississippi River Delta in the Gulf of Mexico. Along with the sediments, there are large amounts of nutrients deposited at deltas, resulting in a high biological productivity in these regions. Deltas are constantly changing, and under natural conditions will change shape or migrate back and forth along the coastline slowly over thousands of years. Human impacts on coastal rivers have dramatically altered these processes and led to coastal conservation issues (see Chapter 4).

Rocky Coasts

Rocky coasts are found where the coastline is steep and there is little sediment accumulation (**Figure 1-9**). Many of these rocky shorelines have little if any area that would be

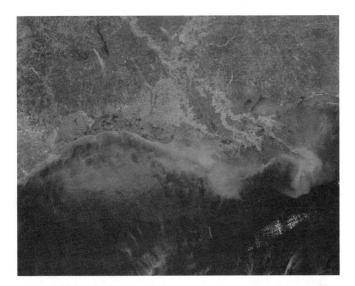

Figure 1-8 Satellite image of the Mississippi River Delta in the northern Gulf of Mexico. (See color plate 1-8.)

considered a beach. Rocky coasts are typical in more geologically active regions along the active continental margins, often in areas with relatively recent volcanic activity. Although erosion and change occur gradually, these coasts are much more stable than beaches. The organisms that live in the **rocky intertidal** zone are specially adapted to withstand a constant pounding from wave action or frequent stranding with the change in the tide. The ecology and conservation concerns about rocky shorelines are covered in Chapter 3.

Estuaries

Estuaries are waters partially surrounded by land where fresh water and salt water meet as a river enters the sea.

These may be associated with deltas, but estuaries are also formed by rivers that do not produce deltas. Because of the mixing of ocean and river waters, the salinity can range from fresh to full strength sea water. In isolated coves or tidal flats, evaporation may result in salinities substantially higher than those of ocean waters. These extreme and variable conditions mean that organisms living in estuaries must be adapted to withstand unique physiological stresses or have the mobility to avoid harsh conditions. Although physical stresses can limit organisms living in estuaries, they gain advantage from being in one of the most biologically productive environments on Earth due to the large input of nutrients and sunlight. Estuaries also provide protection from wave action and a diversity of habitats for shelter and protection. A large biomass and diversity of organisms, thus, are associated with estuaries, at least during some portion of their life.

The living resources of estuaries and access to the sea have attracted human settlements to estuaries for millennia. For example, the Nile Delta in Africa (**Figure 1-10**) and the lower Tigris and Euphrates Rivers have supported some of the oldest continuous civilizations on Earth. The Chesapeake Bay, where large U.S. cities have developed, is an estuary, and the waters flowing through the Mississippi River Delta form an estuary. The impact of activities associated with human settlements and the use of the biological

Figure 1-9 A rocky coastline on the northern shore of Maui, Hawaii.

Figure 1-10 Satellite photo of the Nile River Delta and Estuary. (See color plate 1-10.)

production for human needs results in numerous conservation issues in estuaries. The ecology and conservation of estuaries and associated habitats will be discussed in Chapter 4.

Reefs

Reefs are made by living organisms attached or partially embedded into the bottom. Although reefs can form in temperate coastal regions or even in the deep sea (see Chapter 8), massive reefs are most prominent in clear, warm, shallow waters of the tropics as **coral reefs**. Although corals predominate, other organisms such as algae and sponges help to form the reef complex. The hard calcified skeletons of these organisms form massive structures in shallow waters. In fact, the Great Barrier Reef off Australia—considered the largest structure on Earth—is made by living organisms, comprised primarily of reef-forming corals.

Tropic coral reefs are mostly limited to shallow waters due to the dependence of the reef-forming organisms on sunlight for photosynthesis. Their structure, however, varies with the geologic history of the region in which they occur. These structures can be placed into three basic types based on their general appearance, age, and mechanism of formation (**Figure 1-11**). The first to document and

propose an explanation of the variation in reef structures was Charles Darwin. He noticed in his global travels that while most reefs are associated with coastal areas, some consisted of rings of coral reef isolated in the open ocean. He developed a classification scheme for reefs into three types that are still used today. As scientists learned more about plate tectonics and island formation, Darwin's explanation of reef formation was refined to explain why these different reef formations exist. As a reef begins to form around a relatively new volcanic island, it grows in a fringe adjacent to the coastline. These are called **fringing reefs**. As oceanic plates sink as they move from the spreading center, the islands begin to sink as well. The coral organisms build the reef upward at a rate fast enough to keep the reef alive near the surface. The reef thus now encircles the island but is separated by a lagoon from the portion of the island remaining above the surface; these are **barrier reefs**. When the island eventually sinks below the water surface, the remaining circular reef encloses a shallow lagoon and is now referred to as an **atoll**. In some atolls, small islands form around the reefs. Because corals can only grow at a rate of about one centimeter per year, this evolution of reef formation occurs over an extremely long period of time; it may take tens of millions of years for an atoll to form. Some

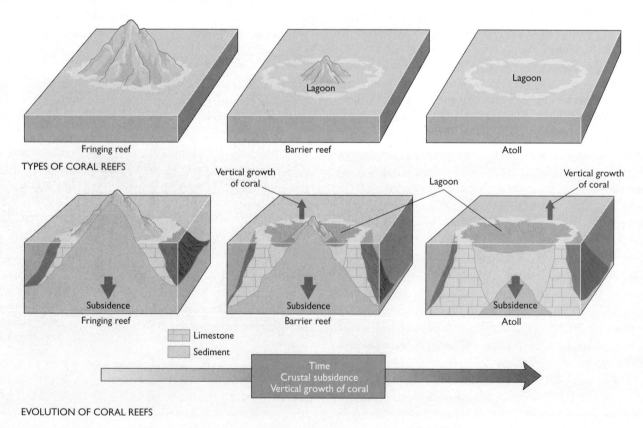

Figure 1-11 Representation of coral reef development. Fringing reefs form around new volcanic islands, developing into barrier reefs and atolls through geologic time in response to land sinking or sea level rising around the island.

scientists are now concerned that rapid global warming could result in the death of many coral reefs if coral growth cannot keep up with the rate of sea-level rise, leaving the corals in waters too deep and dark to survive.

Regardless of the class of reef formation, tropical coral reef habitats attract and support a similar assemblage of diverse organisms. In fact, coral reefs have higher species diversity than any other marine ecosystem. One reason for this high diversity is that organisms are more readily able to evolve specialized adaptations in the stable environment of the coral reef habitat. For example, physical variables such as sunlight, salinity, and temperature are relatively moderate and constant. Because many organisms on the reef have evolved in response to this stability, coral reefs can be very sensitive to disruptions. Human actions disrupt this stability in many ways. By blocking sunlight penetration with excess sediments, modifying temperature through global climate change, or removing major components of the ecosystem by excessive fish harvest, the coral reef ecosystem rapidly begins losing its biological diversity. Tropical coral reef ecology and conservation are discussed in Chapter 5.

Islands

Islands are non-continental land masses projecting above the surface of the water. The size of islands can vary considerably. In the strictest sense continents might be considered islands, and a boulder projecting above the sea surface also might be considered an island. Islands are typically considered as non-continental land masses of intermediate size, however, and often are classified according to the processes by which they are formed.

Barrier Islands

Barrier islands are relatively narrow bars of sand that form parallel to the shore along shallow gently-sloping coastlines. In order for barrier islands to form, there must be adequate sands and the forces of waves, tides, and currents must be strong enough to move and accumulate the sands. They will not form, however, where tides and currents are extreme. The proper combination of physical conditions is present for the formation of barrier islands along about 13% of the world's continental margins. Not only are barrier islands limited geographically, they also are more prevalent during some global climate conditions than others. Although some barrier island are simply accumulations of sand on submerged bars offshore, most are formed under special conditions during periods of rising sea levels. As sea levels rise along sandy shorelines, the waters break through the sands to form a lagoon behind the dunes. Eventually these dunes become islands separated from the mainland by open water. During periods of rising sea level, wave action causes the islands to migrate toward the mainland. Many

current barrier islands actually began forming during the last major sea-level rise about 6,000 years ago.

One of the most extensive series of barrier islands is along the U.S. Atlantic and Gulf of Mexico coasts, where there are almost 300 separate barrier islands, with a combined length of over 4,000 km. These islands are continually changing as a result of current and wave action. Hurricanes or large storms will often overwash smaller islands, form new inlets, or even split them into separate islands. New islands can be formed from **spits** (sand barriers joined to the coast). For example, barrier islands along the Mississippi/Alabama Gulf of Mexico coast are regularly split or combined, especially in association with major hurricanes. In 2005, for example, Asbury Sallenger and colleagues reported substantial changes after a single event, Hurricane Katrina (**Figure 1-12**). Despite this instability, humans have connected many of the barrier islands to the mainland by bridges and developed the islands as cities or resort communities. These include Atlantic City, New Jersey; Miami Beach, Florida; Galveston, Texas; and Pensacola Beach, Florida. It can be a constant battle for residents of these islands to avoid loss of their homes or businesses to the actions of the sea, and no one can predict where the next hurricane might devastate developments on one of these islands. Fortunately, the U.S. barrier islands that have escaped development have been largely protected in the past few decades. Issues concerning conservation of barrier island habitats are discussed in Chapter 3.

Volcanic Islands

Most of the largest islands, as well as those found away from the continental margins, are formed by volcanic activity. As mentioned previously, this volcanic activity is usually associated with the convergence of oceanic plates with continental plates during seafloor spreading or the formation of hot spots in the middle of plates. Many volcanic islands are only a small portion of the volcano, with much more found under the water's surface. Volcanic islands formed at convergence zones are typically arranged in an arc adjacent to ocean trenches. These island chains are easily recognized on maps because of their shape (**Figure 1-13**). They include the Western Antilles in the Caribbean, the islands of Japan and the Mariana Islands in the western Pacific, and many others. These islands include some of the most active volcanoes on Earth; however, many are remnants of long-extinct volcanoes.

Volcanic islands that are formed in the middle of oceanic plates are also found in linear chains but are not near spreading centers or convergence zones. These are created by hotspots that push volcanic material to the surface. The classic example of this type of formation is the Hawaiian Island chain (Figure 1-13). These islands are formed in line

(a)

(b)

Figure 1-12   Satellite photos of barrier islands in the northern Gulf of Mexico **(a)** on October 15, 2004 and **(b)** taken on September 16, 2005 after Hurricane Katrina. Note that Dauphin Island, Alabama (second island from right) has been split in two and that Ship Islands (starting second from left) are substantially smaller. Dauphin Island also migrated landward, leaving some oceanfront homes in the sea. Petit Bois Island, 8 miles to the west, was a part of Dauphin Island 150 years ago. (See color plate 1-12.)

Figure 1-13   Satellite view of the Hawaiian Islands, formed by volcanic activity as the Pacific Ocean plate slides over the mid-ocean hot spot currently located beneath the island of Hawaii (lower right); the oldest islands (Kauai and Niihau) are in the upper left. (See color plate 1-13.)

because the hotspot creating them remains relatively stable while the oceanic plate passes over it. A series of islands thus is created over the millions of years it takes the plate to pass over the hotspot.

The Ocean Water Column

The waters of the oceans comprise vast, largely unexplored realms that continue to fascinate the human mind. Other than the thin layer at the surface, most of the waters of the oceans have been accessible to humans only in the past few decades. Even today much of the deep ocean is still relatively unexplored. The depths of the ocean are extreme from a human perspective, the most extreme being about 11,000 meters below the ocean surface, in the Mariana Trench of the western Pacific. The deepest in the Atlantic Ocean is the Puerto Rico Trench at 8,400 meters. Although these deep trenches compose only a relatively small area of the sea floor, the average depth of the oceans is still remarkable at about 4,000 meters. Most of the biological production in the ocean is limited to the upper 200 meters of the water column, where nutrients, oxygen, and light are most available. Biological production and oxygenation originating in surface waters, however, feed most of the biological activity in the deep sea, down to the deepest basins and trenches. These deep-sea regions have only recently been explored; development of technology that allows us to access these regions means that these are exciting times in ocean research. Still, the deep ocean remains as the last largely unexplored frontier on the planet. It remains to be seen whether increases in accessibility will lead primarily to conservation or to excess exploitation of deep ocean resources. These issues are discussed further in Chapter 8.

Ocean Zonation

For convenience and delineation of habitats, the ocean can be divided into separate zones according to the distance from shore and depth of the water (**Figure 1-14**). The divisions between zones are not precise lines, and there is typically no sharp transition from one zone to the next. The zones do give a useful reference, however, and some general characteristics of these zones are fairly consistent.

Moving horizontally from the coast, the **intertidal zone** (also called the **littoral zone**) is the region above the low tide line that is regularly covered and exposed by the movements of the tides. The **neritic zone** is the region over the continental shelf, from the intertidal zone offshore to the shelf break. This is the region of the highest biological productivity, primarily as a result of nutrient inputs from the continents and exposure to adequate sunlight for photosynthesis. High production and accessibility by humans results in high exploitation of resources from this region. The **oceanic zone** includes the remaining waters of the seas beyond the waters above the shelf break. In terms of biological production, this region is like a desert relative to the neritic zone. But because this ecosystem is larger than any other on Earth, there is a great diversity of organisms and large amounts of biomass. Virtually no region in the oceans is devoid of life.

The ocean waters also can be divided into vertical zones, classified according to their distance below the

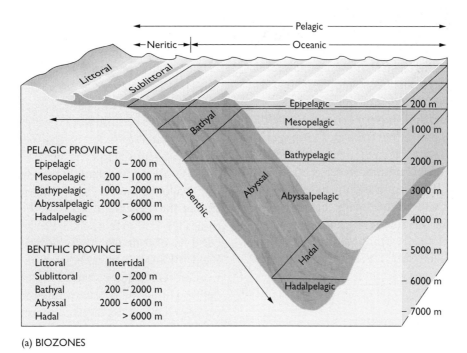

(a) BIOZONES

Figure 1-14   Ocean regions established to discriminate biological zones based on water depth.

ocean's surface. The sea floor is referred to as the **benthic region**, and the bottom habitat itself is called the **benthos**. The water column of the ocean is called the **pelagic zone**. The **epipelagic** generally corresponds to the **photic zone**, the region with adequate light penetration for photosynthesis. This is typically 100 to 200 m depth in the clear open ocean. The **mesopelagic zone** is the mid-region of the open ocean, from the photic zone down to about 1,000 meters depth. This is sometimes called the *twilight zone* because of the low levels of light. Many of the organisms in the mesopelagic depend on the sinking of food particles or nutrients from waters of the photic zone. Other organisms migrate to the surface at night to take advantage of the productive photic zone waters. The deep sea below 1,000 meters is divided into three zones. The **bathyl zone** (or bathypelagic when referring to the water column) is down to 4,000 m; the **abyssal** (or abyssopelagic) zone is below 4,000 meters to about 6,000 meters. The **hadal** (or hadopelagic) zone includes the deepest ocean trenches, down to 11,000 meters depth.

1.3 Water Movements: Wave, Tides, Currents

Coastal and seafaring civilizations have been studying waves, tides, and currents for practical purposes for thousands of years. The knowledge gained enhanced the abilities of humans to travel the oceans and exploit its resources. Scientific advancements over the past century have dramatically increased our understanding of these water movements so that information that used to require the accumulation of more than 100 years of knowledge now can be obtained almost instantaneously (for example, using satellite imagery). An understanding of the water movements not only allows us to navigate the ocean and exploit its resources but also gives us an understanding of how they influence marine organisms and ecosystems.

Waves are the most visible type of water movements. Waves differ from currents in that the water forming the wave does not move at the speed of the wave; it is the movement of energy that produces waves. This is why an object floating just offshore moves only slightly toward shore with each passing wave. The energy that drives the wave can come from various sources including the wind, movements of the Earth (e.g., earthquakes), gravitational forces, and atmospheric pressure. These energy sources are the basis for classifying waves.

Wind Waves

Wind waves are disturbances in the ocean surface formed by the transfer of wind energy to the water (**Figure 1-15a**). Wind waves are typically formed in an undulating pattern with the waters rising to a **crest** (the highest point on the wave) and falling to a **trough** (the wave's lowest point). The size of the wave is defined by the **wave height**—the vertical distance from the crest to the trough—and the **wavelength**—the horizontal distance between adjacent crests. The **wave period** is the time it takes adjacent wave crests to pass a fixed point. Wind waves can vary tremendously in size, depending not only on the wind speed but also the wind **duration** (or how long it blows over the sea), and the **fetch** (the distance over which the wind blows).

Large waves may originate from distant winds; winds of greater duration and speed impart more energy to the water and make larger waves. Until the 1930s it was believed to be impossible for a wind wave to reach heights over 20 meters. This was challenged when a U.S. oceanographer documented waves higher than 30 meters during a long duration storm in the Pacific Ocean. Wave measurements during Hurricane Katrina in 2005, with winds up to 150 mph, showed that large waves are not as uncommon as once believed; waves over 18 meters were not uncommon and 25-meter waves were observed. Oceanographers have predicted that hurricane waves could reach heights as great as 40 meters. Although these waves are caused by storms or other unusual winds, extreme waves could occur miles from the actual storm. Sometimes single unusually large waves occur amongst average size waves in the open ocean. These **rogue waves** are surprisingly large for the sea conditions in which they occur. Though unusual, such waves over 25 meters appear to occur regularly in the open ocean. Their mechanism of formation is still debated, but they seem to often occur in deep water where there is a convergence of physical forces such as strong winds and fast currents, and may be a result of a consolidation of multiple smaller waves.

Regardless of what the maximum height of an ocean wave proves to be, you will not see wind waves approaching these heights near shore, especially in shallow coastal waters. This is because wind waves break as they move into shallow waters (Figure 1-15b). The motion of the wave actually extends below the surface to a distance of about 50% (½) of the wavelength; therefore, as the wave moves into shallower water it "feels" the bottom. As the lower portion of the wave begins to slow, the top of the wave continues forward making the wave steepen until it breaks. A wind wave breaks when its wave height is about 75% (¾) of the depth of the water. For example a 75-centimeter wave would break when it reaches waters of about one-meter depth; a three-meter wave would break in four-meter deep waters. Higher water levels come ashore and cause damage during hurricanes due to the combination of an increased water depth from the **storm surge** with the higher wind waves that form in these deeper waters.

(a)

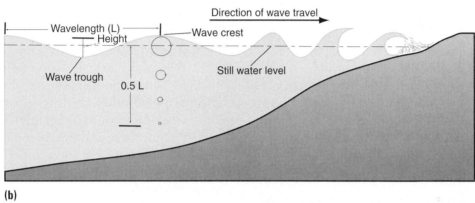

(b)

Figure 1-15   **(a)** A wind wave breaking as it moves ashore. **(b)** The pattern of water motion in a wind wave approaching shore.

As waves break onshore their energy is transferred to the bottom. Because beaches are often composed of loose unconsolidated sands, they are easily moved by waves; this is why beaches are so unstable. Waves usually do not move onto shore directly parallel to the shoreline but approach at an angle, resulting in gradual movement of sediments alongshore down the beach in a process called **beach drift**. Although the movement of sand caused by one wave is not significant, the cumulative effect of continuous wave action can be substantial. Conservation issues resulting from human interactions with beaches and wave activity are discussed in Chapter 3.

▪ Tsunami

The largest and most damaging waves in the ocean are formed not by the wind, but by geologic activity under the sea. These have often been called *tidal waves*, but because they are not a result of tidal activity this is a misnomer. Now the more acceptable term for these waves is **tsunami** (from a Japanese word meaning *harbor wave*). Tsunamis are the result of undersea earthquakes or volcanoes, or coastal landslides rapidly displacing the water. Tsunami waves can be extremely long, with wavelengths of over 160 km, and move extremely fast, at speeds over 640 km per hour. At this speed a tsunami could move from Alaska to Hawaii in about five hours. The height of these waves in the open ocean may only be a meter or two; thus, their steepness is so low that they would hardly be noticeable to a boat under which the tsunami passes. This is the reason that tsunami waves are so difficult to detect in the open ocean, requiring sensitive instruments to monitor changes in the water level. If the wave height is so low, then how can it cause so much damage when it reaches the shore? As the tsunami slows coming to shore, its long shallow waveform is translated into a narrow and much higher wave. Tsunamis can reach heights of over 30 meters as they hit the coastline. The waves appear to an observer as a fast flood of water rushing ashore, not as a breaking wave commonly imagined.

The December 2004 tsunami in the Indian Ocean, caused by one of the largest earthquakes in decades, 240 km off the coast of Sumatra, created waves as high as 20 meters moving onto shore at speeds as high as 800 km per hour (**Figure 1-16**). This tsunami resulted in the deaths of over 280,000 people and damages totaling billions of dollars. Although damaging tsunamis are very infrequent in a single region, globally there is a tsunami that causes major damage and loss of lives on average about every seven years. For example, in the 50 years from 1925–1975 there were at least seven, including tsunamis in Russia, Alaska, Hawaii, Chili, and Newfoundland, Canada. There is nothing humans can do to eliminate tsunamis or predict their occurrence with any accuracy, but we can minimize their impacts by maintaining natural vegetation and structures along coastlines (see Chapter 4). The increased deployment of monitoring systems in the open ocean will increase our ability to warn coastal residents of a tsunami as it develops, but whether there is adequate time to respond depends on the distance from the wave's point of origin. The 2010 tsunami in Japan, resulting from an unprecedented earthquake approximately 70 kilometers off the coast, began moving ashore within about 30 minutes, giving little response time for avoidance; however, advanced earthquake warning systems likely saved the lives of thousands of people. Although there were few natural structures in this heavily developed region of Japan to buffer the wave, tsunami protection walls were in place; however, the unexpected height of the storm surge, over 30 meters in some locations, overtopped walls and inundated tsunami protection shelters.

Tides

Technically speaking, tides are types of waves. They are not typically recognized as such because they rise and fall over a period of 12 or 24 hours and they have extreme wavelengths, as long as half the circumference of the Earth. Tides are periodic waves caused by the gravitational force of the Sun and Moon on the Earth, and the motion of the Earth (**Figure 1-17a**).

Tides vary over a diurnal (daily) cycle due to the Earth's rotation and a monthly cycle resulting from changes in the Earth's position relative to the Moon and Sun (Figure 1-17b). **Spring tides** are periods when the tide range is highest because the Sun, Moon, and Earth are in line (during new and full Moons) and the gravitational attractions of the Sun and Moon are combined. **Neap tides** occur when the Sun, Moon, and Earth form a right angle (first and third quarter Moons). Tides serve an important function in transporting nutrients, wastes, and organisms into and out of coastal inlets. Tidal currents can flow through estuaries as fast as 40 km per hour or more in some regions. Tidal currents differ from wind-driven currents in that they switch directions as the tides move in (**flood tide**) and move out (**ebb tide**) of an inlet or coastal area. In some coastal areas (i.e., the U.S. east coast) there are complex networks of tidal creeks miles inland along the coastline. The biological importance of these areas and conservation issues are discussed in Chapter 4.

Tides are observed as the daily rising and lowering of the sea level on a periodic cycle. Due to a number of factors, such as the shape of the ocean basin, the frequency, and height of tides vary around the globe (**Figure 1-18**). On some coastlines the tides are **diurnal**, with one high and one low tide per day. Regions with diurnal tides include the northern and western Gulf of Mexico, the north Pacific, and southwest Australia. On other coastlines there are two high and two low tides per day. These **semidiurnal** tides occur on the east coast of North America, most of the African and European coastlines, and much of the South American coastline. Along some coastlines the tides are mixed, having two cycles per day, but successive tides are of different heights, such as on the west coast of North America and in much of the western Pacific.

The **tidal range**, or difference in the level of high and low tides, varies substantially around the Earth. This is due to a complex interaction of various physical factors, including the shape and depth of the basin, coastline, or bay where the tide is observed. The typical tidal range is somewhere between 30 centimeters and 2 meters; however, the greatest tidal range is over 15 meters, in the Bay of Fundy on the

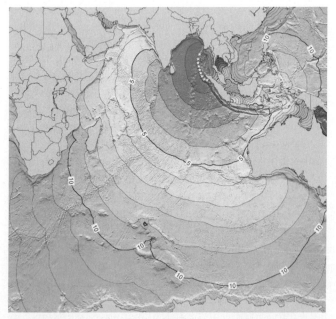

Figure 1-16   Movement of the tsunami wave produced by a magnitude 9.4 earthquake off Sumatra, Indonesia, on December 26, 2004, progressing in time from the epicenter (indicated by stars), from lighter to darker grays. (See color plate 1-16.)

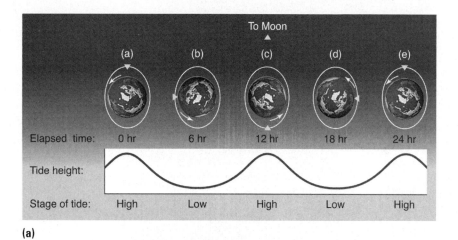

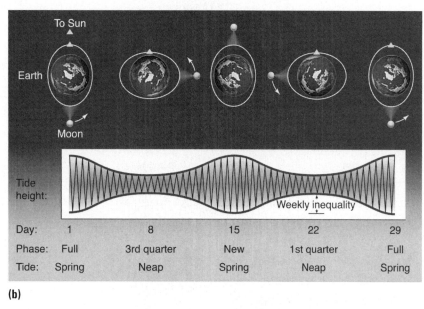

Figure 1-17 Tidal cycles. **(a)** Diurnal cycles in tides caused by the position on the Earth relative to the tidal bulge (arrow indicates a fixed position on the Earth). **(b)** Monthly change in tidal range due to the relative position of the Earth, Moon, and Sun; maximum (spring) tides occur when the Earth, Moon, and Sun are in line.

Atlantic coast of Canada. This variation in tides can have a large influence on organisms that use the tides for movement in and out of coastal areas. It also affects conservation issues through the mixing and transport of materials such as organic matter, nutrients, and pollutants, especially in areas where there is little river flow into the coastal region. These issues are discussed in Chapters 3 and 4.

Circulation Patterns

Surface Currents

Water and energy also can move through the ocean as **currents**. Whereas waves are primarily a movement of energy, currents are masses of moving water. **Surface currents**, driven mainly by surface winds, are layers of water flowing within the surface waters of the oceans; they can reach to depths as great as 400 meters. They serve an important

ecological function by transporting drifting marine organisms, organic matter, and nutrients; they enhance the movements of migratory organisms.

Viewing a map of wind-driven ocean surface currents (**Figure 1-19**), a pattern of circular **gyres** is apparent as flows around the ocean basins of the northern and southern hemispheres. These gyres are primarily due to the **Coriolis Effect**, resulting from the forces imparted on the ocean due to the Earth's rotation. In the northern hemisphere, the Earth's rotation (toward the east) appears to deflect the flow to the right of the wind direction. The water continues to veer to the right around the basin resulting in a clockwise rotation. Not only is this obvious throughout the major ocean basins of the Atlantic and Pacific, but smaller gyres occur in smaller basins such as the Gulf of Mexico. (It is a myth that this rotation can have a significant effect on

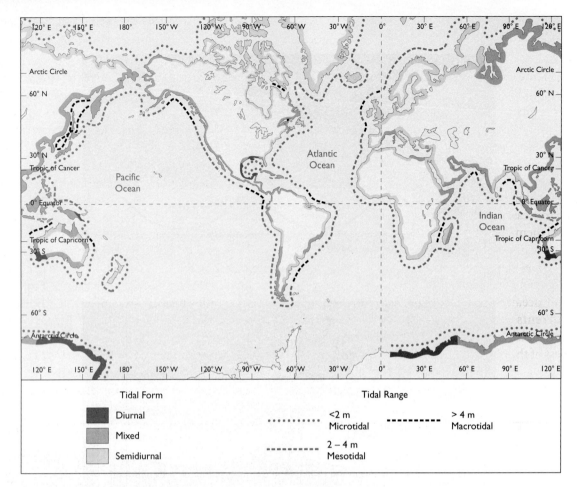

Figure 1-18 Global variation in tidal forms and ranges.

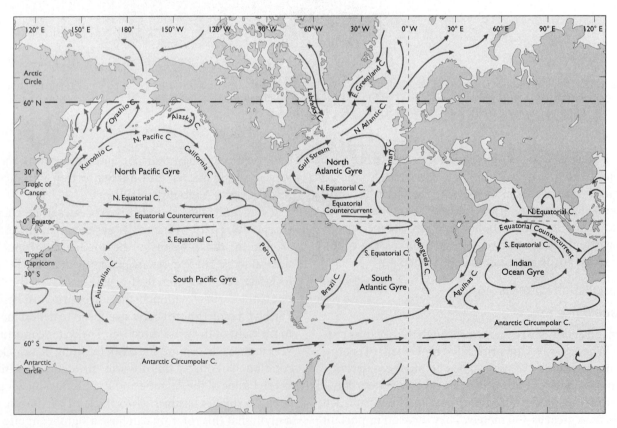

Figure 1-19 Global wind-driven surface currents indicating major circulation gyres.

water sitting in a small basin such as a tub or a toilet; other forces acting on the water greatly overwhelm the Coriolis force at this scale.) In the southern hemisphere the opposite occurs; currents are deflected to the left, resulting in a counterclockwise rotation. Wind patterns and deflection of currents by land masses also contribute to the formation of these gyres. Navigators of ocean vessels have taken advantage of these circulation patterns for centuries. It's not a coincidence that many early explorers followed similar paths around the seas; that's the direction the currents (and winds) carried them. Marine organisms take advantage of this circulation, riding the currents to spawning, feeding, and nursery grounds.

The currents flowing along the coastlines form one component of ocean circulation patterns. The **western boundary currents** (found on the west side of the ocean basins, along the east coast of the continents) are the fastest and deepest of the currents making up the ocean gyres. These currents tend to carry warm waters from equatorial regions toward the poles. Although this has been known

for hundreds of years, only recently have we been able to get detailed surface temperature maps using satellite imagery (**Figure 1-20**). The warm waters flowing northward along the east coast of North America form the Gulf Stream, the largest of the western boundary currents. This current has an average width of 70 km, travels about 8 km per hour, to a depth of around 450 meters. It has important influences on fisheries along the U.S. Atlantic coast. For example, as it flows adjacent to the east coast of Florida it attracts oceanic fishery species into waters close to shore, and continues to influence marine populations as it moves north. The Gulf Stream also has a moderating effect on climate along the U.S. east coast due to its influence on air temperatures. If this current did not exist the climate along the eastern North American coastline would be much colder (see Box 1-1, Conservation Concern: Global Warming and the Ocean).

Eastern boundary currents flow along the east side of ocean basins (along the west coastlines of the continents). These are the opposite of the western boundary currents in

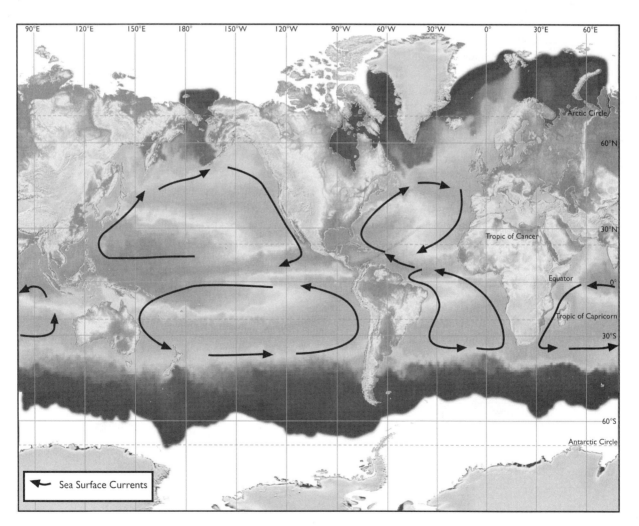

Figure 1-20   Global ocean surface temperatures during summer in the northern hemisphere. Grays indicate temperatures, with lighter gray being warmer and darkest grays being cooler. (See color plate 1-20.)

many ways. They tend to be smaller, weaker, broader, and shallower than western boundary currents. Flowing from the poles toward the equator, eastern boundary currents carry cold water southward in the northern hemisphere. For example, cold waters are transported well down the U.S. west coast. This is why, at comparable latitudes, waters along the U.S. east Pacific coast tend to be substantially colder than those along the Atlantic coast. The Peru Current in the South Atlantic is a southern hemisphere eastern boundary current with similar characteristics (but flowing equatorward from south to north).

Coastal Upwelling

Of course, the ocean is not a one dimensional plain. Along with the surface currents there are vertical components to ocean circulation. Coastal **upwelling** is one component of vertical circulation that influences conservation issues and is discussed in later chapters. Coastal upwelling can most easily be understood by looking at it in a stepwise fashion as it would develop (**Figure 1-21**). A prime example is the Peru Current, an eastern boundary current flowing along the Pacific coast of South America. As the surface currents flow northward along the shore, the waters are deflected offshore away from the coast due to surface winds and the Coriolis effect (to the left of the current direction in the southern hemisphere). As these waters are pushed away from shore, deeper waters are pulled up to replace them. This is the process of upwelling. These waters can be recognized by their temperature signature; the cold **thermocline** (a layer of relatively cold water) is extended to the surface along the coast. The waters upwelling from depth have high nutrient levels, resulting from the accumulation of organic matter that sinks from the productive surface waters toward the bottom. Because there is relatively little biological activity in deeper darker waters to use the nutrients, they accumulate until being drawn back to the surface. The nutrients

upwelling to the surface support one of the most biologically productive regions and one of the most valuable fisheries in the ocean. A similar, but less pronounced pattern can be observed in other coastal upwelling zones, such as off portions of the coast of the western United States and western Africa. Later chapters will discuss the significance of these upwelling zones relative to ocean fisheries.

The pattern described above is the most typical circulation pattern along the western coast of South America. Sometimes this pattern is modified, however. For example, a change in surface winds can interfere with the surface currents, which disrupts the upwelling. In the tropical Pacific such a change occurs about every three to eight years, for reasons that are not completely understood. This change initiates events referred to as **El Niño** (Figure 1-21b). The effects of El Niño events have long been noticed off the coast of South America, because this is when upwelling practically stops, which reduces nutrient levels and leads to dramatic declines in fish populations (see Chapter 11). These events were named El Niño ("the Child") by Peruvian fishers because they are usually initiated around the Christmas season. Although their immediate impact is a change in weather and ocean circulation in the eastern Pacific, El Niño events have many effects on oceans and weather around the world. One of these, discussed in Chapter 5, is the increase in tropical water temperatures that impacts coral reef ecosystems.

Vertical Circulation Patterns

Ocean water circulates vertically not only along the coastlines but also in the open ocean (**Figure 1-22**). This vertical circulation, presented as The Great Ocean Conveyor by Wallace Broecker, is driven primarily by water density differences and is called **thermohaline** ("temperature-salinity") circulation. Saltier or colder water is denser and therefore tends to sink. Waters driven to the poles by surface

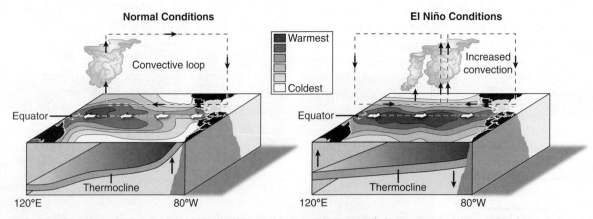

Figure 1-21 Coastal upwelling circulation off the Pacific coast of South America; grays indicate temperature, from darkest gray (warmest) to white (coldest). Left: During normal conditions the thermocline extends to the surface along the coast due to upwelling of deep waters. Right: During El Niño conditions; upwelling stops and the thermocline remains in deep waters. (See color plate 1-21.)

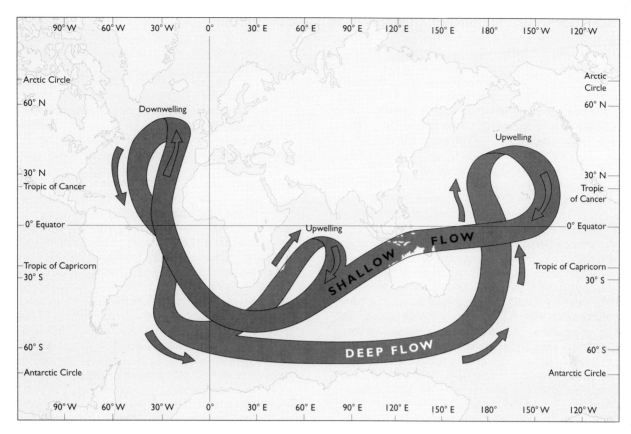

Figure 1-22 Global "conveyor belt" thermohaline circulation pattern between the surface and deep oceans.

circulation will tend to sink toward the bottom of the ocean as they get colder. These waters then slowly move across the deep sea floor toward the equator. Ecologically this is important because these waters carry dissolved oxygen throughout the deep open basins (see Chapter 8). In fact, without this circulation, the deep ocean could not support oxygen-dependent life, including all multicellular organisms. Eventually these bottom waters upwell to the surface at various locations. They are important in bringing nutrients to the surface in equatorial regions, making these tropical waters very productive. This deep-sea circulation is much slower than the surface circulation pattern discussed earlier. Most deep ocean bottom waters take 200 to 300 years or more on average to rise to the surface. Compare this to a bit more than a year it takes water in the North Atlantic surface current gyre to make a complete circuit. Although these water masses generally maintain some integrity, there is much mixing among water masses throughout the ocean. There is also seasonal and year-to-year variability in all oceanic circulation patterns. Many fear that changes in global climate could disrupt or change this circulation pattern and impact water exchange between the surface and the deep sea, with dramatic impact on deep-sea ecosystems (**Box 1-1. Conservation Concern: Global Warming and the Ocean**).

1.4 Sea-Level Changes

For the average person it is difficult to envision dramatic changes in sea levels because few have witnessed noticeable changes along ocean coastlines. It is well documented that in the past sea levels have been much higher and lower than they are at present, however (**Figure 1-23**). For example, during the last ice age, about 18,000 years ago, sea levels were over 120 meters below their current level. In fact, for most of the past 200,000 years the sea levels were at least 30 meters lower. And three million years ago sea levels were 35 meters higher than at present.

Global Sea Level Changes

Variations in global sea level are determined by measuring the mean sea level in relation to relatively stable continental coastlines. The major causes of variability in global sea level are changes in global climate. The most obvious factor is the conversion of ice water in glaciers and the polar ice caps to liquid water that flows into the oceans. When global temperatures are colder, more water is tied up in glaciers; therefore, sea levels are lower. When global temperatures are warmer, less ice is frozen; therefore, sea levels are higher. However, another important factor is **thermal expansion.** At temperatures above 4°C water expands as it is warmed.

Box 1-1 Conservation Concern: Global Warming and the Ocean

Possibly the most debated environmental issue in the world today is global climate change. Debates have centered on questions concerning: whether we are truly in a global warming trend, whether human activities are causing or influencing these changes, our capabilities to reverse recent trends, and what impact climate change will have on the Earth and its ecosystems, including the marine environment. An authoritative source of information on climate change is the Intergovernmental Panel on Climate Change (IPCC; http://www.ipcc.ch/), an international group of scientists established in 1988 through programs of the United Nations. One of the charges of this panel is to assess scientific documents from around the world and provide summaries of these documents. IPCC documents indicate that assessing and predicting global warming trends are complex processes. Because temperatures fluctuate seasonally, annually, and on longer term cycles, a few warmer years cannot be taken as proof of a long-term trend. On the other hand, a few cooler years cannot be taken as proof that global warming is *not* occurring. Decisions must be made concerning where to measure temperatures for documenting climate change, because they could be increasing in one place while decreasing in another. Consistent methods must be applied to enable comparisons of historic and recent temperatures; standard established criteria enable such comparisons over about the past century.

Comparisons of global average air temperatures near the Earth's surface show an increase of about 0.5 to 1 degrees Celsius during the 20th century (**Figure B1-1**). In the past 30 years, however, the temperature rise has averaged about 0.2°C per decade. Although this may appear trivial, it represents a large amount of energy globally and some regions have warmed faster than others. It is estimated that the Earth is within 1°C of

the highest temperature in the past one million years. Although some of the popular news media may give the impression that there is wide disagreement among scientists, a 2008 scientific poll of over 3,000 Earth scientists showed 90% agreement that there has been a significant increase in mean global temperatures in the past 200-plus years.

The next critical question addressed by the IPCC is whether humans are influencing climate change. The conclusion of scientific studies is that most of the observed increase in temperature since the mid-1900s was caused by increased input of **greenhouse gases** (those that have an effect of trapping heat from solar radiation in the atmosphere, primarily water vapor, carbon dioxide, and methane), and that the primary source of the increase was carbon dioxide from fossil fuel burning and deforestation. Since the mid-1700s, carbon dioxide in the atmosphere has increased by over 30%, to levels not experienced on Earth for hundreds-of-thousands or possibly millions of years (**Figure B1-2**) (methane has increased by almost 150%). Carbon dioxide is estimated to contribute more than 60% of the warming attributed to greenhouse gases. Despite these data, scientists and political leaders debate as to the degree of influence these increases have on global warming. However, 97% of polled research climatologists—those scientists who would have the greatest expertise in climate change—agreed that humans have played a role in global climate change (fewer than 60% of the general public agreed).

The next question, whether we are capable of reversing recent global warming trends, is a social and political question as much as a scientific one. Decreasing deforestation and the burning of fossil fuels is physically possible, but politically difficult to initiate when it is viewed as causing economic hardships or as unnecessary. Recent agreements among climate

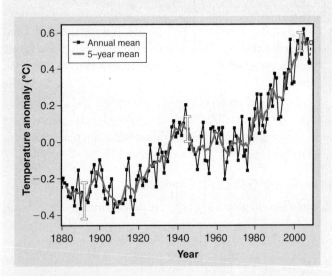

Figure B1-1 Global annual temperature anomaly indicating change in temperature from 1880 to 2009. The time period from 1951–1980 is used as the zero point for reference. The dotted line indicates the annual mean and the solid line is the five-year mean. White bars show the range of uncertainty. Data are compiled from surface air measurements at meteorological stations, ocean ships, and satellite measurements.

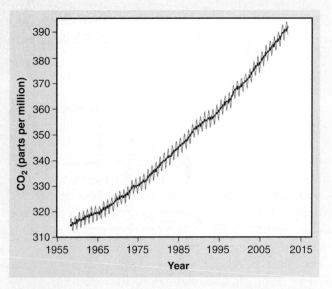

Figure B1-2 Plot of carbon dioxide record at Mauna Loa, Hawaii, by Scripps Institute of Oceanography (1958–1974) and the National Oceanic and Atmospheric Administration (1974–2011). The curves show monthly averages in parts per million (ppm). The solid line indicates annual averages.

scientists have led to increased actions around the world. The primary international agreement dealing with actions to reduce greenhouse gas emissions is the Kyoto Protocol, negotiated in 1977. Over 160 nations have ratified the treaty; the only industrialized nation that has not is the United States. Most nations, including the United States, however, have taken some measures toward moderating emissions. It is widely debated whether these actions will be enough to decrease global warming.

Other methods of reducing carbon dioxide levels in the atmosphere are being researched, including **carbon sequestration**, which is capturing carbon dioxide from burning of fossil fuels and storing it underground or elsewhere. Encouraging increased plant production is a more natural way of sequestering carbon dioxide from the atmosphere. This can be accomplished by planting trees or possibly by encouraging growth of phytoplankton in the ocean (see Box 1-2, Conservation Controversy: Fertilizing the Ocean). Some scientists argue that, even if we reduced greenhouse gas emissions considerably, global warming could still continue until the end of the 21st century. Another response to fears of global warming therefore is to begin adapting to predicted changes. These adaptations would include increased fortification of coastal cities (see Chapter 3) and abandoning islands and coastal areas at low sea levels. Depending on the amount of sea-level rise, it is debatable whether this should be viewed as a valid option, because much of the world's population resides in coastal areas.

The final question of what impact climate change will have on the Earth and its ecosystems may already be partially answered. Glacial retreat in mountainous regions around the globe has been linked to increases in atmospheric greenhouse gases. Examples include the Himalayas in Asia, the Alps in Europe, the Rocky Mountains and Cascade Range in North America, and the southern Andes in South America. Recent warming in the Arctic has resulted in shrinkage of ice cover. Since accurate measurements began in the late 1970s using satellite imagery, record lows in sea ice coverage were observed several times, all occurring in the 2000s. In 2007, the greatest summer decline in Arctic sea ice was recorded. The Northwest Passage, pursued in vain by early explorers for centuries, opened for the first time ever in human memory. Based on projections from recent ice melting, it is predicted that the Arctic Ocean could be free of summer sea ice by the mid to late 21st century. This is noteworthy because there is no scientific evidence of an ice-free Arctic for hundreds-of-thousands of years. Some scientists, however, question our ability to make such predictions with reliability.

With an increase in melting of ice, the percentage of liquid water on the planet must increase; thus, sea-level change is a predictable outcome of global warming. The rate and total amount of change is difficult to estimate precisely, however. Other changes in climate are not as predictable. Single weather events or anomalous weather years cannot necessarily be linked to global climate change. Climate predictions include changes in rainfall patterns causing changes in water distribution patterns, and increased frequency of extreme weather events. Predicting how these changes will affect land use patterns and the abundance and distribution of various organisms is now the focus of many scientific studies.

Other than ice melting and changes in sea level, there are numerous scenarios that might develop in oceans due to global warming and climate change. It is predicted that El Niño events will increase in frequency and intensity. One pattern that develops during El Niño conditions is an increase in water temperatures in shallow tropical waters. Recent extreme temperatures during El Niños have negatively impacted tropical reefs; it is still debated whether this is a direct result of global warming (see Chapter 5 for further discussion of these issues). Increased frequency of hurricanes and other tropical storms since the mid-1980s, especially in the North Atlantic, has been linked to increased water temperatures, but it is still debated whether this trend will continue with global warming.

Carbon dioxide (CO_2) plays an important role not only in global warming but also in ocean chemistry. Increases in atmospheric CO_2 result in higher CO_2 levels in the ocean. The reaction of CO_2 with water to form carbonic acid (H_2CO_3) results in ocean acidification. Ocean pH has decreased (become more acidic) by about 0.1 units since the industrial revolution and is predicted to continue its decline as the ocean absorbs more CO_2. Some climate models predict that the pH of ocean surface waters could decline by over 0.7 pH units due to anthropogenic input of CO_2 over the next few centuries; if so, this would probably be the lowest level experienced by ocean organisms in about the past 300 million years. An analysis by Timothy Wootton and colleagues has shown that calcareous species in regions around the globe are being stressed at current pH levels, and there is growing evidence that higher acidity is impacting coral reef ecosystems (see Chapter 5). Even if the input of CO_2 into the atmosphere were to be returned to normal levels, it could be a thousand years or more before the heat and CO_2 dissipate such that the ocean returns to normal.

One of the most dramatic predictions regarding climate change and the oceans involves changes in ocean circulation. According to one theory, melting of Arctic glaciers could disrupt vertical thermohaline circulation patterns in the North Atlantic as follows. Meltwater from glaciers and the polar ice cap is made up primarily of freshwater, which is lighter than the surrounding seawater (the salts in seawater make it about 3.5% denser than freshwater). The increased meltwater as a result of global warming therefore could lower the overall density of polar seawaters. Less dense polar waters would not sink as they typically do, and this could disrupt the global thermohaline vertical circulation pattern (see Figure 1-22). Because the sinking of North Atlantic waters is the major source of oxygen for the deep Atlantic Ocean, a disruption of this vertical current could result in death of most deep-sea organisms. In addition, upwelling of deep Atlantic waters is a major source of nutrients for North Atlantic plankton. A disruption of the circulation could reduce the plankton to less than half their current biomass, which would have a major impact on ocean food webs and fisheries. Less CO_2 would be taken up by ocean waters if thermohaline circulation is weakened; this would cause further increases in global warming.

Effects of the loss of the thermohaline circulation pattern would not be limited to the oceans. The disruption of near-surface flow of warm waters from the tropics into the North Atlantic would allow colder polar waters to flow equatorward and replace them. These warm currents play a major role not only in warming the north Atlantic but in moderating the climate and temperatures of eastern North America and western Europe; most of the heat that is transported to the atmosphere comes from ocean waters. This scenario therefore would result in major climate changes in these regions with some regions experiencing a paradox of much colder weather as an effect of global warming. Other predicted changes in climate include changes in rainfall pattern and storm frequency around the globe.

Since this hypothetical scenario was developed in the 1980s there has been much discussion of its likelihood, and some scientists have even suggested the pattern is already developing. There is strong historical evidence that such a disruption in thermohaline circulation has happened in the past. The last time was about 12,000 years ago, initiated by large inputs of meltwater at the end of the last glaciations period; temperatures in Scandinavia were lowered about 30°C. These changes probably happened rapidly, in a matter of years (but probably not as rapidly as presented in the 2004 film, *The Day After Tomorrow*, where within days the U.S. citizenry was rapidly being forced toward the Mexican border by an instant ice age). Whether we are in the midst of or near such a scenario is being debated among oceanographers. Mathematical models suggest that a shutdown of this circulation would require a global warming of at least 4°C above current temperatures. Although freshwater input into the North Atlantic is increasing, some models estimate that predicted melting rates will not put in enough freshwater to stop the circulation within this century. The IPCC predicted a possible slowing of 25% by the end of the 21st century, but not a complete shutdown. Some measurements suggest that the circulation may have already begun slowing substantially. This evidence has been questioned because large-scale automated systems have been in place to accurately monitor changes in North Atlantic thermohaline circulation only since the mid-1990s. A better understanding of this thermohaline circulation system that is so critical to ocean life and global climate is gradually being gained. We may soon have better, but not perfect, predictive capabilities. Even if/when we are able to establish the scenario under which large-scale changes in thermohaline circulation will develop, this will not answer the bigger question of how scientists convince the citizens of Earth of the need to make sacrifices to slow or stop the processes leading to global climate changes: will there be the commitment before it is too

late? Once the process of ocean warming begins it is not easily reversed. Susan Solomon and colleagues presented observations and models indicating how persistent the effects of greenhouse gasses are on climate change and ocean warming. Atmospheric warming from carbon dioxide is nearly irreversible for over 1,000 years, even if emissions are stopped. The transfer of heat from the atmosphere to the ocean surface layer can take ten years or less, and centuries are necessary for transfer to the deep ocean. Once the impact of ocean warming has occurred, however, the dissipation of heat will take hundreds of years, even if atmospheric warming is reversed. This suggests that once critical levels of greenhouse gasses have accumulated, we could be dealing with the impacts for millennia regardless of what actions are taken.

The attempt to gain a better understanding of global warming, climate change, and ocean circulation is a good example of the process of science that is often lost to the general public in the hype that can be presented through the news media. Predictive models are generated; hypotheses are developed and tested, supported, or refuted; theories are generated; conclusions are drawn; paradigms become established. Although science may be able to come up with an ideal solution to conservation problems, unfortunately science cannot implement that solution. Conservation depends on human decision-making. What is the "right" answer to a scientist is not always going to be viewed as such to the government leader, politician, industry representative, non-governmental organization (NGO), or the general citizenry. The ongoing challenge thus is not just how to do the best science, but also how to get the science incorporated into conservation. Many scientists are now arguing for application of the **precautionary principle**, which is a responsibility to take cost-effective actions to limit global warming, even if there is a lack of full scientific certainty that those actions are necessary to limit climate change and the associated environmental damage. Even as world leaders become more willing to apply this philosophy, however, there remains the difficulty in defining and balancing what is "cost effective" and how much scientific uncertainty is too much (for example, how much hardship are people willing to accept to take a chance at solving unpredictable climate problems). To get world leaders to listen to these scientific concerns, a consensus must be reached as to what actions are most likely to limit global warming. These are a few of the dilemmas prevalent in marine conservation; many more are presented and discussed throughout this text.

Although the expansion is proportionately small relative to the volume of water, the cumulative effect over the entire ocean can be great. Most of the recent rise in sea level is due to thermal expansion, and at the predicted rate of glacial melting it will probably be the major factor through the 21st century. Thermal expansion is more predictable than glacial contributions to sea-level rise. Projecting anticipated glacial melt is a primary reason for the variation in predictions of future rates of sea level change.

Average sea levels have been gradually increasing at around 2.5 millimeters per year. Although this is a relatively

small change, the cumulative effect, especially if this rate increases, will have serious impacts in some coastal areas. Predictions of future sea-level rise vary greatly among scientists. Most predictions fall in a range from about 10 centimeters to 1 meter through the end of the 21st century. The reliability of these predictions will be important for conservation of coastal ecosystems. Although some debate continues, in recent years it has been well documented and generally agreed upon by experts that we are in a period of global warming. The most important debate from a conservation perspective is over how much humans are affecting

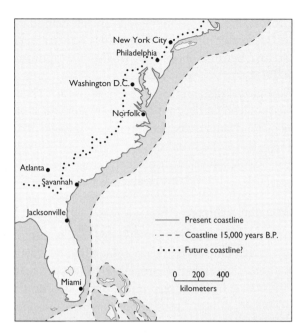

Figure 1-23 Position of the eastern U.S. coastline when sea levels were lower 15,000 years ago during the last glaciation, and predicted coastline if melting of the world's ice sheets continues at the projected rate.

global climate change and sea-level rise through air pollution and other activities, and what effects it will have on ocean currents and habitats (see Box 1-1, Conservation Concern: Global Warming and the Ocean). The effects of climate change on various marine and coastal ecosystems are discussed in later chapters.

Local Sea-Level Changes

Local sea-level changes are perceived changes in sea level relative to some point or structure on land. These can be caused by various factors. Because this is a relative change, either rising water levels or **subsidence** (sinking) of land can result in a local sea level change. One form of subsidence is the compaction of sediments in coastal areas. For example, acres of lands in coastal salt marshes of Louisiana are being lost each year as they literally sink into the sea. A major cause of this subsidence is the channelization of the Mississippi River system. The river normally has provided sediments to the marshes during periods of flooding to replenish the marshes. The channelized river transports these sediments out to sea and off the continental shelf. These issues and possible solutions are discussed in Chapter 4.

Another source of subsidence is on a much longer time scale of millions of years, associated with seafloor spreading. As seafloor gradually spreads and progresses toward the edge of the ocean plate, islands sink lower relative to ocean sea levels and eventually sink beneath the sea surface. For example, the flat-topped guyots are remnants of volcanoes that were exposed to weathering before sinking beneath the sea surface. This phenomenon can be observed in island chains, such as the Hawaiian Islands, where the newer islands rise higher above the sea surface, and the older islands to the west are lower or have disappeared below the surface.

The relative sea-level change at any point is a combination of all of these factors. Relative sea level is continually changing as a result of the ocean's volume, changes in coastal land formations, and processes of plate tectonic and sea floor spreading. From a human perspective these processes are typically slow; however, geologic events, including earthquakes or volcanic activity, or human modifications in the environment, including global climate change, could result in dramatic and even catastrophic short-term changes (see Box 1-1, Conservation Concern: Global Warming and the Ocean).

1.5 Ocean Chemistry

Seawater, Salts, and Trace Elements

The primary factor that distinguishes seawater from fresh water is the presence of relatively high concentrations of **inorganic solutes,** dissolved chemicals not formed by life processes. Within the water solution, these solutes exist as electrically charged atoms called **ions**. For example, sodium chloride (NaCl) disassociates into sodium ions with a positive charge (Na^+), and chloride ions with a negative charge (Cl^-). When these solutes are crystallized as sea water evaporates they form into **salts**. The primary salt dissolved in seawater is sodium chloride. Other ions are present at lower concentrations, including magnesium (Mg^{2+}), calcium (Ca^{2+}), potassium (K^+), sulfate (SO_4^{2-}), and bicarbonate (HCO_3^-). The ratio of these ions to each other in sea water is remarkably constant throughout the ocean. Many other elements are present at even lower concentrations and are considered minor or **trace elements**. The concentrations of these elements can vary dependent on local geological events or the input or incorporation by biological organisms. These include elements important to sea life such as nitrogen (N), phosphorus (P), and iron (Fe). They also include elements that are toxic at excess concentrations, such as lead (Pb) and mercury (Hg). The effects on organism and ecosystems of the increase in trace elements by human activity are discussed in Chapter 7. Some trace elements, such as manganese (Mn), accumulate in rock-like **nodules** around particles such as pieces of bone or algae on the deep sea floor over millions of years (**Figure 1-24**). The nodules are in concentrations great enough that commercial mining is being considered, with potential impacts to deep-seafloor ecosystems (see Chapter 8).

The measurement of the total concentration of the dissolved inorganic solids is called **salinity**. The average

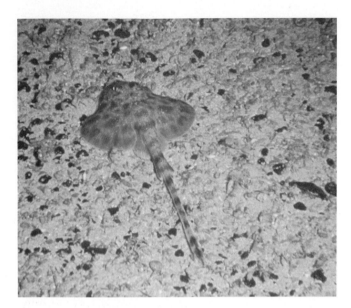

Figure 1-24   A skate rests on a region of the deep-sea floor scattered with manganese nodules.

the process of evaporation the salts are left behind. Much of the freshwater moving through this cycle originates in the oceans, because about 86% of global evaporation is from the oceans.

The global distribution of water is about 97% in the oceans and less than 1% as liquid freshwater (most of that being groundwater). The remaining 2% is in glaciers and polar ice; this is the storehouse of water that controls the large historic changes in absolute sea level, and the reason for concerns about melting of glaciers and the ice caps.

Dissolved Gases

Most gases in the atmosphere dissolve readily into ocean waters, and organisms in the oceans are highly dependent on these gases for biological processes. The oxygen required by almost all animals for respiration originates from near-surface waters either by dissolution from the atmosphere or from photosynthetic activities of plants (the oxygen bound up in water molecules cannot be obtained directly by animals for respiration). Deep-sea animals thus are dependent on vertical circulation of waters from the surface. Marine photosynthetic organisms are limited to water near the surface due to the necessity of adequate sunlight; however, they also need carbon dioxide from the air to support their metabolic processes.

The major gases in the ocean are the same as those in the atmosphere, although the proportions are somewhat different due to solubility differences. **Nitrogen**, about 78% of the volume of the atmosphere, makes up about 50% of the dissolved gases in the ocean. Nitrogen is needed by organisms to make proteins and other organic chemicals; however, most organisms cannot use the dissolved nitrogen directly. They depend on nitrogen that has been converted into organic forms. These processes are discussed below.

Oxygen makes up about 35% of the gas dissolved in the ocean; however, the concentration of oxygen in the ocean is extremely low compared to the atmosphere, which is about 21% oxygen. The average dissolved oxygen concentration in the ocean is 6 milligrams per liter of water (which can also be presented as 6 parts per million or ppm). This requires that most aquatic animals use gills to efficiently extract oxygen from the water. Even with efficient extraction and utilization of oxygen, most fish and large invertebrates cannot survive oxygen concentrations below about 2.5 ppm, a condition referred to as **hypoxia**. When waters reach near zero oxygen concentration they are called **anoxic**. Hypoxic conditions can develop when oxygen depletion occurs without adequate circulation. This can occur in coastal areas where there is excess nutrient input, as is discussed in Chapter 6.

ocean salinity is about 3.5% (salinity is typically given in units of parts per thousand (ppt or ‰) rather than percent (%); for example, 3.5% is 35 ppt). The range of salinities in the open ocean is from about 33 to 37 ppt. In coastal areas where freshwater influence is high the salinities can be much lower. Divisions used to define habitats according to salinity are somewhat arbitrary, but useful, because salinity is an important factor governing the kinds of organisms that can survive in those habitats. Moving from the ocean through estuaries into rivers, there is a gradation in salinities from full strength (greater than 30 ppt) through **brackish** (about 3–30 ppt) to **freshwater** (< 3 ppt). In isolated intertidal pools where evaporation is high, salinities can become higher than those of natural seawater, and are called **hypersaline**.

Sodium chloride contributes the most to the ocean's salinity; the concentration of sodium and chloride ions in seawater is about 3.0% with the remaining salts made up of sulfate, magnesium, and other ions listed above. Minor or trace elements are present in concentrations measured in parts per million or parts per billion. Although some of the salts dissolved in seawater originate from the flow of freshwaters from the continents and coastal erosion, most of the sea salts are from other sources. For example, sodium ions originate from the weathering of rocks of the Earth's crust, but chloride ions originate in the Earth's mantle and are put into the ocean by hydrothermal vents and volcanic activity at the ocean ridges. Other ions originate from various combinations of these sources. Part of the ocean water is continually recycled into freshwater systems through evaporation, condensation, and rainfall; however, during

Although the concentration of **carbon dioxide** naturally occurring in the atmosphere is very low (about 0.04%), due to its depletion by plants as a source of carbon, it is very soluble in seawater and thus comprises 15% of the dissolved gases in the ocean. Once CO_2 dissolves into the ocean it begins movement through a complex cycle (**Figure 1-25**). Carbon dioxide is continually extracted from water by plants and other photosynthetic organisms; however, it is still more concentrated in the ocean than in air. One reason for the high dissolution of CO_2 in seawater is its ability to combine with CO_2 to form carbonic acid (H_2CO_3; see Box 1-1, Conservation Concern: Global Warming and the Ocean). Because of these complex processes, CO_2 and pH levels vary widely among regions of the ocean, and within regions on a daily and seasonal basis. Factors affecting CO_2 and pH include photosynthesis and respiration, horizontal and vertical mixing, and exposure to the atmosphere. Anthropogenic increases in carbon dioxide in the atmosphere increase acidification of the oceans. Increased acidity could have a dramatic impact on ocean ecosystems, and may already be affecting some ecosystems (see Box 1-1, Conservation Concern: Global Warming and the Ocean and Chapter 7). Coral reefs may be especially sensitive because higher acidity can reduce the ability of organisms to build the reef (see Chapter 5).

Some of the dissolved carbon dioxide ends up in the shells and skeletons of marine animals and eventually into sediments. These processes begin with dissolved carbon dioxide forming carbonate ions that combine with calcium in sea water to form calcium carbonate ($CaCO_3$). This is used by marine organisms to build shells and skeletons. (The shells of many invertebrates and microorganisms are comprised primarily of calcium.) When these organisms die in the open ocean their shells gradually sink toward the bottom. On the sea floor above 4,500 meters depth calcium carbonate shells are deposited, mixing with other sediments to form thick deposits of **carbonate oozes**, recently discovered to house a rich assemblage of bacteria and other microorganisms (see Chapter 8). This calcium carbonate may eventually be incorporated into limestone rocks. Below 4,500 meters other materials, such as the remains of **silica** shells, take the place of the carbonates in forming deep sea **siliceous oozes**.

Chemicals of Life

Because living organisms are comprised of chemicals that are mostly derived directly from the ocean, understanding the biogeochemical cycle through which the elements move is critical to conservation. **Carbon** is considered the basic building block of marine (and other) organisms. The

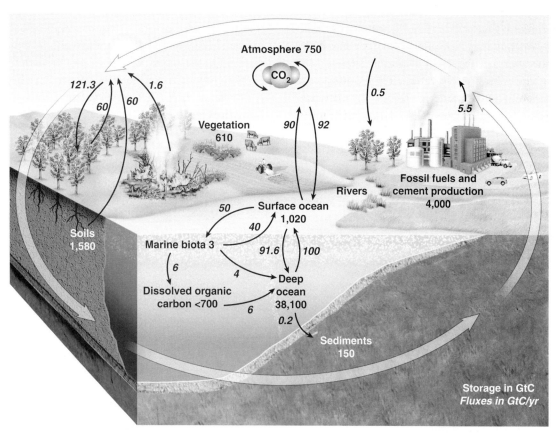

Figure 1-25 Global cycle of annual carbon fluxes and storage. (GtC = gigatons carbon.)

carbon dioxide from the atmosphere dissolves in water and is fixed into **organic molecules** using energy from sunlight during photosynthesis. This carbon is transferred, along with other chemicals, through the food web (see Figure 1-25 and Chapter 2). The carbon from photosynthetic organisms ends up either incorporated into the animal as new tissue (about 45%), expelled from the animal in carbon dioxide as a byproduct of respiration (45%), or excreted as waste into the water as **dissolved organic carbon** (**DOC**; 10%). DOC is used by bacteria and eventually ends up in other organisms. The organisms (or parts of organisms, such as shells) that escape being eaten, sink to the bottom and the carbon enters the geologic cycle as described above.

Many other elements combine to form organic chemicals in organisms; any compound that is used in the production of organic matter is considered a **nutrient**. Most nutrients are not typically considered as of concern in conservation issues because they are relatively abundant in the ocean compared to their need by organisms. The element that is most often considered as a **limiting nutrient** to ocean organisms is nitrogen (**Figure 1-26**). This is not due to a lack of nitrogen-based chemicals but because of limits in **available nitrogen**, that which is in a chemical form that organisms can utilize. Free nitrogen, dissolved in the water as a gas (N_2) in large concentrations, cannot be used by organisms directly. They depend on nitrogen that has been **fixed**, that is, converted from inorganic N_2 to organic forms (such as ammonia, NH_3), primarily by bacteria. Other major sources of available nitrogen are nitrates from rivers and precipitation. These sources are limited enough that ocean waters are frequently nitrogen-limited, especially in regions away from large river inputs.

The other major nutrient that can be limiting under certain conditions is phosphorus. Ocean currents also can have a major influence on the distribution of nutrients, especially in upwelling regions where nutrients that have accumulated in deeper waters are brought to the surface. The distribution of nutrients in surface waters can be mapped indirectly by measures of surface chlorophyll concentration made possible by satellite imagery (**Figure 1-27**).

While many ocean areas are considered nutrient-limited, others have unnatural and undesirable excesses. Marine ecosystems have evolved and adapted to the typical levels of nutrients available in the water, and additional nutrients may support one component of the ecosystem to the detriment of others. This has led to numerous conservation issues, especially in coastal regions. For example, nutrients may support bacterial or algae growth in excess; and excess bacterial activity can deplete waters of oxygen, creating **hypoxic** (low oxygen) zones (see Chapter 6). Excess nutrients in coral reef ecosystems produce blooms of algae that grow over and outcompete the corals (see Chapter 5). The source of excess nutrients in coastal and near-shore ecosystems is typically fertilizers, sewage waste, or excesses of other organic materials flowing into the ocean from rivers. Trace elements in the ocean are typically not limiting because they are used by organisms in such small amounts. Iron concentrations, however, can be so low, due to its insolubility and tendency to adhere to falling particles, that it can be a limiting nutrient to biological production in open ocean waters. Proposals to enhance plankton production in the open ocean as a mechanism for removing carbon dioxide from the atmosphere have been met with controversy (**Box 1-2. Conservation Controversy: Fertilizing the Ocean**).

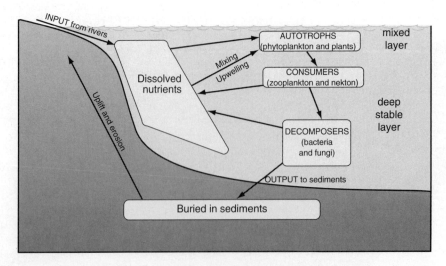

Figure 1-26 Cycle of the major ocean nutrients nitrogen and phosphorus.

>.01 .03 .1 .2 .5 1 2 5 10 20 30
Ocean: Chlorophyll *a*
Concentration (mg/m³)

Figure 1-27 Image taken by the Sea-viewing Wide Field-of-view Sensor (SeaWiFS) satellite. Gray ocean colors represent chlorophyll concentrations, which are indicative of primary production. Note high chlorophyll levels in the North Atlantic, in upwelling zones along the west coasts of South America and Africa and along the equator, and at the mouths of major rivers such as the Mississippi and Amazon. (See color plate 1-27.)

Box 1-2 Conservation Controversy: Fertilizing the Ocean

Iron is a **micronutrient** necessary for the survival and growth of phytoplankton in the open ocean. It is required in concentrations much lower than for macronutrients such as nitrogen and phosphorous (the ratio of iron to nitrogen in ocean phytoplankton is over 100,000 to 1), however. Despite the limited need for iron by phytoplankton, there are areas of the open ocean where it is the major limiting nutrient. This means that production will not increase without additional iron, even if excessive amounts of the other required nutrients are available. Although speculated since the early 1900s, this condition was not documented by scientists until the 1980s, primarily by oceanographer John Martin. This soon led to proposals to fertilize the ocean with iron in order to enhance biological production and reduce global warming. The reduction in global warming would be achieved through the uptake of atmospheric carbon dioxide by the phytoplankton. A large overall reduction in CO_2 is not achieved, however, if the plankton are eaten and incorporated into ocean food webs, because some of the CO_2 eventually will be released by organisms up the food chain during respiration. In order to achieve a net reduction in atmospheric carbon dioxide, the plankton

or their remains would have to die and sink to the bottom. This would result in sequestering of the carbon in deep ocean waters or sediments (in the same manner that sequestering of carbon occurs naturally when carbonate tests of plankton sink into the deep sea).

Ocean experiments began in the 1990s to examine the effect of the iron "fertilizing" on biological production (**Figure B1-3**). Since then experiments have been carried out of various sizes in different areas of the ocean. The results have varied. Some experiments showed a substantial production of phytoplankton but little sequestering of carbon because most of the plankton was eaten and little carbon sank to the sea floor. In other experiments, a relatively large amount of carbon was sequestered and transported to deep waters; however, it is uncertain if this represented a permanent loss of the carbon to the deep sea. None of these experiments were carried out on a scale and in a time frame that would be necessary to see a significant change in atmospheric carbon dioxide. In a natural experiment, George Wolff and colleagues found that areas with elevated concentrations of iron from natural leaching

Figure B1-3 SeaWiFS image of the northeastern Pacific Ocean during a 2002 iron fertilization experiment. Chlorophyll concentration increases from dark to light gray in the open ocean. The fertilized bloom is the small, light area in the bottom center. (See color plate B1-3.)

from volcanic islands supported a larger biomass and density of deep-sea life and a species assemblage with lower evenness than unfertilized areas, and they concluded that large scale fertilization would likely affect deep-sea ecosystems. Supporters of proposals to use iron fertilization of the ocean as a means to reduce carbon dioxide in the atmosphere argue that there is enough evidence to begin large-scale programs. Some industries would like to see this used as a way to mitigate their release of carbon dioxide into the atmosphere.

There is still widespread opposition among conservation biologists to proposals for large-scale application of iron to the oceans, including Ken Buesseler and colleagues, who argued that a stronger scientific foundation of the risks and benefits of such actions are needed. Some doubt the effectiveness of iron fertilization in reducing atmospheric carbon dioxide. Even if it is shown to be effective, the unpredictability of impacts on marine ecosystems may never make it worth the risk. Ocean ecosystems are complex and changing those ecosystems intentionally is considered irresponsible. We do not know what indirect effects might occur. For example, encouraging blooms of certain algae could result

in an increase in species that are detrimental to animals in the current ecosystem, such as toxic organisms or jellyfish, inedible to many fishes and marine mammals. Excess nutrients present the risk of initiating harmful algae blooms or creating hypoxic "dead zones" (see Chapter 4). Environmentalists argue that those pushing for ocean fertilization as a method to reduce carbon dioxide are looking for an excuse to allow carbon pollution to continue, rather than making the tougher decisions to radically reduce carbon emissions. This debate presents a somewhat unique type of conservation dilemma. Rather than suggesting we reduce actions that create climate change (e.g., burning fossil fuels), it is proposed that we increase actions that reduce the effects of climate change (e.g., ocean fertilizing). Following a conservation ethic, with a primary focus of maintaining the health of the natural world, one would argue in support of maintaining the natural balance and focusing on the reduction of carbon emissions. We have learned the hard way too many times that the best of human intentions, when they involve manipulating the environment, often return to haunt us. Other examples will be presented throughout the text.

STUDY GUIDE

Topics for Review

1. Distinguish and define the four major sub-disciplines of oceanography. Discuss how they overlap.
2. Describe the concepts of seafloor spreading and plate tectonics.
3. What is the difference in the subsidence at ocean trenches and subsidence in coastal marshes?
4. Compare the slope of the continental shelf, slope, rise, and deep-sea basin.
5. Compare characteristics and formation processes between active and passive margins.
6. What mineral resource from deep-sea regions is most likely to be exploited by humans?
7. How does the formation of island arcs differ from that of the Hawaiian island chains?
8. Why are guyots an important habitat in the deep sea?
9. What features differentiate beaches, deltas, and rocky coasts as habitats for coastal organisms?
10. What features make estuaries attractive to marine organisms as well as humans?
11. What are the major physical differences in the three basic reef types?
12. Why are barrier islands more vulnerable to erosion than volcanic islands?
13. What factors are responsible for higher levels of biological production in the neritic zone than the oceanic ocean?
14. What factors are responsible for higher levels of biological production in the epipelagic than the mesopelagic?
15. What distinguishes waves from currents relative to movements of energy and water?
16. Explain why wind waves of relatively large size are commonly seen coming ashore on calm, windless days.
17. Describe why the highest surfing waves are more commonly found along active coastlines (e.g., California) than passive coastlines (e.g., the eastern U.S.) even if wind conditions are similar.
18. How does a relatively shallow tsunami wave have enough energy to move ashore with such a great destructive force?
19. Describe the factors responsible for the variability in tidal range on a daily and monthly cycle at a single point along the coastline.
20. How are tides important to coastal organisms?
21. How does the Gulf Stream affect climate along the U.S. east coast?
23. How does coastal upwelling enhance fish populations off Peru?
24. How does thermohaline circulation enhance animal populations in the deep sea?
25. How does thermal expansion influence global sea levels?
26. What are the sources of minerals that make the ocean salty?
27. What chemical effect does excess carbon dioxide have on the oceans?
28. The concentration of dissolve nitrogen gas in the ocean is much greater than that of carbon dioxide. Explain why nitrogen is often a limiting nutrient to marine organisms, but carbon is not.
29. How do biological organisms contribute to sequestering of carbon in deep-sea sediments?

Conservation Exercises

Develop science-based arguments in support of each of the following statements:

1. Coral reef ecosystems are more sensitive than most marine ecosystems to environmental changes.
2. Barrier island beaches are inappropriate places for building homes.
3. Although tsunamis are unpredictable and unavoidable, preparations can reduce their impact on coastal settlements.
4. If the frequency of El Niño events increases it could have a substantial impact on coastal fish harvest.
5. We are currently experiencing a global warming trend, and humans are largely responsible.
6. Without a reduction in carbon dioxide emissions global warming will continue.
7. Because of the potential effects of the disruption of thermohaline circulation we must make sacrifices to stop global climate change.
8. Fertilization of the ocean with iron is not the best answer to resolving climate change problems.
9. Sea-level rise will continue with global warming even without large-scale melting of glaciers and sea ice.

FURTHER READING

Broecker, W. S. 1987. The biggest chill. *Natural History Magazine* 97:74–82.

Broecker, W. S. 1991. The great ocean conveyor. *Oceanography* 4:79–89.

Buesseler, K. O., S. C. Doney, D. M. Karl, P. W. Boyd, K. Caldeira, F. Chai, K. H. Coale, H. J. W. de Barr, P. G. Falkowski, K. S. Johnson, R. S. Lampitt, A. F. Michaels, S. W. A. Naqvi, V. Smetacek, S. Takeda, and A. J. Watson. 2008. Ocean iron fertilization-moving forward in a sea of uncertainty. *Science* 319:162.

Caldeira, K., and M. E. Wickett. 2009. Anthropogenic carbon and ocean pH. *Nature* 425:365.

Doran, P. T., and M. K. Zimmerman. 2009. Examining the scientific consensus on climate change. *Eos, Transactions American Geophysical Union* 90:22–23.

Hansen, J., M. Sato, R. Ruedy, L. Lo, D. W. Lea, and M. Medina-Elizade. 2006. Global temperature change. *Proceedings of the National Academy of Sciences* 103:14288–14293.

Jacobson, M. Z. 2005. Studying ocean acidification with conservative, stable, numerical schemes for non-equilibrium air-ocean exchange and ocean equilibrium chemistry. *Journal of Geophysical Research Atmosphere* 110:D07302.

Komar, P. D., and J. C. Allan. 2008. Increasing hurricane-generated wave heights along the U.S. East Coast and their climate controls. *Journal of Coastal Research* 24:479–488.

McPhaden, M. J., A. J. Busalacchi, R. Cheney, J.-R. Donguy, K. S. Gage, D. Halpern, M. Ji, P. Julian, G. Meyers, G. T. Mitchum, P. P. Niiler, J. Picaut, R. W. Reynolds, N. Smaith, and K. Takeuchi. 1998. The tropical ocean-global atmosphere observing system: a decade of progress. *Journal of Geophysical Research* 103:14,169–14,240.

Pinet. P. R. 2006. *Invitation to Oceanography.* Jones and Bartlett Publishers, Sudbury, Massachusetts.

Sallenger, A., W. Wright, J. Lillycrop, P. Howd, H. Stockdon, K. Guy, and K. Morgan. 2007. Extreme changes to barrier islands along the central Gulf of Mexico coast during Hurricane Katrina. Pages 113–118 in *Science and the storms: the USGS response to the hurricanes of 2005.* US Geological Service.

Schiermeier, Q. 2006. Climate change: a sea change. *Nature* 439: 256–260.

Siddall, M. T., F. Stocker, and P. U. Clark. 2009. Constraints on future sea-level rise from past sea-level reconstructions. *Nature Geoscience* 2:571–575.

Solomon, S., G. K. Plattner, R. Knutti, and P. Friedlingstein. 2009. Irreversible climate change due to carbon dioxide emissions. *Proceedings of the National Academy of Sciences* 106:1704–1709.

Solomon, S., J. S. Daniel, T. Stanford, D. M. Murphy, G.-K. Plattner, R. Knutti, and P. Friedlingstein. 2010. Persistence of climate changes due to a range of greenhouse gases. *Proceedings of the National Academy of Sciences* 107:18354–18359.

Sunda, W. G., and S. A. Huntsmann. 1995. Iron uptake and growth limitation in oceanic and coastal phytoplankton. *Marine Chemistry* 50:189–206.

Turley, C. 2008. Impacts of changing ocean chemistry in a high-CO_2 world. *Mineralogical Magazine* 72:359–362.

Wolff, G. A., D. S. M. Billet, B. J. Bett, J. Holtvoeth, T. FitzGeorge-Balfour, E. H. Fisher, I. Cross, R. Shannon, I. Salter, B. Boorman, N. J. King, A. Jamieson, and F. Chaillan. 2011. The effects of natural iron fertilisation on deep-sea ecology: The Crozet Plateau, Southern Indian Ocean. *PLoS ONE* 6(6):e20697 DOI: 10.1371/journal.pone.0020697.

Wooton, J. T., C. A. Pfister, and J. D. Forester. 2008. Dynamic patterns and ecological impacts of declining ocean pH in a high-resolution multi-year dataset. *Proceedings of the National Academy of Sciences* 105:18848–18853.

Marine Biology, Ecology, and Conservation

2.1 The Conservation Ethic and Biodiversity

A **conservation ethic** has developed gradually through the past century as humans have achieved advances in scientific understanding and the ability to rapidly communicate knowledge, while the number of humans on the planet and their ability to exploit living resources has expanded dramatically. Societies eventually came to realize that virtually all resources, even those in the most inaccessible areas of the ocean, are vulnerable to overexploitation or destruction. They could disappear entirely or be reduced such that humans could no longer benefit from them directly, or they would no longer support functioning ecosystems. Although in reality humans still often operate selfishly in their use of resources, many actions are now guided by a moral philosophy or ethic toward conservation, including protection, allocation, and sustainable resource use. As governments and other ruling entities have adopted a conservation ethic, mechanisms have been developed to establish and enforce rules that can achieve protection and sustainable use of environmental resources. Even if a conservation ethic has become a strong guiding force ruling our use of marine resources, divergent views continue to exist. For example, can conservation be adequately achieved if we attempt to maximum our use of marine resources, or should it be governed by total protection whenever possible? Alternatively, is there a happy medium that can be agreed upon? Should decisions be guided solely by science, and, if so, how do we balance the social, political, and economic concerns that guide human behavior? This text attempts to answer some of these questions; however, in many cases the goal is to provide enough information to develop educated decisions that can guide behavior.

Many conservation biologists now argue that the major focus of conservation efforts should be to maintain biodiversity; this is based on the premise that the healthiest an ecosystem can be is when it is at its highest diversity. This approach to a conservation ethic is a fairly recent development. In the 1800s before the current use of the term *conservation* was adopted, marine animals were mostly considered free for the taking by humans, with the underlying assumption that they would always be there (see Chapter 12 for further discussion of these attitudes). Through the early 1900s even some prominent biologists believed that humans were incapable of substantially affecting populations of marine species by harvest (see Chapter 11). As management evolved as a science and a realization developed that humans are capable of dramatically impacting species and ecosystems, efforts were made to protect and manage the marine species that were considered of some value. This focus, for at least the first half of the 20th century, was based largely on the protection of species that humans desire for food or other uses, a mindset often to the detriment of other species that were considered a nuisance or dangerous. For example, corals were removed to improve ship access to ports, salt marshes were filled to remove breeding habitats for mosquito pests, sharks were killed because of their perceived danger to swimmers, and dolphins were captured and removed because of perceived competition for harvested species. Because only a small percentage of even the larger animals and plants were exploited or used in some way directly by humans, this attitude resulted in declines in ocean biodiversity around the world, concentrated in accessible coastal regions. Eventually it was accepted that all species are dependent on others for survival, and we rarely know what effect the loss of any one species will have on the marine community. Through the last half of the 20th century, a new ethic developed that led to attempts to restrict use of the oceans by humans when it endangers the survival of any species, regardless of its perceived utility to humans (see Chapter 9). This led to efforts to enhance overall biodiversity, which necessitates the protection of habitat along with the organisms that live there, and has led to an ecosystem approach to marine conservation (see Chapters 3–9). To understand the workings of an ecosystem requires knowledge of all interacting components: the living organisms.

The marine environment supports a remarkable **biodiversity**, the assemblage of living organisms that have evolved, diversified, and adapted to live and propagate there over the past four billion years (**Figure 2-1**). The establishment of theories of **evolution** and **speciation** over the past 150 years has provided scientists with tools to develop an increasingly thorough understanding of the biological makeup and species relationships in ocean ecosystems. Research, however, has been limited by the relative inaccessibility of an environment that is foreign to humans. With the advent of modern technologies such as powerful

Figure 2-1 A section of a coral reef ecosystem in Belize showing some of its biodiversity.

motor-driven vessels, scuba, and deep-sea submersibles, there has been a new explosion in the understanding of marine biodiversity. There are still regular discoveries of new species and we can only estimate how many species still remain unknown and undescribed. The first step in describing the known biodiversity of the oceans is to define our unit of interest using standard methodologies for naming and classifying organisms.

Scientists use a standard set of rules for naming organisms first established over 250 years ago that has been modified and refined. This system of **biological nomenclature** allows scientists to quantify biological diversity, establish evolutionary (**phylogenetic**) relationships, and avoid confusion when referring to a species or groups of species (**Figure 2-2**). The standard classification scheme includes

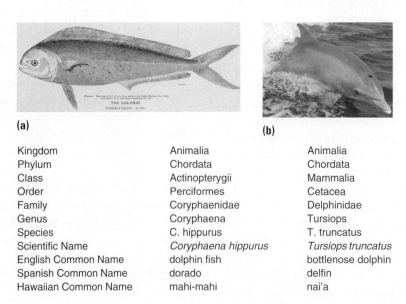

	(a)	(b)
Kingdom	Animalia	Animalia
Phylum	Chordata	Chordata
Class	Actinopterygii	Mammalia
Order	Perciformes	Cetacea
Family	Coryphaenidae	Delphinidae
Genus	Coryphaena	Tursiops
Species	C. hippurus	T. truncatus
Scientific Name	*Coryphaena hippurus*	*Tursiops truncatus*
English Common Name	dolphin fish	bottlenose dolphin
Spanish Common Name	dorado	delfin
Hawaiian Common Name	mahi-mahi	nai'a

Figure 2-2 A comparison of phylogenetic classifications and some common names of the dolphin fish and bottlenose dolphin.

the familiar **monophyletic groupings** (i.e., including all organisms descended from a common ancestral form): kingdom, phylum, class, order, family, genus, and species. Due to the large diversity in some groups, however, subgroupings have been added to this classification scheme (e.g., suborder, superclass, or subspecies). The **species** is the standard unit by which diversity is typically measured. The standard definition accepted by most biologists is: a group of reproductively isolated and closely related individuals capable of interbreeding and producing viable offspring. With recent advances in genetic techniques, a more practical method of defining species is based on the genetic similarity among individuals. As the power of techniques has improved, however, decisions must be made concerning the genetic difference needed to indicate separate species. All populations show some genetic differences.

Genus and species are the accepted Latinized names that are unique for a given species; no two species of organisms have identical genus-species names. This avoids confusion in common names that differ among different languages and different geographic regions. Even within the same language, there may be several common names used for the same species (**Box 2-1. Conservation Brief: What's in a Name?**). This text uses the accepted English common names for most species; however, the first time a species is mentioned in the text it will typically be accompanied

Box 2-1   Conservation Brief: What's in a Name?

In scientific reports and publications, the scientific name (genus-species) must be referenced to avoid confusion. If the report is for a public audience, a common name is often used but after first defining the scientific name. These scientific names are standardized internationally using specific rules. For example, the names are typically taken from Latin or Greek roots and written in a Latin grammatical form; the genus is capitalized, the species is not; and they are written within text in italics. Rules are written out in detail as **nomenclature codes** for animals, plants (including protists), and bacteria. Names are assigned by the scientist who first describes a newly discovered organism, but they must follow these rules and be submitted to an international commission. For example, the code for animals is the *International Code of Zoological Nomenclature* and is managed by the International Commission on Zoological Nomenclature. The species is also placed within the accepted phylogenetic classification scheme (family, order, class, etc.). Once a name is agreed upon it is used in all scientific references to the species. This scientific name is unique for a species—no two biological organisms on earth are given identical names. The value of this system is that the same scientific name is used in all languages and in all parts of the world. For example, *Thunnus thynnus* (Atlantic bluefin tuna in English) is the same fish whether it is caught in the waters of Spain or Florida and whether it is eaten in Japan or Africa.

It would be most practical if all reference to marine organisms used scientific names. Common usage is not regulated by law, however, so names develop to be descriptive in the native language of the region. Names often are chosen to be appealing for marketing purposes. Marine conservationists and fisheries managers, thus, are forced to deal with the confusion rendered by the use of common names. One potential area of this confusion is the monitoring of fisheries harvest, especially at the international level. Around the world a harvested species typically is called different names in different languages. Even within a language a species can go by more than one common name. For example, *Sciaenops ocellatus*, a harvested fish along the U.S. Atlantic and Gulf of Mexico coasts, is variously called drum, red drum, redfish, channel bass, or spottail bass. On the other hand, the same common name can be used for many species; for example, a redfish might refer to

Sciaenops ocellatus, a drum in the southeast United States, *Sebastes marinus*, a deepwater scorpionfish in the northeast United States, *S. mentella* or *S. norvegicus* in Canada, three other *Sebastes* species in Norway, *S. fasciatus* in the United Kingdom, and any of three species of *Lutjanus* snappers or a deepwater squirrelfish *Centroberyx affinis* in Australia. Other names that commonly lead to confusion are sea bass, ocean perch, and rockfish (**Figure B2-1**). To avoid confusion, in most nations names are designated for reporting a harvested species to management agencies. There may still be confusion if the name is not familiar to fishers and when there are similar species that are difficult to distinguish. Internationally, organizations

(a)

(b)

Figure B2-1 **(a)** *Sebastes entomelas* and **(b)** *Sebastes ruberrimus*, two of four fish species marketed in the United States as "rockfish."

that monitor harvest match local common names to scientific names to standardize the data.

Naming confusion can be most problematic when dealing with trade and marketing of seafood. Seafood companies and exporting nations desire to sell their products under a name that makes it appealing to the consumer. After all, which would you rather eat: orange roughy or slimehead, dogfish or rock cod, toothfish or sea bass? (These pairs are common names used for the same fish species). The assignment of names for marketing purposes is considered acceptable in national marketing and in international trade. Most nations have laws against using names that would lead to confusion or misrepresentation. When many fish are marketed by the same generic name, however, consumers may be confused. And seafood companies may be defensive of their rights to a market they have developed under a specific name. For example, dozens of species are commonly called "sardines." Because of the perceived value of the name, international organizations designate what species can be traded under the name "sardine." Still, there have been several international trade disputes, arbitrated by the World Trade Organization (WTO), over naming of seafood products. One dispute resulted in an agreement that the label "sardine" would be reserved for "*Sardina pilchardus*" and that other species of "sardines" (which could include herrings and anchovies) would have to be labeled not simply as "sardines" but as some specific type of sardine.

Assigning a name for marketing purposes can avoid confusion when more than one species goes by a similar common name. For example, the English name "dolphin" can refer to either a fish (*Coryphaena hippurus*) or a marine mammal (*Tursiops truncatus*) (Figure 2-2). The marketing of the dolphin fish as *mahi-mahi* (from the Hawaiian name for this species) enhances its value and avoids the mistaken perception of consuming a marine mammal.

Assigning generic names to several different species for marketing purposes may be a source of confusion to the consumer who may wish to avoid eating certain species because they are overfished or for health reasons. For example, names like "ocean perch" or "sea bass" could apply to many different fish species; and a "shrimp" or "salmon" might be farm-raised using environmentally harmful practices or caught from the wild in a well-managed fishery. Fortunately, there has been increased pressure by consumers for more specific labeling of seafood products, and it is more common for packaged products to at least indicate the location of harvest. Nongovernmental organizations (NGOs) have worked toward developing ecolabeling systems for seafood products; if this becomes more commonplace then some of the naming confusion will be resolved (see Chapter 12).

One additional problem with the use of common names for marketing is the potential for the use of misleading names or selling a product under a false name, despite laws against such practices. When a seafood product gains popularity and value, there is a temptation to market other cheaper and more readily available products by that name. Sometimes there is a thin line between what is considered misleading and what is not. An example is imitation crab meat produced from fish protein and made into a paste called **surimi** (from the Japanese word for minced fish) to which artificial flavor and coloring are added. In the United States it must be clearly labeled as imitation crab or "krab"; in England it is commonly sold as fish sticks or crab sticks.

It is often more difficult to regulate what is sold in restaurants, because seafood names can be changed at any time from the point of harvest through marketing channels up until it reaches the restaurant menu. For example, in the 1980s blackened redfish (in this case *Sciaenops ocellatus*) became a trendy menu item after its inception in famous New Orleans restaurants. Soon after, the largest fishery for these redfish was closed; however, blackened redfish stayed on menus for years following this closure. It was eventually discovered that many fish species were being cooked and sold as "redfish," one of these being a cheaper, less appealing fish, the black drum (*Pogonias cromis*). Even if such name-switching practices are illegal, they are virtually impossible to monitor closely due to the large number of businesses involved and the fact that it is difficult to recognize a species simply by looking at a fillet. With recent advances in DNA technology, however, the identification of animals using a small piece of protein is virtually 100% certain and is becoming more routine. In a study carried out by Dr. Mahmood Shivji, samples were collected from restaurants all over the United States that had grouper or snapper on the menu. His results showed that over 50% of the time the fish were actually some other species, often cheaper imported tilapia or catfish. Although we are a long way from having the Food and Drug Administration (FDA) or local health inspectors walk into a restaurant and analyze the DNA of every item on the menu, there are efforts to increase enforcement. To provide guidance and help enforcement, the U.S. FDA produces a Regulatory Fish Encyclopedia online (http://www.fda.gov/Food/FoodSafety/Product-SpecificInformation/Seafood/RegulatoryFishEncyclopediaRFE) that lists acceptable market names, common names, and scientific classification for over 1,700 species. It also includes photographs of whole fish and fillets, and electrophoresis gel banding and DNA sequence patterns are being added.

Mislabeling products also can be used to circumvent international trade restrictions. For example, the fish *Dissostichus eleginoides* has become a popular restaurant item internationally, which has led to increased regulations and wide-scale illegal pirate fishing (see Chapter 11). You may know this fish as Patagonia toothfish or Chilean sea bass, as it is called in North America, but it also goes by the name *merluza negra* in South America, *mero* in Japan, and *Patagonsky klykach* in Russia. It has become somewhat of a game for unscrupulous international traders to sneak these fish into the market under various names in order that trade restrictions might be avoided. Increased use of genetic "fingerprinting" of trade products will enhance enforcement of trade restrictions.

by the scientific name. In some cases families, orders, or other levels of classification will be given when the species is first mentioned to help place the species in a recognizable category.

If species is the unit of interest in defining diversity, we still must clarify what the term *species diversity* means. Is it simply a count of the number of species or something more complex? **Biological diversity** in its original meaning was considered as a measurement that gives some indication of both the number of species (**richness**) in an area and the relative number of individuals of each species (**evenness**). Various indices have been defined to mathematically quantify diversity; however, for each of these a maximum diversity is reached with the largest number of species and with an equal numbers of individuals in each species. This is most useful when considering similar species for which we would expect an even distribution (for example, diversity of grazing fish species on a coral reef or diversity of protists in the plankton). A simpler and more easily obtained unit of diversity, however, is a count of the total number of species in a given area, used to compare the relative diversity among organism in different ecosystems; this is what the term **biodiversity** is most commonly used to specify. The total number of species on Earth, including the oceans, lands, and all other habitats is believed to be between four million and 100 million. The reason for the broad range in estimates is that our knowledge of biodiversity, especially in the oceans and other poorly explored ecosystems, is far from complete. This is especially true of microorganisms and invertebrates; however, the discovery of a new species of fish, especially in the deep sea, is not unusual. Currently, approximately 1.8 million species have been described and named. Over 160,000 (or about 10%) of the named species are marine, but as many as ten million marine species have not been described.

2.2 Classification and Biology of Marine Organisms

The original groupings of organisms into "animals" and "plants" developed in the 1700s is problematic because it excludes many organisms, especially microscopic organisms, whose characteristics either do not put them in either category or place them into both. By the late 1800s most classification schemes included another group, the **protists**, which incorporates **prokaryotic** organisms (those that lack a nucleus and other organelles). Throughout the 1900s various modifications of the traditional three-kingdom view of biological classification were made. One commonly accepted scheme is a six-kingdom grouping: Archaea (or Archaebacteria), Bacteria (or Eubacteria), Protista, Fungi, Plantae, and Animalia. Each of these

kingdoms is separated into smaller taxonomic groupings including thousands to millions of species. The discussion below will use these groupings, but also describes ecological groupings that recognize similarities in function within marine ecosystems. The goal is not to provide a comprehensive discussion of the biology of life, but to focus on organisms that are ecologically important and that have particular relevance to the conservation issues discussed throughout the text.

The following sections of this chapter provide a general overview of the diversity of marine organisms in some of the major taxonomic groupings and detail some of the ways the organisms are considered integral to the function of healthy marine ecosystems. The discussion focuses on major groups and those that are considered to be important players in conservation issues discussed throughout the remainder of the text.

■ Marine Microorganisms

Although not considered a single phylogenetic group, the term *microorganisms* provides a practical, functional, and ecological category for the discussion of marine conservation issues. Microorganisms include living organisms that are microscopic, that is, not visible individually to the naked eye. Most marine microorganisms are unicellular prokaryotes: bacteria, archaea, or protists. Many of these are found floating in the water column, with most of their movement dependent on movement of the water, a functional grouping called **plankton**. **Phytoplankton** are plankton that are capable of photosynthesis and are thus limited to either shallow areas or the upper portion of the water column.

Every drop of water near the surface of the ocean is teaming with microorganisms. Their importance in the flow of energy was pointed out by Lawrence Pomeroy and others in the 1970s to establish a new paradigm in understanding ocean food webs. Because they cannot be captured with nets used to collect larger plankton, their true abundance and importance has only recently been realized. Microorganisms are also extremely abundant in the sediments at the bottom of all areas of the ocean. Much is not known about the diversity of microorganisms in the ocean, especially the smallest bacteria and protists and those residing deep in the sediments or in the deep sea. Even in well-studied regions of the ocean, genetic analysis has shown that organisms that were considered single species based on morphological characters may actually comprise as many as ten genetically distinguishable species. The following discussion is limited to the major groups of microorganisms and those that play important roles in conservation issues discussed in later chapters.

Bacteria

Bacteria are typically only a few micrometers in length or less, often small enough that individual cells cannot readily be seen even with the use of the most powerful light microscopes (due to the physics of light, the maximum resolution of a traditional light microscope is about 0.2 microns or 0.0002 millimeters). Bacteria make up much of earth's living biomass; one liter of seawater may contain about one billion bacteria. Bacteria are important in the recycling of nutrients, converting them into forms available to other organisms through **fixation** of nitrogen from the atmosphere.

Although many bacteria form symbiotic relationships with other organisms, they also play an important role as free-living organisms in all marine ecosystems. Their critical role in marine food webs has only recently been appreciated, as pointed out in studies by Farooq Azam and colleagues. Bacteria decompose the wastes or remains of dead marine organisms, from microscopic plankton to the largest whales. This decomposition releases nutrients that are cycled through other marine food webs. Bacteria themselves may be an important source of food for other organisms. Other microorganisms may feed on them individually, but some larger invertebrates and fishes use bacteria, along with the decaying organic matter, as a source of nutrition. Bacteria cycle organic matter into marine food webs through the **microbial loop**, consuming **dissolved organic matter** (**DOM**) that cannot normally be assimilated by other organisms (**Figure 2-3**). DOM originates from leakage from cells of algae, excretion of wastes by protists and animals, and the breakdown of organic particles. This organic matter assimilated by the bacteria is then introduced into marine food webs when consumed by protists, which are consumed by small planktonic crustaceans (see discussion of food webs below).

Some free-living bacteria contain photosynthetic proteins and are important primary producers (**autotrophs**) in ocean waters. These are often referred to as Cyanobacteria. Their global importance in this role was only recognized in the later part of the 20th century when their great abundances were discovered. It is now accepted that they account for much of the primary productivity in the ocean. In fact, bacteria, along with other photosynthetic plankton, account for an estimated 40% of global primary productivity.

Archaea

The archaea are similar in size and appearance to the bacteria. Only since the late 1900s have they been recognized as a distinct phylogenetic group, based on differences from bacteria in their genetics and biochemistry. Archaea were first discovered in extreme environments, such as around hydrothermal vents, but more recently they have been found in other of marine environments, in some cases being the most prevalent organisms. For example, species of archaea are found in extreme cold of the polar seas, extreme salinity environments, in the surface plankton, in deep-sea waters, and in marsh and

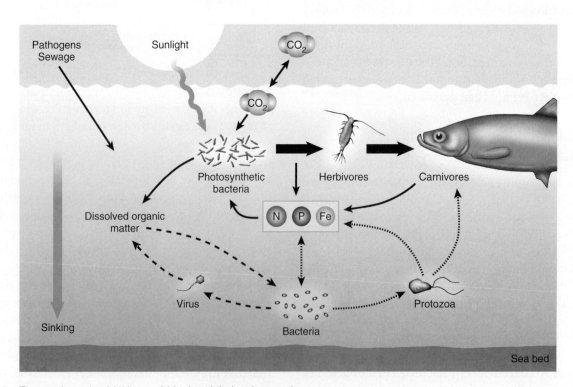

Figure 2-3 The marine microbial loop within the global carbon cycle.

deep-sea sediments. In some regions they comprise as much as 40% of the biomass of microorganisms in the plankton, and they are the dominant organisms in sediments greater than one meter beneath the seafloor. Archaea are important as decomposers in marsh and marine sediments. They also function as important components of nitrogen, carbon, and sulfur cycles in some marine ecosystems, putting these elements into forms available to other organisms. Archaea are more abundant in open-ocean waters than previously assumed and may function in roles similar to marine bacteria, as primary producers and in cycling of organic matter within the microbial loop. Archaea are difficult to study due to their small size (most are less than one micron long) and extreme habitats; therefore, much is still unknown about their importance in marine ecosystem, although they are estimated to comprise as much as 20% of the earth's total biomass.

Protists

Other than the bacteria and archaea, most of the remaining microorganisms in the ocean are protists (**Figure 2-4**). Although classification of these groups is continually being modified and debated, and many taxonomists do not consider protists a valid phylogenetic grouping, the protist

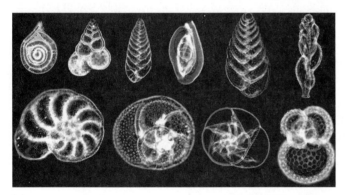

(a)

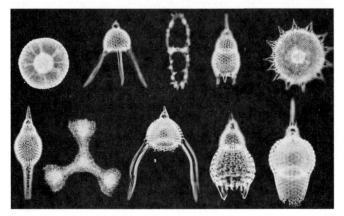

(b)

Figure 2-4   Examples of marine protists: **(a)** foraminiferans and **(b)** radiolarians.

grouping typically includes the unicellular **eukaryotes**. (Eukaryotes are organisms, including animals and plants, with cells containing a nucleus and other membrane-bound organelles.) Most protists exist as independent cells. This broadly defined group is comprised of many organisms; there are about 40 different phyla, each containing thousands of species. Much is not known about the biology and ecology of protists, although they serve many important functions in marine ecosystems. One of the most critical is in the plankton, functioning as primary producers and recyclers of nutrients, or consumers of other microorganisms. They also serve an important role in removal or **sequestering** of carbon dioxide from the atmosphere and production of marine sediments. They thus are considered to be critical in controlling the accumulation of greenhouse gases considered responsible for global warming, and there have been controversial proposals to fertilize the ocean to enhance their populations (see Chapter 7).

Protists, of various types, comprise a large portion of the **phytoplankton**, organisms floating in the water column that are capable of photosynthesis. Structural differences can be used to divide the protists into major groupings. The naked forms (those without a shell covering) include the amoebas. The shelled forms are covered by a shell or **test**, often of an ornate form, and include the **foraminiferans** (sometimes called **forams**) (Figure 2-4a), coccolithophores (coccoliths), and **radiolarians** (Figure 2-4b). Forams and coccoliths have shells composed primarily of calcium carbonate. Radiolarians have shells made of silica (which is also the primary component of sand and manufactured glass; thus, radiolarians literally live in tiny glass houses). After these organisms die, their shells either dissolve in the sea water or slowly sink to the deep-sea floor as a major component of the benthic **carbonate** and **siliceous** oozes, discussed in Chapter 1. Because most sediment particles from land sink to the bottom before reaching the deep sea, over 60% of the deep-sea bottom is covered by oozes made of the remains of these organisms.

Diatoms are single-celled protists that have intricate, ornate shells composed of silica. They are the most diverse of the protist groups with approximately 200,000 described species (though some estimate the true number could be as high as two million). Many species are recognizable by the shape and structure of their tests, called **frustules**. They are a major component of the phytoplankton throughout all oceans. There are also many **benthic** (bottom-associated) species, which often form films on the surface of sediments in shallow or intertidal waters. Diatoms are the major primary producer at the base of many marine food webs and may provide as much as 25% of oxygen in the atmosphere. They undergo dramatic population growth during **plankton blooms** that develop during spring months in temperate coastal waters. These plankton blooms support coastal plankton-feeding fishes that supply the largest fisheries in the world (see Chapter 11).

Dinoflagellates are the second-most diverse group of protists, with over 1,500 described free-living species, most of which are components of the microscopic plankton. Dinoflagellates range in size from five microns to about two millimeters, a 400-fold range. They often have long spiny projections as a defense against predation. Their name derives from Greek and Latin roots meaning "whirling whip," for the whip-like flagella many species possess for propelling them through the water in a whirling motion. They are most prevalent in warmer tropical waters and are an important component of many marine food webs. Despite their microscopic sizes, concentrations of dinoflagellates may be visible to the naked eye. For example, some species are luminescent, emitting light when disturbed. This may result in a greenish glow visible at night in breaking waves or the turbulence from boat propellers. This **bioluminescence** is most dramatic in some tropical bays, some of which are popular destinations where people come to swim at night in the glowing waters. The bioluminescent dinoflagellates are harmless; however, other types produce chemicals that are toxic when consumed by animals, including humans. Heavy concentrations (more than one million cells per milliliter) form **toxic algae blooms**, often called **red tides** based on the appearance of their red **photosynthetic pigments**. These concentrations of billions of dinoflagellates kill some marine organisms with which they come in contact, and their toxins can be passed on to people who eat bivalves or fish in which they have accumulated. Although red tides and other toxic algae blooms may be natural occurrences, the influence human pollutants or climate change have on their frequency is debated (see Chapter 7).

Other species of dinoflagellates are **endosymbionts**, living in the tissues of reef-building corals and some sponges in a prolonged relationship; the dinoflagellates provide energy through photosynthesis while deriving nutrients from the coral or sponge. The dinoflagellates are what give tropical reef-building corals their color. Under stress they are sometimes expelled by the coral resulting in a discoloration referred to as **coral bleaching** that may result in the coral's death (see Chapter 5).

Marine Fungi

Fungi are eukaryotic organisms comprising seven different phyla and about 100,000 described species; they include microscopic (e.g., yeast and molds) and macroscopic (e.g., mushrooms) species. One major structural difference between fungi and plants is that the cellulose in plant cell walls is replaced by chitin in fungi. Fungi are not photosynthetic, but are **heterotrophs**, taking in organic carbon for energy, obtained from the water or substrates on which they grow. Most multicellular fungi grow by forming cells called **hyphae**, which are long, tubular, and filamentous. Reproduction varies among species; many species can produce by either asexual or sexual spores. Ecologically, fungi play an important role in nutrient recycling. There are over 400 known species of marine fungi, some found only in ocean ecosystems and others that also live in freshwater or terrestrial habitats.

Marine fungi are most commonly found on rotting wood or as parasites of seaweeds or animals. At least one species has an **endosymbiotic** relationship with a brown alga, *Ascophyllum nodosum*, which is a common harvested seaweed found in the intertidal zone along North Atlantic shorelines (see Chapter 3). Fungal parasites have been identified in marine fishes, but are relatively uncommon. Fungal infections, however, can result in disease and death in herrings *Clupea harengus*, affecting their population size in the western Atlantic. Other fungi infect marine invertebrates, including the egg and larval stages. Fungi can cause disease in oysters. For example, in coastal waters of Western Europe and India the fungus *Ostracoblabe implexa* burrows through the shell of the oyster, obtaining nutrients from the organic material in the shell; in response, the oyster deposits extra shell material resulting in a deformation of the shell. In corals, fungi can cause diseases but also can be symbiotic. Fungal diseases can cause deaths in large numbers of cultured animals, including bivalves, shrimp, and fishes, where crowded conditions allow easy transmission. The use of fungicides to control these diseases can result in pollution if waters are released into the marine environment. Fungi in marine sediments serve a role as decomposers of decaying organisms. Marine yeasts have been isolated from seawater and some are parasitic on animals.

Marine Plants

The organisms commonly characterized as plants do not necessarily fit within a single phylogenetic category. As a functional category, however, plants can be defined as: multicellular eukaryotic organisms that contain **chloroplasts** to carry out photosynthesis, have cellulose **cell walls**, and lack the power of locomotion. This definition includes the **seaweeds** (multicellular marine algae) and the **vascular plants** (those with specialized vascular tissues for transporting water, minerals, and photosynthetic products), such as grasses and woody plants. The primary vascular plants associated with marine ecosystems are seagrasses, marsh grasses, and mangroves.

Seaweeds

The multicellular algae commonly called seaweeds (**Figure 2-5**) are the most obvious plants in many nearshore areas

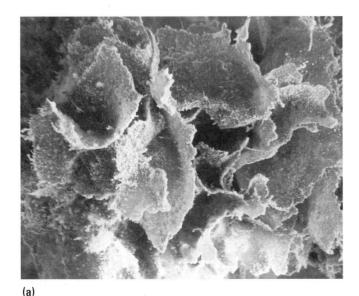

(a)

(b)

Figure 2-5 Examples of seaweed: **(a)** green algae and **(b)** red algae.

of the ocean. Because all seaweeds depend on photosynthesis for energy, they are only found in areas exposed to sunlight, either in the littoral zone or other shallow marine waters, or floating at the surface. Superficially, seaweeds resemble vascular plants. Most have a stem-like structure called the **stipe**, a **holdfast** to attach to a surface, and leaf-like blades called **fronds**. Many also have expanded leaf-like structures called **lamina** that serve as flotation structures. One distinguishing characteristic among seaweed groups is color. Although **chlorophyll** is present as a **photosynthetic pigment** in seaweeds, other **accessory pigments** impart a reddish or brownish color. These colors are often distinctive within a phylogenetic group of seaweeds; thus, they are grouped as red algae (Rhodophyta), brown algae (Phaeophyceae), or green algae (Chlorophyta), based on their

primary accessory pigments. These groups are often difficult to recognize based simply on their color. A relatively small number of seaweed species are free-floating in the ocean; some species (such as the brown algae *Sargassum*) are found at the surface over some of the deepest ocean waters (see Chapter 7).

Seaweeds have important ecological functions at the base of food chains and are used as structure for attachment or concealment by other organisms. They are the primary living structure within some ecosystems (e.g., kelps; see Chapter 6). Seaweeds are important to humans as food products, much more so than many people realize. The use of seaweed-based food products historically has been concentrated in East Asian countries, including Japan, Korea, and China; however, with the popularity of Asian cuisine, the use of seaweed has expanded into other regions. Edible seaweed is most familiar as the *nori* wrapping on sushi, a dried form of red algae *Porphyra*, but many other foods contain seaweed products. **Hydrocolloids**, gelatinous substances extracted from seaweeds, are contained in hundreds of food items, including meat and poultry products, desserts, salad dressings, cheeses, yogurt, mayonnaise, and peanut butter. Hydrocolloids also have industrial applications in paper, adhesives, dyes, and explosives. Other products containing seaweed include toothpastes, cosmetics, paints, and livestock feed. Seaweeds are currently being considered for large-scale use as fertilizers and in the production of biofuels. Seaweed harvest is from a variety of sources. Some seaweeds, including kelp and *Sargassum*, are harvested from natural marine ecosystems (see Chapters 6 and 7). Others, including *Porphyra*, are cultured in coastal waters, primarily in the western Pacific.

Marine Vascular Plants

Vascular plants in the ocean (**Figure 2-6**) are limited to soft-bottom coastal and shallow regions due to their requirements for sunlight exposure and rooting in bottom sediments. They can be a critical component in the transitional region between terrestrial and freshwater habitats and the ocean. Mangroves and marsh grasses dominate marshes in coastal regions and include species tolerant to fresh, intermediate, and full-strength seawater; although parts of the plants can remain submerged, the leaves are exposed to air the majority of the time. Most seagrasses are true marine species, tolerating continued exposure to the ocean in shallow-water habitats.

Seagrasses

Seagrasses (Figure 2-6a) are flowering vascular plants that are capable of living in salt water. They are confined to shallow soft-bottomed coastal waters because of their need to be exposed to sunlight and rooted in the sediments.

(a)

(b)

Figure 2-6 Marine vascular plants: **(a)** seagrass and **(b)** salt-marsh grass.

Although seagrasses are sometimes partially exposed during low tides, most seagrass ecosystems are in subtidal waters. Seagrasses support a productive ecosystem as they provide food and habitat for a diverse assemblage of organisms including microorganisms, invertebrates, fishes, sea turtles, and marine mammals. Human impacts on seagrass ecosystems are severe in some regions, resulting in serious conservation concerns. The ecology and conservation of seagrass ecosystems are discussed in Chapter 6.

Marsh Grasses

Salt-marsh grasses (Figure 2-6b) are also true grasses but are limited to intertidal areas of coastal salt marshes. These grasses are distinguished from most others by their ability to tolerate exposure to salt water (termed **halophytic**). They are not physiologically restricted to salt waters but their distribution is typically limited to salt marshes and **brackish** marshes (those that have intermediate salinities). Although salt-marsh grasses tolerate immersion for a limited time, their stems and leaves are typically exposed to the air. The number of grass species that are considered to be true salt-marsh species are few. In fact most are in one genus—*Spartina*, commonly called cordgrass—with species distributed around the world. Although there are 14 species in this genus, each geographic region is typically limited to one or two dominant species. Because very few plants are able to tolerate the extreme variability in salinity and exposure, there is little competition by other plants in salt-marsh habitats. The high productivity of marsh habitats supports a large plant biomass, however, and thus stands of *Spartina* may dominate many square kilometers of marsh lands. Some the largest areas covered by salt-marsh grasses are found along the U.S. Atlantic and Gulf of Mexico coasts. The muddy sediments of the marshes provide a stable substrate to support the roots, protecting these habitats from storms and tidal current. The grasses also serve an important function by trapping sediments flowing through coastal estuaries. Marsh grasses are most prevalent in temperate areas, being replaced by mangroves in the tropics. Marsh grasses are an important component of salt-marsh ecosystems; their ecology and conservation are discussed in Chapter 3.

Mangroves

Mangroves are woody vascular plants that form shrubs and trees in tropical coastal areas. The term *mangrove* can be applied to any woody plant that is adapted to tolerate salt water conditions; however, botanists often consider only four to five families of shrubs and trees as true mangroves. Because mangroves cannot physiologically tolerate cold winters, they are limited to the tropics, being replaced by salt-marsh grasses in temperate regions. Mangroves serve to protect tropical coastal regions from storm and wave activity. Because of their bushy structure and complex root systems, they weather coastal storms and even hurricanes. They provide a sediment filter for fresh waters flowing into the marine environment. Mangroves are critical habitats for many fishes and invertebrates, especially the young stages that need protection from marine predators. Although few fishes or large invertebrates feed directly on mangroves, they are the base of a productive food web in tropical coastal regions. Impacts of mangrove removal by humans for coastal development and aquaculture activities has received much attention in recent years; these and other conservation issues are detailed in Chapter 3.

■ Marine Invertebrates

Although it is difficult to provide a simple description that includes all of the nearly 1.4 million described animal species, the kingdom Animalia might be defined as a monophyletic group of motile heterotrophic multicellular eukaryotes. Animals are often separated into two groups:

vertebrates and invertebrates. Invertebrates are not a mono-phyletic grouping, and are comprised of 35 separate phyla. Here, the term **invertebrate** follows its common usage to refer to those animals that are not in the subphylum Vertebrata and therefore do not have a vertebral column. Invertebrates comprise about 95% of the named animal species; however, estimates of the total number, including undescribed species, ranges from several million to tens of millions. Most species are in the eight most common phyla: Porifera, Cnidaria, Platyhelminthes, Nematoda, Annelida, Mollusca, Arthropoda, and Echinodermata. Each of these phyla includes thousands to tens-of-thousands of marine species. This section will review the marine diversity in these major phyla, focusing on the groups that are most common and that are most affected by human activities.

Porifera: Sponges

Marine sponges are a group of approximately 9,000 species with diverse body shapes and colors (**Figure 2-7**). They are the simplest of all animals, having no specific organs or tissues but consisting of a large aggregation of specialized cells. Cells within canals in the sponge's body contain **flagella**, the movement of which creates water currents through the sponge. Water enters through surface pores near the base of the sponge and flows through the canals. Food organisms and organic particles are filtered and processed by special-ized cells of the sponge. The water exits the sponge through an opening at the top, the **osculum**. All sponges are **sessile**, that is, attached to the bottom or some structure. Techni-cally speaking, sponges have a **skeleton**, a rigid framework that supports the body. It may be composed of a gelatinous material sandwiched between cell layers, a fibrous protein (**spongin**), and/or hard, transparent **spicules** made of silica or calcium carbonate.

Figure 2-7 Marine sponges.

Sponges are found throughout the oceans in bottom habitats from the poles to the tropics and from the littoral zone to abyssal depths; however, their greatest diversity is in shallow tropical waters including coral reef and mangrove ecosystems. They typically attach to rocks or other hard structures (including coral skeleton or mangrove roots); however, some species attach to soft bottoms with their root-like base. Sponges are often sensitive to physical dis-turbances because excess sediments may block their pores, making respiration and feeding difficult. Spicules serve as a defense against predation, and some sponges shed them to ward off benthic predators. Others produce defensive chemicals to resist predators or keep other animals from attaching. Despite these defense mechanisms, coral reef sponges are an important source of food for some tropi-cal vertebrates. For example, sponges are a primary source of food for hawksbill sea turtles and some butterflyfishes (family Chaetodontidae).

Sponges are a critical component of many marine ecosystems, most notably coral reefs, and thus they are an important consideration in conservation issues. Sponges are not considered edible to humans and are therefore not harvested as food products. They are com-monly used in **bioprospecting**, however, the search for chemical compounds useful to humans, for example as pharmaceuticals. The most familiar use of sponges by humans is as bath sponges; although most bath sponges currently used are synthetic, there is still some harvest of natural bath sponges. Ecology and conservation issues related to sponges are discussed in other chapters, pri-marily Chapter 5.

Cnidaria: Jellyfish, Anemones, Corals

The phylum Cnidaria contains over 9,000 species of ani-mals found mostly in marine environments. It includes cor-als and anemones (Anthozoa) and various types of jellyfish and related animals (Medusozoa) (**Figure 2-8**). Although the shapes and appearances of species in this phylum are di-verse, there are several distinctive common features. Their bodies consist primarily of a gelatinous substance sand-wiched between layers of **epithelium** cells. They have two body forms. The **polyp** is a sessile stage with a mouth sur-rounded by tentacles; this stage is most evident in the anem-ones and corals. The **medusa** is the bell-like stage, floating and oriented upside down compared to the polyp; this stage is most evident in the jellyfishes. Both stages have a ring of tentacles encircling the mouth, the only opening to the body, used in feeding and respiration. These tentacles are swept through the water to capture food organisms. They contain stinging structures called **nematocysts**, which contain small barbed structures that inject toxins when fired into target organisms. The firing of the nematocyst is stimulated by a

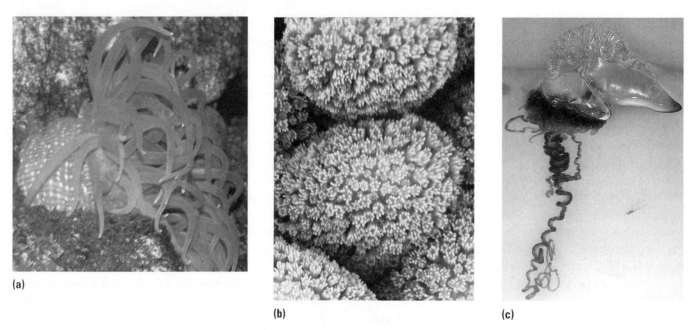

Figure 2-8 Cnidarians: **(a)** anemone, **(b)** coral polyps, and **(c)** Portuguese man-of-war (siphonophore colony).

combination of chemical and physical stimuli produced by potential prey or predators.

Mobility is limited in the sessile forms of cnidarians. At some point in their life history, however, most exhibit a medusa stage. This stage has limited swimming abilities using jets of water squeezed out of the cavity of the **bell**, the body of the jellyfish. They are able to control movements, but swimming is rarely strong enough to travel against currents and, thus, even larger jellyfish are considered planktonic. The importance of jellyfish movements to global ocean mixing might be underappreciated. Kakani Katija and John Dabiri developed models to suggest that, despite limited mobility, the mixing of water resulting from the cumulative movements of billions of these animals is likely important in the circulation of heat and organic material in the oceans, and may even influence global climate. For example, colder nutrient-enriched bottom waters could be dragged toward the surface and initiate a mixing with warmer surface waters.

Cnidarians serve important roles in a diversity of marine ecosystems. Corals and the skeletal materials they produce form the base for one of the most diverse ecosystems on earth, the coral reefs. Impacts on the coral organisms and conservation of the reef ecosystems are the focus of Chapter 5. Anemones are a component of many coastal ecosystems, including coral reefs, mangroves, and seagrasses discussed in later chapters. Jellyfish often comprise most of the biomass in epipelagic ecosystems. They are an important source of food for many animals, including some giant-sized animals, such as the leatherback turtle and the ocean sunfish. Jellyfish may serve an important

role after their death in transporting organic matter from surface waters to the deep-sea floor.

Although jellyfish typically are not considered edible in western cultures, in eastern Asia, particularly Japan and China, there are fisheries in the western Pacific targeting jellyfish marketed for consumption.

Many cnidarians release their nematocysts when they come in contact with humans. The impacts of stings may be barely noticeable, a nuisance, a major health concern, or lethal, depending on the species and the conditions. The Portuguese man-of-war *Physalia* (technically not a Scyphozoan jellyfish, but a colony of siphonophores in Class Hydrozoa) is notorious for its danger to swimmers who become entangled in the long tentacles and are exposed to its toxic nematocysts. They have a gas-filled float with a sail that propels them across the ocean surface for long distances. They commonly wash ashore in some regions, and the nematocysts and toxin can remain potent for days after the animal's death. Thousands of man-of-wars may congregate as a result of their propulsion by winds and currents. Thousands of stings to humans are reported each year; the venom can cause severe pain, rashes, other allergic reactions, and rarely death. The most venomous jellyfish are the box jellies (Class Cubozoa). There are about 19 different species in this group, often called sea wasps. Although toxicity and danger to humans varies, they are sometimes lethal. One of the most dangerous is *Chironex fleckeri,* which typically causes deaths each year off the northern coast of Australia. This species is unique in that it swims well enough to actively pursue its prey. Its tentacles may extend to about three meters length, resulting in further danger to swimmers. Nets are used in swimming areas to exclude these box jellies. Additional conservation concerns regarding jellyfish include local population explosions linked to removal of predators by fish harvest, or increases in plankton prey due to nutrient pollution and climate change (see Chapter 7).

Platyhelminthes: Flatworms

Although organisms commonly called worms are not in a monophyletic group, there are several phylogenetic groups of invertebrates that exhibit similar worm-like body structures. The simplest of the worms are the flatworms, the phylum Platyhelminthes, with a flattened unsegmented body and a single opening to the outside. They do not have respiratory or circulatory organs; their flattened shape allows for nutrients and oxygen to diffuse through their bodies. Most species have a single external opening and must regurgitate undigested materials. The turbellarians are mostly free living, while most other flatworms are parasites, at least as adults. Other groups of flatworms include the cestodes (tapeworms) and trematodes (flukes), which live as adults in the digestive systems of vertebrates (including humans), and the monogeneans, external parasites on aquatic animals. Most parasitic flatworms have complex life cycles that include the production of eggs and intermediate stages that infest secondary hosts.

There are over 3,000 species of turbellarians (**Figure 2-9**), ranging in length from about one millimeter to over a half meter. Some are parasitic, but many are free living, often brightly colored, predators. Most have **ocelli**, light-sensing organs incapable of detailed vision. Many turbellarians are able to reproduce by budding or by cloning themselves after splitting into two parts. They are also capable of sexual reproduction through internal fertilization. All are **hermaphroditic**, with each individual containing both male and female reproductive organs. Some have an interesting reproductive behavior where two individuals will go through a "penis-fencing" ritual. The winner of this dual impregnates the other, who produces eggs, serving the role of the female. The eggs hatch into free-living individuals and there usually is no larval stage. Some turbellarians (in the genus *Stylochus*) are considered a nuisance when they infect oyster beds.

All of the over 3,000 species of tapeworms live as parasites of vertebrates. They have a tiny head with a **scolex** for attachment to their host, but have no mouth or gut, depending on nutrients absorbed from their host through the skin. One common life cycle in marine species is initiated when fish excrete egg-containing segments of tapeworms (**proglotids**). The eggs develop into larvae eaten by small crustaceans (e.g., copepods), which serve as intermediate hosts. When fish eat the crustaceans they are infested by the flatworms. Although many species are microscopic, tapeworms have been found to reach lengths as great as forty meters in whale intestines.

Trematodes include about 20,000 species of flatworms most of which are parasites on mollusks and vertebrates. They have a complex life cycle that includes multiple hosts and stages that reproduce sexually and asexually. For example, in salt marsh habitats several species of parasitic trematodes may infest intertidal *Littorina* snails. A swimming stage is shed into the water where it infests fish, crabs, or bivalves. These are eaten by shorebirds, which then harbor the trematodes as they develop into adult worms. The eggs that are produced are transmitted in feces to snails, completing the cycle. Although infestations by trematodes are not necessarily fatal to the host species, they may play an important role in the ecological community. For example, Chelsea Wood and colleagues found that trematode-infected *Littorina* snails consume 40% less algae than uninfected snails. These changes in algae as food resources or habitat could be a major factor controlling community structure.

Monogeans attach to gills or skin of fish with hook or clamp-like structures. They are hermaphroditic with functional reproductive organs of both sexes in each individual. Eggs are produced that develop into larvae that attach to a new host; they do not have intermediate hosts. They can be problematic in aquaculture facilities where infestations may result in large losses of fish. Some fishes feed on monogean parasites that infest reef fish, which may improve the health of the parasitized fish.

Nematodes: Roundworms

Roundworms, or nematodes, are long, thin, cylindrical organisms with a mouth, gut, and anus. They are one of the

Figure 2-9 A turbellarian flatworm, *Pseudoceros imitatus*.

most diverse of all animals with 80,000 species described, and estimates of as many as 500,000 undescribed species. Nematodes inhabit virtually every habitat on earth where life is possible, including the Antarctic and the deep-sea trenches. They represent approximately 90% of the life of the sea floor. About 20% of nematode species are parasitic; they can be found in most species of marine vertebrates. Pathogenic nematodes are found in both plants and animals. Reproduction in most nematode species is sexual; with internal fertilization by males that are typically smaller than the females. Various reproductive modes exist, however, including self-fertilizing hermaphrodites and forms that are **parthenogenetic** (embryos develop without fertilization by males).

Free-living nematodes species feed on a large variety of materials, including algae, fungus, fecal matter, dead organisms, and other animals. Benthic nematodes play a crucial ecological role in marine sediments as decomposers, recycling nutrients into the ecosystem. Sizes vary, but many are classified as **meiofauna** (animals just visible to the naked eye, between 0.1 and 1 millimeters in size). Due to variability in their tolerance, nematodes serve as a useful tool to conservation biologists for monitoring the impact of pollutants on benthic ecosystems.

Parasitic nematodes serve important functions as disease-causing agents in many marine food webs. Nematode ichthyoparasites are common in marine fish muscle or internal organs, however, and the impacts on the fish's health are typically minimal. Nematode presence can impact the marketability of fish, and, for fish whose muscle tissue is commonly infected, visible nematodes are commonly removed during processing. Humans are not suitable hosts for nematodes that infect marine fish and they are readily killed by cooking or freezing; therefore, they rarely cause human health problems. Consumption of live nematodes or larvae from un-cooked fish that has not been previously frozen can cause abdominal pain and intestinal upset, however. There are cases of nematodes living in human intestines, but the infection is not fatal.

Annelids: Segmented Worms

There are about 17,000 species of annelids, worms with long bodies divided externally into segments. Other distinguishing characteristics include a **cuticle** layer covering the body and hair-like **chaetae** projecting from their body. Annelids include the earthworms (Oligochaetes), leaches (Hirudinea), and polychaetes (Polychaeta). The earthworms are mostly terrestrial burrowers. The leaches, although mostly freshwater and terrestrial, include some marine blood-sucking parasites, primarily on fishes. The polychaetes, most of which are free-living marine species, are the most ocean-dependent of the annelids.

Figure 2-10 A marine annelid, an intertidal polychaete worm.

There are almost 10,000 species of marine polychaete worms (**Figure 2-10**). They include the species that go by the common names ragworm, bristle worm, lugworm, and sandworm. Size varies from one millimeter to three meters length, but most are less than ten centimeters long. Polychaetes differ from other annelids in that they have limbs, called **parapodia**, extending from each segment to assist in burrowing, crawling, and swimming. Many polychaetes have gills for respiration, often as a part of the parapodia. Size and shape of the mouth varies with diet; many species have a pair of jaws and a **pharynx** that can be everted to assist in grabbing food. Reproduction is typically sexual, each individual being either male or female (**dioecious**). There is internal fertilization in a few species, but most are **broadcast spawners**, with males and females releasing gametes concurrently to be dispersed into the water. Fertilized eggs develop into a planktonic larval form that eventually develops, settles to the bottom, and continues developing into an adult.

Habitats of polychaetes include burrows, cracks and crevices, and tubes built in the bottom sediments; some giant forms are endemic to deep-sea hydrothermal vents and cold methane seeps (see Chapter 8). Tube-builders often are **filter feeders** with elaborate feather-like appendages for sieving food organisms from the water. Others are **deposit feeders** that extend their bodies from burrows and pick up food items that have settled on the bottom. Large colonies of polychaete tubes can be seen exposed at the surface of some coastal mudflats at low tide.

Polychaetes form an important ecological role in many marine bottom habitats. Tube-building and burrowing polychaetes can be the most important consumers in mudflats. Some species of lug worms are harvested as fish bait (see Chapter 3). Tube worms around deep-sea vents play an important intermediate role between endosymbiotic chemosynthetic bacteria and consumers around deep-sea vents and methane seeps (see Chapter 8).

Mollusks

There are about 100,000 described species in the phylum Mollusca, with possibly as many undescribed species. Mollusks occupy a large range of habitats around the globe. They represent a large diversity of shapes, sizes, and behaviors (**Figure 2-11**). Although this is a monophyletic grouping, it is difficult to choose defining characters that fit all species, and there are exceptions to almost any general description. Most are soft-bodied animals covered by a hard calcium carbonate shell that is secreted by the **mantle**. Mollusks typically have a head with eyes and other sensory organs, and a muscular "**foot**" used in locomotion. Most feed using a rasping mouthpart called a **radula**. They have a complex digestive system, and typically have eyes and sensors for detecting chemicals and vibrations. Reproduction is sexual, typically with external fertilization of eggs that develop into larval forms. Mollusks inhabit virtually all marine environments from the open ocean to deep-sea vents to rocky coastal zones. There are eight classes of **extant** (currently living) mollusks; however, over 99% of the species are in four classes: Gastropoda, Bivalvia, Polyplacophora, and Cephalopoda.

The largest class of mollusks is Gastropoda with about 70,000 species, comprised of snails, limpets, and abalone, among others. Snails are composed of a crawling foot, topped by a coil of organs and tissues, covered by a coiled shell. Most snails scrape algae and other food from rocks with their radula, others scavenge food off soft bottoms, and some are carnivores on other invertebrates. Snails are an important component of many nearshore and intertidal ecosystems; for example, their interactions with other organisms have important conservation implications in salt marshes (see Chapter 4). Many species are harvested for food; one of the most valuable gastropod fisheries is for the conch, with overharvest problems in shallow tropical waters. The harvest of gastropods and other mollusks for

(a)   (b)

(c)   (d)

Figure 2-11 Examples of marine mollusks: **(a)** limpets, **(b)** oysters and mussels (bivalves), **(c)** chiton, and **(d)** squid (cephalopod).

the ornamental shell market also can impact their populations (see Chapter 5).

Limpets have simpler conical shells, typically not coiled, that are attached to hard surfaces by suction; they often feed by grazing on rocky surfaces. Limpets are important components of rocky intertidal ecosystems. They have been used by coastal cultures for millennia as a source of food and more recently by anglers as bait; overharvest has occurred in some regions (see Chapter 3).

Nudibranchs, commonly called **sea slugs**, are colorful gastropods without shells. The colors are used for camouflage or to indicate distasteful chemicals or toxins to potential predators. Some species function as grazer in coral reef ecosystems.

The second-largest class of mollusks is the Bivalvia, with about 20,000 species, including clams, oysters, mussels, and scallops. The bivalves, as their name implies, have two rounded shells or **valves**, joined at one edge by a **hinge**, within which the living portion of the organism is contained. They do not have a head or radula, as do other mollusks. Gills are used for filtering food as well as for extracting oxygen from the water, which is taken into the shell through a **siphon**. Bivalves either attach to a surface or use a muscular foot for digging into the bottom. Reproduction is sexual and most bivalve species are dioecious. Fertilization is typically external through broadcast spawning. Planktonic larvae develop from the eggs and eventually settle to the bottom and develop into an adult. Bivalve species are present in most bottom ecosystems. In estuaries, bivalves, primarily oysters, serve an important role in filtering organic matter and sediments from the water (see Chapter 4). They are an important source of food for organisms that have mechanisms to overcome the shell covering. Predators include snails that bore through the shell with their radula, crabs with crushing claws, fishes with heavy crushing teeth, birds that peck into the shells, and animals such as octopuses and sea stars that can pry open their shells. Bivalves have been an important resource for coastal human populations throughout history. Commercial harvest of bivalves is still prevalent in coastal areas where they are obtained by scraping from the substrate or digging from the bottom. Overharvest and coastal pollution present conservation problems for bivalve populations (see Chapter 4).

The class Polyplacophora includes about 1,000 species of **chitons**. Chitons have eight adjacent shell plates covering their body. They cling to hard surfaces but can flex upward in order to move over rocky surfaces. When dislodged they can curl up into a ball for protection. Chitons are important grazers on rocky coastlines, and can be found high up in the intertidal zone (see Chapter 3). Predators include sea stars, crabs, and birds. Although not considered a major

human food source, chitons are eaten by some indigenous coastal cultures.

The most atypical of the mollusks are in the class Cephalopoda. Cephalopods include about 900 species, mostly squids and octopuses. In most species the shell is absent or reduced in size and located internally. Octopuses, squids, and cuttlefish have large heads and arms or tentacles (actually a modification of the molluscan foot). They are very active swimmers, especially the squid, moving by forcing water out of the body through a **siphon**. Cephalopods can be efficient predators on large invertebrates and fishes, using tentacles and suckers for capturing prey, transferring the food to their mouths; they have a hard **beak** structure with which they bite the food. Most cephalopods have an **ink sac** from which they can expel a cloud of dark liquid to confuse potential predators. They use **chromatophores** or bioluminescence for camouflage or to produce displays for startling predators, attracting mates, or communication. The largest octopuses can have a four meter arm span and the giant squid, although rarely seen, is up to at least fourteen meters long. Squids and octopuses are commonly harvested for food by humans. Squids support large commercial fisheries in some regions; they are typically caught by a hook and line after being attracted to boats at night with lights. Although harvested from ocean waters around the globe, much of the market for cephalopods is in Japan and Asian countries where they are an important component of *sashimi* (raw seafood) and other traditional foods. Squid dishes are often marketed as *calamari*, from Italian recipes. Problems with overharvest occur that are much the same as those for other fisheries species (see Chapter 11).

Arthropods

Although estimates vary, there are over one million species in the phylum Arthropoda. Many of these are in groups for which there are few, if any, living marine representatives (including insects, spiders and scorpions, and millipedes and centipedes), however. The trilobites are a major extinct marine group. There are three extant groups of arthropods that are mostly marine: the sea spiders (Pycnogonida), the horseshoe crabs (Xiphosura), and crustaceans (Crustacea).

The "sea spiders" are **pycnogonids**, superficially similar to true spiders, with four pairs of legs attached to a small body. There are over 1,300 named species of pycnogonids found in shallow oceans around the world; however, some species live in waters as deep as 7,000 meters. Although some deep-water species are over 90 centimeters across, most in shallower waters are smaller, around 10 millimeters across. They obtain oxygen for respiration through diffusion and there is no respiratory system. Most species are dioecious, with external fertilization; some larvae develop into a parasitic form before maturing into a free-living

Figure 2-12 A marine arthropod, the limulid horseshoe crab.

adult. Sea spiders are usually not obvious, being camouflaged and hiding beneath rocks or among algae. They are mostly carnivorous, feeding on sponges, polychaetes, anemones, and other benthic invertebrates. A common method of feeding for some species is to use their **proboscis** to suck nutrients from the body of soft-bodied invertebrates such as anemones.

Horseshoe crabs (**Figure 2-12**) are a small group of arthropods, with only four species in the family Limulidae, more closely related to spiders than true crabs. They are considered an **ancestral** form of arthropod, little changed over the past 400 million years. The most abundant species, *Limulus polyphemus,* is common in the Gulf of Mexico and along the Atlantic coast of North America. Limulids have a large shell, or **carapace**, consisting of three parts covering the body. The shell appearance results in a common name of "helmet crabs." One of the shell segments forms a tail-like spine called the **telson**. They have five pairs of walking legs and five pairs of **book gills** used in respiration.

Prey is mostly bottom invertebrates, captured with a small pair of pincers. Horseshoe crabs reproduce sexually and the females are typically larger than the males, about 80 versus 60 centimeters length. When mature at about ten years of age, the adults come in to the littoral zone from the continental shelf in spring to spawn. A male latches onto a female, waiting to fertilize the eggs she deposits in a nest excavated in the sand.

Horseshoe crabs serve several important ecological roles. Their carapace provide an area for attachment by a variety of species, including algae, barnacles, and mollusks. Although few fish prey on adult horseshoe crabs, they provide an important source of food for some sea turtles and their eggs are a critical food source for migrating shore birds. Their harvest for bait, used in fishing traps, has removed an important source of food for migratory birds along the U.S. east coast (see Chapter 3). The blood of horseshoe crabs is taken for medical research due to its unique characteristics in clotting reactions (this harvest does not present a conservation concern as the animals are returned to the ocean after bleeding). They have also been used extensively in scientific research of vision, due to their large compound eyes and a readily accessible optic nerve.

Crustaceans

Crustaceans include over 50,000 described species of mostly aquatic arthropods with a hard **exoskeleton** and appendages specialized for feeding, crawling, swimming, mating, or attaching themselves to other objects (**Figure 2-13**). The most familiar marine crustaceans are larger **macroinvertebrates**, such as crabs, shrimp, and lobsters; however, the most numerous are zooplankton or benthic meiofauna. Planktonic crustaceans include copepods, amphipods, ostracods, cladocerans, and other groups. Crustaceans are an important component of marine ecosystems, functioning

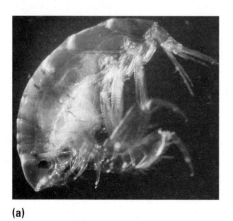

(a)

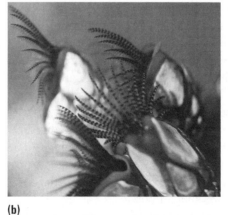

(b)

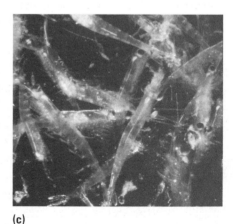

(c)

Figure 2-13 Marine crustaceans: **(a)** copepod, **(b)** barnacles, and **(c)** krill.

in all roles within food webs, including prey, predator, scavenger, and parasite. Reproduction is typically sexual, with eggs developing externally. Most species go through multiple larval stages before transforming into adults. They occupy all types of marine habitats, and some species are adapted to living on land.

Copepods are a large group (over 20,000 described species) of relatively small crustaceans; most are about 1 to 2 millimeters long; however, sizes range from 0.2 to 10 millimeters. They are prevalent in both freshwater and marine environments, including extreme conditions adjacent to hydrothermal vents, on Arctic ice, in hypersaline waters, in deep ocean trenches, and in **hypoxic** (poorly oxygenated) sediments of marshes and the deep sea floor. There are thousands of planktonic, benthic, and parasitic copepod species in the oceans. Copepods are typically the most common animals in the **zooplankton** (animals floating in the water column), and are a critical link between the microscopic plankton and **nekton** (swimming organisms). Like other crustaceans, they are covered with an exoskeleton and have two pairs of antennae on their head. They have specialized appendages adapted for swimming, burrowing, crawling, or obtaining food. Smaller planktonic copepods feed on phytoplankton and protists. They can undergo dramatic population growth in response to seasonal phytoplankton blooms. Larger species feed on zooplankton, including other copepods. Benthic species typically feed on organic **detritus** (decaying organic matter), bacteria, or other microorganisms. Copepods serve an important function in ocean ecosystems. They are possibly the largest source of animal biomass and the greatest source of carbon and nutrients in the deep sea, transferred from near-surface waters through the sinking of the waste products and remains.

Amphipods make up about 7,000 species, including pelagic, benthic, and parasitic form. Some are also terrestrial in marine coastal regions, including *Talitrus*, commonly called the sand flea or sandhopper. Sizes of amphipods range from one millimeter in some pelagic species to about 14 centimeters in some benthic forms. Most species are **ovoviviparous**, with the female producing eggs that develop internally; thus, the offspring that do not go through a planktonic larval stage. Amphipods are not as typical in the zooplankton as copepods; *Themisto* amphipods, however, can be an important part of food webs in some pelagic ecosystems where they prey on copepods and other zooplankton and serve as prey for planktivorous fish. Amphipods are often the most numerous benthic crustaceans in cold waters and the deep sea. Some graze on smaller invertebrates, but others are scavengers and important recycling biomass from decaying animals that sink to the bottom. The deep sea also includes pelagic species that feed on sinking detritus.

Barnacles, although superficially similar to mollusks because of their shell covering, are actually crustaceans. Most of the 1,000-plus named barnacle species attach to the surface of objects or other organisms and are **suspension feeders**, straining small planktonic organisms with feathery appendages they wave through the water. Most barnacle species live in shallow marine waters, many in the intertidal zone. They are especially important members of the rocky intertidal ecosystem (see Chapter 3). Barnacles are considered a nuisance when they attach to ships and are one of the major **fouling** organisms, reducing the performance of the ship by increasing resistance in the water. Scraping them from the boat's hull is time consuming, expensive, and temporary. Antifouling paints can successfully eliminate barnacles and other fouling organisms; however, they have a negative effect on some aquatic species (see Chapter 3). Although most barnacles are too small to be considered edible, the goose barnacle *Pollicipes pollicipes* is considered a delicacy in some regions, especially Portugal and Spain. It is named for the long stalk, typically about 35 millimeters long, with which they attach to rocks.

Krill, also called **euphausiids**, are shrimp-like crustaceans of a relatively small size, about 1 to 2 centimeters length for adults of most species. Krill differ most obviously from shrimp by the krill's externally visible gills. Krill have multiple appendages used in feeding and swimming. The appendages in the head region have comb-liked structures used for filtering small food particles from the water. Their diet is composed of planktonic organisms, largely phytoplankton such as diatoms. They congregate in areas with abundant food, sometimes in swarms of tens-of-thousands of individuals per cubic meter. Krill typically undergo a **diurnal vertical migration**, descending into darker colder waters in the daytime to avoid predators and conserve energy, and rising to the surface at night to feed (see Chapter 7).

Krill form an important component in food webs, especially in Antarctic waters. They function as a link between the plankton and some of the largest organisms on earth. They are a major food source for various whales, seals, penguins, squids, and fishes. Although krill are generally smaller than other crustaceans valued for human harvest, a commercial fishery has developed in some regions, especially in Antarctic waters. Although the current level of harvest is only a small fraction of the krill's total biomass, it remains controversial due to the feared impact on populations of large filter-feeding whales (see Chapter 10). The harvested krill are used for aquaculture feed, frozen bait, or food products, primarily in Japan.

The most familiar and largest crustaceans are the **deca-pods**, including shrimps, lobsters, and crabs (**Figure 2-14**). Decapods have five pair of legs, the first of which usually has claws for defense and feeding. The body is covered by a strong carapace. Shrimp are characterized by their long abdomens, the "tail" portion of the shrimp. They feed by scavenging along the bottom in many types of habitats. Shrimp are one of the most economically valuable fishery species in coastal waters in many regions of the world. Their harvest and rearing have led to several conservation controversies. In tropical countries it is now commonplace to farm them in ponds; this form of aquaculture is controversial because it often requires the removal of natural mangrove habitats and can result in coastal pollution (see Chapter 4). The gears used to harvest wild shrimp are controversial because they also can harvest large numbers of organisms other than shrimp and may destroy bottom habitat (see Chapter 11).

Lobsters are similar to shrimps in appearance, but are usually larger, with adults of most species ranging from about 25 to 50 centimeters in length. Due to their value as food, larger individuals are sought after, with some reaching

(a)

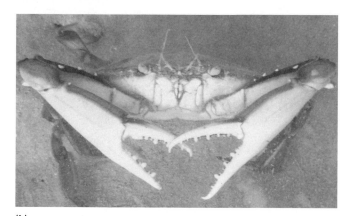

(b)

Figure 2-14 Marine decapod crustaceans: **(a)** clawed lobster and **(b)** portunid crab.

weights well over 10 kilograms. Lobsters are typically found individually, hiding around rocks or crevices during the day and feeding at night. During certain times of the year some species undergo mass migrations into deeper waters. This has been most well documented in the spiny lobsters *Panulirus*, which move across the bottom to deeper waters in single file, possibly to avoid the onset of storms that increase turbidity and turbulence and decrease water temperatures. Lobster species are distributed around the world in bottom habitats ranging from the shoreline onto the continental shelf. Food preferences vary, but most species are opportunistic predators on fish or macroinvertebrates. Some, but not all, species develop large claws used in feeding, defense, and competition.

Lobsters are an important part of marine food webs, but are rarely the dominant animal in an ecosystem. Lobsters currently support valuable fisheries and bring some of the largest prices per weight of any seafood. One of the largest of these fisheries is off the northeast U.S. coast for the American lobster *Homarus americanus* (see Chapter 11). Ironically, prior to the mid 1800s consuming lobsters was considered a sign of poverty in this region. Clawless spiny lobsters (palinurids) are warmer water species that support valuable local fisheries in the Caribbean Sea region (see Chapter 5).

Crabs have a broader body shape than shrimp and often a harder carapace and larger claws. Species of crabs are present in most marine ecosystems around the world. They range in size from pea crabs less than 10 millimeters across to spider crabs with leg spans of almost 4 meters. Crabs are often found in crevices and holes when not actively feeding. They are important predators in intertidal zones (see Chapter 3). Crab claws are used not only in capturing food but often for intraspecific interactions to defend territories or attract mates (e.g., fiddler crabs). Many are opportunistic feeders. Diet can include algae, detritus, or other animals.

Crabs are important in many marine ecosystems; they are a dominant predator and scavenger in some intertidal communities. Humans have a long history of harvesting and consuming crabs, and there are many valuable fisheries for crabs in habitats ranging from estuaries to offshore waters as deep as 1,000 meters. Some crab fisheries are valuable, competitive, and intensively managed (e.g., king crabs; see Chapter 8). Valuable crab fisheries in coastal estuaries have been negatively affected by coastal pollution and other environmental impacts (see Chapter 4).

Echinoderms

The word *echinoderm* means "spiny skin" in reference to the spines or bumps covering the body of many members of the phylum Echinodermata (**Figure 2-15**). These bumps are a portion of the skeleton formed by hard plates or **ossicles** embedded in the skin. Another distinctive characteristic is the five-way or **pentaradial symmetry** of the body, a character most obvious in the sea stars. Echinoderms typically have a mouth opening centrally on one side of the body. They do not have complex eyes, but some have ocelli that allow them to discriminate light and dark. Echinoderms have a unique **water vascular system**, a series of water-filled canals functioning primarily for transport of gases and nutrients within the animal. It is usually most visible in the external extension of this system called "**tube feet**" that are often used in locomotion. Echinoderms are dioecious and reproduce sexually by broadcast spawning, producing free-living planktonic larvae. In some species the larvae are able to clone themselves by splitting or budding, possibly to minimize the chances for predation. As the larvae develop and begin to form the skeleton, they settle and transform into an adult. Despite these similarities in characteristics, the forms that are exhibited in the seven classes (about 7,000 species) of echinoderms are remarkably varied, including star shapes (sea stars and brittle stars), disks (sand dollars), spheres covered with spines (sea urchins), and worm-like tubes (sea cucumbers).

Most **sea stars** or starfish (Asteroidea) have five arms radiating out from a central **oral disk**. They move about by pulling themselves along with tube feet. They also can use the tube feet to attach to other organisms and assist with feeding. Another unique feeding adaptation is the presence of two stomachs, one which can be everted through the mouth to engulf prey. Sea stars have remarkable powers of **regeneration**, replacing lost arms or even replicating new individuals from a single arm attached to a portion of the oral disk. The diet of sea stars varies among species and habitats, and most species are generalists in their food choice. Food items include bivalves, which they slowly pry open using their arms and tube feet, sponges, coral polyps, and dead animal matter.

Ecologically, sea stars play an important role in rocky intertidal ecosystems as a predator on other invertebrates (see Chapter 3). In coral reef ecosystems some species undergo dramatic population explosions, preying heavily on the corals (see Chapter 5). Due to the lack of large discrete muscles and the spiny skin, sea stars are not considered edible to humans and therefore are not desirable for fisheries harvest; however, some are taken and dried to be sold as souvenirs or to rear in aquariums. In some areas, they are considered a nuisance to fishermen due to their predation on clams and other harvestable organisms. Some attempts by fishermen to control sea star populations have backfired when individuals cut into pieces did not die, but regenerated into separate animals, thereby increasing the populations.

The 1,500-plus species of **brittle stars** (Ophiuroidea) are structurally similar to sea stars in that they have five arms radiating from a central disk. Brittle stars' arms are flexible and whip-like, up to 60 centimeters long in some species, and used in locomotion and feeding. The central disk is more sharply separated from the arms, and the viscera (digestive and reproductive organs) are limited in their entirety to the central disk. The tube feet are not modified into suckers or depended upon for locomotion, but are used to transport organic particles to the central mouth. Brittle stars are able to regenerate arms that are

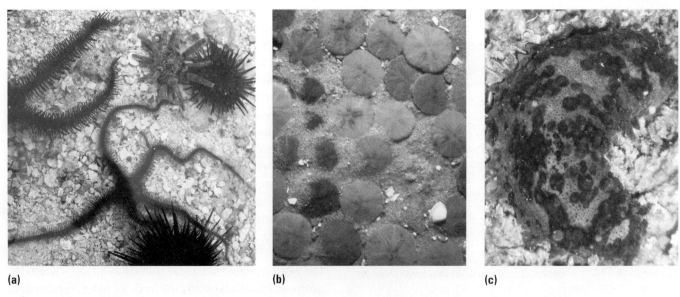

(a)   (b)   (c)

Figure 2-15 Examples of echinoderms: **(a)** brittle stars and sea urchins, **(b)** sand dollars, and **(c)** sea cucumber.

lost or deliberately shed to avoid predation; however, the arms cannot regenerate a new individual. They also can feed on suspended organic matter or small organisms using the tube feet, or small crustaceans or worms transported toward the mouth with the arms. The central mouth is surrounded by a set of jaws composed of five tooth-like plates that meet in the middle.

Brittle star species reside in most marine habitats, including the deep-sea abyssal zone, but they are most diverse in shallow coastal ecosystems. They are not always apparent due to their ability to hide in crevices, under rocks, within coral heads, buried in the mud or sand, or in sponges; and most of them feed nocturnally. Brittle stars are sometimes taken for the aquarium trade, but are not harvested for human consumption. The primary conservation concern is the protection of their shallow-water habitats from destruction or coastal pollution.

Sea urchins (Echinoida) are different in appearance from the other echinoderms in that the long sharp spines covering their bodies are typically the only part that is visible. These spines, used to avoid predations, are typically about 1 to 3 centimeters long, but can be up to 30 centimeters in the long-spined sea urchin *Diadema*. Sea urchins move by extending and shortening the tube feet to pull their body along the bottom. They feed by moving over their food and scraping it from the surface using an apparatus called **Aristotle's lantern**, comprised of a set of five sharp-edge plates. Many species feed on algae, but their diet also can include detritus or other invertebrates. Sea urchins can consume small food particles by picking them off the surface of their bodies with small pincer-like structures (**pedicellariae**) located between the spines. Sea urchin larvae are planktonic; as the hard outer shell begins to develop, they settle to the bottom and transform into adults.

Sea urchins are an important, sometimes dominant, component of many marine ecosystems. Under certain conditions, they may undergo population explosions and form **feeding fronts** as they move across stands of seaweed or seagrass beds consuming most of the vegetation in their path (see Chapter 6). Although their spines limit predation, some fish species have specialized mouths and teeth for consumption of urchins. Sea urchins are also a preferred food of sea otters along the eastern Pacific coast. Although the body of the sea urchin has little food value to humans, the **ovaries**, called **roe**, are considered a delicacy in some regions of the world. Urchin roe is especially valuable when prepared in Japanese sushi as *uni*, imported to Japan from urchins harvested by U.S. and eastern Asian fishers; there have been concerns about overharvest in some regions.

Sand dollars (Clypeasteroida), although closely related to sea urchins, are very different in appearance as they lack long spines and have a flat, rounded body shape. Although most coastal tourists would recognize those that are dead, dried, and bleached, when alive, sand dollars are covered with a purplish or greenish skin with small, short spines. The sand dollar moves over the bottom and buries itself by the action of these spines. The rows of pores that are visible in a star-like pattern on the surface are where the **podia**, extensions of the water vascular system, project for gas exchange. The oral surface is on the underside of the sand dollar; podia lining a food groove transport food particles to the centralized mouth. Food items include detritus, algae, diatoms, copepods, and invertebrate larvae. Sand dollars are most commonly found in shallow, sandy coastal areas on top of or buried just beneath the sand. They are often found in large congregations in preferred habitats.

Although sand dollars do not comprise a large portion of the biomass in coastal waters, they live in areas that are relatively devoid of large bottom organisms or habitat. Despite the hard tests, they are a common source of food for some fishes, such as skates, sea stars, and snails. Sand dollars are popular as a novelty in tourist shops of coastal areas, prepared by bleaching with the skin covering removed. In coastal areas, taking sand dollars may be restricted because of the ease of harvest.

Sea cucumbers (Holothuroidea) have a long tube-like body. The thick skin covers a skeleton made up of small ossicles joined by connective tissue, allowing them to be more flexible than most other echinoderms. This flexibility is enhanced by an ability to loosen fibers within their body wall and transform temporarily into a gelatinous texture and squeeze through tight spaces. Most sea cucumbers are scavengers on the sea floor, grazing on small invertebrates or decaying matter. Tentacles around the mouth are used by some species to filter food from the water. Respiratory organs are located just inside the anus, through which water is taken in and expelled.

Sea cucumbers are sometimes common in shallow near-shore waters. They are so abundant in some deep-sea regions, however, that they make up most of the biomass. Large aggregations of sea cucumbers have been observed moving across the deep-sea floor. Some deep-sea sea cucumber species can regulate their buoyancy and float over the sea floor to move more easily to new areas. Sea cucumbers enhance their defenses by expelling long sticky tubules to entangle potential predators along with toxins when they are disturbed. Most are pelagic spawners producing thousands of drifting gametes.

Despite characteristics that make them unappealing to many humans as a food item, sea cucumbers are sometimes harvested for consumption. In some East Asian countries they are considered a delicacy, and in some regions sea cucumbers are considered to have healing properties, through products taken internally or applied in oils or creams. High prices can support fisheries not only in the west Pacific but in developing regions around the world, where their ease of harvest makes them a valued local fishery. In the Asian Pacific and Caribbean, commercially valuable species are greatly depleted. In a global survey of the sea cucumber fisheries, Veronica Toral-Granda and colleagues discovered that many of the fisheries are not well-regulated, and where regulations are in place they are typically not well enforced.

As a result of the great abundance of echinoderms in large areas of the seafloor, they likely serve an important role as a carbon sink, sequestering carbon dioxide through the decay of dead organisms into marine sediments. Calculations by Mario Lebrato and colleagues indicate that echinoderms transport more carbon into seafloor sediments than do foraminiferan protists, indicating an important role in reducing the effects of excessive carbon dioxide on the marine environment (see Chapter 1).

Marine Vertebrates

Marine vertebrates include marine fishes and marine mammals as well as some birds, amphibians, and reptiles that, although they do not live their entire lives in the ocean, are sufficiently dependent on the ocean that they are commonly considered marine vertebrates and are regulated as a marine resources and protected as marine species.

Marine Fishes

Fishes are the most diverse vertebrates on earth, and most (about 60%) of the approximately 25,000 named fish species live exclusively in the world's oceans and estuaries. As with other groups designated by common non-technical names, the "fishes" are not a monophyletic group, but are composed of several different phylogenetic groups. By far the largest of these is the Actinopterygii, the "ray-finned" fishes, with about 30,000 species that include the most recently evolved (**derived**) fish groups. This includes almost all of the species that are commonly designated as fishes. The other, more **ancestral**, lineages include sharks, lobe-finned fishes, lampreys, and hagfishes (**Figure 2-16**). The largest of these groups is the Chondrichthyes, the "cartilaginous fishes," which includes about 800 species of sharks, rays, and related groups. The Sarcopterygii are the "lobe-finned" fishes, with fewer than ten fish species, including the coelacanth (this group is actually more closely related to mammals and other tetrapods than to other fishes). The other animals that are commonly considered as fishes are the **lampreys** and **hagfishes**; these are in groups considered only distantly related to the other fishes. Lampreys are mostly freshwater species, but include the sea lamprey that comes into fresh water to spawn and has become problematic as an invasive species in the

(a)

(b)

Figure 2-16 Examples of marine fishes from ancestral groups: **(a)** hagfish and **(b)** eagle ray.

North American Great Lakes. Hagfishes include only marine species, a result of their inability to regulate salts in their tissues.

Hagfishes (Figure 2-16a) are mostly associated with the bottoms of deep ocean waters. Although global diversity is low, hagfish serve an important role as scavengers of animals that sink to the sea floor, including some of the largest fishes and marine mammals. A large dead whale on the seafloor is likely to be crawling with hundreds of hagfish soon after it hits the bottom. Hagfishes feed by burrowing into and tunneling through the decomposing carcass. Although hagfishes do not have true jaws, their **dental plates** are covered with teeth to rasp into prey. They produce relatively large eggs (about 3–4 centimeters long) laid on the sea floor, from which their young hatch. Although typically considered to be dioecious, there is evidence that some hagfish species are hermaphroditic. Hagfishes are known for producing copious amounts of mucus or slime that clogs the gills of potential predators. Although slime production and scavenging behaviors are often considered a nuisance, especially from individuals caught in fishing traps, hagfishes are sometimes targeted for commercial harvest, mostly to supply a market for food and to make "snakeskin" leather. Most hagfish populations and fisheries are not closely monitored, but there is concern that some populations are being overharvested.

The Chondrichthyes include a large diversity of fishes of many body shapes and sizes living in most marine habitats. Most species are **elasmobranchs**, the sharks, rays, and skates. There are about 450 species of sharks, including the largest fish species, the whale shark (*Rhincodon typus*) that can reach lengths up to twelve meters. The whale shark and two related species are filter feeders, using comb-like structures called **gill rakers** to filter plankton and small nekton as they swim with mouths open through the water. Most other shark species have well-developed jaws, and tearing or crushing teeth to prey on fish or macroinvertebrates. Sharks inhabit many marine habitats but are somewhat restricted by the lack of a **swim bladder**, present in most other fishes, to maintain buoyancy; therefore, they must either continuously swim or rest on the bottom. Sharks have a wide variety of reproductive strategies. Fertilization is internal; males use **pelvic claspers** as an intromittent or copulatory organ. Large eggs may be released in an egg case to develop externally (**oviparous**), or development can be internal with a placenta-like arrangement (**viviparous**). Most sharks are **ovoviviparous**, however; eggs are produced, but development and hatching are internal and the offspring are born as free-swimming juveniles. A few shark species exhibit a phenomenon called **uterine cannibalism** in which the first-hatched or largest may kill and consume its siblings, resulting in the birth of a single offspring. Shark gestation period is up to 24 months for some species. Sharks reach maturity at a relatively late age compared to other fishes, many at ages from 10 to 20 years. These characteristics result in a relatively low offspring production that limits population growth rate. This is an important consideration for fisheries management, because recovery of reduced populations is likely to be slow. Most sharks are **apex predators**—at the top of food chains—and thus have few natural predators except humans. Historically they were killed primarily for sport or due to exaggerated concerns of their danger to humans. In recent years, however, the increased popularity of sharks as a food item has resulted in a dramatic increase in harvest and resulted in global conservation controversies (see Chapter 11).

Skates, rays, and sawfish (batoids) are also elasmobranchs. They typically have a flattened body with broad wink-like pectoral fins and are associated with bottom habitats. There are a few large pelagic filter-feeding species, such as the manta ray (*Manta birostris*) that can reach lengths greater than 7 meters and weights of more than 2,000 kilograms. Most batoids, however, are predators on bottom invertebrates, with heavy tooth plates for crushing

mollusks, bivalves, or crustaceans. Life history characteristics are variable and similar to sharks. Skates often produce leathery egg cases attached to bottom structures by tendrils at each corner (these egg cases often wash ashore and are commonly called "mermaid purses"). Skates and rays are sometimes harvested in commercial fisheries, often unintentionally in fisheries targeting other species. There are concerns about overharvest in some regions of the world; however, their harvest is not typically monitored closely so much is not known of their population sizes.

The Osteichthyes, including the Actinopterygii (ray-finned fishes) (**Figure 2-17**) and Sarcopterygii (lobe-finned fishes), are by far the most diverse group of fishes, with almost 30,000 species named and on average about 100 new species described each year. Because of this great diversity, it is impossible to present general characteristics that apply to all groups. Sizes range from less than 7 millimeters, for a male anglerfish *Photocorynus spiniceps* (which lives parasitically on the female), to a body weighing over 2,200 kilograms, for the ocean sunfish *Mola mola*. Most fishes have flexible fins supported by **rays** and **spines** (these are the ray-finned fishes, Actinopterygii). These fins provide speed, maneuverability, light-weight protection, and may be used as camouflage or to attract mates. Most Osteichthyes have a swim bladder (absent in the elasmobranchs) that provides buoyancy and allows them to maintain a position in the water column even when not actively swimming; some that live resting on the bottom have lost their swim bladders through natural selection. They typically have well-developed jaws, with numerous modifications and feeding structures, including teeth associated with the jaws, tongue, mouth, throat, and gill cavity used for filtering, tearing, grinding, puncturing, and manipulating food items.

This suite of characters has resulted in a remarkable diversity of bony fishes, with success in adapting to life in virtually all aquatic habitats. Other physiological adaptations make life in extreme physical conditions possible. For example, some fishes contain antifreeze proteins allowing them to live beneath polar ice (e.g., channichthyid ice fishes); others live in heated waters adjacent to hydrothermal vents (e.g., zoarcid eel pouts). Some fishes are able to come out of the water onto land to feed and breathe (e.g., gobiid mudskippers); others tolerate the high pressure and low oxygen of the deepest ocean trenches (e.g., ophidiid cusk eels). Some tolerate living in lagoons where salinities may range from freshwater to hypersaline (e.g., poeciliid mollies). The greatest diversity of bony fishes is in tropical coral reef ecosystems, but the most abundant species are those living in mesopelagic waters of the deep sea (gonostomatid bristlemouths). The most heavily harvested group of fishes, the anchovies and herrings (clupeioids;

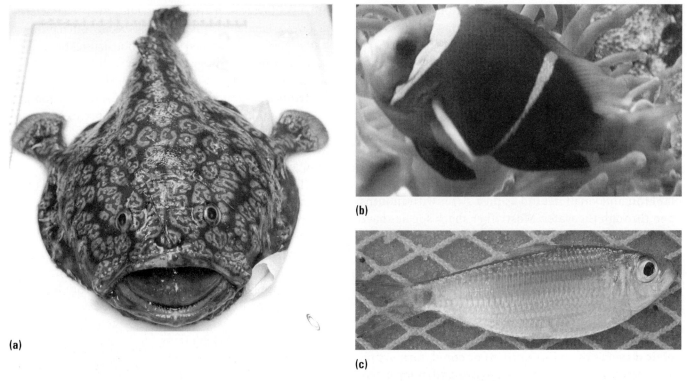

(a)

(b)

(c)

Figure 2-17 Examples of marine ray-finned fishes: **(a)** anglerfish, **(b)** clownfish, and **(c)** herring.

Figure 2-17c) are in near-shore pelagic waters, but the most valuable fisheries species are species of tunas (scombrids), sold as *sashimi*.

Reproduction and life histories of fishes are also too variable to easily generalize. Most pelagic species are broadcast spawners, producing from thousands to millions of small eggs that are fertilized externally and develop into planktonic larvae. These larvae drift in the currents for times ranging from a few days to almost a full year (in the European eel *Anguilla anguilla*). This reproductive strategy allows for a high reproductive rate and thus the potential for greater population growth, leading to a rationalization for harvest of a large proportion of the adult population based on the assumption that a relatively few fish could easily replenish the population. The unpredictability of survival of vulnerable early life stages in broadcast spawners, especially during a **critical period** when they must switch from a reliance on yolk reserves to eating external food sources, however, may result in extremely low survival in some years. The application of the rational of near-infinite reproductive potential in wild fish populations has led to mismanagement of many marine fisheries (see Chapter 11). The broadcast-spawning strategy of marine fishes also may lead to exploitation by the aquarium trade and aquaculture, because it is often easier to capture wild fish than to rear fish from birth in captivity due to the difficulty

of replicating conditions needed for successful broadcast spawning (see Chapters 5 and 11).

Reproductive strategies of some bony fishes include nest spawning (e.g., clownfish *Amphiprion*). Hundreds to thousands of eggs are laid and sometimes guarded by a parent; however, often the larvae hatch into drifting planktonic forms that settle onto reefs before developing into adults. Nest spawners have been reared in captivity more successfully than pelagic spawners, but the popularity of some species has led to their decline from harvest of adults from the wild. A relatively few marine fish are livebearers (ovoviviparous or viviparous), with internal fertilization and larval development. These are mostly estuarine species such as mollies (poeciliids) but also include some open ocean species (e.g., viviparous bythitid cusk eels).

Although most marine bony fishes are dioecious, that is, each individual is only male or female, a wide array of reproductive conditions are common in which an individual functions as both sexes during some time of its life (hermaphroditism). Species that exhibit **simultaneous hermaphroditism** have individuals that function as both males and female simultaneously. Few, if any, of these are self-fertilizing. Typically an individual will function as either male or female at any given time. For example, some species in coral reef ecosystems (e.g., the black hamlet *Hypoplectrus nigricans*) practice **reciprocal spawning** in which, after an elaborate display, partners alternate in the release of sperms or eggs. **Sequential hermaphrodites** function as only male or female at a given time but may switch sex at some point in their life, usually permanently. **Protandrous** hermaphrodites switch from male to female; **protogynous** hermaphrodites switch from female to male. These strategies can be advantageous when larger males are able to dominate a larger territory and service more females (protogyny), or so that larger individuals are females, enabling them to produce more eggs (protandry). It is important to consider these conditions when hermaphroditic species are targeted for harvest. The taking of larger individuals in a population can cause a skewed sex ratio and affect reproduction. In extreme cases, if all the males or all of the females are removed from a population, although population size may appear healthy, no individuals will be produced in the next generation (see Chapter 5).

Bony fishes are a critical component of most marine ecosystems. Because all life stages typically are present in the water column, a single species can serve multiple roles within the food web. Planktonic fish larvae (**ichthyoplankton**) are an important link between the invertebrate zooplankton or phytoplankton and larger organisms, juveniles (and smaller adults) are an important source of food for larger predatory fishes, and large adults are apex predators that may have a controlling effect on organisms below them in the food chain. Migratory fish species can play a direct role in more than one ecosystem. For example, many coastal fishes live and feed in estuaries early in their life cycle but return to the open ocean as adults. Other species may move in and out of multiple ecosystems on a seasonal or daily basis; for example, many coral reef fish move onto seagrass flats each night to feed. This impacts conservation of fishes in numerous ways. For example, fishes may be exposed to pollutants at any life stage, and the ichthyoplankton in particular have little ability to swim away from the polluted waters. Fishes can be vulnerable to impacts at any point in their migration cycle, making them especially vulnerable to habitat destruction in coastal or bottom ecosystems (see *Migrations* below). Apex predators may be exposed to any pollutants to which other organisms in the food web are exposed, especially those that accumulate up the food chain (also see *Food Webs* below).

Because of their abundance, desirability for consumption, and importance as a source of protein and nutrition, most coastal human populations (and many away from the coast) have depended on the harvest of marine fish. As human populations have grown exponentially over the past century and technology has dramatically increased our ability to harvest fish, managing this harvest has become one of the biggest issues in marine conservation. These issues are a primary focus of Chapter 11.

Seabirds and Shorebirds

No bird species has adapted to living its entire life in the marine environment. All must come to land to nest and produce offspring, and even birds that are capable of swimming and feeding underwater must come to the surface to breathe. Most birds classified as **seabirds** (**Figure 2-18**) spend much of their time adjacent to the water, in the water, or flying over the water, and depend on the ocean for food. All seabirds nest on land or in vegetation, and many species form large nesting colonies during mating and nesting. One of the major conservation concerns for seabirds thus is the loss of nesting areas. The term *seabird* is not a phylogenetic grouping, and many of the similarities among groups are a result of **convergent evolution**; that is, they share the same biological traits due to similar natural selection pressure, not necessarily due to genetic relatedness. For example, most have webbed feet and waterproof plumage, and beak morphology is determined by types of food and feeding behavior.

The capture of food organisms in the water by seabirds is achieved using three basic strategies (**Figure 2-19**). **Surface feeding** is accomplished either during swimming, as exhibited by gulls (Laridae), or during flying. For example, skimmers (Rynchopidae) fly over the water with the lower bill in the water probing for fish at the surface. **Pursuit diving** is represented in its greatest extreme by the penguins (Spheniscidae), which have sacrificed the ability to fly and have developed fin-like wings to enable swimming in the pursuit of prey. Other birds that are capable of both flying and pursuit diving include the shearwaters (Procellariidae), which can dive to greater than fifty meters below the surface. **Surface plungers** attack their prey by plunging into the water from flight, using their flight momentum to propel them through the water. These include boobies (Sulidae) and some terns (Sternidae). Some of the most dramatic of the plunge divers are the pelicans (Pelicanidae), which can dive from twenty meters above the surface, plummeting awkwardly toward the water before reorienting their body to plunge into the water. Seabirds also can feed by scavenging or predation out of the water. Frigatebirds (Fregatidae) are **kleptoparasites**; that is, they steal food from other birds. However, they also must feed from the surface to get adequate food. Some birds can feed as scavengers; for example, gulls will feed on dead organisms that wash ashore. Others are predators on other birds, mainly the young and eggs; the giant petrel *Macronectes* will take chicks, which they batter or drown before eating.

Figure 2-18 Some seabirds: **(a)** royal tern, **(b)** African penguins, **(c)** brown pelican, and **(d)** magnificent frigatebird.

Although some seabirds feed along the shore, there are many other **waders** or **shorebirds** (**Figure 2-20**) that do not have webbed feet, but wade along the beach or shoreline searching for food. Most of these are members of one order, the Charadriiformes. Most pick invertebrates from the sand or mud with their bills, which vary in length and thickness depending on the prey on which they are adapted to feed. Some of the most common shorebirds in marine coastal habitats are the **plovers** (Charadriidae) and **sandpipers** (Scolopacidae). Other **wading birds**, such as **herons** (Ardeidae) and **egrets** (Ardeidae), prey on food items while wading in shallow waters. These may be common in estuaries, marshes, or shallow coastal waters.

Seabirds and wading birds are especially vulnerable to environments that have been modified by humans. Most seabird species are **k-selected**; they mature late in life (up to age ten years), and they often produce only one clutch of eggs per year. The recovery of affected populations thus may be slow. Many seabirds and shore-birds are migratory (the Arctic tern *Sterna paradisaea* can

Figure 2-19 Seabird feeding strategies.

migrate from pole to pole in a single year); migrants may be vulnerable to human impacts at any point along the migration route, especially where they feed or breed (see Chapters 3 and 4). Bird populations on islands may be especially vulnerable to predation by invasive species, such as snakes or mammals, because they may not have evolved protective breeding strategies and nests are often built in unprotected open areas on the ground. Some diving birds, such as **albatrosses** (*Diomedeidae*), have been driven to endangerment from death caused

(a)

(b)

Figure 2-20 Coastal wading birds: **(a)** plover and **(b)** snowy egret.

by entanglement in fishing nets or hooks in the open ocean (see Chapter 9). The greatest concerns for penguins are impacts on the Antarctic ecosystem (including global warming) and removal of their food resources by overfishing (see Chapter 9). Predatory birds, such as pelicans, are susceptible to bioaccumulation of toxins up the food chain, which has resulted in their endangerment (Chapter 7).

Marine Amphibians and Reptiles

Marine species comprise only a small fraction of amphibians and reptiles (**Figure 2-21**). Most amphibians (class Amphibia) are not tolerant to submersions in seawater. The crab-eating frog *Fejervarya cancrivora*, however, which lives in mangrove marshes in southeastern Asia, can tolerate submersion in brackish waters (those of intermediate salinities) for long periods, and full-strength seawater for a short time. The legs of crab-eating frogs are considered edible and they are farmed in Indonesia. There is some export of frog legs of these and various species from Indonesia, leading to concerns about possible effects on populations. Ian Warkentin and colleagues argue for more thorough monitoring of international trade to determine the role that harvest plays in population declines.

Few reptiles are adapted to live out their entire life cycle in the ocean; however, three of the four taxonomic orders of reptiles have species adapted to living, swimming, and feeding in the ocean. Most must come to land for nesting. These orders are the Crocodilia (including crocodiles and alligators), Squamata (including lizards, iguanas, and snakes), and Tustudines (including turtles).

(a)

(c)

(b)

Figure 2-21 Examples of marine reptiles: **(a)** saltwater crocodile, **(b)** sea snake, and **(c)** green sea turtle.

Several species of crocodilians may be found swimming in salt or brackish waters, but none are found exclusively in salt water, and nesting and juvenile habitat are typically in fresh waters. Crocodiles tend to be more tolerant of salt water than alligators. For example, the American crocodile *Crocodylus acutus* lives mostly in areas near the coast in Central and South America and the Caribbean, and sometimes may be found in mangrove swamps or even at sea. It is considered endangered throughout part of its range due to habitat loss and pollution, but largely from hunting, primarily for its skin. In the United States, controls on hunting and trade beginning in the 1980s led to the recovery of crocodiles in southern Florida, where they are classified as threatened through the U.S. Endangered Species Act (see Chapter 9). International conservation efforts are focused on establishing protected habitats to enhance recovery.

Although attacks on humans by the American crocodile occur rarely, the saltwater crocodile *Crocodylus porosus* (Figure 2-21a) has a greater reputation as a man killer. It is the largest of all crocodilians, attaining lengths of at least four meters and weights over 400 kilograms. Although attacks, some resulting in mortality, are reported each year, danger of the saltwater crocodile to humans is often overstated. Their geographic range extends from southeastern Asia to Australia. Populations have been reduced or **extirpated** from much of this range by habitat destruction, excessive hunting for the hides, and due to fears for safety, however. Even where saltwater crocodiles are protected, poaching can be problematic. The healthiest remaining populations are in Australia and Papua New Guinea.

Although some lizard and iguana species feed in coastal habitats such as dunes or mangroves, there is only one species that spends long periods of time foraging in salt water. The marine iguana *Amblyrhynchus cristatus* resides on the Galapagos Islands off South America, where it lives primarily on rocky shorelines. It can dive into the water to depths of ten meters or more to feed on submerged seaweed. Its time underwater may last as long as thirty minutes, and is limited by the cold waters and the need to come to the surface to breathe. Marine iguanas spend much of their time out of the water basking in the sun to warm their bodies.

Marine iguanas are protected from harvest under the laws of Ecuador and through international agreements. One conservation concern is predation by domesticated dogs and cats on iguanas. As with many other island-adapted species, the marine iguana has not evolved to avoid terrestrial predators where none existed prior to human settlement.

Most of the sixty-plus species of **sea snakes** (Hydrophiidae; Figure 2-21b), found only in the Indian and western Pacific Oceans, spend their entire life in the ocean. They are the only reptiles adapted to an entirely marine life,

including early developmental stages. They have achieved this through ovoviviparity; eggs are retained by the female and young are born alive and free-swimming in the ocean water. Although sea snakes must come to the surface to take air into their lungs for respiration, many species are able to respire through their skin to obtain a portion of their oxygen needs. They are distinguished from other snakes by their paddle-like tails and laterally compressed bodies to enhance swimming, large lungs for storing air during dives and maintaining buoyancy, and valves over the nostrils to keep out water. Most are **piscivores** (fish-eating predators), and many feed selectively on eels. Most sea snakes are highly venomous. Although some are aggressive and use their bite defensively, most use it to immobilize prey. Sea snakes are typically docile and can be handled by humans with little fear of being bitten, and often little venom is injected with the bite. Venom of many sea snake species is considered more potent than that of most terrestrial species, however, so bites can be fatal. Sea snakes are sometimes hunted for their skin or meat; however, no species are considered threatened or endangered with extinction internationally due to this exploitation.

The most widespread, and the most endangered, of the marine reptiles are the **sea turtles** (Chelonioidea) (Figure 2-21c). At least one of the eight sea turtle species can be found in all oceans of the world, with the exception of colder polar seas. Sea turtles spend almost their entire lives submerged but must come to the surface to breathe air. Routinely they will dive for several minutes to feed, but can remain underwater resting for hours at a time. Despite this ability, they can be drowned in a short time when held under the water by fishing nets because of the activity and stress (see Chapter 9). Diet of sea turtles varies among species. For example, green sea turtles (*Chelonia mydas*) are completely **herbivorous**, feeding on sea grasses and algae; hawksbills (*Eretmochelys imbricata*) typically eat sponges, tunicates, and other invertebrates around coral reef habitats; loggerheads (*Caretta caretta*) and ridleys (*Lepidochelys*) feed largely on mollusks and crustaceans, using jaws adapted for crushing and grinding; and leatherbacks (*Dermochelys coriacea*) feed only on soft-bodied organisms such as jellyfish and tunicates. Most sea turtles undergo seasonal migrations, some for thousands of kilometers, to reach breeding and nesting areas. In all sea turtle species, females must come onto shore on sandy beaches for nesting. Using her flippers, she pulls herself onto shore and digs a circular, narrow cavity about fifty centimeters deep within which she deposits around 100 to 200 eggs. She covers the nest with sand and returns to the sea. After about two months incubation, the eggs hatch and the young dig their way out of the nest cavity and head toward the sea. During this migration to the sea the young are vulnerable to predators; however, they use a **predator swamping** strategy to enhance survival, with all individuals in a nest hatching concurrently. The young swim out to sea feeding in productive waters, such as among floating *Sargassum* seaweed or in the vicinity of upwelling zones. When the turtles mature, after about three years they migrate to primary feeding and breeding regions. Sea turtles do not always nest each year; the interval between nesting ranges from one to nine years depending on species and individuals.

Sea turtles are vulnerable to **anthropogenic** impacts (those resulting from human activities) throughout their life cycle. This has resulted in some degree of endangerment and the need for protection by national and international laws for all sea turtle species (see Chapter 9). Sea turtles historically have been exploited as food, either as adults or as eggs. Nesting success and survival of hatchlings during their short migration from the nest to the sea are affected by development and use of beaches (see Chapter 3). Loss of food sources and feeding habitats in coastal waters has resulted in sea turtle population declines. And, unintentional capture of sea turtles during fishing operations continues to kill numerous sea turtles around the world (see Chapter 11).

Marine Mammals

Mammalians include over 5,000 species of mostly **homeothermic** ("warm-blooded") vertebrate animals that bear live young (**vivipary**) that they nurse with milk produced by **mammary glands**. About 120 species of mammals have returned to the sea, evolving through natural selection from terrestrial ancestors to adopt an aquatic life (**Figure 2-22**). Some marine mammals, seals for example, have retained their connection to land and return for reproduction and rearing of offspring; others (e.g., whales and dolphins) are fully aquatic and capable of bearing young at sea and remain at sea for life. Because mammals must breathe air through lungs, they all must make regular trips to the surface for respiration. Some, however, have remarkable abilities to stay submerged for extended periods and dive to great depths by conserving oxygen. In order to maintain a constant body temperature required for homeotherms, most species retain a thick layer of fatty blubber for insulation. Three of the 29 mammal orders have representatives that are considered marine, either living primarily in the ocean or depending on the ocean for food. These include the Sirenia (manatees, sea cow, and dugong), Carnivora (marine representative include the seals, walruses, sea otter, and polar bear), and Cetacea (whales and dolphins).

Due to internal development of offspring, the life history of marine mammals is not as complex as for most other marine animals. The young typically reside in the same environment as adults and depend on the same food items. Conservation issues concerning habitat protection, thus, may appear to be simpler for marine mammals than for animals with complex life cycles and dependence of multiple habitats and food resources through their life. Most marine mammals are apex predators and, thus, are dependent on the health of all organisms beneath them in the food web. The loss of any of these organisms could have an indirect impact on marine mammals.

The life cycle common to marine mammals also has implications for regulating harvest. All marine mammals would fall near the extreme k-selected end of the r-k continuum relative to other marine organisms. They have relatively few live-born offspring, high parental care, delayed reproduction, and great longevity. For example, the largest whales are not mature until about 5 to 10 years of age, give birth to a single calf every 2 to 3 years, and require a large amount of resources to attain sizes over 100 tons. This makes these animals an unlikely candidate for sustainable harvest. The profitability of whales, historically for their fats and oils, however, resulted in excessive harvest and many species have been unable to recover even after almost fifty years of protection. The life history of marine mammals has contributed to a proportionately large number of species being classified as endangered. Current population declines for unknown or debated reasons could result in long-term effects.

Another factor in conservation is the purported intelligence of cetaceans. Although intelligence is a difficult concept to define, by most commonly-applied measures cetaceans are considered to be the most intelligent of marine animals. This has led to ethical controversies over their protection for aesthetic reasons beyond issues of protection their populations. Although cetacean intelligence results in adaptability and flexibility in behavior that can be used to adjust to changes, they remain vulnerable to anthropogenic changes in their environment. Although cetaceans have developed strategies to largely avoid predation in the marine environment, they are vulnerable to the most efficient and adaptable of land-based predators, humans. Protecting cetaceans from direct harm imparted by humans has been the greatest conservation challenge. These conservation concerns are discussed in more detail in Chapter 10. The discussion below provides an introduction to the major groups of marine mammals and how their biology affects conservation.

Sirenians

The sirenians' closest terrestrial relatives are the elephants. They are fully aquatic and strictly herbivorous, feeding largely on sea grasses that they gather with large prehensile lips. One species, Stellar's sea cow (*Hydrodamalis gigas*),

(a)

(d)

(b)

(e)

(c)

Figure 2-22 Marine mammals: **(a)** manatee, **(b)** Antarctic fur seals—male, female, and pup, **(c)** California sea lion, **(d)** sea otter, and **(e)** polar bears.

was driven to extinction in the 1700s. The four extant species include three manatees (Figure 2-22a) and the dugong, adapted for moving about and grazing in shallow waters with a paddle-shaped tail, reduced forelimbs, and only skeletal remnants of hind limbs. Their bodies are large, heavy, and very muscular. Manatees can reach sizes of up to four meters length and 1,500 kilograms mass. Stellar's sea cows reportedly reached lengths of eight meters. The geographic range of the dugong (*Dugong dugong*) is throughout coastal waters of the Indo-Pacific; however, most individuals are found in waters of northern Australia. There are two marine manatee species (the other lives in freshwaters of the Amazon River); the West Indian manatee *Trichechus manatus* inhabits shallow waters of the Caribbean Sea and Gulf of Mexico, although they may stray up the U.S. Atlantic coast in the summer. The West African manatee *Trichechus senegalensis* is found along the coast of western Africa.

Possibly because sirenians have few natural predators, they did not evolve an avoidance behavior to be used when encountered by humans and our machines. The Stellar's sea cow was easy prey for sailors, and manatee populations in the Caribbean were likely reduced by hunting for meat even in pre-European times. There is evidence that dugongs have been hunted for thousands of years, and continue to be by indigenous groups in some regions, for meat, oil, and other products. The largest impact on dugong populations throughout its range is probably loss of seagrass habitats where they feed (see Chapter 6). Manatees are affected by habitat loss and also by collisions with boats, particularly in Florida waters. Various efforts to reduce such collisions and the injuries and mortality associated with them include listing the Florida manatee as endangered, reducing boat traffic, and providing public education (see Chapter 9).

Marine Carnivorans

The Carnivora includes over 250 species of predatory mammals, most of which are terrestrial. The pinnipeds, which evolved from bear-like ancestors over 20 million years ago, comprise about 35 species of marine mammals that include seals (Figure 2-22b), sea lions (Figure 2-22c), and walruses. Most live in colder waters and have a thick layer of blubber. They are characterized by their wide, flat appendages used as flippers (pinnipeds means "fin-footed") and have excellent swimming and diving abilities; some species can remain underwater for up to two hours. Sizes range from about 30 kilograms (Galapagos fur seal) to over 4,000 kilograms (southern elephant seals). Pinnipeds must come ashore to breed and rear their young. In most species there is a sexual dimorphism, with males attaining larger sizes than females. Associated with this, their mating systems are often **polygynous**: dominant males reproduce with multiple females. The males are often territorial and may defend their rights to breed with a **harem** of up to 50 females.

The eared seals (Otariidae) include sea lions and fur seals. They are better at moving about on land than other pinnipeds with larger foreflippers and rear flippers that can be turned up under their bodies. As their name implies, the otariids have external ears. Protections under the U.S. Marine Mammal Protection Act have given them a refuge from harvest and exploitation in the United States (see Chapter 9). They are kept in marine parks and zoos and the seal used in traditional animal shows is typically a species of sea lion. Some animal welfare groups object to this kind of use. In the wild, special protection under the Marine Mammal Protection Act has resulted in large increases in some populations, such as the California sea lion. As these animals become more accustomed to a human-modified environment, they are often viewed as pests and competitors, invading docks and anchored boats and eating fish from endangered salmon runs. Dealing with these issues has become a conservation dilemma (see Chapter 9).

There are nine species of fur seals, all of which are easily taken when they come onto shores for breeding; this led to excessive harvest for furs and dramatic declines through the 19th century. In some populations, harvest was limited to males without a harem (**bachelor males**) under the premise that the loss of these non-reproductive males would not impact population growth. Such harvest resulted in negative effects on the genetic health of the population, however. For example, when a dominant male dies it is replaced from the bachelor male population; if the fittest bachelor males have been removed, genetically inferior males will be allowed to reproduce. Most populations are now protected but still vulnerable to other environmental impacts and competition with commercial fishers for food resources (see Chapter 9).

The earless seals (Phocidae) are considered "true seals," and differ from the eared seals by their inability to bring their rear flippers forward for moving on land. They still must come onto land or pack ice for breeding and rearing young. Although they are more awkward on land than the eared seals, their streamlined shape results in more efficient swimming and the ability to make longer foraging trips in the water.

Phocids include two species of elephant seals (*Mirounga*), which acquired their name from their large proboscis, used to make loud calls during mating. As for other pinnipeds, the elephant seals are vulnerable to harvest when on land; populations nearly disappeared by the end of the 19th century. Protections have resulted in recovery, however, with some populations showing dramatic growth.

Monk seals (*Monachus*) are smaller than elephant seals but can still reach weights of over 300 kilograms. There are two extant species, one in the Pacific Ocean around the Hawaiian Islands and one in the Mediterranean. These are the most endangered of the pinnipeds with total populations of each species at fewer than 1,500 individuals. A third species, in the Caribbean, was last sighted in the 1950s and has been declared extinct. Monk seals are strictly protected but have been unable to recover after overharvest in the 19th century. They are impacted by overfishing of food resources, entanglement in fishing nets, disease, and anthropogenic environmental effects (see Chapter 9).

There are fourteen other species of phocid seals. Four of these species are limited to waters around Antarctica, where they come onto the pack ice for mating and the rearing of their young. These populations are less impacted due to their isolated habitats; they include the crabeater seal *Lobodon carcinophagus*, with populations that are likely in the tens-of-millions. The remaining species are found primarily in the cold northernmost waters of the Northern Hemisphere. Most of these species have been harvested historically for their coats, sold as furs, but with protection most have exhibited at least a partial recovery. Harvest for some species continues at low levels, largely by indigenous groups. The harp seal *Phoca groenlandica* is the most heavily harvested, which has resulted in high-profile controversies in recent years. The harp seal is distributed in three primary populations across the North Atlantic and they are harvested primarily by Canada, Norway, Russia, and Greenland. The largest and most controversial harvest is in Canada, where several hundred-thousand may be killed legally each year (along with about 10,000 hooded and gray seals). The population has not declined substantially with this harvest and, therefore, the primary opposition is from animal welfare groups. This resistance has been concentrated on harvest of immature seals, in particular the "white coats" that are less than two weeks old. These and other issues with conservation of pinnipeds are discussed in more detail in Chapter 9.

Two species of otters (Lutrinae, in the weasel family Mustelidae) are considered marine mammals. The marine otter *Lontra feline* is **congeneric** with three species of river otters, but rarely inhabits freshwater habitats. They are found mostly along the southwestern coast of South America in rocky intertidal regions. They have been hunted to low levels for their fur and due to perceived competition with commercial fishers. Little is known about their ecology and behavior. They are protected internationally as an endangered species but there are fears that poaching could continue to reduce populations.

The sea otter *Enhydra lutris* (Figure 2-22d) is possibly the most well studied of marine mammals. Its original distribution was from northern Japan through the Aleutian Islands in the North Pacific and down the North American coastline as far south as the Baja Peninsula of Mexico. Sea otters are small relative to other marine mammals (typically less than 40 kilograms), and their primary insulation is not from blubber but with a thick coat of fur. They typically mate and bear young in the water but regularly come onto shore. They feed mostly on bottom invertebrates including sea urchins, mollusks, and crustaceans. Physical adaptations are not as extreme for sea otters as for other marine mammals. They swim by moving the rear end of the body, including a short, thick, flattened tail and webbed feet, up and down in the water. They dive to obtain food in shallow waters, remaining below the surface for up to about four minutes. Sea otters are noted for their tool use; they use rocks to pry abalone from the bottom and to crack open hard shells.

Sea otters can be a critical component of coastal ecosystems where they are abundant. They are often considered a **keystone species**, having a large effect on their environment disproportionate to their abundance. For example, both their absence and abundance have had dramatic effects on kelp ecosystems off California (see Chapter 6). As with other marine carnivorans, harvest of the sea otters for fur in the 1700s and 1800s resulted in extreme population declines. With recent protections they have recovered to high densities in some regions and remain in low numbers in others. Populations were reduced to about 1,000 by harvest for the fur trade, but have recovered to over 100,000. The largest populations are in Alaskan waters; however, recent declines are of concern. Along the North America coastline, through Canada to Washington State, reintroduction and protection have resulted in recovering populations. In California, populations that were originally as high as 16,000 were reduced to 50 individuals by fur trade; with protection as marine mammals and endangered species they have recovered to about 3,000. This recovery is not without controversy because commercial fishers for mollusks view the sea otters as competitors. Chapter 9 includes a further discussion of sea otter conservation issue.

Although the **polar bear** (*Ursus maritimus*) (Figure 2-22e) is not considered an aquatic species, it is often considered a marine mammal because it spends much of its time on sea ice and feeds primarily on other marine mammals. Its distribution is limited to the region around the Arctic Circle. Its primary prey is ringed seals that it drags up onto the ice when the seal comes to the surface to breathe. Polar bear populations have been targeted for hunting by indigenous groups, fur traders, and trophy hunters. This harvest was generally low and sustainable; however, concerns about overharvest increased as the polar bears became more vulnerable to high-powered rifles fired from snowmobiles and airplanes. This led to international agreements among countries bordering the Arctic region to limit harvest. Populations in recent years have been considered stable at about 25,000 bears. Concerns have increased over the impact of global warming on polar bear populations. With observed and predicted decreases in sea ice, populations will likely decline because polar bears cannot feed efficiently without the pack ice. The loss of ice could also force the bears to swim longer distances to reach feeding and denning areas, and increase their interactions with humans as they search for food. *Ursus maritimus* is considered threatened by the United States; however, increasing protection may not help the species in the long run if warming in the Arctic continues at its current pace.

Cetaceans

Of all marine mammals, the cetaceans (**Figure 2-23**) are the most closely linked to the ocean and most well-adapted to living in the aquatic environment throughout their lives. None of the approximately 80 species of whales, dolphins, and porpoises come onto land, other than through strandings; out of water they are prone to overheating and the weight of their body inhibits breathing and can crush internal organs. Archaeological evidence suggests that cetaceans are descendants of terrestrial hoof-footed predatory mammals that lived about 50 million years ago. As ancestral groups evolved into modern cetaceans, characteristics developed through natural selection that made them superficially similar to fish (**convergent evolution**). Hind limbs were reduced to **vestigial** skeletal remnants, and the front limbs were modified into flippers. A flipper-like tail developed at the end of the vertebral column, with horizontal **flukes** (in contrast to the vertical oriented lobes of the **caudal fin** of most fishes). The nostrils (**blowholes**) migrated to the top of the head. And a streamlined muscular body developed, with the retention of a thick layer of blubber for insulation. Mammalian characteristics that were retained include live birth and nursing of young, the necessity for breathing air into lungs for respiration, and the need to maintain an elevated, constant body temperature (**homeothermy**). Most species of marine mammals are predatory, with strong jaws and well-developed teeth. Some of the largest whales have developed specialized structures (**baleen**) for filtering relatively small food items from the water (e.g., krill and other pelagic crustaceans), however. The ability to reach great sizes may have evolved in giant filter feeders (also including whale sharks and manta rays)

Figure 2-23   Examples of cetaceans: baleen whales (**a–e**) and toothed whales (**f–j**).

to increase efficiency of feeding on large swarms of plankton and reduce vulnerability to predation.

Cetaceans are divided into two main groups, the **baleen whales** (Mysticeti) and the **toothed whales** (Odontoceti). There are fourteen species of baleen whales in four families; all use baleen plates to filter food from the water. They include the largest of the whales, with adults ranging from about three tons, for the pygmy right whale, *Caperea marginata*, to over 100 tons, for the blue whale, *Balaenoptera musculus*, the largest known animal to have ever existed. There are six marine families of toothed whales (four dolphin species in four separate families live in fresh water rivers).

The oceanic dolphins (Delphinidae) are classified as toothed whales. There are 37 species, including the bottlenose dolphin *Tursiops truncatus*, the species most commonly used in marine parks and most commonly used in cetacean intelligence and behavior studies. This family includes the killer whale, commonly called orca, sometimes kept in marine aquariums, and an important component of nearshore ecosystems. It also includes the pilot whale *Globicephala*, one of the smaller of the historically harvested whales (commonly called "blackfish") that are well known for their tendency to become stranded on beaches, often for undetermined reasons. Chapter 10 covers more conservation issues regarding dolphins.

The Monodontidae includes two species, the narwhal *Monodon monoceros* and the beluga *Delphinapterus leucas*. The narwhal is distinguished by its long helical tusk extending from upper jaw of males. The tusk was sought after by whalers to be sold for its perceived medicinal value; it is believed to have helped support the unicorn myth in medieval Europe after a story-telling whaler returned with one from a whaling expedition. The function of the tusk has been much debated; it is believed to be used during mating or in fighting for dominance. Recent discoveries by Martin Nweeia and colleagues of nerve endings embedded in the tusk, however, suggest it may also function in sensory perception, such as detecting temperature or salinity changes. The beluga is commonly kept in captivity in marine aquariums, and is distinctive due to its all-white coloration. It commonly comes into bays and estuaries where it is exposed to coastal pollution; high incidences of cancer and accumulation of organic pollutants have been observed in belugas (see Chapter 10).

Although the term *porpoise* is sometimes used for any small dolphin, the true porpoises are in the family Phocoenidae. This family includes six relatively small species that commonly inhabit nearshore waters. Because they feed in areas where commercial fishing is common, the main human impact on porpoises is their entanglement in fishing nets. The most endangered, and the smallest (at about 50 kilograms), is the vaquita *Phocoena sinus*. This species is found in the northern Gulf of California where it is **endemic** (found nowhere else). The vaquita is rarely seen in the wild, and the entire population size may be as low as 100 individuals. The most well known porpoise species, the harbor porpoise *Phocoena phocoena*, can experience thousands of deaths each year as a result of entanglement in fishing nets.

The sperm whale *Physeter macrocephalus* is the only member of the family Physeteridae; however, two smaller sperm whale species comprise the family Kogiidae. The sperm whale is the largest of the toothed whales. Males reach sizes of over 20 meters and 55,000 kilograms; females tend to be about 30% to 50% smaller than males. It was once the most sought after species by whalers and is the source of legends, including the focus of the novel *Moby Dick*. The primary value of the sperm whale was for the **spermaceti oil** contained in a special organ in its large head, which was considered ideal as a heating oil and for many other uses in the 18th and 19th centuries (see Chapter 10 for a discussion of the sperm whaling industry and conservation). There are several hypotheses as to the primary function of the spermaceti organ. It could be used to adjust the whale's buoyancy to assist in diving; the oil's density could be increased by cooling the oil with cold sea water passed through the organ. It may be used to focus the whale's sonar beam, used in **echolocation** of prey. And, it can be used in aggressive interactions for head-butting other whales (there are documented cases of sperm whales ramming and sinking wooden whaling ships, perhaps viewing the ship as a competitor or predator). The other unique adaptation of the sperm whale is its teeth; there are about 20 cone-shaped teeth, but only on the lower jaw. They are apparently not needed for feeding and may be used in aggressive interactions between males. Sperm whales are one of the deepest diving mammals, capable of diving greater than two kilometer below the surface and remaining submerged for over one hour in pursuit of prey, primarily squid in the deep sea.

The remaining approximately twenty species of marine toothed whales are in the family Ziphiidae, the beaked whales. These are moderately sized whales, weighing from one to fifteen tons, most of which feed on the bottom using suction created by movements of their large tongue. They have been documented to dive to depths of about two kilometers and remain underwater for over one hour. Their common name comes from the beak-like appearance created by a lower jaw. They have tusk-like teeth that are not used for feeding, but as a **secondary sexual characteristic** in mate selection and interactions between males that compete for harems of females. Although beaked whales have not been targeted for harvest because of their elusive diving behaviors and remote habitats, there are concerns about accumulation of toxins in their tissues from global pollution. As fishing has increased in their feeding habitats they have been more frequently entangled and killed in fishing nets. Beaked whales appear to be sensitive to sonar used for military purposes when it causes them to surface from deep dives too rapidly, sometimes resulting in physiological injury or death (see Chapter 10).

2.3 Marine Ecology Concepts

In order to establish conservation goals, it is not adequate to simply catalog the diversity of organisms in the marine environment. Understanding conservation needs also requires an analysis of the interactions among the biological components and between the biological and the physical environment. The fields of **ecology** and **environmental science**, thus, are used to establish conservation goals.

Population Ecology

A **population** is a group of interacting and interbreeding organisms of a single species. **Population ecology** focuses on interactions among organisms in a population and their interactions with the environment. Some of the concepts of population ecology are discussed in the descriptions of the taxonomic groups above. This section provides a review and summary of some of these concepts.

Life History Strategies

The **life history** is the changes an organism passes through in its lifetime; this is a composite of the "where, what, and when" in the life of the organism (**Figure 2-24**). The life history includes reproductive interactions, how the organisms develop and change as they grow, periodic movements to get to feeding or breeding areas, and other movements throughout the life of the organism. Invertebrates and fishes tend to have the most complex life histories of marine organisms, going through multiple developmental stages and playing several ecosystem roles throughout their lives. Understanding life histories is critical to conservation, because each of these stages has different needs and is affected differently by outside influences. It is important to consider each stage, not just the adults that might be most obviously affected by human activities.

Life history strategies are the combination of life history characteristics of a species that have evolved through natural selection as a strategy for survival. These strategies are sometimes assigned to the categories of k-selected and

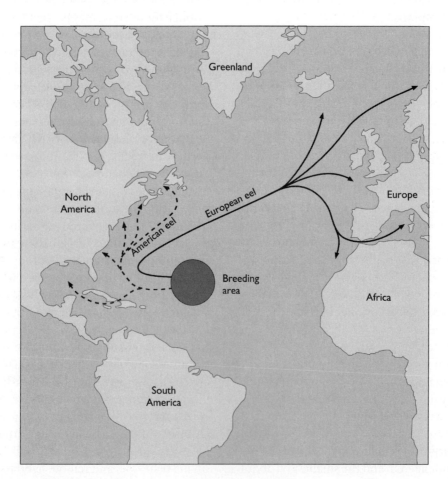

Figure 2-24 Marine migration route from the Sargasso Sea toward river mouths for the American and European eels, demonstrating a catadromous life cycle.

r-selected. K-selected species have a relatively late age at maturity, low **fecundity** (reproductive output), and often reach larger sizes. Sharks, as mentioned earlier, are good examples. Their age at maturity can be 10 to 20 years, and they produce from one to tens of offspring in a year. K-selected species tend to have more stable populations but recover more slowly from population declines, whether those are from natural or anthropogenic affects. For example, unregulated harvest of some shark populations has resulted in dramatic declines and slow recovery (see Chapter 11).

R-selected species have a greater reproductive output, mature earlier in life, and tend to be smaller in size. Their populations tend to be more variable, sometimes undergoing dramatic increases and decreases in a single year. R-selected species can typically recover more rapidly from population declines if the factors causing the decline have been removed. Anchovies are good examples of r-selected fishes. They commonly have a high fecundity as tens-of-thousands of eggs produced in a spawning season, and mature at about two year's age. Anchovies support a relatively large annual harvest, but their populations tend to undergo dramatic changes due to environmental affects as well as harvest (see Chapter 11).

Many organisms are not at either extreme of these r-k strategies and thus it must be viewed as an **r-k continuum**. A given species might be viewed as closer to an r- or k-selected extreme than others. This can be useful in making management and conservation decisions. A more k-selected species may require tighter restrictions on harvest and more careful monitoring. A further difficulty in relying on r-k life history concepts for conservation of marine animals is that many do not fit neatly into an r-k continuum. They may be r-selected for one character and k-selected for another. For example, many large fish have high fecundities, in the tens or hundreds-of-thousands, but a late maturity, ten years or later, and achieve large sizes. One example is the swordfish, which matures at about age six years and produces millions of eggs each year. Despite this large annual offspring production, excessive harvest has caused large population declines and although harvest reductions have resulted in an increase in population numbers, the size of individuals in the population has stayed relatively small (see Chapter 11).

Reproductive Strategies

One of components of the life history most important to conservation of marine organisms is the **reproductive strategy**. This is not simply the offspring production, but the overall approach an organism takes to survive and successfully reproduce, thus passing on hereditary characteristics to its offspring in the next generation. Categories can be assigned according to the methods an animal uses to reproduce. They include, for example, asexual production of offspring by some nematodes, the broadcast spawning of many fishes, the production of large eggs by sharks and birds, and the internal fertilization and development of marine mammals, among many others.

Each strategy has advantages, disadvantages, and tradeoffs, but has evolved as a successful strategy for the organism in which it is exhibited. For example, **broadcast spawning**, common in the marine environment, allows the production of numerous "cheap" offspring. Reproduction is often synchronized, with all individuals in a population releasing eggs and sperm at the same time. This enhances the chance of fertilization of the eggs so that at least some of the eggs and young survive. Only a very small percentage of the young needs to survive to adulthood for the population to sustain itself. This is an r-selected strategy, leading to the potential for a more rapid recovery of populations from declines due to human or natural impacts. Broadcast spawners can take advantage of currents that will transport them to areas that are good for their survival. This comes with a tradeoff; as they drift through the water the larvae are at the mercy of the environment and are prone to predation. **Nesting** provides the opportunity for shelter and protection from the physical habitat or parents. The chance of survival of each offspring is relatively great if the eggs are larger, but because fewer eggs are produced the reproductive potential is less. **Live-bearers** (such as marine mammals, sea snakes, and some sharks) avoid the risk of predation by fertilizing and bearing their young internally. Because of the limitations on space and resources inside the mother, these species bear few offspring; however, because each offspring is larger and more developed it has a much greater chance of survival. This is a k-selected strategy, and because the reproductive potential of live bearers is less, their populations do not recover as easily if reduced by overfishing or other factors.

Migrations

With the assistance of water currents and using efficient swimming capabilities many marine organisms move great distances during their lifetimes. A **migration** is a directed, regular, or systematic movement, typical for the purpose of feeding or breeding. Migrations are often seasonal. For example, many coastal fish species feed in estuary or coastal waters and make an annual movement offshore to congregate for breeding. Large pelagic bluefin tuna migrate from the middle of the Atlantic Ocean into the Gulf of Mexico or the Mediterranean Sea. Some whales migrate from polar waters where they feed into tropical waters for breeding. Some sea turtles feed around islands in the mid-Atlantic Ocean and migrate to the American continent to lay eggs. Seabirds can spend much of the year at sea feeding and move to nearshore islands to nest. The longest migration on record is that of the Arctic tern. This sea bird migrates 35,000 kilometers in one year, round trip between the Arctic and the Antarctic. This long-distance migration allows the tern to experience summer conditions year-round.

Marine migrations are not always seasonal. Larvae of many pelagic spawning fishes drift in the currents to **nursery areas** in estuaries and salt marshes, where they grow to larger juveniles or adults before returning to adult feeding and breeding grounds; these are sometimes call **estuarine dependent species**. Some marine fishes exhibit even more extreme migrations, spending a large portion of their life in fresh waters, a condition called **diadromy**. There are two kinds of diadromy. **Anadromy** is most common in five species of Pacific salmon in the genus *Oncorhynchus* that reproduce in streams of the west coast of North America. They carry out long-distance migrations into shallow headwaters of Pacific coast streams. There they lay their eggs in the gravel and die. After hatching, the offspring swim downstream to the Pacific Ocean where they live and feed for several years before returning to spawn and start the cycle again. The Atlantic salmon *Salmo salar* has a similar life cycle; however, the adults are capable of returning to streams to spawn multiple times. **Catadromous** fishes carry out a migration in the opposite direction. For example, the American and European eels of the genus *Anguilla* are spawned in deep waters of the mid-Atlantic in a region called the Sargasso Sea (Figure 2-24). As the eels develop from egg through the larval stage, they drift toward the coast of North America or Europe. They enter freshwater rivers and swim upstream feeding until they are mature and ready to return to spawning grounds.

Long-distance migrations of marine organisms lead to many conservation dilemmas. During any point along their routes migratory animals are potentially vulnerable to a variety of human impacts. These include exposure to pollutants, removal of food resources (e.g., shore birds; see Chapter 3), harvest as they congregate for spawning (e.g., tropical groupers; see Chapter 5) or egg laying (e.g., sea turtles and sea birds; see Chapter 9), or as they enter the "bottleneck" at the mouths of streams (for example, anadromous and catadromous species; see Chapter 4). Global climate changes that affect ocean currents could affect an animal's ability to complete migrations (see Chapter 1).

Population Dynamics

The description and analysis of marine populations have formed the primary basis for the scientific management of marine animals for much of the past century. The description of animal populations is called **population dynamics**. It includes the description of the number of individual organisms in a population (the **population size**), the relative number of individuals at each age (**age structure**), the gender makeup of the population (**sex ratio**), movements into or out of a population (**immigration** and **emigration**), and the survival and mortality of individuals in the population. These data are obtained by making direct observations or taking samples from the population. Scientific samples are often made using the same methods that fishers use to harvest the animals (see Chapter 11). Ages of individuals are obtained by various methods, depending on the type of organism. Many animals have calcified structures (including shells or skeletons) that form rings counted to estimate ages. For example, hard corals have been aged by taking cores of their skeletal deposits (see Chapter 5). Fishes are often aged by counting rings on bony parts.

Gender can sometimes be determined by **primary sexual characteristics**, those associated with reproduction. For example, sharks have claspers as **intromittent organs** and mammals have external **genitals**. For most other marine vertebrates gender is determined by observing **secondary sex characteristics**, those not used directly in reproduction. For example there can be **sexual dimorphism** between male and female fishes in color patterns or size. Often, however, there are no apparent external differences in males and females. In order to determine gender, invasive procedures must be used to observe reproductive structures internally.

Population dynamics data can be used to assess the overall health of a population or assess impacts of environmental or anthropogenic events. For example, data can reflect effects of environmental pollutants on age-specific survival, or even induced changes in gender that affect sex ratio (see Chapter 3). Biased sex ratios that affect reproductive potential for a population can be detected with population dynamics data (see Chapter 5). Harvested populations of fishes are routinely monitored using population dynamics to determine how harvest might influence population sizes, mortality rates, or age structure, and can be the basis for establishing management policies (see Chapters 11).

■ Community and Ecosystem Ecology

Although population data are considered necessary for making conservation and management decisions, conservation goals cannot be adequately achieved without looking at the overall picture of the environment. This realization has led to a movement towards ecosystem-based management (see Chapter 11). This approach requires an assessment of the **community ecology**, the study of the interactions among populations.

The **community structure**, the individuals of all species in the community, is largely a reflection of **interspecific interactions**, including **competition** or predator-prey relationships. Other environmental components, including the physical habitat, interact with the biological community to form the **ecosystem**.

The simplest way to describe interactions within a marine ecosystem is through the **trophic** (feeding) interactions among species. The linear movement of biomass and energy from one organism to another by feeding, from the primary producers, through intermediate "links," to the apex predators is called a **food chain** (**Figure 2-25**). Primary producers can be any organisms that use sunlight or chemical energy to convert carbon dioxide and water into organic matter, including photosynthetic or chemosynthetic bacteria, some protists, algae, and vascular plants. Some, but not all, of this energy and biomass is transferred through the various **trophic levels**. In fact, on average only about 10% of the energy and biomass is transferred from one trophic level to the next; this is sometimes called "the **ten-percent rule**" (In some marine ecosystems, however, the efficiency of transfer is actually somewhat better than 10%, but the concept still applies). This inefficient transfer produces a **food pyramid** with higher trophic levels, at the top of the pyramid, having a lower biomass than trophic levels below

it. It also results in a limited number of trophic levels in a food chain. For example, with 10% transfer, it would take 1,000 grams of phytoplankton to produce 100 grams of planktivores. Those 100 grams of planktivores would only produce 10 grams of consumers at the next link in the food chain. If there are four links in the food chain from primary producers to apex predators, the biomass of primary producers must be 1,000 times that of the apex predators. This limits the number of levels in a food chain to around four or five; otherwise, the biomass needed to support a population of apex predators would be excessive.

Trophic interactions lead to some important considerations for conservation. One consideration is the inefficiency of harvesting apex predators as food for humans. It takes a tremendous amount of biological production to produce a small amount of food. We can get much more from the ecosystem by eating lower on the food chain, for example, eating anchovies instead of sharks. Farming predatory fish is also a wasteful process. It would be much more efficient to eat the herrings, rather than feed them to predators, such as salmon raised in pens. **Aquaculture as a Solution to Overharvest**, Chapter 11, discusses fish farming controversies.

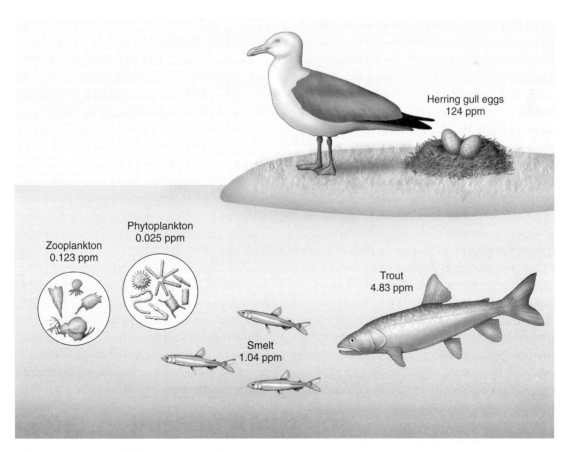

Figure 2-25 Bioaccumulation of PCBs in an aquatic food web.

The transfer of materials up the food chain leads to another important conservation consideration. Although the transfer of biomass up the food chain is relatively inefficient, the transfer of some toxic materials can be extremely efficient. Some toxins are transferred at close to 100% efficiency. This means the transfer of non-degradable toxins may be ten times that of organic materials. This results in a **bioaccumulation** of toxins up the food chain by an order of magnitude (ten times) at each level (Figure 2-25). For example, a chemical at concentrations as low as ten parts-per-billion in the plankton can accumulate to one part-per-million if transferred four levels in a food chain. Not all chemicals are stable enough to accumulate in this manner, but some pesticides (e.g., dichlorodiphenyltrichloroethane [DDT]), organic compounds (e.g., polychlorinated biphenyls [PCBs]), and heavy metals (e.g., mercury) are stable enough to accumulate to dangerous levels in marine organisms, even those far from the source of the pollution. Chapter 7 has a deeper discussion of conservation issues related to bioaccumulation of toxins in the marine environment.

Principles that incorporate the concept of a food chain in natural communities can be oversimplifications because the trophic transfer of energy and biomass is not typically in a linear or chain-like fashion. Any organism might use a number of different food resources and be vulnerable to predation by a number of different predators. This can result in ocean **food webs** (**Figure 2-26**) that are so complex it is difficult to understand all the details. Biologists attempt to gain an understanding of the interactions in marine food webs as well as possible in order to predict what impact the modification or removal of one link will have on other links in the web. This is more difficult than it may appear because many organisms are able to modify their role in the food web as it changes. **Generalists** may be able to take advantage of many available food resources and adjust feeding with changes in their abundance, whereas **specialists** may depend on a single species and decline as a direct result of its decline. We also change the food web by introducing new species (**exotics**). It is typically impossible to predict what impact exotics will have on the food web until it is too late. For example, lionfish from the Pacific have begun showing up in waters of the Caribbean Sea with unknown long-term impacts on the natural marine community (see Chapter 5).

Studying marine systems at the level of the community is difficult. It requires a thorough understanding of each population that comprises the community and how they interact among each other. Basic data that a community assessment would pursue include population dynamics of each species, behavioral and trophic interactions among species, and relative numbers of individuals in each population.

The final and most comprehensive ecological level of organization is the ecosystem. The ecosystem includes the populations that comprise the community, along with the non-living or **abiotic** components. In marine ecosystems these can include the physical habitat of the water, the sea floor, or the beach sands. Living organisms often form a physical as well as biological component of an ecosystem; for example, the coral skeletons that make up a reef, the mangrove plants within a marsh, or the remainders of dead plankton that comprise deep-sea sediments. Anthropogenic materials, including pollutants or physical structures, become an unnatural part of the marine ecosystem. At this level of complexity it is even more difficult to predict responses. It is often not known how the biotic components of an ecosystem will respond to the physical changes within the ecosystem.

Despite the difficulties, marine conservationists and managers attempt to take an ecosystem approach. Conservationists understand the importance of habitat and the environment. For example, in the United States, fisheries management policies now encourage or require that the ecosystem approach to management be considered, and the protection of species endangered with extinction requires that critical habitat be defined and protected. Unfortunately, the lack of manpower, resources, or knowledge means that this approach is often not realized. Issues related to the ecosystem approach will be discussed throughout the text.

Ecosystems are often labeled according to the dominant organisms or physical habitat within the ecosystem, and conservation issues are often designated according to the ecosystems. The text will follow this approach in Chapters 3 through 8, beginning at the coast and continuing through intertidal and near-shore waters to the pelagic zone and the deep sea. It will become clear, however, that even this level of organization is somewhat artificial. No marine ecosystem exists without some degree of interactions with other marine ecosystems, and even the lines we draw between the sea and the land, and the sea and the atmosphere, are not rigid. These components will be drawn into the discussion when appropriate.

The focus of this chapter has been on biological concepts regarding the "natural" marine environment. The next step toward conservation is the incorporation of the human element. If we define *conservation* as "the preservation and management of the environment and natural resources," it is humans that must do the preserving and managing. In the conservationists' ideal world, we would simply have to focus on preservation, keeping the environment in an unaltered condition. This could be achieved by limiting or eliminating human influence on natural

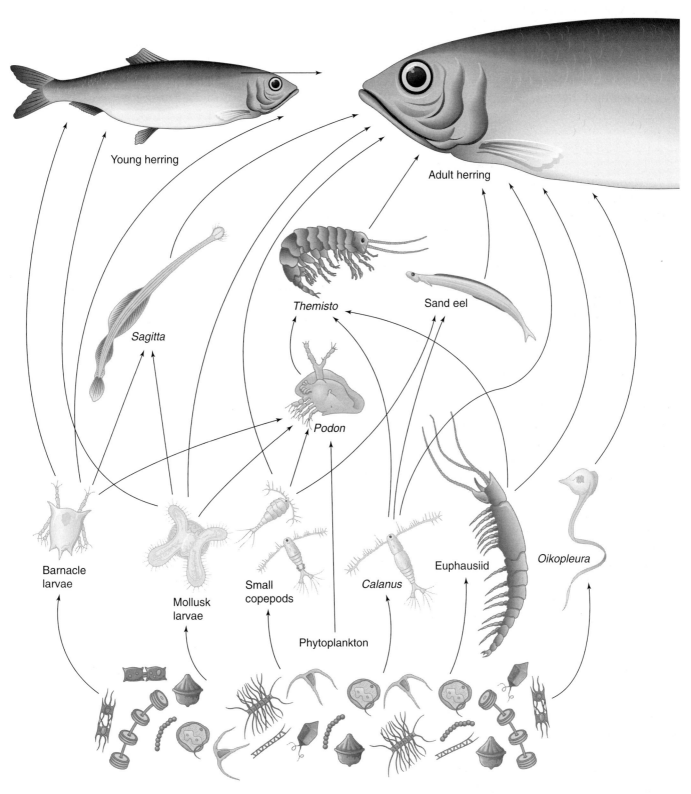

Young herring

Adult herring

Sagitta

Themisto

Sand eel

Podon

Barnacle
larvae

Mollusk
larvae

Small
copepods

Calanus

Euphausiid

Oikopleura

Phytoplankton

Figure 2-26 Diagram of food web that produces herring in the North Atlantic.

marine ecosystems. Many marine ecosystems remain that are minimally impaired by human actions or that can still be returned to unimpaired conditions. It is reasonable for conservationist to fight for the total "preservation" of these ecosystems. To limit ourselves to this approach, however, assumes that human influence is not a component of any marine ecosystems. Humans have become and will, for the foreseeable future, remain as a component of the marine environment (though obviously more so in some ecosystems than others). We must understand and accept that conservation cannot be carried out in isolation from the human factor in today's world. Conservation of many marine ecosystems thus will continue to require "management," sustaining ecosystems to meet both ecological and human needs. The remaining chapters of this text present an ecosystem approach to preservation and management of the marine environment and those elements of the environment that are considered natural resources for humans.

STUDY GUIDE

■ Topics for Review

1. Why has it been more difficult to understand biological relationships in the sea than in terrestrial environments?
2. What is the difference between *biological diversity* and *species richness?*
3. How did attitudes concerning conservation of marine diversity change throughout the 1900s?
4. Describe the problems associated with trying to group all organism into the categories "animals" and "plants."
5. How do bacteria enhance cycling of organic matter in marine ecosystems?
6. Explain why, until recently, primary productivity in the open ocean was greatly underestimated.
7. How could enhancing production by protists help reduce global warming?
8. How do dinoflagellates enhance survival of coral?
9. How are protists important in the formation of deep-sea sediments?
10. What are the primary characteristics that differentiate seaweeds and seagrasses?
11. What are the primary characteristics that differentiate seagrasses from salt-marsh grasses?
12. How do mangroves and salt-marsh grasses differ in geographic location?
13. What are the primary shared characteristics that results in placing corals and jellyfish in the same phylum?
14. Compare regeneration abilities of sea stars and brittle stars.
15. Describe the variability in reproductive strategies in sharks.
16. Describe how k-selected life history traits affect conservation of sharks.
17. What advantages are gained by giant size for filter feeding elasmobranchs and whales?
18. What characteristics of osteichthyan fishes have allowed them to achieve greater marine diversity than chondrichthyan fishes?
19. How has the prevalence of broadcast spawning been used to rationalize large harvest of marine fish?
20. Why are ovoviviparous fishes better candidates for aquarium fishes than pelagic spawners?
21. Compare pursuit diving and surface plunging as used by seabirds in pursuit of prey.
22. Why are sea snakes considered more truly marine than other reptile species?
23. Describe the variability in food items eaten by sea turtles.
24. What major characteristics differentiate the earless seals from the eared seals?
25. Why is the sea otter considered a keystone species along the U.S. Pacific coast?
26. In what ways are cetaceans similar to and different from osteichthyan fishes?
27. Compare possible non-feeding functions of the unusual teeth of the narwhal, beaked whales, and sperm whale.
28. Describe the differences between anadromy and catadromy.
29. Give examples of sexual dimorphism among marine mammals.
30. How do trophic interactions result in bioaccumulation of toxins in organisms high on the food chain?
31. Describe similarities and differences in the body structure, feeding methods, predator avoidance, and movement among the major echinoderm groups (sea stars, brittle stars, sand dollars, sea urchins, and sea cucumbers).
32. Compare the typical mechanisms used to obtain food by each of the following groups: sponges, flatworms, tapeworms, jellyfish, tube-building polychaetes, snails, limpets, bivalves, cephalopods, copepods, barnacles, krill, and baleen whales.
33. Define hermaphroditism as a reproductive strategy and describe its variability in the following groups:

turbellarians, monogeans, nematodes, black hamlet fish, protandrous fishes, and protogynous fishes.

34. Compare mating and spawning methods of the following: mollusks, bivalves, horseshoe crabs, echinoderms, sharks, poeciliid fishes, marine mammals, and sea turtles.

35. Describe the primary reason the following are sometimes considered a danger or nuisance to humans: dinoflagellates, box jellies, barnacles, sharks, nematode ichthyoparasites, hagfish, saltwater crocodiles, sea lions, and sea otters.

■ Conservation Exercises

1. Describe the primary reasons for human exploitation of each of the following organisms:

a. sponges	**b.** jellyfish
c. seaweeds	**d.** polychaete lug worms
e. gastropods	**f.** cephalopods
g. limpets	**h.** horseshoe crabs
i. krill	**j.** sea urchins
k. sea cucumbers	**l.** hagfish
m. crocodiles	**n.** dugongs
o. harp seals	**p.** sperm whales
q. narwhals	

2. Describe how each action or conditions affects conservation of marine organisms.
   a. reporting and marketing fishes using common names
   b. fungal infections in cultured fish
   c. fungal infections in oysters
   d. trematode infections in snails
   e. introduction of exotic predators to islands
   f. boat traffic in waters containing sirenians
   g. melting ice in polar bear habitat
   h. bioaccumulation of toxins in cetaceans
   i. the "ten-percent rule"

3. Describe how each character or behavior affects conservation of the given animals.
   a. k-selected life history strategies in marine mammals
   b. broadcast spawning in fishes
   c. sequential hermaphroditism in fishes
   d. polygyny in fur seals
   e. deep-diving in whales

FURTHER READING

Ababouch, L. 2005. Fisheries Issues. Product identification: trade implications of fish species. *In: FAO Fisheries and Aquaculture Department* [online]. Rome. Retrieved August 2, 2011, from http://www.fao.org/fishery/topic/14807/en.

Azam, F., T. Fenchel, J. G. Field, J. S. Gray, L. A. Meyer-Reil, and F. Thingstad. 1983. The ecological role of water-column microbes in the sea. *Marine Ecology Progress Series* 10:257–263.

Carrier, D. R., S. M. Deban, and J. Otterstrom. 2002. The face that sank the Essex: potential function of the spermaceti organ in aggression. *Journal of Experimental Biology* 205:1755–1763.

Clarke, M. R. 1970. Function of the spermaceti organ of the sperm whale. *Nature* 228:873–874.

DeLong, E. F., and N. R. Pace. 2001. Environmental diversity of bacteria and archaea. *Systematic Biology* 50:470–478.

Herrnkind, W. F. 1985. Evolution and mechanisms of mass single-file migration in spiny lobster: synopsis. *Contributions in Marine Science* 27:197–211.

Hudson, H. J. 1986. *Fungal Biology.* Cambridge University Press, Great Britain.

Katija, K., and J. O. Dabiri. 2009. A viscosity-enhanced mechanism for biogenic ocean mixing. *Nature* 460:624–626.

Lipp, J. S., Y. Morono, F. Inagaki, and K. U. Hinrichs. 2008. Significant contribution of Archaea to extant biomass in marine subsurface sediments. *Nature* 454:991–994.

Lohmann, K. J., J. T. Hester, and C. M. F. Lohmann. 1999. Long-distance navigation in sea turtles. *Ethology Ecology & Evolution* 11:1–23.

Lebrato, M., D. Iglesias-Rodriquez, R. Feely, D. Greeley, D. O. B. Jones, N. Suarez-Bosche, R. Lampitt, J. E. Cartes, D. R. H. Green, and B. Alker. 2010. Global contribution of echinoderms to the marine carbon cycle: a re-assessment of the oceanic $CaCO_3$ budget and the benthic compartments. *Ecological Monographs* 80:441–467.

Lutz, P. L., and J. A. Musick (editors). 1996. *The biology of sea turtles.* CRC Press, Boca Raton, Florida.

Marsh, H., H. Penrose, C. Eros, and J. Hugues. 2002. Dugong status reports and action plans for countries and territories. UNEP/DEWA/RS.02-1. http://data.iucn.org/dbtw-wpd/edocs/2002-001.pdf.

Nelson, J. S., E. J. Crossman, H. Espinosa-Perez, L. T. Findley, C. R. Gilbert, R. N. Lea, and J. D. Williams. 2004. *Common and scientific names of fishes from the United States, Canada, and Mexico.* American Fisheries Society. Bethesda, Maryland.

Nweeia, M. T., N. Eidelman, F. C. Eichmiller, A. A. Giuseppetti, Y.-G. Jung, and Y. Zhang. 2005. Hydrodynamic sensor capabilities and structural resilience of the male narwhal tusk. *16th Biennial Conference on the Biology of Marine Mammals,* San Diego, California.

Pomeroy, L. R. 1974. The ocean's food web, a changing paradigm. *Bioscience* 24:499–504.

Ramaiah, N. 2006. A review of fungal diseases of algae, marine fishes, shrimps, and coral. *Indian Journal of Marine Sciences* 35:380–387.

Thorbjarnarson, J., E. Sanderson, F. Buitrago, M. Lazcano, K. Minkowski, M. Muniz, P. Ponce, L. Sigler, R. Soberon, A. M. Trelancia, and A. Velasco. 2005. Regional habitat conservation priorities for the American crocodile. *Biological Conservation* 128:25–36.

Toral-Granda, V., A. Lovatelli, and M. Vasconcellos. 2008. FAO Fisheries and Aquaculture Technical Paper No. 516. *Sea cucumbers: A global review of fisheries and trade.* Food and Agriculture Organization of the United Nations, Rome.

Vanormelingen, P., E. Verleyen, and W. Vyverman. 2008. The diversity and distribution of diatoms: from cosmopolitanism to narrow endemism. *Biodiversity and Conservation* 17:393–405.

Warkentin, I. G., D. Bickford, N. S. Sodhi, and C. J. A. Bradshaw. 2009. Eating frogs to extinction. *Conservation Biology* 23:1056–1059.

Wood, C. L., J. E. Byers, K. Cottingham, I. Altman, M. Donahue, and A. Blakeslee. 2007. Parasites alter community structure. *Proceedings of the National Academy of Sciences USA* 104:9335–9339.

Conservation of Intertidal and Coastal Ecosystems

This chapter considers the environment and conservation issues in the intertidal zone, the region extending from the level of the lowest to highest tides (**Figure 3-1**). Although this region is clearly defined, its makeup varies widely depending on many variables including the range and nature of the tides (see Chapter 1), and it is strongly influenced by and integrated with neighboring ecosystems, both terrestrial and marine. The discussion is primarily concerned with the region extending from dunes, typically above the high tide line but strongly influenced by ocean processes, to the lowest zone of the subtidal, exposed to air only rarely. Intertidal regions are typically categorized by the primary substrate type as either rocky, sandy, or muddy. Marine intertidal zones differ from other coastal habitats in that they are relatively free of vascular plants, especially on rocky and sandy substrates. The majority of the surface of the intertidal zone of estuaries and marshes is covered by plants, such as mangroves or marsh grasses, growing in muddy substrates. These ecosystems are discussed in Chapter 4.

A continual change in exposure to air and ocean tides and waves, sometimes multiple times per day, creates physically stressful conditions in the intertidal zone. Ecosystems associated with marine intertidal zones have assemblages of organisms uniquely adapted for survival under the stresses of living in these physically dynamic environments. Although some organisms move in and out of the intertidal zone, many are adapted to remain, using adaptations to tolerate or avoid exposure even when the tidal waters recede. Species assemblages vary in part because diverse intertidal ecosystems are found at almost any latitude. Broad similarities occur from region to region in each type of habitat; however, species assemblages are adapted to local environmental conditions. There is much variability in physical factors, such as temperature and the intensity of wave action, and biological factors, such as food availability and exposure to predators.

Conservation issues related to intertidal ecosystems are largely related to direct interactions of humans with the habitat or organisms residing there. This is truer of the intertidal than most other marine ecosystems, because humans have easy access to these habitats and are attracted to these regions to live and recreate. Interest in conservation issues in the intertidal has recently increased as a result of some dramatic oil pollution events, and because changes from global climate change and sea level rise could greatly modify these habitats. This chapter begins with a discussion of the ecology of intertidal ecosystems and follows with a summary of conservation problems and potential solutions. Due to the contrast between the physical characteristics, biology, human accessibility and use, and vulnerability of muddy, sandy, and rocky intertidal regions, each will be discussed separately. Global climate change issues are similar for each type of intertidal regions and are discussed in a closing section of the chapter.

Figure 3-1 Rocky and sandy intertidal region on the north coast of Puerto Rico.

3.1 Dune and Beach Ecosystems

Dunes and sandy beaches are found on all continents (with the exception of Antarctica) and at almost all latitudes. However, they most typically form in areas with low sloping shorelines. Although they are interspersed among other types of coastal ecosystems, sand dunes and beaches are most prevalent along passive margins, and commonly form on barrier islands (see Chapter 1). Some of the most prevalent beach/dune complexes are found along the U.S. Atlantic and Gulf of Mexico coasts, forming an almost continuous band from Maine to the Yucatan Peninsula. Much of the coastline of Africa, Australia, and western Europe are also composed of dune/beach ecosystems (**Figure 3-2**). These regions are not as biologically diverse as most other coastal ecosystems due to the low level of nutrients and the dynamic, changing nature of these habitats. They are critical to the survival of many marine organisms, however, including migratory animals (e.g., sea turtles and sea birds) that use them for nesting. Conservation issues are mostly related to protection of these habitats from physical disturbance by humans because they are attractive to developers and tourists.

The Dune Ecosystem

Coastal dunes will form only if there is an ample supply of sands available to be carried inland by normal winds. Where they occur, coastal dunes occupy an area above the normal intertidal zone, extending to the furthest reach of coastal storms. This region is exposed to winds, sand blast, salt spray, and shifting substrate with low water-holding capacity and low organic matter. Despite these sources of stress, some plants thrive under these conditions and form the base of the coastal dune community.

Coastal dunes can be divided into regions based on their location relative to the ocean (**Figure 3-3**). The predominant vegetation in the **foredune** is grasses that function to stabilize the dunes. Along the southeast U.S. coasts the most prevalent grass is sea oats (*Uniola paniculata;* **Figure 3-4**). These grasses may reach 1 to 2 meters height and withstand exposure to seawater and salt spray; they tolerate partial burial by the sand and some exposure of the root system. The extensive lateral root system and **rhizomes** (horizontal stem with shoots and roots) of the sea oats stabilize the dune, and allow the plant to expand across the dunes. Sea oats also reproduce by seeds, formed in a large head; pollination and seed dispersal is primarily by wind. Other similar grasses, in the genus *Ammophila*, commonly called marram grass, beachgrass, and bentgrass, grow in dense stands on dunes along other North Atlantic coastlines, including Europe, Canada, and the northeastern United States. *Ammophila* grasses have expanded their distribution greatly through intentional introductions along the coasts of the U.S. Pacific, Australia, South Africa, Japan, Argentina, and Chile. Considered invasive in many of these areas, these grasses outcompete the native dune vegetation and modify the ecosystem enough to lower animal diversity. Where these thick stands of grasses do not occur,

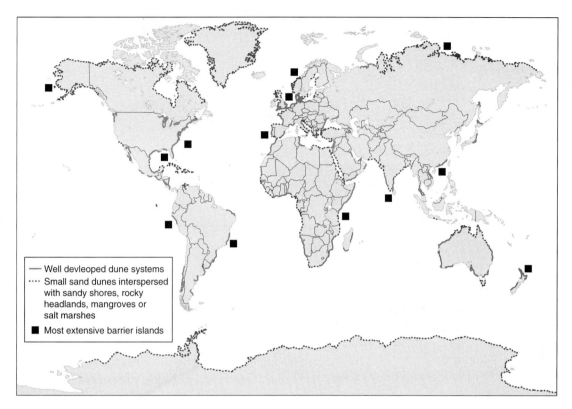

Figure 3-2   Global distribution of coastal dune systems.

they are replaced by more sparse grasses and low-growing trailing plants. Many dune plants tolerate regular immersion in sea water (termed **halophytic**), allowing them to extend onto the beaches. These include succulent plants such as the sea purslane, sometimes called sea pickle, and *Salicornia,* sometimes called pickleweed, found growing on the foredune (**Figure 3-5**); both are considered edible. In some arid areas such as northeast Africa, Saudi Arabia, and northwest Mexico, experimental plots are being raised as livestock food and for the development of biodiesel. Sea pickle has been touted as an environmentally friendly way

of conserving freshwater supplies by using saltwater for agriculture.

Other plants are mostly on the landward side of the dune—the **backdune**—protected somewhat from exposure to salt water and winds. These include spreading grasses, broad-leaved forbs, or low bushes (**Figure 3-6**). For example, the sea rocket *Cakile maritima,* native to Europe and with white-to-purplish flowers, grows in the backdune in clumps in the sand. It has been widely dispersed by humans, and is considered and exotic invasive species along the east and west coasts of North America. Coastal goldenrod

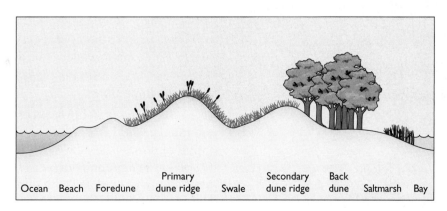

Figure 3-3   Profile of coastal beach and dune habitat.

Figure 3-4 Example of coastal dune grass, sea oats *Uniola paniculata*.

Figure 3-6 Habitat and plants of the backdune.

Solidago simplex and various lupines of the genus *Lupinus* are characteristic of North American west coast backdunes. In northern Europe, many of the backdunes are characterized by the goldmoss stonecrop *Sedum acre*, a creeping ground cover plant that produces small yellow flowers; it has also been introduced into North America. Many of the backdune plants are adapted to survive with low amounts of fresh water in a manner similar to desert plants. For example, the beach elder *Iva imbricata*, a common plant along the southeast U.S. coast, has succulent leaves that retain water. Many of the dune plants are related to or even the same species as inland grasses and forbs that grow along freshwater beaches or in arid areas. There are many more

plants associated with dunes than can be discussed here, and the species assemblages vary with the physical conditions and biogeography of the specific region. Of course, not all dunes are the same, and the presence and size of these zones depends on the width and characteristics of the dunes and the degree of protection from human impacts. Although there typically is not a sharp defining line, the backdune ends as the vegetation changes, most often to shrubs and then forests in unaffected areas.

Biodiversity is relatively low on dunes compared to most other coastal habitats. The stress from such factors as sand movement, high temperatures, drought or flooding, salinity, and low level of nutrients tends to limit the diversity of plants and animals. Many of the animals are of terrestrial origin, including: insects, such as ants, moths, butterflies, and beetles; reptiles, such as snakes and lizards; and mammals such as mice, raccoons, voles, foxes, rabbits, and deer. Some of these move onto the dunes opportunistically from inland habitats; others depend on the dune for survival, many of which have become endangered by loss of habitat. Seabirds may feed or nest around the dunes, and include some species endangered due to habitat loss.

Dune Impacts

The health of natural dune ecosystems varies greatly around the world. The most common reason for exploitation of dunes is to take over the space for other uses. Coastal human population density, thus, is the variable most strongly correlated with dune ecosystems impacts. Humans have dramatically modified and removed dunes for hundreds of years. A large proportion of the world's

Figure 3-5 Example of foredune plant, sea purslane *Sesuvium portulacastrum*.

population lives near the coast, and recent increases in coastal recreation have resulted in further modification and stress. These pressures are especially severe in regions with large areas of dunes and beaches. For example, about 70% of the U.S. population visits beaches when on vacation; over 80% of Australians live near the coast; and over ⅓ of the population of Europe lives within 50 kilometers of the coastline. Since 1900, about 25% of coastal dunes in Europe have been lost and, furthermore, it is estimated that about 50% of the remaining dunes do not function naturally and 85% are threatened.

So what are the specific effects of human actions on coastal dunes? The most dramatic are excavating the sands from the dunes or covering them with roads, buildings, or other structures. Artificial barriers, such as walls or dikes, put in place to protect coastal areas, may destroy or affect the characteristics of the dune or keep them from undergoing natural changes. Dunes and their ecosystems are highly modified when used for agriculture, such as livestock grazing. Pollution of the dune ecosystem is common with industrialization or urbanization of coastal regions. Free access to the dunes for tourism and recreation may cause trampling from foot or vehicle traffic on the dunes, the greatest effects being harm to plants. Plant removal destabilizes the dunes. Even if dunes are not otherwise affected, the low productivity may make colonization of new seeds difficult, slowing recovery. Vehicles may cause harm to bird populations in the dunes by affecting their behavior and destroying nests, eggs, and young. Replacement of native grasses with exotics may affect the dune structure and the overall plant and animal diversity; non-native grasses have often been introduced to stabilize the dunes or for livestock grazing without regard for the ecological function of the dunes. The most substantial of these effects will be elaborated on using specific examples.

One such example is found along much of the southeast U.S. coastline, where the preferred area for tourist resorts and amenities is near sandy beaches, historically on top of the dunes. It is not unusual to see lodgings and roads undercut or even washed into the sea after a strong storm. The location of these structures not only makes them vulnerable to winds and storm surge, in particular during hurricanes, but it also removes the ecological function of the dunes. In most U.S. states, laws have been passed to protect the remaining dunes and require that building be set back from the dunes; however, this typically does not require the return of lost dunes.

Building over the dunes is not as common along steep coasts (e.g., the U.S. west coast) as it is along gradually sloping coastlines (e.g., the U.S. southeast coast). The higher coastal gradient encourages building on higher lands overlooking the coast, and the greater wave activity along steep coastlines discourages building near the shore. Much of the harm to coastal dunes on the U.S. Pacific coast has been

the introduction of non-native *Ammophila* beachgrasses. These grasses, native to Europe and the U.S. Atlantic, were introduced over 100 years ago, probably because they created more stable dunes. *Ammophila* dunes are morphologically different, however, and the replacement of native dune grasses and other vegetation has changed the ecology, resulting in a lower diversity of animals in the dune ecosystems. In many areas efforts are now being made to return the native dune ecosystem (see Box 3-1. Conservation Success Stories: California and Netherland Dunes).

Culture, as well as geography, has resulted in a different set of impacts on coastal dunes of Europe. European coastlines have been attracting large numbers of human settlers for hunting, fishing, and agriculture since prehistoric times. Eventually many dune habitats were converted to farmland or residential areas decades or centuries ago. As populations grew, coastal settlements expanded rapidly onto habitats on or adjacent to the coastal dunes. Many of the dunes were removed or modified to support agriculture or settlements.

Livestock grazing has been carried out on former dunes in some areas of Europe for centuries or millennia. Cattle and sheep are grazed on dunes, and from the Middle Ages through the mid-1900s grazing by rabbits, for food, hunting, and fur was reportedly a primary factor affecting the sand dune community in much of Europe. Livestock grazing removes the natural ecological function of the dunes by impeding the succession process, reducing plant diversity, and lowering animal diversity.

Urbanization in Europe resulted in air, water, and soil pollution, habitat loss, water extraction, and sand excavation. Dunes were often modified by building up berms or dikes for coastal protection. During World War II many unsettled coastal areas of Europe were declared military grounds, prohibiting agriculture and new settlements, which afforded some protection to the dunes. Following the war, some of the dunes were developed into tourist destinations but not necessarily in the interest of protection as natural habitats. Many were made into parks that became so popular that trampling or parking cars on the dunes was prevalent. Grazing continued in some regions but has been abandoned in others due to loss of profitability. In some areas where agriculture was abandoned there is pressure to expand cities and industries. After attitudes began to change in the 1990s, more attempts have been made to convert dunes into natural areas and encourage a natural biodiversity (see Box 3-1. Conservation Success Stories: California and Netherland Dunes).

▪ Dune Conservation

Where there has been large-scale dune removal, recovery of coastal regions to a natural condition is difficult and expensive. In many of these areas the impacts are ongoing.

Few examples of efforts to bring back dunes on a large scale in heavily populated areas thus exist. Even where there have been large losses of structures due to hurricanes, the most typical response has been to rebuild in the same area with stronger structures or larger barriers to protect buildings. Despite commercial pressures from development and tourism, there are still extensive areas of dunes around the world that remain relatively pristine or only minimally impacted. These areas are where the primary conservation focus lies today and is the focus of this discussion.

Where dunes have been spared the negative effects of coastal development, there are usually legal protections in place to save the remaining healthy dune ecosystems. For example, scattered along the U.S. coastlines there are refuges, nature areas, parks, and public beaches where foot traffic is limited to specific areas. Often there are boardwalks elevated above the dunes (**Figure 3-7a**). Along the U.S. southeast coast the removal of sea oats, for example by residents to enhance their view of the ocean, is prohibited by state laws. Endangered species laws (see Chapter 9) provide protections not only for species using the dunes but also for dunes that are considered "critical habitat." For example, at least 15 species of California coastal dune plants are protected by endangered species laws. Dunes also are considered critical habitat for animals such as the endangered Alabama beach mouse and several animal species in California, including the Morro blue butterfly, the shoulder band dune snail, and the California least tern. Many other countries have habitat and protected-species laws that provide dune protection. In Europe, international agreements encourage protection of defined habitats including dunes. For example, the Habitats Directive of the European Union defines several types of natural dune habitats as priority to be maintained for protecting wildlife. Other directives, such as the Birds Directive, define areas to be protected as habitats for certain groups of species.

Education is an important component of dune ecosystem protection. Coastal residents should be informed of the important ecological and physical functions of dunes. Tourists should be informed of the sensitivity of the dunes to plant removal or foot traffic. Educational displays at visitor centers in parks and reserves, or simply signs near public beaches of walkways, serve an education function (Figure 3-7b).

In situations where dune or former dune habitat is protected but has been affected in some way, more active conservation measures may be needed. For example, in areas where plants have been lost due to anthropogenic or natural causes such as storms, replanting of dune vegetation may be recommended to enhance recovery. To replant an entire dune area requires extensive effort, and replanting in patches may stabilize the dune enough that it begins

(a)

(b)

Figure 3-7 Measures to protect coastal dunes: **(a)** walkway through dunes to access the beach, and **(b)** sign educating visitors to avoid walking on dunes.

to recover naturally. Along the eastern U.S. and Gulf of Mexico coastlines, native grasses such as beachgrass, sea oats, and bitter panic grass (*Panicum amarum*) are often used to stabilize dunes because of their high survival and rapid growth. Where efforts are being made to reestablish dune sands, a barrier is put in place to trap the sands prior to planting grasses. A sand fence made of spaced wooden slats is placed several hundred feet behind the high tide line (**Figure 3-8a**). After several months sands will accumulate around the fence that are stable enough to plant grasses.

(a)

(b)

Figure 3-8 Measures to restore coastal dunes: **(a)** dune fences on Dauphin Island, Alabama, and **(b)** dune restoration with dredged sands, East Timbalier Island, Louisiana.

Grasses are often planted by hand, but may be planted with a tractor-drawn planter in large areas (Figure 3-8b). The recovering dune is protected by prohibiting or discouraging walking on the dunes, and walkways are often built over sensitive dune plant communities to keep them from being trampled.

Revegetating of dunes with non-native plants results in a modified ecosystem that may not support native wildlife species. One concern is the potential disruption of biological interactions among the plants and animals that have co-evolved in the ecosystem. Invasive species also may cause physical habitat changes when they do not support the formation of the same type of dunes as the native plants. Because the artificial planting of *Ammophila* grass may result in steeper, more stable dunes and thus modify the ecosystem, these invasive grasses may be removed to enhance dune recovery. On California dunes, *Ammophila* grasses are often removed simply by uprooting by hand (**Box 3-1, Conservation Success Stories: California and**

Netherland Dunes). Herbicides are sometimes used but potential toxic effects to native vegetation must be considered. Burning has been implemented when there are large overgrown areas with little native vegetation that is intolerant of burning.

Dune formations that have been modified by storm surge or strong winds typically will return naturally given time. These are natural occurrences that dune ecosystems have evolved with and adapted to tolerate over the long run. As plants return, either naturally or with human assistance, they begin stabilizing the sands in a manner that will eventually return the dunes to their natural formation. If the loss of the dune removes protection for human dwellings or other structures, earth-moving equipment may be used to re-establish the dune structure. This, however, is not typically needed for long-term conservation purposes. Dunes re-created in this manner will be unstable and vulnerable to destruction by winds or the next storm surge, at least until vegetation is reestablished.

The Beach Ecosystem

Soft bottom habitats where there are substantial waves and currents typically accumulate sands and form beaches in the intertidal zone. Beaches form in the same geographic regions as dunes, along gradually sloping sandy regions of the coast, primarily from sediments eroded from the land or from pulverized hard parts of marine invertebrates (see Chapter 1). The beach typically extends from the seaward edge of the dunes to the water's edge at low tide (**Figure 3-9**). Physically, beaches differ from dunes in that the primary force moving the sands and otherwise affecting the ecosystem is water in the form of waves. Although there is some transport of sands and biological interactions between the beach and the dunes, the transition between the two is fairly abrupt. Beaches are possibly the most naturally physically controlled coastal ecosystem. Beach-associated organisms have to cope with living in one of the most physically dynamic and constantly changing habitats on earth. The most obvious biological effect is the lack of attached plants, largely a result of shifting sands and pounding by waves.

The general lack of large primary producers on the beach means beach food webs depend largely on the import of nutrients and organic matter from the sea to the beach. The low level of nutrients and physical stress also results in a relatively low biodiversity. The animal diversity is higher than it first appears, however, because most of the animals associated with the beach are either hidden in the sands or cryptic to avoid exposure to wave action and predation.

The lack of rooted vegetation on the beach means that vascular plants and macroalgae are found only if they have been uprooted and deposited. Primary producers associated with the sands include microalgae, such as benthic

Box 3-1   Conservation Success Stories: California and Netherland Dunes

Many of the dunes of the U.S. Pacific coastline have been protected from destruction by excavation or construction but are endangered by the spread of exotic grasses. These dunes support a unique type of native grassland that is found only on the foredunes of this region. The beachgrass *Ammophila arenaria,* native to Europe, was introduced to California dunes in the mid-1800s and has largely replaced native American dune grass *Leymus mollis*. This outcompetition is apparently due to *Ammophila*'s greater efficiency in utilizing nutrients and tolerating drought (by tightly rolling its leaves). The consequences of replacement, however, reach far beyond effects on the grasses. *Leymus* grows sparsely, traps relatively little sand, and forms dunes that are gradually sloping, resulting in a series of low mounds or **hummocks**. *Ammophila* tends to grow in more tightly packed bunches, and thus traps more sand and forms a steep dune face. The steep dune face limits the disturbance and movement of sand onto the back dune region. Limiting natural physical disturbance may seem beneficial; however, the native dune ecosystem has evolved with natural disturbance and they enhance biodiversity. Although some mammals and birds benefit from *Ammophila*'s better cover, other species are endangered by the changes; for example, the western snowy plover is considered threatened by the loss of the *Leymus* dune habitat.

As conservation attitudes developed, there was an increased awareness of the unnatural conditions of many of the dune ecosystems on the U.S. Pacific coast. In northern California, efforts to restore dunes were initiated in the early 1980s on the Lanphere Dunes, protected as a preserve and wildlife refuge (**Figure B3-1**). It was discovered that a natural recovery could be initiated by simply removing the invasive *Ammophila* by hand. This simulated a natural disturbance to the dunes that initiated the succession process; these regions are adapted to frequent disturbances by storm surge and winds. Native plants from other areas, including some endangered species, quickly colonized the area. Continued monitoring of the dunes to remove exotic grasses allows the natural ecosystem to remain intact. Since these initial studies, efforts

Figure B3-1 European *Ammophila* beach grass, an invasive that has replaced native grasses in some California dune habitats, grows in tight bunches, trapping sand and modifying the natural dune structure.

to restore coastal California dunes have expanded into many other areas. Efforts involve numerous government and non-government agencies, including the Nature Conservancy, the U.S. Fish and Wildlife Service, the California Conservation Corp, the U.S. Bureau of Land Management, and many local groups, such as Friends of the Dunes. Restoration methods may include burning and herbicides to recover large areas with less expense when deemed safe. This conservation issue is ideal for volunteer involvement. Programs that involve locals in the manual removal of grasses provide a sense of stewardship and result in rapid and obvious results as native grasses, flowers, and wildlife return.

Much of Europe has a long history of human use of coastal areas that has resulted in modification of the dunes and, therefore, restoring natural dunes may require more extensive efforts. For example, in the Netherlands much of the habitable area is in low-lying coastal areas. As populations expanded, many inner dunes were excavated and the sand used to fill lowlands inland to expand cities and towns. Frequent flooding of coastal settlements during storm surges led to extensive dike building to protect these areas from high waters. This system of dikes, currently over 2,000 kilometers in total length, limits the restoration of many regions of coastal dune habitat. Where dunes remain, many of the multispecies plant communities have changed into monospecific stands of grasses in the past several decades. The reasons for this trend are not certain, but appear to be associated with efforts made since the 1950s to stabilize the foredunes against **blowouts** (breaking through the dunes by storm surges) that affect the natural dynamics of the inland dunes. Additionally, some dunes have been managed for livestock grazing.

With an increased environmental awareness, recent efforts have been made to restore natural coastal dunes to some areas of the Netherlands coastline. One such area is the Meijendel Dunes, found north of The Hague, surrounded by one of the most densely populated areas in the Netherlands. This is a relatively small area, about 6 kilometers long and 3.5 kilometers wide, but it has been designated to receive the highest degree of protection by the national government. Protection of this as a natural dune ecosystem is a challenge because it is used not only as a nature reserve, but also as a catchment area for fresh drinking water, a recreation area for about one million visitors per year, and an important barrier to defend low-lying areas inland against the sea. To use the dunes as a catchment area, artificial lakes were created as reservoirs by inundating the dune valleys; groundwater is supplemented with water piped in from rivers. Much of the area was developed into an unnatural park from the 1950s–1980s, including the planting of exotic trees and shrubs, creating recreation areas, and allowing car parking on the former dunes. The barrier function of the dunes has been enhanced over hundreds of years by planting of marram grass, which stabilizes the sand and avoids blowouts.

Priorities for the Meijendel Dunes changed in the 1990s, with an increased focus on nature preservation. Zones have been established within which priorities are given for the conflicting functions of the dunes. There is still a focus on recreation, but car traffic is restricted, bike and walking trails have been established, and people are encouraged to focus more on nature observation than other forms of recreation. Some of the dunes are still used for drinking water catchment, but in a smaller area. Dunes have not been planted with marram grass since the early 1990s but are allowed to be naturally colonized by a more diverse natural assemblage of vegetation; this makes the dunes more mobile and small-scale blowouts are tolerated. Nature conservation is given priority in about 35% of the area; only foot traffic is allowed in designated areas. Surveys have shown that the majority of people living in the regions are willing to pay more for services such as drinking water to support these conservation efforts.

Because of pressures from the large human population in a large vulnerable area, few of the dunes of the Netherlands (and many other European countries) may ever be returned to truly pristine conditions. A balance has to be achieved between the needs of humans and the desire for conservation. How this balance is achieved depends largely on culture and attitudes. Efforts like those in the Meijendel Dunes reflect a change in conservation attitudes and a desire to prioritize the reestablishment of at least some areas to their former ecological function. This desire is reflected by the willingness to pay for changes, both monetarily and through changes in behavior.

diatoms and phytoplankton. Other microscopic organisms include protozoans and bacteria. Meiofauna, including copepods and amphipods hiding among the grains of sand, consume these primary producers and form an important link in the macrofaunal food chains. Macroinvertebrates, especially juvenile crabs and other crustaceans, may prey heavily on the meiofauna, especially at high tides. Macroinvertebrates may be abundant but are well camouflaged or bury themselves in the sand, and rarely form permanent burrows because of the instability of the sands. On beaches with considerable wave action, however, the growth of diatoms attached to sand grains is limited by vertical movement of the sediments (the photic zone is typically limited to about 5 millimeters depth in the sands). Planktonic diatoms may accumulate in large numbers in a film on the surface of the sands; in some cases they are abundant enough to form a line of foam. The diatoms in a given region are mostly of a single species, most in the genus *Attheya* or *Anaulus*. Other photosynthetic organisms include bacteria, cyanobacteria, and flagellates. These too are limited by instability of the sands. Some live as deep as 20 centimeters in the sand, and are capable of vertical migrations, moving toward the surface during daytime and low tides. They may reach densities as high as 1,000 cells per cubic centimeter in beach sands. Some of the same organisms extend into the subtidal sediments. Although seagrasses or other large plants typically don't grow on the beach, they may be an important component of food chains in areas where dead plants wash onto the beach. They accumulate in the **wrack zone**, the region at the upper extent of the tides, and serve as an important source of food for some invertebrates.

Microorganisms and Meiofauna

Bacteria are found mostly attached to beach sand grains, at densities in the hundreds-of-millions of cells per gram of sand. Ciliates and foraminiferans are the most common of the protists and may feed on bacteria, diatoms, and other protists. Intermediate-sized **meiofauna** (organisms roughly ¼ to ½ millimeters in diameter) live mostly interstitially among the sand grains. Nematode worms and harpacticoid copepods are often the most prevalent, with nematodes more common in finer sands. In finer silts, such as found in marsh sediments, these interstitial forms are replaced by burrowing meiofauna. Other common marine meiofauna groups reside in the beach sands in lesser abundances; many beaches have more than 100 meiofauna species. The average densities of meiofauna living in beach sands range from about 50 thousand to 3 million per square meter; they may be found as deep as two meters or more in loose, well-oxygenated sands, but are most prevalent in the top 30 centimeters. They are in greatest densities in the mid- to upper beach. The microorganisms and meiofauna below the top few centimeters of sands are not accessible to macrofauna and, therefore, often form a closed food web.

Macroinvertebrates

Macroinvertebrate diversity is low on beaches compared to vegetated coastal habitats. This is likely due to the lack of structure and habitat variety, and the importance of physical factors. Diversity appears to be controlled primarily by

Figure 3-9 A beach on Turks and Caicos Islands; notice the accumulation of debris in the wrack zone.

physical factors, rather than biological factors such as competition and predation. Typically there are 20 to 30 macroinvertebrate species present on an individual beach. Beaches with finer sands, lower slope, and greater tide range tend to have higher biodiversity as well as an increased abundance of individuals. Even where they are diverse and abundant, invertebrates may not be noticeable due to camouflaged coloration or the ability to bury themselves or burrow in the sand. Due to the mobile nature of the sands, sessile invertebrates are not associated with the beach. Each species has its own forms of locomotion, which may include burrowing, swimming, crawling, leaping, surfing, or running, and is strongly oriented to the rhythm of the tidal and solar cycles, and frequency of waves. Although many specialized groups of macroinvertebrates reside in beach habitats, about 90% are crustaceans, mollusks, and polychaete worms. Although tube-building polychaete worms are much more common on mud flats and sheltered areas, polychaetes such as the lugworm *Arenicola* may occupy lower areas of sandy beaches and be the primary macroinvertebrate in sheltered beaches. Sensitivity to exposure and coarse sands limits their distribution.

Gastropods may be a prevalent and relatively visible component of beach ecosystems. Mollusks, including the prosobranch gastropods (snails), often are predominant in intermediate areas of the intertidal shores (**Figure 3-10**). These include whelks of the genus *Bullia*, found on tropical beaches around much of the world, with the exception of the Americas. *Bullia* are opportunistic feeders, eating any dead animal they can find. They are very efficient at locating new sources of carrion by following the chemicals emitted by decaying organic matter. Upon detection they will emerge from the sand, surf (using an enlarged foot as an underwater sail) to a location near the food, and follow its odor trail by placing their siphon on the surface film of water. Once the whelk reaches the food, it holds onto it by creating suction or folding the food into its foot, attaching with the long tubular

proboscis, or dragging it below the surface of the sand. They can consume their catch rapidly, ingesting up to one-third of their own tissue weight in 10 minutes.

Gastropods also may be important predators on the beach. *Terebra*, found on tropical beaches around the world, preys on other invertebrates, mainly polychaetes, by injecting poison using harpoon-like radular teeth. *Olivella*, sometimes the most numerous gastropod on U.S. Pacific beaches, burrows along under the sands, with only the tip of its siphon exposed, feeding on small mollusks and crustaceans. *Natica* and *Polinices*, commonly called moonsnails, are predatory gastropods, mainly of bivalves on beaches, which bore through shells using the radula to feed on the soft tissues. A single round hole identifies the shells of predated snails when washed up on the beach. Various species of these snails are found on the coasts of Europe, Africa, Australia, and the Americas.

Beach bivalves (**Figure 3-11**) are typically found buried in the sand on the mid to lower portion of the beach, extending into the subtidal because they must be submerged to filter feed on organisms suspended in the water. In beaches with fine sands they often comprise the majority of the biomass. *Donax* is the most common bivalve genus on wave-swept beaches, including 64 species typically a few centimeters in length; species are found along beaches around the world, with the exception of colder sub-polar and polar waters. Intertidal species, as discovered by Olaf Ellers, jump out of the sands and extend their foot and siphons to ride the waves up and down the shore with the tides and then bury themselves rapidly—often in less than 5 seconds—in loose sands as waves recede. The common species on beaches of the U.S. South Atlantic and Gulf of Mexico is the coquina *Donax variabilis,* named for the remarkable variability in shell colors among individuals (presumably a mechanism to keep bird predators from

Figure 3-10 The olive snail *Olivella biplicata,* a mollusk found in intertidal sands.

Figure 3-11 Beach bivalve shells accumulated on lower beach.

learning to recognize the shells by color). When abundant, *Donax* can be easily found by digging in sands in the lower beach exposed to the surf. Other beach bivalves include the surf clams *Spisula* and razor clams *Siliqua,* both of which include species commonly harvested in some locations around the world.

Arthropods living in beach ecosystems are mostly crustaceans. One exception is the horseshoe crab, a chelicerate arthropod that comes into the intertidal zone of some beaches in large numbers to breed (**Figure 3-12**). *Limulus* comes ashore on both the Atlantic and Pacific coasts of North America; there the female deposits several hundred eggs into shallow burrows where they are fertilized by the male.

Crustaceans are typically the primary invertebrates in the upper intertidal area of exposed sands, comprising a broad diversity of species around the world (**Figure 3-13**). Copepods are the typically the most numerous meiofaunal crustaceans. Some live interstitially among the sand grains, others burrow in finer sands, and some are a part of the zooplankton in the surf zone. Amphipods are a common inhabitant of the upper beach, often associated with the wrack zone, feeding on the dead plant material washed up by waves. Others are more aquatic and can be found on the lower beach. Along the U.S. Atlantic and Gulf of Mexico coasts, amphipods such as *Haustorius* may be the predominant beach invertebrate. Isopods can be found from the upper intertidal to the surf zone, with each species having its own specific zonation and activity pattern. Some species are herbivorous, depending on seaweeds or other washed-up plant material, and others are carnivores. *Excirolana* is probably the most widespread isopod and is often the most common macroinvertebrate on the beach. Some species of shrimp use the surf zone as a nursery area

(a)

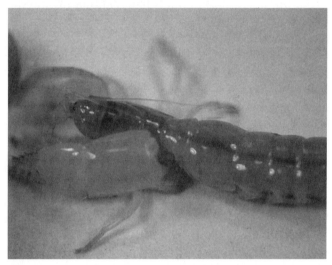

(b)

Figure 3-13 Beach-associated crustaceans: **(a)** a ghost crab *Ocypode quadrata* pulling a dead hermit crab into its burrow; **(b)** ghost shrimp.

Figure 3-12 Horseshoe crab *Limulus polythemus* on the beach of Horn Island in the northern Gulf of Mexico.

and others burrow into beach sands (e.g., *Lissocrangon* in North America). The mole crab *Emerita,* often called "sand flea," and *Hippa,* the "sand louse" with shorter antennae, live in the surf zone and can rapidly burrow into the sands. They move up and down the beach with the tides. These are commonly captured to be used as bait. Hermit crabs, such as *Clibanarius,* occupy abandoned shells of snails such as whelks or moon snails and will move onto the beach to scavenge. Terrestrial hermit crabs *Coenobita* may feed on the upper beach in the tropics. They are commonly caught and sold as pets. Sand crabs or ghost crabs, in the genus *Ocypode,* are fast-moving crabs, living in burrows on the mid to upper beach that they defend, and moving out onto the beach to feed, primarily at night. Some ocypodid crabs are harvested as food in some regions of the world. Beach crabs are typically opportunistic feeders, scavenging on dead animals when available, or preying on smaller invertebrates otherwise. Larval shrimp and small mysid shrimp may be important parts of the plankton in the surf zone, feeding on organisms and organic material that are concentrated by the water circulation. Burrowing shrimp, such as the ghost shrimp *Callianassa*, may be extremely abundant in beach sands (and also in salt marshes), with densities in some areas higher than 400 individuals per square meter. They live in burrows and feed by scraping organic material from the sand grains. These shrimp are easily harvested for bait by locating the opening to burrows and digging them from the sand.

Echinoderms such as brittle stars, sea stars, and sea urchins, though common in the shallow subtidal, do not typically move onto beaches with high wave energy. One exception is *Echinocardium*, a light colored sea urchin that burrows into intertidal sands and is most abundant in the surf zone. Echinoderms such as brittlestars may be abundant in low energy beaches. Sand dollars of the genera *Encope* and *Mellita* can also inhabit the intertidal zone, where they feed on small organic particles and organisms deposited among the sand grains.

Vertebrates

Fish may be an important component of the beach ecosystem at high tide and in the surf zone. These are most typically larval and juvenile forms of many species that feed opportunistically on whatever organisms accumulate in the surf zone or are available on the surface of the sand (including the siphons of clams). The surf zone is an important nursery area for some fish species, including anchovies and flounders, for feeding on zooplankton, avoiding predators, or hiding in plant material washed into the surf zone. Over 100 different fish species have been found in the surf zone of some beaches; many of these are the same species that use estuaries as nurseries (see Chapter 4). Larger adult fishes do not typically reside in the intertidal areas of beaches; however, flounders, skates, and rays commonly move onto the flooded beach at night to feed, their dorsoventrally flattened shape allowing them to move into very shallow waters. Due to the dynamic nature and moving sands of the intertidal region few fish species spawn there. The California grunion *Leuresthes tenuis* is one exception, moving onto the beach at night just above the highest reach of spring high tides and depositing eggs in the sand (**Figure 3-14**). The larvae are then washed out during the next spring tide two weeks later. Atlantic silversides (*Menidia menidia*) lay eggs in protected areas such as crab burrows or in plant matter washed onto shore during daytime of spring high tides.

Terrestrial vertebrates that could be considered a part of the beach ecosystem include mammals, such as mice, raccoons, or coyotes that move from the dunes to the beach to feed. The most common vertebrates on the open beach

Figure 3-14 California grunions *Leuresthes tenuis* spawning on a beach.

are birds, however. Some species (e.g., terns, cormorants, pelicans, and penguins) use the beach for roosting and feed at sea; some (e.g., plovers and terns) nest on the upper beach, but more typically in the dunes; and many species feed on organisms in the intertidal sands of the beach. The species and abundance of birds on the beach vary depending on many factors, such as season, migration periods, food availability, biogeography, beach characteristics, and human impacts. As many as 1,000 birds per kilometer may appear on beaches during peak migration periods, and as many as 100 birds per kilometer may be present throughout the year.

Birds that commonly feed on the beach include gulls and terns, sandpipers and sanderlings, plovers, and oystercatchers (**Figure 3-15**). These birds use three different mechanisms for feeding. **Probers** (e.g., sanderlings and

Figure 3-15 American oystercatcher, one of the many birds that feed in beach ecosystems.

oystercatchers) move along the beach feeling with their bills for invertebrates buried in the sand; **surface feeders** and peckers (e.g., plover) visually search for food on the surface of the sands; and **scavengers** (e.g., gulls) feed on dead animals stranded on the beach. Most of the birds feeding on the beach are opportunistic within these feeding types; the kinds of food consumed are largely determined by what is available. Bird species, however, often specialize in their feeding based on location and size. For example, sanderlings feed at the edge of the water, moving up and down the lower beach with the waves, while plovers search for food higher on the beach. Bills are often adapted to handle certain sizes of organisms. Some birds have developed specialized behaviors to access food from the beach. For example, some gulls will pick up clams, fly above a hard surface such as a rock or roadway, and drop the clam to break it to get to the soft tissue.

Birds may consume a substantial portion of the biomass produced by beach organisms (as much as 40%–50% appears to be common). This organic matter is eaten in forms ranging from bacteria and protozoans to bivalves and crustaceans. Birds also contribute organic matter back to the beach ecosystem in the form of feces, feathers, and carcasses. Predation by birds does not typically affect the long-term health of invertebrate populations. Even at higher levels of predation, such as when large migratory bird populations move onto a beach to feed, prey populations appear to recover rapidly. Beach organisms are adapted to resist predation, not only physically, by developing hard shells or exoskeletons, but behaviorally, by moving vertically in the sands on a solar or tidal cycle to avoid predation.

The vertebrates using beaches that receive the most conservation interest are the sea turtles. All sea turtles require access to sandy beaches for nesting (**Figure 3-16**). Females return from the ocean to beaches where they were hatched to lay their eggs. Because turtles do not feed on the beach, their importance to the beach ecosystem is limited. Although predation of hatchling turtles is common, predators cannot rely on the turtles as a regular source of food because turtle nesting is concentrated over a short seasonal period. This predator swamping keeps predators from maintaining populations large enough to remove a large percentage of the hatchlings (see Chapter 2).

The life history of sea turtles varies among species; however, they all require access to relatively loose, moist sand above the high tide level, or in some species the lower dunes, in which to dig a shallow pit where the eggs are deposited. About 50 to 150 eggs are laid in the burrow, which the female covers before she returns to sea. An individual female will return to the beach to lay eggs several times during the reproductive season at two- to three-week intervals. During the approximately two-month incubation time, the eggs are not commonly exploited by predators (with the exception of some terrestrial mammals, including humans!); however, hatchlings are preyed upon, mostly by crabs and seabirds. The hatchlings dig their way out of the nest at night, using the reflection of ambient light off the ocean's surface to orient themselves as they head to the ocean. There they are vulnerable to sharks and other fishes as they move toward adult feeding grounds. Other aspects of the sea turtle life history and survival are discussed in later chapters.

Beach Impacts

The desires of people to live, recreate, or establish businesses near the beach result in numerous anthropogenic effects on the beach ecosystem. Many of these are due to physical disturbances that modify the beach landscape and geological processes in a manner that affects their stability and habitability by wild organisms. Other effects are caused by introducing pollutants that eliminate or modify the function of beach ecosystems. Taking organisms from the beach for human use or consumption also results in potential harm to the ecosystem.

Beaches are naturally such dynamic habitats that resident organisms are adapted to regular physical disturbances, and are typically able to recover quickly. Many invertebrates respond to storms or cold temperatures by moving into subtidal waters or burrowing deeper into the sands. Although storms may remove sands from the beach, they typically redeposit them elsewhere. However, more frequent storms may result in a permanent modification of the beach, and the transport of sands offshore may be a net loss of beach. Beaches may recover rapidly from human impacts that mimic natural disturbances on a local scale, such as moving sands, moderate organic pollution, or local harvest of clams, if the beach is allowed to recover without continual disturbance. Unnatural or more permanent

Figure 3-16 Loggerhead turtle laying eggs on a beach in Japan.

disturbances, however, may have serious and long-term negative effects on beaches.

Beaches are in constant motion and if one observes a beach over a period of months or years, or hours in the case of storm activity, there may be a noticeable natural change in the shape or size of the beach. A given beach may either decrease or increase in size substantially. This is not problematic to the organisms associated with the beach; they simply move with the sands. Human beings, however, are not as willing or able to move along the coastline to follow the movements of the sands. In order to maintain "our" beach, we are willing to go through enormous efforts and expenses to either stabilize the beach or renew the sands that gradually move elsewhere (**Box 3-2. Conservation Focus: Who Owns the Beach?**).

One way to accomplish desired modifications of the beach is to reshape it using earth-moving equipment. This can be used to create barriers to avoid overwash or flooding of dunes or buildings from storm tides, to create canals or channels, or to modify the beach for recreational purposes by making it broader and flatter. Some of these changes may

enhance the beach and dune ecosystems in the long run if they are done to recover natural conditions. Otherwise beach ecosystems may be affected negatively through mortality or movement of organisms and modification of habitat. If the human disturbances are infrequent and not too extreme, organisms are likely to return naturally. Attempts to modify the natural beach condition may be short-lived, as the beach will likely soon return to its natural state.

The most common way to provide protection from ocean waves and tides on or near the beach is to put an artificial hard structure in place as a protective barrier; this is referred to as **beach hardening** (**Figure 3-17**). One of the simplest but most dramatic methods of hardening the coast is by placing **seawalls** or **revetments** (long mounds or banks) along the upper reaches of the beach parallel to the shore (**Figure 3-18a**). These protect buildings and structures behind the beach from all but the most severe waves and storms; however, they often result in dramatic modification or total loss of the beach. For example, on some California beaches the upper intertidal zone was lost in front of sea walls, wrack accumulation decreased,

Box 3-2 Conservation Focus: Who Owns the Beach?

Living close to the shoreline, near the intertidal zone, is attractive for many reasons. It provides ready access to the ocean, whether it is for harvesting food, transportation, recreational activities, or the aesthetic appeal of the beach-front view. But, what happens when the ocean intrudes on what you consider your private property (**Figure B3-2**)? When societies were more

Figure B3-2 Beachfront eroding away from homes on Dauphin Island, Alabama. Note (in background) sands accumulated around foundations in an attempt to stave off erosion.

mobile and populations were lower the solution was simple: move further inland. But as we have become more sedentary and populations have increased, land further inland is likely already occupied and moving your home is not a simple process. How individuals and nations deal with this will become more and more important as coastal erosion increases and sea levels rise. The response will vary depending on numerous environmental, social, and political factors.

In the United States (and many other countries) the oceans and subtidal waters along the shore are considered **common property** unavailable for private ownership. The states have primary control over regulating activities in coastal waters (see Chapter 12 for a discussion of rights to access ocean resources). The intertidal zone, including beaches, is considered a part of the region to which there is common access. Landowners above the upper edge of the intertidal have personal property rights, however, and conflicts over the transition zone from private property to common property are commonplace. One reason for conflicts is that the legal designation of beach boundaries does not always coincide with the physical limits of the beach or the biological beach ecosystem. The typical legally-defined upper boundary of the beach considered for common access is the mean high tide line. The physical beach typically extends beyond the mean high tide, however, often to the level of highest spring tides or storm tides (the beach above the mean high tide line is legally referred to as the "dry sand beach"), and may be considered part of the private property of land owners. The biological beach/dune ecosystem extends even further inland to the coastal dunes, which are generally considered private property.

Due to disagreements over what comprises the common property zone of the beach, and the fact that beaches are constantly being changed by natural processes, numerous conflicts arise for which there are no simple solutions. For example, can landowners modify the beach to protect their private property? Does the public have rights to cross private property to access the beach? What activities are allowed on the beach and—possibly most controversial—what happens when private property is taken over by the sea or converted into common property (the intertidal zone) due to erosion or sea level rise? The solutions are complicated because each U.S. state has its own set of laws dealing with personal property rights.

One set of conflicts relates to dune protection. Because the dune zone is not considered part of the intertidal zone, it is not considered common property unless specifically designated as protected as a park, reserve, or other public property. Many states, however, have limits on building in the dune region. For example, in Florida there are setback requirements for building along the coast, and in many beach communities there are local laws to prohibit or limit building near the beach. In Connecticut, construction is not allowed within 100 feet of the reach of the mean high tide; however, zoning boards can give exceptions if certain requirements are met. In Texas, setback requirements for building range from 25 feet to 350 feet landward from the edge of dune vegetation, depending on local ordinances; proposals to require at least a 300-foot setback statewide have, predictably, met with resistance by property owners who argue that they will lose property value if they can't build in this region. Other, less direct methods of limiting building on the dunes include high insurance rates or limiting government disaster relief, especially in hurricane-prone areas.

Even if there are limits on building near the shoreline, what happens if erosion or sea level rise forces the shoreline onto your property, or the property is engulfed entirely by the sea? Should you have rights to land further inland; can you attempt to control beach erosion to protect your property? Must the government compensate you for your losses? How these questions are answered legally varies from state to state and in many coastal states the laws available to deal with these issues are unclear. For example, in Texas, coastal erosion, especially in areas affected by recent hurricanes, has resulted in movements of the beach landward into the region that used to be coastal dunes. Private property on which people built houses (typically on stilts to avoid flooding) has turned into beach, which is, therefore, considered public property of the state. Although laws are unclear, in this case the state offered to move affected houses inland; however, landowners are not compensated for the loss of property. Although U.S. laws require that compensation be made for property taken by the state, the state of Texas argued that the loss of these coastal lands to erosion is not their responsibility. Landowners counter by arguing—thus far unsuccessfully—that government projects are at least in part responsible for the increased erosion, due to building ship channels and levies that rob the beach of sources of sands or affect sea levels. Similar situations have occurred along the Florida Atlantic coast and other regions along the Gulf of Mexico as coastal erosion increases and sea

levels rise. In Florida, beach lost suddenly due to storm activity (a process called **avulsion**) can be recovered by nourishment and landowners retain their original property boundary, but if natural erosion shifts the intertidal zone, landowners may lose property without compensations. No doubt legal arguments will accelerate as to who should bear the legal and financial responsibilities of the effects of erosion and climate change.

The natural erosion and movement of beaches, even in the absence of sea level change, may result in local erosion and loss of beachfront property. The legal protection of beaches as common resources has resulted in outlawing the use of hardened structures, such as groins, to protect beaches in some regions. For example, hardened structures have been banned from the North Carolina coastline since the mid 1970s; regulations are based on the belief that it is best to let the beaches change in response to natural processes. Prospective property owners are warned that boundaries of inlets and beaches may change. Recent attempts to obtain permits that would grant exceptions to this ban and allow building of groins to protect beaches adjacent to some large expensive homes has led to resistance by conservation groups.

Access to the intertidal zone of the beach itself for recreational, nondestructive activities is typically given to the public. But when the only access to those lands is through privately owned properties, the definition of "public access" is not always clear cut. For example in New Jersey, Public Access Rules require that cities not limit access to the beach; however, it is debated whether they must provide adequate parking or other facilities. In Hawaii, the upper reach of the publicly accessible shoreline is defined as the highest reach of waves on the shore, identified by the accumulated debris in the wrack zone or the lower edge of vegetation; when these borders are in conflict it is debated which to consider. In North Carolina, public property extends to the mean high tide mark on the beach; however, it is unclear whether the public has rights to access the privately owned area of the upper beach (the so-called "dry sand beach"). New Jersey has similar laws allowing access to the intertidal portion of the beach, and courts have decided that this right includes crossing private dry sand beaches to access the intertidal "wet sand" beach; however, it is unclear under what circumstances and how much of the dry sand beach can be freely accessed and used, and whether parking and restrooms must be provided by municipalities. Texas has some of the most liberal beach access laws. Here, according to the Open Beaches Act, the public has free and unrestricted access to the beach, and the beach is defined as the area between the lower edge of the dunes and the edge of the water.

Court cases have become numerous related to these issues of private and common property rights in coastal intertidal zones; they will continue as long as sedentary humans wish to live in unstable and mobile environments such as the beach. As sea levels rise, flooding currently stable lands, conflicts will undoubtedly accelerate over management and ownership of coastal properties. Decisions should be made now concerning issues such as beach stabilization, compensations to coastal land owners, and movements of current coastal settlements. History tells us that, unfortunately, many of these decisions will be made only when they are inevitably forced upon us.

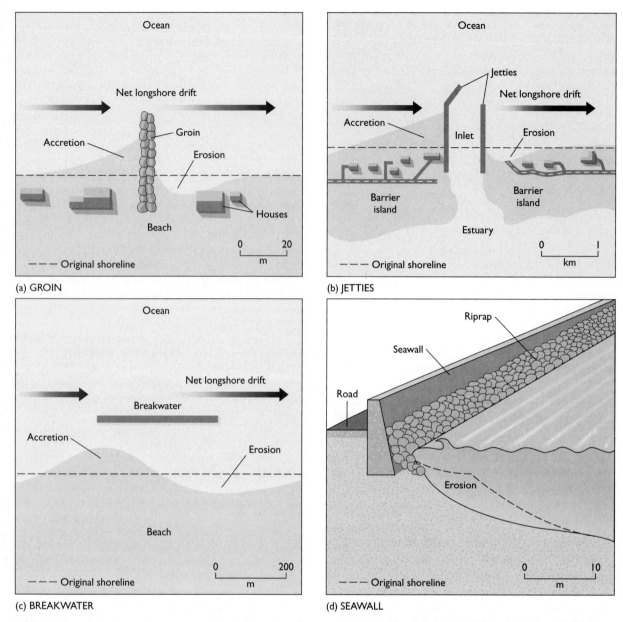

Figure 3-17   Coastal hardening structures, designed to limit beach erosion: **(a)** groin; **(b)** jetties; **(c)** breakwater; and **(d)** seawall.

and shorebird diversity and abundance was less. Natural movements of sand along the beach eventually may result in complete erosion of the beach with time, unless sands are replaced regularly.

One way to minimize the movements of beach sediments is to replace the sands with more stable materials. **Riprap**, a loose pile of large broken stones or boulders put into place along the shore, serves the purpose of protecting the shoreline from effects of waves and tides under typical conditions; however, this is carried out at the expense of covering over all or much of the beach (Figure 3-18b). If some beach remains between the riprap and the shoreline, it will likely be lost in the same manner as beaches isolated by seawalls. Although organisms will inhabit the rocks, they

occur in much lower diversity than in beach habitat and few beach-adapted organisms can live there. Riprap is commonly used along navigation canals and ports, or adjacent to coastal roads, buildings, or industrial facilities built along the coast.

In areas where the desire is to maintain a beach but avoid allowing it to undergo undesirable natural changes (e.g., to protect lodgings built adjacent to the beach), methods have been designed to disrupt the migration of beach sands along the shore. One way of achieving this is through the use of **groins**. Groins are constructed of rocks, wood, or other materials jutting out perpendicular to the shoreline (Figure 3-18c), acting as sediment dams to block the flow of sands carried by along-shore currents and waves, and thus retaining and building up the beach. The sands are retained

(a)

(b)

(c)

(d)

Figure 3-18 Coastlines with hardening structures in place to protect coastal areas from erosion and storm impacts: **(a)** seawall and riprap to protect the city of San Juan, Puerto Rico; **(b)** coastline hardened with riprap to protect a coastal town in Caribbean island of Guadeloupe; **(c)** groin to protect historic fort on east end of Dauphin Island, Alabama; and **(d)** fabricated beach hardening structures on the coast of Kyushu, Japan.

on one side of the groin; however, beaches downdrift from the groin are starved of sand and eventually may be lost from erosion. On coastlines with adjacent private properties, each landowner would be forced to build a groin to protect their beach at the expense of their neighbors, with the process continuing indefinitely down the beach. One way to minimize the effect is through designs that allow some sands to bypass the groin. **Jetties** are similar, but typically larger, structures built perpendicular to the shore at the mouth of inlets or harbors to avoid sediment deposition. They have the same effect as groins in that they stop the movement of sands along the shore, starving beaches down the coast of replenishing sands.

Although beach hardening has affected beach ecosystems for centuries, with technological advances in machinery used for building structures and movements of human populations toward the coast, the use of beach hardening has rapidly increased. Marinas, ports, and coastal developments often include hardening structures and other types of sand barriers that cause similar changes by blocking the movement of beach sands; beaches created adjacent to marinas or developments are typically managed not for natural structures, but to maintain sandy areas for recreation. Along the sandy beaches of the U.S. east coast, construction of groins, seawalls, revetments and other hard structures increased rapidly from the early 1900s through the 1960s.

Eventually, concerns for maintaining a more natural beach environment for recreation and conservation purposes resulted in a decline in beach hardening, and in many areas building such structures is illegal or requires a permit. Still many structures remain and continue to be maintained, especially where cities, industries, and ports have been established along the shore. In heavily populated coastal regions of European and Asian countries, much of the sandy intertidal coastlines have been hardened for coastal protection. For example, in Norway and Japan the majority of the sedimentary coastlines are protected with hard structures (Figure 3-18d). In developing countries, hard structures are still the preferred method of protecting settlements and structures from flooding or storm impacts. Dredging channels may have a similar effect, by removing the sands that would replenish beaches down the coast. For example, maintaining ship channels is considered a primary cause of net erosion of Gulf of Mexico beaches associated with barrier islands (see Chapter 1). These changes have been minimized by returning dredged sands to the beaches, and there are ongoing debates whether such programs should be expanded.

An alternative to building groins or seawalls is to place offshore **breakwaters** parallel to the shore. This is typically achieved by placing large rocks in shallow waters of the subtidal area, extending up to near the water's surface. Breakwaters are designed to reduce the wave energy before it comes ashore and thus help to hold the beach sands in place. They also allow people to build closer to shore because the wave effects are lessened. Breakwaters affect beach ecosystems by limiting the exchange of organisms, organic matter, and sediments between the subtidal waters and the beach and between the beach and the dunes. This also decreases the water circulation and reworking of the sands in the surf zone. For aesthetic reasons, breakwaters can be placed underwater out of site near recreational beaches. But because they reduce the wave action that beach goers consider appealing, they may be considered undesirable.

Problems resulting from adding hard structures to protect the beach have led to a dilemma for coastal residents and tourism facilities: how to maintain beaches in desirable locations without stopping the natural movements of sand and substantially modifying the beach habitat. One compromise, used most extensively in developed countries, is to "nourish" the beach. **Beach nourishment** is simply the addition of sand to the beach (**Figure 3-19**). It is used to protect buildings and man-made structures from wave action, maintain and improve beaches for recreation, and create new beaches. If this is deemed desirable, the obvious question is: where do we get the sands to nourish the beach? In situations where there is an accumulation of sand on an adjacent beach, this **accreted** sand can be moved by

Figure 3-19 Beach nourishment by the U.S. Army Corps of Engineers on Kure Beach, North Carolina by dredging and replenishing sand.

trucking or by pumping a sandy sludge to another beach. This is useful for moving sands from the accreting side of a groin to the eroding side. Transferring sands in this manner, however, may result in death of benthic invertebrates, disruption of feeding and breeding of more mobile organisms, and increase in turbidity and sedimentation. Transfer of sands from dunes to the beach may harm the dunes' physical and ecological function, as discussed earlier. Where sand is not available from an adjacent beach, sand is either pumped or dredged from the bottom offshore and piped or barged to the beach (with the disadvantage of affecting these offshore ecosystems, which will be discussed in Chapter 6). This is especially popular for beaches maintained for recreation. For example, in coastal tourist locations along the U.S. east coast, beaches are routinely nourished using sands dredged from offshore. Beach nourishment may be expensive and is temporary because erosion will simply remove the beach again; however, it is deemed the best alternative to support tourism and to protect tourism-related facilities that desire an "oceanfront view." On beaches of the U.S. middle-Atlantic coast, sand needs to be re-nourished on average every five years.

Possible negative effects of beach nourishment on the beach ecosystem include changes in beach morphology and sand grain sizes and changes in the biotic community. Beaches are more resilient to physical impacts than most other coastal ecosystems and effects are often temporary, with sands added to the beach being colonized rapidly by a native assemblage of invertebrates. Care must be taken, however, that the sands be of similar quality as the natural beach sands and not be unnaturally compacted; otherwise invertebrates may be unable to burrow into the sands and they may not support a natural beach community. Compacted sands may affect sea turtle nesting, both for females digging nesting burrows and for hatchlings digging out the nests.

Although beach nourishment is often primarily to maintain recreational beaches or to protect inland areas, sometimes it is carried out primarily for conservation purposes, to restore the ecological function of beaches. For example, in Florida, beaches and dunes that are severely eroded due to hurricanes can be nourished through the state's Beach and Shore Preservation Act. Nourishment of the upper beach helps to maintain the beach and protect adjacent dunes. The nourished beach often differs physically from the natural beach in terms of morphology (it is typically wider and steeper), sediment composition, moisture, and compaction. This may affect the kinds of invertebrates that colonize the beach. Nevertheless, it protects the dunes and beach from erosion and if done gradually over several years will encourage the eventual colonization by a natural assemblage of organisms.

Beach Pollution

Beach pollution may be a cause of concern for humans and wildlife. There are many regions around the world, especially heavily populated coastal cities of developing countries, where human waste, including sewage, is discharged directly into coastal areas. Stormwater runoff can be a major cause of beach pollution from street runoff or untreated sewage when sewage treatment plants malfunction. Industrial effluent and oil spills also may cause long-term harm to the beach ecosystem. Because sediments tend to accumulate pollutants, effects may continue long after a pollution event.

Sewage is the most widespread source of pollution affecting beaches and intertidal waters. It receives much attention due to aesthetic concerns and effects on human health but may have dire consequences for beach ecosystems as well. Raw sewage is still routinely discharged into coastal waters in many regions of the world. Even popular tourist destinations often have inadequate sewage treatment capabilities; in developing nations, the temptation is great to expand the tourism industry before adequate waste water treatment facilities and laws are in place. Primary treatment of sewage is preferable to releasing raw sewage, and makes the effluent less obvious, but does not solve problems with the release of excess nutrients. Partial treatment methods include macerating the sewage, adding chlorine before it is released, and allowing the solids to settle into a sludge and releasing the remaining liquids into the sea. The sludge is then disposed of on land or deeper

ocean water (see Chapter 8 for conservation issues related to ocean sewage-sludge disposal). Because partially treated sewage effluent is less visible, enforcement of regulations may be difficult and public pressure to tighten regulations is lessened. This low visibility is increased by using long discharge pipes to release wastes further from shore, but doing so transfers the problem from one ecosystem to another (e.g., coral reefs, discussed in Chapter 5).

The effects of sewage on beach organisms vary. Sewage effluent does not typically contain large amounts of toxic chemicals, so the effects are mainly due to **eutrophication**. Excess nutrients and organic matter from sewage or agricultural runoff increases primary productivity and may cause an explosion of algae and bacteria populations, which remove oxygen from the waters. Hypoxic waters support a lower biodiversity. On the beach the increased nutrient load may result in deoxygenation of interstitial waters, affecting organisms residing buried in the sands. Input of other sources of nutrients, such as excess fertilizers or agriculture waste that wash into coastal waters, may have similar effects.

The greatest impetus to reduce sewage pollution on beaches and in intertidal waters is human health concerns. This is an especially great concern in regions of the world where sewage treatment facilities are minimal and water quality is not closely monitored. Even in countries with the strictest pollution regulations, sewage pollution of beaches still occurs due to inadequate laws and enforcement, accidental discharge, inadequate treatment facilities, and extreme flood events.

Heavy rain often contributes to beach pollution, for example, when sewage systems are overwhelmed, causing raw sewage to be flooded directly into coastal waters. Although the reason for most closings is the presence of **fecal coliform** bacteria (used as an indicator of human fecal contamination resulting from a failure in water treatment), often officials are unable to identify the specific source of the contamination. Another problem is that beach monitoring is typically carried out in waters just offshore from the beach; pollutants filtered out by the beach may be retained longer than pollutants in intertidal waters. Measuring contamination of the sands is becoming more routine in some places.

In most developed nations, beach pollution is monitored primarily to protect human health. Although discharge of raw or inadequately treated sewage is typically prohibited, how well beaches are monitored for pollution varies from nation to nation. In the United States, growing concern over the public health hazard of polluted beaches resulted in the U.S. Congress passing legislation referred to as the BEACH program (Beaches Environmental Assessment, Closure, and Health) through the U.S. Environmental Protection Agency (EPA). This program requires consistent national health standards for beach water, along with monitoring and public notification programs by states. It includes funding to states so that they can expand monitoring and public information programs. Monitoring beaches includes testing waters adjacent to the beach for fecal coliform; beaches are closed when levels considered unhealthy are documented. The U.S. EPA and the European Environment Agency (EAA) have developed interactive Web sites through which recent beach water quality and beach closings can be accessed.

Where there is regular environmental monitoring, beach closures can be commonplace. In the United States, beach pollution resulted in over 24,000 beach closings and swimming advisories in 2010. Sewage spills and overflows accounted for about 8% of closings and warnings. About 36% were due to stormwater runoff; heavy rain contributes to beach pollution by washing **nonpoint source pollutants** (those originating from diffuse sources and not from a single identifiable source) into beach waters. Although over half of the closings were a result of fecal coliform contamination, often the source of contamination could not be identified.

Monitoring for **point source pollution** (from a defined source; e.g., industrial chemicals from factory effluent) on beaches is not as routine as is monitoring for organic pollution. However, the release of such pollutants is more strictly regulated at the source in most countries. In the United States, the discharge of wastes from factories into coastal waters require environmental impact studies before a potential source of pollution is allowed. Still, regulations and monitoring are either not present or are unenforced in many countries. With the exception of metals (see Chapter 7), the effect of exposure to specific chemicals on human health and coastal organisms is often not well-documented.

Historically, the effect of pollutants to aquatic organisms has been monitored by toxicity testing; organisms are exposed to varying concentrations of the pollutant in the lab and monitored for survival. But this does not consider the interactions of other environmental factors to which the animals are exposed in the wild or effects on population factors such as reproduction and growth. It has become more commonplace recently to look at general effects of pollutants on the health of the natural ecosystem by comparing animal communities in impacted and unimpacted sites. In beach ecosystems meiofauna populations can be used as indicators of the effects of beach pollution. For example, harpacticoid copepods may be very sensitive to changes in the quality of the water within the beach sediments, while nematode worms are generally insensitive to factors associated with pollution. The proportion of nematodes to copepods thus can be used as an indicator of beach health, with higher values indicating more impacted beach ecosystems.

Even if they contain no pollutants, fresh waters flowing onto the beach may affect the habitat and ecosystem. For example, canals or river modifications often present new sources of water discharge over the beach. This may cause excess beach erosion and reduce biodiversity of beach-associated organisms. Most of the invertebrates associated with the beach are not tolerant of long-term exposure to freshwater or large variations in salinity. Unnatural increase in freshwater input thus typically lowers abundance, biomass, and diversity of beach organisms.

One source of pollution that has been thoroughly studied is oil and other petroleum products. Oil pollution on beaches may come from a number of sources, the most dramatic being tanker spills or well blowouts (see Box 3-3, Learning from History: *Torrey Canyon, Exxon Valdez,* and *Deepwater Horizon*). There are many smaller scale sources of oil pollution on beaches too. For example, groundings of smaller vessels often release petroleum products onto beaches, and relatively small spills at oil terminals may be a source of chronic pollution of some shorelines. Offshore drilling activities also results in routine release of oils that may end up on beaches. Approximately 3 million gallons of oil and petroleum products are spilled into U.S. waters each year on average. Whether the spill reaches the beach depends on many factors. When it does, the greatest harm is to the upper beach, where the oils tend to accumulate as they are washed in by the tides.

The effects of oil on beach ecosystems vary depending on the size of the spill, the distance from the beach, wave action, water temperature, and the chemical nature of the oil products (i.e., whether it is crude oil or refined products). The most toxic hydrocarbons in oils tend to evaporate first, so that they may dissipate from offshore spills by the time they reach the beach. Removal of the oil from the ocean surface before it reaches the beach may reduce the effect of oil spills on beaches dramatically. One method of cleaning oil slicks is to use **chemical dispersants**, which cause the oils to mix with the water so that they will not wash ashore. When combined with the oils, oil dispersants may be more toxic than the oil itself, however, and their use near the shore may seriously harm beach ecosystems. Because of this, many countries regulate the distance from shore within which dispersants may be applied. Cleanup of oil from the beach may physically harm or remove invertebrates. The cleanup is beneficial to their population health in the long term, however, but must consist of removing the oiled sands and not simply burying them out of sight on the beach. Due to the possibility of oil leaching into the sands, spills on beaches may be more persistent than spills on rocky shoreline, but many environmental factors affect the degree of persistence to oil spills and their harm to intertidal ecosystems (see Box 3-3, Learning from History: *Torrey Canyon, Exxon Valdez,* and *Deepwater Horizon*).

The effects of oil on beach-associated invertebrates vary from species to species. Invertebrates residing in the sands may be mostly eliminated by a serious spill, due to physiological effects of the contaminants in the oils. Even when the spill is not lethal, however, effects on feeding, reproduction, or behavior may indirectly cause mortality and inhibit their recovery from the event. For example, oil that is buried in the sand clogs spaces within the sands where meiofauna and microorganisms reside, thereby reducing water circulation and oxygenation of the interstitial spaces. Meiofauna typically recover within a year on high-energy beaches that rapidly dissipate the oils. Even moderate spills may harm filter feeders such as bivalves dramatically due to the effect on their feeding mechanisms. Crustaceans that migrate vertically in the sands may discontinue movements to the surface to avoid the oils. Beach-associated birds can often avoid oil spills, and numbers killed by spills washing onto the beach are not typically enough to reduce populations in the long term.

After the oil is removed or begins to dissipate and decompose, the beach will gradually recover. On sheltered beaches where oil remains trapped in the sands, the recovery may be slower. Tolerant species will be the first to return. Some species of polychaete worms tend to be very pollution tolerant, and may dominate the beach ecosystem for years before the natural biodiversity returns. Conservation issues related to oil spills in other coastal ecosystems will be discuss in later sections and chapters.

Impacts of Recreational Use of Beaches

Although many tourists and recreational users of the beach are environmentally conscious, they can still directly affect beach ecosystems. The tolerance of beach litter varies widely among regions of the world; although there are still places where there is little concern among beachgoers over litter, in others the beaches remain pristine even with heavy use, and citizen groups commonly volunteer for beach cleanups. With increasing environmental awareness globally, most regions strongly discourage or penalize littering of the beach. Education can have a strong influence on attitudes concerning litter. Voluntary beach cleanups raise awareness of beach pollution and improve the condition of the beaches. Although much litter on the beaches is from local sources, in some regions large amounts of litter is washed onto the beach from other sources, including dumping areas or items discarded from boats (**Figure 3-20a**). The non-degradable nature of plastics and other synthetic materials results in its accumulation on beaches around the world. Millions of pieces of trash enter the oceans each day. Currents can carry these long distances where they may be deposited on isolated beaches, sometimes on otherwise pristine uninhabited islands (see Chapter 7). For example,

(a)

(b)

(c)

Figure 3-20 Examples of anthropogenic impacts on beaches: **(a)** marine debris can accumulate on beaches; **(b)** tourist foot traffic compacts beach sands; and **(c)** lodging built close to shore can cover dunes and create light pollution.

a survey of beaches by Tim Benton revealed that the isolated and uninhabited Pitcairn Islands of the South Pacific had comparable amounts of litter as those in Ireland, with some of the Pitcairn litter originating from thousands of miles away.

There is much concern over litter's detraction from the aesthetic values of the beach; however, there is potential for harm to the beach ecosystem as well. The effect of plastics on beach invertebrates is not well known, but its ingestion or entanglement by birds or nesting turtles can be fatal (see Chapter 7). In some areas the litter is buried above the high tide level, where it eventually resurfaces.

Another human impact results from driving on beaches with motorized vehicles. Off-road vehicles driven on the upper beach may disturb nesting shore birds and destroy nests and eggs. The tires pack the sand, affecting the ability of animals, including hatchling sea turtles, to burrow through the sands. Crustaceans such as crabs, isopods, and amphipods may be crushed or have their burrows disrupted. Tire ruts can trap young birds and turtle hatchlings attempting to move down the beach. On most beaches in tourist locations, motorized traffic is prohibited or tightly restricted; however, enforcement of restrictions on isolated beaches is difficult. Driving on the packed wet foreshore of the beach is not as likely to harm burrowing beach organisms, but may temporarily disrupt feeding by seabirds.

Recreational beach-goers can impart heavy foot traffic on the beach or change the ecosystem by other activities (Figure 3-20b). Because the beach ecosystem is adapted to physical stress from wave activity, effects of beach goers are likely minimal and evidence of trampling is usually gone following the next high tide. Delicate crustaceans and juvenile bivalves, however, could be damaged by foot traffic. Comparisons between heavily used and protected adjacent beaches have indicated few differences in crustaceans living in the sands. The feeding behavior of shore birds may be changed due to frightening by human activities, possible resulting in movement to less productive beaches, decreased foraging time, or forcing more foraging at night time when fewer people are on the beach. In the United States, least terns and piping plovers have been affected by coastal development and tourism, altering behavior, reproductive success, and abundance. Birds may become habituated to humans in areas frequented by tourists.

People on the beach at night may reduce the likelihood of sea turtles from coming ashore to nest in that area, and beach furniture, such as beach chairs or umbrellas, may inhibit movements of turtles on the beach. Compaction of the sands in heavily used beaches may result in problems for hatchling turtles digging out of the sands. In beaches with excessive foot or vehicular traffic, fences may be put around nests for protection or eggs are sometimes relocated to more desirable areas of the beach. Relocating nests is discouraged if it can be avoided because it may result in lower hatching success and a skewed sex ratio (the gender of hatchling turtles is partly dependent on temperature, which is affected by where and how deep the eggs are buried).

Although wrack that accumulates naturally on beaches forms an important ecosystem function, it is often considered aesthetically unappealing for its appearance and propensity to create odors and attract flies as it decomposes. Wrack thus is often routinely removed, especially from popular recreational beaches. This removal may affect species that use it for food or shelter and alter the ecology of the beach. The organic matter is an important source of nutrients, and organisms living in the wrack are an important source of food for predators such as crabs or birds. When the wrack is removed by beach-cleaning machines, organisms beneath may also be removed and there is the additional effect of the machine driving over the beach.

A less apparent impact on beaches is caused by **light pollution** from roads, hotels, or other buildings (Figure 3-20c). Although most organisms using the beach are probably unaffected by artificial lights, lights visible from the beach may confuse and result in death of hatchling sea turtles. As the young turtles hatch and dig their way to the surface of the beach, they begin to look for cues to lead them to the ocean. The primary cue is the reflection of moon and star light by the ocean. Hatchling turtles tend to head in the brightest direction, away from silhouettes caused by the profile of the dune and vegetation. If there is a brighter man-made light in the inshore direction, the turtles head toward that light instead of the ocean. In areas where it is not feasible to reduce lighting that is visible from the beach, turtle nests may be moved to other areas, or hatchling turtles may be collected and carried to a darker beach or directly to the water. These actions can impart additional stress on the turtles. An increased awareness of these problems has resulted in restrictions on lighting visible from the beach. For example, in Florida and other tourist areas where sea turtles nest, there are local ordinances that restrict lights that are visible from the beach during times sea turtle eggs are likely to hatch.

Beach Harvest Impacts

On many beaches around the world where large invertebrates are abundant, there is a long history of local harvest for food or bait. Beaches are easily accessible and harvest may be as simple as walking onto the beach at low tide and picking the animal up or digging it out of the sand. Much of the harvest is of bivalves and crustaceans that burrow into the sands, but also may include animals that move ashore to nest, such as sea turtles and horseshoe crabs.

Crab Harvest

Crabs that feed on the beach are often nocturnal and burrow into the sands during the day or when disturbed. Harvest typically involves digging them from their burrows. Much of this harvest is small scale and for local use; for example, ghost crabs and mole crabs are harvested from some African beaches, and populations appear to remain stable at moderate harvest levels. Digging and trampling the beach to access the crabs may be more harmful to the ecosystem in the long run than the removal of the crabs.

Bivalve Harvest

The most commonly harvested organisms on beaches are bivalves (**Figure 3-21**). Commercial harvest has been documented for at least 15 different bivalve species. Along the North American Pacific coast these include the Pismo clam *Tivela stultorum*, a large clam reaching over 150 millimeters in diameter, and the Pacific razor clam *Siliqua patula*. Along the South American coast the yellow clam *Mesodesma mactroides*, the macha *Mesodesma donacium*, and guacuco *Tivela mactroides* are commonly harvested, with harvest of the guacuco extending into the Caribbean. In New Zealand three *Paphies* species, called toheroa, with a maximum size over 150 millimeters, are harvested. And several smaller *Donax* species are harvested around the world, mostly as small-scale local harvest for either food or bait. After peaking in the early 1900s, commercial harvest has declined for many of these species in a pattern similar to many commercial fisheries (see Chapter 11). The largest remaining commercial beach clam fishery is for the macha in Chile, where the clams are harvested from the beach or subtidally by divers from small boats.

One of the most unusual harvested bivalves is the geoduck (pronounced "gooey duck") *Panopea abrupta*

Figure 3-21 Bivalves are harvested by digging into beach sands.

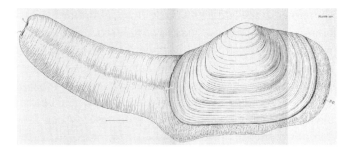

Figure 3-22 The geoduck clam, a valued bivalve harvest from intertidal sands and mudflats.

from the west coast of the United States and Canada (**Figure 3-22**). This is the largest burrowing clam. Although its shell is only about 20 cm (9 inches) long, the long siphon sometimes extends over one meter resulting in weights over 5 kilograms (11 pounds). The geoduck is found from the lower intertidal into deeper waters, but is most easily harvested in the lower intertidal zone of silty-sand beaches and intertidal mudflats. In Puget Sound of Washington State there has been a recreational harvest for over a century. Increased demand, especially in Asia for the sashimi (raw seafood) market, has resulted in commercial harvest from subtidal waters. Harvest is closely regulated and monitored. Aquacultural production in privately owned tidelands (the geoducks are grown in PVC pipes embedded in the sediments), however, has resulted in controversies over property rights and environmental impacts in the intertidal in regions.

Because of the potential for overexploitation, restrictions are typically placed on harvest of bivalves from the intertidal zone, including seasons, sizes, and limits on numbers collected. In no case are the harvested species considered biologically endangered, and the bivalve populations probably would recover rapidly if harvest was discontinued. Beach harvest may have other ecological effects, however, because it typically requires excavation of the beach and may remove most of the macroinvertebrates. Ecological damage is typically low where harvest is considered recreational and clams can only be taken for personal use. When these excavations are localized and do not affect a substantial portion of the beach, recovery of the beach and return of invertebrates may be rapid, as quickly as within a single tidal cycle.

Sea Turtle Harvest

Harvest of sea turtles and their eggs from the beach has been affecting sea turtle populations for hundreds, if not thousands, of years in some regions. This, along with many other impacts at sea, has led to the reduction of most populations to only a small fraction of their original size. Anecdotal accounts from early European explorers of the Caribbean shorelines in the late 1400s, including Columbus, suggest populations so large that every available space on the beach was occupied by a turtle during nesting season, such that many females remained off the coast unable to lay their eggs. Explorers began taking nesting turtles as a source of fresh meat onboard sailing ships; the turtles would be kept alive, but immobile on their backs, until time for their consumption. Eventually international trade of sea turtles for meat was widespread; for example, there were large harvests from the West Indies islands to supply markets in the United States. The turtles were harvested with no restrictions and no regard for conservation. The trade stopped only when nesting sea turtle populations were so low that it was no longer profitable.

Eventually conservation concerns resulted in protections of nesting sea turtles. Through the 1900s protection of nesting sea turtle populations expanded around the world but taking eggs from nests, whether legal or illegal, remained prevalent through the end of the 1900s and continues on some unprotected beaches. Poaching has been curtailed in many regions. For example, pressure from non-governmental organizations (e.g., the World Wildlife Fund, see Chapter 12) enhanced enforcement of the protected status of beaches in Japan in the late 1900s. In Mexico, although laws were passed in 1990 to outlaw taking of sea turtle eggs, poaching eggs remains a problem in some areas. Local traditions include consumption of the eggs based on the myth they are an aphrodisiac. Education campaigns are developed by conservation organizations in an attempt to educate the public in hopes of limiting egg poaching. In some places predation of eggs by dogs, feral pigs, or foxes that come onto the beach threatens survival of eggs and hatchlings.

Despite international agreements to protect nesting sea turtles, enforcement of protective measures can be extremely difficult, especially in isolated areas. For example, on Central American beaches the survival of leatherback turtles (*Dermochelys coriacea*), the largest and one of the most endangered sea turtles, varies depending on local regulations and culture. Large nesting populations were documented by Juan Patino-Martinez and colleagues along 100 km of coastline of the Columbia and Panama Caribbean, where over 5,000 clutches of eggs were laid in 2005 and 2006. The most successful nesting was on beaches where local native Kuna society has established regulations prohibiting taking of nesting turtles or eggs or destruction of nesting sites. There is no coastal construction along these beaches. Here a local culture of respect for the nesting turtles has resulted in a conservation ethic that has been in place for centuries. At other sites of successful nesting, access to the beach is difficult and local organizations encourage beach conservation. At nearby beaches where successful nesting is less common there are problems with organic waste pollution, predation by dogs, compacted sands, cattle traffic, erosion, and egg harvest. Greater community involvement is recommended at these beaches. This exemplifies how important education and a local conservation ethic can be in protecting intertidal habitats, ecosystems, and species, especially in populated areas with access to the coastline. From a scientific perspective, the recent discovery of a large nesting population of these large sea turtles suggests that ecologically important intertidal habitats may be present in areas around the globe, but have not been documented. Broader efforts are needed to document and protect habitats that may be lost before they can be targeted for protection, rather than just focusing efforts on conservation of a few key well-documented sites.

Horseshoe Crab Harvest

The horseshoe crab is another animal that does not rely on the beach ecosystem, but uses the beach habitat for spawning and thus requires protection of intertidal habitat. Although they have no value for human consumption, horseshoe crabs are easily taken when they come onto the beach to deposit eggs. One market is for harvest of their blood, used in the biomedical industry for testing for bacterial contamination and for pharmaceutical exploration. Because these crabs are returned to the wild alive after a portion of the blood is extracted there is little, if any, effect on populations, however. The largest lethal harvest is as bait in conch and eel fisheries in the northeast United States. Here overharvest has led to population decline, although their numbers are not low enough to be totally protected from harvest. Almost one million horseshoe crabs came ashore around 1990; numbers have declined ever since but appear to have stabilized since about 2000. Harvest is controversial, however, because eggs laid on the beach are an important source of food for a subspecies of the shorebird, the red knot *Calidris canutus,* which stops in Delaware Bay before completing its spring migrations from the southern tip of South America to the Canadian Arctic (**Figure 3-23**). Scientific surveys show that the red knot overwintering populations declined by 75% from 1985 to 2007 to about 15,000 individuals; this decline was largely attributed to a 90% decline in horseshoe crabs eggs, upon which they depend for food. Without adequate food the birds may have inadequate energy reserves to make it to breeding grounds. There have been efforts, unsuccessful to date, to have this subspecies of red knot declared as endangered. Although harvest of the horseshoe crabs is still allowed, restrictions on the numbers harvested have been implemented in New Jersey, Delaware, and Maryland. In New Jersey, a harvest moratorium was implemented in

(a)

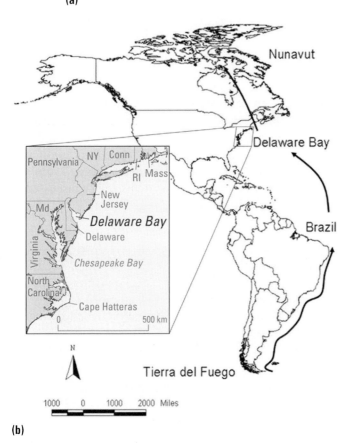

(b)

Figure 3-23 **(a)** Red knot *Calidris canutus* feeding in the intertidal zone. **(b)** Map indicating red knot migration route from the southern tip of South America to the Arctic and its stopover in Delaware Bay where it feeds on horseshoe crab eggs.

2008. In Maryland, a 2:1 male to female ratio on harvest was implemented in 2009 in an attempt to protect the females from overharvest. In Delaware, only the harvest of males is allowed during a defined season. These examples show how complicated conservation of beach and other coastal resources can be when food web interactions must be considered and various user groups, government agencies, and scientists most come to some agreement. As will be reiterated throughout the text, this tends to be the rule rather than the exception.

3.2 Intertidal Mudflats

The term **soft bottom** refers to any bottom land composed of sediment, ranging in size from fine silts to coarser sands. Intertidal soft bottoms bordering the ocean are relatively unstable and constantly shifting because of waves, tides, and currents, and are not typically colonized by vascular plants (e.g., grasses) or macroalgae (seaweeds). In sheltered regions seagrasses are often present but mostly in subtidal regions (seagrass ecosystems are discussed in Chapter 6). Soft intertidal bottoms sheltered in estuaries and marshes are often inhabited by marsh grasses or mangroves (these ecosystems are discussed in Chapter 4). The ecosystems discussed herein are soft bottom intertidal regions exposed to ocean conditions that inhibit the development of the establishment of large plants. These are defined more by physical than biological interactions; one of the defining characters in these ecosystems is the size of the sediments. Calmer, more sheltered areas have finer sediments and form intertidal mudflats. Areas exposed to more intense wave and tide activity accumulate larger sediments and form beaches.

In coastal areas with a low gradient and low wave activity, outside of estuaries, fine sediments accumulate forming **tidal mudflats** where vascular plants and macroalgae are not prevalent (**Figure 3-24**). However, primary producers, largely comprised of diatoms, accumulate on the surface of the sediments. On some mudflats, diatoms and other microalgae are abundant enough to form an obvious greenish film over the sediments. Along with the microalgae, these ecosystems depend largely on organic matter, the remains of plants and animals, or plankton washed in from outside sources, such as diatoms and dinoflagellates or larval invertebrates. The organic matter accumulates in the sediments and, although it can be consumed by animals, much of it is decomposed further by bacteria.

Figure 3-24 An intertidal mudflat on the South Carolina coast.

Microalgae, bacteria, and organic matter function as the base of the food chain and are consumed by other animals and microorganisms.

Below the surface of the sediments, decomposition of the organic matter by bacteria uses up most of the available oxygen, creating **hypoxic** conditions, and the fine silty sediments inhibit the circulation of waters below the surface that replenish the oxygen. Much of the biological activity on tidal mudflats, thus, is in the surface of the sediments. One of primary consumer groups in these upper surface layers are the meiofauna, organism such as copepods and nematodes at sizes ranging from about 0.05 to 0.5 millimeters in size, barely visible to the naked eye. Many intertidal mudflat animals, however, live burrowed into the sediments. These **infaunal** invertebrates avoid low oxygen problems by pumping oxygenated water from the surface. For example, clams burrow into sediments with their foot; crustaceans, such as amphipods and crabs, use appendages to dig; and worms burrow or eat their way through sediments. Some polychaete worms live in tubes they have built with openings at the surface of the sediments. Macroscopic animals living in the soft bottom intertidal can be characterized by their mechanism of feeding. **Deposit feeders** are prevalent in muds and feed on matter and organisms on or in the sediment. For example, some polychaetes take in sediment and organic matter as they burrow. Other polychaetes use sticky tentacles or mucus to pick up detritus. Crabs, such as the *Scopimera* sand-bubbler crabs, feed by removing organic matter and organisms from between sand grains on mud flats. They accumulate the sorted sands into balls, which they then toss back onto the tidal flat. **Suspension feeders** (filter feeders) such as bivalves and tube-dwelling polychaetes filter out plankton and other matter suspended in the water.

Predators on the mudflat include birds, such as sandpipers and plovers, that probe into the muds to feed on invertebrates or herons that feed when the mudflats are covered at high tides. Fishes move onto the mudflats with tidal waters. These include benthic feeders, such as the juvenile spot *Leiostomus xanthurus* along the U.S. Atlantic coast that filters invertebrates from the muds using comb-like **gill rakers**. Juveniles of many fish species feed over the mud flats, including drums (Sciaenidae), herrings (Clupeidae), and flatfishes (Pleuronectiformes).

Environmental impacts on mudflats include pollution from chemicals and oils that readily accumulate in the sediments (see Box 3-3, Learning from History: *Torrey Canyon, Exxon Valdez,* and *Deepwater Horizon*). Organisms that are harvested from the mudflats include clams and crabs. Digging into and trampling the mudflat to harvest burrowing invertebrates may reduce populations and affect the habitat. In some regions worms are dug from the mudflat for bait. Along the British coast, for example, polychaete lugworms (*Arenicola*) and ragworms (*Nereis* and *Nephtys*) can be dug from mudflats in large numbers to be sold as bait (**Figure 3-25**). Although the populations are typically resilient and repopulate rapidly, there are concerns in some regions that local population depletions may occur from such harvest. Harvest of *Arenicola* by dredging in intertidal mudflats has affected benthic populations and habitats. For example, research by Jan Beukema in the Dutch Wadden Sea, where lugworm harvest created 25-centimeter deep gullies in the bottom, revealed a lugworm population reduction of 50% and a near extinction of local populations of *Mya* clams. It took several years after the dredging operations for the ecosystem to recover.

Other impacts on the mudflats, such as those related to global climate change and sea level rise, are similar to those in other intertidal ecosystems and are discussed in the closing section of this chapter. Because mud flats and vegetated muddy intertidal regions are often association with estuaries and marshes formed in shelter areas, the ecology and conservation of these ecosystems will be discussed in more detail in Chapter 4.

(a)

(b)

Figure 3-25 Polychaete worms are sometimes harvested from mudflats for use as bait. **(a)** Tidal mudflat in Ireland with casts of lugworms; **(b)** ragworm *Nereis diversicolor.*

3.3 The Rocky Intertidal Zone

Primarily as a result of geologic factors, sandy beaches are replaced with intertidal areas largely composed of rocks in many coastal regions (**Figure 3-26**). In general, these regions are geologically more recent and too steep to accumulate sands (see Chapter 1). In a given region rarely is the entire coastline rocky; often it is interspersed with sandy beaches or muddy tidal flats. The biology, ecology, and conservation of rocky coastlines, however, is distinctive compared to areas predominated by soft bottoms. The zone dominated by rocky materials—from the lowest tides to the area influenced by sea water spray above the high tides—is commonly referred to as the *rocky intertidal*. Rocky intertidal coastlines are most prevalent along the leading edge of the Pacific Ocean plates, from the western coastlines of the Americas, through the islands of the west Pacific, including Japan, the Philippine and Indonesian archipelagos, and New Zealand. Rocky coastlines are not as prevalent in the Atlantic but predominate on Caribbean islands, along the northern Mediterranean coastline, and in areas with a large tidal range, such as western Europe, the southeast coast of South America, and the U.S. northeast coast.

The Rocky Intertidal Ecosystem

The rocky intertidal ecosystem has been extensively studies by ecologists, in part due to its easy accessibility and in part because of the distinctive interactions within the biological community. The vertical zonation pattern has been particularly well studied and this focus has led to the formation of some major ecological paradigms.

Intertidal Adaptations

Because rock is not easily burrowed into, most of the rocky intertidal organisms live on the rock surface, where they

Figure 3-26 A rocky intertidal coast on the Caribbean island of Guadeloupe.

must deal with great physical stress of exposure to wind, sun, waves, and weather. Rarely, however, is the rocky shore composed of a single slab of rock. More often it is comprised of many rocks with numerous cracks and crevices or boulders of varying sizes and shapes, resulting in a complex habitat. This varied substrate can provide shade, shelter from waves, surfaces for attachment, and areas where water accumulates in pools. Still, rocky intertidal organisms must be adapted physiologically and behaviorally to deal with a suite of extremes that are not prevalent in other intertidal habitats, including desiccation, overheating or freezing, exposure to extreme salinities and other chemical variables, and accessibility to predators.

Many organisms that are permanent residents of the intertidal zone must deal with stress of exposure while emersed out of the water for at least a portion of the day. They do not have the advantage of burrowing into sediments as do invertebrates in beaches or other soft-bottom intertidal areas. Mobile organisms, such as crabs, snails, and sea urchins, however, move into moist shady cavities, crevices, or tide pools. Sessile organisms must use other mechanisms to avoid desiccation. Bivalves can seal themselves within a hard shell and remain inactive until re-covered by water. Other mollusks, such as limpets, seal their shells tightly to the rocks at low tide to retain water during exposure. Fish diversity is limited to species with broad physiological tolerances.

Other physical stresses intertidal organisms must deal with are extreme temperatures due to exposure and extreme salinities, especially in tide pools, resulting from evaporation or rain events. Organisms commonly exposed in the intertidal zone thus are physiologically adapted to tolerate a broad range of physical conditions.

Rocky Intertidal Food Webs

The food web of the rocky intertidal is interlinked with adjacent marine and terrestrial ecosystems. The nearshore marine environment is especially important and there is a strong influence of physical factors originating in nearshore waters, such as wave activity and nutrient input. But many of the biotic interactions are largely confined to the rocky intertidal habitat.

Because of the physical stress of wave action and inability to burrow, many of the plants and invertebrates are sessile organisms attached to the rocks. Although phytoplankton such as diatoms wash into the rocky intertidal, the predominant and most obvious primary producers are macroalgae or seaweeds, attached to the rocks via holdfasts (**Figure 3-27**). Species makeup varies geographically; however, in general, related species can be found in similar rocky intertidal habitats around the world. Examples include palm seaweeds *Postelsia* that colonize bare exposed

Figure 3-27 Macroalgae are the dominant primary producers in the rocky intertidal.

Figure 3-29 Chitons are important grazers in the rocky intertidal.

areas of the rocky intertidal. Other common algae include broad-bladed seaweeds such as *Fucus* and the brown seaweed *Hedophyllum* (sea cabbage), filamentous seaweeds such as *Endocladia,* and turf-forming coralline algae and red algae. The sheet-like sea lettuce *Ulva* is common in the mid to lower intertidal, and kelp-like seaweeds, such as various species of *Laminaria,* occupy the lower intertidal.

Many invertebrates in the rocky intertidal are filter feeders (**Figure 3-28**). Bivalves are typically the most numerous of the attached invertebrates. Mussels avoid displacement by attaching themselves to the rock by **byssal threads;** they remain closed when exposed to air and filter feed when submerged under water. *Mytilus* mussels, such as the blue mussel *Mytilus edulis,* may be especially abundant and often form dense beds. Attached crustaceans include barnacles, such as *Balanus,* that attach themselves to rocks with a cement and filter feed when covered by tidal waters. Other

filter feeders include anemones, such as the giant green anemone *Anthopleura xanthogrammica,* which extend their tentacles when submerged to capture zooplankton and other small crustaceans. They are not as tolerant of emersion and are most prevalent in tide pools that retain water when the tide recedes.

Grazing invertebrates are more mobile, though they commonly remain within a particular zone of the intertidal (**Figure 3-29**). Sea urchins such as the red sea urchin *Strongylocentrotus* graze on algae and invertebrates on the rocks, hanging on by using their tube feet as suckers; they often remain stationary, slowly excavating a cavity in the rocks in tide pools and feeding on drifting algae. Grazing mollusks include limpets and chitons, both of which graze on algae growing on the rocks. They avoid predation and desiccation by attaching to the rocks using suction created by a muscular foot that produces viscous mucus. Some limpets attach with such force that it is almost impossible to physically remove them from the rock without damaging their shells. Species of *Littorina* periwinkle snails are often present as herbivorous grazers, typically feeding on algae on the surface of rocks in the upper intertidal regions.

Predators associated with the rocky intertidal are mostly invertebrates with the ability to overcome the protective shells of mollusks and crustaceans (**Figure 3-30**). Gastropods include predatory snails such as the dog whelk *Nucella,* which use their **radula** to rasp a hole through the shell of barnacles or mussels. Some snails, such as unicorn snails *Acanthina,* also use spines on their shells to pry into barnacle shells. Snails are mobile enough that they can move up and down the shore with the tides. Sea stars, such as three *Pisaster* species on the U.S. Pacific coast, prey on

Figure 3-28 Mussels are dominant filter feeders in the rocky intertidal.

Figure 3-30 Invertebrate predators in the rocky intertidal include sea stars such as the ochre sea star *Pisaster ochraceus* in the eastern Pacific.

various invertebrates, wrapping themselves around the shell of mussels, barnacles, and other mollusks, and using their tube feet to pry open the shells. Sea stars may be an important **keystone species**, having a strong influence on the distribution and abundance of other species. Decapod crustaceans include porcelain crabs, such as *Petrolisthes,* which filter out plankton from the water using setae on their claws. Grapsid crabs move about on the rocks above the tide; they prey on juvenile snails and mussels as well as grazing on algae.

Fishes may move into the rocky intertidal to feed during high tides and some remain in tide pools after the tide recedes (**Figure 3-31**). Tide pool species must be tolerant of extreme temperatures and salinities and include gobies

Figure 3-31 Fishes, such as this puffer, use tide pools for refuge and as foraging habitats.

(Gobiidae), blennies (Blenniidae), sculpins (Cottidae), and clingfish (Gobiesocidae). These are typically opportunistic predators on small crustaceans or mollusks. The rocky intertidal also serves as an important feeding, breeding, or resting area for other vertebrates, including seabirds, reptiles (marine iguanas), or marine mammals, such as seals and sea otters.

Ecological Interactions

The community composition in the rocky intertidal zone depends largely on the ability of organisms to tolerate exposure to wave action and emersion and successfully feed; however, competitive, predatory, and grazing interactions are important regulators of species abundance and distribution. Typically, interactions among the physical and biological factors determine species distribution. Ecologists have carried out numerous studies on interactions in the rocky intertidal. Some examples will suffice to give a general understanding of these interactions and how they influence conservation issues.

One example involves limpets in the upper intertidal region. Where grazing limpets are abundant, they remove much of the algae mat on the rocks and allow invertebrates populations to expand. Limpets do not commonly graze on dense stands of leafy macroalgae in the mid-tidal zone, however, because they cannot firmly attach to the plant surfaces and will be washed away. This results in a dominance of macroalgae in the mid-tidal region exposed to high wave energy, areas avoided not only by limpets but also grazing sea urchins.

Another example, discovered in classic experiments by Dr. Roger Paine in the 1960s and 1970s on the U.S. Pacific coast, is the regulation by keystone species such as sea stars on Pacific coast rocky shores. The sea star *Pisaster* preys on *Mytilus* mussels and other sessile invertebrates. Predation by the sea stars keeps mussel populations in check and allows other invertebrates, such as barnacles, to coexist on the midshore rocks. When sea stars are removed, the community of sessile invertebrates changes dramatically, and mussels and barnacles become predominant, resulting in a decline in species richness by almost 50%. Studies in other regions found similar, but not always as extreme, results when keystone predators were removed. Paine's work led to the development of a principle throughout community ecology illustrating the importance of keystone species. We now realize that generalizations should be applied to diverse ecosystems cautiously, as the keystone species concept may not apply in many other marine ecosystems. Where keystone species are present, however, it has important implications for conservation because the removal of a single species may cause unexpected changes throughout the entire ecosystem.

Distribution Patterns

Complex ecosystems such as the rocky intertidal that are distributed around the globe are difficult to generalize. Even within a given region there may be large seasonal variability in response to physical factors (such as tidal range, storms, or climate change). Most rocky intertidal ecosystems are exposed to similar factors that structure their communities, so some generalizations can be made. The types of patterns most widely documented are **vertical zonation** patterns (**Figure 3-32**). One of the simplest classifications separates the rocky intertidal into four zones from top to bottom: the **splash zone** around the edge of the high tide line, merging into a high zone in the upper intertidal (the **supralittoral**), through a mid zone (the **midlittoral**), to a low zone in the lower intertidal (the **infralittoral**). The demarcation of these zones is somewhat arbitrary and is more of a gradient than a sharp line, and within each zone there may

be large amounts of patchiness. The most obvious prediction, which generally holds true, is that the organisms in the upper intertidal would be adapted to tolerate desiccation, spending long periods of time, sometimes days, without being submerged under water, while those lowest on the shore are more fully marine and can only withstand short periods out of water. Studies have documented that those in the middle zones are intermediate in tolerances and are more likely to be controlled by biological interactions, such as predation and competition (**Figure 3-33**).

Detailed observations reveal more specifics. The splash zone is typically characterized by *Littorina* snails of various species that graze on algae and can withstand long periods out of water; lichens and cyanobacteria growing on the rocks surface; and isopods, such as *Ligia*, commonly called sea roaches, which live in crevasses about the high tide line. Crabs, gulls, and other birds move into this region to feed on the invertebrates. The high intertidal zone is characterized by **foliose algae** such as the red seaweed *Pophyra* (commonly cultured to produce the sushi wrap *nori*) or the brown seaweed *Pelvetiopsis, Littorina* snails, and barnacles such as *Balanus*. The middle intertidal is occupied by encrusting algae or leafy brown algae such as *Fucus,* a diversity of invertebrates including barnacles such as the acorn barnacle *Semibalanus,* oysters, limpets, and *Mytilus* mussels, which are preyed upon by invertebrates such as sea stars or snails. The lower intertidal is typically largely covered by red algae or brown kelp-like seaweeds such as *Laminaria,* and occupied by a diversity of animals, many of which are also found in the subtidal rocky bottoms. These include anemones, sponges, tunicates, bryozoans, and fishes.

Although the species may vary, the patterns described above generally apply to temperate rocky intertidal regions around the world. Another factor leading to variability among ecosystems is the degree of environmental stress. In regions where there is low wave action and less environmental stress, the influence of predation and competition are more important. With greater wave action, physical factors are more important, predation is less influential, and competition for space among sessile organisms becomes more important. In the most stressful high-wave environments, physical factors are more important than all biological factors. To further complicate matters, these generalizations are mostly based on studies in temperate regions, such as the U.S. west coast, and may not necessarily apply to the tropics.

In tropical rocky intertidal habitats with less stress from wave activity and winter temperatures, zonation patterns may not be clear cut and predation may be the primary factor controlling community structure. For example, on areas of the Pacific coast of Panama with low physical wave activity, the rocky intertidal community exhibits high

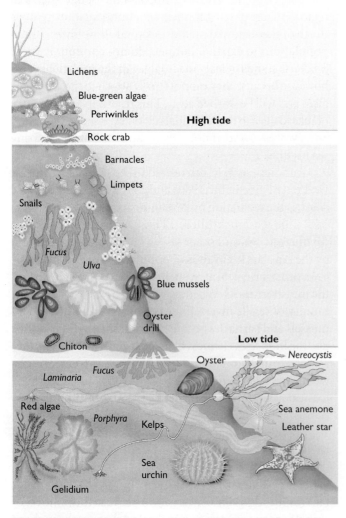

Figure 3-32 Vertical zonation pattern of the benthic community in the rocky intertidal.

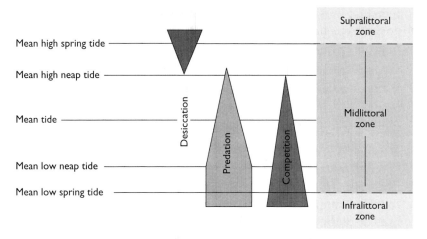

Figure 3-33 Importance of desiccation, predation, and competition in controlling vertical zonation in the rocky intertidal zone.

predation by mollusks, crabs, and fishes. This results in low abundance of sessile organisms, dominance by crustose algae, and indistinct vertical zonation patterns. In tropical areas of the Caribbean with greater wave stress but low tidal range (commonly less than one meter), research by Thomas Good and others revealed that predation is less important. The upper intertidal can be over 50% bare space, with the remainder covered by crustose algae, barnacles, snails, and erect algae; the lower intertidal has higher species richness and is predominated by a diversity of macroalgae, crustose algae, and barnacles. Mobile organisms such as snails and crabs move through the intertidal, but are most prevalent in the lower intertidal.

Even though the rocky intertidal is possibly the most well described of marine ecosystems, it is still impossible to make broad generalizations that apply at most times and in most places. In fact, the more these ecosystems are studied, the less we are able to generalize. This means that it is difficult to predict and understand reactions to human impacts. It is important to attempt to gain an understanding of the relative importance of physical (e.g., wave activity and emersion) and biological (e.g., predation and competition) factors not just for the sake of ecological knowledge, but also from the applied perspective of conservation. Doing so will help develop predictions as to how anthropogenic changes in the physical environment (e.g., through climate change) or biological factors (e.g., through selective harvest) will affect these habitats and ecosystems.

Human Impacts on the Rocky Intertidal Ecosystem

The rocky intertidal zone is not as sensitive to anthropogenic disturbances as most other coastal ecosystems because of the hard rock composition, lack of large exposed biological structures (like trees or grasses) that are easily removed, and their natural adaptations to tolerate extreme events on a daily basis (tides), seasonally (excess cold and heat or storms), or even on a geologic time scale (volcanoes and earthquakes). But because rocky intertidal regions are easily accessed and are often in areas close to large coastal human populations, they are susceptible to various impacts, including nutrient and oil pollution, over-collecting of organisms, siltation, and modifications by artificial structures. Although the effects may be extreme and obvious, more often, separating natural variability from anthropogenic disturbances is not simple. It is especially difficult to identify long-term effects and predict recovery time.

Overharvest

Organisms have been harvested from the rocky intertidal for as long as humans have lived in coastal areas. There is abundant archaeological evidence to show that many organisms in the rocky intertidal zone, including bivalves, snails, sea urchins, barnacles, and seaweeds, were harvested by prehistoric human populations for food, sometimes influencing the intertidal ecosystem. For example, several archaeological studies of shells found in dated layers of **middens** (mounds containing ancient domestic refuse) have provided evidence of a continual decline in the size of harvestable bivalves or limpets, and the switch to smaller species over time. Researchers have found limpets to be one of the major components of middens in many areas of Europe, indicating the harvest of many bivalve species; mussels and limpets were harvested off the coast of the British Isles for thousands of years. Off California, mussels, abalones, urchins, chitons, and crabs have been harvested for millennia with no evidence of harm to the populations. In some areas of the South Pacific, however, limpets apparently disappeared from some regions, and mussels and urchins declined with time in some archaeological sites, possibly due to overharvest. In South Africa, excessive harvest 2,000 to 3,000 years ago appears to have resulted in declines in the sizes of limpets and mussels. One lesson to be learned from this historical evidence is that it does not require large numbers of people with modern harvest technology to have local effects on harvested organisms where they are readily accessible.

Current harvest of "shellfish" such as limpets and bivalves from the rocky intertidal is primarily for local or **subsistence** use, either for human consumption or for use as bait. Rocky intertidal mollusks are typically small and more difficult to harvest on a large scale compared to commercially harvested bivalves taken from subtidal habitats. Harvest, however, can cause local population declines and change the ecosystem makeup of the rocky intertidal. Although climate change and other factors likely have resulted in some declines, many examples point to recent local overexploitation. For example, in France the harvest of sea urchins from the rocky intertidal for their gonads likely contributed to their decline in the late 1900s.

The rocky intertidal organisms with the longest documented history of substantial harvest and current exploitation are limpets. One example is the harvest of large *Patella* limpets in recent times that contributed to local extinctions on the most populated of the Canary Islands and in parts of the Mediterranean, and dramatic declines in the Azores, until a local ban on harvest was implemented. The harvest of the largest limpets can be especially harmful because they are **protandrous**, capable of switching from male to female at larger sizes. Taking the largest limpet can bias the sex ratio and limit reproduction by preferentially removing females from the population.

On rocky coastlines of Chile, limpets are commonly taken for local use and commercial harvest to support one of the largest current limpet "fisheries." The keyhole limpet *Fissurella* is a keystone species in these rocky intertidal ecosystems, so its harvest can cause dramatic changes in the community structure. The creation of a no-take reserve resulted in an increase in size and number of the limpets and *Concholepas* snails (commonly called the "Chilean abalone" because of its appearance), a decline in the algae and mussels on which they prey, and a subsequent increase in barnacles and other organisms. Similar ecosystem recoveries in rocky intertidal marine reserves have been documented in South Africa and other regions. It is apparent that harvest of rocky intertidal organisms can have a large influence on biological communities.

Harvest of invertebrates from the rocky intertidal not only affects populations of the targeted species but also can result in modifications to the ecosystem. The importance of ecological interactions in the rocky intertidal is discussed earlier. Removal of a prevalent species either naturally or by human harvest can affect the ecosystem balance. For example, if sedentary species such as barnacles or mussels are removed, space is opened for colonization by algae. The harvest of predators such as snails or sea stars, whether for food, aquariums, or as **curios** ("souvenirs"), can result in increases in prey species such as barnacles or mussels.

Seaweed probably has been harvested from the rocky intertidal coastlines for as long as invertebrates; however, their remains are not readily preserved in middens. In Japan and China macroalgae has been a staple food and delicacy for over 2,000 years. Beginning in the 1950s, methods of culturing seaweeds were developed and today most commercial production is through aquaculture, primarily in eastern Asian countries. Harvest from the rocky intertidal continues mostly for local or subsistence use and could have local impacts on the rocky intertidal ecosystem.

One region where harvest of rocky intertidal seaweed has been controversial is on the Atlantic shores of Canada and the northeast United States. Here the primary intertidal seaweed is *Ascophyllum nodosum,* commonly called rockweed, a species that does well in relatively sheltered areas such as bays. The intertidal zone in this region is expansive due to the great tidal range, and rockweed can be harvested from the shore or from boats at high tide. Harvest from the Gulf of Maine and Bay of Fundy for fertilizer and fodder in animal feed, and more recently as an herbal supplement for human consumption, has been controversial. Scientists and conservationists are concerned about the ecosystem impact because rockweed is at the base of the food web and functions as a shelter for invertebrates and fishes, many of which are commercially harvested in open waters of the bays and ocean. If all or most of the plant is removed, recovery is slow, and other organisms may move into the space left open and inhibit recruitment of the rockweed. If only the upper fronds are trimmed it may grow back quickly.

Restrictions on harvest of rockweed are implemented by U.S. states and Canadian provinces. In Maine, although 16 inches of the rockweed must remain after harvest so that it can regrow, there are few other harvest restrictions. The harvest has become controversial and has led to conflicts over rights to harvest the rockweed and personal property rights in the intertidal zone. Environmental groups have complained of a lack of enforcement of regulations on harvest and potential effects on the ecosystem. They have called for a moratorium on rockweed harvest so that further studies can be carried out to document its effects. In Nova Scotia, Canada, overharvest of rockweed led to measures implemented in the 1990s to maintain ecosystem sustainability by restricting harvest, distributing licenses, and assigning harvest areas to a few companies. In order to maintain the rockweed, cutting height is regulated and areas are protected after harvest until they regrow. This is an example of how economics, politics, and conservation can clash when valuable resources are easily taken from the intertidal zone. These conflicts also provide examples of how ownership of the intertidal zone and the organisms that reside there is not always clear cut.

Physical Disturbances

Recreational use is not typically as heavy in the rocky intertidal as in sandy intertidal areas. Rocky shorelines are often adjacent to beaches and people may walk onto the rocky shoreline to fish or for nature watching. Trampling of invertebrates is typically minimal because most attached organisms have a hard shell for protection and other organism hide in crevices. Trampling of algae, however, can result in ecosystem changes if foot traffic is heavy. Comparisons have documented that heavily trampled areas have less macroalgal covering and more bare patches that are colonized by grazing mollusks. Daniela Casu and colleagues found that populations of other invertebrates, such as polychaete worms, residing in the algae also can be harmed by trampling.

Taking organisms for bait, aquarium use (e.g., sea stars), or as curios (e.g., mollusks as "sea shells") can have similar effects as other types of harvest if collection pressure is great enough. Although the take by one person may be low, in frequently visited areas the cumulative effect can be substantial. Care also must be taken by researchers and educators to minimize collections from the rocky intertidal. Collection permits may be required for taking some organisms from the intertidal for scientific or education purposes, especially if they are considered endangered or have sensitive populations.

In coastal regions where logging is prevalent, drift logs can be one of the most important physical factors controlling the distribution of sessile species in the rocky intertidal. Along areas of the U.S. Pacific Northwest coastline, for example, logs that move around with tides and storms battering the coastline can have important effects on the distribution of mussel beds. Areas where mussels have been crushed are expanded by the forces of wave action. This can result in colonization of disturbed areas by other sessile species. Those regions battered continuously enough have sessile organisms found only in crevices. Some logs are from natural drift of uprooted trees, but in the most affected areas the majority of the drift logs are from human sources as they show signs of being cut.

Pollution

With rapidly expanding coastal populations come increased impacts to intertidal regions. Rocky intertidal shores are not as vulnerable to habitat destruction as are coastlines with sands or other soft sediments; however, pollution effects can be severe. Organisms may be exposed to any waterborne contaminants, including heavy metals.

Filter feeders are especially prone to accumulation of non-degradable toxicants. These contaminants, however, would not typically accumulate on the shore as they might in soft sediments. Global impacts of contaminants on marine organisms are discussed further in Chapter 7.

Plastics and other non-degradable materials wash onto the rocky intertidal and may accumulate in pools and crevices. Such litter affects this region aesthetically but can also carry contaminants such as oils or other chemicals. The effects on organisms of chemicals that leach from the plastics, such as in tide pools, are not well known.

Much of the pollution documented to harm the rocky intertidal ecosystem is a result of sewage discharge, industrial pollution, or oil spills. Concerns over sewage pollution are similar to those discussed for sandy beaches, although the pollutants from a sewage spill are more likely to dissipate from rocky shorelines, especially if there is a substantial tidal range and wave activity. Chronic nutrient pollution, such as from agricultural runoff or atmospheric deposition, can change the makeup of rocky intertidal communities, however. For example, in regions bordering the Baltic Sea in northern Europe, nutrient pollution apparently has caused a decline in the predominant macroalgae *Fucus* along with an increase in other algae growing on rock surfaces. This has changed the composition of the rest of the rocky intertidal community.

Oil is one type of pollutant that often does not readily dissipate from the rocky intertidal. Especially in areas where there is limited wave or tidal activity, oil can adhere to the rocks and intertidal organisms and wash into crevices. Long-term effects have been observed on the rocky intertidal ecosystem from large oil spills, such as the *Torrey Canyon* spill off the British Isles in 1967 and the *Exxon Valdez* spill off Alaska in 1989 (**Box 3-3. Learning from History: *Torrey Canyon, Exxon Valdez,* and *Deepwater Horizon***). The slow decomposition of oils in the environment suggests that small-scale chronic oil pollution, such as from urban runoff and routine release by the oil industry, may have long-term effects on the rocky intertidal, particularly in areas near oil wells, refineries, or industrialized regions.

Point-source pollution of toxic non-degradable chemicals by industries has been more heavily regulated over the past 50 years, although there are still problems with enforcement in many areas of the world, despite an increased global awareness. A recent concern over the introduction of pollutants has been their function as endocrine disruptors. Certain man-made chemicals, although they cause no noticeable immediate changes in health of organisms, have long-term population effects by disrupting normal endocrine functions that affect development and reproduction. Although endocrine disruptors have been studied most widely in freshwater systems, they are now being ascertained

Box 3-3 Learning from History: *Torrey Canyon, Exxon Valdez,* and *Deepwater Horizon*

Three major oil spills in the past 50 years stand out not only for the damage they caused to coastal habitats and the attention they received, but also from the lessons learned regarding how to avoid and react to subsequent spills. The first of these spills occurred when the *Torrey Canyon* supertanker ran aground off the southwest coast of the United Kingdom on March 18, 1967, spilling 100,000 tons of crude oil into the ocean (**Figure B3-3**). The oil, mostly floating on the surface, was washed rapidly toward the shore. Within a few days, before

(a)

(b)

Figure B3-3 **(a)** The supertanker *Torrey Canyon* off the coast of England in 1967, following the grounding and subsequent release of 100 thousand tons of crude oil. **(b)** The *Deepwater Horizon* drilling rig after the 20 April 2010 explosion. The rig sank the following day, resulting in release of approximately 550 thousand tons of oil at the seafloor over the next 85 days.

cleanup operations could be put into place, one-half of the oil was deposited on the coast of southwest England. By chance, the timing of the spill coincided with a period of extremely high tides, so that much of the oil was deposited high on the shore where it could not be washed away by subsequent tides. The oil remaining on the ship was released when the ship broke up. Northerly winds blew the oil toward the coast of France. Despite efforts to contain the oil, most of it washed ashore on the coast of northwest France, where it covered 90 kilometers of coastline, with deposits up to 30 centimeters thick. The total amount of oil that washed ashore was larger than documented for any other spill (though other spills have released more oil).

As the spill moved shoreward, large cleanup efforts were attempted with varying success. England's response to the spill was to spray dispersants on the oil that had washed ashore to enhance its solubility in water, with hopes that it would eventually by washed away. Dispersants also were sprayed on the oil that eventually washed ashore in France, without much success. The cleanup efforts to remove the oil from the shoreline in France mostly involved crews scraping and scooping it from the surface, though eventually dispersants were used in some tourist areas to quicken the cleanup.

The cleanup operations were carried out mostly without preplanning; methods had not been developed by the time of the accident to deal with such a large spill and minimize the damage to the environment and ecosystems. Decisions had to be made on the spot. Only after extensive studies were carried out years after the spill did scientists gain a reasonable understanding as to the extent of spill's impacts. The greatest lethal harm to organisms was smothering by the oil; the toxicity of the crude oil to most organisms was relatively low. Ironically, the dispersants that were applied to the oil were far more toxic than the oil itself, and resulted in greater mortality of intertidal organisms. Unfortunately, the volume of dispersant chemicals applied to the oil that came ashore in England was almost equal to the volume of the oil itself. Toxicity of the dispersant chemicals had not been tested prior to their use. Intertidal invertebrates were very sensitive, with limpets being most susceptible. In coastal areas near where dispersants were sprayed, virtually all animals were killed, along with much of the algae. Although beaches that were exposed to the spill recovered fairly rapidly, effects on the rocky intertidal were long-lived. Removal of the limpets was especially harmful because they are considered keystone species in this ecosystem. *Fucus* algae were the first to recolonize the shoreline, and thick mats inhibited the recolonization by other animals. Eventually predatory snails returned, keeping barnacle populations from recovering. When *Patella* limpets returned, they underwent a population explosion, eradicating most of the algae. Although the direct effect of the oil and dispersants was long gone, dramatic population fluctuations such as these continued, and it was 15 years before the community finally returned to what were considered normal stable conditions. This is one of many lessons that teach us how unpredictable ecosystem responses can be to anthropogenic impacts in coastal and marine ecosystems.

The second oil spill dramatically affecting rocky intertidal ecosystems, although not as large as the *Torrey Canyon* spill, received much more attention because of its effect on a pristine shoreline and intensive coverage by the news media. The *Exxon Valdez* ran aground in Prince William Sound Alaska in

1989, spilling over 38,000 tons of crude oil that washed ashore onto 1,100 miles of wilderness rocky shoreline (**Figure B3-4**). Storm winds and extreme tides washed the oil high into the intertidal zone. The oil settled onto the rocks and covered them with a black viscous layer. Lessons that had been learned from *Torrey Canyon* and other oil spills resulted in a better preparedness for cleanup; still, the results of the cleanup operation and its effect on the ecosystem were mixed.

Cleanup of oil from the *Exxon-Valdez* spill in the open water was relatively successful. The application of dispersants was avoided due to the potential harm to shoreline ecosystems. Booms were used to contain the oil and keep it from drifting shoreward, especially avoiding salmon-spawning rivers. Skimmers, vacuums, and sorbent materials were used to successfully remove some of the oil from the water's surface. Before the open-water cleanup could be completed, however, many seabirds and sea otters were exposed to oils. Large amounts of money were spent in attempts to clean these animals and return them to the wild. Although this was successful for a relatively small number of individuals, it did little to keep seabird populations from being harmed. Estimates vary, but possibly hundreds of thousands of seabirds died as a result of the spill. Over 1,000 sea otters died as a result of contact with the oil.

As the open water cleanup continued, Exxon was pressured to complete a shoreline cleanup no matter what the

(a)

(b)

(c)

Figure B3-4 **(a)** The *Exxon Valdez* during cleanup following the 1989 spill of over 38 thousand tons of oil into Prince Williams Sound, Alaska. **(b)** High-pressure hot water washing in an attempt to remove oil from intertidal zone. **(c)** Oil containment booms deployed to surround and protect New Harbor Island, Louisiana from oil from the *Deepwater Horizon* spill.

cost, in part because of the heightened awareness by the news media and intense public pressure. This resulted in a push to remove the oil even from shorelines where it might be ill-advised due to the difficulty of oil removal without harm to organisms residing there. Thousands of workers were hired to scoop up the oil with shovels where it was thick enough, and to wipe the oil from the rocks using rags in other areas. Rocks were blasted with high pressure water jets in an attempt to wash the oil into open water where it could be more easily contained and removed. In areas harder to reach or where cold water washing was unsuccessful, heated seawater was used. Dispersants were used along about 70 miles of shoreline; although these dispersants were considered safer than those used for *Torrey Canyon*, use was limited because of fear of ecological impacts. Cleanup efforts continued along the shoreline during summers for two years following the spill. In the end, each of the attempts was only partially successful in removing the oil. In some cases cleanup efforts were more harmful to the ecosystem than the oil itself, especially the hot water to which most coastal organisms were intolerant.

Scientific studies were initiated immediately after the spill and have continued since. There were massive mortalities of invertebrates and algae in the rocky intertidal. Death of *Fucus* algae resulted in the loss of grazing limpets, snails, and predatory whelks. This was followed by colonization by green algae and *Chthamalus* acorn barnacles. The recovery of seaweeds and invertebrate populations became cyclical in a manner similar to the *Torrey Canyon* spill. Normal stable conditions did not begin to return for over a decade after the spill in areas where the oil had disappeared. Surveys 15 years later indicated that some animal populations still had not recovered. In intertidal sediments substantial amounts of oil remain to this day, and effects extend into freshwater and terrestrial ecosystems.

New lessons were learned from both of these spills and cleanup. Of course, the best solution is prevention: avoid the incident before it happens. Although much tighter regulations have been enacted, including structural improvements to the ships and closer monitoring of ships coming into harbor, there is a general consensus that as long as we transport large amounts of oil around the globe, some spills are inevitable. Therefore, we must not only minimize the chance of these spills, but also be ready to minimize the damage when they do occur. Laws have been passed to help achieve this by forcing oil companies to be better prepared to react quickly to future tanker spills. We must be aware, however, that the rush to immediately solve problems created by environmental accidents can result in efforts that do more harm than good. Sometimes the "best available" methods are more harmful than doing nothing at all. For both the *Torrey Canyon* and *Exxon-Valdez* spills, continual monitoring of affected areas have indicated that where there was recovery it occurred naturally, albeit slowly. It is debated whether the massive cleanup efforts in the rocky intertidal considerably enhanced the rate of recovery. Another lesson from these spills is that it may take much longer than anticipated for oil residues to disappear and ecosystems to fully recover. Some shorelines affected by the *Exxon-Valdez* spill are still contaminated with oil over 20 years after the spill, especially in sediments of the upper intertidal regions. Subsurface oil is present in some locations, many of them in the lower intertidal region; in some cases the oil has consistently appeared the same as immediately after the spill. It is clear that affects of the oil spill are more persistent than almost anyone would have predicted.

The most recent major oil spill occurred over a three month period beginning April 20, 2010 following a well blowout, explosion, and sinking of the *Deepwater Horizon* offshore drilling rig in the northern Gulf of Mexico. Over 4 million barrels (780 million liters—roughly 550,000 tons) of oil were released from the blowout at the sea floor at a depth of approximately 1,500 meters below the ocean surface. Although the volume of oil released was over five times that of the *Torrey Canyon* spill and ten times that of the *Exxon-Valdez* spill, the amount reaching the shore was less because much of the oil dispersed into the water, sank to the seafloor, was naturally decomposed, or was skimmed or dissipated from the surface before reaching shore. (The effects of the *Deepwater Horizon* spill on seafloor and open-ocean ecosystems is still uncertain and will be evaluated for decades; see Chapter 8). Based on knowledge gained from previous spills, the use of chemical dispersants was avoided along the coastline and near shore after the *Deepwater Horizon* spill. Various efforts were made to keep the oil from reaching coastal beaches and salt marshes, including the use of skimmer boats, floating containment booms, barriers along the shoreline (see Figure B3-4c), and dispersants near the source of the blowout. Despite these efforts oil began washing onto beaches about one month after the blowout. Oil eventually washed up in areas of coastal beaches and marshes along 200 kilometers of Louisiana's coast, and barrier islands and beaches of Mississippi, Alabama, and western Florida. Cleanup of oiled beaches was carried out throughout this region through the removal of visibly oiled surface sands, much by crews working with hand-held shovels. Removal of oil from salt marshes was more difficult and most successful when pumped or skimmed from the water before the oil had entering vegetated areas. Although long-term impacts will take years to assess, dieback of marsh grasses was documented in some Louisiana marshes up to 30 meters into salt marshes. Recovery from oil impacts has been demonstrated from previous spills to be more rapid in grass-dominated marshes (which are most prevalent in this region) than in mangroves in more tropical regions (see Chapter 4). The direct effect on wildlife of oil coming onto shore was difficult to document. Over 1,500 bird deaths were attributed to contact with oil, and hundreds of dead sea turtles with clear indications of contact with oil were recovered during spill cleanup efforts. Long-term impacts on these populations and on coastal habitats are not known. A more rapid recovery is predicted here than in areas affected by the *Exxon-Valdez* spill due to the relatively low amount of oil reaching shore, the more rapid decomposition of oil in the warmer climate, and the physical makeup of the ecosystems affected. It will likely be decades before a full biological assessment of the *Deepwater Horizon's* impact on coastal ecosystem is completed.

Figure 3-34 The dogwhelk, *Nucillus lapillus*, is sensitive to reproductive effects of exposure to TBT from leachates of paints.

to harm many coastal marine and intertidal organisms. One source of such chemicals is leachates from antifouling paints, used to resist the attachment of algae, barnacles, and other sessile organisms on boat hulls and other structures. One group of rocky intertidal organisms that appears to be particularly vulnerable is mollusks, most thoroughly documented in dog whelks (*Nucella*) in the United Kingdom (**Figure 3-34**). Exposure to the **biocide** tributyl tin (TBT) from leachates of paints can result in the development of male sexual organs in female whelks in a condition called **imposex**. The male organs block the release of eggs, causing female sterility, and in some cases eventual death. This condition was prevalent enough in some bays to cause local extinctions of the whelks. Although TBT was banned by some countries in the late 1980s, it was not until 2008 that an international convention went into effect banning its use worldwide. An increased awareness of similar effects by other chemicals will likely lead to future bans; however, typically there are long time lags between when the effects are recognized and when bans are implemented and enforced.

Introduced Species

Non-native or **exotic species** can become **invasive** and have unpredictable effects on natural ecosystems due to competition, predation, or modifications to the ecosystem. Although intentional introductions are typically illegal or discouraged, unintentional introductions remain problematic. One of the most common ways that aquatic organisms are moved around the world is by ships. Many of these introductions are from the release of ballast water (see Chapter 4). Sessile organisms, such as barnacles or mussels, can attach to the ship hull and then release offspring in coastal areas at the ship's destination. Typically the presence

of exotics is not known until they have become a major component of the rocky intertidal ecosystem. For example, *Elminius* barnacles from Asia and Australia and *Sargassum* seaweed from Japan have been introduced to Europe (**Figure 3-35**). *Elminius* showed up in southeast England in the 1940s and rapidly spread throughout Britain, competing with native *Balanus* barnacles. *Sargassum* has changed the native algae assemblage. In South Africa, *Mytilus* mussels from the Mediterranean Sea have displace the native *Aulacomya* mussels, and competition for space has resulted in declines in algae and limpets. The common periwinkle *Littorina littorea*, as convincingly argued by John Chapman and others, was likely introduced into the rocky intertidal of the Atlantic coast of Canada in the mid-1800s, possibly in rock ballast. This invasive species has caused large changes to the native community through competition with native gastropod species. It is the predominant mollusk along much of the U.S. northeast coast and has colonized the U.S. Pacific coast. The Asian shore crab *Hemigrapsus sanguineus*, native to western North Pacific coastlines, was discovered on the New Jersey shoreline in the late 1980s, likely introduced in ballast water. It has expanded and become well established along much of the northeast U.S. coast, and appears to be outcompeting native crab species, which have declined as the *Hemigrapsus* populations increase. There are concerns that this invasive crab will continue its expansion along the North American coastline within its salinity and temperature tolerance limits.

Although laws are in place in many regions to avoid exotic introductions, it is virtually impossible to eliminate all introductions due to the number of boats travelling from

Figure 3-35 Close up view of *Elminius modestus* barnacles, which are invasive in Europe.

one continent to another. Although most introduced species are rapidly outcompeted or eaten by predators, the few species that successfully reproduce and survive into healthy populations are almost impossible to eradicate.

Rocky Intertidal Protection

Large scale destruction of the rocky intertidal is not as likely as for other coastal ecosystems; and local protection is simpler than for many other marine ecosystems due to the relative ease of limiting access and monitoring the habitat. Anthropogenic changes have actually increased the amount of rock intertidal habitat in some regions through building jetties, dikes, breakwaters, seawalls, and other structure (at the expense of other habitats). Although these are not natural habitats, the ecosystems that become established mimic the rocky intertidal. Marine and atmospheric pollution or global climate change may affect the rocky intertidal as with any marine ecosystem; devising solutions to these problems is complex and must be considered from a global perspective (see Chapter 1). Local protection of rocky intertidal ecosystems is mostly a matter of limiting coastal pollution and protecting organisms from excess harvest.

Protected areas established along coastlines often include rocky intertidal areas. For example, in Australia, Intertidal Protected Areas provide specific protection of rocky intertidal habitats, including prohibition of collecting animals. Southern California has several coastal marine protected areas with rocky intertidal habitats. Removal of organisms from these areas is prohibited, but enforcement and education (i.e., by placing signs) are considered inadequate and illegal collecting is commonplace. For example, the large owl limpet *Lottia gigantea* is commonly collected for food or bait. This is a protandrous species with most large individuals being females; therefore, collecting the largest limpets can bias the sex ratio and affect reproduction. Research by Raphael Sagarin and colleagues revealed that, on inaccessible islands and in areas protected from collecting by monitoring, education, or limiting access, limpets are larger and exhibit a greater range of sizes. Even in Marine Protected Areas a lack of enforcement capabilities, the ease of harvesting rocky intertidal organisms, and the potential for damage from trampling can result in the need for restricted access.

Government and non-governmental organizations (NGOs) work to monitor and protect the rocky intertidal from harm. One such NGO is Conservation and Biodiversity of the Rocky Intertidal of Southern California (CBRISC), which works to assess the health of the rocky intertidal environment and maintain a database for intertidal species, as well as educating the public, possibly the most efficient method of protecting these regions. Possibly the largest organization dealing specifically with the rocky intertidal region is the Multi-Agency Rocky Intertidal Network (MARINe), a collaboration among government agencies, universities, and private groups, which assesses the health of over 100 rocky intertidal habitats along the northeast Pacific coast from Mexico to Alaska and in the New England region of the United States. This network transfers information to resource managers and the public through reports and scientific publications, and provides publications, videos, and education programs for schools.

3.4 Global Climate Change and Intertidal Ecosystems

Intertidal ecosystems could be profoundly affected by climate change because they are heavily influenced by sea levels, storm events, and coastal currents, all of which are predicted to undergo change. Impacts could extend from the dunes through the beach and throughout the rocky intertidal zone. Increased storms will result in greater and more frequent wave activity along the shore. Where beaches are prevalent, some may gain sand from storm erosion; however, most beaches will retreat and become narrower. Under normal conditions there is a balance between the erosion of sands from the beach during storms and the slower return of sands in calmer conditions due to wave action. If this balance is disrupted there will be a net loss of sands from the beaches. Beach erosion also would increase in regions where precipitation and flooding increase as waters flowing to the sea carry sand with them. If beaches are to be retained at their current size there will be a need for more frequent nourishment. Increased storm frequency or increases in coastal winds could increase erosion of dune sands and affect dune development.

Increases in sea level would have the most substantial effect on intertidal ecosystems. It is impossible to develop precise predictions of the degree and timing of sea level increases due to future global climate change (see Chapter 1). If sea levels increase near the 75 centimeters predicted as likely over the next century, not only the intertidal zone, but entire islands and coastal regions will be covered by water (**Figure 3-36**). The amount of beach lost along sandy coastlines could be much greater than one might anticipate, because the horizontal loss of beach averages about 100 times the vertical sea level rise. This means a 75-cm sea level rise could cause a 75-meter horizontal loss of beach. Very few current beach habitats would remain. If intertidal ecosystems convert to open water, human populations will be displaced from low islands and current coastlines, especially where the shore is gradually sloping. If sea level rise is gradual enough, the dune and beach zone will migrate landward where the space is available. Land for human settlements will be at a premium, however, and it is probable

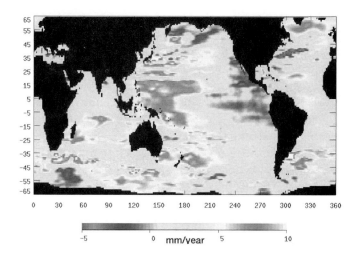

Figure 3-36 Rates of sea level change around the globe from 1993–2008. Note that greatest sea level rise has been experienced in the tropical·western Pacific, a region with numerous populated low-elevation islands. (See Color Plate 3-36.)

that coastlines will be heavily protected by sea walls and other hard structures, keeping expansive dunes and broad beaches from developing in most coastal areas. Even moderate increases in sea level would cause flooding of the current beach zone and result in the loss of barrier islands and other low-elevation islands. Where sea walls or other protective barriers are in place behind narrow beaches, the beach would eventually disappear as the water level approaches the barrier. Beach nourishment would likely be used to at least slow this process. Where dunes are currently still in place, sea level rises would harm the plant communities or convert dunes into beach habitat. Most dunes will not be able to migrate landward due to steep coastlines or hard barriers put in place for coastal protection; thus they will be either disappear be squeezed into a narrow zone. The loss of the dunes will not only affect the ecosystem but remove the protection of inshore regions provided by healthy dynamic dune systems.

The effects of small changes in sea level would not be dramatic in the rocky intertidal relative to other coastal ecosystems; however, larger changes could result in the loss of the lower intertidal. If area is not available above the current high tide line, the rocky intertidal zone would be compacted. If storms become more frequent with climate change, as predicted, there would be an increase in abundance of species that do well in conditions of high wave and wind impact. We, therefore, would expect increases in filter feeders and predators and reductions in grazers and macroalgae. Increases in sea levels and extreme storms could result in increases in the amount of man-made hard structure in the intertidal because we would likely increase the building of hard barriers to hold back the waters and minimize coastal erosion. This could result in increased

abundance in some rocky intertidal species and expand the range of others into regions with little natural rocky coastal habitat (such as the southeast U.S. coastline). Warmer or colder temperatures associated with climate change would result in changes in the species makeup, with the loss of organisms at the extremes of their ranges.

A predicted increase in the frequency and severity of El Niño events could affect the species makeup and nutrient inputs in intertidal areas. The decline in upwelling associated with El Niño would result in lowered productivity as well as temperature changes. Intertidal community responses to recent El Niño can be used to predict future impacts. For example, the 1982–1983 El Niño affected the intertidal zone along North and South American coastlines, a region strongly influenced by upwelling. In the rocky intertidal off Chile, there were large die-offs of brown algae and invertebrates and low recruitment of *Concholepas* gastropods. In California, recruitment of barnacles and tide pool fishes was reduced during the El Niño. The abundance of many organisms declined on Peruvian beaches, but recovered when conditions returned to normal. These responses appear to be associated mostly with nutrient declines.

Intertidal organisms are relatively tolerant to temperature changes and most are not near the upper limit of their temperature tolerance. Temperature changes brought on by global climate change would be gradual enough that many intertidal species would adapt, though species distributions could change as tropical species move into higher latitudes.

The predictions scientists have made of the effects of climate change surely will not prove to be 100% accurate. We do not fully understand the responses of intertidal ecosystems to physical changes, nor do we have a complete understanding of if and when the predicted changes in global climate will occur. Interactions among numerous variables will be important, including temperature, current patterns, sedimentation, erosion, storm events, and human responses. It is crucial, however, for scientists to continue studies of these ecosystems, so that negative effects can be minimized and data can be generated to encourage the public and political leaders to work toward solutions to the climate change dilemma. Although few firm actions are being taken, governments are beginning to discuss preparations for future sea level rise. Taking the attitude that we will deal with it as we have to is not a valid solution. It must be emphasized that not only the environmental, but also the social and economic costs of climate change on coastal ecosystems would be enormous. Ecosystems and species could be lost, a large portion of the human population could be displaced, and the economic costs of protecting and minimizing the impacts on shorelines around the world would be tremendous.

STUDY GUIDE

■ Topics for Review

1. Why are grasses such as sea oats and beach grass important to maintaining dune habitats?
2. What environmental advantages does *Salicornia* have in arid regions over other plants grown for agricultural uses?
3. How do environmental stresses that plants must adapt to differ between the backdune and foredune?
4. What anthropogenic factor has caused the greatest loss of dune habitat along the U.S. Atlantic and Gulf of Mexico coastlines?
5. Other than urban development, historically what was the primary reason for reclaiming dunes in Europe?
6. How do endangered species laws in the United States provide protection of dune habitats?
7. What measures are commonly taken to enhance survival of newly established dune grasses?
8. Why is the planting of *Ammophila* grass discouraged in California dunes, even though it is an excellent way to stabilize the dune sands?
9. What effects do stabilizing Netherland foredunes against blowouts have on grasses in the backdunes?
10. What are the primary forces that move sands along dunes and beaches?
11. How does the lack of large plants affect beach ecosystems?
12. What mechanism of locomotion is used by some bivalves and whelks to move rapidly up and down the beach?
13. Describe the feeding mechanisms of predatory gastropods on the beach.
14. Describe how probing and surface feeding birds differ in the way they recognize food organisms on the beach.
15. Describe how seawalls placed behind the beach affect the structure of the beach.
16. Describe how groins affect the movement of sands along the beach.
17. What are the negative effects of using beach nourishment to maintain a beach?
18. What is the primary reason for declines of beach organism after nutrient pollution?
19. Explain why the proportion of nematodes to copepods provides a good indication of pollution impacts on beach ecosystems.
20. How do oil dispersants assist in the cleanup of oil spills and why are dispersants not recommended for cleaning oil washed onto beaches?
21. How can the removal of the wrack that has washed onto beaches affect beach organisms?
22. How can lights shining onto beaches be harmful to sea turtle populations?
23. Explain how the harvest of horseshoe crabs has affected species using beach ecosystems.
24. How do natural physical stresses limit the kinds of organisms that live in the rocky intertidal zone?
25. Why are macroalgae (seaweeds) more prevalent in the rocky intertidal than on sandy intertidal beaches?
26. Why is the conservation of a single keystone species important to the entire rocky intertidal ecosystem?
27. Why are communities in rocky intertidal regions exposed to extreme wave activity less likely to be controlled by biological interactions?
28. What evidence suggests that mollusks were overharvested by some ancient cultures?
29. How does harvest of the largest limpets from the rocky intertidal affect reproduction?
30. How does harvest of rockweed affect coastal fish populations?
31. How do chemical leachates from antifouling paints affect reproduction in dog whelks?
32. Describe how exotic species of sessile invertebrates are able to travel across ocean basins to colonize rocky intertidal regions.
33. Why are oil dispersants not recommended for use in the cleanup of oil spills that wash ashore into rocky intertidal areas?
34. How would an increase in the frequency of coastal storms likely affect the size of beaches as well as the community makeup of the rocky intertidal?
35. Why would intertidal communities on eastern Pacific coastlines be affected by an increased frequency of El Niño events that could accompany global climate change?

■ Conservation Exercises

Describe how each of the following actions assists in conservation of intertidal ecosystems. Include a description of the terms given in bold.

a. removal of *Ammophila* **beach grasses** from coastal dunes in California.
b. removing **berms** between dunes and beaches in Europe.
c. building elevated **boardwalks** over dunes in Florida.
d. considering the Alabama beach mouse for listing as an endangered species.
e. placing **sand fences** in dune habitat along the U.S. Gulf of Mexico.

f. establishing **priority zones** in the Meijendel Dunes in the Netherlands.

g. establishing restrictions against **beach hardening** in North Carolina.

h. **nourishing** Gulf of Mexico barrier island beaches with sands dredged from shipping channels.

i. establishment of the **BEACH program** by the U.S. EPA.

j. allowing **wrack** to accumulate on the beach.

k. limiting **light pollution** along Caribbean island beaches.

l. relocating eggs from sea turtle nests in Mexico to other areas of beach.

m. regulating the sex ratio of harvested horseshoe crabs in Maryland.

n. protecting **keystone species** from harvest in California rocky intertidal zones.

o. restricting harvest of the largest limpets in the rocky intertidal of Chile.

p. restricting cutting height for **rockweed** along the Nova Scotia coastline.

q. outlawing the use of paints containing **TBT** in the United Kingdom.

r. requiring ships to release **ballast waters** away from the shoreline along the U.S. coast.

s. establishment of the MARINe collaboration in North America.

t. avoiding the use of **chemical dispersants** along the coastline of France during the cleanup of the *Torrey Canyon* oil spill.

u. limiting the building of **sea walls** as sea levels rise due to global climate change.

FURTHER READING

Baeyens, G. M., and L. Martinez. 2004. Animal life on coastal dunes: from exploitation and prosecution to protection and monitoring. Pages 279–293. In *Ecological Studies 71*. M. L. Martinez and N. P. Psuty (eds.). Coastal Dunes, Ecology and Conservation. Springer-Verlag, Berlin, Heidelberg.

Benton, T. G. 1995. From castaways to throwaways: marine litter in the Pitcairn Islands. *Biological Journal of the Linnean Society* 56:415–422.

Beukema, J. J. 1995. Long-term effects of mechanical harvesting of lugworms *Arenicola marina* on the zoobenthic community of a tidal flat in the Wadden Sea. *Netherlands Journal of Sea Research* 33:219–227.

Casu, D., G. Ceccherelli, M. Curini-Galletti, and A. Castelli. 2006. Short-term effects of experimental trampling on polychaetes of a rocky intertidal substratum (Asinara Island MPA, NW Mediterranean). *Scientia Marina* 70S3:179–186.

Chapman, J. W., J. T. Carlton, M. R. Bellinger, and A. M. H. Blakeslee. 2007. Premature refutation of a human-mediated marine species introduction: the case history of the marine snail *Littorina littorea* in the northwestern Atlantic. *Biological Invasions* 9:737–750.

Dayton, P. K. 1971. Competition, disturbance, and community organization: the provision and subsequent utilization of space in a rocky intertidal community. *Ecological Monographs* 41:351–389.

Ellers, O. 1995. Behavioral control of swash-riding in the clam *Donax variabilis*. *The Biological Bulletin* 189:20–127.

Feldman, K. L., D. A. Armstrong, B. R. Dumbauld, T. H. DeWitt, and D. C. Doty. 2000. Oysters, crabs, and burrowing shrimp: review of an environmental conflict over aquatic resources and pesticide use in Washington State's (USA) coastal estuaries. *Estuaries* 23:141–176.

Good, T. P. 2004. Distribution and abundance patterns in Caribbean rocky intertidal zones. *Bulletin of Marine Science* 74:459–468.

Kooijman, A. M. 2004. Environmental problems and restoration measures in coastal dunes in The Netherlands. In *Ecological Studies 71*. M. L. Martinez, and N. P. Psuty (eds.). Coastal Dunes, Ecology and Conservation. Springer-Verlag, Berlin, Heidelberg.

Martinez, M. L., N. P. Psuty, and R. A. Lubke. 2004. A perspective on coastal dunes. Pages 3–10. In *Ecological Studies 71*. M. L. Martinez, and N. P. Psuty (eds.). Coastal Dunes, Ecology and Conservation. Springer-Verlag, Berlin, Heidelberg.

McDermott, J. J. 1998. The western Pacific brachyuran (*Hemigrapsus sanguineus*: Grapsidae), in its new habitat along the Atlantic coast of the United States: geographic distribution and ecology. *ICES Journal of Marine Science* 55:289–298.

McLachlan, A., and A. Brown. 2006. *The ecology of sandy shores*. Academic Press, Burlington, MA, USA.

Pahl, J. W., I. A. Mendelssohn, C. B. Henry, and T. J. Hess. 2003. Recovery trajectories after in-situ burning of an oiled wetland in coastal Louisiana, USA. *Environmental Management* 31:236–251.

Paine, R. T. 1994. *Marine rocky shores and community ecology: an experimentalist's perspective*. Ecology Institute, Nordbruite, Germany.

Patino-Martinez, J., A. Marco, L. Quinones, and B. Godley. 2008. Globally significant nesting of the leatherback turtle (*Dermochelys coriacea*) on the Caribbean coast of Columbia and Panama. *Biological Conservation* 141:1982–1988.

Peterson, C. H., S. D. Rice, J. W. Short, D. Esler, J. L. Bodkin, B. E. Ballachey, and D. B. Irons. 2003. Long-term ecosystem response to the *Exxon Valdez* oil spill. *Science* 302:2082–2086.

Pickart, A. J. 2008. Restoring the grasslands of northern California's coastal dunes. *Grasslands* 18(1):3–8.

Raffaelli, D., and S. Hawkins. 1996. *Intertidal Ecology.* Chapman and Hall. London.

Rick, T. C., and Erlandson, J. M. (eds.). 2008. *Human impacts on ancient marine ecosystems.* University of California Press, Berkeley.

Sagarin, R. D., R. F. Ambrose, B. J. Becker, J. M. Engle, J. Kido, S. F. Lee, C. M. Miner, S. N. Murray, P. T. Raimondi, D. V. Richards, and C. Roe. 2007. Ecological impacts on the limpet *Lottia gigantea*

populations: human pressure over a broad scale on island and mainland intertidal zones. *Marine Biology* 150:399–413.

Thompson, R. C., T. P. Crowe, and S. J. Hawkins. 2002. Rocky intertidal communities: past environmental changes, present status and predictions for the next 25 years. *Environmental Conservation* 29:168–191.

Van der Meulen, F., T. W. M. Bakker, and J. A. Houston. 2004. The costs of our coasts: examples of dynamic dune management from Western Europe. Pages 259–276. In *Ecological Studies 71.* M. L. Martinez and N. P. Psuty (eds.). Coastal Dunes, Ecology and Conservation. Springer-Verlag, Berlin, Heidelberg.

Wiedemann, A. M., and A. J. Pickart. 2004. Temperate zone coastal dunes. Pages 53–64. In *Ecological Studies 71.* M. L. Martinez, and N. P. Psuty (eds.). Coastal Dunes, Ecology and Conservation. Springer-Verlag, Berlin, Heidelberg.

The Estuary and Marsh: Habitat Impacts and Environmental Protection

This chapter covers the biology and conservation of ecosystems in estuaries and associated marshes. An **estuary** is a partially enclosed coastal water body where freshwater river input combines with the salt waters of the sea (**Figure 4-1**). Around the edges of the estuary, large rooted emergent vascular plants (macrophytes), partially exposed grasses (**marsh grasses**) in temperate regions, and woody plants (**mangroves**) in tropical regions, grow in the muddy or flooded marsh sediments. The marshes are characterized by such plants that are capable of tolerating the salinity and environmental conditions of the estuaries. A few species of either marsh grasses or mangroves are visibly dominant in the marshes but support a diversity of other organisms, many of which reside as adults in the open waters of the estuary or near-shore ocean. Estuaries and their associated marshes are among the most productive ecosystems on Earth. There are many conservation issues to discuss because estuaries are a valuable source of exploitable resources and they are adjacent to some of the most populated regions on our planet.

4.1 The Estuary

Estuary Types

By definition, estuaries only occur in places where there is a significant input of fresh water to the sea. The morphology and geologic history of an area, however, help determine the nature and prevalence of estuaries in a given region. The most extensive and most prevalent types of estuaries were formed as sea levels rose at the end of the last ice age, about 10,000 years ago. These **coastal-plain estuaries** are prevalent along gradually sloping coastal margins such as the east coast of the United States; they include Chesapeake Bay and the mouth of the Hudson River (**Figure 4-2a**). **Lagoon estuaries** are formed where sandbars are built up parallel to the coastline, creating peninsulas and barrier islands, behind which lagoons accumulate freshwater run-off that mixes with seawater. These estuaries are common along northwest Europe, parts of Australia, and the southeast U.S. Atlantic and Gulf of Mexico coasts (Figure 4-2b). **Tectonic estuaries** form where the land subsides due to geologic activity along the coastline, allowing seawater to invade. These occur along the California coast; San Francisco Bay is one example (Figure 4-2c). The **fjord** estuary forms as the ocean moves into valleys formed by glacial processes. These are often very deep, but partially cut off from the ocean by a **sill**, a rise deposited at the mouth by the previous glacier. These occur along glaciated coastline, such as Alaska, Chile, New Zealand, and Scandinavian countries (Figure 4-2d).

Physical Characteristics

Most aquatic organisms are adapted to live in either ocean waters, with a fairly constant but elevated salinity, or

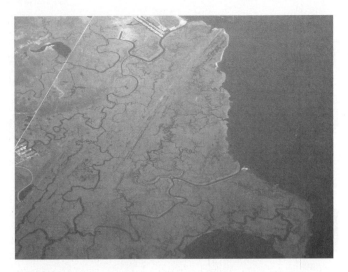

Figure 4-1 Coastal rivers and estuaries on the Gulf of Mexico coast of Texas.

freshwater, with continuously low levels of salts. Estuaries are regions of mixing for the two types of water and the salinity is highly variable on a daily and seasonal basis due to changes in the relative influence of freshwater (e.g., with floods or droughts that affect river input), or the ocean (e.g., due to changes in tidal currents). In general, salinity ranges from fresh at the upper reaches to full-strength seawater (about 35 ppt) at the mouth but can vary considerably within the estuary depending on climate, seasons, tidal cycles, and the geology. In estuaries with substantial freshwater input, a **salt wedge** typically forms at the bottom as the lighter freshwater rides over the incoming sea water (**Figure 4-3**). These are **stratified estuaries**, where a salinity gradient forms from the surface to the bottom as well as from the sea landward. Where there is little freshwater input, estuaries are marine dominated and more **well-mixed** or homogeneous.

The substrate of estuaries also varies depending primarily on the geology and the sediment input. In fjord estuaries, formed at high latitudes by receding glaciers that scrape away sediments, the substrate tends to be rockier; however, accumulating sediments may cover the substrate. Coastal plain estuaries tend to be dominated by thick layer of sediment on the bottom. These sediments are the accumulation of silts and organic matters brought in by the rivers or ocean waters. Silts settle out with decreasing flow as the rivers enter the estuary basin, and are important for the formation of marshes. Other particles suspended in the river waters **flocculate** and sink to the bottom due to chemical processes as the fresh and salt water mix. These flocculants can be an important source of nutrients and organic matter to organism living in the estuary.

Estuaries are physically dynamic and always changing. Not only salinity but also temperature tend to vary more than in surrounding coastal waters. Shallow estuaries can heat up or cool down more rapidly, especially with little tidal flow or river input, and freshwaters entering estuaries have more variable seasonal temperatures than coastal waters. Narrow estuary mouths and the coast adjacent to estuaries tend to dissipate waves coming from the ocean; therefore, wave activity is minimal compared to other coastal regions. Currents in estuaries resulting from river flow or tidal action can be substantial, however, especially in narrow channels. Currents can be important to organisms moving into, out of, or within the estuary. Larval and juvenile fishes, with limited motility that use estuaries as nursery areas, are especially dependent on tidal currents. They coordinate the timing of their migration and depth in the water column with the tidal flow. Turbidity in estuaries tends to be high, resulting from sediments and organic matter in the waters. This is one reason why the majority of the primary production originates from the grasses and woody plants in marshes found along the edges of estuaries, rather than from phytoplankton in the water. Oxygen levels vary greatly in estuaries, depending on water flow and biological activity. In areas with little flow and mixing, oxygen can be rapidly depleted and bottom waters can become extremely **hypoxic**, limiting the organisms that can survive and thrive there. The oxygenated layer in muddy sediments may be very shallow, typically about one centimeter thick, due to low mixing of water and oxygen depletion by organisms in the sediments.

Biological Importance

Estuaries are considered one of the most productive ecosystems on Earth and harbor a rich diversity of organisms. The materials that settle out in the estuary, carried in from the ocean and rivers, are the primary source of nutrients and organic matter supporting this high productivity. The organically rich substrate that is deposited, especially in regions of the estuary sheltered from waves and currents, provides productive sediments where plants can root and animals can burrow and feed. Ecosystems associated with the plants, primarily marsh grasses and mangroves, are discussed below. Most organisms that reside permanently in the estuary are associated with the ecosystems characterized by these plants. The deeper flowing waters of the estuary are an important corridor for diadromous fishes on breeding migrations into freshwaters from the ocean (anadromous species such as the salmons), or from freshwaters to the ocean (catadromous species such as American eels; see Chapter 2).

The majority of fish and swimming invertebrate species living in the estuaries (e.g., crabs and shrimp) are those that migrate from the ocean during some portion

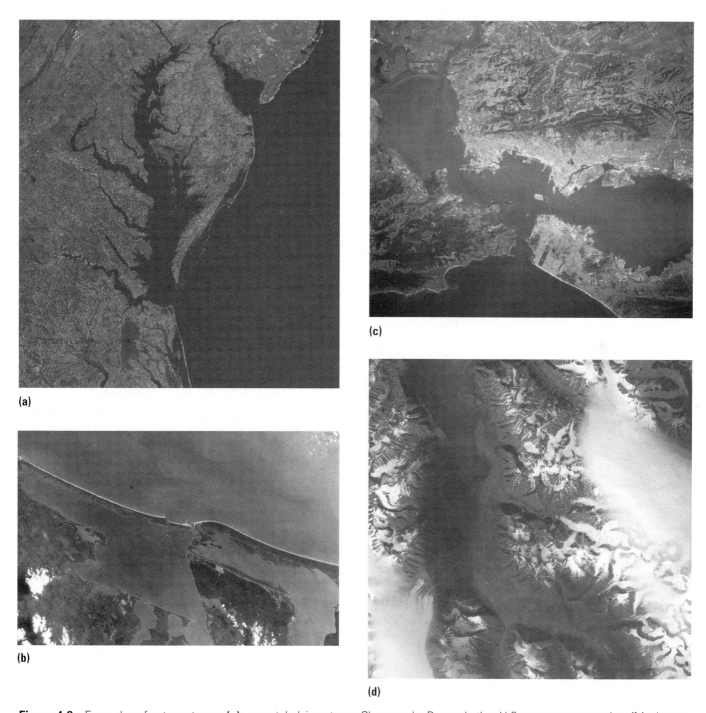

Figure 4-2 Examples of estuary types: **(a)** a coastal-plain estuary, Chesapeake Bay and other U.S. east coast estuaries; **(b)** a lagoon estuary, Matagorda Bay, Texas; **(c)** a tectonic estuary, San Francisco Bay, California; **(d)** a fjord estuary, Boknafjorden fjord in Norway.

of their life history (typically as larvae or juveniles) to use the estuarine habitats for feeding and shelter. In fact, in many regions the majority (about 75% in U.S. waters) of harvested coastal marine species utilizes the estuary, and are thus considered **estuarine dependent**. The use of the estuaries by marine species is highly seasonal and during any time of the year the estuary may be dominated by one or several species. For example, in estuaries of the southeast

U.S. there is a progression of larval fishes that move into the estuary from spawning areas offshore, then feed in the estuary until they near maturity, and migrate out of the estuary to offshore feeding and breeding grounds. These species support some of the most valuable coastal fisheries or are important prey species; they include drums and croakers (family Sciaenidae), anchovies (*Anchoa*); mullets (*Mugil*); and menhaden (*Brevoortia*) (**Figure 4-4**).

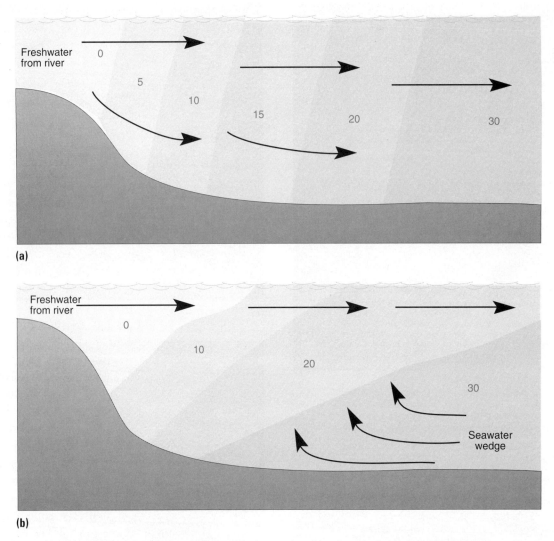

Figure 4-3 Cross sections of **(a)** well-mixed and **(b)** stratified estuaries. Numbers indicate salinities in parts-per-thousand (ppt).

Environmental Impacts

Because of the high biological productivity and presence near mouths of rivers and ports important to humans, estuaries commonly have been exposed to harmful anthropogenic impacts. Important estuarine habitats have been covered for building cities and industrial developments; river runoff can bring in toxic materials and excess nutrients from fertilizers and sewage; and there may be unpredicted effects from factors related to global climate change and human population increases in coastal regions. Around the globe, a proportionately large fraction of the human population lives in the vicinity of estuaries. For example, approximately two thirds of the U.S. coastal population currently lives in counties associated with major estuaries, although these areas comprise less than 6% of the land area along the coast; 70% of the population of southeastern Asia live in coastal or estuarine areas.

Ship traffic through estuaries can be a source of pollution; however, possibly of greater concern is the introduction of foreign aquatic organisms from large ships. Barnacles and bivalves may be introduced when "hitchhikers" come into the estuary attached to boat hulls (see Chapter 3). Many more organisms can be introduced into estuaries from the **ballast** water of tanker ships. When a large tanker ship offloads, it typically takes on water for stabilization during the return voyage to its home port. Small organisms or planktonic larval stages can be taken in with the ballast water. Upon returning to port, the ballast waters are released into estuarine or nearshore waters, along with the exotic organisms they contain. Although most do not survive, only a few individuals may be needed to reproduce and colonize coastal areas, with the potential for long-term negative impacts. One of the most well-documented of such **invasive species** is the European green crab

(a)

(b)

Figure 4-4 The estuarine dependent **(a)** Atlantic croaker, *Micropogonias undulatus* and **(b)** Gulf menhaden *Brevoortia patronus*. Larvae move into estuaries as nursery areas.

Carcinus maenas, a ballast-introduced invasive species that showed up on both the east coast of the United States and the southeast coast of Australia by the early 1800s (**Figure 4-5**). Individuals were documented along the South African and California coastlines in the 1980s, causing declines in native *Hemigrapsus* shore crabs and *Nutricola* clams, and indirectly resulted in increases in polychaete worms. The European green crab has since expanded its range into the Pacific coast of Canada and has showed up

in small numbers on coastlines in several areas throughout the Pacific. It is problematic in estuaries as it can tolerate salinities from 4 to over 50 ppt and temperatures from 0° to over 30°C. New restrictions on the exchange of ballast water may limit future introductions; however, even if this is true, problems controlling the effects of past exotic introductions will likely continue.

Over the past several decades many coastal nations have made progress in establishing monitoring and protection programs targeted toward estuaries. In the United States, the largest such program is the National Estuary Program (NEP), initiated in 1987 through the U.S. Clean Water Act. This program was established to encourage collaboration among agencies and stakeholders to protect the estuary ecosystems; 28 areas have been designated as NEP estuaries. Evaluations carried out through this program have categorized U.S. estuaries overall as in "fair" condition, with northeast coast estuaries in "poor" condition, southeast coast estuaries in "good" to "fair" condition, and Gulf coast and west coast estuaries in "fair" condition (**Figure 4-6**). Impacts identified include poor water quality, commonly indicated by high nutrient levels and low dissolved oxygen; poor sediment quality, including contaminants and toxicity; benthic impacts, such as low community diversity and abundance of pollution-tolerant species; and fish tissue contamination by harmful chemicals. Regions with the highest population densities tend to have estuaries of poorer condition. The most commonly identified environmental concern was habitat loss and alteration, followed in order by declines in fish and other wildlife populations, excess nutrients, contamination by toxic chemicals, presence of pathogens, alteration of freshwater flows, and the introduction of invasive species. A focus of programs such as the

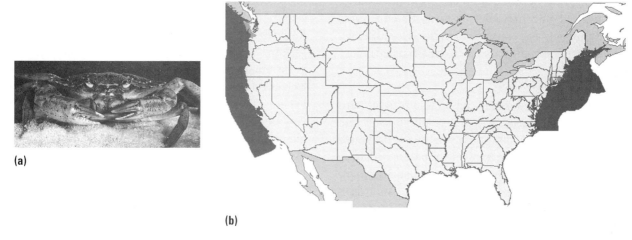

(a)

(b)

Figure 4-5 **(a)** The European green crab *Carcinus maenas*. **(b)** Distribution map indicating the native range (tiny white dots off the east and west coasts) and the invasive range (dark gray).

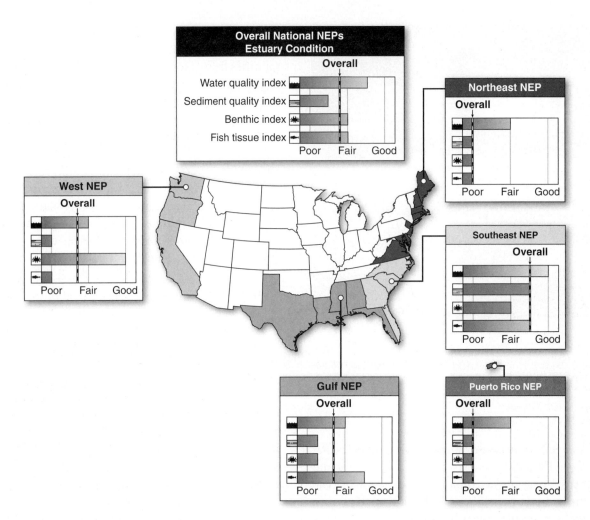

Figure 4-6   Condition ratings assigned by the U.S. Environmental Protection Agency for estuaries in the National Estuary Program (NEP).

NEP is to promote and encourage the protection of estuaries from such impacts. Because the source of these effects can extend all the way from the source of the watershed in the interior of the continents (e.g., agriculture-based pollution) to coastal oceans, simply setting aside estuaries for protection will only partially solve the problems. Some examples of impacts and conservation efforts for estuaries as they relate to the specific ecosystems are discussed below.

■ Fisheries Conservation

The accessibility of estuaries and coastal waters to humans makes estuarine-dependent species especially vulnerable to fishery harvest impacts. Many organisms are not harvested until they mature and move to offshore waters (offshore harvest issues are discussed for some of these, such as menhadens, in Chapter 6). Fisheries harvest in the estuaries is limited because these species are often present only during larval and juvenile stages, when they are either of little fisheries value or are protected from harvest. For example, the harvest season for shrimp (*Penaeus*) in Louisiana waters is typically closed until the young shrimp in the estuaries

reach an average size that is considered adequate to support a profitable fishery without excessive harm to the shrimp populations. In tropical developing countries subsistence fishers are more likely to harvest smaller fishes from mangrove-dominated estuaries. (Fisheries issues in mangrove ecosystems are addressed later in this chapter.) Some species found in temperate estuaries, such as the American and European eels, are valuable enough on the international market at small sizes that young life stages can be overexploited by harvest (**Box 4-1. Current Issue: Eels from the Estuary**). Where commercial fisheries are allowed in estuaries there is typically a limit on the harvest methods; for example, gill nets (see Chapter 11) are so efficient at blocking channels and harvesting a large percentage of the fish that they have been outlawed in most U.S. estuaries. There are many popular recreational hook-and-line fisheries for estuarine-dependent fishes. These include the American shad and striped bass in the northeastern United States, and red drum and spotted seatrout in southeastern U.S. estuaries.

Although anadromous fishes have been harvested from European and North American estuaries for centuries,

Box 4-1   Current Issue: Eels from the Estuary

There are two species of eels in the genus *Anguilla* that have an unusual life history. They are hatched in the deep waters of the mid-Atlantic Ocean, float to the surface, drift toward the coast over a one to three year period, migrate through estuaries up rivers feeding for 3 to 20 years, and finally return to the deep Atlantic where they reproduce and die. These catadromous eels are the American eel (*Anguilla rostrata*), which migrates to North American waters, and the European eel (*Anguilla anguilla*), which migrates to waters of western Europe. During their migration the eels undergo several transformations in body shape. The larvae drifting in the ocean are a leaf-like **leptocephalus**. As they near the coast and reach lengths of about 60 millimeters, they metamorphose into a small, translucent eel-shaped glass eel. As they move into the estuary they transform into an **elver**. Some of the eels (mostly females) enter freshwater rivers and others (mostly males) remain in the estuary. As they feed and grow they develop into an adult form called yellow eels. When the yellow eels reach a size of about 60 centimeters for males or 120 centimeters for females, they begin to mature sexually and transform into a form called silver eels (silvery colored with larger eyes and fins), and stop feeding before they migrate back to spawning grounds (**Figure B4-1**).

Eels are particularly vulnerable to anthropogenic impacts because their survival requires protection all the way from the deep Atlantic into upper reaches of inland rivers. They are vulnerable to harvest from the time they enter the estuary through their freshwater residence until they return through the estuary to the sea.

A popular commercial fishery harvest for elvers using basket traps has been in existence in Europe, including Italy, France, and England, at least since the 1700s. Eels caught in the Thames River and other estuaries were popular, prepared as "jellied eels," until after World War II, when pollution began eliminating the eels from many European estuaries. As these rivers were cleaned up beginning in the 1960s, however, the eel populations began to return. Then from the 1970s to the 2000s the eels underwent another dramatic decline—of about 90%—for unknown reasons. Studies by I. A. Naismith found low recruitment into the population, with recolonization of the upper ends of the estuary to be especially slow. The possible causes of the poor recovery include overfishing, parasite infections, damming of rivers to block migrations, and pollution; some scientists proposed that natural changes in ocean circulation may be contributing to the decline. One pollutant that has been implicated is polychlorinated biphenyl (PCB; see Chapter 7), which may inhibit reproduction and, when transferred to the eggs, result in death of larval eels. Whatever has caused the decline, European eel populations are considered to be facing possible extinction and are now classified as Critically Endangered by the IUCN.

The fishery for American eels in the United States, primarily for export to European and Asian countries, expanded to levels around 500 to 1,000 metric tons per year in the 1950s. The eels were historically harvested at all stages by traps in the estuaries along the northeast U.S. coast. The harvest of glass eels increased dramatically during the 1970s as their value increased on the Asian market. The young stages are vulnerable to high harvest levels because their migration is seasonal and predictable; nets can be placed across narrow channels as the eels move upstream so a large percentage of the migrating population can be caught. Commercial harvest levels hovered around 1,500 metric tons from 1974 to 1981, before declining to under 500 metric tons annually by the late 1990s through the early 2000s. As concern over the species' long-term survival began to be questioned by conservationists, state restrictions were passed by states. Eventually, all U.S. Atlantic states, except Maine and Florida, passed size restrictions that resulted in the stoppage of the glass-eel fishery. In 2004 the American eel was considered for listing as an endangered species in the United States, citing overfishing, blocking of upstream and downstream migrations by dams and other water control projects, and water pollution as possible factors affecting the eels. In 2007, however, the U.S. Fish and Wildlife Service ruled that an endangered listing was not necessary, stating that, although populations had declined in some regions, the overall species was not endangered with extinction.

This is an example of how new, profitable, and practically unlimited markets develop for fisheries that are easily exploited. For the American eel fishery, virtually every individual could be harvested with available methods if restrictions were not implemented. By the time the urgency of the issue is realized and actions are taken by government enforcement agencies and lawmakers to restrict fishing methods, sizes of fish harvested, and limits on amount harvested, it is often too late to continue a sustainable fishery, at least without a protracted recovery period.

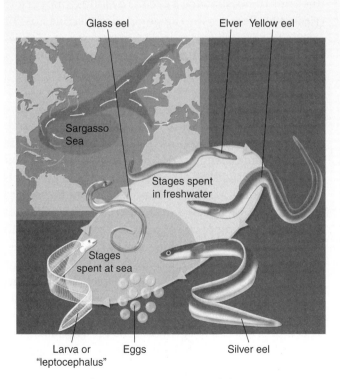

Figure B4-1   Stages of development of the American eel.

currently commercial harvest is tightly regulated because of the ease of catching fish as they pass through narrow channels. Nets placed across these channels during migrations could catch virtually every fish moving into coastal rivers. The use of such nets and fish wheels, devices powered by the water flow that scoop up the fish as they swim upstream, have been outlawed in most estuaries and coastal rivers, except for limited use by some indigenous groups.

Invertebrates that spend much, or all, of their adult life in estuaries have historically supported valuable local fisheries. These can include numerous species of crustaceans and mollusks around the world. Two groups that can be of considerable commercial value are crabs, such as the blue crab in U.S. Atlantic and Gulf of Mexico estuaries, and bivalves, such as clams and oysters.

Blue Crabs

Many crab species are harvested from estuaries around the world. One of the most abundant, well studied, and commercially profitable is the blue crab *Callinectes sapidus* (**Figure 4-7**), an estuarine and nearshore resident of waters of the West Atlantic extending from Nova Scotia, Canada to Argentina. The genus *Callinectes* is somewhat unique in that, even as an adult, they are capable of swimming using oar-like rear legs (*Callinectes* means "beautiful swimmer"), and are thus more mobile than most other crabs. Blue crabs live, feed, and mate in estuaries, but females undergo a seasonal migration into higher salinity waters for egg laying; however, this migration does not typically extend beyond the mouth of estuaries or far into coastal waters. For example, in Chesapeake Bay, the crabs release eggs near the mouth of the bay where the larvae float and swim for several weeks as they mature, undergoing transitions through several stages of development before becoming juveniles. The juvenile crabs (about 2.5 millimeters wide) move toward

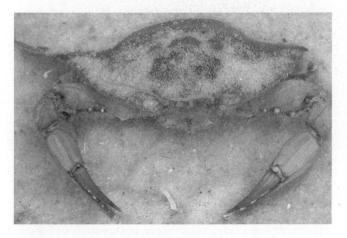

Figure 4-7 The blue crab *Callinectes sapidus,* commonly harvested from estuaries of the west Atlantic.

fresher waters of the estuary, where they settle to the bottom and begin feeding. In order to continue growing, the blue crabs must molt, or shed their shell (exoskeleton), and form a new one (if harvested just after this molting they are sold as "soft-shell crabs" and bring a higher value). Juvenile crabs undergo about 20 molts before maturing when they are 1 1/2 to 2 years old. After mating, females store the sperm for future fertilization of eggs, and move into higher salinity waters where they release their eggs and remain; they reside most commonly in waters with salinities from about 15 to 30 ppt (**Figure 4-8**). After mating, males typically stay in the upper reaches of the estuary, most common in salinities from about 3 to 15 ppt. During winter they often bury themselves in the muddy bottom. Because of the complex mobile life cycle for many estuarine invertebrates, there is a need for conservation efforts throughout the estuary, rather than focusing exclusively on the adult stages.

In many estuaries along eastern U.S. and Gulf of Mexico coastlines, blue crabs support valuable local fisheries. Typically they are caught in baited wire-meshed traps called **crab pots**. Historically one of the most important of the blue crab fisheries has been in Chesapeake Bay. This is the largest estuary on the U.S. east coast and the blue crab is its most valuable commercial fishery. Through the 1980s and early 1990s, commercial harvest hovered near 45 million kilograms per year, but then began declining, and fluctuated around 23 million kilograms from 2000 to 2008. Population estimates followed this same trend, showing a decline over this time period. (**Figure 4-9**). Reasons for the population decline are not certain but are believed to include nutrient pollution resulting in eutrophication and hypoxic conditions, the loss of seagrasses that serve as shelter for the young crabs, and excess fishing pressure.

As the fishery declined, management agencies were pressured to reduce levels of harvest; however, many crab fishers resisted, complaining that the problems were the result of pollution and the EPA had failed to limit nutrient input into the watershed from fertilizers and sewage. Efforts in 2001 to reduce harvest, including setting minimum size limits, were not adequate and the crab populations did not recover. This prompted the U.S. Secretary of Commerce to declare the Chesapeake Bay blue crab fishery a "commercial fishery failure" in 2008, releasing financial resources for crab fishers and for the development of recovery and management plans in the region. Management goals included protecting about 50% of the population from harvest and reducing harvest of females by over 30%. In Virginia waters, the number of traps each fisher could use was limited and the practice of dredging up crabs burrowed in the sediment during winter was stopped. In Maryland waters the number of licenses were limited, daily limits were placed

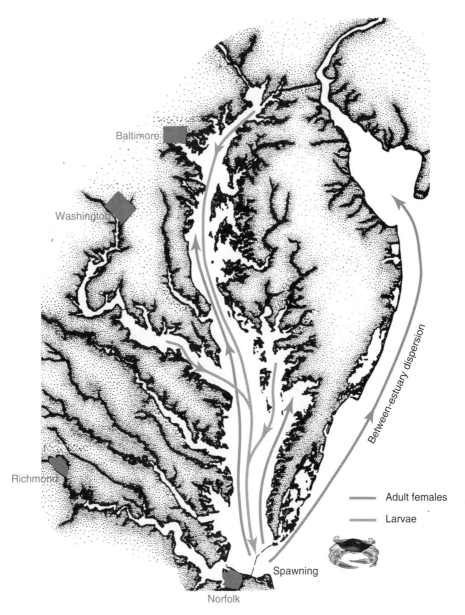

Figure 4-8 Map of Chesapeake Bay, indicating routes taken by adult female blue crabs during spawning migrations and the return routes of the larvae out of the Bay.

on harvest, and the fishery was closed during the peak of female migration in the fall. Some of the fishers were hired to recover lost "ghost traps" that continued to catch crabs. By the spring of 2009 the population of female crabs had increased by about 70%, and by 2010 numbers had more than doubled from the 2008 population size.

This example shows how complicated solutions to estuarine conservation issues can be. When it is difficult to determine exactly why the ecosystem is changing, the tendency is to lay blame on someone else and to resist change. From a social, economic, and political perspective, it is difficult to implement changes even if the cause of the problem appears to be obvious to scientists. When

meaningful change is implemented, however, success can be achieved.

Oysters

Various types of bivalves are harvested from estuaries around the world, with oysters supporting some of the most valuable bivalve fisheries. Oysters can be a good indication of the health of estuary waters, because they feed by filtering organisms from the water and are thus susceptible to waterborne diseases and toxicants. These become a human health issue when transferred by consumption of oysters, especially when eaten raw. Oysters can benefit water quality by removing sediments, nutrients, and algae

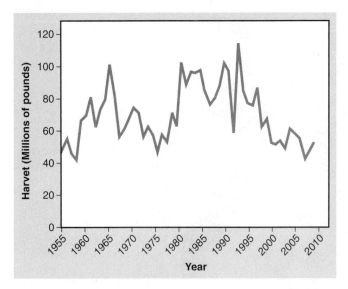

Figure 4-9   Chesapeake Bay blue crab harvest from 1956 to 2009. (Note that the Maryland reporting system changed in 1981, which increased harvest estimates.) Source: NMFS Fisheries Statistics of the United States.

from the water. They provide important bottom habitat in estuaries, providing a hard substrate within the muddy sediments. Many animals live on and among the oysters, including barnacles, anemones, fishes, and crustaceans, and other animals feed on these organisms or the oysters themselves. Some drums (Sciaenidae), such as the black drum, can pulverize the oyster and consume it shell and all.

In estuaries of the eastern United States, the oyster *Crassostrea virginica* typically matures at one year of age and produces sperm that it spews into the water for broadcast spawning. As the oyster grows larger during the following year, it switches to producing eggs (protandry). The fertilized eggs develop into floating, swimming larvae that must find a suitable substrate on which to settle; in estuaries this substrate is typically old oyster shells (in some regions fishers return the shells to encourage settlement of the next generation of oysters). These settled spat, about 25-mm long, accrete a shell and develop into an adult oyster.

Oysters have been harvested from estuaries for millennia. This is evidenced by oyster middens found around the world. For example, spectacular middens located on Dauphin Island, Alabama, a barrier island in the northern Gulf of Mexico, are protected as Indian Mound Park. There are six oyster shell middens, the largest of which is 50 meters across and up to 7 meters high, that were created by Native Americans from 1100 to 1550 AD. The oysters were probably taken during low tide from oyster banks, steamed and eaten, with the shells discarded to form the mounds. There is no indication of overharvest, and the oysters probably provided a reliable source of food throughout this time period.

Oysters continue to be harvested today—but not always sustainably—and are popular seafood items in many coastal areas, eaten either raw or in various preparations. Environmental impacts and overharvest have affected the edibility and the availability of oysters. The list of local problems that have affected oysters are numerous. They include viruses (e.g., herpes) resulting from inadequate sewage treatment, bacterial infestations that cause human illnesses and result in closing oyster beds to harvest, excess nutrients from fertilizers affecting water quality, accumulation of toxins from harmful algae blooms, and contamination by oil or chemical spills. In U.S. waters, there are valuable oyster fisheries throughout the Gulf of Mexico and the Atlantic coast (**Box 4-2. Learning from History: Oysters from Chesapeake Bay**). Over 55% of the U.S. oyster harvest is from Gulf of Mexico waters, about 10 million kilograms (approximately 550,000 bushels) annually.

4.2 Salt Marsh Ecosystems

Salt Marsh Development

Estuaries in temperate regions are characterized by tidal creeks, shallow pools, and mudflats, and are typically bordered by salt marshes (**Figure 4-10**). A salt marsh develops in areas sheltered enough for the accumulation of muddy sediments. Salt tolerant grasses colonize the sediments and spread by developing leaves and roots from extensive horizontal stems (rhizomes) that spread out underground. These plants enhance sediment accumulation, thus promoting further marsh development. Areas with a more extensive tidal range typically have broader salt marshes, especially those on gradually sloping coasts. These conditions are typified on the mid-Atlantic and Gulf of Mexico coasts of the United States, regions with gradually sloping coasts, broad estuaries, and shallow bays that are ideal for the development of salt marshes (**Figure 4-11**). Because the Pacific coast is steeper and rockier, it does not support the development of large salt marshes.

Salt Marsh Ecology

Grasses of the salt marsh typically extend from the edge of tidal flats to the height of the highest tide. The degree of exposure of the marsh varies daily with the tides; however, the grasses are tall enough that even at high tide their tops are typically not covered by water. The extreme and variable physical conditions limit salt marsh ecosystems to a low diversity of plants; typically a few species of grasses dominate in a given geographic region. These extremes include low oxygen levels in waterlogged soils, due to water filling of air spaces around the soils and roots while bacteria consume the remaining oxygen. Waterlogged low-oxygen sediments are not tolerated by most plants because the roots must

Box 4-2   Learning from History: Oysters from Chesapeake Bay

Based on estimates of pre-1600 pristine populations in Chesapeake Bay, Roger Mann and colleagues calculated that oysters could have filtered the entire bay's waters in three to four days; at current population levels it would take almost a year. Oysters populations were heavily harvested by European settlers in the United States far earlier than were many other coastal fishery populations because motor-driven boats and modern technology are not needed. Oysters are simply scraped from the bottom using various types of rakes and dredge devices (see Chapter 11), and at low tide many oyster banks are exposed above the water. The work was hard but profitable.

Oysters were already overharvested in the northeastern United States by the late 1800s. The "watermen" who harvested the oysters then moved south to Chesapeake Bay to continue their business. In Maryland waters, the annual harvest from 1871 to 1878 was ten to fifteen million bushels (the exact definition of a bushel varies from place to place, but the U.S. standard is about 2,750 cubic inches, about 45,000 cubic centimeters; a New Jersey study found that, on average, there were about 270 oysters in a bushel). Annual harvest declined by over 10% in three years as the oysters began disappearing. During this time period, there were no limits on harvest, but it stabilized at about two to three million bushels from the 1930s through the 1950s, and then declined to just over one million bushel by the mid 1960s. The decline of oysters in Chesapeake Bay was apparently due to one cause: excessive harvest (**Figure B4-2**).

By 1960, as it became apparent that oysters would not recover on their own especially as harvest continued, some proactive measures were taken. Oysters need hard bottom for settling of the spat, but removing oyster shells over the past century had destroyed critical habitat from the bay. The state of Maryland began repletion programs whereby oyster shells were dredged from areas where they were abundant and placed into waters where the oysters shell habitat was gone. This was supplemented by developing oyster "seed bars" where young

oysters were allowed to grow until being transplanted into the new habitat. These programs appeared to be successful, because oyster harvests increased to two to three million bushels annually from 1970 through the mid 1980s.

Just when it appeared that management efforts were going to at least stabilize the populations, if not totally recover them, another impact hit Chesapeake Bay. Diseases and parasite infections that killed the oysters began to appear with a greater frequency, probably as a result of more polluted waters. Some infections were documented as early as the 1960s but in the late 1980s and early 1990s three major diseases associated with parasitic infections resulted in oyster death rates as high as 90% in some areas of the Bay. Annual harvest declined to levels under 100 thousand bushels in 1994 and stayed under 500 thousand bushels annually through 2006.

This historical account might suggest that conservation of oyster populations in Chesapeake Bay will continue to be a losing battle. But beginning in the late 1980s, Maryland began restoration programs not only to restore the oysters for harvest, but to enhance their ecological roles, including filtering and providing habitat and food. Sanctuaries were established, some of over 5,000 acres, where oysters have been planted and no harvest is allowed. Other areas are set up as reserves, where oysters are planted and protected for five years before managed harvest can begin. Still, commercially harvested oyster populations remain low and new restrictions continue to be applied to limit harvest.

It remains to be seen whether restoration efforts will be able to restore the oyster populations in Chesapeake Bay to anything near pristine levels of the mid-1800s, especially if disease outbreaks continue and pollution in the Bay cannot be managed. Many entities have come together to work on the oyster problem as well as other issues in the Bay, and educate and encourage the cooperation of the public. These include Federal government agencies, state agencies, nongovernmental organizations (NGOs), and community groups.

One way to provide commercial oysters while avoiding the problems of overharvest is to move toward aquaculture. The aquaculture industry for oysters (as well as many other marine species, see Chapter 11) has grown dramatically. For example, Virginia has continued a multi-million dollar clam and oyster industry by leasing beds to individuals where they plant spat to be harvested later. This has become commonplace in other regions. In Louisiana it has been common practice since the mid-1800s to lease bottoms to oystermen who plant the "seed oysters" for later harvest.

The limited success of oyster restoration efforts through 2011 has led to more creative ideas for oyster restoration in Chesapeake Bay. One proposal, suggested in the early 1990s, was to use the non-native Asian oyster *Crassostrea ariakensis,* commonly called Suminoe oysters, to assist with recovery of the oyster fishery. This recommendation is based on the Suminoe oyster's excellent growth in estuaries and resistance to the diseases that are harming native *C. virginica* populations. Those who support the introduction, including the harvesters and the seafood industry, hope that it would rapidly populate and reduce plankton blooms resulting from eutrophication of the Bay. Oppositions to introduction of the Suminoe oyster are based on fears that it could displace or cross-breed with

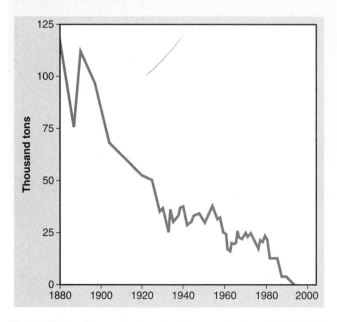

Figure B4-2   Oyster landings in Chesapeake Bay, 1880–2000.

native oysters (possibly producing sterile offspring), damage the ecosystem and fisheries, bring new exotic species or diseases into the Bay, or end up in other waters outside of Chesapeake Bay. Scientific studies were carried out and numerous meetings of state agencies and public hearings were held to assess these alternatives for oyster recovery. State and U.S. federal agencies, including the U.S. Fish and Wildlife Service, NOAA, and the EPA, eventually came out in opposition to any use of the non-native *C. ariakensis*. The State of Maryland Department of Natural Resources (DNR) concluded that only native oysters (*C. virginica*) should be used in ecological restoration and revitalization of the oyster industry. Management alternatives should include restoration of native oysters, implementing further restrictions on harvest, rehabilitating habitat and creating sanctuaries, and expanding

aquaculture with native triploid oysters and those bred for disease resistances. The reasons for the stand taken by the Maryland DNR include observations that methods used to sterilize native Asian oysters are not 100% successful. History with other sterilization programs for aquatic species (e.g., carp in U.S. freshwaters) has taught us that despite precautions, it is likely that eventually there will be an introduction of fertile individuals into the wild.

The battle to recover oysters and the ecosystem of Chesapeake Bay undoubtedly will continue into the foreseeable future, involving science, conservation, politics, and economic considerations. Not everyone will be satisfied with all of the decisions but, hopefully, lessons from past mistakes will be considered and guide decision-makers in the direction that is in the interest of long-term health of the ecosystem.

take in oxygen for respiration. Marsh grasses tolerate these conditions with special adaptations. Anaerobic respiration is possibly for many of these plants; however, this results in lower growth rates and possible death due to the accumulation of toxic chemicals. Marsh grasses can enhance aerobic respiration by conveying oxygen taken in by the leaves through air spaces (**aerenchyma**) to the roots. Succulent plants such as *Salicornia* take in oxygen through aboveground shoots that is transferred to the roots. Although salt marsh plants are tolerant of tidal submersion, their reliance on atmospheric oxygen means that constant submersion is stressful and can be lethal. Another source of stress is highly variable salinities. Marsh grasses are **halophytes** (salt-tolerant plants) that tolerate high salinities by excreting salts through specialized salt glands. Succulent salt marsh plants (e.g., *Salicornia*) avoid excessive salt concentrations in their tissues by minimizing salt uptake and transport to the growing shoots of the plant.

The variability among plant species in the mechanisms for dealing with the physiological stresses of living in the salt marsh environment results in a well-defined distribution along gradients of salinity and exposure. Along the seaward edge of salt marshes the grasses are mostly in the genus **Spartina**. *Spartina* grasses occur worldwide; in salt marshes of the eastern United States, two species of *Spartina* dominate: the cordgrass *Spartina alterniflora* along the shore, and marsh hay *Spartina patens* in the higher marsh. In the highest portions of the marsh another grass, black rush (*Juncus roemerianus*) dominates. Other salt marsh grasses include the saltgrass *Distichlis spicata*, native to the Americas but introduced onto other continents. Glassworts

Figure 4-10 A Mississippi salt marsh.

Figure 4-11 NASA satellite image of the eastern United States showing numerous bays and estuaries along the coast.

in the genus *Salicornia* are succulent, halophytic plants that can be found in salt marshes as well as in beach and mangrove habitats. At the inland edge of the marsh there may be other succulent plants and a line of woody shrubs (**Figure 4-12**).

The salt marsh is one of the most productive ecosystems on Earth. The primary productivity is provided mostly by the grasses, and the muddy bottom is covered with bacteria, diatoms, and algae. Bacteria play an important role by decomposing the dead grass leaves. This is a substantial amount of organic matter because much of the leaf biomass dies during the winters. Because many animals cannot digest the grasses directly, decomposition makes the nutrients available for the rest of the salt marsh food web.

The bacteria themselves also provide a significant amount of nutrition to organisms that consume detritus.

Burrowing **macroinvertebrates** are common in the soft muddy bottoms of the salt marsh. The most obvious are the polychaete worms and bivalves (**Figure 4-13**). Other invertebrates live among the marsh detritus, including **meiofauna**, organisms barely visible to the naked eye, such as copepods and amphipods. Crabs are common salt marsh inhabitants. One of the most conspicuous is the fiddler crab *Uca*, which builds burrows along the edge of the mudflat and feeds on detritus in the mud. Male fiddler crabs attract females to their burrows by waving their enlarged left claw in the air. It is common to see hundreds of these fiddler crabs desperately waving their claws on the exposed

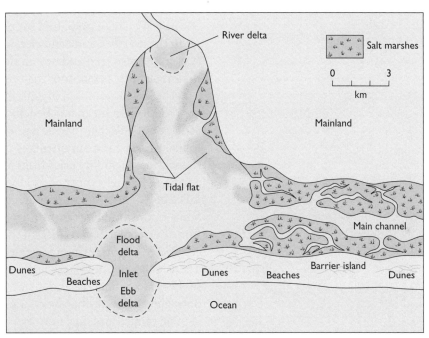

(a) SALT MARSHES AND OTHER COASTAL ENVIRONMENTS

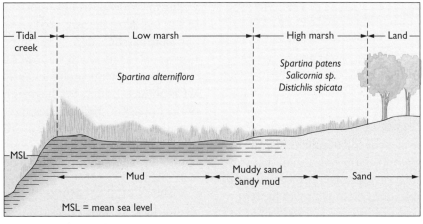

(b) SALT-MARSH PROFILE

Figure 4-12 **(a)** Locations of salt marshes and associated habitats within a coastal estuary. **(b)** Profile of a U.S. east coast salt marsh, indicating dominant vegetation and substrate type along a low to high marsh gradient.

(a)

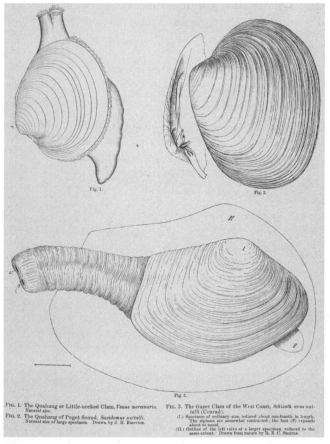

FIG. 1. The Quahaug or Little-necked Clam, *Venus mercenaria*.
Natural size.
FIG. 2. The Quahaug of Puget Sound, *Saxidomus nuttalli*.
Natural size of large specimen. Drawn by J. H. Emerton.
FIG. 3. The Gaper Clam of the West Coast, *Schizothærus nuttalli* (Conrad).
(I.) Specimen of ordinary size, reduced about one-fourth in length.
The siphons are somewhat contracted; the foot (F) expands about as usual.
(II.) Outline of the left valve of a larger specimen, reduced to the same extent. Drawn from nature by R. E. C. Stearns.

(b)

Figure 4-13 Salt-marsh macroinvertebrates: **(a)** a polychaete worm *Glycer* from a South Carolina salt marsh, and **(b)** quahog and gaper clams common in coastal marshes.

mudflats at low tide. Another conspicuous invertebrate in the salt marsh is the periwinkle *Littorina*, a snail that can live and breathe out of the water, and is commonly found slowly climbing up the plants as the tide moves in. *Littorina* feed on organisms attached to marsh grasses and can have a significant impact on the salt marsh under certain conditions (**Box 4-3. Geese and Snails, Top-Down Killers of the Marsh**).

Salt marshes support relatively few subtidal attached organisms, such as sponges, barnacles, or tunicates, compared to mangroves; this is probably due to the lack of hard substrate (with the exception of oyster beds) and the extreme temperatures in temperate shallow waters. Some fish species can tolerate the extremities of the marsh throughout their lives. These include **resident** species such as killifishes (Fundulidae) and silversides (Atherinidae), that feed and reproduce in the marsh. Juveniles of many **transient** fishes and crustaceans use the productive marshes as **nursery areas**, using the tidal creeks and pools at low tide and moving into the marsh grasses at high tide to escape predators that move into the estuary with the tide. Many of these organisms are important fisheries species as adults in coastal waters. Predators include larger fishes, such as the seatrouts (*Cynoscion*) and drums (family Sciaenidae) (**Figure 4-14**). These predators support recreational fisheries in the marshes and estuaries. Many birds such as rails (**Figure 4-15**) also feed and nest in the salt marsh, and small mammals such as raccoons may visit the marsh to feed.

The large amounts of organic matter and nutrients that are produced in the marsh benefit not only the marsh and estuary but adjacent marine habitats as well. It is difficult to make precise measurements of production export from the salt marsh and it varies from marsh to marsh. Generally there is a net export of production through a process called **outwelling** (the outflow of nutrients from an estuary). Some of this transport is through organisms that migrate from the salt marsh into adjacent coastal habitats, and this is considered important in supporting coastal marine fisheries. There is also an outwelling of nutrients and organic matter in the form of detritus and dissolved organic matter. The outwelling from salt-marsh estuaries occurs in pulses dependent on rainfall and tidal flow, and the degree of outwelling depends on the amount of production, the geomorphology of the estuary, and tidal amplitude. The importance of outwelling provides a strong argument for the conservation of salt-marsh ecosystems.

Salt marshes serve other important ecological functions, such as filters to remove sediments and a limited amount of nutrients and pollutants from the water. Marshes act as physical buffers for the mainland by absorbing much of the impact of storm surges and reducing erosion of the coastline. Salt marshes typically recover from the impact of storms and hurricanes by accumulating sediment and regrowing where they have been destroyed.

Salt-Marsh Destruction

As modern human settlements were established along temperate coastlines around the world, many of the salt marshes were destroyed or covered over. Most of the salt marsh habitats in Europe were lost hundreds or thousands

Box 4-3   Research Brief: Geese and Snails, Top–Down Killers of the Marsh

Most of the research on salt marsh impacts has focused on bottom–up effects, that is, how inputs of nutrients, changes in soil chemistry, or physical factors affect the growth of marsh grasses. Research indicated that biomass of the marsh grasses entered the food web only as dead plant material that was turned into detritus and decomposed. It was even debated whether the bacteria that consumed the dead grass were as important as a source of nutrition as the grass itself. The common paradigm was that top–down effects from predation of marsh grasses were not important. Recent studies of two marsh ecosystems have begun to change this paradigm, however.

The first of these is in the marshes of the New England region of the northeast United States. Snow geese (*Chen caerulescens*) and Canada geese (*Branta canadensis*) populations feed and breed in these marshes during the summer months. At historic population densities the geese actually benefitted the salt marshes through their activities. Snow geese would return from overwintering in southern marshes and migrate to breed in the New England marshes. Although the geese fed on the marsh grasses, they also defecated in the marshes, putting nutrients back into the ecosystem. Once they left, the marshes recovered. Over the past 30 years, however, Canada geese and snow geese populations have grown rapidly, in part due to decreased use of harmful pesticides but also due to increased access to farm crops. From winter to spring the snow geese migrate from croplands in the southern United States to salt marshes in the northeast by the millions, where they begin feeding on the roots and rhizomes of the marsh grasses. They can denude millions of square meters of marsh in an hour. Once the grasses are gone, evaporation can increase the salinity in the marshes and eventually dry out the soils, conditions under which the marsh grasses cannot recolonize. Once the soil qualities change in the exposed mudflats, it could take decades for the grasses to return, even if the grazing pressure from the geese is removed.

Another series of studies was carried out in salt marshes along the southeastern United States and Gulf of Mexico coasts by Brian Silliman and colleagues. Die-offs of *Spartina* marsh cordgrasses totaled more than 250,000 acres over a six-year period following severe drought years in 1999–2001.

These die-offs were initially attributed solely to the drought's effect on salinity and factors related to soil moisture. Evidence indicated, however, that other interactions were involved. The marsh periwinkle snail *Littoraria irrorata* feed on fungi that grow on the *Spartina*. During feeding, the snails damage the *Spartina* and facilitate additional fungal infection. At high snail densities, feeding on these infected areas stresses and can kill the *Spartina* plant. Through field observations, experiments, and modeling, Silliman and colleagues found that the drought and snail predation were working together to cause the marsh diebacks. The intense droughts resulted in stressful soil conditions that, either alone or in combination with snail grassing, caused die-offs of the *Spartina* in some areas. The snail populations increased dramatically in the stressed areas. As the stressed area was denuded, the snails concentrated at the border of healthy marshes adjacent to the die-off area. These snail "consumer fronts" then moved into healthy marshes, destroying grasses as the fronts progressed. Eventually the fronts subsided as the snails dispersed; however, the snail fronts persisted in some marshes for as much as one year after the impact of the drought subsided. By 2003 much of the marsh affected by the drought had begun recovering as *Spartina* recolonized the mudflats; however, many areas affected by the snail fronts still had not recovered by 2005. There is a concern that with climate change severe droughts may become more severe and these events may become more frequent. To add to this concern, blue crabs, which are potential snail predators, have recently declined by as much as 40% to 80% in southeastern U.S. salt marshes. There appears to be a natural **trophic cascade**, whereby the crabs prey on the snails that prey on the marsh grasses. Loss of the crabs releases the snail populations that can destroy the marsh grasses.

These studies have shown that ecological interactions in the marsh are complex and factors initiated far from the marsh or even on a global scale may have unexpected effects. These all must be considered in protecting the remaining salt marshes. Scientists are striving to develop a better predictive understanding of the salt marsh ecosystem so that environmental groups and government officials will have the knowledge needed to protect what remains of one of the most productive ecosystems on earth.

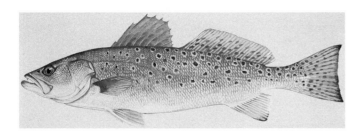

Figure 4-14 Spotted seatrout *Cynoscion nebulosus,* common predatory fish in salt marshes of the southeastern United States.

of years ago. These marshes were modified to support settlements, livestock grazing, and farming, and many are still maintained for agriculture. Early European settlers to North America used the salt marshes for livestock grazing and often modified the marsh with canals and dikes to enhance hay production. More than 50% of the marshes were gone in southern New England before large permanent European colonies were established. Most of these salt marshes were never restored to their original function, and many were eventually filled for development. The greatest effect of urban sprawl on salt marshes has been in the heavily populated areas; for example, around large cities

Figure 4-15 A clapper rail *Rallus longirostris,* which feeds and nests in temperate and tropical coastal salt marshes throughout much of the Americas. Populations have declined in some regions due to marsh loss.

most of the marshes were destroyed by dredging, channel deepening, and port construction.

Through the 1900s, marshes were filled and canals were dug for control of mosquitoes, based on the perception that they harbored diseases and not realizing their importance to coastal fisheries. The most recent major effect on salt marshes was total destruction to create areas for residences, industry, and agriculture, primarily from the 1950s to mid 1970s. In some areas marshes were filled to create tourist accommodations and beaches. Typically the beach is maintained with sands dredged from offshore, a process referred to as **beach nourishment** (see Chapter 3; **Figure 4-16**).

Figure 4-16 A view of the Mississippi Gulf of Mexico coast where salt marshes have been covered to maintain beaches and accommodations.

Even without filling, the marshes can be affected by development or changes in the tidal flow. Too much fresh water can flow into the marsh when woody vegetation is removed from along the landward edge of the salt marsh. Salt water tidal flow into the marsh can be restricted by sea walls or embankments. Without the seawater influence, freshwater vegetation such as cattails (*Typha*) or reeds of the genus *Phragmites* will outcompete the cordgrass. Roadways and railroads affect the flow of water in the marsh, causing much of the marsh to be drained and other areas to be overrun with freshwater. The intrusion of seawater into freshwater marshes can destroy vegetation that is intolerant of salt water, including trees such as cypress.

Along the U.S. Atlantic and Gulf of Mexico coasts, the importance of the salt marsh is now generally appreciated and the remaining marshes are mostly protected from destruction by state and federal laws. Indirect effects of flow modification for flood control or canals built for boat access, however, still affect the salt marshes. For example, the dredging of the Intracoastal Waterway in northeast Florida resulted in the loss of over 35% of the marshes in regions around the canal. Some salt marshes have been designated as Marine Protected Areas and function more like parks, receiving near total protection from direct harm.

Impoundments

During the settlement of the United States, salt marshes were generally considered useless to humans and they were often filled, drained, or impounded. Vast tracts of salt marshes along the U.S. east coast were diked and drained for rice or other forms of agriculture, or impounded for waterfowl. After the Civil War, the abandoned rice field often continued to be maintained for attracting waterfowl (**Figure 4-17**).

Impoundments are constructed by building earthen dikes around an area of marsh to control the tidal flow.

Figure 4-17 A coastal marsh impoundment south of Charleston, South Carolina.

Although some water flow may be allowed, this affects the natural ecosystem by limiting the tidal flow. These impoundments have a more constant salinity, sometimes near zero percent. In summer the oxygen level in impoundments can be severely depleted due to the stagnation of the water, lack of tidal exchange, and decomposition of organic matter. This kills many of the marsh residents that use impoundments. Impoundments managed for waterfowl are controlled at low salinity, because plants preferred by waterfowl will not tolerate higher salinities. Water levels are regulated and the impoundment bed may be cultivated, reducing or eliminating salt marsh species.

In the late 1900s the modification of the salt marshes and the right of private ownership of impoundments became controversial. Many believed that the marshes should be returned to their natural function of supporting coastal ecosystems and fisheries and protected as a common resource. Conservationists argue that the value of the marsh should be in supporting natural ecosystems and fishery species as nursery areas, and that natural salt marsh habitat should be considered a public resource. Laws have been passed in most states to eliminate the impounding of salt marshes and some impoundments have been returned to natural marshes. Still, many impounded areas remain; for example, in South Carolina approximately 15% of the coastal marshes are still impounded to some degree. These impounded areas are managed in a variety of ways. Some are maintained as freshwater wetlands, others retain their estuarine function through openings to coastal inlets, and some are still managed for primarily for migrating or wintering waterfowl.

Pollution

Even away from large cities or impounded areas the cumulative impact on salt marshes can be substantial. Pollution can accumulate in marshes that receive water from various sources. **Point-source pollutants** (those from a single source like an industrial factory) are typically regulated by permits. **Non-point source pollutants** (those not from a defined point, such as street or agricultural runoff), however, can be a significant problem because of the difficulty in monitoring and regulating their input. Excess nutrients from agriculture and other sources entering the marsh through river input or local runoff can result in eutrophication. Even at more moderate levels, an increase in nutrients can modify the species makeup of the marsh. For example, in northeastern United States, nutrient pollution from agriculture runoff has resulted in lower plant diversity in some salt marshes, making marshes less useful as a nursery area for fishes and invertebrates. Salt marshes can be vulnerable to accumulation of heavy metals or pesticides that are deposited into the marsh along with sediments and

organic matter that originate inland. Herbicides can affect salt marsh plants. Studies by Chris Mason and colleagues found that, even at sublethal concentrations, herbicides in England's marshes have been shown to lower growth and production of grasses and diatoms, and play a role in increased erosion of the marshes.

Impacts of River Channelization

Large coastal rivers, such as the Mississippi River flowing into the Gulf of Mexico, provide an important source of sediments to coastal salt marshes. When the rivers flood onto the coastal plains, accumulated sediments support marsh development. But after the marsh lands are formed they gradually compact, causing the level of the marsh to sink or **subside**. Therefore, without a source of sediment to replenish the marsh it will eventually convert to open water. This is exactly what is happening in much of coastal Louisiana.

Over about the past century there have been extreme efforts to keep the lower Mississippi River in its current channel within its banks by building levees and channelizing the river to enhance boat traffic and protect cities such as New Orleans (**Figure 4-18**). (Even New Orleans has subsided below sea level but is normally kept dry by a massive assemblage of pumps.) The completion of river modification projects from the 1920s through the 1960s resulted in a 67% decrease in the sediments delivered to the Louisiana coast. The only times the river and estuary overflow the banks are in times of severe flooding or when hurricanes, such as Katrina in 2005, move up the Mississippi River. Damming of the rivers in the Mississippi River watershed also has reduced the sediment load by over 50%,

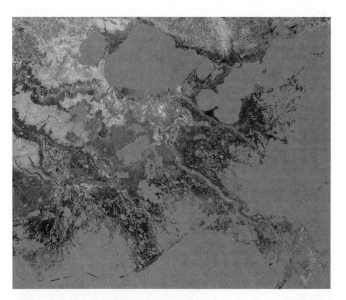

Figure 4-18   The lower Mississippi River and Delta. (See Color Plate 4-18.)

reducing the amount of sediment transported down the river. Most of those sediments are being transported out onto the continental shelf in the Gulf of Mexico.

Not only are the marshes in the vicinity of the Delta hurting, but marshes to the west are also gradually disappearing. These marshes also depend in the long term on sediments from the Mississippi River. After the Delta builds over several millennia, the river eventually switches its location along the coastline and thus provides sediments to a new area. Over several tens-of-thousands of years the Mississippi River has changed its course several times. The current tendency is for more water to flow down **distributaries** (streams that branch off and flow away from the main river channel) to the west of the current river channel, which would produce a new delta and replenish the marsh sediments in that region. Allowing the Mississippi to switch courses, however, would dramatically affect shipping traffic on the Mississippi and the economy of New Orleans and other areas along the lower Mississippi.

As a result of these effects, much of the marshlands of coastal Louisiana have literally sunk into the sea, becoming open water where salt marshes used to be. To add to the impact, canals have been dug through the marshes themselves. During oil exploration this is one of the easiest ways to gain access to the marsh. With erosion, the channels expand into areas of open water in the marshes.

In total, Louisiana has lost almost 5,000 square kilometers of coastal marshes and wetlands in the 20th century (**Figure 4-19**). An increase in tropical storms and hurricanes could increase the rate of marsh loss. The numerous hurricanes that hit the Louisiana coast in the first decade of the 21st century resulted in the loss of over 500 square kilometers of wetlands, and without the sediments to replenish them they will be unable to return on their own. It is predicted that over 1,600 additional square kilometers could be lost in the next 50 years if adequate preventive measures are not taken. If that occurs, one-third of coastal Louisiana will have been lost.

In some marsh systems, hurricanes can enhance the marsh by moving sediments from coastal beaches and dunes into the marsh. For example, this has been documented in coastal areas of Texas. Building over the dunes, however, has inhibited this natural erosion and movement of sediments to the marshes.

What effect does marsh loss have on coastal ecosystems? As described above, salt marshes are an important nursery area for many coastal fish species, and loss of the marshes removes critical habitat. Another effect of this land loss is the loss of habitable land along the coast. Towns can be literally disappearing into the sea. Without the marsh as a buffer, coastal areas are more vulnerable to flooding and erosion from storms. To further exaggerate the problems, freshwater marshes are being destroyed by **saltwater intrusion** as ocean water moves further inland and kills vegetation intolerant to high salinities.

Salt Marsh Restoration Methods

With the realization of the important functions of salt marshes have come increases in efforts to protect the healthy salt marshes that remain. A hands-off approach can be best in some situations. But when the marsh has been destroyed or severely impacted more active restoration efforts may be necessary. Of course, each situation must be approached independently; however, research has established a set of standard methods that can be adapted to a given situation.

In the simplest situation, restoration involves simply removing the fill that covers a former marsh and allowing the marsh grasses to recolonize the area. Changing the drainage and tidal exchange (for example, by adding

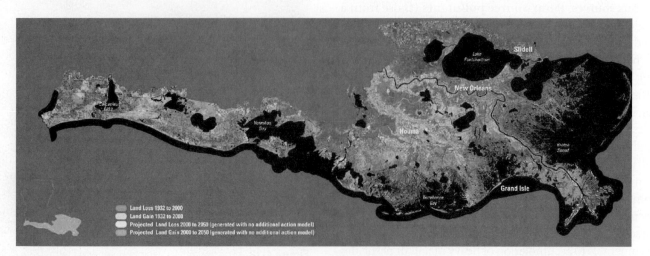

Figure 4-19 Historic and predicted land loss in coastal Louisiana. (See Color Plate 4-19.)

culverts in the proper location during road construction) may enhance the recovery. At times salt marshes are created or restored unintentionally. If land use changes cause changes in river flow and drainage that result in the creation of a new mudflat, this mudflat may be colonized naturally by marsh grasses and eventually develop into a healthy salt marsh. Manmade dikes or levees may naturally erode, resulting in the building of a new mudflat that eventually establishes marsh grasses and develops into a salt marsh.

A natural recovery may be a long, uncertain process as the sediments need to accumulate and marsh plants become reestablished. For severely damaged marshes, more active efforts are needed. In some areas along the U.S. Atlantic coast, recovery methods include pumping dredge spoils onto the marsh and replanting marsh grasses. The removal of invasive plants from marshes may be necessary before replacement by native *Spartina* in some marshes. One such invasive is a European strain of the common reed *Phragmites australis,* introduced into U.S. marshes in the late 1800s. This grass has a low tolerance for high salinity waters, and invades marshes when the salinity regime has been altered by changes in tidal water flow. It outcompetes *Spartina,* forming dense colonies that are barriers to the movement of marsh animals, including shorebirds and wading birds. Eventually open water areas of the marsh begin filling in, raising the marsh elevation. If *Phragmites* is removed (with herbicides and mowing) and salt water tidal flows restored, *Spartina* may eventually return on its own. Because *Phragmites* is intolerant of salinities over 18 ppt, the restoration of tidal flow alone may result in *Spartina* recovery. Because natural replacement can take 10 to 20 years, however, replanting may be desirable. In China the opposite situation exists, where *Spartina* is the invasive species, taking over marsh habitat once dominated by *Phragmites*. *Spartina* was intentionally introduced into estuaries in China to enhance land accretion and enhance biological production; however, as Shuqing An and colleagues report, this has resulted in a loss of native biodiversity in regions where it has been introduced, including a loss of insect species that depend on *Phragmites* and bird species that depend on the natural marsh habitats.

Some of the most active research on salt marsh restoration is in coastal Louisiana. Because of the large area involved restoration will require massive efforts. There are currently over 100 separate restoration projects in Louisiana, sponsored by various federal, state, and NGOs. These include the National Marine Fisheries Service of NOAA, the U.S. Fish and Wildlife Service (USFWS), the U.S. Environmental Protection Agency (EPA), and the Louisiana Department of Natural Resources. Many collaborative efforts have been initiated; for example, the Barataria-Terrebonne National Estuary Program, established by an agreement between the State of Louisiana and the EPA.

Some of the methods used to restore salt marshes in coastal Louisiana include spreading dredge materials across the subsidized area until the level is high enough for marsh development, constructing terraces to protect areas from further erosion, placing breakwaters offshore to minimize wave impact, and replanting grasses in areas being reestablished. Even placing discarded Christmas trees into open waters can assist in the accumulation of sediment and reestablishment of the marsh. One of the most ambitious plans is to harvest sediments from the bottom of the Mississippi River and deliver them as a slurry through pipelines to be deposited into the marshes. It is estimated that this would build up the marshes around the Mississippi River Delta region of Louisiana to their 1956 conditions in 50 years. The price tag, in the hundreds-of-millions of dollars per year, however, will make this project difficult to implement.

One of the most productive long-term solutions would be to allow the Mississippi River to run its natural course and perform its normal function in restoring marsh sediments. Removing levees along the river would allow natural flooding of the marshes and changes in flow would result in redistribution of sediments to other areas along the coast. Allowing this natural progression is not considered acceptable due to our dependence on the river for barge and boat traffic, and flood control is necessary for protection of agriculture lands and cities, including New Orleans. A compromise has been reached to allow some flow into distributaries to the west of the current channel. Thirty percent of the Mississippi River flow is now allowed into the Atchafalaya River using control structures at the junction of the two rivers. This has resulted in a substantial increase in marshlands and delta sediments in this basin. Allowing the diversion of river flow is also controversial because of effects the freshwaters may have on the salt marshes (such as harming oyster beds).

Many of the current and historical impacts on salt marshes reflect a misunderstanding of their function and importance to coastal marine ecosystems, the production of commercially valuable fishery resources, and protection of human settlements in coastal regions. Education, therefore, can be a key to salt marsh protection. With education, the perception of salt marshes as wastelands is slowly being replaced with the understanding that these are critical components of ecosystems, both aesthetically and economically. Programs are in place around the North American coast to educate children, fishers, developers, political leaders, and other citizens while studying, protecting, and preserving salt marshes and other components of estuarine ecosystems. These programs include the National Estuary Program established by the U.S. EPA. NOAA's National

Estuarine Research Reserve System protects 27 areas in different biogeographic regions of the United States and supports long-term research, monitoring of water quality, and education. One of the goals of this program is to work with local communities and regional groups to establish management policies, restore habitat, and address pollution and invasive species issues.

In establishing restoration programs, care must be taken to consider the landscape surrounding the marsh. For example, research by Melissa Partyka and Mark Peterson found that the simple presence of salt marsh habitat within an altered landscape was not adequate to ensure a healthy ecosystem. Marshes located within a natural landscape exhibited healthy conditions, demonstrating the importance of habitat protection along the entire gradient from terrestrial through freshwater ecosystems, through estuaries to the sea.

4.3 Mangrove Ecosystems

Mangrove ecosystems replace salt marshes in comparable environments of tropical regions; the most pronounced difference is the woody vegetation that replaces marsh grasses as the dominant emergent vegetation. Mangroves are the most prevalent of the woody plants that can tolerate exposure to a broad range of salinities and sediments with low oxygen levels. This tolerance allows them to take advantage of the productive silty habitats in tropical coastal areas associated with estuaries. Mangroves are not limited to estuaries, however, and tend to dominate along any tropical coastline that is sheltered and shallow enough to accumulate sediments to support emergent plant growth. They are excellent colonizers of small oceanic islands, where they grow on shallow banks or in lagoons. The roots are exposed to full-strength seawater but can tolerate an extreme range of salinities, influenced by heavy rainfall at one extreme and evaporation at the other. Variations in the hydrology and the topography where mangroves can colonize lead to variations in the animal species that utilize them; this results in a diverse assemblage of organism in mangrove communities. As with other estuarine and coastal ecosystems, increasing human populations and the associated impacts result in many mangrove-associated conservation issues and conflicts.

Mangrove Distribution

The term **mangrove** is not a formal taxonomic term but is used to refer to woody plants, in the form of shrubs or trees, that have a set of physiological adaptation that allow them to thrive in coastal areas exposed to seawater. The ecosystem associated with the mangroves (**Figure 4-20**) is often termed a **mangal;** however, in this chapter "mangrove" will typically

Figure 4-20 Underwater view of mangrove roots and their associated ecosystem.

refer to either the plants or the ecosystem. About 60% to 70% of the Earth's tropical coastlines are lined with mangroves, typically in sheltered areas away from the ocean's wave action. Mangroves follow a similar broad distribution to coral reef ecosystems, and the two are often associated with each other; however, mangroves occur in some regions not tolerated by corals because of excess sediment input (e.g., the west coast of Africa or Amazon coastal region of South America), or cool areas of upwelling (e.g., off the west coast of South America and Africa; **Figure 4-21**). Mangroves are missing from some isolated coral islands in the central Pacific, presumably because the mangrove seeds cannot disperse the distances necessary to colonize these islands. There has been much debate as to why mangroves are limited to tropical regions. One primary reason is that they cannot tolerate freezing well. The full physiological explanation is still uncertain; however, studies by Stephanie Stuart and colleagues indicated that the physical characteristics of salt water result in excess tension in the mangrove xylem during freezing, limiting the mangrove's ability to supply water to the leaves. In temperate areas, typically above about 25 degrees north and south latitude, mangroves are replaced by salt marshes. The mangroves' range is extended farther in some regions due to movement of warm currents along the coast (see Chapter 1), for example, off the east coast of South America, Africa, and Australia, and in the Gulf of Mexico. In the continental United States, mangrove marshes are limited primarily to southern Florida and the southern tip of Texas. The limits on mangrove distributions result in a remarkable contrast, with woody vegetation dominating most of tropical coastlines but almost totally absent from temperate coastlines.

The largest areas of mangroves are in estuaries associated with deltas of large rivers, such as the Indus Delta of Pakistan and the Amazon Delta of South America. Sediments deposited in these areas provide new habitat for mangrove

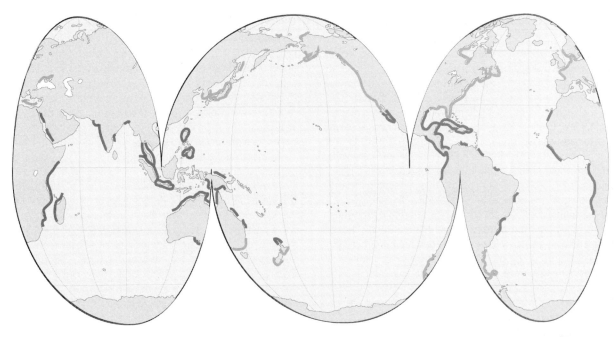

Figure 4-21   The global distribution of mangroves (dark gray), compared to that of salt marshes (light gray).

settlement (**Figure 4-22**). Damming thus can affect these ecosystems by removing a source of sediments in a manner similar to that discussed above for salt marshes. Mangroves are able to trap sediments and can increase sedimentation rate and slow down erosion. Apparently they do not actually create new land by accumulating sediments, however. It is more likely that they take advantage of the accumulation

Figure 4-22   The Amazon River mouth. Mangrove habitat extends along the seaward edge of the river's mouth. Note the sediment plume extending into the ocean. (See Color Plate 4-22.)

of new sediments, or disappear as shores erode. A primary ecological function of mangroves in estuaries thus is to turn what would be a uniform mud flat into a complex productive environment supporting a rich and diverse ecosystem.

Less extensive mangroves are found fringing some coastlines away from river influence, especially where tidal currents transport enough sediment to support the mangrove roots. If lagoons form behind this fringe, mangroves may form. **Overwash mangroves** form away from the coast in some areas of the Caribbean. Here they grow with no substantial source of sediments, but are supported by peat accumulation on small islands (**Box 4-4. Conservation Focus: the Belizean Reef Mangroves**).

Because mangrove habitats are in soils either covered by water or inundated periodically by the tides, the soils are continuously waterlogged. Waterlogged soils are typically low in oxygen for two primary reasons. For one, water that fills the spaces in the soil does not supply nearly as much oxygen as air, and, in addition, the oxygen that is present is rapidly used by bacteria in the soil. Because the roots of plants require oxygen for respiration, waterlogged sediments are not readily tolerated by many plants. Some mangroves species, such as the black mangrove *Avicennia germinans*, tolerate these hypoxic sediments by using porous upward extensions of shallow roots (**pneumatophores**) adapted to exchange gases for respiration (**Figure 4-23**). Other species, such as the red mangrove *Rhizophora mangle*, have long, thin prop roots extending from above the ground or aerial roots extending from branches (**Figure 4-24**). Mangroves also use roots for obtaining nutrients, primarily nitrogen

Box 4-4   Conservation Focus: Belizean Reef Mangroves

Most mangroves are components of estuarine ecosystems; however, mangroves can grow in areas far from significant freshwater input, even on small islands limited almost entirely to exposure to full-strength seawater. One of the most prominent of such ecosystems is associated with coral islands off the coast of Belize, Central America. The coral reefs here receive much attention as the longest continuous barrier reef in the Western Hemisphere. The Belizean Reef Mangrove Ecoregion, covering or fringing most of the undeveloped cayes (islands), also houses a large, diverse, and productive ecosystem. Many of these islands are completely covered with mangroves, mostly red mangroves, which can better tolerate continuous exposure to water than other species (**Figure B4-3**). These offshore mangroves are unique in that they are supported by the largest mangrove peat deposits in the world. There are mangroves over 12 kilometers offshore that have peat deposits as thick as 8 meters. Peat cores taken by Matthew Wooller and colleagues document that these deposits accumulated as sea levels rose over the past 8,000 years.

Many birds are associated with the mangroves, including important breeding and nesting populations of various egrets, herons, ibis, and the magnificent frigate bird *Fregata magnificens*. These birds contribute to the productivity of the mangrove ecosystem by depositing nutrient-rich guano. The underwater portion of the mangrove ecosystem is rich in organisms typical of Caribbean mangrove ecosystems. Much of this mangrove ecosystem is included in the Belize Barrier Reef Reserve; however, limited monitoring and enforcement have led to some conservation problems. These problems include illegal hunting of birds or egg collection during nesting season, disturbance of nesting colonies by tourists, and introductions of rats to some islands.

Tourism development on the islands off Belize has resulted in conflicts between developers and conservationists. Because mangroves naturally cover or fringe these islands, the sandy beaches that many tourists expect are uncommon; therefore, some developers remove mangroves to establish resorts and beaches. One of the most popular tourist destinations and one of the larger cayes in Belize is Ambergris Caye, where past development has resulted in much mangrove removal. The Belizian government, however, has tried to put a halt to future mangrove removal with recent moratoriums and laws requiring permits for any mangrove removal. Much of the southern tip of Ambergris Caye is still covered with mangroves, but there are proposals for development of a portion of this area into a resort.

If approved, this project would require the removal of large areas of mangroves on the southern tip of the island and could also affect the reefs offshore that are included in the Hol Chan Reef Reserve. This situation will be a test for the conservation movement in Belize and could counter complaints by NGOs that politically-motivated environmentally-harmful decisions have been common in the past.

On the island just south of Ambergris Caye lies one of the better protected island mangrove habitats, the 100 acre Caye Caulker Forest Reserve. Despite protection of the mangroves there are still problems with rats, feral dogs, and cats. Man-O-War Caye is a much smaller island protected in the South Water Caye Marine Reserve. This island has one of the 10 largest magnificent frigate bird colonies in the Caribbean and the only nesting site in Belize for the brown booby *Sula leucogaster*. This caye is well protected from direct harm but has been damaged as a result of nearby tourist development. Sand dredging from underwater near the island to build up the foundation and beach around tourist resorts resulted in erosion and undercutting of the mangroves as sand slumped away from the island to fill in the depression left by the excavation (**Figure B4-4**).

Much effort is being put into developing a conservation ethic that will encourage the protection of Belize's mangrove ecosystems by its citizens and leaders. Progress is being made through education. For example, in 1998, a Coastal Zone Management Plan was established to develop policies and strategies for managing Belize's coastal resources for conservation, involving government and NGOs. The Coastal Zone Management Authority and Institute of Belize has recently promoted education by establishing workshops on sustainable mangrove management. It remains to be seen whether compromises can be reached that allow development for tourism along with long-term conservation of the mangrove ecosystems.

Figure B4-4   Evidence of erosion on Man-O-War Caye, Belize, an important breeding site for brown-footed boobies and magnificent frigatebirds.

Figure B4-3   Man-O-War Caye, mangrove island off Belize.

Figure 4-23 Pneumatophores of black mangroves.

and phosphorous, which tend to be low in mangrove soils. Mangroves recycle nutrients using two strategies: the roots penetrate decaying parts of dead mangroves, and the trees resorb most of the nutrients from dead leaves.

Mangroves are not physiologically limited to salt-water habitats. Most species grow well in freshwater but are typically outcompeted in areas that are exposed only to freshwater. In estuary and coastal habitats, mangroves dominate because of their ability to tolerate exposure to brackish water and saltwater. The mechanisms of salt tolerance vary among mangrove species but are limited to several possibilities. First, mangroves tend to be more tolerant than other plants of salt in the tissues, but salt may still need to be eliminated. The main mechanisms mangroves use to deal with this issue are exclusion of the salt by the roots, extrusion of salt from the leaves through glands (if

you look closely you may see salt crystals on the mangrove leaves), or deposition of salt into bark or leaves that are subsequently dropped. Mangrove roots also appear to be capable of selectively using freshwater sources that are accessible, such as at the surface after a rainfall. Mangroves conserve water with succulent leaves covered by a waxy cuticle on the upper surface and a dense layer of hairs on the underside that minimize evaporation losses.

Although there are about 55 species worldwide that are considered true mangroves; about 35 species in four families are most prevalent. One to several species typically dominate the plant community in a given region; for example, four species are common throughout the Caribbean. Mangroves function in the ecosystem not only as a source of energy at the base of the food web, but also as a filter for terrestrial runoff, a sediment trap, structure and shelter for a diversity of animals, and a nursery areas for young fish and crustaceans.

Mangrove Reproduction and Growth

Mangroves reproduce by forming flowers and seeds. Pollination is by wind, insects, birds, or bats depending on the region and mangrove species. All mangroves disperse their **propagules** by water. In red mangrove, the seeds are unusual compared to other plants in that they germinate and sprout while still on the plant; the seedling remains on the plant for several months, and can grow as a spindle-shaped structure 25 cm or longer before dropping onto the ground or into the water (**Figure 4-25**). Those that land in the water can remain alive floating for up to one year, and may even

Figure 4-24 The prop root system of red mangroves.

Figure 4-25 Propagules on a red mangrove.

sprout roots and leaves while floating. Absorption of water at the tip makes the mangrove propagule float point-down. Those that contact the bottom begin to form roots in about ten days. After taking root in the sediments the mangrove can grow rapidly. Growth varies by species and depends on environmental factors; however, intertidal species such as the red mangrove may grow to one meter in one year, into a maze of prop roots in three years, into a small forest in five years, and eventually may reach heights of ten meters or more. Mangroves in areas with low nutrient input (e.g., on small islands) typically exhibit extremely slow growth rates; small stunted trees can be decades old. The reproductive characteristics of mangroves provide an excellent dispersal mechanism and are critical factors for recolonizing areas where mangroves are lost due to severe storms, including hurricanes, or human activities.

The Mangrove Ecosystem

In a given geographic region, the level of the land relative to the water is the primary determinant of which species of mangrove will dominate. The reasons for this are complex, but at least in part it is determined by tolerance to salt water or exposure. A gradient thus is formed from the seawater's edge inland. For example, in the Caribbean, as you move landward away from the water, mangrove stands change from red mangroves (*Rhizophora mangle*) at the water's edge, to black mangroves (*Avicennia germinans*), and then white mangroves (*Laguncularia racemosa*) growing in the shallow intertidal waters and mudflats. Red mangroves are adapted to live in the shallow waters along the coast by elevating themselves with stilt-like prop roots. Black mangroves are most tolerant to high salinities that may occur in the upper tidal regions, while white mangroves are less tolerant of tidal flooding and high salinities and are restricted to higher ground. In river estuaries there may be a gradient of mangrove species up the river, and in some regions there is no clear pattern of zonation, for example, in much of Australia's mangroves.

Although few animals eat the mangrove plant or its leaves directly, the nutrients in the mangrove support a highly productive ecosystem. Although the mangroves do not drop their leaves seasonally, they can produce tons of leaf litter per hectare each year. For example, in Australia, branch and leaf-fall averages about ten tons per hectare per year. The dropped leaves are decomposed by fungi, bacteria, and other organisms, making the nutrients available to other organisms, such as mangrove crabs or small shrimp, which in some regions consumes the majority of the **detritus**. These organisms are an important source of food for larger animals and support a complex food web. Detritus also can be flushed out of the mangrove marshes to support other coastal ecosystems. For example, on average over 3,000 kilograms of particulate organic matter is transported from each hectare of Australian mangrove habitats into marine waters each year. In areas with a substantial tidal range, the mud flats around mangroves can be exposed at low tide and become accessible to wading birds, such as ibis, that feed on worms, mollusks, and crustaceans living in the sediments.

Mangrove Community Diversity
The Mangal Habitat

The importance of mangroves is not limited to the estuarine and marine organisms they support (**Figure 4-26**). Mangroves are used by migratory or roosting birds. For example, the mangroves in many regions serve as breeding and nesting sites for egrets, herons, cormorants, ibis, boobies, frigate birds, and eagles. Surveys by Gaetan Lefebvre and colleagues found that numerous other bird species use mangroves as shelter or for roosting, feeding, or nesting. The mangroves are important habitats for these species because they may be the only woody vegetation along the coastline. Mangroves not only play an important role in the health of these bird populations, but the birds can also contribute to the productivity of the mangrove ecosystem through nutrient input via guano. Wading birds such as herons and egrets also can be important predators on invertebrates and fishes living among the mangroves.

A large diversity of insects and other invertebrates live on and in the exposed portion of the mangal. They are not typically active or apparent during the daytime, probably due to stress from the lack of freshwater and hot sun; many of the insects feed at night or remain inside the

Figure 4-26 Brown boobies (l) and magnificent frigatebirds (r) are dependent of mangroves as roosting habitat in the Caribbean, here on an island off Belize.

plant. Consumption of the leaves by insects is typically low, averaging around 5% of leaf production, possibly due to toxic chemicals or salts in the leaves. Wood borers such as moths and beetles can attack the propagules and branches. They create hollow tubes within the mangrove branches that can be inhabited by scorpions, spiders, moths, termites, ants, and other insects. Some of these insects, for example, the root boring beetle *Coccotrypes rhizophorae,* may be an important consumer of mangrove propagules and young roots. Other invertebrates damage the mangroves by burrowing into the roots. These include shipworms (actually a teredinid bivalve) and isopod crustaceans. Wood borers can be very harmful to the mangroves; extreme infestations can destroy the roots and result in death of the mangroves.

Relatively few amphibians and reptiles reside permanently within the mangrove habitat. A few species of frogs can tolerate the brackish water, including the crab-eating frog *Rana cancrivora,* which may be locally abundant enough to be harvested and eaten. Lizards and snakes of terrestrial origin come into the mangroves to feed. Some crocodiles tolerate brackish or salt water sufficiently to be an important predator in some mangrove ecosystems; these include the American crocodile *Crocodylus acutus* in Central America, the Nile crocodile *Crocodylus niloticus* in western Africa, and the estuarine crocodile *Crocodylus porosus* in tropical southeastern Asia and Australia.

Few terrestrial mammals depend solely on the mangroves, although they may visit the mangroves to feed. These include mongooses, raccoons, deer, rodents, otters, monkeys, rhinoceros, water buffalos, and bats. In Bangladesh, the mangroves are considered critical habitat for the endangered Bengal tiger, although their preference for this habitat is largely due to the loss of critical habitat elsewhere. Grazing domestic camels and buffalo have harmed mangroves in Arabia and Pakistan.

The intertidal area, including the mangrove roots and the mudflats, is often inhabited by a species assemblage similar to that of temperature salt marshes. This includes crustaceans, gastropods, bivalves, and polychaete worms. The species assemblage in a given area is affected largely by the tidal exposure or freshwater influence. The most numerous animals are meiofauna in numerous phylogenetic groups, including copepods, amphipods, nematodes, oligochaetes, and flatworms. These typically live in the upper layers of sediment, feeding on algae, bacteria, and detritus. They are apparently not a major food source for many of the larger organisms but are important to some fishes and crustaceans that feed in the muds.

Crustacean representatives include a diversity of crabs that can be seen feeding at low tide on the mangroves or mudflats (**Figure 4-27**). Many of these are grapsid crabs that forage at low tide and return to their burrows as the

Figure 4-27   Fiddler crabs (*Uca*) in a Jamaican mangrove marsh.

mudflats are covered by high tides. On some mangrove mudflats there may be over fifty crabs per square meter. The grapsid crabs tend to be generalist feeders. For example, the mangrove tree crab *Aratus pisonii* is common in Caribbean mangroves, living on the red mangrove roots above the waterline and feeding primarily on mangrove leaves and seeds as well as insects. Many of the grapsid crabs are primarily herbivorous. Some scrape algae or diatoms from the mud or the plants, but most prefer decaying fallen mangrove leaves that are easier to digest, and they often will store leaves in their burrows before consuming them. Ocypodid mangrove crabs of the genus *Ucides* feed on mangrove leaves and other plant material, detritus, and algae. Research by Inga Nordhaus and Matthias Wolff documented that *Ucides cordatus* serve an important role in mangrove ecosystems of northern Brazil, because their feces makes organic matter from mangrove detritus available to other species. Crabs can affect mangroves' propagule survival. For example, in Malaysia and Australia, most (sometimes over 95%) of the propagules produced are destroyed within days of falling from the trees, but in Florida only about 5% of propagules are taken by crabs.

Uca fiddler crabs can be commonly seen out of burrows on intertidal mudflats around the mangroves feeding on detritus particles and associated diatoms and bacteria (Figure 4-27). Populations of fiddler crabs can be very dense, as many as 70 per square meter are common on mudflats associated with southeastern Asia mangroves. Burrowing by these and other crabs benefits the mangrove ecosystem by aerating the muds and allowing deeper oxygen penetration. Without this mixing, bacterial action results in hypoxia even at shallow depths below the sediment surface. Hermit crabs, mostly *Clibanarius* species, also can move onto the mudflats or into trees to forage.

Many crustaceans live in the muds of the mangroves, and barnacles are often found attached to the intertidal portion of the mangrove roots. These and other **encrusting** organisms are filter feeders that depend on the organisms present in the water at high tide. They can harm the mangroves if present in large numbers by inhibiting gas exchange through the aerial roots and pneumatophores.

Mollusks in the intertidal zone include gastropod snails and bivalves, such as oysters, attached to roots. Snails are conspicuous foragers on the mangroves and mudflats. These include many species that feed by scraping organic particles, diatoms, or algae from the surface of muds, roots, stems, or leaves; few feed directly on the mangrove leaves. The most abundant of these are congeneric with the *Littoraria* periwinkles so important in the salt marshes.

A few fish species can come out of the waters onto the mudflats to feed. The mudskipper, a type of goby, is associated with mangrove ecosystems, primarily in tropical southeastern Asia; it can crawl out of the water and pull itself across the muds, spending most of its time at low tide feeding on organisms living on the mangroves or surrounding mudflats. It retreats into burrows when the mudflats are covered at high tide. The mangrove rivulus, *Rivulus marmoratus,* found in some Caribbean mangrove ecosystems, can tolerate long-term isolation in pools, and has even been observed in standing water on decaying mangrove trunks. This is one of the few species of fishes that are **simultaneous hermaphrodites,** capable of producing both eggs and sperm and self-fertilizing, an adaptation to the likelihood of becoming isolated in pools without access to potential mates.

Subtidal Mangrove Ecosystem

Animals living in the subtidal habitat associated with the mangroves are diverse and abundant due to: the high biological productivity of the mangrove ecosystem, the availability of shelter in the submerged root system, and the nearness of other productive marine ecosystems such as coral reefs and seagrasses. For example, experiments by Pia Laegdsgaard and Craig Johnson found that juvenile fish are attracted to mangroves primarily to take advantage of food availability and shelter from predation. Larger fishes are more likely to move out of the mangroves onto adjacent mudflats to feed.

The region around the subtidal roots exhibits the highest diversity and abundance of organisms associated with the mangroves. A diversity of attached organisms (**epibionts**) cover the roots, including algae, barnacles, sponges, and tunicates (**Figure 4-28**). These organisms typically do not harm the mangroves; in fact, they can benefit the roots by protecting them from root-boring animals.

Some of the most visible animals attached to the mangroves are the sponges. Sponges are mostly limited to the subtidal portion of the mangrove roots; however, some species, for example, as documented in Belize by Klaus Rutzler,

(a)

(b)

(c)

Figure 4-28 Marine organisms living attached to mangrove roots (epibionts): **(a)** barnacles, **(b)** sponges, and **(c)** *Acetabularia,* a single-celled green algae called the "mermaid's wineglass."

can tolerate several hours' exposure during very low tides. Colorful sponges can cover a large fraction of the surface of the mangrove roots. For example, in Key Largo, Florida almost 75% of the available root space is covered by sponges, a total of ten different species. Sponges have a mutualistic relationship with mangroves in the Caribbean, where the mangroves gain protection from root-boring isopods and a source of nitrogen from the sponges, while the sponges obtain carbon from the submerged roots. Sponges can grow up to ten times faster on mangrove roots than other surfaces, and mangroves produce small roots that penetrate through the sponge. Studies by Sebastian Engel and Joseph Pawlik found that there is a competition for space on the roots among the sponge species, with some sponges capable of overgrowing others, and some able to produce **allelochemicals** to resist overgrowth by other sponges. Some of the sponges found on mangroves are restricted to mangrove habitats, but others are the same species as present on nearby coral reefs. Sponges on the mangroves may grow to larger sizes because sponge predators are uncommon around the mangroves. Abiotic factors can be an important determinant of sponge diversity. Such factors as sedimentation, temperature extremes, hypoxia, currents, storms, and anthropogenic factors affect the diversity and distribution of sponges on the mangroves.

Tunicates are another group of conspicuous animals sometimes attached to mangrove roots. Mangrove tunicates have been used for bioprospecting in a search for potential pharmaceutical chemicals, as have coral reef organisms (see Chapter 5). For example, the mangrove tunicate *Ecteinascidia turbinata* has shown potential in reducing tumor growth in cancer research.

The bottom sediments around and near the mangroves are typically partially covered with seagrasses, algae, and other organisms. Crustaceans, bivalves, and polychaete worms also live in the subtidal area around the mangrove roots or burrowed into the sediments (**Figure 4-29**). The upside-down jellyfish *Cassiopea xamachana* is a conspicuous animal around some mangrove ecosystems in the Caribbean, named for its behavior of resting on the bottom with its tentacles reaching upward. Symbiotic **zooxanthellae** reside in the tissues of the jellyfish (a relationship similar to the one found in many corals; see Chapter 5) and provide much of its nutrition; therefore, they are typically found in shallow sunlit waters. The jellyfish can also take dissolved nutrient from the water or use the stinging nematocysts on its tentacles for feeding and protection.

Mangroves as a Nursery

Many tropical shrimp and fish species inhabit the mangrove ecosystem at some stage in their life. Some species remain in the mangrove marshes as adults (e.g., centropomid

Figure 4-29 Upside-down jellyfish *Cassiopea* in a mangrove ecosystem in Belize.

snooks); however, few spawn there and many depend on the mangroves primarily when they are larvae and juveniles. The planktonic larvae of many invertebrates and fish are moved into mangrove habitats by currents and are retained there by the structures, channels, and inlets that reduce the water flow. Up to nine different shrimp species utilize mangrove habitats, especially those associated with estuaries, as nursery areas. The most commercially valuable are the penaeid shrimps and *Macrobrachium*, commonly marketed as freshwater prawns. Strong correlations have been documented between coastal shrimp production and the extent of mangrove habitat. It is estimated that, on average, over 150 kilograms of shrimp are produced for each hectare of healthy mangrove habitat. These shrimp are of high commercial value, but have been displaced in many regions by removal of mangroves for shrimp farms, discussed below.

Crabs are of important local value for fisheries harvest in many regions. The mangrove mud crab *Scylla serrata*, a portunid crab related to the blue crab found in temperate estuaries, is harvested in the Indo-West Pacific region. The ocypodid crabs can be important for local harvest along Atlantic and Pacific coasts of South America. Various species of bivalves (e.g., oysters, mussels, and cockles) and gastropods (e.g., conch) are harvested from and near mangrove ecosystems around the world.

A large number of tropical marine fish species also use the mangrove ecosystem, many as a nursery during juvenile life stages (**Figure 4-30**). For example, mangrove-dominated estuaries in India and Australia are used by almost 200 species of fish. There are two important benefits of the mangroves for these juvenile fish. One is the availability of structure and shallow waters as a refuge for

Figure 4-30 A diversity of fishes use the mangroves for shelter and as a foraging location; here a foureye butterflyfish *Chaetodon capistratus*, a young mangrove snapper *Lutjanus griseus*, and a bluestriped grunt *Haemulon sciurus*.

protection from predation, a major source of mortality for small fish and shrimp. Many juvenile fish and shrimp move substantial distances to gain access to the refuge of the mangrove, especially during high tides. Not only does the structure provide protection, but the abundance of large carnivores is also lower in the mangroves compared to other nearshore habitats. The other benefit of the mangroves as a nursery area is for feeding; mangroves are typically a more productive source of food than adjacent coastal ecosystems. Zooplankton, including crab larvae, are the most valuable food source for larval and juvenile fishes using the mangroves. Small shrimp and mangrove crabs can be the most important food link from the detritus to the larger fish in the mangroves.

The importance of mangroves as a nursery for harvested fishery species has recently been emphasized in encouraging conservation. For example, the proportion of harvested species that use mangrove ecosystems is estimated at over 75% in southern Florida, approximately 60% in Fiji waters, and over 65% off eastern Australia. In some regions (e.g., on some small islands) the harvest supported by mangroves is by subsistence fisheries, which often are not included in harvest statistics. Some of the most common harvested species are detritivores such as mullets (Mugilidae), scavengers such as catfish (Ariidae), and predators such as groupers (Serranidae), snappers (Lutjanidae), tarpons (Megalopidae), snooks (Centropomidae), and sharks and rays. Smaller plankton feeders, such as herrings (Clupeidae) and anchovies (Engraulidae), which are important food for marine fish predators, also use the mangroves as a nursery. The movements of fishes in and out of the mangrove ecosystem tend to integrate the mangroves with other tropical marine ecosystems, including coral reefs, mud flats, and sea grasses. The average biomass of fish within the mangroves, however, is much higher than in adjacent coastal habitats (with the exception

of coral reefs). Mangroves have been shown to support from 4 to over 30 times the number of fish compared to adjacent seagrass beds.

Some marine mammals and reptiles also move into the mangroves to feed, including dolphins, manatees, and sea turtles. Observations by Colin and Duncan Limpus indicate that mangrove leaves can comprise a significant portion of the diet of some sea turtles. Green sea turtles feed not only on the mangrove leaves but also on tunicates, invertebrates, seagrass, and algae in mangrove ecosystems, and the mangroves can be an important nursery for young green sea turtles. Hawksbill sea turtles are primarily sponge eaters but can feed on bark, leaves, and fruit of the mangroves.

Mangroves, in a manner similar to salt marshes discussed earlier, can provide a source of organic matter and nutrients to adjacent ecosystems. A portion of this is from animals that feed in the mangroves and migrate to other ecosystems. Other sources of export are detritus, particulate organic matter, and dissolved organic matter and nutrients. Outwelling of organic matter in the form of particulate matter and leaf detritus can be substantial (e.g., as much as 30% of leaf production from an Australian estuary) but varies depending on such factors as production and tidal flow. For example, a tidal mangrove is more likely to be a source of export than a mangrove on higher land.

Mangroves as a Buffer and Filter

Mangroves are an important buffer for marine ecosystems offshore as well as terrestrial ecosystems inland. Mangroves reduce coastal erosion by stabilizing the shoreline and river banks. The retention of fresh waters in the mangroves protects less tolerant marine ecosystems, such as coral reefs, from harmful salinity fluctuations. Mangroves also protect coral reefs by retaining excess sediments and assimilating nutrients that are input along with the fresh waters. Other pollutants may be retained or decomposed in mangrove sediments, keeping them from harming more sensitive ecosystems such as coral reefs. Mangroves also provide a buffer to terrestrial and freshwater ecosystems especially against the effects of storms, including hurricanes or typhoons, or tsunamis, as discussed below.

Natural Impacts on Mangroves

Organisms associated with the mangrove community have evolved and adapted to survive or recover from most natural environmental events. Infrequent extreme natural events may have substantial local impacts on mangroves, however. Cold weather can kill mangroves at the edge of their distribution range, as occurred in southern Florida in January 1997; however, more cold-tolerant species survived the low temperatures. Flood events originating inland may actually increase the area available to mangroves by transporting sediments into the river delta.

The broad, extensive prop roots and aerial roots of mangroves help them absorb the impact of most storms hitting the coast, and in regions frequented by cyclonic storms the mangroves tend to be shorter and more tolerant of high winds and storm surge. The hurricanes in the Caribbean Sea and Gulf of Mexico, and typhoons in the western Pacific, however, can severely damage or destroy mangroves with high winds, storm surges, and heavy rainfall. Caribbean islands and regions of southeastern Asia are particularly vulnerable. For example, Vietnam is hit by eight to ten typhoons per year on average. Tsunamis, though much less frequent, can cause damage similar to hurricanes.

One of the most well-documented hurricane events affecting mangroves was Hurricane Andrew, which struck south Florida in 1992, severely damaging about 150 square kilometers of mangrove habitat with 240 km per hour winds and a 2-m storm surge (**Figure 4-31**). About 60% of the mangroves were uprooted or broken, and many of the surviving trees eventually died. Of those that remained, the red mangroves survived better than other species. Recolonization by seedlings was rapid; however, mortality was high and growth slow, resulting in a slow recovery. The species distribution of the mangroves was also modified, at least initially, due to differential ability to recolonize and grow. Evidence indicates that even after major hurricanes, recovery of the mangrove ecosystem is likely if the area is protected from other impacts.

Human Impacts on Mangrove Ecosystems

The location of mangroves in areas close to human populations has resulted in severe harm to the mangrove ecosystems. Approximately 20% of the world's mangroves

Figure 4-31 Mangrove forest in Biscayne National Park, Florida in September 1992, three weeks after being crossed by the eye of Hurricane Andrew.

were lost due to human actions from 1980 to 2005, and many more have been seriously impaired. The estimated annual rate of loss was about 1% during the 1980s. The greatest losses were in Asia, Central America, and Africa. Although mangroves are still in decline, the loss rate has slowed somewhat, to 0.7% annually from 2000 to 2005. In a few countries, such as Bangladesh, there has been an increase in mangroves due to protection in forest reserves, and the removal of coastal shrimp farms has allowed for the recolonization by mangroves in some countries, such as Ecuador.

There is a long list of human factors impacting mangroves. Mangroves are removed for agriculture, aquaculture, human settlements, and industrial or tourism development. The trees are harvested for many uses in some regions of the world, including for wood products or firewood, leaves to make baskets or mats, propagules for consumption, sap for producing soft drinks or alcoholic beverages, and various parts for medicinal use. Fishes and other organisms can be overharvested from the mangrove ecosystem. Excess siltation due to deforestation and coastal pollution, and changes in freshwater input through damming, channelization, or irrigation all can cause harm to mangroves.

Changes in adjacent marine ecosystems also can affect mangroves. For example, coral reefs serve as a buffer to the mangroves against strong currents or waves; without them the sediments can erode from around the mangrove roots and seedlings can be prevented from taking root.

Changes in Freshwater Input

There is a certain amount of natural variability in the amount of freshwater entering the mangrove ecosystem on an annual, seasonal, and daily basis. This variability is dependent on climate changes, rainfall, seasonal effects, and whether or not the mangrove ecosystem is associated with an estuary. These variables affect the species composition of the mangroves and other associated communities, as discussed above. The ecosystem, however, is tolerant and can adjust to most natural variability in freshwater input or salinity.

Mangroves may not be able to tolerate extreme conditions that result in a rapid decrease or increase in freshwater river input, such as with damming, channelization, or irrigation. Without adequate river input, the morphology and salinity of the estuaries can change dramatically. Estuarine mangrove species cannot tolerate extreme salinities or drying out of the sediments. In areas of southeastern Asia, for example the Indus Delta of Pakistan, irrigation has reduced river flow and caused a reduction in mangroves.

Sediment input is important to maintain stability for the mangrove plants and may increase the area available for mangrove colonization, but a large influx of sediments can result in death of the mangroves. Increased deforestation inland can cause erosion and such extreme increases in sediment input to the mangrove ecosystem that the mangrove roots are smothered, inhibiting their ability to exchange gases through the roots or pneumatophores.

Coastal Pollution

Pollution can enter the mangrove ecosystem locally or from sources far from the coast through river input. Metal pollution affects mangrove ecosystems in areas close to mining operations or those exposed to industrial waste. Although no noticeable effect on the mangrove plants has been documented, heavy metals are highly toxic to crab larvae and can bioaccumulate in predatory organisms. Increases in agriculture along the coast result in excess inputs of chemicals and nutrients. Some degradation of herbicides and pesticides occurs in the anoxic sediments, so mangrove ecosystems may not be as sensitive to these pollutants as ecosystems that lack anoxic sediments. Low levels of nutrient input from sewage effluent may increase productivity of mangroves, and studies have shown that mangroves might be used for waste water treatment. Higher nutrient levels are likely to cause excess algae growth, however, creating anoxic conditions due to bacterial decomposition. Algae can cover aerial roots and pneumatophores, inhibiting gas exchange, or cover seedlings.

Toxicants applied directly to the mangroves can cause extreme damage. For example, during the Vietnam War in the 1960s the United States destroyed over 100,000 hectares of mangrove forests, over one fourth of the estimated mangrove area, through the application of herbicides and defoliants. Many of these mangroves have not recovered, in part due to inadequate protection and a lack of resources for recolonization efforts.

Ongoing chemical pollution problems include oil or chemical spills in regions near oil terminals, refineries (e.g., the Middle East and Central America), industry, or boat traffic (e.g., the Panama Canal region). Mangroves are especially vulnerable to oil and other pollutants released accidentally because they are exposed to chemicals that float on the water's surface. Heavy oils carried into the mangroves by the tides can coat pneumatophores and aerial roots and inhibit gas exchange. Mangroves appear to be able to eventually recover from moderate impacts from oil spills if protected or assisted. The complex structure of the mangroves make the clean-up of oil spills difficult, however, and it may take decades for an area to recover. Mangroves have been severely damaged in the Niger Delta of Nigeria where oil exploration activities have not been well regulated. Oil spills also have affected mangroves in countries of the Middle East and eastern Africa.

One of the most well-studied, oil-damaged mangrove habitats is the Bahia las Minas on the coast of Panama. A spill in 1968 resulted in deaths of about 4% of the mangroves in the bay, but by 1979 mangroves had returned to most of the area. In 1986, a second large oil spill from a ruptured refinery storage tank occurred in the same area. About half of the mangroves in the bay that were initially covered by the oil died within a few months. Eventually other areas were affected as the oil washed into the mangroves with the currents and tides. Mangrove roots were covered and organisms in the intertidal area showed massive mortality; over 40% of the mangrove habitat in the bay was affected. Natural recovery of the mangroves was slow, in part because residual oil remained in the sediments and the spill killed most of the mangrove seedlings. Some areas were converted into open water due to erosion before the mangroves could recover. Replanting efforts were eventually used to assist in the recovery. Five years after the spill, however, oil was still washing out of the sediments and affecting the ecosystem. Bivalves still had high levels of oil in their tissues. Based on studies of this and other spills, it appears that about 20 to 30 years are needed for mangroves to recover from a major oil spill.

Wood Harvest

Mangroves do not typically grow into trees that are valuable as timber, but many other uses have been discovered for the wood and other parts of the mangrove plant. In some areas mangroves have been used locally to build dwellings or boats. Mangrove wood can be a good source of fuel because of its density and hardness. For example, in some African countries a main reason for removing the mangroves is for smoking fish, and in Pakistan the wood has been used in the boilers of trains. Overharvest for charcoal production has affected mangroves in Central America and Indonesia. In Central America and Asia, the wood is used to extract tannins for tanning leather or fish nets. Other products made from mangroves include roof shingles, fish traps, traditional masks, paper pulp, matchsticks, and household utensils. Materials from mangroves are used to produce beverages, local medicines, or foods. Mangroves are farmed in some areas for some of these uses; however, the harvest from natural mangrove stands is still common in many countries. Most of the wood harvest is a local industry; however, in Indonesia and Malaysia extensive areas of mangroves have been cleared for the international wood chip market.

Sustainable harvest of the mangroves for wood is possible if closely regulated. This is rare, but has been successful in some regions, such as the Matang mangroves in Malaysia. If not regulated and monitored, however, these activities can result in a non-renewable use of the mangroves and cause ecosystem degradation or destruction. Still, wood removal is rarely the main impact on mangroves around the world; the major problems are more typically the removal of mangroves for other uses of the habitat.

Fisheries Harvest

As discussed earlier, mangroves are important to many coastal fisheries species, especially as a nursery for larvae and juveniles, not typically harvested in the mangroves. Many regions have a substantial harvest of organisms directly from the mangroves, including shrimp, oysters, clams, and other invertebrates. For example, the large spiral-shaped marine snail *Telescopium* is harvested from mangroves in the Indo-Pacific for consumption, and chemicals have been extracted from it as potential pharmaceutical products and antibiotics. It is difficult to document mangrove fisheries and their effect on mangrove ecosystems because many are subsistence fisheries for which harvests are not routinely reported. One documented fishery is in the Sarawak mangrove of Borneo, where trap fisheries catch more than thirty fish species, ten shrimp species, two jellyfish species, and at least one species of crab.

Coastal Agriculture

Agriculture for livestock or crops in coastal areas can result in the removal of mangroves and the construction of dykes and embankments to protect the farmland from salt water intrusion. For example, mangrove habitats have been converted to sugar cane farms in many regions. Conversion to rice farms in Africa and Asia is sometimes supported by governments to encourage self-sufficiency of food production. In southern China, farmers built levees across the mouths of inlets and converted the area behind the barrier to rice farms or shrimp ponds. These levies were continually destroyed by typhoons until farmers began planting mangroves on the seaward side of the levies to protect the paddies and ponds. In Guinea, on Africa's west coast, farmers cut through the mangroves and build mud dikes to limit the tidal flow into rice paddies; this removes the connection between the mangrove ecosystem and the ocean and eventually kills the mangroves (**Figure 4-32**). In Indonesia, conversion of mangrove habitat to farmland has been one of the major causes of mangrove loss. In some regions mangrove habitat has been converted to grazing lands for livestock. For example, in arid countries in Africa and the Middle East, grazing camels, goats, or cattle have reduced the quality of mangrove habitats.

Coastal Aquaculture

Over the past 30 years, shrimp farming has probably received more attention globally than any other anthropogenic factor impacting mangroves. In southern China,

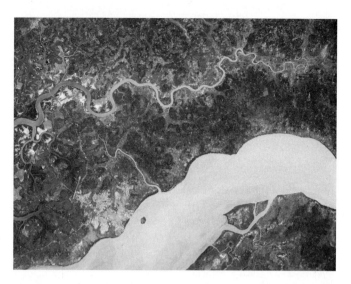

Figure 4-32   Area along the Mansoa River in Guinea-Bissau. Dark gray regions adjacent to the river are mangroves; lighter gray regions are mostly rice paddies. (See Color Plate 4-32.)

mangrove habitats have been used for farming shrimp for centuries. The ancient methods of farming, however, did not require removing the mangroves; wild shrimp were harvested from impounded mangrove habitats. These methods are inefficient for current commercial enterprises, and recent practices are much more destructive to the mangroves.

Since the 1970s, shrimp aquaculture practices have probably damaged or removed more mangrove ecosystems worldwide than any other activity. Methods involve clearing the mangroves, allowing the tides to flood the area, and then building dikes and levees to turn the area into small ponds. Young shrimp were netted from local waters, placed into the ponds, fed naturally occurring organisms, and raised to harvest size over several months. The ponds were typically used for two to five years, and then abandoned (**Figure 4-33**). The farmers moved to another area and repeated the cycle. As the market grew, pond sizes grew, and entire coastlines were being cleared of mangroves for shrimp farming (**Figure 4-34**). When mangroves are cleared, the entire mangrove-based ecosystem is lost or dramatically harmed. To compound the affect, removal of the mangroves for shrimp farming results in a decline in populations of wild mangrove-dependent shrimp. Even when the farms have been abandoned, the wastes can continue to wash into surrounding ecosystems and the ponds are slow to recover. The recovery of abandoned farms to healthy mangrove ecosystems typically takes about 30 years. Although laws have been passed in many countries prohibiting the removal of mangroves, many former mangrove habitats continue to be used as shrimp farms. It is estimated that about 800,000 hectares of mangrove habitat, mostly in Asia and Latin America, were lost to shrimp aquaculture.

(a)

(b)

Figure 4-33 Mangrove marshes are sometimes converted to aquaculture ponds; here shrimp culture ponds are adjacent to coastal mangroves, north of Belize City, Belize: **(a)** flooded in 2006, **(b)** abandoned in 2009.

Figure 4-34 False-color satellite images of the Gulf of Fonseca region on the Pacific coast of Honduras in 1999. Dark grays indicate areas covered in water. Light grays indicate vegetation including mangroves adjacent to the water. Shrimp ponds appear as rectangles. (See Color Plate 4-34.)

As shrimp farming developed into a global industry in the 1980s and 1990s, methods were refined and modernized to provide a greater production of shrimp. Higher production requires replacing the more traditional **extensive** methods with **intensive** aquaculture. In intensive farming, shrimp are produced in hatcheries and raised at higher densities on artificial feeds in aerated ponds away from the coast, using water pumped from coastal areas (**Figure 4-35**). Effluent from the ponds, containing excess nutrients, antibiotics, pesticides, and shrimp feed (30% of feed may remain uneaten) ends up in the surrounding waters, affecting other coastal ecosystems and inhibiting the recovery of the mangroves. Some nutrients and organic matter from farming can be tolerated by mangroves and adjoining ecosystems; however, the area of habitat needed to process the effluent is 30 to over 100 times that of the area of the intensive shrimp farm.

Because the development of intensive aquaculture methods has focused on a few species, it is often not native shrimp species that are being farmed. The two most common cultured species are the giant tiger prawn *Penaeus monodon* and the Pacific white shrimp *Litopenaeus vannamei*.

Figure 4-35 Hatchery ponds used in intensive shrimp farming in Japan.

Color Plates

Plate 1-2 Relief map of the earth. Colors indicate depths and elevations; ocean regions are indicated from light to dark blue, shallow to deep. Note the mid-ocean ridges and other seafloor features.

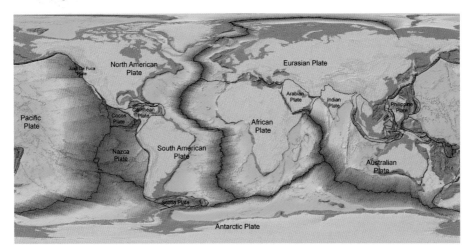

Plate 1-3 Map of the sea floor with evidence of seafloor spreading based on age of ocean sediments. Newer areas of the seafloor near the spreading centers are indicated by reds.

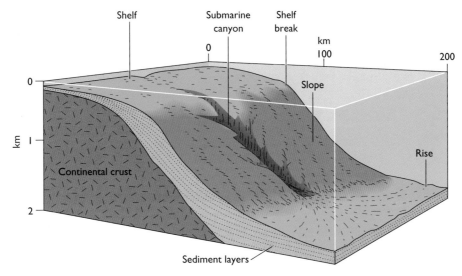

Plate 1-4 Representation of a continental margin, consisting of the continental shelf, slope, and rise. Submarine canyons cut through the continental margin in some regions.

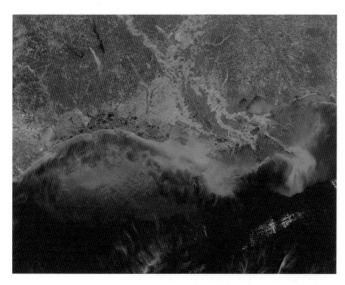

Plate 1-8 Satellite image of the Mississippi River delta in the northern Gulf of Mexico.

Plate 1-10 Satellite photo of the Nile River delta and estuary.

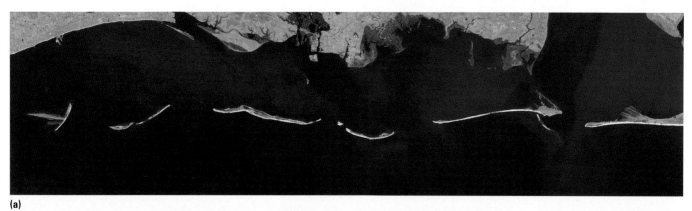

(a)

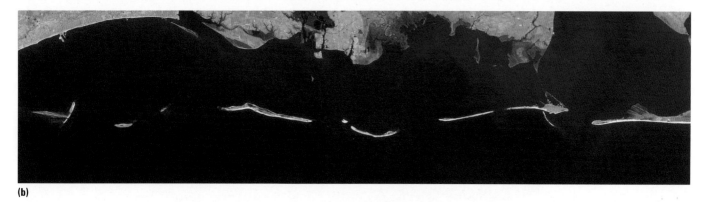

(b)

Plate 1-12 Satellite photos of barrier islands in the northern Gulf of Mexico **(a)** on October 15, 2004 and **(b)** on September 16, 2005 after Hurricane Katrina. Note that Dauphin Island, Alabama (second island from right) has been split in two and that Ship Islands (starting second from left) are substantially smaller. Dauphin Island also migrated landward, leaving some oceanfront homes in the sea. Petit Bois Island, 8 miles to the west, was a part of Dauphin Island 150 years ago.

Plate 1-13 Satellite view of the Hawaiian Islands, formed by volcanic activity as the Pacific Ocean plate slides over the mid-ocean hot spot currently located beneath the island of Hawaii (lower right); the oldest islands (Kauai and Niihau) are in the upper left.

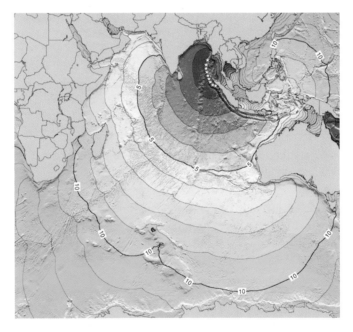

Plate 1-16 Movement of the tsunami wave produced by a magnitude 9.4 earthquake off Sumatra, Indonesia, on December 26, 2004, progressing in time from the epicenter (indicated by stars), through reds, greens, and blues.

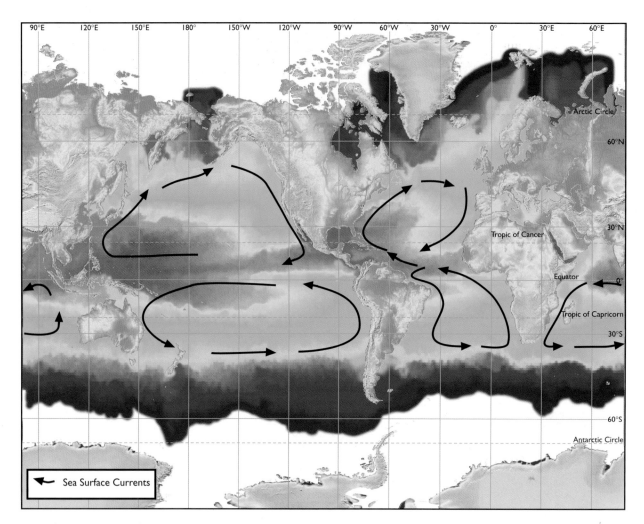

Plate 1-20 Global ocean surface temperatures during summer in the northern hemisphere. Colors indicate temperatures, with red being warmest and decreasing through yellow, green, blue, and purple.

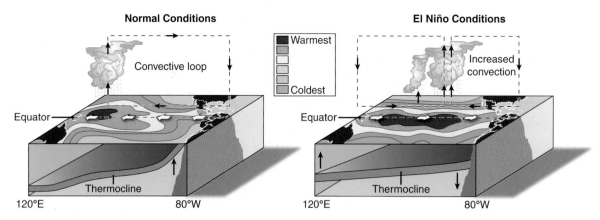

Plate 1-21 Coastal upwelling circulation off the Pacific coast of South America; colors indicate temperature, from red (warmest) through orange, yellow, green, and blue (coldest). Left: During normal conditions the thermocline extends to the surface along the coast due to upwelling of deep waters. Right: During El Niño conditions; upwelling stops and the thermocline remains in deep waters.

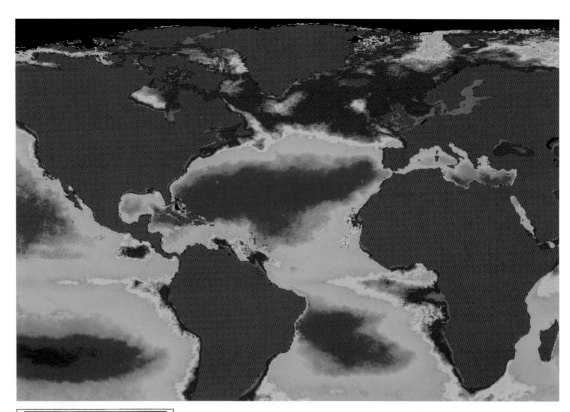

>.01 .03 .1 .2 .5 1 2 5 10 20 30
Ocean: Chlorophyll *a*
Concentration (mg/m³)

Plate 1-27 Image taken by the Sea-viewing Wide Field-of-view Sensor (SeaWiFS) satellite. Ocean colors represent chlorophyll concentrations, which are indicative of primary production. Note high chlorophyll levels in the North Atlantic, in upwelling zones along the west coasts of South America and Africa and along the equator, and at the mouths of major rivers such as the Mississippi and Amazon.

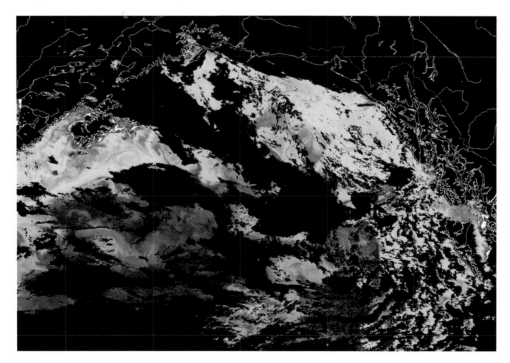

Plate B1-3 SeaWiFS image of the northeastern Pacific Ocean during a 2002 iron fertilization experiment. Chlorophyll concentration increases from blues through yellow to orange. The fertilized bloom is the yellow-orange area in the bottom center.

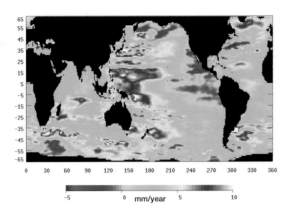

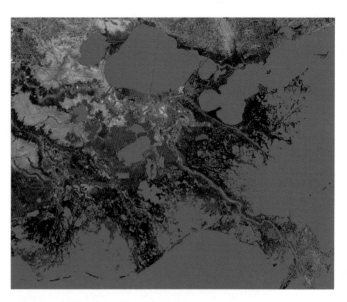

Plate 3-36 Rates of sea level change around the globe from 1993–2008. Note that greatest sea level rise has been experienced in the tropical western Pacific, a region with numerous populated low-elevation islands.

Plate 4-18 The lower Mississippi River and Delta.

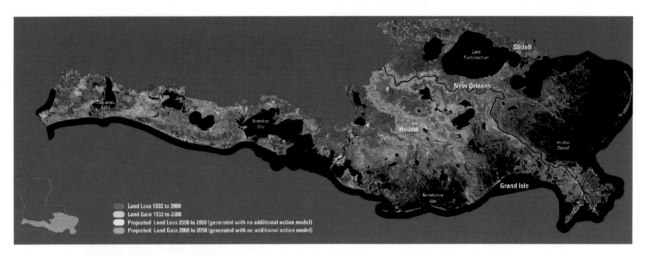

Plate 4-19 Historic and predicted land loss in coastal Louisiana.

Plate 4-22 The Amazon River mouth. Mangrove habitat extends along the seaward edge of the river's mouth. Note the sediment plume extending into the ocean.

Plate 4-32 Area along the Mansoa River in Guinea-Bissau. Dark green regions adjacent to the river are mangroves; lighter colored regions are mostly rice paddies.

Plate 4-34 False-color satellite images of The Gulf of Fonseca region on the Pacific coast of Honduras in 1999. Blues indicate areas covered in water. Green is vegetation including mangroves adjacent to the water. Shrimp ponds appear as rectangles.

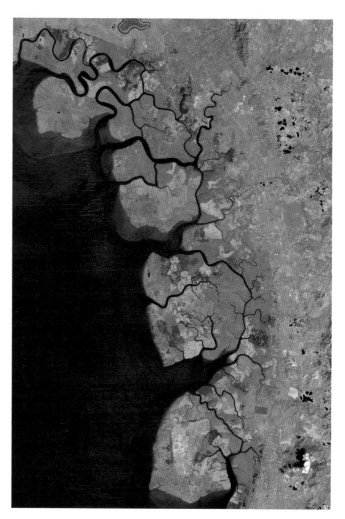

Plate 4-38 False-color satellite photo of the Matang Mangrove Forest, Malaysia. Dark greens are mangroves, blues are rivers flowing through the estuary to the sea, pinks are cleared areas (including towns and fishing villages within the mangroves), and light greens are agriculture or other vegetation.

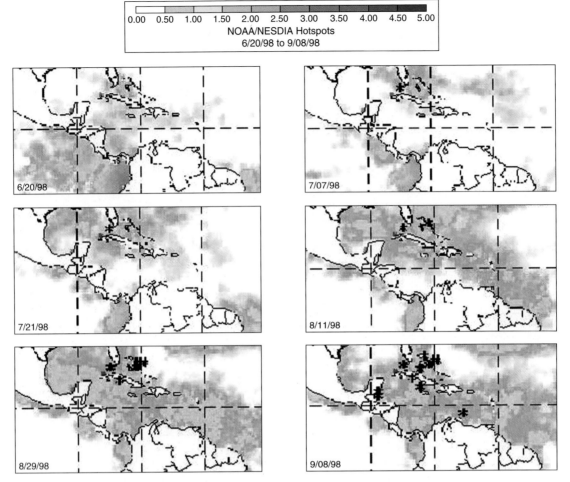

NOAA/NESDIA Hotspots
6/20/98 to 9/08/98

0.00 0.50 1.00 1.50 2.00 2.50 3.00 3.50 4.00 4.50 5.00

6/20/98

7/07/98

7/21/98

8/11/98

8/29/98

9/08/98

Plate 5-9 Temperature anomalies in the tropical western Atlantic Ocean during the 1998 El Niño. Asterisks indicate hotspots vulnerable to coral bleaching.

Plate 6-15 Shark Bay in western Australia, which contains the largest seagrass bed in the world. The blue-green coloration in the bay is the result of a large phytoplankton bloom.

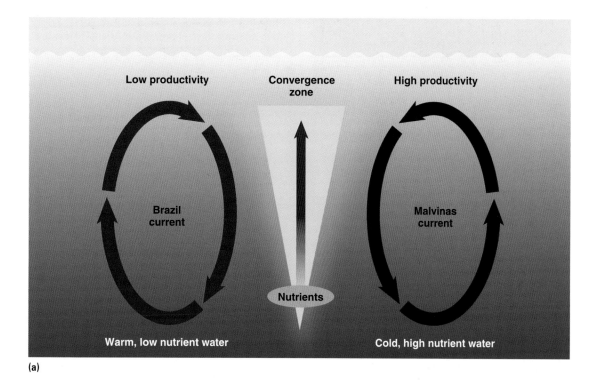

(a)

(b)

Plate 7-2   Upwelling of nutrients into the epipelagic at a convergence zone. **(a)** Convergence of the Malvinas and Brazil currents in the southwest Atlantic. **(b)** Surface image demonstrating elevated chlorophyll concentrations at the convergence zone.

Plate 7-3 Dust blowing from arid regions over the open ocean, such as from the Sahara to the Atlantic Ocean, can provide an important source of nutrients such as iron and phosphorous.

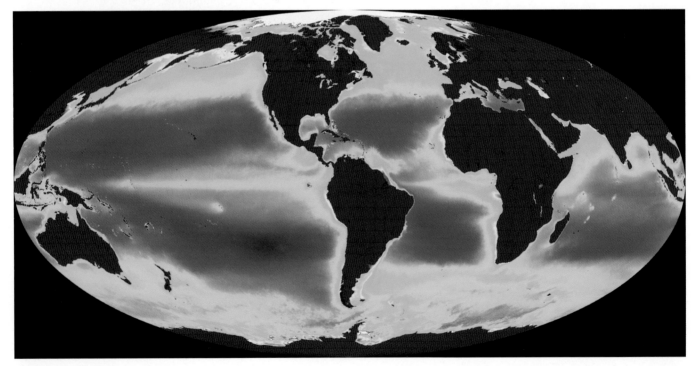

Plate 7-4 Image obtained by SeaWIFS (Sea Viewing Wide Field of View Sensor) carried by the NASA SeaStar spacecraft. A measure of absorption of blue and green light gives an indication of the concentration of chlorophyll (and thus, photosynthesizing organisms) in the water. In this image, purples indicate chlorophyll-a concentrations of < 0.1 mg/m^3, light blues are 0.2–0.5 mg/m^3, reds are 20–50 mg/m^3. Note that the regions of lowest productivity are in the center of major ocean basins.

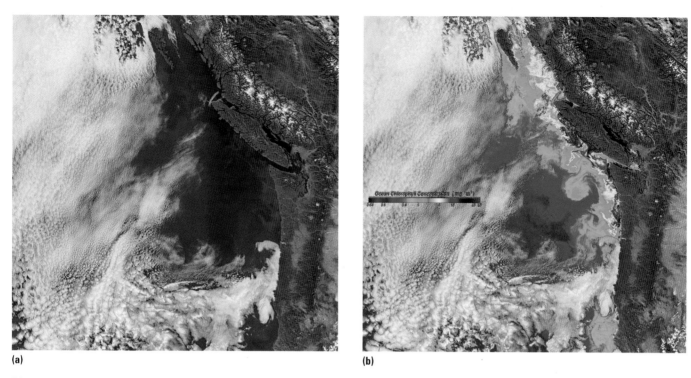

(a)   (b)

Plate 7-13   October 1, 2004 image from the SeaWiFS Sensor during a *Pseudo-nitzschia* toxic algae bloom off the coast of Washington State and Vancouver Island: **(a)** true color; **(b)** phytoplankton chlorophyll concentration (reds indicate highest concentrations).

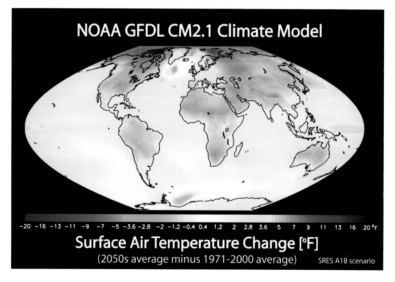

Plate 7-16   Predicted temperature changes from 2000 to 2050 from models developed by the Intergovernmental Panel on Climate Change. Reds and blues indicate warmer and colder temperatures, respectively. The average predicted increase in global temperature is 5.2°F.

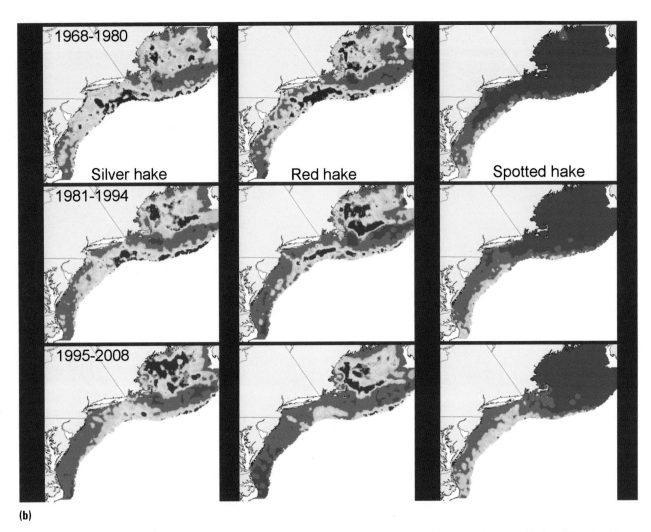

Plate 7-17 **(b)** Distribution of merlucciid fishes, the silver hake (*Merluccius*), and red hake and spotted hake (*Urophycis*) over the same time period. Reds indicate the highest fish concentrations. Patterns indicate a general shift in distribution pattern to the north in association with an increase in water temperatures.

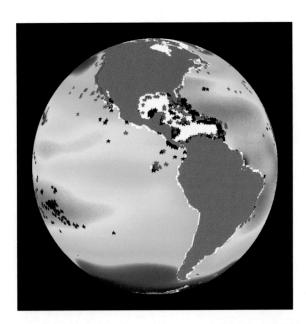

Plate 7-18 Predicted changes in ocean acidity from 1765 to 2100 from models developed by the Intergovernmental Panel on Climate Change. Blue areas are more acidic (lower pH) and red areas are more basic (higher pH). Black stars indicate shallow corals, magenta stars deep corals.

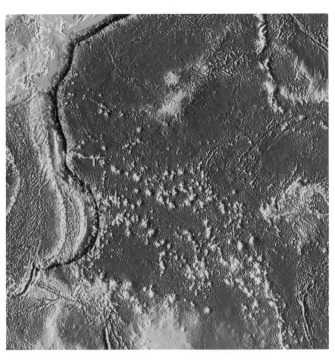

Plate 8-2 Image of western Pacific Ocean with a high concentration of seamounts. Darker colors are deeper regions. The Mariana Trench is indicated by the dark blue region on the left.

Raising non-native species may protect local shrimp populations, but presents other problems, including the potential introduction of a new invasive shrimp species and diseases that may accompany the non-natives. To minimize these problems, laws have been introduced that require the shrimp sold as broodstock to be certified as pathogen-free.

As public pressure to limit the harmful effects of shrimp farming increased in the 1990s, efforts were made to encourage the shrimp farming industry to develop practices that were less harmful to mangrove and coastal ecosystems. One of the first efforts was to encourage consumers to ask stores not to stock shrimp unless they come from sustainable sources. It has been difficult to track and label the source of all shrimp products and to decide what should be considered ecologically sound practices. A more inclusive program was established in 1999 through the World Bank, Food and Agriculture Organization of the United Nations (FAO), World Wide Fund for Nature (WWF), and aquaculture organizations to develop improvements in farming practices and establish educational programs. The international shrimp farming industry has begun adopting many of the recommended practices. In a 2006 meeting organized by the FAO, 50 countries worked out a set of international principles for responsible shrimp farming. In many countries it is now illegal to remove mangroves to build new shrimp farms; instead, new intensive farms are built outside of the mangrove areas. Waste water treatment methods have been developed, and there is some effort by the industry to develop mangrove reforestation projects that use the treated wastes to establish new mangrove stands.

Despite these efforts, most of the mangroves that were cleared for farming still have not recovered, and small farming operations in developing countries still use mangrove habitat to farm shrimp. Many local NGOs continue to discourage the expansion of shrimp farming in mangrove habitats. For example, Friends of Earth Indonesia works to stop the expansion of shrimp farming, and the Network of Aquaculture Centres in Asia-Pacific (NACA) organizes meetings and provides education on sustainable shrimp aquaculture practices.

By 2007, the worldwide annual farmed shrimp production had increased to over 3 million metric tons. Asia was responsible for over 85% of this total, with China being by far the largest producer (1.3 million metric tons), followed by Thailand, Vietnam, and Indonesia. Most of the remaining 15% is from Latin American countries, Ecuador and Mexico being the largest producers. Although farmed shrimp production in the United States comprises only about 0.1% of the world total, over 15% of the production is imported into the United States (European Union nations import comparable amounts). The low price of these imports has driven down the value of wild-caught shrimp in the United States, affecting the profitability of the shrimp trawling industry.

Development and Mangrove Loss

Coastal development is another source of destruction and degradation of mangrove ecosystems. Development can lead to mangrove removal and filling in of the habitat, modification of water flow and circulation, input of toxins, and inadequate treatment of sewage and other wastes. Oil exploration and drilling not only results in removal of the mangroves, but associated canals, roads, pipelines, and other structures can alter the flow and drainage. Urban development has become widespread in tropical coastal areas and often leads to mangrove removal. In some regions, such as south Florida, Central America, and some eastern African and South American countries, urban development has been the primary factor impacting mangroves. The direct loss of mangroves to urban development is relatively permanent; however, control of urban wastes that reach the remaining mangroves can assist in their protection and recovery.

In many regions, development for tourism has been the major cause of recent destruction or harm to mangroves. Mangroves are removed and covered to build hotels, resort areas, marinas, beaches, or golf courses, and the resulting pollution can affect the remaining mangrove ecosystems. Increasing the size of ports to accommodate cruise ships has resulted in the loss of coastal mangrove and reef habitats. Development for tourism can be very attractive to small tropical countries as a quick way to improve the economy by attracting developers, businesses, and tourists. Often times, the long-term impacts of mangrove removal are not adequately considered in rushing to develop. Tourism can be, but often is not, done in a manner that minimizes the environmental effects. Development for tourism is the main factor resulting in recent loss of mangroves in the Caribbean Islands. In Asia and Latin America only shrimp aquaculture has resulted in greater loss of mangroves than tourism.

Social Impacts of Mangrove Loss

Coastal human communities throughout the tropics depend on mangroves both directly and indirectly, especially in subsistence cultures. If the ecological function of the mangrove is compromised, a major source of sustenance and livelihood will be affected. An impacted mangrove ecosystem may not support the fisheries species on which coastal residents rely, and aquaculture or tourist developments can displace coastal residents, especially if they are not employed by these industries.

Even if people remain in coastal settlements near where mangroves have been removed, the indirect impact can be even more severe. Loss of the mangroves removes the buffer against storm-induced waves. There are numerous examples where healthy mangrove forests could have avoided or minimized the effects of tropical storms and cyclones. For example, in Bangladesh, devastation of coastal regions by cyclones has increased dramatically since the removal of the mangrove habitats. Residents of Bangladesh and India realized the importance of the mangroves and resisted their replacement with shrimp farms; at times this even resulted in violent confrontations. It is believed that the four-meter storm surge produced by Cyclone Nargis that destroyed low-lying coastal settlements and killed tens-of-thousands of people in Myanmar (Burma) in May 1982 could have been minimized if coastal mangroves were still in place.

Tsunamis are infrequent and unpredictable but potentially devastating to coastal communities, and areas closest to the source of large tsunamis may not be able to avoid large-scale damage. Even with warning systems in place, there may be inadequate time to move coastal residents to higher ground. Nevertheless, mangroves can provide substantial protection from much of the damage from tsunami waves that move onshore in tropical regions. For example, following the Indian Ocean tsunami that hit regions of southeastern Asia in 2004, an assessment by Finn Danielsen and colleagues indicated fewer human deaths and less property damage in regions that were near healthy mangroves. In one region of India, regions where mangroves had been removed were adjacent to those with healthy mangrove forests still in place. Villages adjacent to the coast without mangrove protection were totally destroyed; those behind the mangroves experienced minimal damages. Models indicate that mangrove stands can reduce the intensity of tsunami waves by at least 90%.

Fishery harvest in the mangroves or of mangrove-dependent species is important in many tropical regions, and these local fisheries provide food security in many African and southeast Asian coastal regions. A decline in fishery harvest has been linked to mangrove loss in various regions, and this is particularly well documented in Jamaica and southern Florida.

Even though coastal shrimp farming results in enormous economic benefits for some individuals, these are often outweighed by the negative economic impacts of mangrove loss and water pollution, especially if the long-term effects are considered. Economic losses resulting from shrimp farming practices can be five times the potential earnings, and in many countries the locals are not the people who benefit from the shrimp farming. For example, in Bangladesh, the shrimp farming industry moved into coastal areas, denied access to the coast by local fishers, and destroyed mangrove ecosystems that were considered common public resources depended on by locals for food and livelihood.

The shrimp farming industry generates about 10 billion dollars annually in export value. Most of the shrimp farmed in Asian countries ends up in the United States, Europe, and Japan, where shrimp consumption tripled over a decade after shrimp farming boomed beginning in the early 1990s. In some cases the profits from shrimp farming end up in the hands of large international conglomerates; however, in other countries, including Thailand, many farms employ and are managed by locals. By the late 1990s some shrimp farming operations were collapsing in Hong Kong, Thailand, China, and other areas due to diseases, overuse of chemicals, and other poor farming practices, leaving the former mangrove habitats in a devastated condition. A large investment or a long time will be needed for these to recover.

In Vietnam the loss of mangroves has resulted in a cascade of effects. Most of the mangroves that were damaged or destroyed during the Vietnam War were converted to shrimp farms before they were able to recover. Now, coastal areas are more vulnerable to storm damage and intrusion of salt water that damages farm crops. Larval shrimp (needed to stock ponds) and edible mud crabs, both of which depend on mangroves, have declined, and an increase in standing pools of water has resulted in an increase in malaria-carrying mosquitoes.

Mangrove Protection Methods

In order to conserve and protect mangrove ecosystems, political leaders, users of the mangroves, and other citizens need to understand the importance and value of the mangroves ecologically, socially, and economically. This has been achieved increasingly over the recent decades, in part as a result of education campaigns involving locals, fishers, government, and NGOs. For example, the NGO Greenpeace has organized peaceful protests in over 15 different countries to demonstrate against harmful practices by the shrimp farming industry.

Education has led to action to protect an increasing percentage of the remaining healthy mangrove habitats. The removal of mangroves for aquaculture is now banned in most countries, and the large-scale removal of mangroves for any purpose often requires an environmental impact statement, though politics and enforcement are often problematic. Many conservation biologists are proposing the increased protection of the remaining mangroves by creating a network of Marine Protected Areas around mangrove ecosystems.

If done properly, replanting mangroves can be an excellent way to recover areas that have been lost due to natural causes, such as hurricanes or typhoons, or man-made causes, such as pollution or aquaculture. This is especially true if the habitat is protected but natural recovery is unlikely or deemed too slow (for example, if there are not enough healthy mangroves nearby to provide propagules). Some mangrove species, such as the red mangroves, can be replanted simply by taking propagules from healthy mangals and inserting them in the mud. Jurgenne Primavera and Janalezza Estaban, however, found that rehabilitation programs in the Philippines were generally carried out in areas not normally colonized by mangroves (mudflats, sandflats, and seagrass meadows) because mangrove habitat was occupied by fish ponds; this resulted in low 10% to 20% long-term survival. Mortality can be also be high from predators such as crabs or toppling of the seedlings by other organisms. Many replanting efforts therefore are enhanced by raising the propagules in nurseries for a few months before replanting. A method called **encased replanting** provides a means of protecting the seedling, for example, with a section of PVC pipe, until aerial roots develop to stabilize the plant (about 3 years; **Figure 4-36**). Studies have shown that local involvement can increase the success of replanting programs over more costly government or international programs (in one study in southeast Asia survival increased from as low as 10% to as high as 97% with local involvement). One probable reason for the success is that locals have a vested interest and will spend the time and energy needed to maintain the plants.

■ Regional Mangrove Status

Summary statistics can give an indication of how mangroves are fairing, on average, globally. The status and protection levels for mangroves vary considerably around the world, however. In order to get a feeling for the variety and extent of mangrove conservation, some specifics for various regions are reviewed below, as reported by the Food and Agriculture Organization of the United Nations (FAO).

Figure 4-36 Mangrove recovery area using encased replanting in southwest Puerto Rico.

Asia has the largest area of mangrove habitat in the world (estimated at over 6 million hectares). Although the extent of mangroves has declined by 1% to 1.5% annually since 1980, there are several large well-protected reserves in the region. The reserve that has possibly seen the greatest long-term protection of any mangrove system is the Sundarbans Forest on the border between Bangladesh and India. Protections of this 542,000 hectare reserve began in 1875, and have remained in place in much the same area ever since. In other parts of India, local residents have donated money to buy mangrove areas to be set aside for protection from destruction for shrimp farming. Mangrove seedlings are collected to be transplanted into areas designated for mangrove replenishment. In these situations education plays a major role in enhancing protection of the mangroves.

The Matang Mangrove Forest Reserve in Malaysia has been managed since 1902 (see below). The Ranong Mangrove Forest in Thailand is a protected mangrove reserve supported by ecotourism and a scientific research center; local fishers are allowed to live in the reserve. Brunei, on the island of Borneo in the western Pacific, has some of the most well-preserved mangroves in southeastern Asia. In Thailand, one of the countries hit hardest by mangrove loss due to shrimp farming, a creative method is being used to protect mangrove habitat. The WWF is working with Thailand's army to develop a nature park in a mangrove forest, including tours and educational programs, with a goal of educating the public of the importance of mangroves. Many countries have established laws to protect mangroves outside of parks and reserves; however, enforcement is often hampered by a lack of resources.

In South America some of the tallest mangrove forests in the world survive as a result of protection in reserves and a lack of accessibility. In areas of northeast Brazil, including the Amazon River Delta, mangrove trees can reach heights of 40 to 50 meters and extend up to 40 kilometers inland.

In the Caribbean, mangroves occur in both estuaries and along exposed coastlines as well as covering small islands that might be covered at high tide (see Box 4-4, Conservation Focus: the Belizean Reef Mangroves). These are particularly vulnerable to destruction for tourism development, so the establishment of protected areas and reserves is important. The Central Mangrove Wetland in Grand Cayman is the largest area of protected inland mangroves in the Caribbean; it includes 4,000 hectares of mangrove habitat protected under the Marine Parks Law (**Figure 4-37**). Protected estuarine mangrove areas, some with mangroves as tall as 30 to 40 meters, are found in Mexico, Costa Rica, Panama, and Belize. In Cuba, mangroves are protected under habitat laws, and major mangrove plantation efforts were begun in 1980 and continue today.

Figure 4-37 A portion of the Central Mangrove Wetland in Grand Cayman (foreground). The city of George Town is visible across North Sound in the background.

The remaining mangroves in south Florida are heavily protected within a system of protected areas. Mangrove recovery is achieved by first assessing and removing the cause of mangrove loss, and then either allowing the mangroves to recolonize on their own or replanting mangrove seedlings. In this region, the mangrove ecosystem will typically replenish itself naturally in 15 to 30 years. In Florida, the Mangrove Trimming and Protection Act prohibits the removal, trimming, or disturbance of mangroves without a permit. In the Bahamas, educational programs have been designed to increase awareness, such as a Coastal Awareness Month, initiated in April 2005. In Central American countries insufficient enforcement of legislation hinders mangrove protection; however, reserves and parks have been established in Costa Rica and Honduras.

In much of Africa, laws are considered inadequate to protect and conserve the remaining mangrove ecosystems; however, legal protections are in place in Congo, Egypt, Kenya, and South Africa. Rehabilitation programs are increasing in some nations, including Mauritius, and in Tanzania all mangroves are legally protected. Even in areas where legal protection is lacking, education programs are increasing to inform locals of the benefits of conserving mangroves, for example, in Guinea, Mauritius, and Sierra Leone. A replanting program was organized by NGOs in Senegal involving thousands of youths from over 100 villages, providing education as well as recovery.

Throughout the world, further legislation and enforcement are needed to protect the remaining mangroves. Education and local citizen actions will serve crucial roles in determining the fate of mangrove ecosystems. Aaron Ellison points out that the mixed results of mangrove recovery efforts calls for increased international cooperation, greater sharing of information among developing countries, and further application of ecological theories to improve the success rate of restoration projects.

Ecotourism has provided an incentive to protect mangroves in some regions, as the income generated both protects the mangroves and employs locals. For example, kayak tours of the mangroves are organized in Florida and Honduras. Wildlife watching attracts tourists in some mangroves. For example, the Kuala Selangor Nature Park in Malaysia has developed paths and walkways, bird-watching blinds, and tours to view the synchronous flashing fireflies. Bird watching attracts ecotourists in mangroves of Trinidad to view the scarlet ibis, and in Belize to visit brown booby roosts. Many ecotourism agencies now include mangrove tours on their agenda, and some snorkeling guides include visits to mangrove habitats. It is idealistic to propose that the majority of mangrove ecosystems around the world be protected as sanctuaries untouched by humans, especially in countries where they are depended upon for subsistence. Managing the mangroves sustainably is better than destroying them for some other use, however. Probably the best example of a middle-ground solution is the Matang mangroves of Malaysia (**Figure 4-38**). This region is comprised of 40,000 hectares; about 5% (2,000 hectares) is protected as a "virgin-jungle reserve" and some smaller areas are protected as bird sanctuaries or for ecotourism and education. The remainder has been managed for over a century for wood production much as a commercial forest. Small areas are clearcut on a 30-year rotation, leaving the adjacent mangroves, including a three-meter strip next to the water, to produce propagules for recolonizing the clearcut area. The harvested wood is used mainly for charcoal production. Replanting is done after one year when necessary, using seedlings from local nurseries. As the mangroves grow, they are thinned, and the mangroves that are removed are used for fishing poles and materials for building village houses. The managed forest differs from that of an untouched mangrove ecosystem; however, there is a diversity of birds and other wildlife in the managed mangroves, and healthy fish populations in the associated marine habitats. The fisheries in the coastal region appear to be managed sustainably, with a harvest of more than 50,000 tons per year. Although this is not an extreme conservation solution to mangrove protection, it is an excellent model to consider when total protection is not a practical alternative.

Mangrove Conservation Prognosis

So, what is the future prognosis for mangrove ecosystems? As discussed above, these ecosystems are vulnerable to numerous impacts around the globe because of their accessibility and location near heavily populated coastal areas. There is no one simple solution to the problems of mangrove degradation and loss because impacts vary with

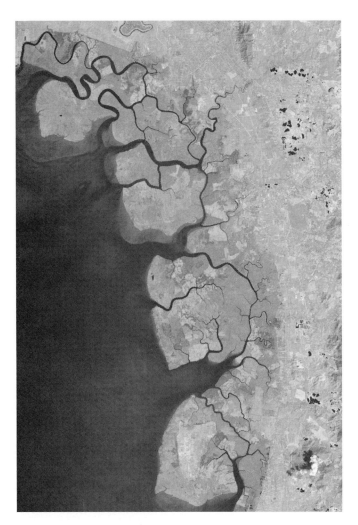

Figure 4-38 False-color satellite photo of the Matang Mangrove Forest, Malaysia. dark grays are mangroves, blacks are rivers flowing through the estuary to the sea, very light grays are cleared areas (including towns and fishing villages within the mangroves), and medium grays are agriculture or other vegetation. (See Color Plate 4-38.)

geography, climate, history, human culture, politics, and economics. The first step is education of the importance of mangroves, tailored to the specific region, and targeted to children and adults, political leaders, businesses and industries, and local residents. Each of these groups must realize that destruction of the mangroves is in no one's long-term interest. Although conservation should not be driven solely by economic value of ecosystems—and these values are difficult to generate—an awareness of the great economic value of mangroves can help to convince political leaders that mangroves are worth protection. For example, an analysis by Patrick Ronnback estimates the market value of fisheries supported by mangroves could be over $10,000 per hectare.

Resources are needed to assist in the recovery of damaged mangrove ecosystems, but involvement by locals can make recovery remarkably inexpensive. More protected areas are needed in parks or reserves to preserve some of the remaining healthy mangroves. Acceptance will be more readily achieved if locals are involved in decision making and can benefit from employment in park management or ecotourism. It is impractical to expect all mangroves to be protected as sanctuaries because many people depend on the mangroves for the natural resources they produce. It is in the interest of these people to develop sustainable management plans, however, which may include harvest of the mangroves for wood or taking mangrove-dependent organisms for food. Larger issues that affect mangroves that involve protecting watersheds and ocean ecosystems or minimizing effects of climate change are not as simple, but need to take the mangrove ecosystems into account. Much scientific progress has been made in recent years in understanding what needs to be done and how to do it. It remains to be seen how seriously people will take this information and apply it to mangrove conservation.

4.4 Global Climate Change and Marsh Ecosystems

Because of the unpredictability of global climate change it is not known when or how much estuary and marsh habitats will be impacted. Changes in temperature, carbon dioxide levels, and rainfall patterns could have effects on salt marsh and mangrove ecosystems that are difficult to predict. Sea level change may have the most predictable impacts. Higher sea levels could result in the inability of salt marsh grasses and mangroves to survive in much of their current habitats. In theory, the marsh ecosystems could "migrate" inland and be re-created in new regions. Many of these areas are currently occupied by human settlements or other developments, however, or blocked by sea walls or levies. If sea levels rise, habitable areas along the coast will be at a premium and it is doubtful they will be abandoned to allow for new salt marshes and mangroves. Coastal communities will be forced to continually replenish the coastline with sediments dredged from offshore or the estuaries.

If sea level rise is within the ranges that many scientists predict (see Chapter 1), many coastal regions will be covered with water within 50 to 100 years or sooner. Other changes that are predicted include: an inundation of brackish waters with sea water, inhibiting the growth of marsh grasses, and greater wave activity and increased erosion due to rising water levels. An analysis by Rusty Feagin and colleagues indicated that salt marshes could be naturally maintained, or even expanded, where there is adequate sediment deposition and accretion of organic material, especially if rates of sea level rise are at the lower bound of current estimates. Urbanized areas, however, could limit the migration or expansion of marshes. At upper bounds of estimates, the rate of accretion may not be adequate in most estuaries to maintain the level needed for the survival of the marshes. Studies by Patty Glick and National Wildlife Federation scientists, for example, estimated that, for Chesapeake Bay, a sea level increase by 0.6 meters would result in the loss of 652 square kilometers of brackish marsh and more than half of the region's salt marshes.

The most obvious way to minimize the impact of sea level rise is to minimize climate change, a global problem discussed in Chapter 1 (see Box 1-1, Conservation Concern: Global Warming and the Ocean). Considering that efforts may not be adequate, there needs to be a consideration of ways to minimize the local effects. For example, in Chesapeake Bay there has been a call for planning by government and NGOs. These plans include: prioritizing sites for protection based on their ecological importance and vulnerability, expanding protected areas to account for migration of the marshes with sea level rise, restoring and protecting other coastal ecosystems, identifying areas that might be enhanced by replenishment of sediments, and expanding monitoring programs. Similar methods are being discussed and applied in other regions; for example, the Mississippi River Delta of Louisiana, as discussed above. Both of these areas are already seeing the effect of sea level rise due to global warming in combination with marsh subsidence. Efforts to replenish the marshes include using dredge spoils to restore islands and marsh lands; however, a continual replenishment as sea levels rise may be cost prohibitive.

Mangroves will be exposed to similar impacts as sea levels rise. If deposition of sediment and detritus is great enough and rates of sea level rise low enough, the ecosystem may be able to maintain itself. This would likely only apply to mangroves associated with some estuaries and river deltas with large sediment inputs, however; those in arid areas and on offshore islands or coastlines away from substantial river input would eventually be exposed to rising water levels. Fossil records from the Caribbean have shown that the mangroves have been able to deposit sediments rapidly enough to keep up with increases in sea level of about 10 centimeters per century. Much higher rates of sea level rise, as have been predicted, would result in the loss of many island mangrove habitats.

As mangrove habitats are covered by rising sea levels, the species zonation pattern discussed above would shift inland. Beyond the current upper extension of the marshes inland, however, the availability of habitat for their migration is low. Many of the coastal regions where mangroves thrive are heavily populated or the habitats are already occupied by rice paddies or shrimp farms. But in Australia, rising sea levels appear to have already resulted in the expansion of mangrove habitat into certain drowned river valleys.

Mangroves are better adapted to tolerate higher temperatures than coral reefs; therefore, a temperature increase would probably not substantially affect their survival or productivity. In fact, increases in temperature may allow mangroves to extend their range into higher latitudes, and habitat is potentially available for expansion along coasts in the Pacific and Atlantic. In the United States, mangroves could replace salt marshes in the northern Gulf of Mexico and northward along the Atlantic coastline beyond the south Florida coast.

The biological community dependent on estuaries, mangroves, and salt marshes will be harmed by the loss of habitat. Many populations—and possibly some species—would be lost. Some species that use the estuaries and marshes as nurseries might survive without them. It is debated whether some species using the estuaries for a portion of their life are truly "estuarine dependent" or whether they might be considered "estuarine opportunist." Although a decline in estuaries may not drive species to extinction, the loss of these optimal habitats could undoubtedly have a dramatic effect on their populations.

The human-caused loss of mangroves and salt marsh ecosystems could be severe in coastal areas, in particular those that depend on these ecosystems for subsistence fisheries or for the protection they provide from violent storms. If global climate change increases the frequency of tropical storms, as some predict, this would further exacerbate the impact of global climate change.

4.5 The Future of Estuary and Marsh Conservation

Solving estuary and marsh conservation problems are complex due to a myriad of reasons. Estuaries and marshes are biologically complex and variable; we may never completely understand the ecological interaction and predict biological responses within these ecosystems. They are also critical, often to an unknown degree, to the survival of other marine and inland ecosystems. They can be affected by impacts originating all the way from headwater streams to the open ocean, due to their position at the intersection of where the freshwaters from rivers meet the tides and currents from the sea. Because they are in regions of the largest human population densities in the world, they are vulnerable to human actions; rising human populations will only exacerbate many of the current conservation problems. Their high productivity will continue to attract exploitation by humans living in coastal regions, continually creating new conservation challenges. Although the habitats themselves are not the most appealing tourist destinations, estuaries and salt marshes are in the vicinity of some of the most popular tourism-dependent regions in the world. Continual work is needed to balance tourism development with conservation. Finally, these ecosystems are managed and regulated by hundreds of governments around the world with different needs and conservation ethics; it is impossible to avoid political and social issues in protecting these ecosystems.

Despite these complexities, much progress has been made in recent years to understand what is needed to protect estuarine and marsh ecosystems. Although there remains much to be learned, we understand the biology of these ecosystems better than ever. Although human actions can be devastating, these ecosystems are more tolerant of moderate impacts than some other marine ecosystems (e.g., coral reefs). Scientists, managers, and conservationists now generally accept that we must consider scientific,

social, political, and economic factors if these ecosystems are going to be protected.

Education programs at the local, national, and global levels are being implemented around the world; positive results are allowing the establishment of model programs to be implemented in other regions. Methods are being developed to minimize harm and still exploit resources from these ecosystems for aquaculture and harvest of renewable resources.

It is important that conservation efforts and outcomes continue to be monitored and that lessons are learned from mistakes. More than for most ecosystems, this will continue to require coordination among scientists, political leaders, government organizations, NGOs, and the public as we move forward in a continually changing world.

STUDY GUIDE

■ Topics for Review

1. Describe how the four types of estuaries differ in size and shape.
2. Why do estuaries tend to be more physically variable than other aquatic ecosystems? Describe the sources of this variability.
3. Why are estuaries and marshes attractive to fishes as nursery areas for young live stages?
4. What are the major impacts from rivers that flow into estuaries?
5. What special considerations are used in U.S. estuaries to protect fisheries species from harmful exploitation?
6. What are the possible reasons for the decline in blue crabs in Chesapeake Bay? What factors have kept these populations from recovering?
7. How do healthy oyster populations benefit estuarine ecosystems?
8. Describe the setbacks that have slowed efforts to recover oyster populations in Chesapeake Bay.
9. List the biological factors that make American and European eels particularly vulnerable to anthropogenic effects.
10. Describe the physical factors that govern the distribution of salt marshes and mangroves.
11. What factors limit the plant diversity in salt marshes and mangroves?
12. How and why do the animal communities of salt marshes and mangroves differ?
13. How does the destruction of salt marshes and mangroves affect inland ecosystems and human settlements?
14. Other than direct destruction, how does coastal development affect the function of salt marshes?

15. How does the function of an impounded salt marsh differ from a healthy salt marsh ecosystem?
16. How have studies of grazing geese and snail consumer fronts changed paradigms about salt marsh ecological functions?
17. How has channelization of the Mississippi River caused the loss of salt marshes in coastal Louisiana?
18. What factors have limited restoration of river flow as an acceptable method of marsh restoration?
19. What reproduction and growth characteristics of mangroves enhance their ability to recolonize after physical impacts such as hurricanes or cyclones?
20. How do the nutrients and organic matter from mangrove leaves reach predatory organisms that feed in mangrove ecosystems?
21. How do mangrove habitats function to protect coral reefs and freshwater marshes?
22. How do mangroves and salt marshes function to protect human settlements?
23. Why are mangroves and salt marshes more tolerant to moderate levels of organic pollution than coral reefs?
24. How would damming of rivers inland potentially affect mangroves?
25. How can wood products be taken from mangrove in a sustainable manner?
26. Describe how extensive aquaculture methods are particularly harmful to mangrove ecosystems as well as coastal fishers.
27. What are the major impacts of coastal tourism on mangrove ecosystems?
28. Why is mangrove replanting sometimes preferable over allowing a natural recovery of destroyed mangals?
29. How is traditional tourism more harmful to mangrove ecosystems than ecotourism?

▨ Conservation Exercises

1. List the methods that you would recommend to most practically and efficiently achieve each of these conservation objectives. Consider biological, social, and economic factors in developing your lists.

a. Recover oyster populations in Chesapeake Bay, while developing a sustainable harvest.

b. Protect and recover endangered populations of the European eel.

c. Develop a strategy for managing coastal impoundments in South Carolina to enhance coastal fisheries.

d. Protect New England marshes from destruction by snow geese.

e. Recover coastal Louisiana salt marshes.

f. Recover mangrove habitats in coastal Vietnam.

g. Establish protections for mangroves in Africa without destroying the livelihood of coastal communities.

h. Recover mangroves in Thailand without economically destroying the shrimp aquaculture industry.

i. Protect reef mangroves in Belize while minimizing the impact on the tourism industry.

j. Protect mangrove ecosystems in Indonesia without totally eliminating harvest for wood products.

FURTHER READING

An, S. Q., B. H. Gu, C. F. Zhou, Z. S. Wang, Z. F. Deng, Y. B. Zhi, H. L. Li, L. Chen, D. H. Yu, and Y. H. Liu. 2007. *Spartina* invasion in China: implications for invasive species management and future research. *Weed Research* 47:183–191.

Beck, M. W., K. L. Heck, K. W. Able, D. L. Childers, D. B. Eggleston, B. M. Gillanders, B. Halpern, C. G. Hays, K. Hoshino, T. J. Minello, R. J. Orth, P. F. Sheridan, and M. P. Weinstein. 2001. The identification, conservation, and management of estuarine and marine nurseries for fish and invertebrates. *BioScience* 51:633–641.

Bertness M., B. R. Silliman, and R. Jefferies. 2004. Salt marshes under siege. *American Scientist* 92:54–61.

Chen, J., B. Zhao, W. Ren, S. C. Saunders, Z. Ma, B. Li, Y. Q. Luo, and J. Chen. 2008. Invasive *Spartina* and reduced sediments: Shanghai's dangerous silver bullet. *Journal of Plant Ecology* 1:79–84.

Cooper, A. 1982. The effects of salinity and waterlogging on the growth and cation uptake of salt marsh plants. *New Phytologist* 90:263–275.

Danielsen, F., M. K. Sorensen, M. F. Olwig, V. Selvam, F. Parish, N. D. Burgess, T. Hiraishi, V. M. Karunagaran, M. S. Rasmussen, L. B. Hansen, A. Quarto, and N. Suryadiputra. 2005. The Asian tsunami: a protective role for coastal vegetation. *Science* 310:643.

Ellison, A. M. 2001. Mangrove restoration: do we know enough? *Restoration Ecology* 8:219–229.

Engel, S., and J. Pawlik. 2005. Interactions among Florida sponges. II. Mangrove habitats. *Marine Ecology Progress Series* 303:145–152.

Feagan, R. A., M. L. Martinez, G. Mendoza-Gonzalez, and R. Costanza. 2010. Salt marsh zonal migration and ecosystem service change in response to global sea level rise: a case study from an urban region. *Ecology and Society* 15(4): 14. [online]URL: http://www.ecologyandsociety.org/vol15/iss4/art14/.

Field, C. D. 1999. Rehabilitation of mangrove ecosystems: an overview. *Marine Pollution Bulletin* 37:8–12.

Food and Agriculture Organization of the United Nations (FAO). 2007. The world's mangroves: 1980–2005. *FAO Forestry Paper* 153. FAO, Rome.

Glick, P., J. Clough, and B. Nunley. 2008. Sea-level rise and coastal habitats in the Chesapeake Bay region. *National Wildlife Federation Technical Report.* National Wildlife Federation, Reston, VA.

Hogarth, P. 2007. *The biology of mangroves and seagrasses.* Oxford University Press. New York.

Keithly, W. R., and P. Poudel. 2008. The southeast U.S.A. shrimp industry: issues related to trade and antidumping duties. *Marine Resource Economics* 23:459–483.

Laegdsgaard, P., and C. Johnson. 2001. Why do juvenile fish utilize mangrove habitats? *Journal of Experimental Marine Biology and Ecology* 257:229–253.

Lefebvre, G., B. Poulin, and R. McNeil. 1994. Temporal dynamics of mangrove bird communities in Venezuela with special reference to migrant warblers. *The Auk* 111:405–415.

Limpus, C. J., and D. J. Limpus. 2000. Mangroves in the diet of *Chelonia mydas* in Queensland, Australia. *Marine Turtle Newsletter* 89:13–15.

Mann, R., J. M. Harding, and M. J. Southworth. 2009. Reconstructing pre-colonial oyster demographics in the Chesapeake Bay, USA. *Estuarine, Coastal and Shelf Science* 85:217–222.

Mason, C. F., G. J. C. Underwood, N. R. Baker, P. A. Davey, I. Davidson, A. Hanlon, S. P. Long, K. Oxborough, D. M. Paterson, and A. Watson. 2002. The role of herbicides in the erosion of salt marshes in eastern England. *Environmental Pollution* 122:41–49.

Mendelssohn, I. A., K. L. McKee, and W. H. Patrick. 1981. Oxygen deficiency in *Spartina alterniflora* roots: metabolic adaptation to anoxia. *Science* 23:439–441.

Naismith, A., and B. Knights. 1993. The distribution, density and growth of the European eel, *Anguilla anguilla,* in the freshwater catchment of the River Thames. *Journal of Fish Biology* 42:217–226.

Nordhaus, I., and M. Wolff. 2007. Feeding ecology of the mangrove crab *Ucides cordatus* (Ocypodidae): food choice, food quality and assimilation efficiency. *Marine Biology* 151:1665–1681.

Odum, E. P. 2000. Tidal marshes as outwelling/pulsing systems. Pages 3–7. In *Concepts and Controversies in Tidal Marsh Ecology.* M. P. Weinstein and D. A. Kreeger (eds.). Springer, Houten, Netherlands.

Partyka, M. L., and M. S. Peterson. 2008. Habitat quality and salt-marsh species assemblages along an anthropogenic estuarine landscape. *Journal of Coastal Research* 24:1570–1581.

Primavera, J. H. 1997. Mangroves as nurseries: shrimp populations in mangrove and non-mangrove habitats. *Estuarine, Coastal and Shelf Science* 46:457–464.

Primavera, J. H. and J. M. A. Esteban. 2008. A review of the mangrove rehabilitation in the Philippines: successes, failures, and future prospects. *Wetlands Ecology and Management* 16:345–358.

Rinehart, K. L. 1999. Antitumor compounds from tunicates. *Medicinal Research Reviews* 20:1–27.

Ronnback, P. 1999. The ecological basis for economic value of the seafood production supported by mangrove ecosystems. *Ecological Economics* 29:235–252.

Roberts, H. H. 1998. Delta switching: early responses to the Atchafalaya River Diversion. *Journal of Coastal Research* 14:882–889.

Rutzler, K. 2008. Low-tide exposure of sponges in a Caribbean mangrove community. *Marine Ecology* 16:165–179.

Rutzler, K., and I. C. Feller. 1996. Caribbean mangrove swamps. *Scientific American* 274:94–99.

Samson, M. S. and R. N. Rollon. 2008. Growth performance of planted mangroves in the Philippines: Revisiting forest management strategies. *Ambio: A Journal of the Human Environment* 37:234–240.

Shepherd, G. 2006. Status of Fishery Resources off the northeastern U.S.: American eel (*Anguillis rostrata*). Northeast Fishery Science Center. National Oceanic and Atmospheric Administration. Retrieved August 10, 2011, from http://www.nefsc.noaa.gov/sos/spsyn/op/eel/.

Silliman, B. R., J. van de Koppel, M. D. Bertness, L. Stanton, and I. Mendelsohn. 2005. Drought, snails, and large-scale die-off of southern U.S. salt marshes. *Science* 310:1803–1806.

Stuart, S. A., B. Choat, K. C. Martin, N. M. Holbrook, and M. C. Ball. 2006. The role of freezing in setting the latitude limits of mangrove forests. *New Phytologist* 173:576–583.

Tufford, D. L. 2005. *State of Knowledge Report: South Carolina Coastal Wetland Impoundments.* South Carolina Sea Grant Consortium, Charleston, S.C.

United Nations Environment Programme. 2008. *Africa: atlas of our changing environment.* Division of Early Warning and Assessment, United Nations Environment Programme, Nairobi, Kenya.

Wooller, M. J. 2007. A multiproxy peat record of Holocene mangrove palaeoecology from Twin Cays, Belize. *Palaios* 24:650–656.

Tropical Coral Reefs: Environmental Impacts and Recovery

There has been a practical interest in coral reefs, primarily for access to food resources, as long as humans have lived in tropical coastal areas. As humans began to move about the world in boats, there was another source of increased interest in reefs, born of frustration from ship groundings as attempts were made to establish ports on tropical coastlines. The first widely distributed publications to consider the nature of coral reefs scientifically were by Charles Darwin in the early 1800s, a product of a curiosity that led not only to his theory of biological evolution, but also to theories that might explain the origin and physical evolution of coral reefs. Scientific studies of coral reefs continued through the 1800s and early 1900s as interest in global exploration increased, but these were mostly limited to the description and naming of organisms associated with the reefs.

The explosion of ecological studies of coral reefs began in earnest as technology was developed to better access the reefs. This included the development of scuba (an acronym for self-contained underwater breathing apparatus), invented in the 1920s, but not readily available until Jacques Cousteau's "Aqualung" was marketed in the late 1940s. Scuba diving enhanced the ability of scientists to study the reefs while also increasing the public's interest in recreation and exploration of the tropical coral reef ecosystem. Through the mid-1900s, there were dramatic increases in both scientific studies and recreational use of the coral reef ecosystem. These explorations established

much about the geology, biology, and ecology of the reefs. Less effort was put into establishing the impacts of anthropogenic activities on coral reef ecosystems, however, and methods and rules to conserve these sensitive ecosystems were only slowly being established. By the 1980s more scientists were looking at conservation issues and serious debates were initiated to establish what effects humans were having on the reefs. This interest exploded as scientists and recreational divers began to see increases in coral disease in the 1980s and mysterious widespread deaths of corals in the 1990s. As we have moved into the 21st century, scientists have reached a consensus that human activities are affecting coral reefs in a dramatic fashion. Unfortunately, it has been difficult to establish clear links between specific causes and effects that could more easily be dealt with through rules and regulations. Because of this uncertainty, the best, most practical, conservation solution has not always been clear; and when scientists are uncertain, encouraging political and social action is more difficult.

In the first decade of the 21st century, we have experienced a dramatic increase in attention to conservation of coral reef ecosystems relative to that received for other coastal ecosystems. The public's interest is, in part, a result of the aesthetic appeal of the clear tropical waters and colorful organisms. Conservation biologists argue that this attention is warranted because of the potential for impacts on the extreme biological diversity of reefs, recent losses of extensive areas of coral reefs, and dramatic unexplained ecosystem changes. Tropical reef ecosystems are ecologically unique and extremely diverse, in part because they occur in relatively stable tropical environments. As humans change the environment and affect that stability, we change the ecology and cause dramatic shifts in species composition. The specific causes of these changes are often hard to discern and solutions may be difficult and expensive to implement. Scientists, resource managers, and the public, therefore, are in a constant struggle to determine how measures might be enacted to conserve tropical coral reef ecosystems around the world.

5.1 The Tropical Coral Reef

Biodiversity

Because clear tropical waters where reef-building corals grow are typically low in nutrients, it might seem unlikely that they would support one of the most diverse biological communities on Earth within a highly productive ecosystem (**Figure 5-1**). This is largely due to the efficient recycling of nutrients by the coral, and this efficiency is enhanced by the symbiotic relationship between the coral and dinoflagellates of the genus *Symbiodinium,* the **zooxanthellae**, incorporated into their tissues. The relationship is considered symbiotic because the corals depend on the zooxanthellae as a source of energy through photosynthesis, and the zooxanthellae depend on the corals as a source of

nutrients through food consumption. As the zooxanthellae in the coral photosynthesize and convert sunlight into biological energy, they take up waste products from the coral as nutrients. The result is that these nutrients do not make it into the surrounding water to support large amounts of primary production by phytoplankton. Water currents bring in other nutrients and zooplankton that are efficiently used by the coral and recycled by other reef organisms. Even though the waters surrounding a healthy pristine coral reef are low in nutrients, the reef ecosystem is highly productive. Because of this high productivity and the warm stable environment, there is a great diversity of specialized organisms living in association with the coral reefs.

Reef-Building Organisms

The structure of the reefs found in clear tropical waters largely comprises coral skeletons deposited over long time periods, millions of years for some reefs. These **calcareous skeletons**, made up primarily of calcium carbonate, are produced and deposited by the living part of the coral over many generations. This thin living layer of the coral consists of many identical coral **polyps** (**Figure 5-2a**). It is the accumulation of billions of these small skeletons that form the framework of the reef. The zooxanthellae within the polyps are an integral part of this reef-building process because, without the zooxanthellae, the coral—if it survives—cannot typically obtain enough energy from feeding to deposit adequate calcium carbonate to build the

Figure 5-1 A coral reef ecosystem off the Turks and Caicos Islands.

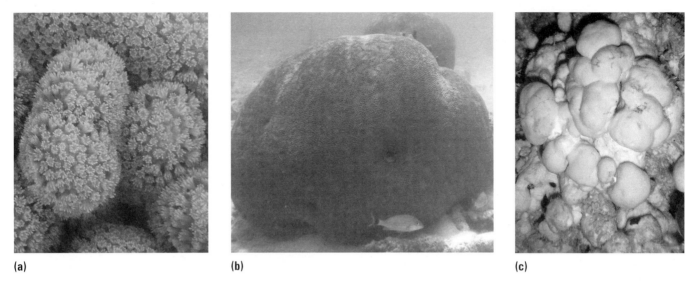

(a)   (b)   (c)

Figure 5-2   Reef building organisms: **(a)** finger coral; **(b)** brain coral, a type of stony coral; and **(c)** encrusting coralline algae.

reef. As sea levels change, reefs need this growth to maintain levels close enough to the sea surface to receive adequate sunlight, and adequate sunlight is needed to support the zooxanthellae.

The stony or scleractinian corals (order Scleractinia) that are most responsible for reef building are typically described by their growth forms (Figure 5-2b). These are mostly "massive corals," including branching elkhorn and staghorn corals, brain corals, platelike corals, and columnar corals. (Other stony corals are found in deeper colder waters; see Chapter 8).

Other corals are less massive and more flexible because they do not produce the rigid reef-forming skeleton; these are commonly called soft corals. The gorgonians (Gorgonacea) are soft corals that are long, branched, and attached to the bottom by a **holdfast**. Gorgonians have common descriptive names such as sea whips, sea fans, sea rods, and sea feathers. Their skeleton consists of flexible protein rods and embedded **spicules**. The true soft corals (order Alcyonacea) are more thick-trunked, branching structures. These are softer than the stony corals because their branches are composed primarily of a gelatinous matrix embedded with skeletal spicules. The soft corals are more common in Indo-Pacific than Atlantic reefs. They are an important part of the reef ecosystem in most tropical reefs but are not as limited by temperature and other physical factors as stony corals. Soft corals, therefore, are found around tropical reefs as well as in colder shallow waters, and there are species without zooxanthellae in deep-sea regions (see Chapter 8).

The **hermatypic corals** (those that contain the symbiotic zooxanthellae) are the major, but not the only, reef builders. The hydrocorals are also a part of the reef but are not technically corals because they comprise colonies of hydroids (class Hydrozoa) that secrete a calcareous skeleton; these include the fire corals with stinging nematocysts that can cause a painful sensation when touched. Encrusting algae (also referred to as coralline algae) also may cover much of the reef, serving as a sort of glue consolidating coral and the calcium carbonate remains of other reef organisms, holding the reef together and protecting it from wave action (Figure 5-2c).

Conditions for Reef Growth

The geographic distribution of coral reefs is greatly limited because the conditions required for reef-building corals to survive long enough to build the reef structure are very specific (**Figure 5-3**). Hermatypic corals depend on both autotrophic nutrition, provided by the photosynthetic zooxanthellae, and heterotrophic nutrition, from food obtained by the coral polyps (mainly small planktonic crustaceans). Neither can be missing for long-term survival of the corals. Because of the zooxanthellae's need for sunlight for photosynthesis, the water must be clear and relatively shallow. Limitations to sunlight penetration mean that typically reefs cannot develop below about 50 meters depth, even in the clearest waters; most hermatypic coral growth is in the upper 25 meters. This limits this type of coral reef mostly to regions near the coast of continents or islands. The water also must be warm, with an average temperature above 20°C for long-term growth and survival of coral. Optimal growth of most corals is at about 25°C; growth and survival decline at lower temperatures. For example, skeletal growth of some branching hard corals declines by about 50% with every three degrees drop in temperatures. Globally, this limits coral reefs mostly to regions below 30° latitude. Cool water temperatures also

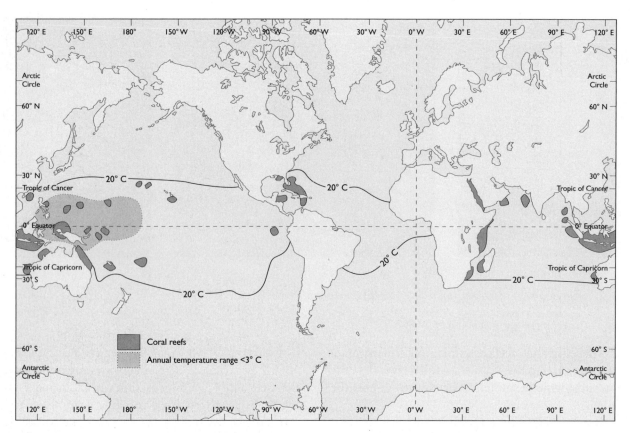

Figure 5-3 Global distribution of coral reefs. The 20°C isotherm is considered to be a major factor limiting the distribution of coral reefs.

keep coral reefs from developing in regions influenced by upwelling, such as along the west coasts of South America and Africa. The water also cannot be too warm. Although some reefs tolerate temperatures above 35°C, if average temperatures are above about 30° to 35°C the coral polyps often become stressed and eventually die. Such high temperatures do not typically occur consistently even in regions near the equator; however, as discussed below, this situation may change with global warming. Excess turbidity is another factor limiting coral reef growth. Sediments and other particles suspended in the water not only limit sunlight penetration but also inhibit feeding activity by the coral polyps. This means that corals are not found where there are substantial inputs of sediment. Finally, most corals cannot tolerate waters that deviate substantially from full-strength seawater (32–36 parts-per-thousand [ppt]). Corals thus are not found associated with estuaries. The combination of turbidity and freshwater results in their absence from large areas on the east coast of South America influenced by the Amazon River, and locally around smaller rivers.

The result of the limitations on coral growth is that the largest areas conducive to growth of tropical reefs are in the central and western Pacific Ocean. This includes regions of the tropical Southeast Asia coastline, the western Pacific island complexes of Indonesia and the Philippines,

and the largest barrier reef system, the Great Barrier Reef off Australia. In the Atlantic Ocean the greatest concentrations of tropical coral reefs are associated with islands and coastal areas of the Caribbean Sea and the Gulf of Mexico. This includes the islands of the Greater and Lesser Antilles, the Florida Keys and Bahamas, and portions of the Central and South American coastline.

5.2 Reef Community Interactions

As with any marine ecosystem, tropical coral reefs can be divided into **functional groups** of organisms, each of which is composed of various species, often representing several phylogenetic groups, which carry out similar ecological functions. Organisms in these functional groups have co-evolved such that the reef maintains itself as a highly diverse community. Although natural perturbations, such as storm events, can at least temporarily disrupt the stability of the reef community, anthropogenic impacts have had dramatic long-term effects on the interactions among these functional groups in the majority of the world's tropical reef ecosystems. Scientists continue to struggle in understanding the interactions within and among functional groups on the reefs so that reef conservation and management can be carried out to maintain high diversity and continue to provide benefits to humans.

Primary Producers

Primary producers in the biological community associated with tropical coral reef ecosystems are not limited to zooxanthellae in the coral tissues; other primary producers include photosynthetic bacteria and protists, macroalgae ("seaweeds"), and sea grasses. Algae may compete directly with the coral for space on the reef, but on a reef with a healthy population of grazers the algae are typically kept in check. Some encrusting coralline algae are important in reinforcing the reef and helping maintain its stability, especially in the presence of wave action, and serve an important role in precipitating calcium carbonate from the water.

Filter Feeders

Many of the filter feeders on the reefs are members of a diverse assemblage of sponges (phylum Porifera). Other than the corals, sponges are often the most visible component of the coral reef ecosystem (**Figure 5-4**). The most obvious are large colorful and column-shaped, named descriptively as tube, pipe, barrel, vase, and rope sponges. Other sponges are encrusting in nature and are found covering hard surfaces or filling cracks in the reef. Sponges can be very colorful with bright red, orange, yellow, or purple colors. This spectrum of colors serve as an advertisement to potential predators of the toxic chemicals contained in the sponges. Sponges vary in abundance from reef to reef, and on some reefs the biomass of their soft tissues may be greater than that of the corals. Especially in areas where corals have declined, sponges may be much more obvious than the corals. Sponges play an important role on a healthy reef as filterers, and help maintain the water clarity that is so important for the corals and other reef organisms. The sponges, especially the encrusting types, also help to consolidate the reef and avoid its collapse or destruction from waves, especially during storm events.

Tunicates (commonly called sea squirts) also can be an important component of coral reefs and are sometimes confused for sponges by divers due to their superficial similarities and bright colors (**Figure 5-5a**). Some tunicates have photosynthetic zooxanthellae incorporated in tissues in a relationship similar to that in corals. Although tunicates are typically a relatively minor component of tropical reef ecosystem, Bernardo Vargus-Angel reported regions in American Samoa in the Pacific where there were population explosions of colonial tunicates from unknown causes. These colonial tunicates appear similar to encrusting sponges and overgrow the reef, causing coral mortality.

Other filter feeders on the reef include tube-building polychaete worms. These include the feather duster worms (also called fan worms) that live in tubes attached to the

(a)

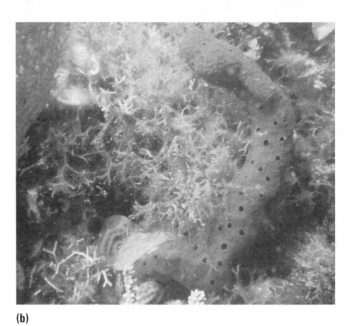

(b)

Figure 5-4 Sponges are a major component of coral reef ecosystems. **(a)** Stove pipe sponge and **(b)** rope sponge.

reef. They are easily recognized by their brightly colored feather-like appendages (**radioles**), extending from the tube for capturing plankton (Figure 5-5b). Other filter feeders include bivalves such as oysters and clams.

Grazers

The coral reef ecosystem has a great diversity of mobile grazing invertebrates, many of which are inconspicuous or camouflaged to avoid predation. On a healthy reef there is intense grazing pressure by fishes and invertebrates on the algae, coral, and any other attached plants or invertebrates (**Figure 5-6**). Sea urchins are sometimes the most important grazers on reefs, feeding primarily on algae but also on seagrasses and various invertebrates. Sea stars can play an important role as grazers; some species depend on the coral

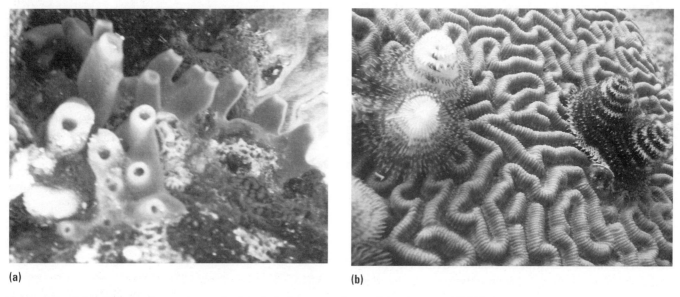

(a)

(b)

Figure 5-5 Other filter-feeding reef organisms include **(a)** encrusting social tunicate and **(b)** Christmas tree worms.

(a)

(b)

(c)

(d)

Figure 5-6 Reef grazers depend on a diversity of food organisms: **(a)** long spined urchins *Diadema antillarum* feed on algae and detritus; **(b)** the crown-of-thorns starfish *Acanthaster planci* eats coral polyps; **(c)** the stoplight parrotfish *Sparisoma viride* grazes primarily on algae attached to the reef; and **(d)** French angelfish *Pomacanthus paru* feeds on sponges and other reef invertebrates.

tissues. Grazing polychaete worms and sea cucumbers feed primarily on detritus and small crustaceans.

Of the thousands of species of gastropods associated with coral reefs, many are grazers on plants, algae, and invertebrates, including the coral tissue for some species. These include snails, such as conchs (*Strombus*), cone snails (*Conus*), cowries (*Cypraea*), and many others, some of which are taken from the reefs for the shells to be sold as curios (see discussion below). Other grazing gastropods are shell-less, including sea hares (Anaspidea) and sea-slugs (Sacoglossa), which feed on algae, and nudibranchs (Nudibranchia), which feed on various invertebrates, including sponges.

Many fishes on the reef are grazers. Herbivorous grazers include triggerfishes (Balistidae), squirrelfishes (Holocentridae), butterflyfishes (Chaetodontidae), and damselfishes (Pomacentridae). Grazing intensity by these animals can be so great that almost 100% of the algae production is removed. Other animals graze on the corals themselves; these grazers include parrotfishes (Scaridae) and surgeonfishes (Acanthuridae). And some angelfishes (Pomacanthidae) are able to overcome the chemical defenses to graze on sponges.

At healthy densities, grazers serve the role of maintaining a stable coral ecosystem and minimize the abundance of algae on the reef. If the makeup of the grazing community is disrupted, however, the reef ecosystem can be changed dramatically and may not return to its original composition for many years. This was demonstrated most dramatically after a mysterious die-off of the long-spined sea urchin *Diadema antillarum* in the early 1980s. This unexplained mortality affected most urchin populations on tropical Western Atlantic reefs, including the Caribbean Sea and Gulf of Mexico, resulting in mortality of 90% to 99% of the sea urchins throughout the region. Following this die-off, there was an immediate increase in the growth of filamentous macroalgae over many of the reefs. A few years later in some regions there was an overall increase in herbivorous fishes, such as surgeonfishes, although the algae's dominance continued through the early 2000s. There are some areas where sea urchin populations have begun to return, but it is still not known what is inhibiting their recovery. Possibilities include outcompetition by herbivorous fishes and predation. For example, Joshua Idjadi and colleagues found that the return of *Diadema* to reefs along the north coast of Jamaica has resulted in reduced macroalgal cover and an increase in new coral, suggesting that the trend is reversible if the reefs remain healthy.

The removal of predators of the grazers by overfishing also can have dramatic effects on the reef ecosystem. When predators and other reef grazers are removed, urchin populations may increase. At high densities urchins reduce algae cover but also prey on other invertebrates, including living coral. Sea urchins are even capable of burrowing into the hard reef structure and, when in great abundance, have been known to destroy large areas of reef. For example, in some reefs in the eastern Pacific, annual rates of reef erosion are as high as 10 kilograms per square meter. In areas of the Caribbean where both herbivorous fishes and sea urchins are gone, macroalgae is likely to dominate the reef. These changes in the reef ecosystem are often unpredictable, but all result in a decline in overall species diversity.

Possibly the most dramatic and well-studied direct impact by a grazing predator on coral reefs was of the crown-of-thorns starfish *Acanthaster planci* (see Figure 5-6b), which feeds on living coral tissue on Indo-Pacific reefs. Although populations of *Acanthaster* typically remain at low densities due to predation and other ecological factors, during the 1960s populations exploded on some reefs in the western Pacific, where tens-of-thousands of starfish caused near-complete destruction of the reef. Interestingly, such outbreaks have not been observed in other areas where *Acanthaster* resides. Evidence suggests that these sorts of outbreaks may be a part of the long-term ecology of the Indo-Pacific reefs, but could be enhanced by human actions. One such action is the taking by shell collectors of the giant triton (*Charonia tritonis*), one of the few natural predators of adult crown-of-thorn starfish. Outbreaks are also associated with increased levels of nutrients, which can be enhanced by coastal clearing or agriculture, especially after heavy rains.

■ Predators

The species diversity on the reef can support many generalist predators, most of which are fishes (**Figure 5-7**). Smaller predators, feeding largely on crustaceans and bivalves, are prevalent on all healthy reefs. Some of the most common of these are grunts (Haemulidae), which, as Richard Appeldoorn reports, serve an important role in linking other shallow-water ecosystems such as mangroves with the reefs through migrations. Some of the largest predators commonly associated with coral reefs include groupers (Serranidae), snappers (Lutjanidae), barracudas (Sphyraenidae), and sharks (Elasmobranchii). The importance of predators is reflected by the numerous defense mechanisms prevalent among reef creatures. Many invertebrates are camouflaged or able to burrow or hide. Many organisms produce or incorporate toxins in their flesh or in spines; these include the sponges and sea cucumbers, and fishes such as scorpion fishes (Scorpaenidae, with a venom contained in their spines), parrotfishes and surgeonfishes (which produce toxic secretions), puffers (Tetraodontidae, which contain a deadly neurotoxin in their tissues), and barracudas (which incorporate a toxin produced by a dinoflagellate that causes ciguatera, a neurological disease in humans; see Chapter 7). However, each of these organisms

(a)

(b)

(c)

Figure 5-7   Reef apex predators include barracudas, sharks, and groupers: **(a)** the great barracuda *Sphyraena barracuda;* **(b)** sand tiger shark *Odontaspis taurus;* and **(c)** tiger grouper *Mycteroperca tigris.*

is vulnerable to predation by species that have co-evolved with them and can overcome their defenses. When non-native species are introduced into the reef ecosystems, the dynamics of predator-prey relationships can change. For example, lionfish native to the Pacific Ocean have recently been introduced into the western Atlantic and Caribbean

with unknown long-term impact (**Box 5-1. Current Issue: Attack of the Lionfish.**)

5.3 Environmental Impacts on Corals

Because corals are so sensitive to water conditions that are continually changing from natural and manmade causes, it is important for conservation biologists to understand not only the tolerance level of the corals, but also how they respond to stress when conditions approach that tolerance level. This is not a simple task, as the effects of extreme conditions are somewhat unpredictable and are dependent on many factors, including the geographic region, species involved, and current health of the ecosystem.

Bleaching

Extreme events such as hurricanes or boat groundings may result in immediate destruction of the coral. Recovery is likely if the damage is not too extensive, there are healthy neighboring reef systems, and adequate time is allowed. Less extreme conditions may result in stress and possibly a slower death; recovery is possible, but only when the stressors are reduced or removed. Stress to hermatypic corals can often be readily recognized due to an event called **bleaching**. Bleaching occurs when the corals expel their zooxanthellae under stressful conditions; one such stressor is excessive water temperatures, typically above about 30°C. Because the color of the coral is primarily from the photosynthetic pigments in the zooxanthellae, in their absence corals appear white or "bleached," a result of light reflecting off the calcium carbonate skeleton below the translucent polyps (**Figure 5-8**). If more than about 70% of the zooxanthellae are expelled the coral will appear bleached. Corals that are bleached do not die immediately and they may recover over weeks or months; however, their survival is tenuous and, if conditions causing the bleaching do not improve, the coral organism will eventually succumb. It is unclear why the corals expel the zooxanthellae, and many factors may be involved. One possible reason is that the zooxanthellae become damaged or lose the ability to produce adequate nutrients to assist with coral growth and thus become a burden to the coral. Another possibility is that the stressed zooxanthellae release chemicals that signal the corals, leading to expulsion. Angela Douglas proposed that bleaching may not be advantageous to coral survival, but could result from a reaction that was advantageous to coral under different environmental conditions.

El Niño

Although localized bleaching of coral has probably always occurred, recently scientists have documented large-scale bleaching events on a scale that appears to be unprecedented

Box 5-1   Current Issue: Attack of the Lionfish

Reefs of the northwestern Atlantic and Caribbean are currently undergoing the most serious invasion of an exotic invasive predator ever documented on coral reefs. That predator is the red lionfish *Pterois volitans* (Scorpaenidae), a tropical salt-water fish native to Indo-Pacific coral reefs (**Figure B5-1**), easily

(a)

(b)

Figure B5-1 **(a)** An invasive red lionfish *Pterois volitans* on a Jamaica coral reef. **(b)** Spearfisher assisting with removal of lionfish from a Jamaica coral reef.

recognized by long feather-like spines and striped body (turkey fish and dragon fish are other common names applied to lionfish). The long spines serve as a protection and warning of the venom contained in the dorsal spines. Despite the awkward appearance, the lionfish is an efficient predator on other smaller fish; it feeds by cornering its prey and lunging, entrapping the fish with the long pectoral fins before swallowing it whole. Lionfish typically hide among corals or other structures at day and feed nocturnally in deeper waters.

Lionfish were initially discovered off Florida in the early 1990s. Although some were documented to have been released into a Florida bay as a result of Hurricane Andrew in 1992, it is uncertain whether this is the origin of the current population. An aquarium release is a likely source because they are popular among fish hobbyists and in show aquariums. In the early 2000s, sightings were becoming commonplace in coastal waters from Miami, Florida to Cape Hatteras, North Carolina, including the coral reefs of the Bahamas. Juveniles were reported as far north as Rhode Island during the summer, although northern populations likely won't remain due to a low tolerance of colder waters. Lionfish populations continued to expand. In 2004, Paula Whitfield and colleagues documented lionfish as second in abundance among fish observed at several locations along the U.S. coast from Florida to North Carolina in waters from 35 to 100 meters depth. Juveniles found off Rhode Island presumably were transported northward along the U.S. coast by the Gulf Stream. By 2005, lionfish had spread throughout the Bahaman Islands. Stephanie Green and Isabelle Cotes found densities of almost 400 per hectare on some Bahama Reefs by 2008, five times the densities observed on native reefs in the Red Sea. On these reefs, the lionfish consume native fish at an average rate of almost 1.5 fish per hour. By 2011 the lionfish had expanded throughout the Caribbean, to the northern coast of South America, and into the Gulf of Mexico (**Figure B5-2**).

It is still too early to know the impact of the lionfish invasion on Atlantic and Caribbean coral reef ecosystems. History tells us that exotic invasions have the potential for serious harm to ecosystems, however. Scientists are working in earnest to measure and predict the effects of the invasive lionfish on coral reef ecosystems. A study by Mark Albins and Mark Hixon found that lionfish can reduce numbers in juvenile reef fish populations by almost 80% in five weeks. In some regions where lionfish are prevalent there has been a documented reduction of various types of reef fishes. The change in fish community structure has resulted in increases in algae coverage at the expense of corals and sponges, likely by similar mechanisms as when coral reef fish communities are affected by other causes. Studies by Michael P. Lesser and Marc Slattery have shown this impact to extend to reefs at depths of 30 to 150 meters that are typically less affected by anthropogenic disturbances than shallower reefs.

So the big question for managers of Caribbean and Atlantic coral reefs now is: can we get rid of the lionfish? One suggestion is to capture the lionfish to be sold to fish hobbyists. Lionfish can be collected readily by divers with hand nets, but profitability would be limited as the market would quickly be saturated. As a method of population reduction, such collections

Figure B5-2   Geographic distribution of lionfish sightings in the northwest Atlantic Ocean and Caribbean Sea as of 2011.

would likely be a losing battle due to the rapid rates of reproduction and ability to inhabit waters over 100 meter depth. Because lionfish are broadcast spawners and eggs are released into the water column, they can easily be dispersed long distances on currents to colonize new areas. Introducing exotic predators to control the lionfish is not considered a viable option due to the potential for harm to native fish populations. The best hope might be that natural predators would control population growth and expansion. Native grou-

pers have been documented to eat lionfish; however, grouper populations have been reduced by excessive fish harvest, and it is unlikely that lionfish would be their preferred prey. It is ironic that there is now a concern over a predatory fish being too abundant, when humans have reduced the populations of most predators from reefs through overfishing and ecosystem effects. Efforts are being made to encourage consumers to eat lionfish. Lionfish can be readily taken by spearfishing, but in some regions local fishers are hesitant to target lionfish for fear of contact with the spines while attempting to harvest the fish or when handling it afterwards. The venom is contained only in the spines and is decomposed with heating; therefore, it does not make the fish dangerous for consumption, and lionfish are considered palatable, healthy, and are commonly sold as food fish in their native range in the Pacific. Although targeted fishing could affect population growth, it is not considered as a realistic means of eradicating lionfish even from local areas due to various factors, including their ability to reproduce throughout the year, efficiently disperse, and occupy habitats in deep waters and around structures such as reefs, reducing efficiency of fishing. Currently, most of the effort dealing with the lionfish invasion is targeted toward monitoring. Sport divers report new sightings and even contribute to their removal.

A general lesson to be learned from this incident is that the effects of moving species around the world are unpredictable. Most exotic fishes that are released or escape from aquariums do not survive long enough to cause an impact or successfully reproduce; however, it only takes one "successful" introduction to have devastating effects on the reef ecosystem. Elimination of the tropical fish trade is not currently a practical option; therefore, hobbyists must take precautions and be educated of the possible dangers of releasing nonnative fish into the wild.

Figure 5-8   Coral bleaching has been a major conservation concern in many tropical regions. Bleached staghorn coral off Jamaica.

and result in large-scale damage over broad areas. Many bleaching events have harmed coral that appeared not to be exposed to physical disturbances or pollution events. In the late 1990s evidence began linking these bleaching events to El Niño events. It is known that, even though these events originate in the eastern Pacific Ocean, they can influence ocean conditions around the globe (see Chapter 1). One influence is on ocean temperature, with temperatures increasing in tropical regions during El Niño years.

During the extreme El Niño of 1998, temperatures around some tropical reefs increased to dangerous levels long enough to cause extensive bleaching and death of previously healthy corals (**Figure 5-9**). Reefs in shallows were the most susceptible; however, reefs at over thirty meters depth were affected in some regions. Even if the corals survived the bleaching events, many did not reproduce for a year or two after the return of normal water temperatures. Overall, more than 15% of the world's coral reefs were

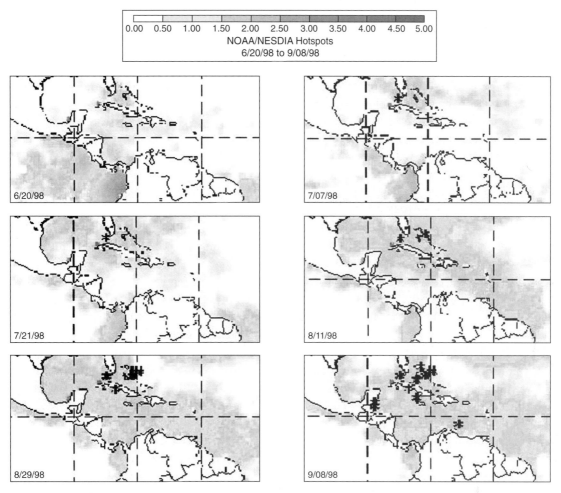

Figure 5-9 Temperature anomalies in the tropical western Atlantic Ocean during the 1998 El Niño. Asterisks indicate hotspots vulnerable to coral bleaching. (See Color Plate 5-9.)

heavily affected by the 1998 El Niño. The greatest damage was in the western Pacific and Indian Oceans, where some reefs lost almost 99% of their structure.

Not all coral reefs are affected equally by elevated water temperatures associated with El Niño events, and some corals have recovered from bleaching events. The reasons for this variability are not always known. Corals in different geographic regions are affected differently, in part because of differences in temperature regimes; however, some coral species appear to be more susceptible to bleaching events than others. For example, during the 1998 El Niño the stony corals *Porites* (a genus that includes finger coral) tended to tolerate the higher temperatures and survive better than the branching staghorn and elkhorn corals (genus *Acropora*) in some regions. In other areas *Porites* were heavily impacted; for example, in some broad regions of the Indian Ocean virtually all *Porites* corals were bleached. When branching corals were bleached the structure of the coral initially remained; however, over a period of two to three years the corals were reduced to rubble. When the percentage of coral loss is high, there may be no source of new recruits to replenish the corals after water conditions have returned to normal.

Despite the clear link between bleaching and elevated temperatures, effects of climate change and El Niño events are not as predictable as once thought. Some species of coral are more tolerant to high temperatures, and, for some coral species, colonies residing in regions with elevated water temperatures (such as enclosed lagoons) are more tolerant than other colonies. This suggests that, at least in some areas, elevated temperatures might not eradicate the coral ecosystem, but could cause a change in the composition of corals and other species in the reef community.

Events that harm the reefs as severely as the 1998 El Niño appear to be historically very rare; for example, some corals that died from bleaching were up to a thousand years old. But even if such extreme events are rare, we do know that El Niño and extreme storm events that eliminate large areas of coral are naturally recurring. Through natural selection, one would expect the corals to be adapted for long-term survival and eventually recolonize once conditions have returned to normal. This expectation is being borne out on some reefs, where corals have begun to recover; globally as many as 40% of the reefs that were bleached have recovered or are on their way to recovery.

The question remains as to whether these extreme events will increase in frequency with global climate change, as many scientists predict. More recent El Niño events (in 2000 and 2003) have caused bleaching, but damage was more localized and not as expansive as the 1998 event. The overall trend is still one of decline. Worldwide, it is estimated that coral reef cover has been reduced on average about 5% per year since 1997. It remains uncertain how much of this decline is due to local human actions, such as coastal pollution, and how much is due to larger scale environmental events, such as El Niño events and hurricanes.

Climate Change and Coral Loss

Even without extreme El Niño events, there is concern that gradual global warming could cause large-scale and permanent loss of corals. Global warming could increase average and extreme water temperature in the tropics, with the potential loss of reefs. Although it is difficult for scientists to predict future water temperatures, a rise in global sea surface temperatures has been documented since the early 1970s (see Chapter 1). Some regions will be more vulnerable to global warming impacts than others, and the typical water temperatures in some areas are already approaching levels that can be lethal to corals. For example, in some areas of the Indian Ocean where corals were recovering from the 1998 El Niño, they have begun dying from temperature increases, apparently due to global warming.

If sea level rise due to increasing temperatures is rapid enough, water levels could increase over reefs more rapidly than the corals can grow (on average about 4 millimeters per year). As water levels increase exposure to sunlight decreases, eventually reaching levels at which the zooxanthellae in the coral cannot survive, at which point massive loss of reefs could result. The sea level relative to the reef also increases if the reef is eroded away following a bleaching event. These factors along with other anthropogenic impacts could result in a permanent loss of corals to some areas.

The loss of coral would not only affect the reef ecosystem, but it could also exaggerate the effect of seal level rise in coastal regions. The coral reefs serve to protect the coast from the impact of waves because they absorb much of the wave energy. As sea levels rise in the absence of coral, wave activity is more likely to increase erosion and damage buildings and roads near the shore. Sandy beaches are especially susceptible to these erosion effects (see Chapter 3); the horizontal loss of beach to an area can be over a hundred times as much as the vertical sea-level rise.

Another way that coral reefs could be threatened by global climate change is through an increase in tropical storms and hurricanes. Hurricanes can cause a local devastation of coral reefs; however, reefs are adapted to infrequent storm disturbances and will recover eventually if the habitat is otherwise healthy. Although there is some evidence for a recent increase in the frequency of large tropical storms, a consensus has not been reached as to whether this is actually a long-term trend.

The effects of increasing concentrations of atmospheric carbon dioxide are not limited to climate change (see Chapter 2); when the excess CO_2 dissolves in sea water it alters the chemical balance and lowers pH (makes the water more acidic). This can cause a reduction in calcification rates for corals (and other organisms with calcium carbonate shells, such as mollusks). Climate change and reef scientist Ove Hoegh-Guldberg and colleagues have shown experimentally that carbon dioxide levels predicted to occur over the next 50 years can measurably reduce calcification rates such that corals will become increasingly rare in tropical reef ecosystems.

A decreased rate of calcification has already been documented in some Great Barrier Reef corals. By looking at coral growth bands (**Figure 5-10**), Glenn De'ath and colleagues found that, from 1990 to 2008, calcification rates in the stony corals *Porites* declined by 14% and linear growth of corals decreased by 13%. These declines are greater than have been seen in the past 400 years; they could be due to changing temperatures and increased acidification as a result of carbon dioxide increases from human influences. If

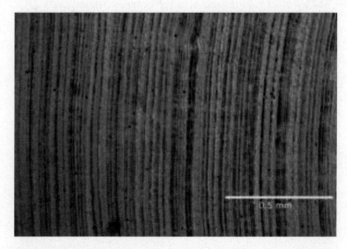

Figure 5-10 A cross section of black coral showing annual bands used to estimate age.

this trend continues, it is predicted that coral growth could stop altogether in the next 50 years. Even before the direct impacts of ocean acidification result in loss of coral, it could add enough stress to enhance coral bleaching or diseases.

■ Coral Diseases

Coral diseases are another cause of coral declines presumably enhanced by human impacts on ocean water quality. Diseases have been associated with increased temperatures, sedimentation, airborne pathogens, bacteria introduced by sewage, eutrophication, and pollution. More than 30 different coral diseases have been described, many documented only in the past three decades. Reasons for the increases in diseased corals are often unclear and difficult to pinpoint. It is difficult to know how much of the apparent increase is due to an actual increase in the number and prevalence of diseases, and how much is due to an increase in awareness and observations. Coral diseases are typically classified by their appearance; some of the best documented and most devastating are the white-band disease, white plague, white-pox disease, and black-band disease (**Figure 5-11**).

White-Band Disease

White-band disease, first reported in the 1970s and resulting from unknown causes, is commonly found in staghorn and elkhorn corals and can be recognized by degradation of the coral tissue along a band that moves toward the tip of the coral, exposing the coral skeleton. White-band disease is most prevalent in the Caribbean but has been documented in Indo-Pacific reefs, including Indonesia, the Philippines, and the Great Barrier Reef. Although the cause has not been definitively identified, bacteria have been found that are associated with the disease; however, their role is uncertain. There has been an increase in white-band disease in stressed corals. Losses of over 90% of staghorn corals have been documented in some regions.

White band has caused major changes in the coral communities in many Caribbean reefs, and some coral species have been driven to near extinction in the region. As with many changes in reef communities, the loss of staghorn coral has had unpredicted effects through complex species interactions. For example, William Precht and colleagues discovered that the death of the staghorn coral resulted in increased mortality of other coral species and greater algae coverage on reefs throughout the Caribbean through complex interactions of fish, coral, and algae. The fish species driving these changes is the threespot damselfish *Stegastes planifrons*. This fish establishes territories where it cultivates dense stands of algae on which it feeds. In order to establish these "gardens," the damselfish must first kill the corals by biting off the living polyps. On healthy reefs the damselfish limit their territories to regions around the staghorn coral. But when their preferred microhabitat in the branching structure of the staghorn coral is missing, the damselfish switch to other areas to establish their gardens. This reduces the area covered by other coral species and increases the area covered by algae. Although threespot damselfish populations may benefit, Precht's studies suggest that there is a high mortality of corals, such as the boulder star coral *Montastraea annularis*, along with the increase in algae covering the skeletal remnants of the coral. Other species associated with the corals are likely to be harmed as well, changing the composition of the reef community.

White Plague

White plague, caused by the bacterium *Aurantimonas coralicida*, from unknown sources, was first discovered in the Florida Keys in the 1970s. It is similar in appearance to white-band disease but affects different species of coral. Biologists have identified three different types of white plague, differentiated by differences in appearance and growth, affecting over 30 different species of corals, including some of the largest reef-building brain corals. One type can kill bands of coral tissue at rates as high as 2 centimeters per day, wiping out small colonies in just a few days. White plague is found throughout much of the Caribbean; for example, it is one of the major diseases affecting reefs of Puerto Rico.

Black-Band Disease

Black-band disease can be recognized as a blackish band up to 3 centimeters wide that moves up the coral, consuming the tissue at rates of up to 2 centimeters per day. It is

Figure 5-11 Brain coral with black-band disease.

apparently caused by a species of cyanobacteria along with several types of bacteria, although their functional roles are not known. Black-band disease was first discovered in Belize and Florida reefs in the early 1970s but has since been documented on many reefs around the world. In the western Atlantic and Caribbean it affects mainly the massive reef-building corals, and staghorn and elkhorn corals appear to be resistant. This is not the case in the Indo-Pacific, however, where it can affect over 25 different coral species. Black-band disease usually only causes partial mortality of a coral colony, but it does open up the affected coral for colonization by filamentous algae and other organisms. Its prevalence is often associated with corals that are stressed by high water temperatures or the input of sediments, nutrients, or toxins.

White-Pox Disease

White-pox disease is the first documented case of a bacterium associated with the human gut serving as a marine invertebrate pathogen. Kathryn Patterson and colleagues documented the association with fecal bacteria from human waste entering the Florida Keys ecosystem. White pox can cause lesions that kill coral at the rate of over two square centimeters per day. This disease has caused a dramatic change in the appearance of the coral reefs where it occurs. It is believed to be a major factor in the loss of more than 70% of the elkhorn coral in the Florida Keys reefs. The loss of staghorn and elkhorn corals in U.S. waters due to diseases and other factors has been so dramatic that they have been listed as threatened under the Endangered Species Act (see Chapter 9). This requires that critical habitat be defined for the coral and that, once it is defined, no activities authorized by the federal government are to harm that habitat.

Studies will continue to describe coral diseases and define their specific causes; however, reducing the impact of these diseases is going to require efforts to achieve an overall reduction of pollutants into tropical coastal areas. Other stresses, such as excess warming, also could make the corals more susceptible to these diseases. Reefs are likely to recover from the effects of diseases if and when this is achieved.

■ Excessive Nutrients and Organic Matter

Even when pollutants do not cause bleaching or disease in corals, their input into reef ecosystems can be harmful. Because coral reefs are **oligotrophic** systems, living in nutrient-poor waters, they are not adapted to tolerate the excessive input of nutrients by humans into coastal ecosystems. Intensive agriculture practices have resulted in a sixfold increase in fertilizer use over the past 50 years around the world, much of which ends up in rivers and eventually coastal waters. As well, the rapid rise in coastal populations in tropical regions often has been accompanied by poor sewage treatment. Together these result in a large input of dissolved nutrients—primarily nitrogen and phosphorous—and particulate organic matter (POM; see Chapter 2 for a discussion of marine nutrient cycles).

Moderate increases in nitrogen levels may increase the growth of zooxanthellae in some corals, but ironically this can reduce the growth rate of the coral itself. Because there are more zooxanthellae, their use of carbon dioxide is greater, which potentially limits the availability of carbon dioxide for calcification and growth of the coral skeleton. At higher concentrations of dissolved nitrogen, the effect on the corals is due to ecosystem changes as the water becomes more eutrophic (nutrient-enriched). There is an increased production of phytoplankton, increase in growth of attached macroalgae, and increased growth of herbivores such as sea urchins (**Figure 5-12**). This can change the balance of the reef ecosystem as algae begin to replace coral and water clarity declines. The system changes from one dominate by nutrient-recycling symbiotic organisms (the coral with their zooxanthellae) to one dominated by macroalgae and organisms that feed on that algae. This is referred to as a **"bottom-up" effect**. The input of nutrients has a direct impact on organisms at the bottom of food chains, but the effect moves up the food chains to affect the entire food web.

Excess POM, for example from sewage input, also can affect corals and the reef ecosystem. A moderate increase in POM can increase growth of types of coral that are able to feed on particulate organic matter; however, this growth may be accompanied by a more fragile skeletal structure in branching corals, resulting in more breakage

Figure 5-12 A coral head being overgrown with algae.

during turbulent conditions. Clear-water species may be outcompeted by more tolerant species as suspended POM increases. The result is a change in the species composition of the reef, often resulting in lower overall biodiversity. At higher levels of POM, the costs outweigh the benefits to the reef ecosystem as a result of decreases in light penetration. In addition, excess suspended POM is usually associated with increases in dissolved nutrients, leading to addition stresses from eutrophication, as described above.

Sedimentation

Many coral reefs have been exposed to unusual levels of sediment from human actions such as deforestation and poor agriculture practices at least since the 1980s, and possibly much longer. Sediments, primarily originating from river input, end up in the waters around coral reefs. The greatest effect is near river mouths, as greater than 95% of the larger sediment particles settle out within a few kilometers and fine sediment particles can be distributed much farther. This increase in turbidity results in decreased photosynthesis by the zooxanthellae and increased polyps incur increased metabolic costs because they must expend energy to remove sediments during feeding. In extreme situations these sediments can smother bottom organisms; for example, coral tissue exposed to high levels of sediments can be killed within just a few days. The degree of impact depends on where the corals live, and different species of coral are affected differently by sedimentation. The effect is greater in sheltered lagoons or bays, where sedimentation rates are greater. Branching corals do better than massive corals with flat surfaces; and corals in deeper waters, below about ten meters, can be affected more by the reduced light penetration. Even moderate levels of sedimentation thus can change the species composition of the reef, lowering diversity and extent of coverage by corals.

In the Indo-Pacific and Caribbean, the decline in coral reefs associated with increased sedimentation from deforestation, agriculture, and land clearing has been well documented, and still continues in many areas today. Some of the hardest hit regions in recent years are in southeastern Asia and the Philippines. In some areas where reefs have been missing for decades, their absence has been taken for granted and assumed to be the norm, but their historical presence is evidenced by large amounts of rubble and other remnants of dead coral in the area (**Figure 5-13**).

Toxic Pollutants

Although nutrients and sediments are considered to be the most widespread pollutants of coral reefs on a large scale, other toxic materials such as metals, hydrocarbons, and pesticides can cause serious harm to coral reefs on a local

Figure 5-13 Reef area in the Caribbean covered with coral rubble.

scale. For example, metals such as copper and zinc have been shown to reduce reproduction and growth of coral. Herbicides from agricultural runoff are known to stress corals, although the long-term effects are not known. Corals can incorporate many types of water-borne pollutants into their tissues, but the long-term impacts are not well documented.

The effects of exposure to hydrocarbons from oil spills have been more thoroughly studied. Effects vary depending on the coral species, the level of exposure, and the chemical composition of the oil. Corals can be killed by short exposure to high concentrations or long-term exposure to lower concentrations. Chronic oil exposure can affect reproduction, feeding, growth, and development of corals; oil also can bioaccumulate in zooxanthellae and be deposited in the coral skeletons. Branching corals are more sensitive to oil exposure than massive corals. During an oil spill, exposure to the oil varies because some oils remain largely floating at the surface while other chemicals sink.

Likely the most harmful to tropical coral reefs was the 1986 Bahia las Minas crude oil spill in Panama, which released about 100,000 barrels (about 15 million liters) of crude oil from a ruptured storage tank. Coral reefs in this region had already been harmed by over a century of excavation, dredging, landfill, deforestation, and erosion; however, some relatively healthy reefs remained that had been surveyed prior to the spill. Researchers Jeremy Jackson and colleagues found that three months after the spill one heavily damaged reef had lost over 75% of its cover in reefs shallower than 3 meters and over 50% in deeper reefs; elkhorn coral totally disappeared. Cover decreased less on moderately impacted reefs, but growth was significantly lower in surviving corals that had been exposed to the spill. Studies have shown that recovery of these reefs from the oil spill could be achieved in a decade. The spill made these reefs more vulnerable to subsequent natural and anthropogenic impacts, however, limiting their recovery. Ten years after the Bahia las Minas spill, oil seeping from a refinery landfill and mangrove sediments continued to be common in the area. U.S. NOAA scientists estimate that it could take a century for these reefs to recover.

Combined Impacts

It is often difficult to attribute coral reef decline to a single factor because multiple impacts often happen concurrently. For example, nutrients, organic matter, and sediments may originate from a single source, and it is hard to piece out which of these factors are the most deleterious. Another complication is that corals stressed due to pollution are more likely to be affected by environmental events like ocean warming or storms. Care must be taken not to assume that a strong correlation between a single factor and coral decline always reflects a cause-and-effect relationship.

Reef ecologists have established a long list of effects that may be due to multiple factors. For example, reproduction and larval survival in coral is dependent on good water quality. The types of inputs discussed above can affect the production or fertilization of eggs, settlement and survival of larvae, and/or growth and survival of juveniles. Because coral reproduction and recruitment of young tend to be more sensitive to pollutants, the adult corals may survive the pollutants. The effects are not noticed until corals are unable to re-establish after a physical disturbance.

Katharina Fabricius has pointed out how pollution factors can harm organisms other than the coral that are critical components of the reef ecosystem, resulting in an indirect effect on the corals. For example, in areas exposed to sewage input there may be an increase in organisms that burrow into and erode the reef skeleton (**bioeroders**), such as the boring sponge and some bivalves. This can harm the integrity of the reef and make it more sensitive to wave action during storms. In extreme situations the rate of erosion can be greater than the rate of calcification, thus inhibiting coral growth.

Excess growth of macroalgae on the reefs might be caused by a number of ecological interactions, such as grazing, discussed below, and in a given situation it may be difficult to discern the most important factor. There is also a direct effect of excess nutrients on algae growth and coral decline, however. Macroalgae are not as tolerant of low-nutrient conditions in which corals thrive and, on a healthy reef, corals dominate because the zooxanthellae and feeding polyps are usually more competitive in using the limited nutrients that are available. Excess nutrients encourage the growth of large colonies of macroalgae that may inhibit the growth and survival of corals. Large dense mats of algae can overgrowth corals and inhibit access to oxygenated waters or shade them from adequate sunlight. When the algae die they can create eutrophic conditions, further limiting coral survival. A review by Renaud Fichez and colleagues points out numerous situations where nutrient input has being strongly correlated with increases in macroalgae and declines in coral. For example, in an area of the Red Sea a 50% increase in nutrients from 1990 to 2004 was linked to a dramatic increase in algae and a 50% reduction in coral cover. A reef in the Florida Keys showed an increase in nitrogen and phosphorous (but not siltation) after an increase in agriculture runoff from the Everglades in the early 1990s. This was followed by an increase in algae cover and a decline in coral cover on the reef, along with a doubling in the relative abundance of herbivorous fishes.

Even when herbivorous fish are present they may avoid large dense stands of macroalgae, resulting in increased algae growth and persistence. Andrew Hoey and David Bellwood discovered that herbivorous fishes avoided dense stands of the large fleshy *Sargasum* algae, preferring to feed in more open areas with more dispersed algae cover, possibly because it limits their view of predators or obstructs movements. These studies suggest that, when grazing fish populations are low, seaweed can reach a critical density beyond which grazers can control seaweed growth, further enhancing the seaweeds' ability to dominate areas of the reef.

Corals and filter-feeding organisms such as sponges and bivalves are often the dominant bottom species on the reef, and their relative abundances can vary for a number of reasons. In some reefs there is an apparent correlation between low coral abundance and a high abundance of filter-feeding organisms. Ecological and physiological studies have shown that this is probably not due to competitive interactions. The coral are most likely absent from these areas due to other factors that do not harm the filter feeders. For example, inputs of organic pollutants may enhance growth of filter feeders due to increased availability of organic matter, while inhibiting growth of corals due to low light penetration. Another possible indirect effect of terrestrial inputs is an increase in coral predators. Outbreaks of crown-of-thorn starfish appear to be linked to increases in terrestrial runoff, possibly because of increases in nutrients that support plankton that feed the starfish.

The vulnerability of reefs to the combined effects of the environmental events will vary depending on a number of factors that may be hard to pinpoint. Generally, however, the most vulnerable coral reefs appear to be those closest to sources of terrestrial runoff: those that are in shallow bays or lagoons with poor water circulation or low wave activity to flush out warmer waters or pollutants, those in deeper waters where sunlight penetration is critical, and those surrounded by shallow sea floor where water movements can keep resuspending sediments. Many reefs appear to be resilient to natural impacts that have always occurred unpredictably on the reefs; however, it remains to be seen whether reefs in regions prone to extreme conditions, such as tropical storms or El Niño events, can recover when exposed to additional anthropogenic stresses.

The effect of multiple impacts is demonstrated in coral reefs in the French Polynesian area of the Pacific. Since 1980, these reefs have undergone seven separate bleaching events, five separate cyclones, and one extensive outbreak of the crown-of-thorns starfish *Acanthaster*. A second outbreak of *Acanthaster,* beginning in 2006, had nearly wiped out the living coral on many of the reefs; however, much of the coral structure remained in place. Then, in 2010, tropical cyclone Oli hit the already-stressed French Polynesian reefs, reducing much of the coral structure to rubble and dramatically affecting invertebrate and fish communities. There are concerns that the degree of destruction, along with continued exposure to pollutants, could inhibit recovery indefinitely, leading to a replacement of the diverse reef ecosystem with one dominated by benthic algae. It is likely to be at least a decade before the future of the ecosystem will be known. If a new reef forms, it will likely be with a new assemblage of corals supporting a different community.

5.4 Harvest Impacts

People living in tropical coastal regions have depended on organisms taken from the reefs for food or as a source of income for as long as there have been substantial human populations residing there (**Box 5-2. Research Brief: Historical Ecology of Reefs**). It is idealistic to think that we can immediately stop all fisheries and other kinds of harvest. Scientists and managers, however, are working with user groups to come up with ways to minimize this harvest to make it sustainable and to establish alternative conservation-friendly employment for fishers.

Sea Urchin Harvest

Fisheries harvest from coral reefs includes more than catching fish. Many invertebrates found around the reefs, including lobsters, conchs, and sea urchins, support valuable fisheries. Sea urchins are harvested mainly for the gonads, referred to as **roe**, or *uni* when sold in Asian markets and

restaurants. A common method of harvest is by divers in areas around and adjacent to reefs, including seagrass beds. Urchins can be vulnerable to overharvest because of the ease of collection; they are highly visible and move slowly, depending on their long, sharp spines for protection. In some regions where there are no local restrictions or enforcement, sea urchins have been driven to near extinction. Much of the current harvest of sea urchins is not from tropical waters, but in cooler waters dominated by kelps, discussed in Chapter 6.

Organisms as Curios

The demand for tourism curios, such as shells, sea stars, sand dollars, dried sea horses, and dead coral, has affected populations around some reefs, especially where harvest is not properly regulated (**Figure 5-14**). It is estimated that over 5,000 species of mollusks (primarily as ornamental seashells) plus unknown numbers of corals, echinoderms (sea urchins, sand dollars, and sea stars), sponges, and sea fans have been harvested as curios, many of them around coral reefs. Although shells are often collected from organisms that have died from natural causes, for commercial harvest most individuals are taken alive to obtain them in the best condition. Fishes taken from reefs as curios and sold dead and dried include seahorses, triggerfishes, and porcupinefishes.

Much of the harvest of animals for the curio market is from developing countries, in particular from regions of the west Pacific. Indonesia is possibly their greatest source, and the United States is the greatest importer. Precise estimates of the number of animals collected and sold as curios is unknown because, when reported, they are not typically tracked by species. By most estimates, however, tens-of-thousands of individuals are collected per year from many of the invertebrate groups given above. The overall effect on the populations and ecosystems is not known because their harvest is not thoroughly monitored or regulated. Although no species is documented to be threatened with total extinction due to collection for curios, populations are often severely reduced where collections occur, and some species that are imported into the United States are considered threatened.

Traditional Medicine

Another market that supports the taking of many tropical reef species is traditional medicines (defined as medical prescriptions rooted in ancient indigenous healing practices). There are believed to be hundreds of marine species harvested for a myriad of uses in traditional medicine. For example, seahorses (Syngnathidae) are purported (but not documented scientifically) to be useful in treating ailments

Box 5-2   Research Brief: Historical Ecology of Reefs

Scientists typically are hesitant to draw conclusions from historical information considered to be anecdotal or that is not validated through scientific studies. This presents a dilemma, however, when such information is all that is available. This is especially true when anthropogenic impacts occurred long before scientific studies were commonplace. Some scientists and historians have moved beyond their hesitancy to develop a field of study called *historical ecology* that looks at the historic interaction between humans and the environment. Although historical accounts not obtained by trained scientists might be used by historical ecologists, the validity of such accounts is taken into consideration. For example, the source is considered and multiple concurring accounts from different sources are considered more valid.

Jeremy Jackson of the Center for Tropical Paleoecology and Archaeology of the Smithsonian Tropical Research Institute has argued that coral reef ecologists should not be turning their backs on history. He points out that ecologists have often defined *healthy* as the conditions observed when scientists first began to study reefs. He argues that greater efforts should be made—using historical evidence—to determine what reefs really were like without human effects. Defining this so-called *pristine* condition has been very elusive to ecologists not just for reefs but for many ecosystems.

Jackson took a historical ecology approach to assess what effects humans might have had on coral reefs in the Caribbean. He considered paleontological evidence of the abundance and distribution of reef organisms as well as historical accounts from journals, reports, and other publications going back as far as the journals from Columbus' voyages in the 1490s. He uses accounts from independent observers in an attempt to piece together what a pristine coral reef would look like. For example, evidence from the 1700s through early 1900s indicates that under pristine conditions there were great abundances of the long-spined sea urchin *Diadema*. This suggests that large sea urchin populations are the natural condition, and that overfishing of predators and grazers was not necessarily the reason for the large sea urchin populations on some reefs, as assumed by many ecologists (see discussion above).

There is evidence that humans were directly affecting coral reef ecosystems even in pre-Columbian times. For example, human populations on the large islands in the Caribbean (Hispaniola, Jamaica, and Cuba) were probably in the hundreds-of-thousands. The coastal waters around these islands appear to be unable to support large populations of local fishers, even when they are fishing using simple methods like canoes and handlines. Overfishing might have resulted in the loss of large fishes around these islands for centuries. There is strong evidence that humans had depleted the Caribbean reefs and coastal ecosystems of many large animals prior to 1800, long before ecologists began studying reefs, and this probably had caused substantial changes to coral reef ecosystems. For example, manatees and sea turtles eat large amounts of sea grasses and disturb the bottom, hawksbill turtles tear sponges from the bottom (**Figure B5-3**), large parrotfishes break up large pieces of coral, and large predatory fish affect the composition of the prey community. The elimination or great decline in these animals probably had profound impacts on the reef ecosystem.

Although arguments based on evidence without scientific documentation are not without detractors, even early in recorded history we know that indirect, but unobserved, effects were occurring long before ecologists knew what the reef ecosystem looked like. Deforestation, and the associated siltation, was occurring on Caribbean Islands as early as the 1600s as sugar cane plantations were being established by European settlers. What has been considered a healthy ecosystem along these coastlines based on recent scientific observations therefore actually may be very different from pristine pre-human conditions.

Jackson argues that we cannot continue to ignore historical evidence and focus only on what we see here and now; we are kidding ourselves if we think it is possible to have a sustainable use of the reefs for everyone, and what we have considered pristine coral reef ecosystems are in reality substantially affected by humans. These conclusions do not mean we should give up on conservation of reefs, however. Re-evaluations may be needed, and some tough decisions made to limit effects of terrestrial activities and harvesting and work to recover damaged reefs. One recommendation is to establish more very large ("hundreds to thousands of square kilometers") marine protected areas. We have lost few species to total extinction, so if we act now, we can save coral reef ecosystems and at least have some areas that approach a pristine condition.

Figure B5-3   Decline of species such as this hawksbill sea turtle that feed on the reef have undefined impacts on the entire reef ecosystem.

(a)

(b)

Figure 5-14 Corals such as **(a)** gorgonian sea fans and fishes such as **(b)** seahorses are sometimes collected from reefs to be sold as curios.

such as asthma, impotence, arteriosclerosis, thyroid disorders, and heart disease, among others. Many of these organisms are taken from coral reefs because of the great species diversity as well as the accessibility and nearness to profitable Asian markets. Numbers on the volume of harvest are not known, but this type of exploitation is possibly the most severe for seahorses, collected mainly in the Indo-Pacific and used locally or exported to countries in eastern Asia. The seahorse harvest and trade is poorly monitored, but knowledgeable estimates are that tens-of-millions of

seahorses were captured in some years during the 1990s for the traditional medicine trade. Their value provides a strong incentive for harvest as some seahorses sell for hundreds of U.S. dollars per pound in Hong Kong. Many species of seahorses that are collected for traditional medicine (as well as for the aquarium trade) are now considered endangered with extinction, and trade is restricted for some species through international agreements (e.g., Convention on International Trade in Endangered Species [CITES]; see Chapter 9). Unfortunately, the pressure on marine species for traditional medicine probably will continue to increase due to its great foothold in Asian countries and its recent increase in popularity in other regions of the world.

Bioprospecting on Reefs

Another non-food use of harvested coral reef organisms is **bioprospecting**, the scientific search for chemical compounds, primarily for pharmaceutical products. Bioprospecting is not limited to coral reef organisms; however, tropical reefs are a popular source of these chemicals, in part because of the accessibility of a great diversity of species. Also, the compounds reef organisms produce to avoid predation or for competitive interactions are considered good prospects for potential drug use.

It is difficult to determine the number of species that have been collected for bioprospecting because there are few reporting requirements and a desire for confidentiality within the pharmaceutical industry. Information from sources such as the U.S. Cancer Institute, however, suggests that chemicals from tens-of-thousands of species have been collected. Most species considered for bioprospecting are echinoderms (e.g., sea stars, sea urchins, and sea cucumbers), urochordates (tunicates), cnidarians (e.g., jellyfishes), and poriferans (sponges). Sponges are the most common source, and probably comprise more than half of the species collected. The pharmaceutical industry already has extracted many promising chemicals from sponges, some being considered for potential cancer and asthma treatment. Most of the chemicals end up in the United States, Japan, or Europe.

The amount of material collected from one species for initial screening is small, less than 1 kilogram and, therefore, conservation concerns are unlikely for the bulk of the species collected. Follow-up collections of promising compounds can take hundreds to thousands of kilograms of material, however, and currently the greatest concern is that collections could cause a decline in local populations. It is unknown if any organisms with bioprospecting potential are species that are endangered due to other factors; if so, they already could be listed as endangered by International Union for the Conservation of Nature (IUCN) for other reasons, leading to trade restrictions (see Chapter 9).

Sponge Harvest

Although it includes a very small fraction of species, the most familiar market for sponges is for the bath sponge industry. Most materials currently marketed as "sponges" are synthetically produced, but there continues to be a limited market for natural sponges, some of which are harvested commercially from coral reef regions. The primary traditional source of sponges was the Mediterranean Sea, but a substantial commercial sponge harvest developed in the Caribbean and off southern Florida in the late 1880s through early 1900s (**Figure 5-15**). In fact, sponges were the most profitable fishery in the Florida Keys until numbers declined from overfishing in the early 1900s, a disease in the 1930s, and a red-tide algae bloom in the 1940s. When sponge populations recovered, the market was depressed from the introduction of synthetic sponges and competition from the Mediterranean. This is reflected in a decline in Florida sponge harvest from about 270 thousand kilograms per year in the 1930s to about 27 thousand kilograms per year in the 2000s. Currently, sponges are harvested from Florida waters primarily for specialty bath shops and as paint applicators.

Despite reductions in the sponge fishery, there is still concern about excess harvest. In order to avoid overharvest, in some regions sponges must be cut from the seafloor by divers. This allows the sponges to regrow. Traditional methods of pulling the sponges from the bottom with hooks from a boat reduce the chance of regrowth. This method also tends to disturb the bottom, increasing turbidity. Commercial sponges are only about 2% of the total sponge biomass in the Florida Keys and, therefore, their harvest probably has not caused substantial harm to the reef ecosystem.

Conch Harvest

Conch, the large spiral-shelled mollusk of the genus *Strombus* or *Cassis*, are harvested from shallow waters, often in the vicinity of tropical reef ecosystems, to support important local fisheries (**Figure 5-16**). Conchs typically supply a local market for food; however, the shells from this harvest are sometimes sold as curios. The fisheries are typically small scale, but conchs can be important economically as well as for export. Most often they are harvested by divers from small boats because the conchs live in shallow waters and

(a)

(b)

Figure 5-15 Sponges supported a large commercial industry in southern Florida through the early 1900s: **(a)** sponge auction wharf in Key West, Florida in 1902. Currently, tourist and specialty shops support a small market for natural sponges; **(b)** sponges sold at a tourist shop in Tarpon Springs.

Figure 5-16 Conch support local fisheries throughout the Caribbean. Here, a Turks and Caicos fisher extracts conch meat from the shells.

are visible from the surface. Even a small-scale fishery's take can lead to overharvest.

Queen conchs (*Strombus gigas*) have been harvested in many areas of the Caribbean and Gulf of Mexico for thousands of years. Even in pre-Columbian times it appears there was local overharvest in some regions of the Caribbean, as evidenced by a decline in shell size found in middens. In recent times, the queen conch population in the Florida Keys was overharvested by the 1970s, resulting in a total ban on harvest in the 1980s. Although some populations in the Caribbean and Gulf of Mexico are healthy enough for a continued harvest, declines throughout their range have resulted in an Appendix II listing by CITES, meaning trade is restricted to regions where harvest is documented to be sustainable (see Chapter 9). Each country in this region has harvest restrictions, including minimum sizes, prohibiting the use of scuba, harvest seasons, closed areas, and harvest quotas. Poor enforcement of regulations has led to unrestricted harvest and poaching in many regions of the Caribbean, however, resulting in continued population declines.

Lobster Harvest

Although, globally, the commercial market for lobsters is dominated by species from temperate waters away from reef ecosystems, tropical lobster harvest can support one of the most valuable fisheries on reefs in some regions. The most commonly harvested reef species are the spiny lobsters (Palinuridae; **Figure 5-17**). The obvious difference between spiny lobsters and clawed lobsters (Nephropidae) is that spiny lobsters have large antennae and do not have large claws. They can be readily taken by hand by divers from crevices in the reefs or by using traps. For example, the

Figure 5-17 Spiny lobsters are a valued species harvested from coral reefs. A Caribbean spiny lobster *Panulirus argus* captured in a trap.

Caribbean spiny lobster *Panulirus argus* supports a valuable fishery in coral reef habitats of the Caribbean and Gulf of Mexico. Regulations governing harvest vary throughout the lobsters' range, but in many areas there are seasons and restrictions.

One of the more strictly managed fisheries for the spiny lobster is in Florida where it was one of the most valuable in the state in the last half of the 1900s. Florida's annual harvest has varied from a high of over 2,200 kilograms in the 1970s to under 900 kilograms in the 2000s. Regulations include seasons of recreational harvest (by diving) and commercial harvest (with traps), restrictions on harvest methods, minimum sizes, and harvest quotas. Overharvest and loss of habitat appear to be the major factors affecting spiny lobster populations, however, and monitoring and regulating fisheries continue to be difficult.

Destructive Fishing Methods

Fishes of all sizes and species are harvested from many tropical reefs for local consumption or to be sold on commercial markets. Larger predatory fish tend to be the most valuable, but almost any species is a potential food item in tropical regions where subsistence harvest is common. Overharvest is always a potential concern. Coral reefs are particularly sensitive to destructive fishing methods and, when fishing methods result in destruction of critical habitat, the fishes may never have a chance to recover. Degradation of reef ecosystems can result in severe economic losses to communities and nations dependent on these resources.

Hook-and-Line and Traps

To minimize physical harm to the organisms that form the structure of the reef, including corals, sponges, and other stationary organisms, fishing on reefs is often restricted to either **hook-and-line** or **trap fishing**. Hooks can be lowered into the water without contacting the reefs, and traps can be set gently onto the bottom near the reef with minimal damage (**Figure 5-18**). These methods also can allow for the release of non-target species if they are handled carefully and not allowed to remain on the hook or in the trap too long before release.

Trap fisheries, as with other methods, can lead to overharvest if not properly regulated. Traps, can be particularly harmful to fish populations and other organisms if they are lost (e.g., during a storm) and not retrieved. These so-called **derelict traps** continue catching and killing organisms and potentially harm the reef itself as they are carried by water movements across the reef. To avoid catching organisms after being lost, degradable escape panels are often required on traps.

Figure 5-18 Fish traps are commonly placed on flat bottoms next to the reef to attract fish. A fish trap used in waters off Jamaica.

Net Fishing

Other methods that are more destructive than hook-and-line or traps are usually outlawed or restricted around coral reefs. If fish can be harvested more easily using these methods and restrictive laws or enforcement are not in place, however, some fishers will continue to use them. Most types of net fishing are potentially destructive when used around the reefs. **Trawling**—pulling a net over the reef to capture fish (see Chapter 11)—can be very destructive to the reef and its associated organisms. It damages the coral reef directly and catches and kills large numbers of all kinds of organisms unintentionally. It may take decades for re-growth of the coral after this type of damage, and as long as trawling continues the reef will never be allowed to re-grow. Trawling in tropical coral reef habitats is illegal in most regions of the world, but restrictions are not always enforced and, if trawling is allowed in areas adjacent to the reef, it will result in destroying patches of coral and other habitats.

Blast Fishing

The most destructive method of harvesting fish from coral reefs is **blast fishing**. This method involves throwing dynamite or homemade explosives onto the reef. The blast from the explosion kills or stuns the fish, which then are collected from the surface. Damage to the reef can be long term because the coral in the blast zone is pulverized by the explosion (**Figure 5-19**). The coral needs a stable substrate on which to form, which the coral rubble may not provide. If the habitat is protected after blast fishing, coral may begin recovery in as little as five years in areas where blast craters are dispersed; however, extensively blasted areas may not recover as long as there is no stable substrate. Blast fishing is so destructive that it is not considered legal in any regions; however, enforcement and education are necessary for the total elimination of this method. For example, despite laws

Figure 5-19 Blast fishing with explosives, although illegal, continues to destroy reefs in some regions. Remnants of blast fishing in Indonesia.

making blast fishing illegal, it has been especially problematic and difficult to control in Indonesia, where blast fishing has been widespread since World War II, and off of coastal Africa, where it is a more recent development.

Cyanide Fishing

Another destructive method of fishing has been problematic around some coral reefs of Southeast Asia—particularly in the Philippines—since the 1960s. The use of cyanide makes capturing fish easier but at the risk of causing long-term harm to the coral reef ecosystem. Cyanide is used to stun or kill the fish, which are then easily taken without having to invest in nets, hooks, or traps. Cyanide tablets are crushed into plastic squirt bottles and mixed with seawater. Divers, usually without scuba, dive down and squirt this solution at fish around the coral. Although the cyanide can be lethal to fish, at the concentrations used about 50% survive, but are stunned and can be easily taken before being allowed to recover in a tank and sold. To avoid the cyanide, some fishes escape into crevices in the reef and the divers may destroy the reef heads to get to these fish. Fish collected by cyanide fishing end up mostly in Asian, European, or North American aquarium shops.

Even if the cyanide is not immediately lethal to the fish caught this way, it can affect their health and result in eventual mortality. Cyanide causes impairment to enzyme systems and damage to internal organs, contributing to up to 80% delayed mortality of wild-caught reef fish. This exacerbates the problem because it provides an incentive to divers to return and capture more fish to make up for this loss. Not only are the fish affected, but the corals and other reef organisms are also harmed by exposure to the cyanide. It can cause loss of zooxanthellae by the coral, bleaching, and possibly death.

Much effort has been put into eliminating cyanide fishing, in particular in the western Pacific around southeastern Asia where it is most prevalent. Community-based management has been recommended as the best solution, which can lead to self-enforcement of regulations by locals. Several non-governmental organizations (NGOs) have participated in education campaigns, providing training to fishers in alternative collection methods such as hook-and-line capture for food fish and net collection for live fish (see below). Still, collectors are slow to switch from cyanide without some incentive if it is a more profitable method. NGOs are working with governments, for example in the Philippines, to establish fish inspections for cyanide residues, and certifications that fish are cyanide-free. A global certification system would help to eliminating cyanide fishing by enabling enforcement of laws regarding the use of cyanide but also eliminating the market for cyanide-caught fish. Distributors and consumers would be less likely to purchase a fish if they knew it was poisoned by cyanide. Fish buyers would be able to pay more for net-caught fish or refuse to purchase cyanide-caught fish, providing an incentive to the fishers to change capture methods. This sort of system is difficult to implement on a large scale, however, because once the fish leave their capture location they are difficult to track or label. The U.S. government is working on developing quick, inexpensive methods to identify fish that have been caught using cyanide; however, this is difficult due to the rapid decomposition of cyanide in the fish. The most efficient way of eliminating cyanide fishing through enforcement would be to focus on the middlemen, the people who provide cyanide to the fishers. But even with laws in place, enforcement is difficult due to the small-scale widespread nature of the fishery.

Spear Fishing

Another efficient way to harvest fish from coral reefs for consumption is by spear fishing. Especially since the widespread use of scuba, divers can easily approach fish within the range of a spear gun. Although spears can damage the reef, this is not as destructive to the corals and non-targeted organisms as other fishing methods such as cyanide or trawl fishing. Because the target species can be picked out easily, this method eliminates unintended incidental catch common with net or trap fishing. Unless strict regulations are enforced, however, fish can be vulnerable to overharvest because of the ease to select certain species and sizes of fish (**Figure 5-20**). Large groupers, for example, depend on their size and camouflage for protection and are thus not always wary of potential predators, including humans. In many coral reef habitats, large predators have now become a rarity because of selective harvest of the largest fish. Targeting the largest fish also can affect the population makeup by removing the fittest individuals or those capable of the greatest egg production. Another less apparent effect is on the sex ratio of the population. Groupers, for example, are protogynous hermaphrodites (they change sex from female to male at some point in their life; see Chapter 2), resulting in all of the largest individuals being males. Targeting the largest fish thus removes males from the population preferentially and could dramatically reduce reproductive potential for the population.

Live Fish Markets

Some food fish are so valuable alive that it is worth the expense and effort to capture them in a manner that does not kill or harm them substantially. There are two types of markets for these fish, the aquarium market and a specialized restaurant market.

Figure 5-20   Spearfishers are able to target the largest or most marketable fish. Spearfishers with a catch of parrotfish in Maui, Hawaii.

The Aquarium Trade

The largest market for live reef fish is the aquarium market. Over one thousand marine fish species are purchased for display in aquariums around the world (**Figure 5-21**). Most tropical marine fish that are bought in the marine aquarium trade are caught from the wild due to the difficulty of rearing most marine fish species in tanks. Because many marine fish species release planktonic eggs that develop into larvae that float in the currents until ready to settle on a new reef, it is hard to duplicate an environment conducive to their survival. It is easier and cheaper to simply capture the fish from the wild. The aquarium trade also includes harvested live coral (so-called "**live rock**") and reef invertebrates (e.g., shrimp, clams, and anemones). Over 90% of the collections are from the western Pacific, mostly by small-scale harvesters from developing countries; for example, there are over 2,000 collectors in the Philippines. Most collection is by simple methods, usually by using small nets (including cyanide fishing, discussed earlier). Unfortunately, even with nets, the mortality of fish collected from reefs is high because of poor handling, maintenance, and transport techniques; this leads to the need for excessive collections to supply the market. The Aquarium Council in conjunction with NGOs (one of these is the Marine Aquarium Council Certification Program) are making efforts to establish procedures and guidelines for the collection and transport of fish and invertebrates collected from the reef.

Most of the live coral export comes from the western Pacific as well. Much came from the Philippines until laws banning it were enforced in the 1980s; however, this only transferred the market to other countries, such as Indonesia. Although the United States bans collection of corals from most of its reefs, since the 1990s imports have

increased by over 10% per year. The United States now imports about 80% of the live coral sold, hundreds-of-thousands of pieces per year. As much as 90% of the coral of some of the most popular collected species are dead before or soon after they reach the aquarium owner. Other marine invertebrates, millions of individuals per year, have been taken and exported, primarily to the United States. Although the proportion coming from coral reefs is unknown, it is presumed to be large.

The overall effect of collecting organisms for the aquarium trade is not known. No species has been documented as endangered with extinction by the trade, and some countries (e.g., Australia) appear to have developed a sustained harvest. Local depletions of certain species are common, however, especially where regulations are inadequate or unenforced and harvest is poorly monitored. For example, Allson Jones and colleagues discovered that popularization of clownfish and anemones for the aquarium trade following the release of the Disney animated movie *Finding Nemo* led to reduced clownfish populations where collections were allowed. Clownfish in protected areas of the Great Barrier Reef were up to 25 times as abundant, likely as a result of reduced collection and less coral bleaching.

The Restaurant Market

Some restaurants also provide markets for live fish. The fish are put in a large aquarium outside the restaurant to attract customers who choose the desired fish for their meal. Live fish can bring a premium price due to the traditional value in markets in East Asia, where the freshest fish are highly valued. It is most common in Hong Kong and mainland China, but the market has recently expanded into other areas of Asia and regions where there is a large ethnic Chinese population. The most popular species are various groupers, snappers, and wrasses. Value varies widely by species; some of the more valuable species, such as the humphead wrasse *Cheilinus undulatus,* can bring over 80 U.S. dollars per pound. Most of the live food fish are taken directly from the ocean and transferred to restaurants, others are captured as juveniles and grown to a larger size before marketing, and others are raised from eggs in hatcheries. Most of the fish come from western Pacific coral reefs.

This type of harvest presents several conservation problems. Harvest methods can cause habitat impacts, especially if cyanide fishing is used. (The fish are held in tanks before being sold to allow the cyanide in the fish to decompose enough that the fish is not harmful for consumption.) The most valuable fish for this market tend to be the larger fish; the effects on the population structure would be similar to those described above for spearfishing. Mortalities, which can be greater than 50% before the fish reach the market, increase the incentive for more

Figure 5-21 Anemones and clownfish, popular marine aquarium fish following the success of the movie *Finding Nemo.*

harvest. Harvest for the live food trade has led to the consideration of some species (e.g., the humphead wrasse) for endangered status, which would lead to trade restrictions. Unfortunately, whenever there are fisheries of such large values (in Indonesia, fishers harvesting live fish can make up to 10 times the profit of other fishers) to supply a large and growing market (tens-of-thousands of tons of fish per year are estimated to be harvested for this market), there is tremendous pressure to continue harvesting, even as new restrictive laws are passed.

Ecosystem Impacts of Fish Harvest

The grazing and predatory fish community on coral reefs is comprised largely of fishes likely to be targeted for harvest in regions where they are not protected. Overharvest can lead not only to the loss of individual species but also to an imbalance in the ecosystem. As described earlier, overharvest of grazing fish species can result in an increase in algae on the reef and a concurrent loss of the coral. Alternatively, when large predators are removed from the ecosystem, smaller grazing species can increase, resulting in a potential loss of coral. This is referred to as a **"top-down" effect**, because the effects of the loss of a top-level predator cascade down the food chains to affect the rest of the ecosystem.

The impacts of fishing on large predators have become apparent in many tropical reef ecosystems. Because few large reefs are totally protected from fishing and many have been fished for hundreds if not thousands of years, it has been difficult to establish baseline data concerning what the biomass of predators should be on a pristine reef. Relatively few marine protected areas have been in place long enough to provide a baseline as to what a pristine reef should look like (see Box 5-3, Research Brief: Contrasting Fish Communities on Hawaiian Reefs).

Spawning aggregations have become a recent focus for protection of large reef fish species. Many reef fisheries species, such as snappers and groupers, form large aggregations during reproduction. Fish can migrate from a broad home range and concentrate for days or weeks each year in areas easily located by fishers, where they are vulnerable to intensive fishing pressure. The availability of modern technologies such as sonar and global positioning systems (GPS) has increased the ease with which fishers can locate these areas. Fishes that form spawning aggregations tend to be large, have a late age at maturity, and a long lifespan. These life history factors tend to make them vulnerable to overexploitation, possibly leading to local extinctions. There have been large population declines in fish species that are predictable in their spawning timing and location, and some migrate for hundreds of miles to form spawning aggregations. For example, Stephania K. Bolden found that the Nassau grouper *Epinephelus striatus* can migrate over 640 km to aggregate at spawning sites in the Bahamas; this species has undergone dramatic declines throughout much of the Caribbean, in some cases greater than 80% in 10 years or less. This has led to endangered and overfished classifications that protect the species from harvest in many areas, including all U.S. waters. Fisheries managers have begun to react to overharvest concerns by limiting fishing on spawning aggregations or creating sanctuaries in which no fishing is allowed; however, enforcement of regulations has been problematic in some regions.

5.5 Development and Tourism Impacts

Coastal Development

Because of the sensitivity of coral reefs to coastal influences such as siltation and nutrient input, coastal development can be a serious threat to survival of the tropical reef ecosystems. Development activities that can directly destroy reefs include dredging of channels and harbors, covering reefs for construction projects, or mining the reef for construction materials, such as for making cement, bricks, road-fill, or barriers. In most regions these actions are tightly restricted in areas where healthy coral reefs remain; however, as tropical nations developed there was a substantial, but often undocumented, loss of corals reefs. Coastal development also may target areas that have lost reefs for other reasons and keep them from ever recovering.

Boat Groundings

Physical collisions with the coral from boat groundings can cause severe local damage to corals. In the days of sailing ships, coral reefs were considered a major nuisance and groundings thwarted the plans of many ocean explorers. Unintentional groundings of large ships on reefs still occur on a regular basis around the world (**Figure 5-22**), sometimes accompanied by spills of oil or other toxic materials. For example, ship groundings off the Florida coast have been a regular occurrence since the 1700s, in part due to the narrow passage between the reefs of the Florida Keys and those off Cuba and the currents of the Gulf Stream. In the 1700s, the U.S. government even considered whether corals could be eradicated in the Florida Keys to avoid ship grounding. As recently as 1989 there were three ship groundings on the Florida Keys within a time period of a few months. Soon after, the Florida Keys Marine Sanctuary was established and groundings have been less frequent, but still several grounding occurred through the 1990s. Groundings of large ships on reefs around the world are frequent enough that they typically receive attention only from the local news media. For example, a large cargo ship was grounded on the Belize barrier reef in January 2009, destroying about 10,000 square meters of reef.

(a)

(b)

Figure 5-22 **(a)** The *M/V Wellwood* grounded on Molasses Reef in the Florida Keys National Marine Sanctuary. **(b)** A portion of a Puerto Rico reef impacted by a ship grounding (left) compared to an adjacent unimpacted area (right).

Grounding of smaller boats on the reef cause less damage but are more frequent and more difficult to monitor; the cumulative effects can be significant. On reefs in the Florida Keys there are, on average, over 50 boat groundings reported each year; many more undoubtedly occur that are unreported. Even if boats avoid grounding, simply anchoring on the reef can cause extensive local damage to the coral. Anchors from large ships can weigh several tons and cause serious damage to the corals and, if dragged over the reef, this damage may extend for hundreds of meters. In many protected reefs, permanent anchoring buoys are provided to avoid this problem. For example, popular dive sites are often provided with buoys anchored to the bottom, which dive operators are required to use.

Development for Tourism

Pollution as an indirect result of development has affected an unknown amount of coral reef ecosystems around the world. Some pollution events are single occurrences and recovery may be possible; however, reefs that are already stressed due to other factors may not recover, especially if the pollution continues. The main chronic pollutants deriving from terrestrial origins are nutrient pollution and siltation from such sources as sewage and runoff from agricultural and development. An increase in land clearing in some tropical regions has resulted in erosion and siltation that has caused a dramatic decline in reef communities. Siltation also can increase when habitats that filter out sediments before they reach the reefs, such as sea grasses and mangroves, are removed.

Tourism has become a major industry around the world, and the majority of that tourism is in coastal areas.

Coral reefs attract tourists for snorkeling and scuba diving, but also because of the aesthetic appeal of clear tropical waters. This makes development for tourism attractive to tropical developing countries. Regions that used to be considered isolated adventure destinations are becoming more accessible to tourists, but along with the tourism amenities come pollution and habitat loss inherent with rapid development.

Although tourism's impact is typically not as severe as destructive activities such as mining or excessive fish harvest, the indirect effect of the construction of roads, airports, beaches, and resorts can be extreme. For example, caution must be taken to avoid an increase in sedimentation during construction. As discussed above, the reef is adapted to a low sediment environment, and sediments affect the corals and the reef ecosystem at many levels.

The population increase from hotel and resort tourists also can result in problems related to the disposal of sewage and other wastes. There is a great temptation in developing countries to encourage tourism because of the economic benefits, but to a level that treatment facilities cannot support. Nutrient input from sewage has resulted in eutrophication in many regions as a direct result of tourism and resort development. Tourism can also encourage excessive fisheries harvest to support an increased popularity in local seafood and for sport fishing.

The cruise ship industry presents another potential tourism-related effect on the reef ecosystem. For example, over 50% of the global cruise industry is in the Caribbean, where accidents such as collisions with reefs or anchors can cause extensive damage. Cruise ships also add an additional sewage burden in ports, where facilities for handling wastes generated on cruise ships are needed. In some regions of the Caribbean tours organized from cruise ships are restricted due to the potential effect of large numbers of people visiting dive or snorkel sites. Large cruise ships may require the building of large ports that require destruction of coral reefs and coastal habitats. For example, in Jamaica, a port completed in 2011 to accommodate large cruise ships required dredging and the removal and transplanting of live coral. Remediation of the impacts also was provided through the placement of ceramic reef modules in other areas to mimic the structure of branching corals. Although measures were taken to minimize the environmental damages, conservationists are concerned over the unknown survival success of the transplanted coral, the removal of natural coastal protections provided by the reef, and suspension of sediments resulting from dredging activities.

Snorkeling and Diving

Although most snorkelers and divers are among the most conservation minded of tourists, their damage to the reefs can be significant if proper precautions are not taken. The greatest harm is from contact with the reef, which can result in breakage, especially of branching corals, or abrasion of living coral tissues, making them more susceptible to disease. Heavy diving intensity may result in decreased coverage by reef building corals or an overall change in the makeup of the coral community.

Damage to the reefs occurs most frequently by divers with inadequate training due to poor buoyancy control, carelessness, and loose equipment and gauges. Scuba certification programs provide some training and conservation education, which is helpful in avoiding reef damage by divers. Certification is typically not required for snorkeling without scuba equipment, however, and snorkelers often go out with minimal training or environmental education. In shallow waters, disturbance by kicking or standing on the bottom or on the coral itself can result in extreme local damage of the reef (**Figure 5-23**).

A common goal of many divers is to view and get close to as many fish as possible. One way to achieve this is for divers to bring food into the water to attract and feed fish. This may not be viewed by divers as affecting the reef negatively; however, feeding has been documented to disrupt the natural behavior of the fish and affect the composition of fish populations and their prey species. Feeding is not allowed in many areas popular for diving; however, resistance by some divers and dive operators can make enforcement of these rules difficult.

Even though the harm from a single diver may be minimal, the cumulative impact can be large at popular dive sites where thousands or tens-of-thousands of divers visit each year. In some areas a reef **carrying capacity** has been

Figure 5-23 Careless sport divers and snorkelers can impact the reef ecosystem by physically disturbing the reef habitat.

determined, indicating the number of divers per year that a site may tolerate; for example, Julie Hawkins and colleagues estimated that coral reefs in the Caribbean can tolerate 4,000 to 6,000 divers per year. A study by Anthony Rouphael and Graeme Inglis on the Great Barrier Reef indicated that reef damage did not accumulate over time in newly established diving areas, but that branching corals recovered quickly from scattered damage by divers. They suggested that managing divers to avoid the damage is more efficient than setting limits on the number of divers visiting a site.

Restrictions have been difficult to implement in most areas because of economic incentives to maximize the number of divers and difficulties with enforcement.

5.6 Approaches to Coral Reef Conservation

Assessment and Monitoring

Numerous efforts are being made around the world to protect and recover coral reef ecosystems. In order to approach coral reef conservation on a global scale, however, the current condition of reefs must first be assessed and a baseline of ecosystem health established that enables the measurement of future progress. Modern remote-sensing and imaging technologies have provided a convenient way to monitor coral coverage on a broad scale, but direct observations are needed to obtain information on species distributions, ecosystem health, and other details concerning the reef ecosystem. There are numerous government agencies and NGOs involved in such assessments of coral reef ecosystems. Hundreds of direct observational studies on coral reef health have been carried out by scientists over the past several decades. Unfortunately, the results of these studies are often difficult to locate and many are not published in the available scientific literature. In recent years, regional databases and informational Web sites have been developed by NGOs (e.g., http://www.ReefCheck.org and The Coral Reef Alliance, http://www.Coral.org). The World Fish Center has established a global reef database (http://www.ReefBase.org) containing information on the status and management of coral reefs; these data are used by the Global Coral Reef Monitoring Network (http://www.gcrmn.org) to improve management and conservation by publishing manuals and databases and providing training, funding, and equipment for monitoring. One central location for the accumulation of all coral reef-related scientific data would be useful toward achieving conservation goals.

Education

Although it is difficult enough to define the harmful impacts on coral reef ecosystems and discover what conservation efforts might be needed, it is even more difficult to implement solutions to conservation problems. Efforts to curb pollution simply to protect the coral reefs for aesthetic and biological reasons may not be successful without educating the public and user groups, such as fishers and the tourist industry, of the sensitivity of reefs to pollution and their importance to the health of the oceans.

Obviously destructive activities such as fishing methods that destroy the reef must be restricted. Simply enacting laws may not be adequate, however; they must be implemented and enforced. This requires a financial investment to hire people for monitoring and enforcement. Education of the public can enable some degree of self-enforcement and protection beyond that enabled by laws. Education can include: informing divers and snorkelers of the harm of contacting coral; educating fishers in non-destructive methods and of the importance of the reef to future fishery populations; educating consumers on how to choose sustainable seafood; educating locals of the effects of pollution on the reef; and educating tourists of the potential indirect harm from tourism activities.

One example of the benefits of education can be seen in a popular snorkeling area near Honolulu, Hawaii. Hanauma Bay was developed to attract tourism beginning in the early 1900s; however, building facilities and a road to the beach made the Bay such a popular destination that the reef ecosystem began to suffer. In order to provide better access to the water for swimmers much of the fringing reef was blasted away. The area became more popular for swimming and fishing and there were few fish remaining in the Bay by the 1960s. In 1967 the Bay was declared a no-take conservation area. Management for recreation continued

to receive precedence, however. Boulders were placed off the beach to reduce erosion and blasting and dredging were used to create swimming areas. The boulders created an artificial reef that attracted fish, making snorkeling popular, and visitors to the bay began feeding the fish with bread and human snacks. By the late 1980s as many as 10,000 people per day were visiting the Bay, and the remaining corals were being killed as a result of trampling. Although fish feeding was attracting many fish, they were of only a few species. Finally, beginning in 2000, the health of the area began to turn around, largely due to a project that included a large education campaign. Bay use is still heavy, at about 3,000 visitors per day, limited by fees and smaller parking lots, but the way the Bay is used has changed. An education center was created away from the beach, and novice snorkelers are required to view a training video before snorkeling. Volunteers on the beach provided conservation education, and fishing, taking animals, and fish feeding are not allowed. Use of the beach and reefs can continue due to increased awareness by users of conservation, endangered species, and effects of human actions on coral reef ecosystems.

Marine Protected Areas

One of the simplest solutions to reef conservation and recovery is to set aside areas for protection and allow them to recover on their own. Total protection can be difficult due to social and political factors, however, because coral reefs are integrated into, dependent on, and affected by other ecosystems. Marine Protected Areas (MPAs; also called Marine Parks or Marine Sanctuaries) are viewed as one of the most desirable means of protecting and recovering coral reef ecosystems (**Figure 5-24**). There are varying levels of protection provided through MPA designation. For

Figure 5-24 A portion of Australia's Great Barrier Reef, much of which is protected as the Great Barrier Reef Marine Park.

example, it may include a total ban on fishing (i.e., no-take areas [NTAs]) and limits on the number of divers who can visit the reef. If there is a long history of fishing on the reef, economic and cultural reasons can make it difficult to convince lawmakers to establish MPAs. Sometimes there is strong resistance from user groups to protect the large areas needed to conserve all components of the reef ecosystem. Even if a large area of the reef is protected, there still may be external factors influencing the ecosystem; for example, pollution from terrestrial sources or movements of harmful organisms onto the reef. When designing MPAs, consideration should be made for interactions with other ecosystems, including recruitment of larvae from surrounding areas.

Enforcement is another necessary consideration. It is easier to declare a reef as an MPA than it is to establish adequate measures to enforce this status. For example, if fishers have historically fished an area and believe they need or deserve to harvest these resources, they may continue fishing illegally. Establishing successful MPAs requires a thorough consideration of scientific, political, and social factors. The region must be assessed biologically to determine the current status of the reefs and what might be the optimal area to establish an MPA, and these assessments need to be continued after the MPA is established. One way to fund enforcement is to require a user fee for those visiting the reef, including snorkelers and divers. Proceeds also can be used to further benefit the reef by employing locals, taking them away from fishing and other activities harmful to the reef.

It is important to educate users and lawmakers of the importance of MPAs. For example, fishers may actually benefit from MPAs even though it means fewer areas to fish. MPAs can function as a source of fish that recruit from the sanctuary to outside areas where regulated fishing is allowed. This is especially true if protected areas include regions where spawning aggregations occur.

Users of the reef and its resources also must be aware that results will not be immediate and that it may take years before damaged reefs begin to return to healthy conditions. In recent years the establishment of MPAs has become one of the preferred methods of coral reef conservation, and there has been a steady increase in their numbers since the 1980s so that protected areas have been established in almost all large reef ecosystems. For example, no-take reefs are now scattered along the coast of Jamaica in regions historically heavily overfished; off Australia, NTAs were expanded into 33% of the Great Barrier Reef Marine Park in 2004; and the largest continuous MPA was recently established in the northwestern Hawaiian Islands (**Box 5-3. Research Brief: Contrasting Fish Communities on Hawaiian Reefs**). The degree of success has been mixed in recovering reefs even in protected areas due to some of the complicating factors mentioned earlier, and many have not been in place long enough to witness results.

Reef Restoration

Direct damage to the reef due to anthropogenic activities, such as coastal development, dredging, boat groundings, or anchoring on the reef, can be so severe that the reef is unable to recover on its own. Re-establishment of the reef thus may require active restoration efforts. In areas where dredging or other destruction is imminent, corals, sponges, and other organisms can be taken by divers or grown in the lab and relocated into other areas (**Figure 5-25a**). If corals have been reduced to rubble, it is likely that a stable base for future coral growth cannot be established. In this case the rubble is removed and dislodged living coral and other organisms can be reattached (e.g., using cement or plastic ties) to remaining stable structures or to artificial reef structures. This has been successful for many types of coral but less successful for sponges, especially when they have been detached from the bottom.

Because coral larvae are planktonic, they often return to an impacted area by natural recruitment if a stable structure is available for settlement. If severe damage to the reef has occurred (e.g., from dredging, ship grounding, or a hurricane), it may be necessary to put large structures in place to encourage a natural return of reef organisms. These structures can include limestone boulders, concrete, metal frames filled with rocks, or preconstructed "modular reef units" (Figure 5-25b). Although artificial reefs have been established from a wide array of salvaged structures, including railroad cars, sunken ships, airplanes, and even household appliances, this is not recommended for replacing natural reefs because of the unnatural appearance and the possibility of smaller items being dispersed by water movements, especially during storms.

The natural return of coral to a protected area is most desirable but may not occur rapidly enough, especially in areas that have had extensive loss of corals and where new larvae are not present to settle on the reef. In these situations it may be necessary to reintroduce corals to the reef. Entire coral heads are rarely available to be "transplanted" in areas that have suffered coral loss without harming the reef from which they are taken. Coral fragments can be taken from a healthy reef with little long-term harm, however. These fragments are obtained from healthy coral colonies and allowed to grow for several years. These methods were used successfully for recovery of areas of the Great Barrier Reef following an outbreak of the crown-of-thorn starfish.

Rather than using coral fragments, coral larvae also can be introduced into a damaged area. Reproducing coral can be taken from the wild and allowed to release their

Box 5-3   Research Brief: Contrasting Fish Communities on Hawaiian Reefs

The northwest islands of the Hawaiian archipelago and the associated coral reefs have seen relatively little human influence, in contrast to the heavily populated main Hawaiian Islands to the southeast. This protection has historically resulted from its isolation and lack of human inhabitants; however, more recently legal protections have been provided by the U.S. government through the establishment of a National Wildlife Refuge and Critical Habitat for Endangered Species. In 2006, the Northwest Hawaiian Islands Marine Coral Ecosystem Reserve was established as a U.S. national monument, the single largest marine conservation area in the world, encompassing over 137,000 square miles (**Figure B5-4**). This region has provided reef ecologists with an unplanned field experiment, comparing a region that has been largely protected from fishing with a comparable region with few protected areas that has also experienced significant anthropogenic impacts from terrestrial influences and heavy fishing pressure for over a century (only 0.3% of the marine area around the main islands is totally protected from harvest). Currently there are thousands of commercial fishers and tens-of-thousands of recreational fishers in Hawaii—and sometimes poor compliance with fishing regulations.

In the 1970s, U.S. agencies, including the National Marine Fisheries Service, the Hawaii Department of Fish and Game, and the University of Hawaii, initiated a five-year study of the marine community of the northwest Hawaiian Islands. A portion of this study, as reported by Alan Friedlander and Edward DeMartini, found a remarkable contrast between the reefs of the northwestern and main Hawaiian Islands. The biomass of large apex predators, including sharks, jacks, and groupers, was greater than 54% around the northwestern Hawaiian islands but only 3% around the main Hawaiian Islands. The evidence suggests that the large predators have nearly been eradicated by fishing from the waters around the main Hawaiian Islands and even fish at lower trophic levels have been greatly affected. Currently the reef ecosystem of the northwestern Hawaiian Islands is one of the few remaining large continuous predatory-dominated tropical reef ecosystems. Coral reef ecologists have begun to realize that a human-caused loss of predators is probably the norm rather than the exception around the remainder of the world. For example, Simon K. Wilson and colleagues discovered that when fishing around coral reefs of Fiji declined, the number of larger predatory fish began to increase substantially. Octavio Aburto-Oropeza and colleagues documented an 11-fold increase in biomass of top predators (and a 4-fold increase in total fish biomass) in 14 years after the establishment of a no-take MPA in the Gulf of California, Mexico; the recovery was largely attributed to effective enforcement of no-fishing restrictions through strong community leadership. On Australia's Great Barrier

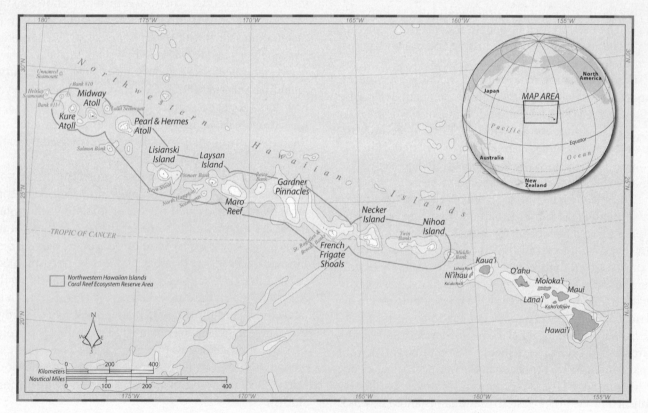

Figure B5-4   The Northwest Hawaiian Islands Coral Reef Ecosystem Reserve, the largest reef ecosystem protected as a Marine Protected Area.

Reef some regions have been protected from fishing since the mid-1970s; studies by Nick A. Graham and colleagues found that biomass of targeted predators is 60% higher on unfished than fished reefs. These studies support the contention being made by some marine ecologists for decades: that human actions are changing coral reef ecosystems and that they would be much more predator-dominated without anthropogenic impacts.

Studies on northwestern Hawaii reefs also show that the abundance of large predators does not lower the diversity of other species on the reef; in fact the opposite appears to be true. The marine ecosystem of the northwestern Hawaiian Islands has a large species diversity (more than 7,000 species) and over 1,700 endemic species, those found nowhere else in the world. Smaller predators, such as goatfish (Mullidae), scorpionfishes, and squirrelfishes, are also abundant on these coral reefs, and the biomass of smaller carnivorous and herbivorous fishes on shallow reefs is double that found around the main Hawaiian Islands. The total fish biomass is over 2½ times greater in the northwestern Hawaiian Islands. In the main Hawaiian Islands there is a notable absence of food fish and colorful aquarium fish species, presumable due to overfishing and collecting (**Figure B5-5**).

Friedlander and DeMartini came to several general conclusions concerning coral reef conservation: (1) large no-take marine reserves are needed to keep some coral reef ecosystems in pristine conditions; (2) natural coral reef ecosystems include many large predators that are vulnerable to rapid overharvest by fishers, so that there are few of these natural, healthy reef ecosystems remaining in the world; (3) the value of marine reserves is more than aesthetic, because protected reef ecosystems can provide fish recruits to neighboring areas and enhance fishing; and (4) management of coral reefs is needed at the ecosystem level rather than focusing on individual species.

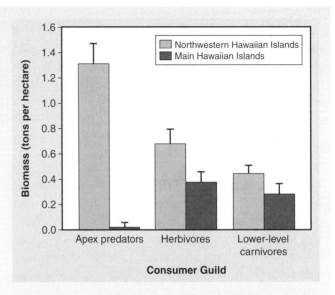

Figure B5-5 Comparison of biomass by trophic level for reef sites surveyed in the main Hawaiian Island and the northwestern Hawaiian Islands, indicating a remarkable divergence in the number of apex predators.

Protection of the reefs of the northwestern Hawaiian Islands provides an outstanding opportunity for a large-scale experiment in coral reef conservation. If this ecosystem can be protected from local anthropogenic impacts, such as excessive fishing and coastal pollution, continued monitoring will provide valuable information concerning effects of global impacts, such as climate change and pollution, on the health of reef ecosystems. The benefits of these protections will hopefully lead to the expansion of tropical reef protected areas around the globe.

(a)

(b)

Figure 5-25 Coral restoration efforts: **(a)** staghorn coral planted on the reef after being raised from sperm and egg in the laboratory; and **(b)** modular reef units being put into place to replace damaged coral in the Florida Keys.

gametes into a tank. The larvae are kept until they are mature enough to settle and then are released into the affected area. This method has been successfully implemented in Australian reefs. Taking this process one step further, programs have been developed to assist the recovery of elkhorn coral to Florida's reefs in which coral egg and sperm are collected from the wild, the resulting offspring are allowed to settle and grow in the lab. The living coral head is then transplanting it onto a reef (see Figure 5-25a).

Education and Community Action

Even if scientists and managers are able to develop the best possible plans to conserve coral reefs in a specific area, these plans are useless unless they can be implemented. Education and local support are critical; this can be especially true in island nations, such as the Philippines, where, on some islands, a large fraction of the population depends on local coral reef resources and there are few other options for their livelihood. Marine protected areas can be most successfully implemented if locals are involved in the planning, and understand the reasons for the park and how it might benefit them (e.g., by providing employment in tourism, enforcement, and management, or improving fisheries outside the park).

The need for local involvement was learned the hard way recently in Belize. Government officials established a protected area around coral reefs, with little communication with those who depended on the reef for fishing. A management station built within the protected area was soon vandalized and burned. This delayed enforcement within the protected area to reconsider regulations on uses of the reef and to communicate with locals, educating them of the potential effects on fisheries and development, and employment alternatives if fisheries were taken away.

Another successful way to implement coral reef protection is through collaborations. This is a fairly new idea in coral reef management, but around the world, partnerships recently have been established among government agencies, NGOs, local communities, and academic institutions. Collaborative efforts are often successful not only in enhancing the health of the reef ecosystem, but also in improving the livelihood of locals and minimizing conflicts among user groups (e.g., fishers, tourism industry, environmentalists, and managers).

For example, a coral reef ecosystem adjacent to Apo Island in the Philippines has been successfully managed by local government and community organizations. A portion of the coral reefs associated with the island was protected as a small unfished reserve. Fishing outside the reserve was allowed to continue, but the area was protected from destructive fishing methods such as trawling and blast fishing. Locals were educated of the benefits of the reserve and volunteers participated in enforcing fishing restrictions. A study by Angel Alcala and colleagues demonstrated that the outcome of this program was a substantial improvement in the health of the reefs and numbers of fish. Fish populations adjacent to the protected area were also enhanced, providing a stable fishery and employment related to ecotourism improved the living standard of the island community.

The lesson learned from collaborative efforts toward coral reef conservation is that the human dimension to conservation must be considered. We will not be able to return to a world where coastal human populations were low and reefs were inaccessible enough that effects from harvest and pollution were negligible. The human factors of education, protection, enforcement, and involvement must be considered if conservation efforts are to be successful.

5.7 Overview and Future of Coral Reef Conservation

Because it is impractical to expect, in the foreseeable future, a return to a world in which most coral reefs are in the pristine conditions they were centuries ago, decisions must be made that include humans as interacting components of reef ecosystems. The result is that the ideal goals of conservation may be out of reach when they are not perceived as being in the best interest of the public. Conservation decisions will have to consider both biological and socioeconomic factors, requiring answers to some difficult questions. For example, what economic hardships will be tolerated to protect a reef from harvest? Where can we get the economic resources required to establish conservation programs? When options are limited, what conservation measures will provide the most benefit with the least cost? How can we balance the desire for tourism development with conservation of the ecosystems? How can we convince lawmakers and political leaders, in the presence of what may be considered more pressing issues, of the importance of coral reef conservation? How do we convince people to act in the long-term interest of conservation rather than focusing on short-term personal gains? A concerted effort will be necessary to answer these questions, and what is feasible in one region may not be possible in another, depending on culture, politics, economics, and history.

Because of where they are—in accessible coastal waters in climates where people like to live and visit—and what they are—extremely diverse and sensitive ecosystems—coral reefs may be one of the most difficult of marine ecosystems to protect from harmful anthropogenic impacts. In order to be successful, both conservation failures and successes must be analyzed and used by scientists, management agencies, and political leaders to establish future directions.

There are some bright spots to provide cautious optimism for future success in coral reef conservation. For example, there has been an increase in public interest in coral reefs and their protection for aesthetic reasons, due to the wide availability of photos and videos, and increased accessibility to the reef with air travel and diving technology. Efforts are being made for international cooperation that could limit pollution and global climate change that harm coral reefs. The number and size of marine protected areas that include reefs has increased dramatically in recent years. We have limited the use of extreme destructive methods for the harvest of fish (e.g., explosives and poisons) in some regions due to international pressure and collaboration among government and NGOs. Finally, we have seen the recovery of many of the coral reefs after the extreme El Niño events of recent years and discovered corals that are tolerant to high temperatures. Obviously, extreme efforts are needed to expand the conservation of one of the most biologically diverse and economically important ecosystems on the planet.

Coral reef ecologists such as Nancy Knowlton have provided recommendations of what realistic actions scientists and citizens can do to enhance coral reef conservation efforts in the future. Scientific studies of reefs need to be continued to understand and document the biological diversity and ecological interactions. This will help us to avoid extinctions and establish a baseline to strive for when we gain the social and political will to make conservation a priority. Scientific studies also will give us a better understanding of cause-and-effect relationships between natural and anthropogenic events and reef conditions. More large no-take marine protected areas need to be established; however, there also needs to be a focus on protecting coral reefs outside of protected areas to achieve a balance between human resource needs and conservation. A better understanding is necessary of how climate change will affect coral reef ecosystems; for example, how will changes in acidity affect coral growth, and can coral reefs adapt to warmer temperatures? And finally, the human side of coral reef conservation must be considered. Over 100 nations have coral reefs in their coastal waters and each of these has its own cultural and political differences that affect the approach taken to conservation. It must be understood that in order to understand, educate, and encourage the cooperation of coastal residents, developers, fishers, divers, tourists, and environmentalists, human behavior will be a key factor determining the future direction of coral reef conservation.

STUDY GUIDE

▣ Topics for Review

1. Describe how corals are able to more efficiently use and recycle nutrients than most other animals.
2. What factors keep coral reefs from forming at the mouths of tropical rivers?
3. What factors resulting from global climate change could affect coral reef growth and survival?
4. What kinds of pollutants have been associated with diseases in coral?
5. Describe the differences in "bottom-up" and "top-down" effects on coral reefs.
6. Describe the different ways that excess sediment can affect coral growth and survival.
7. Describe the various factors that can result in the growth of excess macroalgae on coral reefs.
8. Discuss how coral reef ecosystems have changed due to overfishing of large predatory fishes.
9. What are the various markets for seahorses that have resulted in population declines? Where are the primary sources and markets located?
10. Compare how methods used to harvest live fishes impact coral reefs. What are the effects on the fish populations?
11. Why are coral reef animals considered good prospects for bioprospecting for pharmaceutical products?
12. What characteristics make conch and lobsters vulnerable to overharvest even by locals using primitive fishing methods?
13. Compare the following fishing methods regarding their effects on coral reef ecosystems: hook-and-line, net fishing, traps, blast fishing, cyanide fishing, and spear fishing.
14. Discuss the various dangers to coral reefs when tourism expands too rapidly into tropical coastal areas.
15. Discuss how establishing Marine Protected Areas around coral reefs can benefit residents of tropical islands.
16. How do fish communities around coral reefs that are protected from fishing compare with those in unprotected areas?
17. What are some of the positive trends and developments that may provide hope and optimism toward coral reef conservation?

Conservation Exercises

1. Prepare a presentation to the following groups arguing their roles in promoting coral reef conservation and how doing so would benefit them in the long run:
   a. live fish collectors on a reef in the Philippines
   b. tourists scuba diving on the Great Barrier Reef
   c. global political leaders regulating emissions of greenhouse gases
   d. local regulators promoting tourism development on an island in the Caribbean
   e. distributors of traditional medicines derived from coral reef products in Hong Kong
   f. local fishers on coral reefs in Indonesia

FURTHER READING

Aburto-Oropeza, O., B. Erisman, G. R. Galland, I. Mascarenas-Osorio, E. Sala, and E. Ezcurra. 2011. Large recovery of fish biomass in a no-take marine reserve. *PLoS ONE* 6(8):e23601. Doi:10.1371/journal.pone.0023601.

Aiken, K., A. Kong, S. Smikle, R. Appeldoorn, and G. Warner. 2006. Managing Jamaica's queen conch resources. *Ocean and Coastal Management* 49:332–341.

Albins, M. A., and M. A. Hixon. 2008. Invasive Indo-Pacific lionfish *Pterois volitans* reduce recruitment of Atlantic coral-reef fishes. *Marine Ecology Progress Series* 367:233–238.

Alcala, A. C., G. R. Russ, and P. Nillos. 2006. Collaborative and community-based conservation of coral reefs, with reference to marine reserves in the Philippines. Pages 392–418. In *Coral Reef Conservation.* I. M. Cote and J. D. Reynolds (eds.). Cambridge University Press, Cambridge, UK.

Appeldoorn, R. S. 2009. Movement of fishes (Grunts: Haemulidae) across the coral reef seascape: a review of scales, patterns and processes. *Caribbean Journal of Science* 45:304–316.

Barker, N. H. L., and C. M. Roberts. 2004. Scuba diver behaviour and the management of diving impacts on coral reefs. *Biological Conservation* 120:481–489.

Bellwood, D. R., T. P. Hughes, C. Folke, and M. Nystrom. 2004. Confronting the coral reef crisis. *Nature* 429:827–833.

Bolden, S. K. 2000. Long-distance movement of a Nassau grouper (*Epinephelus striatus*) to a spawning aggregation in the central Bahamas. *Fishery Bulletin (U.S.)* 98:642–645.

Centre National de la Recherché Scientifigue (CNRS). 1 March 2010. Polynesian coral reefs wiped out by cyclone Oli. Retrieved August 15, 2011, from http://www2.cnrs.fr/en/1708.htm

Cote, I. M., and A. Maljkovic. 2010. Predation rates of Indo-Pacific lionfish on Bahamian coral reefs. *Marine Ecology Progress Series* 404:219–225.

De'ath, G., J. M. Lough, and K. E. Fabricius. 2008. Declining coral calcification on the Great Barrier Reef. *Science* 323:116–119.

Douglas, A. E. 2003. Coral bleaching—how and why? *Marine Pollution Bulletin* 46:385–392.

Fabricius, K. E. 2005. Effects of terrestrial runoff on the ecology of corals and coral reefs: review and synthesis. *Marine Pollution Bulletin* 50:125–146.

Fichez, R., M. Adjeroud, Y.-M. Bozec, L. Breau, Y. Changerelle, C. Chevillon, P. Douillet, J.-M. Fernandez, P. Frouin, M. Kulbicki, B. Moreton, S. Ouillon, C. Payri, T. Perez, P. Sasal, and J. Thebault. 2005. A review of selected indicators of particle, nutrient and metal inputs in coral reef lagoon systems. *Aquatic Living Resources* 18:125–147.

Fox, H. E. 2004. Coral recruitment in blasted and unblasted sites in Indonesia: assessing rehabilitation potential. *Marine Ecology Progress Series* 269:131–139.

Fox, H. E., and R. L. Caldwell. 2006. Recovery from blast fishing on coral reefs: a tale of two scales. *Ecological Applications* 16:1631–1635.

Friedlander, A. M., and E. E. DeMartini. 2002. Contrasts in density, size, and biomass of reef fishes between the northwestern and the main Hawaiian Islands: the effects of fishing down apex predators. *Marine Ecology Progress Series* 230:254–264.

Gillett R., and W. Moy. 2006. *Spearfishing in the Pacific islands: current status and management issues.* Secretariat of the Pacific Community, Noumea and Food and Agriculture Organization of the United Nations, Rome. Retrieved August 15, 2011, from http://www.fao.org/docrep/009/a0774e/a0774e00.htm

Graham, N. A. J., R. D. Evans, and G. R. Russ. 2003. The effects of marine reserve protection on the trophic relationships of reef fishes on the Great Barrier Reef. *Environmental Conservation* 30:200–209.

Green, S. J., and I. M Cote. 2009. Record densities of Indo-Pacific lionfish on Bahamian coral reefs. *Coral Reefs* 28:107.

Grey, M., A.-M. Blais, and A. C. J. Vincent. 2005. Magnitude and trends of marine fish curio imports to the USA. *Oryx* 39:413–420.

Hawkins, J. P., C. M. Roberts, T. Van't Hof, K. De Meyer, J. Tratalos, and C. Aldam. 1998. Effects of recreational scuba diving on Caribbean coral and fish communities. *Conservation Biology* 13:888–897.

Hoegh-Guldberg, O., P. J. Mumby, A. J. Hooten, R. S. Steneck, P. Greenfield, E. Gomez, C. D. Harvell, P. F. Sale, A. J. Edwards, K. Caldeira, N. Knowlton, C. M. Eakin, R. Iglesias-Prieto, N. Muthiga, R. H. Bradbury, A. Dubi, and M. E. Hatziolos. 2007. Coral reefs under rapid climate change and ocean acidification. *Science* 14:1737–1742.

Hoey, A., and D. R. Bellwood. 2011. Suppression of herbivory by macroalgal density: a critical feedback on coral reefs? *Ecology Letters* 14:267–273.

Idjadi, J. A., S. C. Lee, J. F. Bruno, W. F. Precht, L. Allen-Requa, and P. J. Edmunds. 2006. Rapid phase-shift reversal on a Jamaican coral reef. *Coral Reefs* 25:209–211.

Idjadi, J., R. Haring, and W. Precht. 2010. Recovery of the sea urchin *Diadema antillarum* promotes scleractinian coral growth and survivorship on shallow Jamaican reefs. *Marine Ecology Progress Series* 403:91–100.

Jaap, W. C., J. H. Hudson, R. E. Dodge, D. Gilliam, and R. Shaul. 2006. Coral reef restoration with case studies from Florida. Pages 478–514. In *Coral Reef Conservation*. I. M. Cote and J. D. Reynolds (eds.). Cambridge University Press, Cambridge, UK.

Jackson, J. B. C. 1997. Reefs since Columbus. *Coral Reefs* 16S:23–32.

Jackson, J. B. C., J. D. Cubit, B. D. Keller, V. Batista, K. Burns, H. M. Caffey, R. L. Caldwell, S. D. Garrity, C. D. Getter, C. Gonzalez, H. M. Guzman, K. W. Kaufmann, A. H. Knapp, S. C. Levings, M. J. Marshall, R. Steger, R. C. Thompson, and E. Weil. 1989. Ecological effects of a major oil spill on Panamanian coastal marine communities. *Science* 243:37–44.

Jobbins, G. 2006. Tourism and coral-reef-based conservation: can they coexist? Pages 237–263. In *Coral Reef Conservation*. I. M. Cote and J. D. Reynolds (eds.). Cambridge University Press, Cambridge, UK.

Jones, A. M., S. Gardner, and W. Sinclair. 2008. Losing 'Nemo': bleaching and collection appear to reduce inshore populations of anemonefishes. *Journal of Fish Biology* 73:753–761.

Knowlton, N. 2006. Coral reef coda: what can we hope for? Pages 538–549. In *Coral Reef Conservation*. I. M. Cote and J. D. Reynolds (eds.). Cambridge University Press, Cambridge, UK.

Lesser, M. P., and M. Slattery. 2010. Ocean exploration: consequences of lionfish invasion to stability of mesophotic coral reefs. *Proceedings from the 2010 AGU Ocean Sciences Meeting*; February 22–26, 2010: Portland, OR.

Maljkovic, A., T. E. Van Leeuwen, and S. N. Cove. 2008. Predation by the invasive red lionfish, *Pterois volitans* (Pisces: Scorpaenidae), by native groupers in the Bahamas. *Coral Reefs* 27:501.

McClanahan, T. 2006. Challenges and accomplishments towards sustainable reef fisheries. In *Coral Reef Conservation*. I. M. Cote and J. D. Reynolds (eds.). Cambridge University Press, Cambridge, UK.

Milazzo, M., F. Badalamenti, T. Vega-Fernandez, and R. Chemello. 2005. Effects of fish feeding by snorkelers on the density and size distribution of fishes in a Mediterranean marine protected area. *Marine Biology* 146:1213–1222.

Patterson, K. L., J. W. Porter, K. B. Ritchie, S. W. Polson, E. Mueller, E. C. Peters, D. L. Santavy, and G. W. Smith. 2002. The etiology of white pox, a lethal disease of the Caribbean elkhorn coral, *Acropora palmata*. *Proceedings of the National Academy of Sciences USA* 99:8725–8730.

Precht, W. F., and R. B. Aronson. 2006. Death and resurrection of Caribbean coral reefs: a palaeoecological perspective. Pages 40–76. In *Coral Reef Conservation*. I. M. Cote and J. D. Reynolds (eds.). Cambridge University Press, Cambridge, UK.

Precht, W. F., R. B. Aronson, R. M. Moody, and L. Kaufman. 2011. Changing patterns of microhabitat utilization by the threespot damselfish, *Stegastes planifrons*, on Caribbean reefs. *Plos One* 5(5): e10835.

Riegl, B. M., and R. E. Dodge (eds.). 2007. *Coral Reefs of the USA*. Springer Publishers, Berlin.

Rouphael, A. B., and G. J. Inglis. 2002. Increased spatial and temporal variability in coral damage caused by recreational scuba diving. *Ecological Applications* 12:427–440.

Russell, M. 2003. Reducing the impact of fishing and tourism on fish spawning aggregations in the Great Barrier Reef Marine Park. *Gulf and Caribbean Fisheries Institute* 54:681–688.

Sadovy, D. Y., A. Cornish, M. Domeier, P. L. Colin, M. Russell, and K. C. Lindeman. 2008. A global baseline for spawning aggregations of reef fishes. *Conservation Biology* 22:1233–1234.

Sutherland, K. P., J. W. Porter, J. W. Turner, B. J. Thomas, E. E. Looney, T. P. Luna, M. K. Meyers, J. C. Futch, and E. K. Lipp. 2010. Human sewage identified as likely source of white pox disease of the threatened Caribbean elkhorn coral, *Acropora palmata*. *Environmental Microbiology* 12:1122–1131.

Sheppard, C. 2006. Longer-term impacts of climate change on coral reefs. Pages 392–418. In *Coral Reef Conservation*. I. M. Cote and J. D. Reynolds (eds.). Cambridge University Press, Cambridge, UK.

Tratalos, J. A., and T. J. Austin. 2001. Impacts of recreational SCUBA diving on coral communities of the Caribbean island of Grand Cayman. *Biological Conservation* 102:67–75.

Vargas-Angel, B., J. Asher, L. S. Godwin, and R. E. Brainard. 2009. Invasive didemnid tunicate spreading across coral reefs at remote Swains Island, American Samoa. *Coral Reefs* 28:53.

Vincent, A. C. J. 2006. Live food and non-food fisheries on coral reefs, and their potential management. Pages 183–236. In *Coral Reef Conservation*. I. M. Cote and J. D. Reynolds (eds.). Cambridge University Press, Cambridge, UK.

Wilkinson, C. 2006. Status of coral reefs of the world: summary of threats and remedial action. Pages 3–39. In *Coral Reef Conservation*. I. M. Cote and J. D. Reynolds (eds.). Cambridge University Press, Cambridge, UK.

Wilson, S. K., R. Fisher, M. S. Pratchett, N. A. J. Graham, N. K. Dulvy, R. A. Turner, A. Cakacaka, and N. V. C. Polunin. 2010. Habitat degradation and fishing effects on the size structure of coral reef fish communities. *Ecological Applications* 20:442–451.

Witzell, W. N. 1998. Origin of the Florida sponge fishery. *Marine Fisheries Review* 60:27–32.

Nearshore Ecosystems: Community Ecology and Habitat Protection

The nearshore zone begins at the lower edge of the intertidal zone and extends into coastal waters (**Figure 6-1**). The extent of this zone is highly variable and there is no precise point where the "nearshore" becomes "offshore." The nearshore is generally considered to be the subtidal region within which there are substantial influences from coastal processes and freshwater runoff. This chapter focuses on environmental biological processes and conservation issues that are unique to ecosystems within this region.

Important physical factors controlling the environment of the subtidal were introduced in Chapter 1. Temperature is more variable here than in deeper waters and thus is one of the most important physical factors controlling the types of organisms present. In the shallow subtidal there can be substantial seasonal changes and water currents can greatly influence temperature. Subtidal shallow bottom communities are also physically affected by currents and waves. And, even though these regions are below the intertidal zone, they are still affected by water movements generated by the rise and fall of the tides. Wind waves can affect waters of moderate depths, sometimes as deep as 200 meters or more. Although the subtidal is not as unstable as the intertidal, turbulence at the bottom may cause instability. The mixing of nearshore waters is also important to subtidal organisms in that it provides adequate oxygen for respiration.

In the shallow waters of the subtidal, even well beyond estuaries, there is typically adequate light penetration to support photosynthesis and substantial nutrients originating from terrestrial sources to support high levels of primary production. A portion of this production is by phytoplankton in the water column or algae on the sediment surface. Often, however, the dominant primary producers in these ecosystems are seagrasses and attached submerged macroalgae, including kelps and other seaweeds. These can extend from the substrate well into the water column, often reaching the surface. In soft sandy bottoms sea grasses are common. In cool waters with rocky bottoms kelp are more prevalent. In relatively warm waters, other kinds of macroalgae are commonly associated with hard bottoms, including coral reefs.

This chapter focuses on the ecosystems that form in these shallow subtidal regions as commonly defined by the dominant primary producers. Outside of coral reef habitats (discussed in Chapter 5), the most diverse of these ecosystems are where seagrasses and kelps occur; however, many subtidal regions, whether dominated by soft or rocky bottoms, are without large multicellular primary producers, less diverse but often times supporting high productivity.

6.1 Soft Bottom Ecosystems

The nearshore subtidal region is dominated by soft, sandy or muddy bottoms, even extending into regions offshore from the rocky coastlines. Organisms that live

Figure 6-1 A nearshore subtidal ecosystem dominated by turtle grass.

(a)

(b)

Figure 6-2 Epifauna living on shallow soft bottoms: **(a)** cushion sea star *Oreaster reticulatus;* **(b)** southern stingray *Dasyatis americana.*

on the surface of soft bottoms, the **epifauna**, are typically small or flattened and camouflaged with a color pattern to match the substrate (**Figure 6-2**). Because the bottom of subtidal regions is more stable and exposed to less stress than the intertidal, there are more epifauna in the subtidal. Predators include sea urchins, sand dollars, sea stars, crabs, and shrimps. Many fishes scavenge on the bottom or prey on bottom organisms. These include cods, pollocks, drums, flounders, and rays. As fisheries, these are classified as **groundfish** (see Chapter 11). Most organisms of the subtidal bottom are **infauna**, that is, they live burrowed under the sediment. These are similar to the organisms of the intertidal (see Chapter 3) and include many species of burrowing worms, clams, and other invertebrates.

The most important factor determining the distribution of organisms in subtidal benthic community is the substrate type. In muddy bottoms, oxygen does not penetrate far below the surface, limiting the ability of organisms, other than bacteria and other anaerobes, to live buried far beneath the surface. Organisms that live buried in the sediments are either more tolerant of low oxygen or transport oxygenated water down into their burrows. For example, clams often live buried beneath the surface but have a siphon extending to the surface to take in water for food and oxygen. In sandy bottoms, organisms can live deeper in the sediments because sands are more porous and are thus oxygenated to a deeper level. An entire community of organisms resides in these sandy bottoms. Most are meiofauna that can move between sand grains; these include crustaceans, such as copepods and rotifers, nematodes, kinorhynchs, polychaete worms, and hydroids.

Where subtidal soft bottoms are unvegetated the primary producers are mainly microscopic protists such as diatoms. Most of the organic matter is produced in the water column and is transported by currents or sinks to the bottom. Most of the nutrients are contained in decaying organic materials (detritus), transported from estuaries and other coastal communities, or present as feces or the remains of dead animal that sink to the bottom.

Anthropogenic Impacts on Soft-Bottom Ecosystems

As a result of habitat overlaps and migratory movements, fisheries in nearshore subtidal ecosystems are largely extensions of those focused in adjacent estuaries, intertidal regions, or offshore habitats. For example, harvest of benthic invertebrates such as oysters and clams extends from estuaries and the intertidal region (discussed in Chapters 3 and 4). Bottom-associated fishes and crustaceans utilizing the subtidal, such as shrimp or groundfish, are the focus of bottom trawl fisheries that extend into offshore waters. These nets can have varied impacts on the ecosystem including habitat modification and destruction (Chapter 8) and overharvest of fisheries (Chapter 11). Much of the conservation focus in shallow subtidal regions is on habitats with substantial large plants or seaweeds, such as seagrasses and kelp, due to the high biological diversity and production, and concerns over habitat loss. Pollution also can extend from the estuaries and intertidal zone into the subtidal, impacting these ecosystems, sometimes in unpredicted ways. For example, Jonathan Puritz and Robert Toonen discovered that plumes of polluted urban runoff in the vicinity of Los Angeles served as a barrier to dispersal of larval sea stars, and thus genetic mixing of populations, along the California coast. Presumably, similar effects could occur with other species and in other coastal regions, resulting in a reduction of genetic mixing among populations that are fed by drifting larvae transported among habitats along coastlines. Other conservation issues regarding nearshore habitats were addressed in Chapters 3 and 4. Issues specific to the subtidal region are discussed in the following section.

Coastal Hypoxia and Dead Zones

Coastal subtidal regions that are strongly influenced by river input can be very productive ecosystems. When nutrient levels are excessive and water circulation is low, however, these regions can develop low oxygen (**hypoxic**) conditions. As nutrients, detritus, and dead organisms sink to the bottom, oxygen is consumed as they are decomposed by bacteria. If mixing is inadequate to renew bottom waters with well-oxygenated surface waters, dissolved oxygen levels can drop to levels critical to the survival of bottom organisms (**Figure 6-3**).

Hypoxic conditions are well-documented in enclosed bays or estuaries, such as Chesapeake Bay, where there is little current flow or wave action and thus little mixing. These conditions are more likely to occur in summer months when the water is warmer (as water temperature increases oxygen retention declines), winds die down, and there is a high level of organic material in bottom waters. This is enhanced by salinity stratification, a result of the isolation of salty, heavier waters near the bottom below fresh waters coming in from rivers (see Chapter 1).

For example, in Mobile Bay, Alabama in the northern Gulf of Mexico, summer hypoxia events become evident when water circulation forces low-oxygen bottom waters into shallow areas along the shore. These conditions drive marine animals into shallow nearshore waters along stretches of the shoreline ranging from a 100 meters to about 20 kilometers (**Figure 6-4**). These hypoxia events vary from year to year in intensity and frequency, but have been documented to occur about five times per year on average when wind and tide conditions are favorable. Locally they are referred to "jubilees," when locals rush to net or gig the

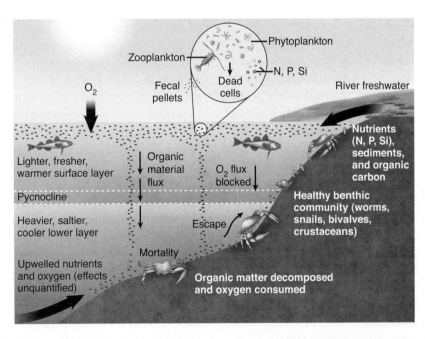

Figure 6-3 How hypoxia develops in coastal waters. Excessive nutrients and organic matter entering from rivers leads to a reduction of oxygen in bottom waters as a result of bacterial decomposition.

Figure 6-4 During "jubilee" events—such as this one in Mobile Bay, Alabama—bottom organisms concentrate along the shoreline when deeper waters become deoxygenated.

oxygen-starved (but otherwise healthy) bottom-dwelling fishes (such as flounder) and crustaceans (such as crabs and shrimp).

Jubilee events have been documented in Mobile Bay since at least the mid-1800s, leading to debates as to whether they are natural phenomena. Studies by Edwin May in the 1970s suggest that these hypoxic conditions have resulted from, or at least been enhanced by, the Mobile Bay ship channel. Digging of the channel was initiated in the mid-1800s and it has been widened, deepened, and maintained ever since. As organic material is washed into the bay from rivers, it accumulates in the channel. As the material is decomposed by bacteria, oxygen levels drop. The waters are trapped in the channel and not allowed to circulate and become re-oxygenated. There also has been a loss of production of fisheries species in the Bay, such as crabs and oysters, due to such factors that enhance hypoxia. Bays channelized for shipping around the world can experience similar conditions, which, when considered along with additional stresses of pollution and overharvest (see Chapter 4), result in substantial stresses to the ecosystem.

Unfortunately, hypoxic events are not limited to bays and estuaries. Conditions resulting in oxygen depletion of coastal bottom waters can develop beyond river mouths also. These conditions typically develop during summer months as organic matter originating from spring plankton blooms in surface waters sink to the bottom. As the surface waters warm they become lighter and remain above the cooler, denser bottom waters. During this seasonal stratification there is little vertical circulation of the coastal subtidal waters, and conditions are further enhanced by calm wind conditions. As bacteria decompose the organic

matter at the bottom, the limited oxygen that is available is depleted. Although this is a natural phenomenon, when excess nutrients are put into coastal waters, the degree of oxygen depletion is enhanced. Major sources of these excess nutrients include agricultural fertilizers and sewage wastes.

Beginning in the mid-1900s, there was a noticeable increase in the formation of oxygen depletion zones in coastal waters associated with population centers and river systems with large nutrient inputs. Fishers began to refer to these regions as "dead zones" due to the dramatic declines in fisheries harvest of benthic animals attributed to high levels of eutrophication. Robert Diaz and Rutger Rosenberg identified dead zones in over 400 coastal marine ecosystems, encompassing a total area of more than 245,000 square kilometers (**Figure 6-5**). Efforts to encourage a reduction in nutrient input into these systems have had limited success as fewer than 5% have shown signs of improvement; however, there has been a virtual elimination of dead zone associated with the Thames River in England and the Hudson River in the United States.

One of the largest dead zones has been documented and followed by Nancy Rabalais and colleagues in coastal areas of the northern Gulf of Mexico. This dead zone has been linked to nutrients from fertilizer runoff and other agricultural activities throughout the Mississippi River Basin. Its coverage ranges from less than 500 square kilometers, during years when river flow is lowest, to over 15,000 square kilometers during high-flow years. Exacerbating the impact of agricultural runoff is the channelization of the Mississippi River, which has removed the broad floodplains from the system that would normally trap the sediments and nutrients flowing down the Mississippi River and recycle them into estuary and wetland ecosystems. Restoring the Mississippi system to its normal flow patterns has been debated in recent years but is not currently practical (see Chapter 4). Dead zones in the northern Gulf of Mexico and many other areas are seasonal. As summer ends, temperature changes and wind and waves result in mixing of the coastal waters. This brings oxygen to the bottom waters and removes the dead zone. Storms and hurricanes also can result in mixing of coastal waters, removing dead zones.

Breaking the cycle that leads to the development of these hypoxic dead zones will take a large effort because they can originate far from the regions that are impacted; for example, 40% of the waters of the United States drain into the Mississippi River system. Recent plans by the U.S. EPA, NOAA, and state agencies are to fund programs to reduce excess nutrients going into the Mississippi River by 30%. These plans are not as strong as environmentalists had pushed for but do set goals and methods for nitrogen reduction. One of the most controversial aspects of proposed plans is to get farmers to use less fertilizer on crops.

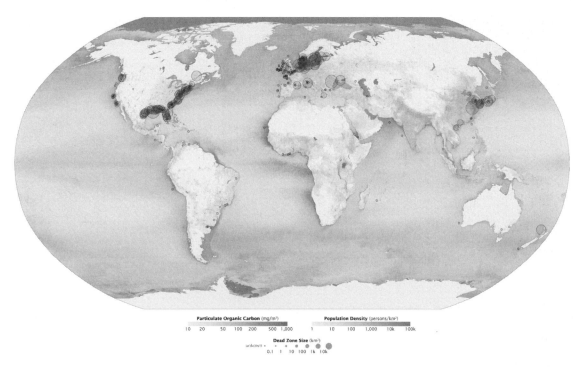

Figure 6-5 Hypoxic or "dead" zones are most prevalent in coastal regions near high human population densities.

Although there is some debate as to whether hypoxia has always been a seasonal occurrence in some coastal areas, there is little doubt that it has increased dramatically along with the increase in industrialization and the use of fertilizers around the world. Many rank coastal hypoxia with habitat loss and overfishing as major global marine conservation problems, emphasizing the need for worldwide initiatives to reduce nutrient inputs into coastal waters.

6.2 Seagrass Ecosystems

Seagrasses

Seagrasses are distinctive in that they are an extremely productive shallow water marine ecosystem that can be found associated with soft bottom habitats at almost any latitude. Seagrasses are found in brackish to full-strength seawater, from the intertidal to 50 meters depth, and from the Arctic to the equator (**Figure 6-6**). Typically a single species dominates in a given habitat; however, worldwide there are about 60 different species of seagrasses. They characterize possibly the most widespread shallow marine ecosystem on Earth, but have not been thoroughly surveyed in many regions where they are prevalent. Almost all seagrass species must remain submerged nearly continuously; only a few species can survive in intertidal waters where they are regularly exposed at low tide. Seagrasses are most typically found in sheltered shallow water, but in clear waters they can be found much deeper. The maximum depth at which adequate sunlight penetrates to support photosynthesis in seagrasses, even in the clearest of waters, is about 50 meters; however, few seagrasses occur at depths greater than about 5 meters. The size of exposed leaves varies by species and with depth. Intertidal species are quite small, with leaves about 1 centimeter long. For some species in deeper waters, long slender leaves can reach over 5 meters in length. The longest sea grass leaves recorded were from *Zostera caulescens* growing in Funakoshi Bay, Japan. For some seagrass species, the roots can extend as much as 5 meters in the sediments.

Seagrasses are the only group of vascular plants that thrive submerged in seawater. Most species are exposed to fairly constant salinities of around 35 parts-per-thousand (ppt) in full-strength seawater, and most species, especially those living in estuaries, can tolerate salinities from about 10 to 45 ppt. Salinities beyond this range are stressful and may kill the seagrasses. Seagrasses avoid desiccation in seawater by maintaining a cell osmotic concentration that matches that of the sea water (in contrast to marsh grasses and mangroves that exclude salts from their cells).

Seagrasses grow in soft muddy or sandy bottoms (only one species is able to cling to rocky shelves). They do best with moderate current flow. Rhizomes extend horizontally and branch into the sediments and thus function to accumulate and hold sediments in place. The leaves dampen the wave and current energy, enhancing sediment deposition. Deposition of fine sediments makes the water around the seagrass beds clearer.

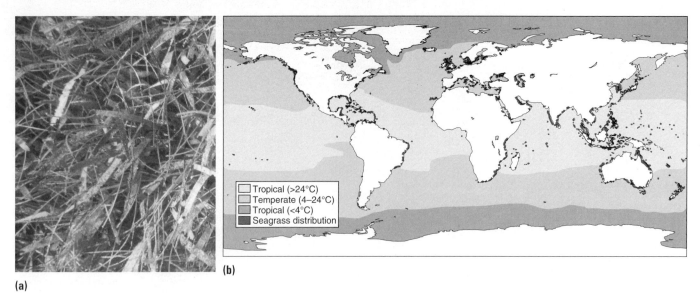

(b)

(a)

Figure 6-6 **(a)** A mixed stand of turtle grass (broad blades) and manatee grass (narrow blades). **(b)** Global distribution of seagrasses, prevalent in both temperate and tropical regions.

Seagrasses are the most productive marine community on subtidal soft bottoms, despite the low nutrient levels in the waters and sediments where they often grow. In organic soils the oxygenated layer can be very shallow. Root penetration and nutrient availability can be limited by hypoxia; however, the plants are efficient at nutrient uptake, which is enhanced by nitrogen-fixing cyanobacteria in the sediments and on the leaves. The ability to take in nutrients from the sediments, through roots, and from the water, with leaves, allows seagrasses to live in more nutrient-poor environments than other primary producers such as macroalgae. This is enhanced by the recycling of nutrients within the seagrass plant as nutrients are often transferred from older leaves to younger growing leaves. Excessive nutrients, such as in eutrophic environments affected by anthropogenic inputs, can result in macroalgae outcompeting seagrasses.

Gas exchange also presents a special problem in submerged habitats due to the poor ability of water to hold dissolved gases. Carbon dioxide is taken in through a porous cuticle covering the seagrass leaves. Gas-filled chambers (**lacunae**) in the leaves transport oxygen to the roots for respiration to compensate for the lack of oxygen in the soils. The lacunae also enhance the buoyancy of the leaves, allowing them to extend toward the surface, enhancing photosynthesis.

The diversity of seagrasses is greatest in tropical and subtropical regions; however, some species are found as far north as the Arctic and as far south as South Africa and southern New Zealand. In the tropics, seagrasses are often associated with mangroves and coral reefs. It is common to find seagrasses in open areas adjacent to mangrove roots

and to find small patch reefs as "islands" within seagrass meadows.

Seagrass Growth and Reproduction

Seagrasses grow by extending horizontal **rhizomes** near the surface through the sediments. Vertical rhizomes are formed at intervals, extending **adventitious roots** into the sediments and new shoots and leaves into the water. This allows the seagrass to expand into open space. A single clone thus can occupy a large area, growing and expanding for years, sometimes growing thousands of separate shoots. These growth patterns explain why many seagrass meadows are occupied by a single species of seagrass. Seagrasses also can expand by **vegetative propagation** because fragments of the plant that are broken off may form roots; in some species fragment plants can remain viable for several weeks, providing a mechanism for longer distance dispersal.

Seagrasses can reproduce sexually with flowers also, but only rarely for most species. The flowers are very small and inconspicuous. Pollen is transported by water, typically over distances of several meters from a male plant to the flower of a separate female plant (most seagrasses are **dioecious**, having separate male and female plants). Seeds produced from sexual reproduction are not buoyant for many species and thus disperse a short distance from the plant. For species with floating seeds, dispersal can be tens of kilometers. Even though sexual reproduction is relatively rare, it serves as an important mechanism for seagrasses to colonize new areas. The extent of time the seed can remain dormant varies considerably among species, and in some species the seeds can remain dormant for several years. This can assist in recolonizing an area where the plants have been eradicated, for example by a hurricane or typhoon.

The Seagrass Community

Because seagrasses occur in tropical and temperate shallow waters around most of the world, the species associations can vary considerably. A species' distribution is affected by temperature, current flow, turbidity, depth, nutrients, and soil type, among other factors. The number of species in a given geographic region also varies, ranging from 17 species in the tropical Indo-Pacific to just one species in the South Atlantic off the coast of South Africa. Although a single seagrass meadow is often dominated by a single species, the number can range from one or two in most temperate regions to as many as seven in the tropics. Although the diversity of species is fairly consistent in a given region, a disturbance can result in changes to the dominant species as a result of superior ability to colonize an open area by some species and subsequent competition for space.

Despite the diversity of species and the habitats they occupy, some generalizations can be made about the biological community associated with the seagrasses. Seagrasses provide a structure to which organisms can attach or within which organisms can hide. Primary production by the plant supports a diversity and abundance of organisms. It directly supports grazing organisms that consume the living leaves or detritivores that use decaying organic matter that was produced by the plants. These organisms support a complex food web of other consumers. The production within seagrass beds also supports adjacent communities through the transport of organic matter out of the seagrasses either directly or through fishes, marine mammals, and reptiles that move in and out of the seagrasses. Although many seagrass-associated species are the same as those found in adjacent habitats, abundance is typically much higher in the seagrasses.

Epiphytes include bacteria, fungi, and algae that grow directly on the seagrass leaves (**Figure 6-7**). The productivity of epiphytes can be greater than 50% of that of the seagrasses, and may be more important to consumers because they are more readily available to grazers. There can be a negative impact of epiphytes on seagrasses through a reduction of photosynthesis, and therefore leaf production. Higher nutrient levels can result in unnatural increases in epiphytes.

Bottom invertebrates can be very abundant in the seagrass beds due to the availability of organic matter in the detritus of decaying leaves and feces of other organisms (**Figure 6-8**). Another important factor is the high settling rate of organic matter resulting from the reduction of water flow by the seagrass leaves.

Figure 6-7   Sea grasses, such as this Caribbean turtle grass, support a productive ecosystem. Note the leaves covered with epiphytes.

(a)

(b)

Figure 6-8   Two invertebrates residing in seagrass ecosystems: **(a)** a giant anemone *Condylactis gigantea*, and **(b)** a giant hermit crab *Petrochirus diogenes*.

Bivalves in the seagrass beds are usually found burrowed into the sediments. The rhizomes can provide protection from excavation by predators and the abundant organic matter enhances their growth. Gastropods often feed on the epiphytes, primarily algae, growing on the seagrasses. This may reduce the negative effects of epiphytes on seagrass photosynthesis but harm the leaf itself through damage to the epidermis. Other gastropods, including conch, feed on the detritus, much of which adheres as small particles to the seagrass leaves. Crustaceans in the seagrass ecosystem are numerous, including copepods, isopods, amphipods, shrimps, and crabs. Many smaller crustaceans graze on epiphytes, but some isopods eat by burrowing small holes into the leaves. Many shrimp and crabs rely on detritus or leaf material; some species harvest the leaves and store them in their burrows to consume later. Although less common, some crabs eat the root and rhizome material. Some burrowing shrimp are important in disturbing bottom sediments by their burrowing activities, affecting distribution of seagrass patches as a result of the differential colonizing abilities.

Echinoderms such as sea cucumbers and sea urchins can be important grazers in the seagrass beds. Some sea cucumbers consume sediments and extract detritus, playing an important role in disturbing the substrate. Sea urchins consume epiphytes, live leaves, and detritus. Sea urchins in some regions, such as the Gulf of Mexico purple urchin *Lytechinus variegates,* can strip enough leaves to leave large patches bare. Once an area is stripped bare, the urchins become more vulnerable to predators. This reduces the urchin populations, allowing the seagrasses to recover. The cycle may then start over. In some regions, sea urchins move out of coral reefs at night to graze on seagrasses. This results in an obvious halo of barren sands around the reefs.

Sea turtles, especially the green turtle *Chelonias mydas* (**Figure 6-9**), are historically an important consumer of

seagrasses around much of the world. Although the green turtle has declined considerably from its former abundance, it is still important in many regions. Turtle grass *Thalassia testudinum* is the preferred food of the green turtle, though the turtles will eat other species as well as algae. Although grazing can be intense, it does not typically cause long-term harm to the seagrass ecosystem. The turtles do not uproot the seagrasses, but bite off the lower portion of the leaves, which are younger and more nutritious. The leaves immediately begin regrowing and the turtles may return to graze on the new leaves. Turtles may remain in one area for more than a year, cultivating the patch through their grazing pattern. As with many large herbivores, green turtles have microbes in their gut that assist with digestion of the plant material. This not only makes the nutrients available to the turtles, but also releases nutrients into the feces, which becomes an important component of the detritus, increasing availability of nutrients to detritivores.

Other important herbivores of seagrasses are the dugong *Dugong dugong* in the Indo-Pacific and the West Indian manatee *Trichechus manatus* in the Caribbean and Gulf of Mexico (**Figure 6-10**). Both species have been dramatically impacted by human activities (Chapter 9) but remain an important component of the seagrass ecosystem in some regions. For example, in Australia foraging groups of dugongs can be as large as 100 individuals. Dugongs feed primarily on sea grasses and are less cautious than sea turtles in their foraging, sometimes eating not only the leaves but also the rhizomes (**Figure 6-11**). Dugongs can reach sizes as large as 250 kilograms and consume as much as one quarter of their body weight per day, removing large areas of seagrass. This does not typically result in long-term damage to the seagrass habitat, however. Dugongs graze in strips as much as 25 centimeters wide, creating paths weaving through the seagrass beds; they always leave a strip of seagrass between

Figure 6-9 The green turtle is one of the few species that can survive by consuming seagrasses.

Figure 6-10 A West Indian manatee feeding on seagrasses.

Figure 6-11 Dugongs are important grazers of seagrasses in shallow waters of the Indo-Pacific.

Figure 6-12 Though few directly consume living seagrasses, fishes, such as these French grunts, are an important component of seagrass ecosystems.

paths and the paths do not cross. The seagrasses between paths thus can recolonize grazed areas quickly by rhizomal growth or vegetative propagation. Dugong grazing can enhance species of seagrasses that easily colonize these paths. Although manatees feed on a broader range of plant materials than dugongs, seagrasses are an important part of their diet. Manatees were probably an important part of the seagrass ecosystem throughout the Caribbean and Gulf of Mexico prior to their decline. The remaining manatees graze extensively on sea grasses, primarily in shallow waters. Lynn Lefebvre and colleagues found that Florida seagrasses were able to recover rapidly from manatee grazing.

Birds, such as swans, geese, and ducks, can be important consumers of sea grasses in shallow waters of some regions and may consume both leaves and rhizomes. For example, in the Gulf of Mexico, much of the population of redhead ducks *Aythya americana* feeds on shoal grass *Halodule wrightii*. Brant geese *Branta bernicla* migrate along the Pacific coast of North America feeding on eelgrass *Zostera marina* in Alaska, California, and Mexico. Much of the biomass consumed by the birds is returned to the seagrass ecosystem as guano; however, in some situations ducks remove as much as 50% of the seagrass in one season.

Few fish species graze directly on living seagrass leaves, though many depend on the production of seagrass plants (**Figure 6-12**). Pinfish *Lagodon rhomboides*, however, consume and derive nutrition from seagrass leaves along the southeast U.S. coast. Parrotfish (Scaridae) and leatherjackets (*Oligoplites*) consume seagrass material along with epiphytes. Parrotfish in the Caribbean can move from coral reefs to graze in adjacent seagrass beds with an effect similar to that caused by sea urchins, clearing an area to create a bare zone around the reef. Many fishes, such as grunts (Haemulidae) in tropical regions, graze on organisms from the seagrass plants or the substrate. Other fishes, such as mullets (Mugilidae), feed on the detritus formed from decaying leaves. Most fishes and many macroinvertebrates using the seagrasses are carnivores, including crabs and predatory fishes. Those that are permanent residents of the seagrasses are typically small, including seahorses and pipefishes (Syngnathidae). Larger predators, such as moray eels (Muraenidae) and snappers (Lutjanidae), sometimes move onto seagrasses at night to feed.

Defining food webs for seagrasses (as well as other coastal or estuarine ecosystems) is difficult because many organisms do not fit discreetly into traditional trophic categories due to the variable foods they consume. In part this is due to feeding changes at different life history stages. For fishes, the same species may subsist on plankton when young and as a predator as an adult. Or, a fish that consumes the seagrasses and appears to function as an herbivore also may depend on detritus or organisms attached to the leaves. Many fishes are opportunistic, even as adults, and may be able to feed on many different types of food items.

The advantage to organisms living in the seagrasses is not only for feeding, but also as a refuge from predators. Many animals seek shelter in the seagrasses for a portion of their life history, especially young life stages. The importance of seagrasses as a refuge for larval fishes has probably been underestimated. For example, 30 families of fish larvae were found on a seagrass bed in Thailand, and their abundance was almost double that over nearby open sandy bottoms.

Seagrass Interactions with Other Ecosystems

Because of the broad geographic range of seagrasses, they serve a critical role in shallow marine and estuarine ecosystems around the world. They are such an integral component of some estuary, coral reef, mangrove, or mudflat ecosystems that it is somewhat artificial to separate them from these ecosystems. Few invertebrates or fishes are limited only to seagrass beds. Many of the species discussed throughout this chapter are opportunistic in their use of seagrass ecosystems. Almost all species can utilize other habitats, depending on geographic region and availability. Many individuals may reside in several different habitats within their lifetime, or sometimes even within one year or one day. Depending on the geographic region, fishes that use seagrasses seasonally or at some point in their life history, especially as juveniles, may use one or more of these other coastal habitats at other times. Common temporary inhabitants of seagrasses include drums (Sciaenidae), surgeonfishes (Acanthuridae), grunts, snappers, and parrotfishes. Some of these use seagrasses as a nursery area for larval and juveniles in a manner similar to salt marsh and mangrove habitats (see Chapter 4).

Emigration out of the seagrass beds by organisms that feed there can serve to enhance other ecosystems through an export of organic matter and nutrients produced in the seagrasses. The importance of connections between these habitats varies, depending on where they co-occur. In tropical regions the net movement of organic matter is generally from mangroves to seagrasses to coral reefs. Access to seagrasses is important enough for some coral reef fish species that they are found in higher abundance when seagrass beds are nearby. Many grunts and snappers move onto seagrass beds to feed at night and then return to coral reefs for shelter during the day. Others, including pinfish and mullets, feed on seagrass beds only during the day. Barracudas and various ray and shark species may move from deeper waters onto the seagrass beds looking for prey. The presence of mangroves can enhance the survival of some seagrass fishes, but this varies regionally. For example, J. E. Jelbart and colleagues found that juveniles from seagrasses in the Indo-Pacific commonly move into mangroves at high tide to feed, influencing fish assemblages in both ecosystems. In contrast, based on studies by Martijn Dorenbosch and colleagues, few fishes move from seagrasses to mangroves to feed in the Caribbean, where tidal range is less. The interconnectedness of these habitats presents a strong argument for conservation efforts that take a broad ecosystem approach.

Physical interactions between seagrasses and adjacent shallow or intertidal habitats also can be important. The seagrass' ability to trap sediments improves water clarity and also minimizes erosion in shallow coastal areas. This can be important to coral reefs in the vicinity of seagrasses due their low tolerance of turbidity (see Chapter 5). The suspension of sediments after the loss of seagrass reduces the water clarity and thus inhibits seagrass recolonization. Removal of seagrasses, therefore, can have long-term impacts on coastal ecosystems.

Human Benefits

Direct use of seagrass leaves by humans is uncommon, though leaves have been used as fertilizer or livestock food, and for weaving into mats, baskets, roof thatch, and fishing nets. Seeds and rhizomes are eaten in a few localized areas, such as in Kenya, and dried leaves and rhizomes are considered useful for treatment of diarrhea in Thailand. Seagrasses were commonly used for food and medicine by Native Americans on the Pacific Coast of North America. Dead leaves are used as compost in some regions, including Korea. The greatest benefit of seagrasses to humans is indirect as it is related to their support of invertebrate and fish populations that provide food and habitat for many marine animals. Maintenance of this diversity is considered important in many ways that benefit production of species exploited by humans. Larvae and juveniles of some species of shrimp depend on structure and food available in seagrasses and other shallow ecosystems. Seagrasses are important in maintaining diversity of coastal fish populations as nursery, refuges, or feeding grounds. Many of these species are important prey for fisheries species or support fisheries at later life stages. In some regions, fishes, bivalves, and gastropods are harvested directly from seagrass meadows; these are especially important as subsistence harvest in Africa and Asia. In Thailand, fishes such as groupers (Serranidae), mackerel (Scombridae), mullets, snapper, and invertebrates such as sea cucumbers and crabs are harvested from the seagrasses. Seagrasses provide habitat and food for marine mammals and endangered species for which most countries have provided special protection, including manatees, sea turtles, and seahorses.

Sea grasses provide physical protection to coastal areas from the influence of storms by reducing waves and currents, although this is probably a more important function for mangroves and salt marsh grasses (see Chapter 4). The ability of seagrasses to encourage settling of sediments can be important in improving water quality for adjacent coral reefs.

Anthropogenic Impacts on Seagrasses

The World Atlas of Seagrasses, prepared by the United Nations Environmental Program (UNEP), estimates that there are more than 175,000 square kilometers of seagrasses remaining worldwide; but that this area declined by 15% from about 1990 to 2000 due to direct and indirect human impacts. In some areas, including Canada and Australia, the loss has been less; but in other areas, including parts of Europe, the Mediterranean, and the U.S. Atlantic coasts, the loss has been more severe.

Most of the documented losses of seagrasses are from a combination of effects, such as increases in turbidity, excess nutrients, and physical damage. Probably the primary indirect cause of seagrass loss is increase in sediment load. Higher sediment levels in the water increase turbidity and reduce water clarity, reducing the ability to photosynthesize, while potentially covering and smothering the seagrasses. Turbidity can decrease the maximum depth at which sea grasses survive. For example, in some Australia bays seagrasses occur to depths as great as 15 meters in unpolluted areas, but are limited to 9 meters depth in areas with increased turbidity. Land-based activities that can cause siltation include agriculture, land clearing, or construction. It is often difficult to pinpoint the cause of the losses. For example, Westernport Bay, Australia has seen several dramatic declines in seagrass coverage since the 1950s. Although these are likely attributable to the input of sediments and nutrients from aquaculture in the watershed that feeds the Bay, it has been difficult to come up with conclusive evidence of a single factor causing the losses. On the U.S. Atlantic, high turbidity totally eliminated seagrasses from Delaware Bay and limits seagrasses in Chesapeake Bay to no deeper than one meter. Much of the turbidity and excess nutrients are from non-point source pollution that is hard to identify and eliminate.

Increases in nutrient input, resulting in excessive nutrients (**nutrient loading**) can reduce photosynthesis due to overgrowth of the seagrasses by epiphytes, plankton blooms, or competition with macroalgae (**Figure 6-13**). Excess nutrients typically originate from agriculture runoff or sewage input. This is probably the primary reason for loss of seagrasses in temperate, industrialized regions, where the seagrasses habitat itself may be protected, but inputs from terrestrial sources affect the ecosystem (see Box 6-1, Conservation Efforts: The Wadden Sea and Chincoteague Bay). Even if wastewater treatment facilities are in place,

Figure 6-13 Macroalgae can overgrow seagrasses when excessive nutrients are input to coastal waters.

increases in nutrients and suspended matter from effluent have been associated with seagrass declines. For example, eelgrass beds in bays in Massachusetts and New Jersey declined 40% to 60% in regions impacted by housing developments. Losses of seagrasses along the south Florida coast have been attributed to increases in sewage waste and turbidity resulting from the rapid increase of residences and hotels in the past several decades. In more isolated areas, however, seagrasses have fared much better, and with improved waste treatment some areas have begun to recover. Seagrass coverage has declined throughout much of the northern Gulf of Mexico due to numerous impacts: increased nutrients from waste waters, increased sediment input in coastal rivers, dredging of ports and canals, scarring from boat traffic, and filling for coastal development. Wastes from terrestrial sources are not the only source of nutrient loading. In regions where fish are raised in pens suspended in estuaries excess feed and waste cause eutrophication and shade the bottom where seagrasses would grow (see Chapter 11).

Toxins, such as those originating from oil or chemical spills or industrial waste, may result in the rapid death of seagrasses. For example, one of the worst oil spills in the Mediterranean, by the oil tanker *Haven*, killed beds of seagrass off Italy in the early 1990s. One argument against offshore drilling on the U.S. west coast is the potential for spills that could harm seagrass beds. In many regions of the world, raw or partially treated sewage is piped directly to the ocean or released as sewage sludge, and there may be inadequate restrictions on industrial waste disposal. If aquaculture activities occur near seagrass beds, antibiotics present in some feeds can impact the seagrasses by changes in microbial activities associated with the seagrass ecosystem.

Introductions of exotic invasive species have harmed seagrasses in some regions. For example, the algae *Caulerpa taxifolia* (**Figure 6-14**) was unintentionally introduced in the Mediterranean in the 1980s and smothered and killed large areas of seagrass beds; it has spread to other regions since and was observed off the coast of California in the late 1990s, where it eventually could have similar impacts.

Physical disturbances are another source of harm to seagrass beds. Seagrasses are adapted to recover rapidly from natural physical disturbances such as grazing and cold temperatures because these usually leave the root system intact. Unpredictable weather events such as storms or floods can cause substantial harm to seagrass beds, but in areas where they are protected recovery is likely. For example, in 1992 a cyclone and two major floods affected seagrasses in eastern Australia. Shallow-water and intertidal seagrasses were uprooted by heavy seas and deep-water seagrasses died from increased turbidity in the waters. The impacted seagrasses eventually recovered in the subtidal areas below 5 meters,

Figure 6-14 A stand of *Caulerpa*, an invasive species found in the Mediterranean and off California.

however, beginning in two years. The intertidal seagrasses took four to six years before they were fully recovered.

Continued disturbance of the bottom by human activities such as dredging, or otherwise disturbing the bottom to develop shipping channels, harbors, ports, or marinas, destroys seagrass beds and they are rarely allowed to recover. It is important that these activities be minimized in areas with sea grasses and that the total amount of protected seagrasses be considered before embarking on new projects. Mining of sand for land reclamation can devastate seagrass beds. For example, sand dredging in Malaysia to build condominiums and stabilize shorelines has resulted in sediment deposition at levels great enough to smother seagrass beds. In some areas of Belize and Jamaica, sands have been dredged to build up beaches, removing seagrass beds. Of course, if the land being "reclaimed" to build a beach is the seagrass bed itself, the losses are permanent. This is especially problematic in mountainous island nations where the availability of flat lands for roads, airports, or commercial developments is low. Japan lost over 30% of the seagrasses from several large bays between 1978 and 1991, primarily due to land reclamation. Some small western Pacific islands, such as Micronesia, have lost many of their seagrass beds to land reclamation for building airports and causeways. Around large coastal cities in the U.S. northeast that have seen over four centuries of development, there is no documentation of the amount of sea grass that has been lost.

Boat propellers and anchors can gouge scars in the seagrass beds, destroying the roots that may take years to regrow, and turbulence from boats can increase turbidity. This is especially problematic in shallow areas around the U.S. Florida Keys, where over 15% of the seagrass meadows have been scarred by boat propellers, despite education campaigns encouraging boaters to avoid shallow seagrass beds.

Fishing activities can harm sea grass beds also. Fishing trawls and seines (see Chapter 11) pulled across the bottom can disrupt seagrass beds, and clam dredging, historically common on the northeast U.S. coast, can remove seagrasses and keep them from recolonizing. Although the focus of conservation efforts related to cyanide and dynamite fishing is on coral reefs (see Chapter 5), they also have impacted seagrass ecosystems in some regions. Easy access to seagrass beds makes them attractive to subsistence fishers without the means to purchase large boats and expensive fishing gear. If not properly regulated, heavy fishing in the seagrass beds can result not only in excess removal of fish, but also harm to the plants. For example, in Mozambique, locals commonly harvest bivalves from the seagrass beds at low tide. This requires digging up muddy sediments from the bottom, including the seagrass plants. Over a decade, beginning in the mid-1990s, most of the seagrasses disappeared from some shallow bays, presumably due to this type of activity. Dredges used to remove clams from the bottom are especially destructive to seagrass habitats.

Coastal activities requiring land reclamation can cover over seagrasses, lead to excavation of the substrate where they are rooted, or negatively impact water flow. Construction of docks and piers can destroy plants or reduce primary production due to shading. Tourism activities also can impact seagrasses, if seagrasses are removed to make the waters appear cleaner for swimmers.

Diseases have been documented to impact seagrasses in some regions. Large die-offs of eelgrass in the North Atlantic and Florida waters have resulted from a wasting disease. This disease was documented by Peter Ralph and Frederick Short to be caused by a protist that dramatically reduces photosynthetic efficiency in affected leaf tissue. Eelgrass was severely impacted along the U.S. Atlantic coastline in the 1930s such that it was almost eliminated from much of the region. Its return has been limited in many areas due to anthropogenic nutrient and sediment input. Seagrass loss in Virginia and New Jersey's coastal bays resulted in a total disappearance of the bay scallop, an important coastal fishery (the larval scallops attach to seagrass leaves). Neither the seagrasses nor the scallops have returned. In New Jersey the loss of eelgrass due to coastal development and pollution also may be harming striped bass, blue crab, and clam populations. Another region affected by sea grass disease is the Wadden Sea region off northern Europe, where extensive seagrass habitat was reduced to a narrow strip around the water's edge in the 1930s. Efforts to recover and maintain seagrasses in the region has had limited success (**Box 6-1. Conservation Efforts: The Wadden Sea and Chincoteague Bay**).

Box 6-1   Conservation Efforts: The Wadden Sea and Chincoteague Bay

Because seagrasses are in shallow waters, they are vulnerable to coastal pollution and readily accessible to humans, making ecosystem protection challenging. This has been especially true in areas affected by the seagrass wasting disease that first occurred in the 1930s. Two areas where conservation efforts have had some success are the western Wadden Sea, on the coast of the Netherlands, and Chincoteague Bay, on the U.S. coast of Virginia and Maryland (**Figure B6-1**). Both areas were largely covered by *Zostera* seagrasses before the 1930s.

After the wasting disease dissipated, seagrasses in Chincoteague Bay began to recover gradually along the eastern shore. This shoreline is adjacent to an unpopulated barrier island (currently protected as a National Seashore) with healthy marshes along the edge of the Bay adjacent to a broad area of

dunes. Despite healthy marshes and low human populations on the western shore of the Bay, the seagrasses did not recover.

Through the 1980s, seagrasses continued their gradual recovery along the eastern shore. Increased awareness of the importance of coastal shallow-water ecosystems led to increased efforts to recover the seagrass habitats. Eventually laws were passed to prohibit most activities that would damage seagrasses and initiate recovery programs. Construction of sewage treatment plants and limiting fertilizer use along the western Bay area resulted in reduced nutrient loads and is believed to have contributed largely to a threefold increase in seagrass coverage in the Bay through the 1990s.

In the 21st century, new problems began to develop to set back the seagrass recovery in Chincoteague Bay. Certain

(a)

(b)

Figure B6-1   Efforts to recover seagrasses after a wasting disease in the 1930s have been met with some success in **(a)** Chincoteague Bay along the U.S. east coast and **(b)** the western Wadden Sea along the northern coast of the Netherlands.

fishing activities continued to be allowed. For example, a fishery for clams developed in the late 1990s used dredges that scarred the bottom, removing seagrass cover. Resource managers were quick to react in implementing laws to prevent dredging in seagrass beds; however, as is often the case when new conservation problems develop, it took several years for the laws to be fully implemented. In the meantime, clam dredging increased, scarring as much as 15% of the seagrass area in 1999.

After a slight decline in 2000, there were record levels of seagrass coverage in Chincoteague Bay in 2001 (over 66 square kilometers), suggesting that conservation efforts might be successful in restoring the ecosystem to its former health. The increase was short lived, however, as seagrasses declined in following years such that the 2007–2008 coverage was less than half that of 2001. Concurrent with this decline, there has been an increase in nutrient concentrations in the Bay. Eutrophication has been indicated by dominance of macroalgae and/or phytoplankton in some areas, suggesting that factors associated with development in the watershed feeding the Bay have begun to have negative impacts on the health of the seagrass ecosystem. The focus of conservation efforts now must be expanded to include issues related to agriculture, sewage treatment, and runoff that eventually reach the Bay.

The Wadden Sea encompasses 10,000 square kilometers along the border of Denmark, Germany, and the Netherlands. It contained large beds of *Zostera* seagrasses until the wasting disease of the 1930s removed much of the seagrass coverage. The southwestern portion bordering the Netherlands (the Dutch Wadden Sea) was particularly hard hit. Seagrasses covered 150 square kilometers prior to the wasting disease; afterward only remnant populations remained in a small portion of the intertidal zone. The bordering region of the Wadden Sea is much more densely populated than the Chincoteague Bay coastline, and increased turbidity and nutrients, dredging for

clams, and coastal construction activities resulted in little or no recovery of the seagrasses.

By the early 1970s only 5 square kilometers of seagrasses remained in the Dutch Wadden Sea. Recovery efforts were initiated in the 1980s as monitoring programs were developed and transplanting experiments were initiated; however, these took several years to be implemented. Eventually clam dredging was prohibited, and conservation efforts began to show some success by the end of the century as seagrass coverage increased to almost 100 square kilometers.

Early in the 21st century, Dutch laws were passed to protected *Zostera* eelgrass as a threatened species and the Wadden Sea was proclaimed a Conservation Area. Habitat protections were initiated to include a no-activity zone around eelgrass plants. A transplanting program was initiated in 2002 when it was realized that seagrasses would be unable to recolonize naturally. Even with transplanting, recovery has been slow. In the Wadden Sea, as in Chincoteague Bay, excess nutrient input is considered a major problem limiting ecosystem recovery and needs to be addressed at the drainage basin level. Proclamation of the Wadden Sea as a World Heritage Site by UNESCO in 2009 hopefully will bring further conservation efforts to the region.

In both the Wadden Sea and Chincoteague Bay on opposite sides of the Atlantic, we see a pattern and a dilemma that are common to marine conservation. In what used to be relatively healthy ecosystems, without the help of extensive conservation efforts, a dramatic decline (in this case from disease) has been experienced. Before these populations could recover on their own, humans had dramatically changed the ecosystem (in this case by increased siltation and nutrient inputs) so that even with extensive local efforts, complete population recovery is doubtful. Only with broader change in the way we use our resources and protect the environment will conservation become a matter of simply protecting what we already have.

Biotic interactions can result in a decline in seagrasses. For example, if sea urchin populations are large, they can overgraze the seagrasses, which causes a dramatic impact on extensive areas. Even though the sea urchins are a natural consumer of seagrasses, their populations may be affected by human activities. Overharvest of sea urchin predators, wrasses, and triggerfish off Haiti and the U.S. Virgin islands resulted in an explosion in sea urchin populations and a loss of seagrass beds.

Seagrass Protection and Recovery

Although only a small fraction of the seagrass habitat remaining globally is given total protection, there is an increasing trend in the protection of seagrass ecosystems through the establishment of Marine Protected Areas (MPAs). These do not necessarily provide protection from all human activities; for example, fishing may be allowed. The seagrasses and most of the associated ecosystem are

provided some degree of protection, however, because one of the goals of MPAs is to maintain healthy ecosystems. Few, if any, MPAs have been designated solely for the protection of seagrasses. The strong association of seagrasses with other shallow ecosystems, such as coral reefs, mangroves, and salt marshes, has resulted in over 250 MPAs in over 70 countries that include seagrasses as one of the protected habitats. Seagrasses are not always designated for specific protection in these MPAs, in part because of a lack of awareness of their importance in comparison to other shallow marine ecosystems, especially coral reefs. Protection of seagrasses is also affected by ineffective management and enforcement in some MPAs; for example, trawl fishing may not be restricted and boat propellers may destroy patches of seagrass through carelessness.

One of the most influential MPAs for protecting seagrass ecosystems is in Shark Bay, Australia (**Figure 6-15**). The Bay contains over 4,000 square kilometers of sea grass

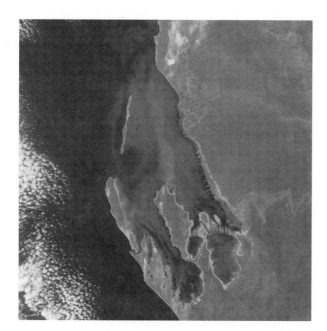

Figure 6-15 Shark Bay in western Australia, which contains the largest seagrass bed in the world. The blue-green coloration in the bay is the result of a large phytoplankton bloom. (See Color Plate 6-15.)

beds, with a quarter of that area in a single bed, the largest in the world. The largest stable population of dugongs, largely dependent on the seagrasses, resides in this MPA. It contains 12 species of seagrasses, 2 that are endemic to southern Australia, one of the most diverse assemblages in the world. Parts of the Bay support the growth of stromatolites, structures formed by consolidation of sediments by cyanobacteria, uncommon today, but important in the fossil record over 2 billion years ago. Many smaller MPAs in Australia protect seagrasses as parks or fish habitat areas.

Specific protection of seagrasses outside of designated areas has been implemented by some countries. These protections typically include restrictions on trawling or dredging or introducing sediments or pollution around the seagrass beds. Some of the strongest protections are provided in Australian waters, where federal fisheries and native vegetation protection laws give specific protection to seagrasses. For example, in Queensland, activities cannot be carried out that damage seagrass unless a permit is issued. In the United States, the Clean Water Act prohibits dredging and filling activities around seagrass beds without a permit. They are also protected as Essential Fish Habitat through the Fishery Conservation and Management Act (see Chapter 12), and various states have protective laws in place. Other regions and countries have implemented monitoring and protection measures aimed specifically at seagrasses.

Simply protecting impacted seagrass populations from further harm sometimes results in their recovery. For example, some seagrass populations harmed by wasting disease on the U.S. Atlantic coast have recovered well. A recurrence of the disease in New Hampshire in the 1980s was followed by natural recovery via seed production, and by 1996 coverage was greater than before the disease.

There have been some promising recoveries of seagrasses where water quality has been improved by reducing nutrient input and turbidity in shallow coastal regions. For example, seagrasses covered over 300 square kilometers in Tampa Bay, Florida prior to the 1900s. Through the 1960s, poorly treated sewage, industrial wastes, and dredging in the Bay caused the loss of most of the seagrasses. After an enhanced environmental awareness and stricter water quality laws beginning in the 1970s, a turnaround in the health of Tampa Bay and a continued increase in seagrasses have resulted, so that they are now back to about 40% of the original coverage. NGOs such as the Tampa Bay Watch combine education with recovery efforts by training high school and college student interns in techniques for replanting seagrasses. A similar program is in place in Rhode Island.

Transplanting of seagrasses can be time-consuming and expensive because it typically involves manually placing individual plants and may involve divers (**Figure 6-16**). Plants are taken from healthy seagrass stands and then transplanted to the recovery area. Studies by Marieke van Katwijk and colleagues found it crucial to consider habitat and source of donor plants; transplants should be collected from comparable habitats and placed into areas confirmed to be historical seagrass habitat.

Figure 6-16 Replanting of seagrasses may be recommended if water and sediment conditions are healthy. Preparations for replanting seagrasses in Santa Rosa Sound, Florida.

Planting and restoration efforts have only been about 30% successful around the world, but some areas and species have had better success than others. Some areas impacted by wasting disease in the U.S. Atlantic have been transplanted successfully with eelgrass. Other methods that have been attempted include attaching plants to a wire mesh frame. Seeds can be spread over the water surface for planting, but predation on the seeds can lead to low survival. Mechanical planters also have been developed to sow seeds in the bottom. In Florida, methods have been developed where boat propeller scars are filled with gravel and then silt, into which seagrasses are planted. One of the largest and most successful seagrass restoration efforts is being carried out in Virginia waters. It involves volunteers collecting millions of seeds from healthy seagrass beds in the spring; these seeds are raised in tanks and then transplanted as seedlings in the fall. Over 77 hectares of eelgrass have been recovered using these methods.

One of the major challenges to restoring seagrasses ecosystems is education and public awareness. Communities are more likely to work to minimize the impact on seagrasses if they understand their importance in supporting local subsistence fisheries. Locals and tourist are more likely to tailor their activities to avoid harm to seagrasses when they are aware of their importance to endangered animals such sea turtles, dugongs, or manatees. For example, in Thailand, an education campaign, including placing dugongs and seagrass on a postage stamp, has increased awareness and appreciation of seagrasses. After enforcing measures to protect seagrasses, fish harvest began to increase. Employing locals to enforce protection in an MPA also educated other local people of the importance of the seagrass habitat.

In industrialized countries there has been an increased awareness of the importance of limiting nutrient pollution and siltation. For example, after a new sewage treatment system was installed near Boston Harbor, seagrasses returned naturally after being missing for 200 years. In Chesapeake Bay, policies and plans have been developed through cooperative efforts among non-governmental, state, and federal agencies, born of the knowledge that seagrass health is linked to water quality and important to fisheries species. Similar success stories are not as common as they could be. Stricter laws and increased enforcement are needed in most regions to protect seagrasses.

Limiting the physical destruction from recreational activities is best achieved by educating boaters to stay out of seagrass beds. If boaters do enter the beds, they should be encouraged to stop the engine and paddle out of the bed to avoid further destruction. Laws are needed to restrict such activities as clam dredging around seagrass beds, such as those initiated in Chincoteague Bay on the Virginia coast (see Box 6-1, Conservation Efforts: The Wadden Sea and Chincoteague Bay).

Seagrass Restoration

Protecting seagrasses from the negative effects of human activities is typically the most practical way to conserve and recover seagrass ecosystems; however, there are some situations when restoration of seagrass beds is advisable. Scientists are improving techniques for transplanting seagrasses into sites with appropriate conditions such as high water clarity, low siltation and nutrient input, and adequate sediments. If pollution problems are resolved, replanting presents a means to accelerate recovery of the seagrass ecosystem. As mentioned above, planting seagrasses can be time-consuming and expensive because it often involves divers transplanting individual plants.

In the U.S., funding has been provided through governmental organizations for recovery of seagrass ecosystems. Because most seagrasses are in state waters in the United States, much of this funding is through state agencies. As an example, in New Jersey the State Department of Environmental Protection has provided funds for transplanting seagrasses onto barren bottoms.

A part of the problem with protecting and recovering seagrass ecosystems is a lack of knowledge of the severity of the problem. Scientists have begun to monitor and assess the status of seagrass ecosystems through a global monitoring program called Seagrass Net (http://www.SeagrassNet.org). This initiative has standardized methods of assessment so that seagrass habitat can be documented around the world and used as a baseline for future comparisons. It has encouraged monitoring of many seagrass habitats in regions that had not been previously monitored. In the Caribbean, the CARICOMP network (Caribbean Coastal Marine Productivity Network) monitors seagrasses (along with coral reefs and mangroves) through a data management center in Jamaica, involving the cooperation of marine labs and conservation organizations in 17 different countries. As with other ocean resource issues, research and education can go a long way towards avoiding the further loss of seagrass ecosystems, protecting those that remain, and working to recover at least some of those that have been lost.

6.3 Rocky Bottom Ecosystems

The habitat and ecosystems that develop in shallow subtidal regions depend on many physical and biological factors, and rocky bottoms persist in a minority of coastal regions.

Even in geologically active areas where rocky shorelines dominate, the subtidal region may be covered in sediments, especially where there is substantial river input. In tropical regions, much of the hard bottom is covered by the living organisms of the coral reef community (detailed in Chapter 5). In temperate ecosystems, many of these subtidal areas exist as extensions of adjacent rocky intertidal ecosystems and have similar communities as the lower intertidal zone. These regions are often covered in large attached macroalgae such as kelps.

Where open rocky subtidal regions exist, the biological community is typically more diverse than surrounding soft bottoms. The rocks provide a substrate to which organisms can attach, something that may be hard to find in areas of sand or sediment accumulation. Some of the hard bottom in temperate regions is formed by living organisms, such as algae, polychaete worms, and oyster shells that form a reef structure.

The most conspicuous residents of the rocky bottom are seaweeds. There is a remarkable diversity of colors, shapes, and sizes that may cover almost every square inch of a rocky bottom area. Some seaweed species are the same as those found in the rocky intertidal, but the red and brown seaweeds are most prevalent in the rocky subtidal. Factors that regulate the seaweed community include wave action, currents, temperature, and light. Low light penetration limits the depth at which algae can thrive in turbid waters, sometimes to less than one meter below the surface. In clear waters, however, seaweeds can survive to depths as great as 200 meters.

Invertebrates compete with seaweeds for space in the rocky subtidal (**Figure 6-17**). Attached **sessile** invertebrates dominate in regions that have little sunlight penetration. These animals live attached to the surface because burrowing into the rock is much more difficult than into soft bottoms; they include sea anemones, sponges, barnacles, and soft corals.

Other invertebrates in the rocky subtidal include slow-moving grazers, sea urchins typically being the most prevalent. Other grazers include mollusks and gastropods, such as abalones and various snail species. Some of the invertebrates are predators on other invertebrates. These include the sea stars, brittle stars, and other snails. In the rocky subtidal much of the competition is for the limited space, and animals compete by predation or growing over organisms already present on the rocks.

Fishes also may function as grazers or predators (**Figure 6-18**). Fish grazers in the rocky subtidal of tropical waters are similar to species found around coral reefs (e.g., parrotfish and surgeonfishes), and in temperate waters they are similar to species found in kelp ecosystems.

(a)

(b)

Figure 6-17 Organisms such as anemones and fish **(a)**, sea cucumbers, urchins, brittle stars, and macroalgae **(b)** compete for space in the rocky subtidal.

Crevice-dwelling fishes such as sculpins (Cottidae) and blennies (Blenniidae), however, are more persistent in the rocky subtidal. Grazers on invertebrates often have specialized teeth developed for scraping (e.g., sparids) or crushing mollusks and gastropods (e.g., drums). Predators include crabs and many fish species similar to those in other shallow coastal ecosystems. The ability of hard structures to attract a diversity of organisms is taken advantage of by placing them in shallow water as artificial reefs, often with a goal of enhancing fish abundance and diversity for angling or sport diving (**Box 6-2. Conservation Controversy: Artificial Reefs**).

(a)

(b)

Figure 6-18 Fishes such as rockfish **(a)** and sculpins **(b)** feed on invertebrates associated with the rocky subtidal.

Box 6-2   Conservation Controversy: Artificial Reefs

Biodiversity in subtidal ecosystems tends to be greater around structure, such as rocks, reefs, or vegetation, than in comparable areas covered with relatively flat sandy or muddy bottoms. For hundreds of years, humans thus have been attempting to enhance fish communities by placing hard artificial structures in the water to create **artificial reefs**. Any structure that is stable, hard, and heavy enough to stay in place might be used as an artificial reef. Materials include rocks and boulders; appliances such as washers and refrigerators; concrete and steel structures such as culverts and bridges; vehicles including automobiles, railroad cars, ships, and aircraft; and retired oil rigs (**Figure B6-2**).

In many ways, artificial reefs function similarly to natural reefs. Sessile organisms such as barnacles and clams settle on the reefs, attracting fishes that feed on them, which attract other predators. Eventually a diverse ecosystem is established around the reef consisting of organisms that otherwise might be limited to natural rock outcroppings or rocky shorelines. Research studies have shown that, at least in some situations, artificial reefs enhance fish populations and benefit ecosystems.

Artificial reefs are most often created to enhance fishing opportunities mainly for predatory species like snappers and groupers. Artificial reefs also can be attractive to sport divers, especially in regions where coral reefs do not develop. It is often argued in support of creating artificial reefs, that they enhance fish populations that are overfished or depressed for other reasons. There is an ongoing debate as to whether reefs put in place to enhance fishing actually increase the size of fish populations or whether they simply concentrate the fish for harvest. There is even concern that artificial reefs placed to enhance fishing may actually be harmful to fish populations by concentrating the fish, making them more vulnerable to harvest, and thereby leading to overfishing.

Although reef-based rigs provide an obvious benefit to fishers, the assumption cannot necessarily be made that fishery enhancement (greater fish harvest) is a result of higher fishery production (an increase in fish biomass). Improved fishing could be due to attraction of fish to the structure away from other habitats and not greater reproduction, survival, and growth that would increase population size. This argument

of "attraction versus production" is likely the most volatile for snappers, a highly valued commercial and recreational fishery species in the northern Gulf of Mexico. James Cowan and colleagues present strong arguments that artificial reefs, including oil and gas structures, in the Gulf that enhance red snapper fishing do so through attraction. The aggregation of red snappers to easily located artificial structures closer to shore (from ledges, cliffs, and rock outcroppings further offshore) could increase vulnerability to fishing, and predators could be attracted into nursery areas of juvenile red snappers. The combination of such factors could offset any increased production resulting from artificial reef placement.

Even if artificial reefs are not encouraged as a conservation measure, their economic impact provides another argument for their placement in some coastal waters. For example, off the U.S. South Carolina coast the economic impact of artificial reefs is estimated to be almost $20 million, mainly due to its enhancement of the sport fishing industry. The state of Florida has spent over $15 million on artificial reef programs, creating reefs from used or fabricated concrete structures, military equipment (e.g., tanks), steel vessels, and scrap steel or limestone. A further advantage is that materials placed as reefs do not have to be disposed of elsewhere.

One of the largest programs for creating artificial reefs is in the northern Gulf of Mexico off the Louisiana and Texas coasts. Here many of the artificial reefs are retired oil rigs. Through "Rigs to Reefs" programs, oil companies donate the material from oil and gas rigs, either toppling the rigs in-place or transporting them to other areas to create structures to "enhance living marine resources" and create fishing and diving opportunities. These programs are touted as a win–win situation. Oil companies save money and limit scrap metal that has to be disposed of otherwise, and anglers and commercial fishers benefit through enhancement of fish harvest. These artificial reefs are especially popular in the charter boat fishery, which typically target snappers, grouper, triggerfish, sheepshead, and other species. Commercial fishers also target fishes around the artificial reefs, the red snapper being one of the most valued.

Many smaller structures are commonly placed as reefs in coastal waters. Off most U.S. coastal states, artificial reef pro-

(a)

(b)

Figure B6-2 Artificial reefs: **(a)** a sunken ship off the U.S. Georgia coast; **(b)** fabricated reef modules in the Bluefields Bay marine sanctuary, Jamaica.

grams are run by state agencies that control the placement of reefs but allow regulated fishing around the reefs. Typically, the structures are marked by a buoy and may be identified on maps so that they can be located by anglers or divers and avoided by trawlers. Where private citizens are allowed to place artificial reefs, they typically cannot claim ownership of the reef due to common access laws in coastal waters (see Chapter 12). For example, in Alabama waters, private citizens can create artificial reefs, but the state doesn't defend their ownership. If someone discovers a privately placed reef they can also use it. Fishers off Alabama guard information on the location of "their" reefs very closely; for example, charter boats may not allow anglers to carry GPS units to identify reef locations. Limitations also are placed on the types of structures that can be placed as reefs. One of the most common objects used to create these reefs used to

be old cars (with requirements for the removal of toxic materials), but sinking washing machines, shopping carts, toilets, and other junk was common. After scraps began being displaced by currents and storms and interfering with fish trawls, the state of Alabama tightened restrictions on the types of materials used for artificial reefs. As these restrictions have increased, businesses have been created to prefabricate artificial reefs. At least one company has even developed a cemetery where ashes from human cremations are incorporated into concrete "tombstones" to form the reef's structure.

In other countries, rules concerning the placement of artificial reefs vary. For example, in Japan, fishing communities can establish cooperatives that have ownership of artificial reefs. These communities have the incentive to protect and enhance the productivity of these resources.

6.4 Kelp Ecosystems

Kelps

Kelp is a type of brown seaweed (Laminariales) of various species, known to form some of the largest living structures growing from the seafloor. These cool-water ecosystems over coastal hard bottoms are referred to as kelp ecosystems or kelp forest ecosystems (**Figure 6-19**). The kelps are attached to the bottom by **holdfasts** and the blades (**fronds**) extend through the water, sometimes to the surface, attached to the stalk (**stipe**). Kelps can survive relatively deep waters because the fronds are exposed to light at the surface, while the holdfasts attached to the bottom keep them in place.

The diversity of kelp species is relatively low, with a region typically being dominated by one or two species. The U.S. Pacific coast is especially diverse and includes about 20 species, with 5 being prevalent (**Figure 6-20**). Kelps can be grouped into three categories based on size. The largest of the kelp are referred to as canopy kelps. These occur primarily along the Pacific coasts of North and South America and produce canopies that float at the surface. These include bull kelp *Nereocystis luetkeana* and giant kelp *Macrocystis pyrifera* (**Figure 6-21**) that can extend up to 45 meters from the bottom and grow at rates up to 50 cm per day. Canopy kelps can reach heights of 15 meters in just three years, and individual kelps can live about 25 years. Stipate kelps do not grow as large, up to 10 meters length, and are held up in the water by rigid stipes. Stipate kelps are found in various locations in the Pacific and Northwest Atlantic. Prostrate kelps are the smallest, growing over the substrate, and are common on both sides of the North Atlantic. All three types of kelp can co-occur in one area, and other

Figure 6-19 Kelp ecosystems support a diverse assemblage of fishes including these rubberlip seaperch in the eastern Pacific off California.

bottom plants and encrusting algae can be found in the kelp ecosystem. This results in a great diversity of habitats for other organisms.

The life history and reproduction varies among kelp species, and an understanding is necessary when

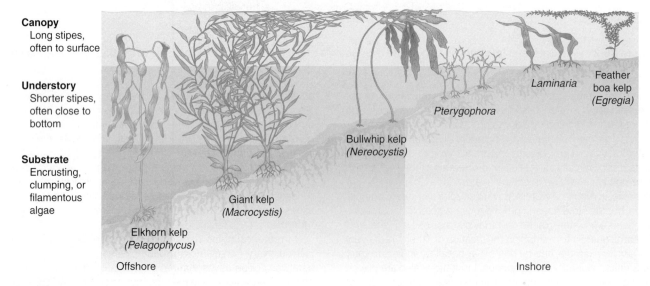

Figure 6-20 Kelp forests off the U.S. Pacific coast are dominated by *Macrocystis* or *Nereocystis,* but include a diverse assemblage of plants in the understory.

Figure 6-21 Macrocystis kelp growing off the coast of California.

areas impacted by natural or man-made disturbances. One unusual aspect of reproduction is that it is not seasonal but is linked to high nutrient levels or cool water temperatures. This means that kelp can rapidly adjust to optimal survival and growth conditions, suggesting that they are adapted to a variable environment. This also suggests that the dramatic fluctuations observed in kelps may be a normal condition in some areas.

Temperature is important in limiting the distribution of kelps, which, in contrast to mangroves, seagrasses, and coral reefs, cannot tolerate warm water. Coastal current gyres have a strong influence on kelp distribution. For example, kelps can be found scattered along much of the western coast of North America because of the cold waters transported down the coastline by the California Current. Upwelling of nutrients also supports the high productivity of those kelp ecosystems. In the western Pacific the distribution of kelps is much more restricted, limited to the northern Asiatic coast off Russia and northern Japan. In the southern Pacific, extensive areas of kelps are found along much of the South American coast. In the Atlantic, kelps are restricted to scattered regions at higher latitudes where colder waters are available. In lower latitudes kelp disappear due to their low tolerance to high temperatures, and are replaced by other large fleshy brown macroalgae such as sargassum. Kelps are less tolerant to low light levels than are most other algae. Kelps thus are uncommon above 60° latitude due to the relatively low amount of sunlight available and extended periods of darkness. This light limitation restricts kelps to coastal waters shallow enough that they can get adequate sunlight for photosynthesis.

considering reintroducing or transplanting kelp. For the giant kelp, there is an alternation between an asexual **sporophyte** stage and a sexual microscopic **gametophyte** stage (**Figure 6-22**). Fertilization by the male and female gametophytes results in the leafy sporophyte. Spores can remain swimming and viable for several days, an adaptation for long-distance dispersal, important for recolonization of

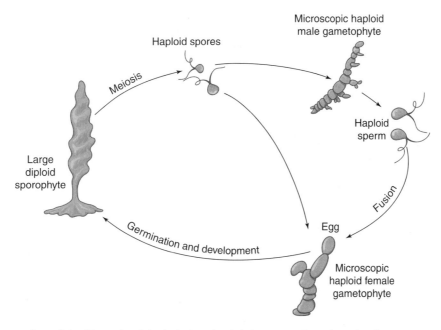

Figure 6-22 An understanding of the life cycle of the kelp *Laminaria* is important for reintroduction programs.

Kelp Communities

The forests that kelps create are home to a large diversity and abundance of organisms (**Figure 6-23**). The kelp canopy creates productive protected habitats by dampening waves, altering water flow, enhancing productivity, and creating various light conditions. The kelps provide a substrate for the attachment of invertebrates and algae and habitat for animals that live and feed on the kelp and other organisms. The kinds of organisms present depend in part on the nature of the kelp habitat. Those that live in associated with the kelp holdfasts include various polychaete worms, small crabs, shrimp and other crustaceans, and brittle stars. Animals living on the leaves include tube-dwelling polychaetes and many sessile invertebrates. Inhabitants of the rocky bottoms around kelps include crabs, sea urchins, sea stars, sponges, octopuses, and abalones. Kelps are also an important source of biomass for coastal herbivores and detritivores that feed on the plant materials littering the coastline.

Although there are few fishes that are exclusively limited to kelp ecosystems, a large diversity of fishes inhabit the kelp forests and feed on organisms residing there (**Figure 6-24**). For example, off California about 50 species of coastal fishes are common around kelp beds. These species tend to group based on habitat and feeding preferences. One group is associated with rocky areas and dense stands of kelp. Many of these are in the same families as those found in tropical reef habitats, such as wrasses (Labridae) and damselfishes (Pomacentridae). One of the most important predators is the California sheepshead wrasse (*Semicossyphus pulcher*), which uses its strong teeth to feed on invertebrates such as barnacles, crabs, mollusks, and sea urchins. Another obvious kelp inhabitant is the bright orange garibaldi damselfish (*Hypsypops rubicundus*), which defends its nest among algae that it tends on the rocks, feeding on the algae and bottom invertebrates. The blacksmith damselfish (*Chromis punctipinnis*) swims above the rocks to feed on zooplankton. The feces it deposits among the rocks are an important source of nutrients for benthic invertebrates.

Other fishes are associated with the kelp fronds in the canopy, where they graze on small invertebrates. Juvenile rockfishes (Scorpaenidae) of various *Sebastes* species also graze among the kelp fronds. Other groups of fishes associate with the rocks or kelp on the bottom; these include small cryptic fishes such as blennies, sculpins, and kelpfish (Clinidae) that commonly feed on small invertebrates. Adult rockfishes and lingcod (*Ophiodon elongatus*) feed on fishes and larger invertebrates. Surfperches (Embiotocidae) live at the shoreward margin of the kelps and feed on invertebrates hiding in algae beds. Finally, there are fishes that commute among the various kelp habitats searching for fishes or invertebrates on which to feed. These include surfperches, rockfishes, the piscivorous kelp bass (*Paralabrax clathratus*), and the rubberlip seaperch (*Rhacochilus toxotes*), a nocturnal invertivore.

Associations among plants, invertebrates, and fishes can be very important to maintain the kelp ecosystem. For example, if a stand of kelp is destroyed by storm activity, recolonization can be inhibited by urchins and fishes grazing on new kelp plants. But if algae recolonize the area before the kelp, they provide some protection to the young kelp from grazing fishes, such as the señorita wrasse *Oxyjulis californica*, which clips off the kelp in order to feed on attached bryozoans. In another important interaction, the señorita feeds on isopods that, if left unchecked, can destroy mature kelp plants.

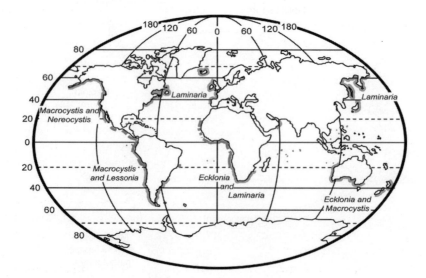

Figure 6-23 Kelp forests are largely restricted to cooler waters at higher latitudes or in regions of coastal upwelling. The darker gray represents forests in deep water, and the lighter gray represents kelp forests on the surface.

(b)

(a)

(c)

Figure 6-24 Some components of the complex kelp ecosystem: **(a)** kelp, other seaweeds, and invertebrates; **(b)** garibaldi damselfish and urchins; and **(c)** sea otter, eating an urchin.

In other regions of the world, associations within the kelp ecosystem are similar but involve different species, and many of these associations are not yet well documented. It is important that the kelp forests be viewed from an ecosystem perspective. When any component of the food web is removed, the response of the ecosystem is rarely predictable but can result in devastating effects to the kelp or other organisms.

■ Impacts to the Kelp Ecosystem

Kelp can be lost due to a variety of factors, including disease, excess herbivory, and physiological stress. Natural losses are more common at the edge of the kelps' geographic range, and are usually due to anomalous temperatures or changes in salinities or nutrients. Although large areas may be affected, the kelp will typically recover over a few years time. For example, large die-offs of kelp have been documented off the coast of Japan since at least the early 1900s, attributed to changes in salinity or current patterns affecting water temperature. El Niño events can cause kelp

die-offs when the upwelling of cold nutrient-rich waters is interrupted, as can occur along the South America Pacific coast (see Chapter 1). Physical stress also can make the kelps more susceptible to diseases. Kelps off Tasmania have undergone dramatic fluctuations and shifts in distribution linked to El Niño events, changes in coastal current patterns, and storm events. When the environment returns to favorable conditions, however, recovery has been rapid.

In the center of the range of kelp ecosystems, in midlatitudes between about 40° and 60°, many of the documented die-offs of kelp have been due to herbivory by sea urchins (**Figure 6-25**). This has been most prevalent in the northern hemisphere, for example, in Alaska, the Gulf of Maine, Canadian Atlantic, Iceland, and Norway. These die-offs appear to be fairly recent phenomena—since the 1960s in most areas—suggesting some anthropogenic cause. The most well-documented effect is from excess harvest of sea urchin predators. When the predators decline, the sea urchin populations increase dramatically. Sea urchins are a native component of the kelp ecosystem that normally feed on

Figure 6-25 Sea urchins feeding on giant kelp.

detached, drifting kelp; however, when the urchins become overpopulated, they begin to eat live kelp fronds and hold-fasts. When the holdfasts are gone the kelp drifts away and dies. Large areas of kelp can be cleared completely by the sea urchins, leaving a "barren" containing virtually no kelp. The development of kelp barrens has been associated with overharvest of cod in the northwest Atlantic and California sheepshead off southern California. In each of these cases, in areas or times when predatory populations are healthy, sea urchin populations are kept in check; it is only with the decline of predators that the urchin populations expand dramatically. In regions off the Aleutian Islands of Alaska kelp loss has been attributed to declines in urchin-eating sea otters; however, this may not be a simple predator–prey response, but the result of a chain of interactions (**Box 6-3. Learning from History: Phase Shifts in Kelp Ecosystems**).

Nutrient input from anthropogenic sources does not appear to be a major source of harm to most kelp ecosystems. Kelps are more tolerant and, in fact, need high levels of nutrients compared to nearshore ecosystems such as coral reefs; however, sewage pollution can result in a cascade of events that negatively affect kelp ecosystems. For example, in California a coastal sewage spill was followed by an increase in sea urchins and a decline in kelp canopy. It has been suggested

that the input may have resulted in stress to the kelp, making it more vulnerable to grazing by sea urchins. These kelps recovered quickly when healthy water conditions returned.

Kelp harvest is another potential source of harm to kelp ecosystems. Harvest is carried out for various uses, including extraction of alginates as thickening agent for products such as ice cream, jelly, toothpaste, cosmetics, and pharmaceuticals, as a health food, and as a smoothing agent in paints. In waters off both California and Norway over 100,000 tons of kelp are harvested each year on average using ships with machinery designed for cutting the fronds (**Figure 6-26**). Although kelp harvested in this manner will regrow, there is concern over potential negative ecosystem effects.

Stressed kelp ecosystems may be more vulnerable to exotic invasive species. For example, in the western North Atlantic, where biodiversity is low in the kelp ecosystems, several exotic species have become problematic. These include encrusting bryozoans that cover the kelp and cause loss of fronds during the summer, and crabs that sometimes dominate the ecosystem. These kinds of invasions have not occurred in areas where kelp ecosystems with a high species diversity.

Kelp Protection and Recovery

Because many kelp populations undergo dramatic natural fluctuations, it is sometimes difficult to know when active methods would be beneficial to assist in recovery. Minimizing impacts on the system and maintaining natural animal communities, including predators, may allow the kelps to recover when optimal conditions return. In today's world this is rarely simple. Analyses of the current status of kelp ecosystems by Robert Steneck and colleagues led to the conclusion that much of kelp conservation is related to issues of overfishing and climate change, with overfishing being the most manageable of these problems. Many of the issues discussed in this chapter are not unique to kelp conservation and will require a global concerted effort to resolve, a theme to which we will return numerous times.

Restoration of biodiversity—including key predator species—to kelp ecosystems would probably be the most beneficial focus of conservation efforts. Sea otters currently are protected in U.S. waters, but negative impacts on other components in the ecosystem may be affecting otter populations, as discussed earlier. Restoration of North Atlantic cod and California sheepshead could renew the healthy function of these kelp ecosystems.

Harvest of sea urchins in areas where they are overabundant may assist the recovery of kelp; however, as described for U.S. Atlantic kelp, it would not likely restore a normally functioning ecosystem. In reality, biodiversity would likely decline when both the predators and the herbivores are removed from the ecosystem. We have to ask the question: do we want the kelp without its ecosystem?

Box 6-3   Learning from History: Phase Shifts in Kelp Ecosystems

In ecology, a **phase shift** refers to a rapid transition in the living organisms (biota) in an ecosystem. For example, the dominant organisms in a kelp ecosystem can switch predators to herbivores as a result of natural events, such as storms or diseases, or man-made changes, such as overfishing or pollution. When dramatic and frequent phase shifts occur in an ecosystem, anthropogenic causes are always suspect. Kelp ecosystems are ideal ecosystems to look for factors causing phase shifts. They are fairly easily monitored compared to other marine ecosystems and tend to undergo dramatic changes in the prevalence of the dominant organism, the kelp itself.

There is some evidence for phase shifts in kelp ecosystems that predate recent history. Although scientific proof of these events or their causes is often unavailable, it can be useful to develop "educated guesses" that can apply to recent events and assist in the development of conservation plans. Analyses by Robert Steneck and colleagues provide likely phase-shift scenarios. Kelp forests have likely been present in the region of the Aleutian Islands of Alaska for tens-of-millions of years. The dominant players in this ecosystem included sea urchins, sea otters, and the Steller's sea cow *Hydrodamalis gigas*. The ecology of the sea cow is not known because the species was driven to extinction in the mid-1700s by European fur traders (see Chapter 9); however, it is likely that it specialized on eating the canopy kelp. Because the sea cows probably could not dive, they would have fed on the kelp fronds in the canopy and not caused kelp die-offs. Sea otter predation probably kept sea urchins from overgrazing the kelps. Other animals were present, but were probably not major players, and this was likely a relatively stable ecosystem for millions of years.

With the movement of humans into this region, evidence of changes becomes apparent, suggesting a decline of sea cows long before their extinction. One possible scenario is that the first settlers of the Americas about 10,000 years ago overharvested the relatively defenseless sea cows, reducing their populations before moving on to the harvest of whales. The reduction in sea cows probably did not have a dramatic influence on the health of the kelps. Sea otters would have kept the sea urchin populations in check and prevented overgrazing on the kelp. Harvest by Native Aleuts began diminishing sea otter populations about 2,500 years ago, however, resulting in increases in sea urchins (as indicated by the larger sizes of sea urchins in remains from this period), and presumably increasing urchin grazing on the kelp. The remaining sea otters were eventually hunted to near extinction by North American fur trades in the 1700s and 1800s. This resulted in a phase shift as sea urchin populations exploded, overgrazing the kelps and causing a massive die-off (**Figure B6-3**).

The next phase shift was a return to the original kelp-dominated ecosystem. This resulted from legal protection of the sea otters initiated in the 1960s and 1970s (Chapter 9). The recovered sea otter populations once again controlled the sea urchins, allowing the kelp to return to a healthy state. Finally in the 1990s a new and unexpected phase shift occurred. Sea otters began declining, despite legal protection, and sea urchin overgrazing returned. This appears to be due to a chain of events, as follows:

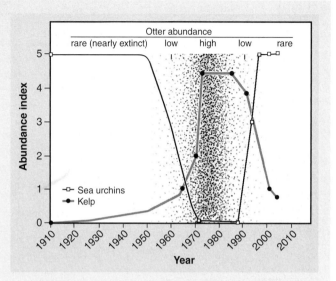

Figure B6-3   Abundance trends indicating interactions among kelp, sea urchins, and sea otters in an Alaskan eastern Pacific kelp ecosystem.

1. Something caused a decline in sea lions and seals. This could either be the human overharvest of their fish prey or other environmental changes.
2. Killer whales, which usually prefer to eat sea lions and seals, began to prey more heavily on sea otters.
3. The sea otter populations declined, resulting in an increase in sea urchins and overgrazing of kelp.

Similar phase shifts have been noted in kelp ecosystems in the western North Atlantic. Although these ecosystems may not be as ancient as those in the eastern North Pacific, they appear to have been fairly stable, until recently, over at least the past 5,000 years. In this region, the Atlantic cod and other predatory fish control the sea urchin populations and keep them from overgrazing the kelp. Historical accounts suggest that kelp dominated coastal ecosystems from Greenland to Cape Cod at least until the 1930s. Cod have been harvested from this region for hundreds of years, but it was only in the 1930s that fishing pressure became great enough to dramatically reduce cod populations (see Chapter 11). The removal of most of the predatory fish by fishing had predictable impacts on the kelp ecosystem, a phase shift to an urchin-dominated ecosystem and loss of most of the kelps (**Figure B6-4**). Development of an urchin fishery resulted in a rapid decline in sea urchins, and resultant recovery of the kelp. The ecosystem once again appeared, superficially, to be a normal kelp-dominated ecosystem; however, the species diversity had declined because fishing had now removed both the predators (cod) and the herbivores (urchins). In the Gulf of Maine this has opened up a niche for a new type of predator, crabs (*Cancer*), which moved into the kelps in massive numbers. Attempts to reintroduce urchins have been unsuccessful because they are quickly attacked by large swarms of crabs. This latest phase shift, to an herbivore-free ecosystem of lower diversity, appears to be stable, at least for the time being. Interestingly, the dominant predators

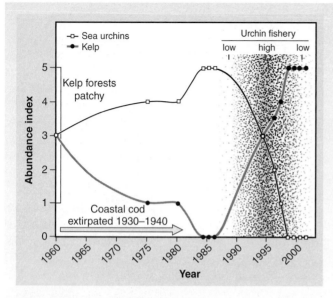

Figure B6-4 Trends in abundance of kelp and sea urchins in a Gulf of Maine, western Atlantic, kelp ecosystem.

of sea urchins in European kelp ecosystem are crabs. It is possible that removal of fish predators by European fishing over hundreds of years resulted in this current situation.

These examples emphasize several points generally applicable to marine conservation. One is the danger of focusing management on individual species rather than working toward protecting entire ecosystems. Another is the unpredictability of human impacts on complex ecosystems; any effect on one link in the food web can cascade throughout the food web in unpredictable ways. And finally, humans are very capable—with or without modern technology—of modifying populations and entire ecosystems, sometimes in ways that may be virtually impossible to mend.

Active methods of restoring kelp have been attempted with variable success (**Figure 6-27**). These include lab culturing of zoospores (reproductive cells) and sporophytes (vegetative structures) onto an artificial substrate (e.g., plastic plates or twine) and attaching these to the natural substrate in the field. This method has worked with variable levels of success. Problems with mortality result from grazing by invertebrates or being covered by sedimentation. If attached to twine, the transplants can be elevated above the bottom to avoid most grazers. Other methods involve collecting healthy juvenile or larger sporophytes or drifting detached pieces of kelp. These are attached by splicing their

holdfasts among the strands of a rope, which is then attached to the substrate. When healthy kelp forests are near the recovery area, it may be sufficient to clean the substrate of sediment or other attached organisms, or to add rocks as attachment areas, and allow natural kelp recruitment. In most cases it is expected that other organisms natural to the kelp ecosystem will move into the area on their own.

Several large-scale efforts have been made in recent attempts to recover kelps along the U.S. Pacific coast. An intensive introduction was carried out to establish a 61-acre kelp reef on the U.S. Pacific coast for mitigation of environmental impacts of a nuclear generating station.

Figure 6-26 Products from harvested kelp fronds are used in the food, cosmetics, pharmaceutical, and paint industries.

Figure 6-27 A diver examines a kelp planting during a restoration project.

Quarry rock was added as substrate to encourage natural recruitment, which occurred in some areas from kelp as far away as 3.5 kilometers, presumably from dispersal of spores. Another large restoration effort was carried out off the southern California coast in 2001–2007. NGOs, in co-operation with federal agencies, used various methods to restore about 20,000 square meters of kelp bed. Kelp were grown in aquariums and outplanted by attachment to rocks using rubber bands, drifting kelp were re-attached to the bottom, and urchins were relocated to other areas to promote natural recruitment of kelp. Expense was minimized with this program by involving volunteers, and education was incorporated by involving students in the program.

Laura Carney and colleagues at San Diego State University carried out experimental studies to determine feasible methods for smaller scale, less expensive restoration of bull kelp. The most successful transplanting method was achieved by taking juvenile sporophytes from natural populations and weaving the holdfast into nylon ropes, which were then attached to natural substrate with epoxy. Mortality was 70%, but 30% of the kelp that survived were healthy. Another method attempted was to transplant newly settled zoospores and lab-cultured microscopic sporophytes onto natural substrate in Petri dishes; however, survival was low due to overgrowth by other organisms and sedimentation.

Kelp restoration programs may become more necessary in the future if coastal development or climate change continues to affect these ecosystems. It is recommended that experiments continue to develop useful transplant methods. Conservation efforts should be incorporated into transplanting programs if a natural ecosystem is to be established. As with any re-introduction efforts, care must be taken to maintain the genetic integrity by establishing kelps that are of genetic stock native to the region. Where natural recruitment is possible, transplantation may not be necessary at all if impacting factors can be removed.

It may be unrealistic to think that we will be able to return all kelp beds to pristine conditions and protect them from future human impacts. But the more areas that are designated as marine sanctuaries or protected areas, the more successful kelp conservation will be. Along the U.S. Pacific coast, four marine sanctuaries protect kelp ecosystems. These include the Monterey Bay National Marine Sanctuary, a 560-mile strip off the central California coast protecting giant kelp ecosystems. Despite protection of the kelp ecosystem within the sanctuary, such issues as pollution from outside sources and nearby dredging continue to stress the ecosystem. A marine preserve off the Channel Islands of California contains kelp ecosystems and is one of the largest designated no-fishing zones in U.S. waters. The Gulf of Farallones and Olympic Coast National Marine Sanctuary protect bull kelp ecosystems.

6.5 Climate Change and Nearshore Ecosystems

As with most marine conservation issues, we must consider what effects global climate change could have on efforts to recover and protect shallow subtidal ecosystems. For example, some climate change models predict that ocean stratification will increase with climate change, resulting in increased chances for the development of hypoxic zones. Rainfall patterns could result in changes in nutrient input to coastal systems with increased river flows. On the other hand, increases in coastal storms would decrease stratification and reduce the occurrence of nearshore hypoxic zones.

Seagrasses could be affected by rising sea levels, circulation patterns, tidal regimes, ultraviolet (UV) radiation, salinities, and the occurrence of extreme events. Sea level rise increases marsh erosion, which increases turbidity and thereby reduces seagrass coverage; this may lead to further erosion because the grasses reduce wave action. Increased hurricanes could destroy more seagrasses. Climate change impacts would not necessarily be negative; for example, higher carbon dioxide levels could increase seagrass productivity.

As climate changes, seagrass distributions is likely to change. This is especially true at the edge of the ranges of seagrass species. For example, North Carolina is at the southern edge of the range of eelgrass and at the northern edge of the range for shoal grass, but widgeon grass (*Ruppia maritima*) tolerates a wide range of temperatures and ranges north and south of North Carolina. Changes in water temperatures would likely result in shifts in the species makeup in this region.

Due to the sensitivity of kelp to warm temperatures, global warming would likely result in a range reduction. Kelp ecosystems closest to the tropics may be lost. An increased frequency of El Niño events could stress kelp ecosystems in upwelling regions due to the increases in temperature and reductions in upwelling of nutrients. An increase in intense storms may affect kelp; however, kelp ecosystems tend to recover from storm events fairly rapidly. Sea level rises could increase sedimentation and turbidity, which would reduce kelp abundance.

Although many questions remain concerning the severity and time frame of global climate change (see Chapter 1), it is useful to develop predictions of potential effects of global climate change in coastal and shallow marine ecosystems, which are sensitive to temperature changes and coastal influences. Even if it is difficult to reach a consensus over what the predicted impacts could be, knowledge currently being obtained and methods currently under development will be needed to react to changes as they occur. Incentives to protect these ecosystems for the long term include the benefits they provide as habitat and reservoirs of biodiversity as well as for recreation and the production of resources for human use.

STUGY GUIDE

Topics for Review

1. Explain why temperature is one of the most important factors affecting the types of species living in subtidal ecosystems.
2. Describe the conditions that lead to hypoxia in coastal bottom waters.
3. Describe why hypoxic conditions are more prevalent during summer months.
4. Describe the conditions that lead to "jubilee" events in Mobile Bay, Alabama.
5. Explain how channelization of the Mississippi River has enhanced the development of dead zones in the northern Gulf of Mexico.
6. What factors contribute to seagrasses being more broadly distributed than other shallow marine ecosystems?
7. What mechanisms to seagrasses use to thrive in environments where low nutrients limit the growth of macroalgae?
8. How are seagrasses adapted to live in soils containing low concentrations of oxygen?
9. How are seagrasses adapted to rapidly recolonize patches lost due to local disturbances?
10. How are seagrasses adapted to recolonize large areas lost due to large scale environmental impacts?
11. What factors may lead to the decline of seagrasses with excess nutrient inputs?
12. Discuss why seagrass ecosystems support such a large diversity of invertebrates despite the fact that few species can digest seagrass leaves.
13. Describe how grazing by sea turtles and dugongs affect seagrasses differently.
14. Describe biological interactions that lead to haloes around patches of coral reefs that are bare of seagrasses.
15. Describe how seagrasses enhance adjacent coral reef ecosystems.
16. Describe how increases in turbidity can affect sea grasses.
17. Explain how an increase in coastal tourism can lead to the physical destruction of seagrass beds.
18. Describe how commercial fishing activities can physically harm seagrass beds.
19. Describe how the loss of seagrass can affect scallop populations.
20. What activities have contributed to recovery of seagrasses in Tampa Bay, Florida?
21. Present arguments in opposition to placing artificial reefs to enhance fishing in coastal regions.
22. What aspect of kelp reproduction is adapted for surviving annual variations in environmental conditions?
23. What are the primary factors that limit and control the natural distribution of kelp?
24. How does the senorita fish both help and harm the kelp by feeding on invertebrates in kelp ecosystems?
25. What factors impact kelps during El Niño events?
26. Describe the ecological interactions that presumably kept kelp ecosystems stable in pre-human times.
27. What factors have resulted in kelp ecosystems in the northwest Atlantic having healthy kelp populations but low species diversity?
28. What are some of the predicted impacts of global climate change on kelp ecosystems?

Conservation Exercises

Discuss the factors that have hindered the success of the following conservation efforts.

a. reducing the number and size of dead zones in coastal waters.
b. reducing turbidity that has caused declines in seagrass populations.
c. reducing nutrient loads that result in the decline of seagrass coverage.
d. protecting seagrass habitats in Marine Protected Areas.
e. limiting destruction of seagrasses by boating activities.
f. restoring seagrasses in the western Wadden Sea.
g. restoring kelp ecosystems around the Aleutian Islands.
h. transplanting kelp off the U.S. west coast.

FURTHER READING

Carney, L. T., J. R. Waaland, T. Klinger, and K. Ewing. 2005. Restoration of the bull kelp *Nereocystis luetkeana* in nearshore rocky habitats. *Marine Ecology Progress Series* 302:49–61.

Claudet, J., and D. Pelletier. 2004. Marine protected areas and artificial reefs: a review of the interactions between management and scientific studies. *Aquatic Living Resources* 17:129–138.

Cowan, J. H., C. B. Grimes, W. F. Patterson, C. J. Walters, A. C. Jones, W. J. Lindberg, D. J. Sheehy, W. E. Pine, J. E. Powers, M. D. Campbell, K. C. Lindemna, S. L. Diamond, R. Hilborn, H. T. Gibson, and

K. A. Rose. 2011. Red snapper management in the Gulf of Mexico: science- or faith-based. *Reviews in Fish Biology and Fisheries.* 21:187–204.

Diaz, R. J., and R. Rosenberg. 2008. Spreading dead zones and consequences for marine ecosystems. *Science* 321:926–929.

Dorenbosch, M., W. C. E. P. Verberk, I. Nagelkerken, and G. van der Velde. 2007. Influence of habitat configuration on connectivity between fish assemblages of Caribbean seagrass beds, mangroves and coral reefs. *Marine Ecology Progress Series* 334:103–116.

Green, E. P., and F. T. Short (eds.). 2003. *World atlas of seagrasses.* UNEP-WCMC, University of California Press: Berkeley.

Herrier, J. L., J. Mees, A. Salman, J. Seys, H. van Nieuwenhuyse, and I. Dobbelaere (eds.). *Proceedings 'Dunes and Estuaries 2005.'* International Conference on Nature Restoration Practices in European Coastal Habitats, Koksijde, Belgium. VLIZ Special Publication 19.

Hogarth, P. 2007. *The biology of mangroves and seagrasses.* Oxford University Press, New York.

Jelbart, J. E., P. M. Ross, and R. M. Connolly. 2007. Fish assemblages in seagrass beds are influenced by the proximity of mangrove forests. *Marine Biology* 150:993–1002.

Jones, G. P., and W. J. Seaman. 1997. Do artificial reefs increase regional fish production? A review of existing data. *Fisheries* 22:17–23.

Lefebvre, L. W., J. P. Reid, W. J. Kenworthy, and J. A. Powell. 2000. Characterizing manatee habitat use and seagrass grazing in Florida and Puerto Rico: implications for conservation and management. *Pacific Conservation Biology* 5:289–298.

Luczkovich, G., P. Ward, J. C. Johnson, R. R. Christian, D. Baird, H. Neckles, and W. M. Rizzo. 2002. Determining the trophic guilds of fishes and macro-invertebrates in a seagrass food web. *Estuaries* 25:1143–1163.

Marsh, H., H. Penrose, C. Eros, and J. Hugues. 2001. *Dugong: Status report and action plans for countries and territories.* United Nations Environment Programme, Nairobi, Kenya: UNEP/DEWA/RS.02-1.

May, E. B. 1973. Extensive oxygen depletion in Mobile Bay, Alabama. *Limnology and Oceanography* 18:353–366.

Moyle, P. B., and J. J. Cech. 2004. *Fishes: an introduction to ichthyology.* Prentice-Hall, New Jersey; 726 pages.

Nagelkerken, I., and G. Van der Velde. 2004. Are Caribbean mangroves important feeding grounds for juvenile reef fish from adjacent seagrass beds? *Marine Ecology Progress Series* 274:143–151.

Orth, R. J., T. J. B. Carruthers, W. C. Dennison, C. M. Duarte, J. W. Fourqurean, K. L. Heck, A. R. Hughes, G. A. Kendrick, W. J. Kenworth, S. Olyarnik, F. T. Short, M. Waycott, and S. L. Williams. 2006. A global crisis for seagrass ecosystems. *Bioscience* 56:987–996.

Puritz, J. B., and R. J. Toonen. 2011. Coastal pollution limits pelagic larval dispersal. *Nature Communications* 2:226 doi:10.1038/ncomms1238.

Rabalais, N. N., R. E. Turner, R. J. Diaz, and D. Justi. 2009. Global change and eutrophication of coastal waters. *ICES Journal of Marine Science* 66:1528–1537.

Ralph, P. J., and F. T. Short. 2002. Impact of the wasting disease pathogen *Labyrinthula zosterae,* on the photobiology of eelgrass *Zostera marina. Marine Ecology Progress Series* 226:265–271.

Reed, D. C., C. D. Amsler, and A. W. Ebling. 1992. Dispersal in kelps: factors affecting spore swimming and competency. *Ecology* 73:1577–1585.

Rick, T. C., and J. M. Erlandson (eds.). 2008. *Human impacts on ancient marine ecosystems.* University of California Press, Berkeley.

Steneck, R. S., M. H. Graham, B. J. Bourque, D. Corbett, J. M. Erlandson, J. A. Estes, and M. J. Tegner. 2002. Kelp forest ecosystems: biodiversity, stability, resilience and future. *Environmental Conservation* 29:436–459.

van Katwijk, M. M., A. R. Bos, V. N. de Jonge, L. S. A. M. Hanssen, D. C. R. Hermus, D. J. de Jong. 2009. Guidelines for seagrass restoration: importance of habitat selection and donor population, spreading of risks, and ecosystem engineering effects. *Marine Pollution Bulletin* 58:179–188.

Wazniak, C., L. Karrh, T. Parham, M. Naylor, M. Hall, T. Carruthers, and R. Orth. 2005. Seagrass abundance and habitat criteria in the Maryland coastal bays. In *Maryland's coastal bays: ecosystem health assessment 2004.* Wazniak, C. E., and M. R. Hall (eds.). Maryland Department of Natural Resources, Tidewater Ecosystem Assessment: Annapolis, MD.

The Open Ocean: Anthropogenic Inputs and Environmental Impacts

CHAPTER

7

The open ocean beyond the littoral zone holds over 99% of the water on earth (**Figure 7-1**). Because there are very few structures above the seafloor that can be used as shelter or to which organisms can attach, most either swim or are free-floating in the water column. As primary production in the open ocean is relatively low compared to most nearshore waters (see Chapter 2), animals are more widely dispersed; however, because there is such a massive amount of habitat, they can be extremely abundant. Historically, the waters in the open ocean were not as heavily impacted by anthropogenic pollutants or large-scale fish harvest. This has rapidly changed as pollution and climate change have become prevalent on a global scale, and capabilities have been developed to harvest animals efficiently from remote ocean regions. Regulation of pollutants is difficult due to their origin far from the area of impact, and regulating harvest is problematic because of the difficulty in monitoring regions far from shore and the international nature of the fisheries.

This chapter addresses environmental and conservation issues in the upper reaches of the open ocean realm: the epipelagic or photic zone. The focus is on the direct and indirect impacts of anthropogenic inputs into the ocean, many of which are transferred from coastal waters or the atmosphere. Effects of the removal of fishes, primarily those harvested for food, are discussed extensively in other chapters, primarily Chapter 11.

Marine mammals, including cetaceans and pinnipeds, and seabirds are important components of many epipelagic ecosystems. Conservation issues concerning these species are discussed in other chapters, primarily Chapters 9 and 10.

7.1 The Photic Zone

Nearly all the primary production in the ocean pelagic region is in the *photic zone,* the mass of water from the ocean's surface down to depths at which there is adequate sunlight penetration to support photosynthesis (see Chapter 1). How deep the photic zone extends in a given region depends on the water clarity and the amount of sunlight. The maximum depth is about 200 meters below the surface in clear waters of the open ocean. The majority of the epipelagic in the open ocean is **oligotrophic** (has relatively low nutrient levels). This is largely a result of low nutrient input from outside sources relative to the neritic zone, where nutrients and organic matter are enhanced by river input or coastal upwelling. Significant areas of upwelling also exist in the open ocean, however. For example, in equatorial regions thermohaline circulation brings nutrient-laden waters to the surface. Nutrient-enriched water is also forced upward as an upwelling zone in association with smaller-scale vertical circulation gyres at convergence zones where two ocean currents meet

Figure 7-1 A moon jellyfish drifts in the epipelagic zone.

(**Figure 7-2**). These upwellings supply nutrients to other areas of the epipelagic when distributed by surface currents. Another source of nutrients is dust blown from the continents, especially from dry desert regions (**Figure 7-3**). Although the quantities of nutrients from this source are small compared to those entering the ocean by river input, they can be important source of iron and phosphorous that are limited in nutrient-starved reaches of the open ocean (see Chapter 1).

Production in the epipelagic zone also serves as a source of nutrients to most deep-sea ecosystems, transported downward in water masses through thermohaline circulation or the sinking remains of organisms. These sources are critical to the deep sea because primary production is largely limited to isolated patches where chemosynthesis is possible, such as hydrothermal vents (Chapter 8). Production in the photic zone is the source of about one-half the carbon export to the sea floor. This is an important factor in the control of atmospheric carbon dioxide and global warming (see Chapter 1).

Oceanic Epipelagic Ecosystems

Open Ocean Production and Fisheries

Recent developments in the use of satellite imagery have provided a rapid mechanism for analyzing production patterns in ocean surface waters (**Figure 7-4**). Prior to the late 1900s when this technology became available, primary production was estimated largely by extrapolating from phytoplankton samples taken from boats. These methods were not only time consuming, but, as we now realize, tended to underestimate photosynthesis in open-ocean surface waters by as much as 50%. This was primarily due to an underestimation of the importance of plankton

less than 20 micrometers in diameter (**nanoplankton**, 2 to 20 micrometers in diameter, and **picoplankton**, less than 2 micrometers). These planktonic organisms (single-celled algae, bacteria, archaea, and small protists; see Chapter 2) are so prevalent that almost every drop of water in the top 100 meters of the ocean contains thousands of plankton cells. They account for more than 95% of the photosynthesis in the ocean, which equates to about half of the primary production on the entire Earth. The open ocean phytoplankton thus play a critical role not only in supporting ocean ecosystems, but also in global oxygen production and sequestering carbon from the atmosphere (see Chapter 1).

Although there is a new appreciation of the importance of primary production in the open ocean, the mid-ocean epipelagic is considered analogous to a biological desert because of its low production relative to the neritic regions. For example, the concentration of chlorophyll a (the predominant form in most photosynthetic organisms) in the center of the major ocean basins is typically 0.02 to 0.05 milligrams per cubic meter (mg/m^3), as compared to about 0.2 to 0.3 mg/m^3 in equatorial upwelling regions, 1 to 2 mg/m^3 in most coastal waters, and as high as 10 to 40 mg/m^3 at river mouths and nearshore upwelling regions.

It is difficult to generalize about primary production within any region of the epipelagic because it is highly variable with time and place. Much of this variability is governed by oceanographic processes (see Chapter 1). The typical low productivity at the center of ocean basins is a result of isolation from coastal input and upwelling that are the origin of much of the ocean's nutrient input. This is especially pronounced in temperate and tropical areas. In these regions solar heating creates warmer, lighter waters that remain at the surface, resulting in little mixing between the low-nutrient surface waters and the colder high-nutrient waters at greater depths. The separation between these water masses is marked by a **thermocline**, a thin layer with a sharp vertical temperature gradient located between 100 and 1,000 meters below the surface (**Figure 7-5**). For example, in regions of the Indian Ocean surface temperatures can be over 25°C at the surface and 10°C at 200 meters depth.

The depth of the thermocline is influenced in part by the amount of mixing of surface waters by wave activity. The thermocline is more permanent in the tropics, and in polar regions it is very shallow or nonexistent due to the low temperatures at the surface. In temperate regions, however, the thermocline is often seasonal. It is disrupted when cooling during winter increases the density of surface water, causing it to sink. This thermal mixing, in

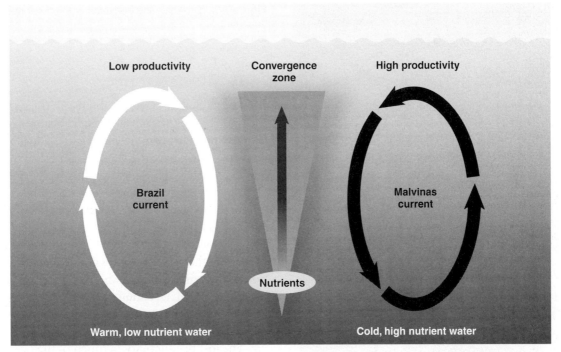

(a)

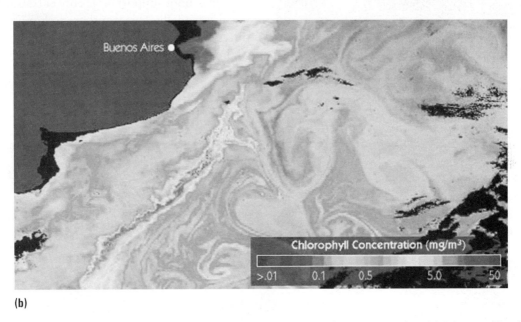

(b)

Figure 7-2 Upwelling of nutrients into the epipelagic at a convergence zone. **(a)** Convergence of the Malvinas and Brazil currents in the southwest Atlantic. **(b)** Surface image demonstrating elevated chlorophyll concentrations at the convergence zone. (See Color Plate 7-2.)

conjunction with wind mixing, causes an overturn of the water column, removing the thermocline. Nutrient-rich waters are brought to the surface, increasing nutrient levels by about an order of magnitude (i.e., a factor of 10). Low light and cold temperatures limit the phytoplankton growth until spring, when sunlight and surface water temperatures increase as days become longer. Phytoplankton respond to the high nutrient levels and reproduce rapidly, forming plankton blooms. This response cascades up the food chain as higher trophic levels respond by moving into the area and increasing growth and reproduction. As planktivorous fish increase, predators move in to feed. Finally, human predators—the fishers—respond by moving in to harvest the fish. Blooms are especially obvious during spring months in regions of the North Atlantic and North Pacific Oceans and are evident by the green color produced

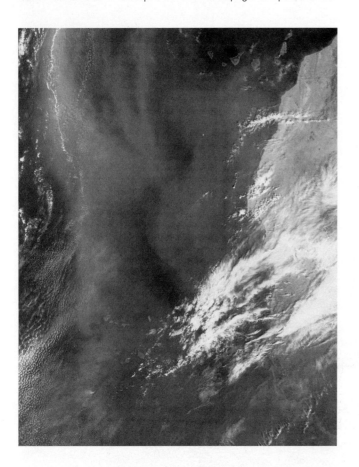

Figure 7-3 Dust blowing from arid regions over the open ocean, such as from the Sahara to the Atlantic Ocean, can provide an important source of nutrients such as iron and phosphorous. (See Color Plate 7-3.)

from the billions of phytoplankton cells (**Figure 7-6**). As summer approaches, the surface waters become calmer, with less mixing to bring in new nutrients. The planktonic organisms, which have depleted the waters of nutrients, die and are consumed by larger organisms. The high production, thus, typically does not last beyond the spring. As the bloom dissipates, some of the nutrients and organic matter from dead organisms and wastes sink below the thermocline and accumulate there. This sets up conditions for a renewal of the cycle the following spring. Although a portion of the sinking organic matter ends up on the bottom or in deeper waters, enough remains in waters just below the thermocline to be circulated to the surface again during the following winter and spring.

The low productivity in much of the oceanic pelagic region cannot support large concentrations of plankton, even seasonally. This translates to low densities of small filter feeders, such as herrings and anchovies, and, thus, there are relatively few fisheries for small pelagic species

Figure 7-4 Image obtained by SeaWIFS (Sea Viewing Wide Field of View Sensor) carried by the NASA SeaStar spacecraft. A measure of absorption of blue and green light gives an indication of the concentration of chlorophyll (and thus, photosynthesizing organisms) in the water. The regions of lowest productivity are in the center of major ocean basins. (See Color Plate 7-4.)

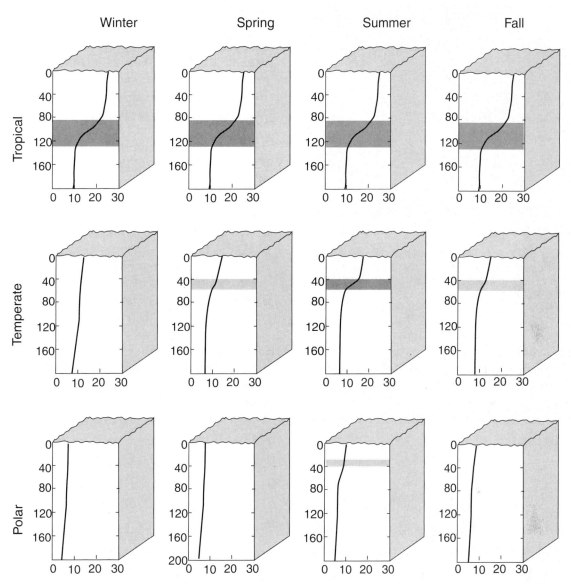

Figure 7-5 Seasonal development of thermoclines in tropical, temperate, and polar ocean waters. Depth (meters) is on the vertical axes and temperature (degrees Celsius) is on the horizontal axes.

compared to more productive waters. Although the low productivity also translates into relatively low production of large predators, their large size, great value, and tendency to congregate for spawning or feeding make them profitable to harvest. Migratory predators, such as tuna and swordfish, are capable of moving great distances in search of food organisms. As humans from many nations have gained access to the fishes of the open ocean, overfishing has become a major conservation issue, as discussed in detail in Chapter 11. The fish harvest biomass could be increased, in theory, by enhancing productivity of open ocean regions. Although adding nutrients such as iron to open ocean waters has been considered as a means to accomplish this fertilization, it is highly controversial (see Chapter 1). The uncertainty of the outcome of ocean

fertilization and laws to ban such practices (e.g., in U.S. waters in 2010) suggest that they are not likely to be applied on a large scale in the near future.

Open Ocean Food Webs

Despite the relatively low production and biomass in much of the open ocean, the massive amount of area allows for a great biodiversity and abundance. Components of epipelagic food webs are similar to those in many other marine ecosystems (**Figure 7-7**). The food webs differ from those of coastal ecosystems, however, in that they are not as closely linked to terrestrial components and, because few multicellular plants can survive floating in the open ocean, the primary producers are mostly unicellular plankton. (One major exception is the floating

Figure 7-6 Surface waters of the North Atlantic southwest of Iceland. This image was taken from NASA's Aqua satellite by the MODIS imager. Swirls indicate plankton blooms. Note plumes of dust blowing off Iceland.

seaweed *Sargassum,* whose distribution is centered in the central region of the North Atlantic, called the Sargasso Sea; **Box 7-1. Conservation Success Story: Protecting the Creatures of the Sargassum.**) The major contributors of primary production are the diverse bacteria, archaea, unicellular algae, and protists (see Chapter 2). Most

bacteria and archaea depend on dissolved organic carbon, derived from the waste and remains of other organisms. These, along with unicellular algae of picoplankton size, often dominate the phytoplankton in regions of low production. For example, flagellated *Micromonas* algae are found in most ocean waters, from the tropics to the Arctic. Other species are most prevalent in the presences of higher nutrient levels; for example, *Ostreococcus* may reach concentrations greater than 100,000 cells per milliliter of seawater. Coccolithophores are common protists in the phytoplankton of the open ocean, especially in tropical and subtropical regions. Small coccolithophores of nanoplankton size are often the major component of open ocean plankton blooms. Coccolithophores are sensitive indicators of ocean temperature, salinity, and acidity, and are considered important indicators of current ocean conditions and historic ocean changes, as reflected in seafloor sediments. Dinoflagellates are more common in the tropics due to their tolerance of warm waters, and, although they may not be as important as other primary producers globally, they can be a major component of harmful algae blooms under certain conditions (discussed below). Diatoms are more abundant in shallower waters nearer to the coast; in the open ocean they are most abundant in temperate and polar regions.

Figure 7-7 A simplified Arctic pelagic food web.

Box 7-1   Conservation Success Story: Protecting the Creatures of the Sargassum

Most multicellular marine plants are limited to shallow waters where there is substrate to which they can attach with roots (vascular plants) or a holdfast (benthic seaweeds). One prominent exception to this is the floating seaweeds (technically a brown macroalgae) in the genus *Sargassum*. This genus has over 200 species, most of which live attached to rocks and reefs in shallow tropical waters but are able to survive floating when detached by wave activity. Two species, however, *Sargassum natans* and *S. fluitans,* are adapted to float at the surface for their entire life cycle using berry-like air-filled bladders to enhance buoyancy (**Figure B7-1**). These "sargassum weeds" are concentrated in mats that float in the center of the major North Atlantic current gyre, a region referred to as the Sargasso Sea. They also can be found in other clear waters of the North Atlantic Ocean and Gulf of Mexico, especially where surface currents converge, and commonly wash up on adjacent shorelines.

In the middle of the Atlantic Ocean where there are few structures to serve as habitat, sargassum can be a magnet to creatures looking for shelter and food. Several species live much of their lives hiding in the sargassum and many have evolved appendages and coloration that mimic the fronds of the sargassum. These include fishes (e.g., the sargassum fish *Histrio histrio* and planehead filefish *Stephanolepis hispidus*), the sargassum crab *Portunus sayi,* sargassum shrimps

(a)

(b)

Figure B7-1   Sargassum seaweeds serve as a point of aggregation in the open ocean otherwise devoid of structure. **(a)** Smaller fish, such as filefish and triggerfish, swim beneath the sargassum, attracting **(b)** predators such as dolphin fish.

Latreutes, and the sargassum snail *Litiopa melanostoma.* Epibionts, such as colonial hydroids, grow on the sargassum fronds and serve as an important source of food for grazing fishes such as the filefishes. Some fish species use the sargassum as juvenile nursery habitat, hiding and feeding among the fronds; small crustaceans, such as amphipods, serve as an important food source. Larger predatory species are attracted to the sargassum ecosystem to feed on the residents. These include jacks, dolphin fishes, tunas, and marlins. Dolphins and loggerhead sea turtles may associate with the sargassum also. Over 80 different fish species and over 100 invertebrates have been observed in association with sargassum habitat off the U.S. coast where Gulf Stream currents carry the sargassum mats. Allan Stoner and Holly Greening found a broad diversity of animals living in association with sargassum throughout its range. The diversity and abundance of animals in and near the sargassum is much greater than in the surrounding epipelagic waters. Sargassum also contributes to productivity of the epipelagic by releasing organic material and nutrients into the water.

Fishers have long known of the importance of the sargassum as habitat in epipelagic ecosystems. The sargassum weed itself became an exploited resource in the 1970s, however, when harvest was initiated off the U.S. coast for use as a livestock feed supplement. Soon, about 2½ metric tons were being harvested per year. Even though this was a small fraction of the total biomass (estimated as 4–11 million metric tons throughout the North Atlantic), there was concern over the local impact on the sargassum as a habit and base of food webs supporting important fisheries. Because sargassum had not been considered of value for harvest, there were no legal restrictions in place to keep the harvest from being expanded. Conservationists, scientists, and fishers banded together to encourage government agencies to develop a management plan that would protect the sargassum from extensive harvest off the U.S. coast. By declaring the sargassum as essential habitat for fishes and sea turtles, regulations were implemented in 2004 to limit total annual harvest to 1.1 metric tons in U.S. waters and restrict the locations of harvest. To limit harming or catching animals, harvest methods were restricted to 4-by-6 foot nets with 4-inch mesh openings. Protection of sargassum habitat in international waters of the Sargasso Sea has been considered through resolutions by the International Commission for the Conservation of Atlantic Tunas (ICCAT; see Chapter 12), to recognize the importance of sargassum as a critical habitat for tunas and other species in the North Atlantic.

Issues focused around the sargassum provide an example of a positive outcome to a conservation controversy. Although it took several years to put protective measures into place, actions were taken before the harvest industry developed to a point that economics and political pressure could stop the implementation of protective measures. Rather than focusing on a single species, habitat was considered as a critical factor needing protection. There are still problems with protecting harvested fishes associated with the sargassum (see Chapter 11), but habitat protection is a major step in the right direction.

Protists also can serve the role of **heterotrophs** in the open ocean, relying on other organisms, primarily other microbes, for nutrition. Some are "**mixotrophs**," functioning as both autotrophs and heterotrophs, containing chloroplasts but also ingesting prey. Much is still unknown about interactions among protists and other microbes at the base of open ocean food webs. Without this information it will remain difficult to understand their function in the community and to predict how they will respond to environmental changes.

Various types of planktivorous meiofauna (0.1 to 1 millimeter in diameter), mostly crustaceans, function as consumers of the primary producers and other microscopic plankton in the open ocean. Copepods are the most prevalent crustaceans in most ocean regions. Copepods feed by creating water currents with their paddle-like appendages to draw food organisms within reach. They then trap them on net-like bristles and scrape them into their mouth. Cnidarians, mostly jellyfishes, can also be important consumers of plankton in the open ocean. Jellyfishes serve as an important source of food for some sea turtles and fishes. These invertebrates spend their entire life as plankton (called **holoplankton**). Other animals spend only a portion of their life functioning as plankton (called **meroplankton**). These include the larval stages of many invertebrates (including shrimp and crabs) and pelagic fish species, most of which produce floating eggs that hatch and develop into larvae that drift in the currents as **ichthyoplankton.**

Mid-sized predators feed at intermediate trophic levels in epipelagic food chains. Many of these are residents of deeper waters, to several hundred meters or more during the day, migrating to the surface to feed at night. This vertical migration is an adaptive advantage for energy conservation and predator avoidance. Energy conservation is achieved by taking advantage of lower metabolic requirements in colder waters below the thermocline during the day. Feeding takes place at night when surface waters are relatively cool. Predators are avoided by staying in darker waters, at the surface at night and at depth during the day.

Waters below the photic zone are of limited accessibility to many surface predators due to light limitations and other factors. For example, fishes with swim bladders are unable to rapidly move up and down in the water column due to physiological effects of rapid pressure changes (however, sharks, and tunas—without swim bladders—are capable of rapid dives). Birds and mammals are limited in their time away from the surface due to oxygen needs (however, some marine mammals are capable of dives of great depths and durations; see Chapter 2). Diurnal vertical migrations to and from the surface require little energy; a slight change in buoyancy or active swimming allows the animal to slowly float up or sink down in the water column. Diurnal vertical migrations can be exhibited by plankton and planktivores up to about 10 centimeters length. The concentrations of these plankton are so great that they form a "deep scattering layer" in much of the open ocean, detectable by sonar equipment used to map the sea floor (**Figure 7-8**).

Larger juvenile and adult fishes and macroinvertebrates also feed on the plankton in the epipelagic. They typically filter organisms from the water using specialized structures (e.g., comb-like appendages or feeding structures in invertebrates, gill rakers in fishes, baleen in whales). Many fishes that migrate from the mesopelagic (some of the most abundant are gonostomatid bristlemouths), however, can either pick individual plankton from the water or filter organisms with gill rakers. Although anchovies and herrings are mostly concentrated in neritic regions that experience large plankton blooms, they also can be important in oceanic areas of high productivity. Krill are the predominant plankton

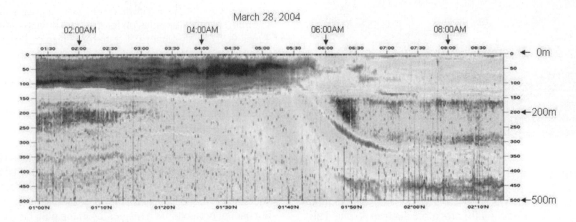

Figure 7-8 An acoustic transect obtained from 1:30 to 8:30 AM in pelagic ocean waters. Dark regions indicate a dense congregation of planktonic organisms moving from shallow to deeper waters at dawn.

feeder in colder oceanic regions such as the Antarctic. They concentrate in large swarms and serve as an important food source for fishes, seabirds, and even the giant baleen whales.

Apex predators in the epipelagic include fishes, marine mammals, and birds (**Figure 7-9**). The diversity of predatory animals in the open ocean is relatively low due to the low biological production, and many migrate long distances in search of food. Most large pelagic fishes are categorized as "highly migratory predators" and include billfishes (Xiphiidae), jacks (Carangidae), dolphin fishes (Coryphaenidae), mackerels and tunas (Scombridae), and sharks. These fishes support some of the most valuable fisheries in the world and are especially difficult to monitor and manage due to their highly migratory nature and position at the apex of ocean food chains (these fisheries are discussed in Chapter 11). Squids and jellyfishes also are important predators in the epipelagic, primarily on smaller fishes. Marine mammals in the oceanic epipelagic include dolphins, whales, and seals. These are given special legal protection but are still influenced by indirect impacts of global pollution,

overharvest of food resources, and unintentional harm by other human activities. Marine mammal protection is discussed in Chapters 9 and 10. Birds that feed in the oceanic epipelagic include penguins and diving birds such as albatrosses. Although these birds are protected from intentional harvest, many are endangered by anthropogenic impacts, including entanglement in fishing gears (Chapter 9), loss of food resources (Chapter 11), and loss of shore nesting areas (Chapter 3).

Food chains in the epipelagic can be relatively simple. There may be only a single trophic link from the zooplankton to the largest fish, the whale shark, and from the plankton to the largest mammals, the baleen whales. But for most large marine vertebrates there are two to four or more links from the plankton to the adult carnivores. Epipelagic food chains tend to be more efficient than most and are an exception to the *ten percent rule* discussed in Chapter 2. Epipelagic herbivores have been documented to convert more than 20% of the energy obtained from their food into growth, and even the carnivores convert more than 10%. This allows for long and complex epipelagic food chains.

(a)

(c)

(b)

Figure 7-9 Open ocean epipelagic predators include birds, cetaceans, and fishes, such as **(a)** emperor penguins *Aptenodytes forsteri,* **(b)** killer whales *Orcinus orca,* and **(c)** bluefin tunas *Thunnus thynnus.*

7.2 Open Ocean Pollution

Throughout most of human history the open ocean away from coastlines was virtually unaffected by our actions. Pollutants released into coastal waters reached the open oceans only in very low concentrations, and atmospheric pollutants were not released on a scale to cause global impacts. Pollutants that were released from ocean-going vessels rapidly decomposed or were diluted such that they caused little long-term harm to populations of marine organisms. Until about the past 50 years it was widely presumed that the oceans were so vast that humans were incapable of having anything more than local impacts in waters near the coast. As populations have exploded and accessibility to the sea has increased, however, humans have learned to access and develop chemicals that persist for decades or centuries in the environment, and anthropogenic influences in the open ocean have become more and more apparent.

Of the numerous tons of hundreds of chemicals that have ended up in the ocean due to human actions, for only a fraction can we hope to reliably estimate the potential long-term impacts on ocean ecosystems. The greatest danger is from chemicals that do not decompose rapidly and become dissolved or mixed into the ocean waters, potentially to be dispersed throughout the ocean. Even when the massive volume of ocean waters dilutes these chemicals to concentrations that are barely measureable, they can accumulate to harmful concentrations in biological organisms, especially those at upper trophic levels of food webs.

As discussed in previous chapters, much of the pollution focus in coastal ecosystem is on excessive input of nutrients and naturally-occurring organic chemicals. Because nutrients are typically limited in the open ocean, other categories of pollutants are often of more concern. Pollutants can include elements that occur naturally but at extremely low concentrations in the ocean environment, such as metals, as well as materials produced by humans for industrial or agricultural purposes, such as pesticides and plastics. The remainder of this section focuses on harmful impacts of these persistent pollutants in the open ocean.

Metals

Metals are naturally occurring elements with specific chemical ionization and bonding properties in common. These properties make metals useful in various commercial applications and industrial processes, leading to wastes that can eventually end up in the environment. Metals do not decompose under normal environmental conditions, and thus can accumulate in the environment and in certain living tissues. Some metals, including iron, copper, manganese, and zinc, are necessary for the function of many organisms; however, because these are typically micronutrients needed only in very low concentrations, they may be harmful in excessive amounts. Other metals, such as mercury and lead, are considered toxic to most organisms, often even at low concentrations. (The term *heavy metal* is often applied to these metals, but the term is poorly defined, so this text refers to these simply as toxic metals.) As a result of their propensity to bioaccumulate, toxic metals are especially problematic to animals feeding at upper trophic levels or that are long-lived.

Lead

Lead has received much attention in recent years due to its toxic effects on humans and common occurrence in fuels, paints, and other industrial products. Effects on humans and other vertebrates include interference with nervous system development and various physiological processes. Lead is introduced naturally to the ocean from freshwater input of eroded materials. This results in concentrations that are extremely small, however, and most of this lead becomes trapped in sediments in coastal waters, never reaching the open ocean. The majority of the lead introduced into the open ocean since the Industrial Revolution has come from atmospheric pollutants. Ocean lead pollution, as estimated by analysis of snow and ice layers and coral deposits, began increasing gradually in the late 1800s. Levels had increased to almost five times the natural concentrations by the early 1900s in some regions of the North Atlantic. A further increase in lead pollution resulted from the introduction of leaded gasoline in the 1930s (**Figure 7-10**). As gasoline usage increased (primarily in North America) from 1950 to 1970, lead deposited on the surface of the oceans increased threefold. About one-half of the lead introduced into epipelagic waters from the atmosphere is transported to the seafloor within a few years, after being incorporated into the remains and wastes of organisms. The other half remains in upper water column for a few years to several decades, depending on circulation patterns, where it remains available for bioaccumulation in marine animals. Ocean lead levels began declining after restrictions on the sale of leaded gasoline were imposed by the U.S. Clean Air Act in the 1970s and in Europe in the 1980s. In the 1990s, however, the rate of decline began to slow in the North Atlantic, leaving lead concentrations in the pelagic at about double those in preindustrial times. The probable sources of most lead deposited into the oceans since about 1980 is from air pollution from industrial activities such as coal burning and smelting. The impact of lead accumulation in marine animals has not been well documented; however, lead levels have not resulted in restrictions on seafood consumption as has occurred for other metals, especially mercury.

Mercury

The fear of toxic effects of mercury has led to its being possibly the most investigated contaminant of open ocean marine

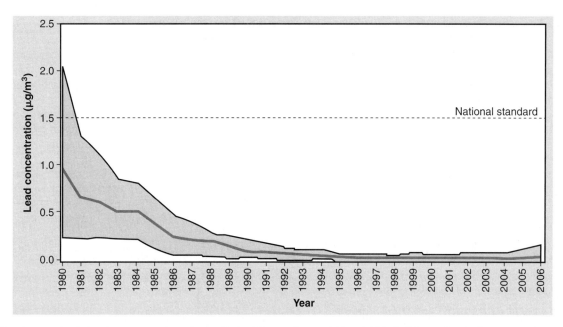

Figure 7-10   Atmospheric lead concentrations in the United States from 1980 to 2006. Solid gray line indicates average values for all sites tested; the gray region encompasses 80% of observations.

species. Mercury is a naturally occurring element in the ocean, with a steady input into the ocean through erosion. Because mercury concentrations in the earth's crust are very low, less than 0.1 parts per million (ppm) on average, the amount of mercury eroding naturally into the ocean is small. Mercury is also introduced into the atmosphere from geologic processes such as volcanic eruptions, but these are typically minor sources as well. Natural sources of mercury are diluted to sub-toxic levels in the ocean. The addition of anthropogenic sources can lead to bioaccumulation to levels that are harmful to many animals, however, including humans. These sources include thousands of tons of mercury mined from concentrated deposits or released during burning of fossil fuels.

Elemental mercury is relatively harmless to animals because it is not easily absorbed and assimilated. But mercury entering rivers and oceans is rapidly converted by bacteria and other microorganisms into methyl mercury, a form easily assimilated by organisms, not readily decomposed, and thus prone to bioaccumulation in ocean food webs. Concentrations in the ocean water that average about 1 to 2 parts per trillion (ppt) can be taken in by planktonic organisms and, after transferring through several levels in the food web, accumulate in the tissues of predatory fishes to concentrations around 1 ppm. That is a bioaccumulation of six orders of magnitude (i.e., the concentration is 10^6 or 1 million times greater in the predatory fish than in the surrounding water)! At these concentrations the mercury is not immediately lethal to fish; however, high levels of mercury may affect reproduction, growth, development, and behavior. Sublethal effects of moderate mercury concentrations

on animals have not been well defined. The major focus of concern has been mercury's effect on health of humans from seafood consumption. Although the lower levels considered to be harmful to humans are debated, consumption of contaminated seafood has led to chronic medical conditions, illness, and death, and numerous warning about seafood consumption (**Box 7-2. Learning from History: Mercury Scares and Seafood Consumption**).

Currently, most of the mercury entering the open ocean is transported from the atmosphere in rainwater. The total annual input of mercury into the atmosphere from all sources is estimated to be over 8,000 metric tons. About 3,000 tons is believed to be from anthropogenic sources; the largest source, about 1,400 tons is from coal-burning power plants. Even though coal contains only small amounts of mercury (about 1.5 grams per metric ton or less), about 7 billion tons of coal are burned annually. The majority of the mercury emissions globally are from Asia, much from coal-fired power plants in China. Europe and the United States each contribute about 8% to global atmospheric mercury pollution.

There are continual efforts to develop technologies to reduce mercury emission, and pollution-control devices designed to remove sulfur dioxide also can remove about 35% of the mercury from coal. Reductions in atmospheric mercury emissions have been achieved by many nations, 70% in Europe, and about 40% in North America since the passage of the 1990 Clean Air Act. Control of atmospheric pollutants containing mercury is truly an international issue because much of the mercury ends up in the global atmospheric cycle with the potential for reaching the sea.

Box 7-2 Learning from History: Mercury Scares and Seafood Consumption

Mercury has many unusual chemical characteristics for a metal, most notably its normal liquid state in the elemental form, that have made it beneficial to humans for a myriad of uses. The most familiar of these is in mercury thermometers and batteries. Much larger volumes of mercury have been used in industrial processes. For example, mercury was commonly used as an electrode in an electrolysis process to extract chlorine from sodium chloride salt. Mercury is used in plastics production and in small-scale mining operations to extract gold from its ores (currently the second largest source of anthropogenic mercury emission into the atmosphere). Mercury compounds have been used in numerous cosmetic and medical products, including mascaras, antiseptics, diuretics, laxatives, syphilis treatments, and tooth fillings. Many, but not all, of these uses have been phased out.

Because liquid mercury is not absorbed easily by ingestion or contact—less than 0.1% of swallowed mercury is absorbed by the intestines—mercury's toxicity has not always been apparent. Mercury is volatile, however, especially when heated, and inhalation of mercury fumes during mining and refining from ores has been a historic cause of mercury poisoning. Mercury-containing compounds that are soluble in water are also toxic and readily absorbed into the body. One widely cited case of mercury toxicity was from exposure by hat makers to mercury compounds used to soften leather in the manufacture of felt hats in the 1700s and 1800s in Europe (this is a possible origin of the phrase, "mad as a hatter"). As more evidence of mercury's health effects became available, exposure to mercury vapors during mining and industrial processes became a concern. These types of exposures are primarily limited to people working in these industries, and alternate procedures were eventually developed to limit exposure; however, they still present health problems in some regions of the world. The most publicized concern about mercury toxicity eventually became exposure of the general citizenry to mercury in the environment. Even though people were rarely exposed to quantities in the environment that were toxic upon direct exposure, mercury's persistence and ability to bioaccumulate led to concentrations in some seafoods great enough to cause health problems. The primary concern is neurological problems, reflected by impairment of the senses, emotional and memory problems, and lack of coordination. Children born to women who are exposed to mercury during pregnancy may have serious birth defects.

The potential impact of mercury in seafood came to attention of the public around the world in the 1960s, when thousands of people living in coastal regions of western Kyushu, Japan became seriously ill for unknown reasons. Many of these people exhibited permanent brain damage or paralysis, children were born with birth defects, and hundreds may have died. It was eventually determined that the cause of these maladies was mercury poisoning, and that the source of the mercury was the effluent of a fertilizer and petrochemical company in operation since the 1930s. The resulting condition became known as Minamata disease after one of the bays in Japan where the condition was documented. Seafood consumption was high in this region, and the toxic mercury compounds that had accumulated in fish and clams were the major source of

contamination. This scare led to the elimination of mercury input into Minamata Bay by 1968, and was a major impetus for a nationwide environmental movement in Japan. These incidents also had a global effect, as publicity of the incident led to major cleanups of mercury pollution around the world. Nations began closely monitoring input of mercury-containing wastes and phasing out industries that were responsible for mercury pollution in rivers and bays.

Controlling industrial mercury pollution did not end the concern over mercury in seafood. As the health effects of mercury began to be more thoroughly documented and mercury in aquatic species began to be monitored more closely, a new awareness developed of mercury's ability to bioaccumulate in predatory species far from any source of mercury pollution. Although industries had begun to tightly control the release of mercury, small amounts still were escaping into the atmosphere and waters. Incineration of municipal and medical waste was putting mercury into the atmosphere, and in some countries small-scale gold mining operations were using mercury with little monitoring or control. Mercury from these sources has been greatly reduced, but not eliminated. Currently, however, the largest source of the release of mercury into the environment—some of which eventually ends up in the open ocean—comes from burning of coal for heat and power production. Enough of this mercury eventually reaches animals harvested as seafood in regions far removed from the source that health effects are a serious concern. For example, it is estimated that 40% of the mercury consumed by U.S. citizens comes from seafood caught in the Pacific Ocean, and the primary source of this mercury is from pollution originating in Asia.

A total elimination of ocean mercury pollution is not currently practical, so governments and consumers are forced to make decisions over restrictions on seafood consumption for species with mercury contamination. This is the most common way most people are exposed to mercury (**Figure B7-2a**). The primary focus has been on the consumption of species such as shark, swordfish, king mackerel, and tilefish, for which average mercury concentrations range from about 0.75 to 1.5 ppm. These are all long-lived predatory fishes.

Factors other than longevity and trophic level also elevate levels of mercury in fish. Anela Choy and colleagues found the highest mercury levels in fish that feed at greater depths, including swordfish and some tunas, where prey have relatively high mercury levels. Lowest mercury values are found in fish feeding at lower trophic levels; for example, herrings and sardines, anchovies, cod, oysters, and crabs typically have mercury concentrations below 0.1 ppm. Other fishes with moderate levels of mercury, averaging about 0.4 to 0.5 ppm, include marlin, orange roughy, groupers, and Patagonian toothfish. A 2003 study by Tetsuya Endo and others found mercury concentrations that were above levels considered safe by the Japanese government (0.4 ppm) in the meat of all kinds of predatory toothed whales being marketed for consumption in Japan (**Figure B7-2b**). Values ranged from 0.5 to 81 ppm; the majority of species tested had mercury levels of 5 to 10 ppm. (Almost all whale meat from baleen whales was below 0.4 ppm, however, reflecting their diets low on the food

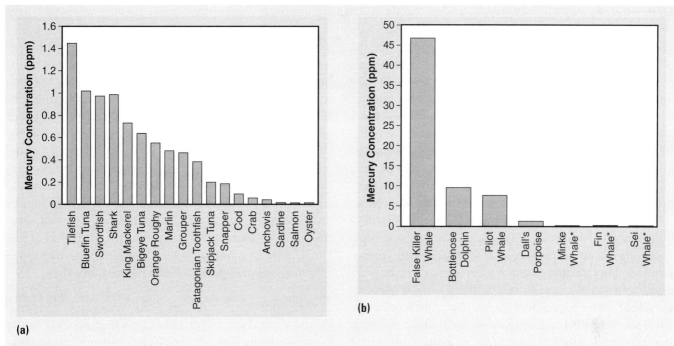

(a)

(b)

Figure B7-2 Mercury concentrations in muscle tissues of marine animals. **(a)** Mercury in commercially harvested fishes and invertebrates. (Data are from USFDA www.fda.gov, and Storelli and Marcotrigiano, 2001.) **(b)** Mercury levels in cetaceans harvested for human consumption in Japan. *Values for minke, fin, and sei whales are below 0.1 ppm each. (Data are from Endo et al., 2003.)

chain.) It is not yet known how these observations will affect the whaling industry in Japan, which is already under attack over ethical and ecological concerns (Chapter 10).

It is generally understood that consuming too much mercury-tainted seafood is dangerous to human health. But unsafe levels are difficult to define because health effects at low exposure are hard to monitor, and mercury exposure depends on frequency of seafood consumption and types of species consumed. Studies in several regions where seafood consumption is high, including the Faeroe Islands, Seychelles, and New Zealand, indicate that mothers who had consumed mercury-containing seafood bore children with noticeable, but subtle, impairments to nerve function (however, this has been a controversial conclusion because other studies have not always supported these findings). The levels of mercury designated acceptable in seafood range from 1 ppm in the United States to 0.5 ppm in Europe and 0.4 ppm in Japan (however, seafood is not routinely monitored for mercury levels or removed from the market when high levels are discovered). Another measure of safety for seafood consumption is the amount of mercury consumed on a daily basis, measured as micrograms mercury per kilogram body weight (µg/kg). These levels vary from 1.5 µg/kg as set by the World Health Organization, to 0.4 µg/kg designated by the U.S. Food and Drug Administration (FDA), and 0.1 µg/kg by the U.S. Environmental Protection Agency (EPA). The variance in recommendations by different countries and different agencies as well as different conclusions from different epidemiology studies have

led to consumer confusion over the past decade over what should be considered safe. Avoiding seafood is not an option in many regions where it is the primary source of protein, and the health benefits of consuming fish are considered to outweigh the dangers of consumption of small concentrations of mercury in fish. To balance safety concerns and health benefits, the U.S. FDA recommends that women and children eat up to 12 ounces of fish per week, limited to those with lower mercury concentrations. For canned tuna, the most commonly consumed fish in the United States, the lowest mercury levels are typically found in "light tuna" (this is skipjack tuna *Katsuwonus pelamis*, the smallest of the commercially harvested species). Mercury levels in tuna vary widely by species and labels do not typically indicate the species of the tuna (see Chapter 11 for a discussion of tuna fisheries). Consumers eating fresh fish should consider that older, larger fish are likely to have the highest levels of mercury or other contaminants. For example, sushi from bluefin tuna sold in the United States has been commonly documented to contain mercury levels above the 1.0 ppm that is considered unsafe for consumption. For those who eat large amounts of fish, fears of the consumption of excess mercury can be avoided by eating primarily seafood with lowest concentrations of mercury (e.g., anchovies, herrings, cod, salmon). There are additional benefits of consuming these species because many of them also support some of the most sustainable fisheries (Chapter 11) and are, thus, considered the most environmentally friendly to consume.

For example, about 25% of the mercury released from U.S. coal-burning plants falls to the Earth locally; the remainder is transported into the global atmospheric cycle.

Despite progress in reducing mercury pollution, anthropogenic sources have increased mercury levels in the atmosphere by 250% since pre-industrial times, leading to mercury concentrations 25% higher in the surface ocean and 10% higher in the deep ocean. Of particular concern is the unexpected accumulation of mercury in polar regions far from its source. Although the processes that led to this accumulation are still uncertain, it was discovered in the mid-1990s that Atmospheric Mercury Depletion Events (AMDEs), resulting from reactions with sea salt and sunlight, convert elemental mercury in the atmosphere into reactive forms that are easily converted to toxic methyl mercury. It had been assumed that meteorological conditions resulting in AMDEs were limited to spring in polar regions when the Sun reappears over the horizon. Ann Steen and colleagues recently discovered, however, that conversion of atmospheric mercury into reactive forms continues through the summer months in the Arctic (likely from oxidation by ozone in the atmosphere), increasing concerns over the impact of global mercury pollution and the resulting bioaccumulation on sensitive polar ecosystems.

International scientific and political debates continue over how to deal with mercury pollution issues. There has been much discussion about establishing further limits on sources of mercury and other atmospheric pollutants. It remains politically difficult, however, to establish meaningful agreements.

Persistent Organic Pollutants

Many toxic chemicals that end up in the marine environment are synthetically produced organic chemicals. Those that do not decompose readily and thus persist in the environment are called Persistent Organic Pollutants (POPs). The source of many POPs in the oceans is from pollutants released into rivers or the atmosphere. Many of the POPs are of low solubility in water but highly soluble in lipid; thus, they are easily retained in the fatty tissues of animals and can bioaccumulate in ocean food webs. Top predators like tuna and swordfish can have high levels of these toxins even though the concentrations in the ocean water are relatively low. Documented harm to wildlife includes damage to the nervous, reproductive, and immune systems or death. Some have been shown to cause cancer in humans.

As with metal pollutants, POPs often show up in high concentrations in waters and animals found in the open ocean far from the source of the pollution. Transport of these chemicals from warmer to colder regions is especially problematic, resulting in their accumulation in waters and ice near the poles. This accumulation occurs through a process called **global distillation**. Because many POPs have a relatively high volatility (i.e., they vaporize easily), they are readily transported from the land to the atmosphere at relatively warm temperatures. Once they enter the global atmosphere cycle, they are transported by winds and then condense when they reach cooler temperatures. After condensation, the chemicals fall to Earth in precipitation. This results in a net transport from lower to higher latitudes, where they can bioaccumulate in animal tissues. As a result, predators, including humans, eating marine animals at higher latitudes are exposed to some of the highest concentrations of POPs despite being far removed from the sources of these pollutants. The concentrations are especially pronounced in humans that eat high amounts of animal fats, such as indigenous groups for which blubber from marine mammals is an important component of their diet.

Pesticides

Many of the chemicals classified as POPs initially were designed to be used as pesticides. The most environmentally harmful of these have been insecticides used for agricultural purposes or for controlling disease-carrying insects. Those that do not decompose readily can enter the environment and potentially end up in ocean ecosystems through erosion, aquatic applications, and vaporization or suspension in the atmosphere. Pesticide concentrations that are lethal to aquatic life are most prevalent in fresh or coastal waters near the source of application. Persistent pesticides behave as POPs and can accumulate in the marine environment and organisms, however, affecting ecosystem health and causing physiological and behavioral changes that are often difficult to document.

The pesticide that has received the most attention and possibly has caused the most harm to wildlife is dichlorodiphenyltrichloroethane (DDT). DDT, a member of a synthetically-produced group of chemicals called organochlorines, was first widely used to control transmission of insect-borne diseases such as malaria and typhus during World War II. After the War, DDT was applied as an agricultural pesticide in massive amounts (about 40,000 metric tons per year from 1950 through 1980) until evidence indicated its potential harm to wildlife and humans. This evidence was compiled by Rachel Carson in the book, *Silent Spring*, in which she questioned the large-scale use of this and other chemicals in the environment with little concern or monitoring of the long-term impacts on the health of wildlife and humans. This book was an important landmark not only in the control of pesticide use but also in the global environmental movement. Carson's book and the resulting public outcry led to the ban on the use of DDT in the United States in 1972 (although U.S. companies continued producing it for export until 1985), and an eventual ban on agricultural use worldwide. DDT is still produced and used on a smaller scale (about 5,000 metric tons per year) in other parts of the world for control of malaria and other diseases, mainly in Asian countries.

DDT is slow to degrade in the environment, and its breakdown products also can be persistent and toxic. The degradation rate varies depending on environmental conditions; but it can remain over 50% intact in the environment for up to 30 years under some conditions. Although DDT does not readily dissolve in water, it is quickly absorbed by organisms and is stored in body fats. It thus tends to bioaccumulate at higher trophic levels in food webs (**Figure 7-11**). Much of the DDT that reaches the open ocean arrives from atmospheric sources. Once it reaches predatory animals it is slow to be metabolized; the majority of the DDT and its breakdown products remain intact in humans for up to a decade. In the marine environment DDT is less toxic to mammals than it is to crustaceans, fishes, and birds. Its most severe impact has been on predatory birds. DDT was considered to be the major reason for declines of bald eagles, pelicans, and osprey, all of which often rely on fish as a major portion of their diet. These impacts were slow to be recognized because death of adult birds did not typically occur. Its greatest effect was on reproduction. DDT's interference with calcium metabolism can result in the production of fewer eggs and thinner egg shells that crack prematurely, resulting in death of the young. Banning the use of DDT

was considered a major reason for recovery of bald eagles and pelicans and their eventual removal from the U.S. Endangered Species list. In some regions, including areas of the U.S. Gulf of Mexico coast, entire populations of brown pelicans disappeared, requiring re-introductions from other regions. By 1982 the brown pelican had recovered to levels considered healthy enough that it could be removed from the Endangered Species List. On the Louisiana coast, where the bird had been virtually eradicated, there were consistently over 100,000 successful nestings by the 1990s. But even though restrictions on the use of DDT in much of the world have resulted in fewer lethal impacts on marine animal populations, because of its persistence and effects of global distillation, DDT has stayed with us. Heidi Geisz and colleagues discovered that DDT levels may actually be increasing in Antarctic waters, coming from glacial meltwaters as an effect of global warming. The original source was the historic accumulation of DDT transported from the atmosphere by snowfall. This increase is reflected in DDT levels in the Antarctic Adelie penguin *Pygoscelis adeliae*, which have not declined since the reduction in global use of DDT after the 1970s.

Although the global impact of DDT has been the most well documented, other pesticides are considered POPs, including aldrin, dieldrin, chlordane, endrin, heptachlor, and toxaphene. Almost all uses of these were phased out in the United States by the mid-1980s, and usage has been restricted globally.

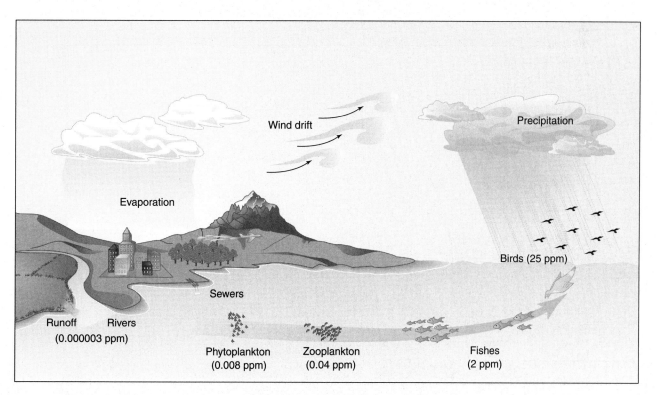

Figure 7-11 Modes of transfer of DDT into the ocean environment and an example of bioaccumulation through a marine food chain. Numbers indicate DDT concentrations in parts-per-million (ppm).

Polychlorinated Biphenyls

Polychlorinated biphenyls (PCBs) are a group of synthetically-produced POPs, comprised of a mixture of over 200 separate compounds that are used primarily in industrial processes, including as coolants and lubricants in electric transformers and in the manufacture of paints, plastics, adhesives, and other products. Although the toxicity of PCBs was recognized by the 1930s, production was not banned in the United States and other industrialized nations until the 1970s and 1980s. Hundreds of tons of these chemicals are still in use in electric transformers and other equipment, however, and at least half of the PCBs produced are either still in use or in storage. Because PCBs are very stable and do not degrade easily, they cannot readily be eliminated from the environment. PCBs are present in large quantities in some landfills in items such as in old transformers. Many PCB toxins have already accumulated in aquatic ecosystems from past releases before environmental restrictions were enacted. For example, over 500,000 kilograms were released into New York's Hudson River from the 1940s to 1970s, resulting in fishing bans and fish consumption warnings over fears of effects on human health. Efforts to clean up contaminated river sediments have been ongoing since the 1980s. Lethal effects on marine species were first documented in the 1970s when dead seabirds with high concentrations of PCBs in their tissues from an unknown source washed up on beaches. Since this incident, PCB ingestion has been documented to result in various health effects on humans and other animals, including liver damage, immune system effects, and neurological problems.

Biomagnification of PCBs has been documented in the tissues of predatory marine species far removed in time and place from the sources of the pollution. Toxic levels of PCBs have been found in marine mammals including seals, dolphins, whales, and seals (**Figure 7-12**). These include effects on the immune system, reproductive impairment, and the development of tumors. Peter Ross and colleagues discovered that killer whales in the eastern Pacific off North America have some of the highest PCB levels documented in any marine animal. Concentrations of over 200 ppm have been found in the blubber of some individuals, especially those that fed on other marine mammals. For example, a concentration of 1,000 ppm was measured in a dead beached killer whale. Despite a decrease in the level of PCBs in the environment, Brendan Hickie and colleagues estimated that lifetime exposure could continue to affect killer whales for an additional 50 years or more due to persistence in the marine environment, bioaccumulation, and the whales' great longevity.

Polycyclic Aromatic Hydrocarbons

Polycyclic aromatic hydrocarbons (PAHs) are another group of POPs, comprising over 100 chemicals formed from incomplete burning of coal, oil, gas, and other organic

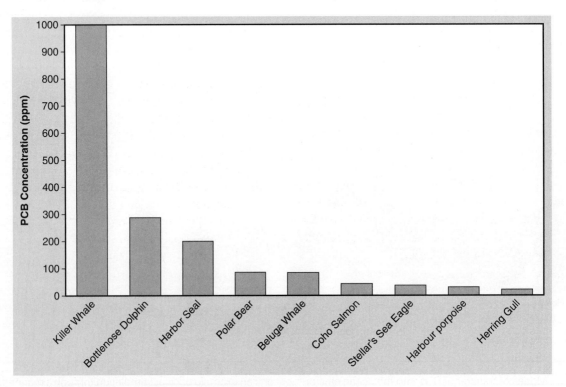

Figure 7-12 Examples of PCB levels (ppm) found in selected marine animals.

materials. They are contained in products such as coal tar, crude oil, and roofing tar also. Other PAHs are produced for use in dyes, plastics, and pesticides. PAHs can be transported in the atmosphere, mainly from burning coal and other fossil fuels, and deposited in the ocean. Many of the PAHs are toxic to and can bioaccumulate in marine organisms. Exposure has resulted in various effects on fish growth, survival, and health, including genetic mutations in young. This has been most commonly documented in estuaries and coastal areas near the source of definable pollution sources.

POP Regulation

A thorough documentation of the harmful impact of POPs in the environment is expensive and time consuming and thus is not always be feasible. Governments typically move cautiously when attempting to eliminate the use of these chemicals. Even when agreements are reached, economic considerations often result in a slow phasing-in period rather than an immediate ban on chemicals discovered to be harmful. More testing is required than ever before of chemicals on the environment, wildlife, and humans, but thorough tests are not carried out on all chemicals before they are released into the environment, resulting in the release of many chemicals with unknown effects on living organisms.

Recent efforts have been made to develop agreements among nations to ban the production and use of POPs considered to be the most harmful. The Stockholm Convention on Persistent Organic Pollutants, organized by UNEP (United Nations Environmental Program) in 2001, established bans or restrictions on 12 POPs internationally that went into effect in 2004. Nine chemicals were identified for elimination, including PCBs. DDT was designated as restricted for use only for disease vector control, primarily because it is the most efficient means of controlling mosquito-borne diseases like malaria in tropical countries. Even when banned, many of the chemicals will likely persist in the environment for decades to come. New environmental impacts of chemicals are recognized each year, and these can be considered for addition to the Convention; this resulted in nine new chemicals added to the list in 2010. Over 160 nations have ratified the Stockholm Convention treaty. Although the United States bans many of the chemicals listed, it is the only large industrialized nation that has yet to ratify the treaty.

The current global impact of pesticides, PCBs, and other POPs on marine ecosystems is unknown. Although the toxic effects of DDT and PCB have been well documented, there are still many questions about their sublethal impacts to organisms. Other POPs are less well studied, and in many situations multiple POPs accumulate in the environment, hindering the ability to distinguish the relative effect of each. Adverse effects associated with POPs have been well documented in species living near the poles, apparently due to global distillation of various POPs in this region. For example, stress has been indicated in polar seabirds with relatively high contamination rates. Killer whales in North Atlantic waters near Norway show high levels of various pesticides and PCBs, apparently from consuming herring with high contamination levels. Evidence that contaminants are reaching deep-sea ecosystems is indicated by high levels of various POPs found by Michael Unger and colleagues in squid and octopus collected from depths of 1,000 to 2,000 meters in the North Atlantic Ocean. This also might explain the high contamination levels in whales that feed in these deep waters.

■ Harmful Plankton Blooms

Harmful Algal Blooms

Not all toxic chemicals in the ocean environment are from anthropogenic sources; some are produced naturally by marine organisms. These chemicals are produced to serve many physiological functions or may be produced for protection from predation. Organisms that commonly produce such chemicals include sponges, jellyfishes, and some fishes. But the broadest environmental effect of natural toxins in the pelagic ocean environment is from production by some planktonic protists, most commonly diatoms and dinoflagellates. The environmental effect of the toxins is minimal under typical conditions when the plankton cells are at relatively low concentrations. When conditions are optimal for growth and reproduction, however, the plankton populations can rapidly increase into blooms. When these blooms are of toxin-containing species they are called **harmful algal blooms** (**HABs**; here *algae* is used broadly to include photosynthetic protists such as dinoflagellates and diatoms). HABs also can include blooms of plankton species that cause mechanical damage to animal tissues such as fish gills, or they can create hypoxic conditions from excessive respiration or bacterial decomposition. The term **red tide** is commonly used for dinoflagellate blooms that impart a reddish color to the water, and is sometimes used to include all HABs.

Harmful algae blooms are typically composed of a single species that grows and reproduces rapidly, outcompeting other planktonic organisms. Concerns over HABs include the direct lethal effects to fishes and other marine animals and the harmful effects to humans who consume seafoods contaminated with these toxins, especially filter feeding animals such as oysters, clams, and other bivalves. Exposure of humans to algae- or protist-produced toxins is not limited to those associated with HABs. For example, dinoflagellate-produced ciguatoxin produces the most commonly reported marine toxin disease in the world, ciguatera (**Box 7-3. Consume with Caution: Ciguatera Poisoning**).

Box 7-3 Consume with Caution: Ciguatera Poisoning

The algal toxin that most commonly affects humans is not from a species that produces large HABs. The dinoflagellate *Gambierdiscus* lives on the surface of calcareous algae, seaweeds, and dead coral skeletons on tropical reefs. It produces **ciguatoxin**, which can be taken up by algae-grazing reef species and bioaccumulate in fish-eating predators. When humans or other mammals eat these predatory fish, the toxin can result in illness or even be lethal. The disease associated with ciguatoxin poisoning is called **ciguatera**, the most commonly reported marine toxin disease in the world. The species most commonly associated with ciguatera are tropical reef-associated predators, including barracuda, groupers, and snappers; however, ciguatera has been documented in a wide array of species that feed in tropical waters, including pelagic predators such as tunas and mackerels. Although *Gambierdiscus* dinoflagellates are found throughout the tropics, the primary concentrations of ciguatera outbreaks have been in three regions: the tropical Pacific and Indian Oceans and the Caribbean Sea. However, within each of these regions the toxicity of *Gambierdiscus* varies. Toxicity also varies with physical conditions (e.g., higher temperatures can increase the level of toxins), seafood consumption patterns, and other factors, leading to large variability in reports of ciguatera throughout the tropics.

Scientists and consumers are frustrated by the difficulty in predicting the likelihood of exposure and illness from ciguatoxin poisoning. Under-reporting and misdiagnosis have hampered predictions and resulted in underestimates of the prevalence of ciguatera. For example, in the United States the Centers for Disease Control (CDC) estimates that only 2% to 10% of cases are reported. In coastal regions of the southeast United States, the number of incidences has been impossible to estimate reliably because many doctors do not recognize the symptoms. For example, although only about 50 ciguatera cases were reported annually in southeast Florida in the early 2000s, the true number of cases could have been as high as 5,000. In regions of the Caribbean where seafood consumption rates are high, the best estimates are that 3% to 5% of individuals in the population experience ciguatera each year. On small islands in the South Pacific and Caribbean, over 50% of the population may have been exposed at some time. It is possible that an average of 50,000 people or more contract ciguatera around the world each year.

Although over 70% of those exposed to ciguatoxin by ingesting heavily contaminated fish are affected, lethal reactions are rare, probably less than 1%. Lethal effects usually require that the fish's internal organs be consumed, which is not a common practice in developed nations, but common cuisine on small islands in the South Pacific, where the entire fish may be used for making stews. These islanders often continue consuming fish until symptoms are noticed, then avoid fish while being treated by local remedies and, when symptoms subside, resume fish consumption.

There is some evidence that the prevalence of ciguatera is expanding with increasing ocean temperatures associated with climate change. For example, the first reports of ciguatera came only recently in the western Gulf of Mexico and eastern Mediterranean. It has been impossible to develop a baseline of historic incidence rates, however, due to poor reporting and misdiagnosis. Although a blood test has recently been developed as a diagnosis tool, it not yet in common use. In subsistence fishing communities, even when symptoms are recognized, they are rarely reported to physicians or health officials. Factors affecting the incidence of HABs (i.e., excessive nutrient inputs) or reductions in coral reef health also could result in increases in *Gambierdiscus* populations. There is an additional concern that artificial reefs and other structures (e.g., oil drilling rigs) in near-surface waters could increase areas for colonization.

Symptoms of ciguatera are many and vary widely among regions of the world and among infected individuals; they include a large variety of gastrointestinal (e.g., vomiting and diarrhea), neurological (pain, headaches, tingling, respiratory distress, and anxiety), and cardiovascular (arrhythmias, reduced heart rate) symptoms. Some of the most unique symptoms, though not always present, are a sensation of loose teeth and temperature sensation changes, whereby cold items can feel extremely hot and hot items extremely cold. Neurological symptoms, including itching of the skin, headaches, muscular aches, and weakness can persist for weeks or months after exposure to the toxin, and a recurrence of symptoms sometimes persist for several years. Consumption of fish, alcoholic drinks, caffeine, and nuts are sometimes associated with a recurrence of the symptoms. There is no medical cure for ciguatera; however, various drugs help to control the symptoms. There are over 50 purported herbal medicines and other local remedies from regions where the disease is prevalent; some have been shown to be effective in reducing symptoms, but controlled studies of active agents in the remedies have not been carried out.

Except for abstaining from the consumption of predatory fish, there is no certain way to avoid ciguatera-tainted fish in regions where the toxin is prevalent. Although the odds of any individual fish containing ciguatera toxin are low, the chances of being exposed increases with frequency of consumption of certain species from particular tropical regions. One of the most dangerous fish for consumption is the barracuda (**Figure B7-3**), which is often considered inedible in

Figure B7-3 The barracuda is a fish commonly associated with ciguatera poisoning.

ciguatera-prone regions. Grouper and snapper should be considered with caution, and when cleaning fish for personal consumption, caution should be taken to avoid contaminating the flesh with matter from internal organs. Ciguatoxin does not affect fish tissue in any visible way, and infected fish are not recognizable by appearance, odor, or taste. There is no simple method of neutralizing the toxin, as neither cooking nor freezing affects the toxicity. Although laboratory bioassay techniques have recently been developed to test for presence of ciguatoxin, there is no simple, rapid method for reliably determining toxicity of fish to humans. Traditional methods of determining which fishes were safe for consumption was feeding the fish to other animals, including cats, mongooses, guinea pigs, and rats. Such attempts were only sometimes successful, and there are ethical concerns of routinely using domesticated animals in this manner. Consumer groups argue that international seafood suppliers should make greater efforts to document that groupers and snappers are not from a region known to have high incidence rates for ciguatera.

Public awareness and medical and scientific research to address concerns over the prevalence and future status of ciguatera has increased substantially over the last decade. Robert Dickey and Steven Plakas summarized recent advances, pointing out that, although much progress has been made concerning environmental risk and public health concerns, there is much we still do not understand about the ecology and physiology of the dinoflagellate *Gambierdiscus* and ways of dealing with its toxic effects on humans.

Reasons for occurrences of HABs vary depending on the species, time, and place of the bloom. Many HAB species are native to the region where blooms occur. The organisms are capable of surviving at low concentrations when conditions for growth are not ideal; for example, with low levels of nutrients. When exposed to high nutrient levels or other conditions conducive to growth, they undergo massive population explosion leading to concentrations so dense there can be thousands of organisms in a single drop of water. Although some HABs are apparently natural occurrences, others appear to be a result of excessive nutrients put into the oceans from anthropogenic sources.

HABs often occur seasonally under conditions similar to those supporting other plankton blooms; for example, in regions of upwelling (see Chapter 1). The occurrence and toxicity of algae blooms may depend on very specific conditions that are difficult to define and predict, however. For example, *Pseudo-nitzschia* diatoms bloom and produce neurotoxins only under very defined conditions including specific currents, temperatures, and nutrient concentration (**Figure 7-13**). These conditions co-occur only during some time periods in the Pacific Ocean off the coast of North America, and appear to be influenced by natural variation in factors such as estuarine outflow, upwelling, and El Niño events. In other regions, HABs are predictable because they occur seasonally as a response to factors such as increased nutrient input, warm temperature, and calm seas. Historic accounts of HABs prior to large anthropogenic effects give further support to their natural occurrence; for example, red tides were reported in the Gulf of Mexico in the accounts of 18th century explorers.

Contact with waters containing harmful algal blooms does not always cause harm to humans or other animals, and the toxins often are not lethal to the animals that consume them. Consumption of organisms in which the toxins have accumulated, however, can cause illness or death in fishes, marine mammals, or humans. For example, humans exposed to toxins in bivalves can get shellfish poisoning, as classified into four syndromes: paralytic, neurotoxin, diarrheal, and amnesic. Some of the most well known toxin-producing protists are the dinoflagellates *Alexandrium* and *Karena* and the diatom *Pseudo-nitzschia*.

One source of HABs that is of particular concern in U.S. waters is *Alexandrium fundyense*. Saxitoxin formed by this dinoflagellate accumulates in bivalves that, when eaten, can cause paralytic shellfish poisoning in humans. Symptoms include nausea and vomiting, tingling and burning sensations, shortness of breath, lack of coordination, and sometimes death. Marine mammals, such as sea otters, humpback whales, and monk seals, also have died apparently as a result of consuming clams or fish contaminated with saxitoxin. These toxins do not bioaccumulate to the same degree as POPs because they are decomposed or eliminated by most organisms within a few weeks. Some bivalves retain the toxins, however, apparently as a chemical defense mechanisms against predation; for example, butterclams (*Saxidomus giganteus*) can retain the toxins for up to two years. In the Atlantic Ocean off South Africa, *Alexandrium catanella* is the species commonly causing harmful algal blooms; the toxins accumulate in bivalves harvested as seafood. In the U.S. Gulf of Mexico, the dinoflagellate *Karenia brevis* produces a neurotoxin that accumulates in bivalves and can cause a neurotoxic shellfish poisoning and other environmental impacts (see Box 7-4, Conservation Controversy: Are Red Tides Natural or Man Made?).

Not all toxic dinoflagellates impart a color to the water that makes the HAB readily visible. For example, *Pfiesteria piscicida* becomes apparent only because of its effects. This is because *Pfiesteria* is incapable of producing chlorophyll, and it survives by ingesting other phytoplankton and using their chloroplasts. Although this species was hardly

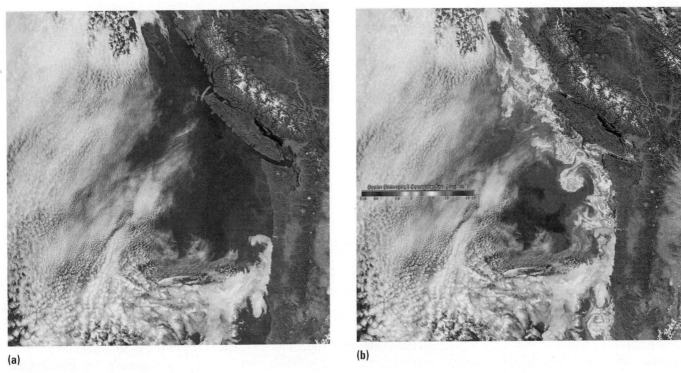

(a) **(b)**

Figure 7-13 October 1, 2004 image from the SeaWiFS Sensor during a *Pseudo-nitzschia* toxic algae bloom off the coast of Washington State and Vancouver Island: **(a)** grayscale; **(b)** phytoplankton chlorophyll concentration (gray tone corresponding to 10 on the scale indicate highest concentrations). (See Color Plate 7-13.)

known prior to these fish kills (it was not discovered until the 1980s), it is naturally present along much of the U.S. Atlantic and into the Gulf of Mexico. This late recognition is a result of *Pfiesteria*'s unique life cycle. It can remain in a dormant cyst stage in the sediments for long periods, and only becomes a free-swimming flagellated cell when slime and excretions from fish, primarily the menhaden *Brevoortia,* are present in large amounts in the water. This is the toxic stage, when the *Pfiesteria* produces a neurotoxin that is lethal to fish and another toxin that initiates the formation of deep lesions in the fish (**Figure 7-14**). The fish tissue released into the water from the sores is fed upon by the dinoflagellate, which then encysts and returns to the sediment. *Pfiesteria* can tolerate a broad range of temperatures and salinities, but the toxic forms are most prevalent at intermediate salinities, around 15 ppt, and warm temperatures, over 26°C. A high level of nutrients in the water also enhances the likelihood of a bloom. The combination of these conditions is most likely in late summer months in coastal estuaries. Excess nutrients flowing into U.S. coastal estuaries appear to have increased the likelihood of *Pfiesteria* blooms. Because the *Pfiesteria* organism was only recently identified, however, it is uncertain whether historic cases of these events went unrecognized.

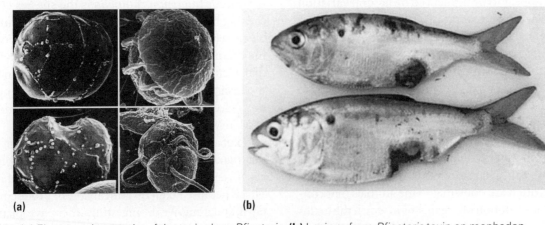

(a) **(b)**

Figure 7-14 **(a)** Electron micrographs of the toxic algae *Pfiesteria*. **(b)** Lesions from *Pfiesteria* toxin on menhaden.

The few large *Pfiesteria* blooms along the U.S. mid-Atlantic coast in the 1990s caused massive fish kills and affected human health. The most serious effects on humans resulted from handling exposed fish that exhibited sores. There were also effects from exposure to the toxins in the water or air during the blooms. *Pfiesteria* toxins do not appear to accumulate in animals, including humans and, thus, illnesses are not likely a result of eating contaminated seafoods as is true for other HAB toxins. Symptoms of exposure include disorientation, memory loss, nausea, and skin lesions. The broad range of symptoms and high level of news reporting and sensationalism made it difficult to assess the overall health impact of blooms. After the recognition of *Pfiesteria* as a source of these blooms, extensive research has been carried out to determine causes and effects; however, there is still wide ongoing debate of these issues. As with other HABs, if we wish to reduce their likelihood in the future, it is important to understand how important the input of excessive nutrients into coastal regions is in initiating the blooms.

The most debated conservation questions concerning harmful algae blooms are: have they increased in frequency, and, if so, why? In part, the increase in reported HABs could be the result of an increase in awareness because more people are spending time on or living near the ocean and the waters are more closely monitored. There is strong evidence that nutrient enrichment of coastal waters increases the intensity of at least some HABs, but it is often difficult to prove (**Box 7-4. Conservation Concern: Are Red Tides Natural or Man Made?**). Although many HABs involve native species, some are comprised of exotics transported around the globe in ballast water of ships, and global climate change could enhance blooms of warm-water plankton.

How or whether we can minimize the impacts from HABs to humans and the marine environment is a much-debated topic. The feasibility of preventing blooms depends on knowledge of factors that cause the blooms. Where an excess of nutrients is a factor, reducing nutrient input is the most advisable way to prevent HABs, and there is the additional benefit of further environmental improvements (see previous chapters). For example, this has been successful in areas of Japan and Hong Kong where reductions of nutrient input into the ecosystem was followed by a substantial decline in occurrences of harmful algae blooms. The addition of foreign chemicals is rarely desirable as a means of controlling blooms because the damage to the environment is unpredictable and could be worse than that resulting from the bloom. Biological control could be feasible if bacteria or other organisms are identified that specifically target the organism causing the HAB, especially if that organism is natural in the environment. This ideal has

rarely been achieved, and unpredictable effects of adding organisms to the ecosystem are always possible. The addition of natural clays to control a bloom that is in progress has received much attention; however, this does not prevent the bloom from occurring and the harmful toxins are transferred to the sediments where they may continue to harm the ecosystem. The impacts on humans can be minimized by monitoring blooms and the presence of toxins in organisms likely to be consumed and warning the public of potential dangers of seafood consumption.

Jellyfish Blooms

The term *plankton bloom* usually refers to an explosive growth of microalgae and protists populations, and the response of meiofaunal crustaceans to these events. One level further up the food chain, macroplankton, mostly jellyfish, also can respond to bloom conditions. Large aggregations of jellyfish can occur naturally; however, as with other plankton, recent increases in frequency and size of the blooms around the world suggests that human influences are having an effect. Although human deaths have resulted from contact with the jellyfishes in theses blooms, this is a relatively rare occurrence (see Chapter 2). Other impacts include reduction of fish abundances through competition and predation, and interfering with fishery harvest by clogging nets or dominating catches.

Anthropogenic effects that potentially influence the frequency and severity of jellyfish blooms include overfishing, eutrophication in nearshore regions, climate change, and translocations of exotic species. Overharvest of fish can remove competitors for food resources. For example, sardines feed on the same planktonic organisms as jellyfish. Areas where sardines have been overharvested off the western coast of southern Africa are now dominated by jellyfish. Overharvest of jellyfish predators could remove controls over jellyfish population growth also. Jellyfish blooms are often of native species; however, in many cases exotic species that undergo population explosions were translocated from other regions by ship ballast water or boat hulls to which polyps can attach. For example, the large (average of about 50 centimeter bell width) spotted jellyfish *Phyllorhiza punctata* from the western Pacific Ocean became a nuisance in the Gulf of Mexico. It is not toxic enough to be considered harmful to humans, but shrimp fisheries are affected when jellyfish clog trawl nets. Jellyfish may affect fish populations by consuming eggs and larvae also.

Jellyfish have several attributes that sometimes allow them to survive better in disturbed marine environments than other plankton consumers. Jellyfish prey more readily on a broad range of plankton than do fishes and other plankton feeders. If nutrient enrichment increases blooms of dinoflagellates, which are smaller than diatoms and other

Box 7-4 Conservation Controversy: Are Red Tides Natural or Man Made?

When the dinoflagellate *Karenia brevis* reaches high concentrations in ocean waters, its photosynthetic pigments impart a distinct red coloration to the water (**Figure B7-4**). As these blooms expand into coastal waters they appear as a red water mass, which gives the blooms their common name. The algal blooms associated with the red tide phenomenon are of interest to conservationists and seafood consumers because *Karenia brevis* produces a neurotoxin, called brevetoxin, that can reach dangerous levels in the waters and animal tissues during red tide events (although no fatalities have been documented in humans). Large die-offs of millions of fish from hundreds of species have been documented. Fish can be affected by consumption of the dinoflagellates directly, eating other organisms containing the toxin, or by absorption of the brevetoxins in the water through the gills.

In order to counter common perceptions that the impact of red tides is largely limited to fish kills and warnings about seafood consumption, Dr. Jan Landsberg and colleagues reviewed the evidence that effects on the natural environment were more far reaching than had been assumed. They found that toxins from *Karenia brevis* red tide blooms can bioaccumulate in the food web. Because bivalves typically are not killed by the red tide and some crustaceans and fish consume *Karenia brevis* without lethal effects, brevetoxins that can be transferred up the food chain. Clams and oysters can retain toxins for at least several weeks, and planktonic crustaceans and fishes containing sublethal concentrations also can transfer the toxins. When these organisms are eaten by predators the effects can be lethal. For example, the death of large numbers of birds such as cormorants, mergansers, and ducks, and marine mammals, including dolphins, has been associated with accumulation of red tide toxins. Predators exposed to sublethal concentrations are often able to recovery, however, when removed from the source of the toxin. Animals also can be affected by toxins remaining in the waters or in bottom sediments after the bloom has dissipated. This is one possible reason for increased sea turtle stranding and manatee deaths following red tides. In humans as well as marine

mammals, respiratory irritation is associated with inhaling sea spray associated with red tides. Concerns over visiting the beach during red tide events for fear of exposure, as well as aesthetic reasons when there are fish kills, can have a serious impact on tourist industries. Fear of seafood consumption also impacts tourism as well as commercial fishing industries. During severe red tide events, fish kills may be massive enough to affect the health of the ecosystem as water quality deteriorates and oxygen levels decline due to bacterial decomposition. For example, a red tide event along the Texas coast in 2005 was associated with the death of tens-of-thousands of planktivorous animals, followed by mortalities of predatory fishes and, two weeks after the red tide, a die-off of benthic invertebrates such as shrimps, crabs, bivalves, and polychaetes. Indirect impacts are difficult to ascertain; however, in coastal Florida, by some estimates, each large bloom event has resulted in about a $20 million economic impact.

Red tides occur during late summer and autumn months during most years, but number of incidences, their size, and duration vary from year to year. Their development is influenced by numerous factors, including nutrient levels, temperature, and current patterns. In the Gulf of Mexico, *Karenia brevis* red tide blooms are typically initiated in offshore waters over the continental shelf in response to an upwelling of deeper nutrient-laden offshore waters onto the shelf. The blooms concentrate in regions where water currents converge, and can be transported toward shore, where they have the greatest impact. Visible red tide blooms have been observed in the Gulf of Mexico from the Yucatan peninsula to Texas and through the northern Gulf of Mexico. The most common, extensive, and harmful red tides have occurred off the west coast of Florida, where currents transport the blooms toward shore and along the coastline.

From a conservation perspective, one of the biggest questions concerns the extent to which human actions affect the size, frequency, and impact of *Karenia brevis* red tide blooms. There is little doubt that the red tides can occur naturally, because Native Americans, Spanish explorers, and early settlers reported such blooms long before the time of significant anthropogenic influences. The apparent increase in the frequency and impact of red tides in recent times, however, has initiated questions concerning the degree of human influence. For example, red tides were reported in the Gulf of Mexico about once every two years from the mid-1850s to mid-1950s, but in about 80% of the following 50 years. Recently, there has been an increase in reported deaths of animals associated with red tides; for example, mortalities of bottlenose dolphins and manatees were associated with red tides in each year from 2002 to 2006. This evidence suggests that anthropogenic factors such as nutrient runoff or climate change could be resulting in an increase in red tides. Another possible explanation is that a rise in coastal populations, ocean use, and awareness has increased reporting of red tide events, rather than there being an actual increase in the events themselves. The increase in animal deaths associated with red tides may reflect scientific advances that allow for better documentation. The truth is undoubtedly some combination of these proposals. Algal blooms that cause red tides are a regular event in the

Figure B7-4 Red tide blooms of the dinoflagellate *Karenia brevis* can result in large fish kills and bioaccumulate up the food chain.

Gulf of Mexico and can develop in any given year under certain natural ocean conditions. An input of land-based nutrients is needed to sustain the blooms in coastal waters, however, and anthropogenic sources of nutrients may increase their size, duration, and impact. One proposed mechanism for increased harmful algal blooms is called the *silica deficiency hypothesis*. According to this hypothesis, because anthropomorphic nutrient sources are low in silica, they support blooms of harmful dinoflagellates (with cellulose coverings) over diatoms (with siliceous coverings).

Without a firm knowledge of the degree of human influences on the occurrence of these red tides, it is difficult to know whether it is feasible or desirable to develop methods for controlling the blooms. Although the influence of nutrients is often unclear, it is a reasonable goal to reduce inputs into coastal waters for the overall benefits to the marine environment, as discussed in previous chapters. Proposals to apply chemicals to control the blooms are controversial due to the likelihood of impacts of the chemical on other components of the marine ecosystem. Prior to the existence of current environmental regulations, experimental attempts were made to control red tides with copper sulfate. The application of about 100 tons of copper sulfate over a 40 square kilometer area off west Florida in 1957 resulted in a rapid elimination of the bloom; however, the *Karenia brevis* populations returned to high concentrations after about 10 days. This was the only attempt for large-scale control using copper sulfate. It is not considered a valid option because it is a costly temporary solution and—more importantly—because of the toxic effects of the copper sulfate on the environment. Treatments used for drinking water have been attempted also; ozonization and chlorination of waters were successful in controlling *Karenia brevis* experimentally. Methods of biological control have been considered, too; for example, application of naturally occurring aligicidal bacteria. Another method being considered that does not require the use of unnatural chemicals is application of natural clays to the waters where the bloom is in progress. The clay particles cause aggregation and possible death of the dinoflagellate cells, which sink to the bottom. Although this may successfully remove the cells from the water column, they could still accumulate in the sediments and possibly bioaccumulate in benthic organisms. No wide-scale efforts to specifically control these blooms are currently in place. The goal of any program to control the impact of harmful algae blooms such as those causing red tides needs to include a consideration of any environmental damage that may result.

phytoplankton, jellyfish would survive more readily than consumers that require large plankton to feed efficiently. Oxygen levels are another important consideration. Jellyfish are more tolerant of low-oxygen conditions than are fishes; thus, hypoxia resulting from increased eutrophication in coastal waters would allow jellyfish to take advantage of high-nutrient conditions during hypoxic events. Nutrient pulses may enhance jellyfish because they have fast growth rates when conditions are ideal, but can tolerate starvation after nutrient sources are used up by shrinking their body size. Overfishing in the presence of eutrophic conditions may start a cycle of interactions that enhance jellyfish. This cycle begins when jellyfish numbers become excessive after fish competitors and predators have been removed. Then, by preying on the fish eggs and larvae, they inhibit the return of overharvested fish populations. This sort of feedback mechanism makes it difficult for jellyfish-dominated ecosystems to return on their own to natural healthy conditions.

As research uncovers more evidence that some regions of the ocean are being driven into jellyfish-dominated ecosystems by anthropogenic impacts, scientists are beginning to consider possible ways to avoid these situations and return ecosystems to their normal function. Anthony Richardson and colleagues have reviewed the research on jellyfish blooms and presented recommendations and problems. They suggest that long-term strategies should be to reduce eutrophication, overfishing, and global warming. These are worthy goals but, as discussed throughout this text, are difficult to achieve. Other local management responses could include developing fisheries and markets for jellyfish products as food; however, this depends on the marketability of jellyfish products in new regions (jellyfish are currently harvested for consumption primarily in Southeast Asian regions to supply markets in China and Japan), and might encourage the harvest of non-problem species. Another possibility would be to destroy the jellyfish using nets designed to cut through the animal; this is currently still experimental, and it is uncertain whether the jellyfish could not regenerate or simply return through natural reproduction. Jellyfish could be readily destroyed during the benthic polyp stage; however, this would require extensive efforts for physical removal or research on environmentally safe chemicals that would selectively kill the polyps. Biocontrol agents might be used to selectively kill jellyfish during the pelagic stage; however, no such chemicals are known and the ability to develop such chemicals is highly uncertain. Finally, there could be greater attempts to eliminate exotic introductions by controlling ballast water transfer, establish hull-cleaning regulations, and eliminating the aquarium trade in jellyfish. Such protocols are being developed for control of other exotics but are politically difficult to implement and enforce.

It is now apparent that a combination of anthropogenic environmental impacts is a major factor in the recent increases in blooms of macroplankton, such as jellyfish, as well as microplankton, such as HAB-forming organisms. The complexity of interactions among these factors makes it difficult to establish cause-and-effect relationships with certainty and, without this certainty, changing human behavior and implementing new restrictions will remain difficult. Each of these factors is of concern regarding not only the ocean pelagic region, but the overall health of the global ecosystem. The evidence provided by scientific studies, thus, should provide further impetus for political leaders, environmental managers, and the public to act locally to produce a cumulative positive effect globally in resolving the myriad environmental problems we now face.

Plastics in the Open Ocean

As the world has become increasingly industrialized, the open ocean has become a reservoir not only for chemical pollutants but also for larger items that are commonly classified as litter. The largest portion, about 60% to 80%, of this marine litter is plastics. Although using the oceans as general garbage dumps has been outlawed through international agreements for about half a century (see Chapter 12), humans continue to put tons of plastics and other trash into the sea each year unintentionally or illegally. Although plastics break down into small pieces when exposed to ultraviolet (UV) radiation in sunlight, the polymers that make up the plastics are not readily biodegraded. Due to the persistent nature of plastics, these materials can remain afloat at or near the surface of the ocean for decades, being washed far out to sea. Even though only a small fraction of plastic produced reaches the ocean, hundreds of tons are produced annually; for example, each U.S. citizen uses over 100 kilograms of plastic each year on average. The United Nations Environmental Program (UNEP) estimates that on average there are 46,000 pieces of plastic per square mile floating in the ocean. Much of the plastics in the ocean that do not eventually wash up on shore accumulate in ocean circulation gyres. For example, the so-called Great Pacific Garbage Patch is trapped within the North Pacific Gyre current, containing debris extending for over 2,000 kilometers across the ocean. Plastics comprise about 90% of this floating litter, about 80% of which was introduced from land. The rest comes from ships, including trash thrown overboard, remnants of fishing gears, and materials lost at sea during shipping accidents. It is misleading to refer to this patch as a *garbage island,* as it is sometimes portrayed by the news media, because most of the plastic materials have been broken up or shredded and thus are sizes of a less than 3 millimeters diameter. This patch was discovered by scientists pulling nets through surface waters to capture plankton; in the North Pacific Gyre the mass of plastic material netted was six times that of the plankton. A major source of plastics floating in the open ocean is small pea-sized thermoplastic resin pellets, called *nurdles,* produced to be used in the manufacture of plastic products.

A chemical analysis of these nurdles by Lorena Rios and colleagues revealed that they can cause additional harm to wildlife by absorbing other toxic chemicals, including POPs such as DDT and PCBs. Ways these pellets get into the ocean ecosystem include spills from trucks or at manufacturing facilities that wash into rivers or from container ships. This is a problem with an easy solution if greater care were taken in handling and disposal.

Plastics are pollutants not only in mid-ocean patches, but in open ocean and coastal waters around the world. Many of these plastics end up washing onto beaches of islands far from the continents (see Chapter 3). Others are picked up by sea birds, mistaken as a source of food. Over 40% of seabird species ingest floating plastic taken from the ocean surface. Ingestion of plastics is especially problematic for albatrosses nesting on isolated islands (**Figure 7-15**). In their search for food, the adults bring numerous plastic articles back to the nest to feed to their young. Consuming plastics appears to affect albatross survival because young with the most plastic in their guts tend to have higher mortality rates. Larger plastic items, such as plastic bags floating near the ocean surface, present problems to marine animals that feed in the ocean pelagic. The leatherback turtle *Dermochelys coriacea,* for example, feeds largely on jellyfish, for which floating plastic can be easily confused. If these plastics obstruct the passage of food into the gastrointestinal tract (GI) of the turtle, they can cause death. A survey of reported leatherback deaths by Nicholas Mrosovsky and colleagues found that 37% of the turtles had plastics in their gastrointestinal tracts and some of these plastics had blocked the GI tract in a manner that could have caused death.

Although plastic materials are very persistent, recent studies by Katsuhiko Saido and colleagues have shown that, under certain conditions, slow decomposition in ocean

Figure 7-15 Plastic debris accumulated in the stomach of a dead albatross. Adults mistake the debris for food and bring them back to feed to their young in the nest.

waters releases potentially harmful chemicals. For example, polystyrene, one of the most commonly used plastics, begins to decompose after one year. This is concerning because the products of decomposition are chemicals that are potentially harmful to ocean life and have been shown to disrupt hormone functions, affect reproduction, and are suspected carcinogens.

Removal of the majority of the plastic materials from the ocean environment is not practical, beyond efforts to remove large remnants of fishing gears that could entangle animals (see Chapter 11). The simple way to limit the future effects of plastics is to reduce their introduction into the marine environment. This would include tighter restrictions on the manufacturers of plastic products to control the release of wastes, further recycling of plastic products, containing plastic wastes in dumps and landfills, reducing litter, and developing and replacing plastics with materials that are biodegradable into natural byproducts.

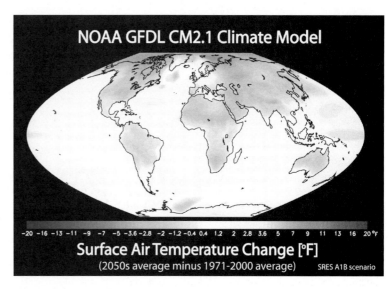

Figure 7-16 Predicted temperature changes from 2000 to 2055 from models developed by the Intergovernmental Panel on Climate Change. The average predicted increase in global temperature is 5.2°F. (See Color Plate 7-16.)

7.3 Climate Change and the Ocean Pelagic

The physical and chemical effects of global climate change and controversies over causes and solutions were discussed in Chapter 1 (see Box 1-1, Conservation Concern: Global Warming and the Ocean). Issues concerning sea level rise are not of as great concern in the open ocean as they are in coastal regions (see Chapters 3 and 4); however, changes in circulation patterns, temperatures (**Figure 7-16**), and ocean chemistry could have significant effects on the organisms and ecosystems of the open ocean.

Ocean warming would result in the extension of warm waters from the tropics, thus expanding the range of tropical open-ocean species while contracting the range of temperate species. A disruption of ocean circulation gyres could lead to warmer waters on the eastern side of ocean basins due to a disruption of flow of cold water away from the poles. Waters on the western side of ocean basins would be cooler if circulation of warm waters from the equatorial region is reduced (see Chapter 1). These conditions would change the ranges of oceanic species, where some species would become more abundant while others decline, and the survival of exotics could be enhanced. Julie Roessig and colleagues analyzed trends in fish distributions and predicted that even small temperature changes could affect the distribution and abundance of fishes. Some predicted effects of warming ocean temperatures include changes in

spawning times and growth rates for polar fish species such as cod and haddock, and northward shifts in distribution for temperate fishes such as salmon and tuna. Changes in distribution patterns have been documented in many ocean regions. Along the North American coastline, Janet Nye and colleagues found that some fish species have already begun to shift northward and into deeper waters (**Figure 7-17**). Allison Perry and colleagues found that over the past 25 years in the North Sea, distributions of the majority of the marine fish species have shifted closer to the poles or into deeper waters, and the populations that have made this shift mature earlier and grow to smaller maximum sizes.

Ocean temperature changes could affect growth and survival of species that remain within their native geographic range as conditions change. Temperatures at the extremes of a fish's tolerance range can lower growth due to increased physiological stress. Sudden shifts in temperature, especially excessive cold conditions, commonly result in fish kills in the U.S. Atlantic and Gulf of Mexico. Because higher temperatures result in higher metabolic rates, less energy would be available for growth and reproduction. This already has been documented in some regions for Atlantic cod and Atlantic salmon *Salmo salar*. Larval and juvenile fishes (e.g., cod) can be very sensitive to changes in temperature, influencing survival and growth of an entire year class due to physiological effects. The frequency and distribution of plankton blooms are affected by water temperatures, impinging on food availability throughout

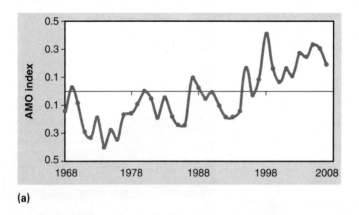

(a)

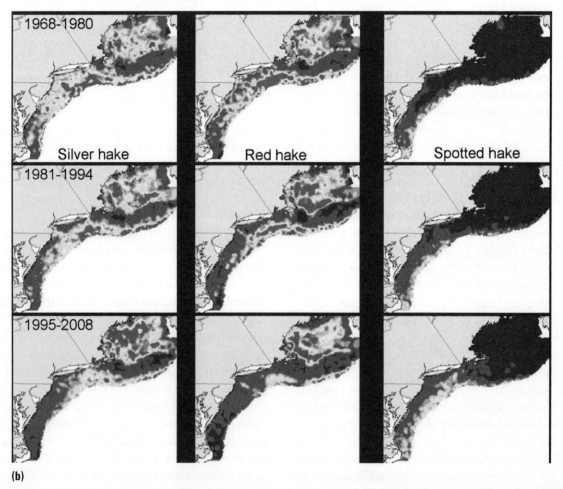

(b)

Figure 7-17 **(a)** Atlantic Multidecadal Oscillation Index (AMO; an index of North Atlantic temperatures) from 1968 to 2008 in the northwestern Atlantic. **(b)** Distribution of merlucciid fishes, the silver hake (*Merluccius*), and red hake and spotted hake (*Urophycis*) over the same time period. Mid-tone grays indicate the highest fish concentrations. Patterns indicate a general shift in distribution pattern to the north in association with an increase in water temperatures. (See Color Plate 7-17b.)

entire food webs and influencing migratory patterns of fishes feeding within bloom regions.

Higher temperatures could be especially harmful to marine animals living in polar oceans. They may not be able to physiologically tolerate warmer waters or could be unable to compete with temperate species that expand their range poleward. Preferred food items may become less abundant. For example, warming in the Antarctic has been associated with declines in krill, potentially affecting populations that depend on them as a food resource, including the baleen whales. A 30% decline, since 1987, in Antarctic populations of the Adelie penguin, a krill consumer, could be a direct result of the loss of krill. Warmer waters also have been shown to enhance some jellyfish populations, allowing them to outcompete planktonic fishes.

Changes in carbon dioxide levels in the atmosphere influence not only global temperature changes but also ocean chemistry, in particular the pH of ocean waters (**Figure 7-18**). It is difficult to predict how severe these changes will be and how they will influence open ocean ecosystems. Shell-forming organisms are those most likely to be impacted because at low pH it becomes more difficult to develop calcium carbonate shells, due to the higher dissolution of carbonates in the water. The first organisms to be affected are those that use aragonite as the form of calcium carbonate in their skeletal structures. These include primarily bottom organisms such as mollusks and corals (see Chapter 5). Predicted pH levels eventually could harm pelagic organisms that form shells of calcite (another form of calcium carbonate). These include foraminiferans, coccolithophores, and crustaceans. Studies by Luc Beaufort and colleagues provide evidence that these effects may already be occurring. They found that coccolithophores in more acidic waters, resulting from elevated CO_2, typically form thinner shells. Thus, strongly calcified species may be outcompeted, resulting in a shift in species composition. A predominance of more weakly calcified species could result in less carbon uptake in the marine plankton, reducing the ocean's ability to sequester carbon.

Higher CO_2 levels in the open ocean could have other physiological effects on organisms. For example, some photosynthetic protists and algae could benefit from increased CO_2 concentrations if they are able to tolerate modified pH levels. For other species elevated CO_2 could have negative effects. There is evidence of impacts on development, growth, photosynthesis, digestion, and

Figure 7-18   Predicted changes in ocean acidity from 1765 to 2100 from models developed by the Intergovernmental Panel on Climate Change. Gray areas near the poles are more acidic (lower pH) and grays areas surrounding the equator are more basic (higher pH). Black stars indicate shallow corals; gray stars indicate deep corals. (See Color Plate 7-18.)

mortality for some organisms. Another potential effect of pH change is an alteration of the chemical form in which nutrients occur, which could reduce their availability to living organisms. For example, changes in nitrogen availability could affect rates of photosynthesis and growth and alter the entire food web. A lower pH could increase the solubility of iron and make it more available, however, thereby increasing primary productivity in iron-limited regions (see Chapter 1). There remains much uncertainty in not only the future levels of carbon dioxide in the atmosphere, but also what effects a given CO_2 concentration will have on ocean ecosystems. Better predictions will be possible only with a more complete understanding of physiological impacts of carbon dioxide and pH on marine organisms and interactions within marine ecosystems. Once again, the precautionary principle is the recommended approach. The increasing evidence of numerous negative impacts of elevated CO_2 in the world's ocean environments should encourage continued efforts to reduce these inputs. Chapter 1 (see Box 1.1, Conservation Concern: Global Warming and the Ocean) has a discussion of ways this could be achieved.

STUDY GUIDE

■ **Topics for Review**

1. Why is most of the primary production in the epipelagic zone by unicellular phytoplankton?
2. How do sands blowing from continents affect biological production in the ocean epipelagic?
3. Why is primary production typically lower in the center of ocean basins than in coastal regions and upwelling zones?
4. How did technological developments result in increases in estimates of open ocean primary production?
5. How do winter conditions cause a disruption of the open ocean thermocline, and why does this enhance production in surface waters?
6. What are the adaptive advantages of diurnal vertical migrations of plankton in the open ocean?
7. Why are large apex predators vulnerable to overharvest even when taken in relatively low numbers?
8. Why do open ocean food chains tend to be longer than those in other ecosystems?
9. What was the primary source of lead in the ocean environment in the 20th century?

10. What kinds of fish can be consumed to avoid excessive mercury contamination?
11. What characteristics of POPs lead to their high rates of accumulation in ocean organisms?
12. How does global distillation lead to the accumulation of POPs in polar waters?
13. What arguments support contentions that HABs are primarily natural occurrences?
14. What arguments support contentions that HABs are mostly due to human activities?
15. What characteristics distinguish *Pfiesteria* blooms from other HABs?
16. Why are biological and chemical control methods not widely used to control HABs?
17. How do red tides impact coastal areas economically?
18. What factors have led to poor documentation of incidences of ciguatera?
19. How does overfishing affect the occurrence and persistence of jellyfish blooms?
20. Why do jellyfish tend to survive better than other planktivores in disturbed marine environments?
21. What chemical effects do plastics have in the open ocean?

■ **Conservation Exercises**

1. Describe effects of each of the following on the given open ocean species or characteristic:
   a. DDT — seabirds
   b. *Pfiesteria* blooms — fish
   c. red tides — fish
   d. jellyfish blooms — fish populations
   e. plastics — seabirds and sea turtles
   f. elevated carbon dioxide — calcite-shell forming plankton
   g. warming waters in the North Atlantic — fish populations
2. Describe how each of the following actions could enhance the health of open ocean ecosystems or consumers of marine organisms:

   a. outlawing the use of sargassum in livestock food supplements
   b. reducing the amount of coal burned for power generation
   c. producing alternatives to DDT for control of malaria
   d. encouraging the consumption of planktivorous fishes instead of large predators as seafood
   e. replacing consumption of marine mammals with other seafood products
   f. educating consumers of the symptoms of ciguatera
   g. reducing nutrient inputs into waters prone to HABs
   h. increased use of plastics that biodegrade into natural byproducts

FURTHER READING

Arctic Monitoring and Assessment Programme (AMAP)/ United Nations Environmental Program (UNEP). 2008. *Technical background report to the global atmospheric mercury assessment.* Arctic Monitoring and Assessment. Programme/UNEP Chemicals Branch, 159 pages. www.chem.unep.ch/mercury/

Bosch, D. F., D. A. Anderson, R. A. Horner, S. E. Shumway, P. A. Tester, and T. E. Whitledge. 1997. *Harmful algal blooms in coastal waters: options for prevention, control and mitigation.* NOAA Coastal Ocean Program, Decision Analysis Series No. 10, Special Joint Report with the National Fish and Wildlife Foundation.

Beaufort, L., I. Probert, T. de Garidel-Thoron, E. M. Bendif, D. Ruiz-Pino, N. Metzl, C. Goyet, N. Buchet, P. Coupel, M. Grelaud, B. Rost, R. E. M. Rickaby, and C. de Vargas. 2011. Sensitivity of coccolithophores to carbonate chemistry and ocean acidification. *Nature* 476:80–83.

Choy, C. A., B. N. Popp, J. J. Kaneko, and J. C. Drazen. 2009. The influence of depth on mercury levels in pelagic fishes and their prey. *Proceedings of the National Academy of Sciences USA* 106:13865–13869.

Dickey, R. W., and S. Plakas. 2010. Ciguatera: A public health perspective. *Toxicon* 56:123–136.

Endo, T., Y. Hotta, K. Haraguchi, and M. Sakata. 2003. Mercury contamination in the red meat of whales and dolphins marketed for human consumption in Japan. *Environmental Science and Technology* 37:2681–2685.

Geisz, H. N., R. M. Dickhut, M. A. Cochran, W. R. Fraser, and H. W. Ducklow. 2007. Melting glaciers: a probable source of DDT to the Antarctic marine ecosystem. *Environmental Science and Technology* 42:3958–3962.

Hickie, B. E., P. S. Ross, R. W. MacDonald, and J. K. B. Ford. 2007. Killer whales (*Orcinus orca*) face protracted health risks associated with lifetime exposure to PCBs. *Environmental Science and Technology* 41:6613–6619.

Landsberg, J. H., L. J. Flewelling, and J. Naar. 2008. *Karenia brevis* red tides, brevetoxins in the food web, and impacts on natural resources: decadal advancements. *Harmful Algae* 8:598–607.

Mason, R., and N. Pirrone. 2009. *Mercury fate and transport in the global atmosphere.* Springer, New York.

Mrosovsky, N., G. D. Ryan, and M. C. James. 2009. Leatherback turtles: the menace of plastic. *Marine Pollution Bulletin* 58:287–289.

Nishimura, M., S. Konishi, K. Matsunaga, K. Hata, and T. Kosuga. 1983. Mercury concentration in the ocean. *Journal of the Oceanographical Society of Japan* 39:295–300.

Nye, J. A., J. S. Link, J. A. Hare, and W. J. Overholtz. 2009. Changing spatial distribution of fish stocks in relation to climate and population size on the Northeast United States continental shelf. *Marine Ecology Progress Series* 393:111–129.

Patterson, C. 1987. Global pollution measured by lead in mid-ocean sediments. *Nature* 326:244–245.

Perry, A. L., P. J. Low, J. R. Ellis, and J. D. Reynolds. 2005. Climate change and distribution shifts in marine fishes. *Science* 308:1912–1915.

Rhiannon, L. M., S. E. Reynolds, G. A. Wolff, R. G. Williams, S. Torres-Valdes, E. Malcom, S. Woodward, A. Landolfi, X. Pan, R. Sanders, and E. P. Achterberg. 2008. Phosphorus cycling in the North and South Atlantic Ocean subtropical gyres. *Nature Geoscience* 1:439–443.

Richardson, A. J., A. Bakun, G. C. Hays, and M. J. Gibbons. 2009. The jellyfish joyride: causes, consequences and management responses to a more gelatinous future. *Trends in Ecology and Evolution* 24:312–322.

Rios, L. M., C. Moore, and P. R. Jones. 2007. Persistent organic pollutants carried by synthetic polymers in the ocean environment. *Marine Pollution Bulletin* 54:1230–1237.

Roessig, J. M., C. M. Woodley, J. J. Cech, and L. J. Hansen. 2004. Effects of global climate change on marine and estuarine fishes and fisheries. *Reviews in Fish Biology and Fisheries* 14:251–275.

Ross, P. S., G. M. Ellis, M. G. Ikonomou, L. G. Barrett-Lennard, and R. F. Addison. 2000. High PCB concentrations in free-ranging Pacific killer whales, *Orcinus orca*: effects of age, sex and dietary preference. *Marine Pollution Bulletin* 40:504–515.

Shen, G. T., and E. A. Boyle. 1987. Lead in corals: reconstruction of historical industrial fluxes to the surface ocean. *Earth and Planetary Science Letters* 82:289–304.

Sherr, B. F., E. B. Sherr, D. A. Caron, D. Vaulot, and A. Z. Worden. 2007. Ocean protists. *Oceanography* 20:130–134.

Steen, A. O., T. Berg, A. P. Dastoor, D. A. Durnford, L. R. Hole, and K. A. Pfaffhuber. 2010. Natural and anthropogenic atmospheric mercury in the European

Arctic: a speciation study. *Atmospheric Chemistry and Physics Discussion* 10:27255–27281.

Stoner, A. W., and H. S. Greening. 1984. Geographic variation in the macrofaunal associates of pelagic *Sargassum* and some biogeographic implications. *Marine Ecology Progress Series* 20:185–292.

Storelli, M. M., and G. O. Marcotrigiano. 2001. Total mercury levels in muscle tissue of swordfish (*Xiphias gladius*) and bluefin tuna (*Thunnus thynnus*) from the Mediterranean Sea (Italy). *Journal of Food Protection* 64:1058–1061.

Torres-Valdes, S., V. M. Roussenov, R. Sanders, S. Reynolds, X. Pan, R. Mather, A. Landolfi, G. A. Wolff, E. P. Achterberg, and R. G. Williams. 2009. Distribution of dissolved organic nutrients and their effect on export production over the Atlantic Ocean. *Global Biogeochemical Cycles* 23(4):GB4019.

Turley, C. 2008. Impacts of changing ocean chemistry in a high-CO_2 world. *Mineralogical Magazine* 72:359–362.

Unger, M. A., E. Harvey, G. G. Vadas, and Michael Vecchione. 2008. Persistent pollutants in nine species of deep-sea cephalopods. *Marine Pollution Bulletin* 56:1498–1500.

Wooton, J. T., C. A. Pfister, and J. D. Forester. 2008. Dynamic patterns and ecological impacts of declining ocean pH in a high-resolution multi-year dataset. *Proceedings of the National Academy of Sciences, U.S.* 105:18848–18853.

Seafloor and Deep-Sea Ecosystems: Resource Harvest and Habitat Protection

Prior to the development of technology that allowed the mapping and exploration of the seafloor, little was known about bottom ecosystems beyond the intertidal zone. During the last century, however, much progress has been made in exploring the deep sea. Most of the exploration has been from indirect observations, such as mapping of the seafloor using sonar, methods that do not readily discriminate smaller physical features or organisms on the seafloor, or provide detailed information such as species identifications. They thus provide only limited information for conservation of marine communities. Nets are pulled through deep waters or traps are set on the bottom to give us a glimpse of life in the deep sea and on the seafloor. This type of exploration is difficult and samples only a very small area of the seafloor, and organisms that are pulled up from the deep sea do not provide information on habitat, behavior, or interaction of organisms within a community. In recent years, visual observations of the deep sea have become more feasible with technological advancements. Although scuba diving is mostly limited to water less than 50 meters depth, submarines and remotely operated submersible vehicles have given us a direct access to deep-sea and seafloor communities (**Figure 8-1**). Due to the vast area encompassed by the deep sea and the difficulty and expense of working in these environments, however, only a small fraction has been explored directly by humans. Despite these limitations, we have accumulated a large amount of knowledge in the past

50 years about the deep ocean environment and the organisms residing there.

As the ability of humans to explore the seafloor and deep sea has increased, so has the ability to exploit the resources there. While it is not yet feasible or profitable to harvest animals or extract resources from many of the deepest regions of the sea, through time we have developed the capability to exploit ocean ecosystems at greater and greater depths. Even if much of the deep sea remains inaccessible to fishing and mining, we now realize that pollutants we put into the atmosphere and surface waters of the ocean have the potential to reach the seafloor and affect deep ocean waters.

8.1 Continental Shelf and Slope Ecosystems

As one moves away from the coastline into deeper waters over the continental shelf and slope, there is a rapid decline in primary production due to low sunlight penetration. Life in regions beyond the photic zone depends on organic matter originating elsewhere; sources include particles sinking from near-surface waters, or dissolved or suspended organic matter circulating in waters from the epipelagic or transported offshore from coastal regions. Along the inner regions of the continental shelf, the influence of coastal rivers can be especially important in transporting nutrients into coastal bottom ecosystems. Oxygen depletion can be

Figure 8-1 The remotely operated vehicle (ROV) *Hercules* working at the deep seafloor.

a biologically limiting factor in waters over the shelf and slope, especially in regions near coastlines where there is an excessive input of nutrients from anthropogenic sources (see Chapter 6 section on Coastal Hypoxia and Dead Zones); however, in most ocean regions there is adequate circulation from waves and currents to supply enough oxygen to these bottom habitats to support life. Wind waves can enhance mixing of bottom waters to as deep as 200 meters or more, and vertical thermohaline circulation gyres result in the sinking of surface waters in many regions (see Chapters 1 and 7).

Most of the continental shelf and slope seafloor is smooth and soft, covered with sand, silt, and fine organic particles. Food webs here are supported largely by benthic detritus and particulate organic matter suspended in the water. Benthic and epibenthic invertebrates can be numerous, including a diverse assemblage of mollusks, crustaceans, and worms. River canyons and gullies extending onto the continental shelf can accumulate organic matter and provide relief and structure to attract organisms and some protection from predation and harvest by fishers.

The density of fishes and large invertebrates tends to be lower over the continental shelf and slope than in comparable shallow habitats. For example, most of the fish biomass typically comprises a few species, though there may be as many as 50 species within a slope/shelf ecosystem (in contrast, a kelp forest and coral reef complex may have over 100 and 1,000 fish species, respectively). Some of these fishes (e.g., gadids, such as cod, and flatfishes, such as flounders) are commonly targeted as important fisheries species because the biomass is in a few species that can be easily captured by pulling nets over a smooth sandy substrate. These are caught mostly in regions over the inner shelf, down to about 150 meters depth. Avoiding excessive

harvest of these species has been problematic, and is discussed in Chapter 11.

As depth increases along the shelf/slope gradient, the biomass and diversity of living organisms shows a decreasing trend; however, structure-oriented fishes, such as rockfishes (*Sebastes*), may be concentrated around bottom structures, including seamounts or gullies. Scavengers, such as hagfishes, also can be abundant in these ecosystems. The species assemblage in a given region is governed not only by depth but by other physical factors, such as temperature, currents, and bottom type. Although average biomass can be low in deeper waters, congregation of fish and crustaceans for reproduction can make them more vulnerable to harvest. The biology of animals in deeper waters of the continental shelf appears to be similar to comparable species in shallower coastal waters; however, the greater the depth, the less is known of the ecology and life histories of organisms. Fishes and crustaceans that characterize deeper bottom regions of the shelf typically remain in deep waters throughout their lives, making them difficult to study. Although some species migrate to surface waters or into shallower continental shelf regions as young, they typically return to deep waters as adults.

As depth increases across the outer shelf toward the continental slope, the bottom sediments become dominated by silts and clays, and species assemblages gradually change. More deep-sea fish groups are present, including cuskeels (Ophidiidae) and eelpouts (Zoarcidae). At the break between the continental shelf and slope there is a moderate increase in production due to the movement of nutrient-rich deep bottom waters onto the shelf. Onto the slope beyond the shelf break, beginning at about 400 meters depth, the ecosystems change gradually into something very similar to those found in the deep sea. Production is substantially lower, in part due to the low primary production in surface waters over the slope and deep sea relative to the continental shelf. Deep-sea bottom fishes, such as grenadiers (Macrouridae) and longnose eels (*Synaphobranchus*), become more prevalent. Animals below about 1,500 meters are not currently taken by fishers as their value does not warrant the effort and expense of fishing at such depths; however, harvested species can be impacted even to depths beyond the range of fisheries, for reasons discussed below.

8.2 Seamount Ecosystems

Seamounts are scattered around the oceans, many projecting up from the deep seafloor in regions far from coastlines (**Figure 8-2**). Although seamounts are formed by volcanic activity, a few are currently active volcanoes, and explosive eruptions of undersea seamounts are uncommon due to

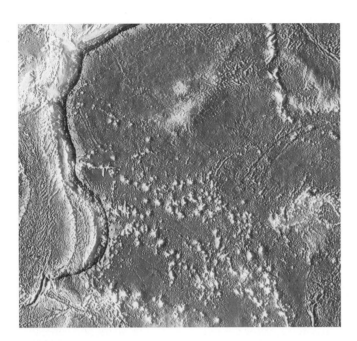

Figure 8-2 Image of western Pacific Ocean with a high concentration of seamounts. Darker areas are deeper regions. The Mariana Trench is indicated by the black region on the left. (See Color Plate 8-2.)

detect with current technology. The total area encompassed by seamount biomes worldwide was recently estimated by Peter Etnoyer and colleagues. They used satellite data indicating vertical gravity gradients to identify and estimate the total area covered by seamounts over 1 kilometer height above the seafloor. Their estimate of total seamount coverage of about 30 million square kilometers is 20% greater than the total area of the continental shelves and almost double the combined area of coastal habitats (marshes, seagrasses, coral reefs, mangroves, and beaches). This indicates the importance of seamount ecosystems in the deep sea, in concurrence with recent studies indicating their biodiversity and biological production, as discussed below.

Possibly the most important factor affecting the type of ecosystems that form on seamounts is water depth. For example, those that have been eroded just below the surface may support tropical reef formation. Below the photic zone, seamounts differ from the open seafloor at comparable depths, resulting in unique habitats and ecosystems. Major differences result from the enhanced productivity around seamounts due to unique physical processes. For example, circulation gyres (called *Taylor cones*) may form over the top of the seamount as currents flow over and around the seamount; these gyres entrap organic materials and plankton (**Figure 8-3**). Currents are amplified as they flow over the seamounts, enhancing the recruitment of larvae and supplying additional food resources to planktivores. Seamount productivity also can be enhanced by a process called *bottom trapping*. This occurs when plankton migrate from deep waters to the surface at night (part of the deep scattering layer described in Chapter 7) and attempt to return the next morning. Those that have drifted over the seamount are trapped on top as they sink, thus enhancing

effects of water pressure. Chains of seamounts occur along the boundaries of ocean plates, including island arcs, or at mid-ocean ridges (e.g., the mid-Atlantic ridge); however, many seamounts form in the middle of ocean plates and project above the abyssal plains of the deep sea. These are formed over hotspots, such as those along the Hawaiian Island chain (see Chapter 1); the tallest island peak on Earth (measured from the seafloor) is the Hawaiian island of Mauna Kea. The height of seamounts above the seafloor and their depth below the sea surface vary widely. They range from volcanic islands, seamounts that project above sea level, to low hills in the deep sea or on continental margins that may be completely covered by sediments.

The area of highest concentration of seamounts is in the Pacific Ocean (about 60% of the global total); many of these are beyond the continental shelf and slope (about half of large seamounts are more than 320 km from shore). Reasons for this distribution are not completely understood; however, the distribution is important from not only a scientific perspective, but also relative to conservation because regions beyond 320 km are in international waters and require international agreements to regulate exploitation of resources (see Chapter 12).

The total number and sizes of seamounts are difficult to estimate. Adrian Kitchingman and colleagues estimated the number of large seamounts (greater than 1.5 kilometer height) to be, conservatively, over 45,000. There could be hundreds-of-thousands of large seamounts, with many more smaller mountainous features that are difficult to

Figure 8-3 Circulation gyres called Taylor cones may function to retain larvae over seamounts.

food for plankton-feeding organisms. Another potential source of enhanced productivity is nutrient-rich deep-sea currents flowing up and over the seamounts.

The enhanced productivity of seamounts supports a complex food web, including many organisms that inhabit the seamounts throughout their life history (**Figure 8-4**). The rocky substrate of seamounts can provide shelter, but much of the habitat is provided by living benthic organisms. These organisms include sponges and corals, which comprise much of the seamounts' benthic biomass (see Box 8-1, Conservation Concern: Deep-Water Coral Reefs). The structure they provide attracts bottom invertebrates, including a diverse array of mollusks, crustaceans, and echinoderms. These provide food for numerous fish species, some of which remain on the seamounts and others that

(a)

(b)

Figure 8-4 Photos taken from a submersible in the Davidson Seamount ecosystem off California, a portion of the Monterey Bay National Marine Sanctuary. **(a)** Starfish, coral, and shrimp; and **(b)** anemone, sea cucumber, and starfish.

move in to feed. Much is not known about the biodiversity of seamount habitats because most have not been thoroughly surveyed; those that have exhibit a remarkable biodiversity. The proportion of endemic species appears to be relatively high due to the geographic isolation of many seamounts. Many species produce planktonic drifting larvae that enhance their chances of colonizing other seamounts. Initial surveys of most seamount ranges discover many new species. For example, surveys of Norfolk Ridge seamounts (south of New Caledonia Island in the southwest Pacific) found over 50% of the 730 animal species to be new to science. The most diverse groups of macrofauna were mollusks, crustaceans, and fishes.

The life history of most species living in association with seamount habitats is poorly documented. Research has found that many species exhibit great longevity and slow growth rates relative to related species in shallower ecosystems. Some corals live over 1,000 years, and some fish species can live over 100 years. Average age of maturity for fishes that aggregate on seamounts is over 12 years (compared to 4–6 years for fish species from most other oceanic regions). The orange roughy (*Hoplostethus atlanticus*), one of the most well studied and heavily harvested seamount fish species, can live up to 150 years; they do not reproduce until over 20 years of age, even though a typical length is only about 40 centimeters. Some *Sebastes* rockfish species can live up to 200 years. This great longevity may be an evolutionary response to the unpredictability of recruitment and early survival in seamount ecosystems. Changes in current patterns could dramatically decrease production and larval transport into the ecosystem. By reproducing for many years, animals enhance their potential for passing on offspring to the next generation.

An important aspect of ecology and behavior of abundant seamount fish species is their propensity to congregate, in part as a response to the limited size of their habitat but also to enhance reproductive success. This behavior also enhances harvest of species such as orange roughy, alfonsinos (Berycidae), Patagonian toothfish (*Dissostichus eleginoides*), oreos (Oreosomatidae), pelagic armorhead (*Pseudopentaceros richardsoni*), grenadiers, and rockfishes (**Figure 8-5**). Some migratory fish species also aggregate around seamounts, including sharks, tuna, and billfishes, for feeding, mating, or to enhance navigation. On shallower seamounts, large reef-associated fishes such as seabass (Serranidae) and jacks (Carangidae) sometimes congregate for spawning. Common continental shelf species (e.g., zeiform dories) may also be common on seamounts. Many of the species documented to congregate on seamounts are also found in other areas of the deep sea, but typically at lower densities.

(a)

(b)

Figure 8-5 Fisheries species that are commonly associated with seamount ecosystems include **(a)** orange roughy and **(b)** alfonsino.

The biological nature of seamount ecosystems make them vulnerable to direct anthropogenic impacts, currently limited primarily to fish harvest. Populations of species with slow growth rates, great longevity, delayed reproduction, and unpredictable reproductive success and recruitment do not recover rapidly after removal of large numbers from their populations. Seamount fish populations appear to go through time periods of 10 to over 20 years within which recruitment of new individuals is extremely low, limiting their ability to recruit new cohorts after excessive harvest. Unfortunately, seamounts have become an attractive target for fish harvest in the past 50 years as improved fishing technology has made them more accessible, and harvest has moved into deeper water to replace fisheries overharvested from shallower waters. The schooling of resident fishes and aggregation of migratory fishes on seamounts makes them prone to overharvest. Today most of the species targeted for harvest from the deep sea are associated with seamounts.

8.3 Deep-Sea Ecosystems

The deep sea away from photic zone and isolated from substantial influences of continents, islands, and seamounts is a remarkable, largely unexplored region. Despite the fact that the majority of the biologically inhabitable environments on Earth are found here, only in about the past 20 years have resources and efforts been devoted to exploring this region. Prior to the 1900s it was largely assumed that little life could survive below a few hundred meters depth in the ocean due to the extremes of pressure, absence of primary production, and low temperatures, oxygen, and light. Many scientists in the mid-1800s still believed that life disappeared totally below 500 meters (ignoring the fact that collections in the early 1800s had found basket stars [*Asterias*] at depths of 2,000 meters). Eventually, studies carried out in association with the laying of trans-oceanic telegraph cables and scientific expeditions in the 1860s and 1870s accumulated irrefutable evidence that life was present on the seafloor even in waters below 4,000 meters. The most famous of these cruises was the *Challenger* expedition organized by England to explore the physical, chemical, and biological nature of the deep sea. This sailing ship circumnavigated the globe from 1872 to 1876 collecting plankton samples from the water column and dredge sample from the deep seafloor. The *Challenger* expedition verified that diverse life was present at all depths in all ocean basins. A major biological conclusion of the expedition, which has been bourne out in more recent surveys, was that diversity and abundance of animals on the seafloor decreases with depth.

Extreme conditions continue to limit human exploration of the deep sea, so relatively little is known about the ecology and ecosystem interactions there. A great deal of knowledge has been accumulated, however, concerning the makeup of deep-sea life through the capture of animals by dredges or nets deployed from the sea surface and dragged along the bottom. Information concerning physiology and ecology through these surveys is limited because the physical stress and change in pressure from bringing animals to the surface often kills and deforms them. Manned and remotely-controlled submersibles have increased our ability to study deep-sea ecosystems directly, but their capabilities are still extremely limited, both physically and financially, considering the expansive geographic area of the deep sea. Only a very small fraction of the deep seafloor thus has been explored or biologically sampled.

The Mesopelagic Zone

The ocean pelagic region below the photic zone but above the sea floor is by far the most extensive habitat on Earth; however, this was the last region to be verified to have abundant life. It was only after nets could be sent to depths and opened and closed before being brought to the surface that explorers could be certain that captured animals actually originated in this zone. By the end of the 1800s it was finally accepted that there was a diversity of life at all depths of the ocean, both at the seafloor and in the water column.

The mesopelagic region, as described in Chapter 1, is located immediately beneath the photic zone where there is still some light penetration but not enough to support photosynthesis. The depth of the mesopelagic varies depending on water clarity; in clear waters it extends from about 200 to 1,000 meters below the surface. This midwater region contains a diverse assemblage of animals with adaptations for living in a cold, dark or dimly lit, low-nutrient environment.

One of the primary factors controlling the numbers and types of organisms present in the mesopelagic is the low availability of food. There is about 10 times the biomass in near-surface waters as there is at 1,000 meters depth. Much of the biomass in the mesopelagic is from zooplankton, mainly krill and other crustaceans. But the diversity of fishes is relatively high (about 850 species, compared to 250 in the epipelagic), though as much as 90% of those may be in two families: the Myctophidae (lanternfishes) and Gonostomatidae (bristlemouths). The low food availability results in small sizes for mesopelagic fishes. The most abundant of these (and the most abundant vertebrate on Earth) is the bristlemouth *Cyclothone*, 3 to 6 centimeters in length, a size typical of many of the mid-water fishes. These fishes are opportunistic feeders, a characteristic common in a region with scarce food resources. They feed on small plankton by filtering with fine gill rakers, but can also feed on organisms that are large (relative to their body size) with their many sharp teeth and large mouths.

Larger predatory fishes in the mesopelagic tend to be long and eel-like, with large mouths, unhinged jaws, fang-like teeth, and distensible guts, enabling them to take advantage of prey organisms even larger than themselves in a region sparsely populated with prey items. The muscles of these fishes tend to be reduced, implying lower activity to reduce energy use. Species include the dragon eels (Saccopharyngidae), dragonfishes, and loosejaws (Stomiidae; **Figure 8-6a**). Larger invertebrates also can be common predators in midwater ecosystems, including squids and jellyfishes. Lengths of the giant squid *Architeuthis* and colossal squid *Mesonychoteuthis* are up to 10–12 meters or more. These large sizes could be due to their association with highly productive seamount and ridge habitats. These large squids are a primary source of food for sperm whales, frequent visitors to the mesopelagic.

Many mesopelagic animals have photophores with light produced by symbiotic bioluminescent bacteria. Photophores arranged in rows along the underside of the body camouflage mesopelagic fishes predators swimming

(a)

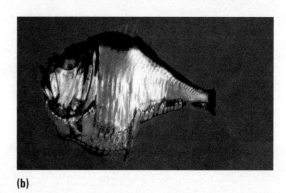

(b)

Figure 8-6 Mesopelagic fishes are often of bizarre shapes and have photophores scattered on their body: **(a)** a loosejaw with a photophore beneath its eye; and **(b)** a hatchetfish with photophores lining the ventral side for counterillumination.

below, because the light breaks their silhouette against the relatively light background of surface waters. For example, hatchetfishes (Sternoptychidae), the second most abundant fish group in the oceans, have semicircular bands of photophores lining the underside of their laterally flattened bodies (Figure 8-6b). Photophores are also used in mate selection and attraction in an environment where mates may be widely dispersed and difficult to locate. For example, lanternfishes (Myctophidae) have photophores arranged in rows in species-specific and gender-specific patterns. *Malacosteus* dragonfishes produce red lights, visible to conspecifics but invisible to most other deepwater predators. This is because many deep-sea organisms have lost their ability to see reds and other low wavelength colors due to its absence in these regions; the red portion of the color spectrum is refracted by the water from any sunlight that might penetrate the ocean depths.

One way that many mesopelagic animals respond to the limited food resources is daily vertical migrations to and from the surface to take advantage of the relatively abundant food available there. Many planktonic invertebrates and small fishes congregate in midwater regions by day (as part of the deep scattering layer) and float or swim to the surface at night to feed. Feeding in the epipelagic at night also helps to avoid surface predators while taking advantage of waters with lower nighttime temperatures that require less metabolic energy. These migrations play an important ecological role in the mesopelagic by transporting large amounts of biomass and energy from the surface to the midwater.

The Deep-Sea Pelagic

In the deep pelagic regions (the bathyl, abyssal, and hadal; see Chapter 1), below about 1,000 meters, is a dark, cold, relatively constant environment. Although 75% of the world's liquid water is found in this region, the density of life here is very low due to the lack of sunlight and low amounts of nutrients and organic matter. This region is almost entirely dependent on the fall or circulation of organic particles from the surface waters as the source of biological energy. Because only about 5% of the organic matter produced in the photic zone makes it to the deep sea, and much of that ends up on the seafloor, the scarcity of food organisms is the most critical limiting factor controlling the ecology of this region.

Deep-sea animals do not migrate to surface waters from such depths to feed, as do those in the mesopelagic, because of the great distances—and thus, time—involved and possibly the great pressure changes. At the base of deep-sea food chains are invertebrates similar to those near the surface. Many species have bodies that are darker or reddish with little other coloration. Pelagic invertebrates near the seafloor include sea cucumbers, a group normally limited to benthic habitats (**Figure 8-7a**). Fish species are less than 20% as diverse as in the mesopelagic (about 150 vs. 850 species). The most common fishes are bristlemouths, in the same family as those in the midwater regions but usually slightly larger (up to 10 centimeters long).

Approaching greater depths closer to the seafloor, the assemblage of animals gradually changes and their density becomes very low. These very deep waters are limited not only by food but also by low oxygen levels, especially in regions where there is little circulation between the surface and the deep. For larger animals such as predatory fishes, energy conservation becomes very important. Most are sedentary, floating in the water waiting for prey to come to them. Because there are relatively few large predators to avoid, they can save energy by being less mobile. Muscles are flabby, skeletons are weak, organs are reduced, and body water content is higher. This limits energy expenditure but also enhances buoyancy in the water. Maintaining air in a swimbladder, the primary mechanism by which most fish remain buoyant, is difficult due to the high pressures and low oxygen levels. (Otherwise, high pressure in the deep sea is not considered a major limiting factor, as animal tissues are composed primarily of water, which is virtually uncompressible even at highest pressures.)

In regions where oxygen is not limiting, fishes exhibit fast growth rates and greater egg production on average than those in the mesopelagic, even though food resources are lower. This is probably due to the low energy needs of sedentary deep-water fishes relative to mesopelagics that undergo regular vertical migrations. This allows the

(a)

(b)

Figure 8-7 Some deep-sea pelagic animals: **(a)** a pelagic sea cucumber, unusual in that it can swim above the bottom; and **(b)** an anglerfish, which lunges at prey it attracts with a lure at the end of a modified fin on its head.

deep-water pelagic fishes to quickly grow to sizes at which they are less vulnerable to predation, further enhancing their survival rate and ability to live to great ages.

Deep-water pelagic fishes also have specialized adaptations to detect and capture prey in a dark sparsely populated environment. They have excellent hearing and sense of smell (which also helps to locate mates). These fish are typically uniformly black, with few photophores on their body. Some species, such as the anglerfishes (Figure 8-7b), however, have photophores on the tip of modified fins used as lures to attract prey organisms. These lures are designed to mimic the photophores found on some invertebrates, thus enabling them to attract smaller predators. With large mouths, sharp teeth, and distensible guts, these predators can eat relatively large animals. They also can capture smaller prey using well-developed gill rakers.

The Deep Seafloor

The ecology of the deep-sea changes near the seafloor. As discussed above, there are remarkably diverse ecosystems along the continental margins and on seamounts and ridges. Even on the plains of the abyss, however, life is more abundant than in the overlying waters of the abyssopelagic. This is directly related to the distribution of food in the deep sea. Deep-sea animals depend on the food particles that rain down from the surface waters, and those that are not captured by animals in the water column end up on the seafloor. Accumulation of these particles enhances biological production; however, food availability is very low relative to benthic nearshore regions or the epipelagic (**Figure 8-8**). Most of the food is comprised of particles such as fecal pellets that sink rapidly but have limited nutrition, and much of it is not easily digested, such as the skeletal remains of crustaceans or phytoplankton (e.g., diatoms, coccolithophorids, or foraminiferans). In fact, much of the sediment on the plains of the deep sea is biogenic in origin, formed from the accumulation of calcified shells from plankton (see Chapter 1). Bacterial decomposition plays an important role in making indigestible organic matter available on the seafloor.

Despite the relatively low abundance of living organisms in most places on the seafloor, this life may be highly diverse. Although the fraction of the deep seafloor that has

been sampled is too small to enable a precise estimate of the actual number of species, samples of small areas of the deep sea suggest that as many as 1 to 10 million species could live on the deep seafloor (**Figure 8-9**). Many of these are related to shallow water species, sometimes with bizarre adaptations. For example, there are anemones, brittle stars, lobsters, and eel-like fishes. Some invertebrates exhibit gigantism, being much larger than their shallow-water relatives. These include long-legged pycnogonid "sea spiders," spider crabs (*Macrocheira*) over 4 meters across, and giant isopods (*Bathynomus giganteus*) over 40 centimeters long. The reasons for this phenomenon of deep-sea gigantism are unclear.

Most fishes on the deep seafloor are blind, the loss of vision occurring through natural selection in a region with no sunlight. Those species with vision probably use it to detect bioluminescent organisms. Tripod fishes *Bathypterois* use long fin rays to elevate themselves off the bottom, snatching small organisms that pass by in the current. Many fishes have long sensory rays and whisker-like barbels to feel and taste in the water. In regions of the deep sea where there is little vertical circulation of waters from

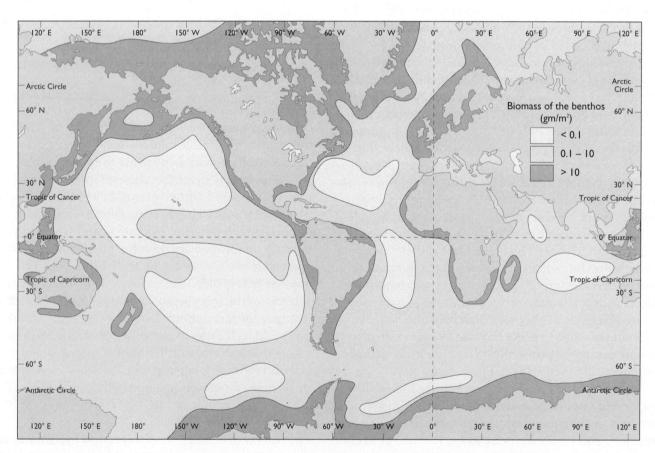

Figure 8-8 Biomass of benthic organisms on the seafloor. Note that regions of high and low concentrations generally correspond to those regions in pelagic waters, with low biomass in the center of ocean basins, and high biomass near coastlines.

(a)

(b)

Figure 8-9 Deep-sea benthic organisms are related to those found in shallower waters but often have bizarre specializations and are giant sized. **(a)** An arrow crab and **(b)** an echiuran spoon worm, cnidarian sea pen, stalked crinoids, and brittlestars.

the surface, oxygen and nutrient levels can be limiting to animals living there and, therefore, they tend to be smaller but more abundant.

Numerous scavengers in the deep sea appear to rely on larger carcasses of fishes and whales that make it to the seafloor, arriving in a sequence to colonize and feed on the decaying carcass. First to arrive are scavengers such as grenadiers, hagfishes, sharks, crabs, and amphipods. Hagfishes scavenge on dead bodies by burrowing into the animal and eating their way outward; they produce a noxious slime that helps to lubricate passage into the carcass and also resist predation. It may take up to two years to remove the soft tissue from the bones of a large whale carcass. Polychaete worms then move in to feed on organic matter in the bones and surrounding sediments. They can reach densities of tens-of-thousands of worms per square meter and remain for two years, at which time all that is left is the bones. As described below, bacteria continue to decompose the

bones of a large whale for about another 50 years and attract other grazers.

Many of the animals on the deep seafloor live within the fine, muddy sediments found there. These include the meiofauna, animals barely visible to the naked eye and similar to those found in shallow water sediments. They are the links in food chains that transfer the energy from the microscopic bacteria and dissolved organic matter to the larger bottom animals. Many of these animals are hidden buried in the sediments. The dominant species varies from region to region of the deep sea. In most areas it is polychaete worms, crustaceans, or bivalves; in others it is brittle stars, sea stars, or sea cucumbers. Predators on these invertebrates include crabs, brittle stars, and starfishes.

Animals of the deep sea are similar to shallow-water relatives but several aspects of their life histories are distinctive. For example, they typically produce fewer and larger eggs than their shallow-water relatives and tend to grow slower and live longer. One proposed explanation is that cold water slows metabolism and the low food supply slows growth. The animals need to achieve large sizes to produce large offspring, and large offspring are necessary because they must contain enough food reserves to survive with an unpredictable source of nutrition. The lack of large predators makes predation a less likely source of mortality, allowing longer lives.

Vent and Seep Ecosystems

Hydrothermal Vents

One of the most remarkable biological discoveries in the sea in the past century was the diverse ecosystems found in association with hydrothermal vents. In 1976, divers in submersibles unexpectedly found large mussels and clams up to 26 centimeters long at densities as great as 10 kilograms per square meter in regions of the deep sea where nothing comparable was known to exist. The source of energy supporting this biomass was a mystery, but it appeared they were thriving in areas with toxic levels of hydrogen sulfide and metals. Eventually, worms with long feather-like red plumes extending out of two-meter long tubes were discovered (**Figure 8-10**). These polychaete worms, named *Riftia pachyptila*, were discovered to have no mouth or digestive system but grew at rapid rates of almost one meter per year. A diverse community of microorganisms, bivalves, crustaceans, echinoderms, and fishes were discovered later in association with these ecosystems.

Several years after the discovery, the mystery of the existence of this diverse ecosystem was solved. The energy supporting the rich food web was found to be chemical based, primarily stored in hydrogen sulfide, rather than originating from photosynthesis. Although

(a)

(b)

Figure 8-10 A complex and unique ecosystem is supported by the chemical energy provided by hydrothermal vents, including **(a)** giant vent tube worms *Riftia pachyptila,* and **(b)** hydrothermal mussels and shrimps.

chemosynthesis by microbes was known since the 1880s, this is the first time it had been documented to be ecologically significant in supporting a complex ecosystem of larger organisms. The tube worms, giant clams (*Calyptogena*), and vent mussels (*Bathymodiolus*) house symbiotic bacteria in their tissues. The animals provide chemicals needed for chemosynthesis (hydrogen sulfide, carbon dioxide, oxygen, and nitrogen) and the bacteria nourish the animals with their biological production. This symbiotic relationship is most highly evolved in the tube worms, which are unable to feed and thus rely entirely on this energy source. There are also free-living chemosynthetic bacteria that provide a food source for filter feeding invertebrates such as copepods and other crustaceans, anemones, polychaete worms, barnacles, and limpets.

One reason for the excitement over these discoveries is that the energy stored in the chemicals supporting this

ecosystem is not derived from the Sun's energy but from the geological energy beneath the Earth's crust. This energy is reflected in the high water temperatures associated with these hydrothermal vents. For example, hydrothermal chimneys called *black smokers* emit waters at 350°C (**Figure 8-11**). The black plume that is emitted forms as iron and other metals precipitate out when the hot water meets the cold deep-sea water. Sulfides and other chemicals precipitate out to form chimneys up to 50 meters high, growing at 5 meters per year. White smokers form at cooler vents from whiter-colored minerals such as silica, anhydrite, and barite. Vent animals typically live in waters surrounding these vents, and some, such as the Pompeii worm *Alvinella pompejana,* can tolerate temperatures up 100°C.

Another important way that hydrothermal vents differ from most of the deep seafloor is that they are unstable and unpredictable. A single vent may endure for only a few decades, although vent fields (which typically range in size from 10 to over 1,000 square meters) stay active for longer, centuries or millennia. Once the vent field expires, the sessile vent animals that live there soon die due to the loss of their energy source. The giant tube worms are some of the first to go due to their total dependence on the chemosynthetic energy. Other animals can feed on

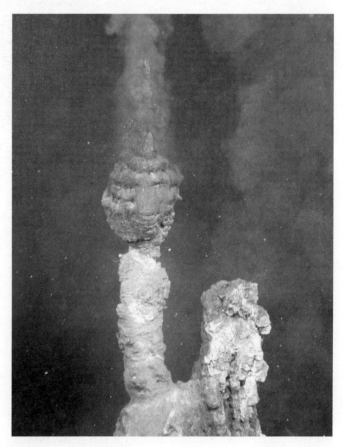

Figure 8-11 A black smoker hydrothermal chimney.

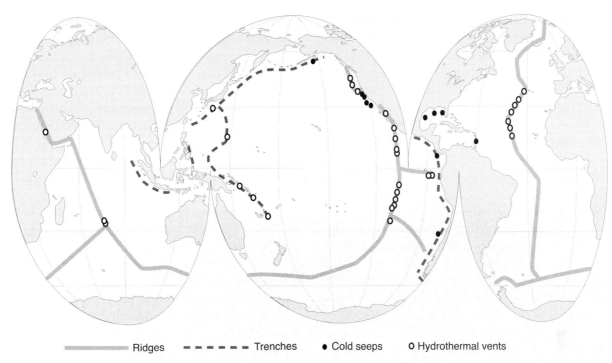

Ridges ———   Trenches – – – – –   Cold seeps ●   Hydrothermal vents ○

Figure 8-12   Confirmed locations of hydrothermal vents and cold seeps, mostly associated with deep-sea trenches and geologically active ridge systems.

free-living bacteria and other organisms for awhile but eventually die as well. Lava flows in the geologically active areas around the vents also can smother the vent animals. This unpredictable duration of the vents has important ecological implications. Animals at the vents must have mechanisms for colonizing new vent regions or they would become evolutionary dead ends. This colonization can be observed when a new vent site is established. Within a year, tube worms, giant clams, and other components of the vent ecosystem have become well established. Although few, if any, of the adult organisms at the vents can travel long distances to colonize a new vent ecosystem, their larval stages can. Larvae of the sessile animals at the vent are pelagic, and drift with currents. Because the currents typically follow the deep-sea ridges, they carry the larvae to other areas where there may be vents (**Figure 8-12**). Most of the vent animals cannot settle and survive away from vents; therefore, the only ones to survive are those that are able to find and settle on another vent. Where vents systems are widely dispersed (such as across broad ocean basins), biogeography barriers are evidenced by the presence of ecologically similar, but different, vent communities in separate regions.

Cold Seeps

Upon their discovery and description, hydrothermal vent ecosystems were assumed to be unique in the ocean environment in their dependence on chemosynthesis as the major source of primary production. Soon, however,

other chemosynthesis-based communities were discovered when regions thick in clams, mussels, snails, and tube worms were observed during submersible explorations in the Gulf of Mexico far from hydrothermal energy sources (**Figure 8-13**). These newly discovered ecosystems were in regions where energy-rich chemicals such as hydrogen sulfide and methane gasses seep from the seafloor. These are commonly referred to as **cold seeps**. These conditions are found mostly in areas where there are large accumulations

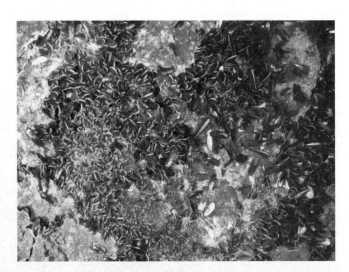

Figure 8-13   Mats of bacteria and a bed of mussels are part of the cold seep communities, supported by energy-rich chemicals such as hydrogen sulfide and methane.

of organic sediments from river input, primarily at the edge of continental margins (see Figure 8-12). Any region where sulfides, methane, or other hydrocarbons come to the surface of the seafloor can support cold seep ecosystems; however, some are even found in ocean trenches where sediments have accumulated. Although seep ecosystems are not directly supported by photosynthetic production, they are indirectly supported by photosynthesis as their chemical energy is derived from sulfides and hydrocarbons that originate from long-dead organic matter originally produced through photosynthesis.

Organisms at these sites are similar to, but rarely the same, species as those found in the hydrothermal vent communities. Prominent components of the seep ecosystems include large bacterial mats, crabs, bivalves, tube worms, anemones, and soft corals. As in hydrothermal vent ecosystems, many of these species have symbiotic chemosynthetic bacteria in their tissues; for example, this is true of about 30% of over 200 species identified in cold seeps in the Atlantic and Pacific. Visitors to seeps include fishes, octopuses, and other predators.

Other than the obvious variance in temperatures, other differences between cold seeps and hydrothermal vents are reflected in ecological differences. Seeps tend to be longer-lived than vents but are less productive. This appears to have resulted in physiological differences in animals living in these environments as seep animals tend to be slower growing but longer lived. For example, the seep tube worm *Lamellibrachia* has been documented to live over 200 years to attain a size of 2 meters (a size reached in just 2 years by the vent tube worm *Riftia*). The greater stability of seeps results in more specialization and thus a greater species diversity. As many as 15 macroinvertebrate species can be common around a seep. Methane is the primary source of energy at seeps (it is hydrogen sulfide at the vents); however, most seep symbiotic bacteria use hydrogen sulfide, released by free-living bacteria that use the methane.

Methane Hydrates

Other potential sources of methane that can support biological production in the deep sea are structures called **methane hydrates** or **gas hydrates**, solid masses of frozen water containing pockets of gas molecules, mostly methane but also some hydrogen sulfide. These hydrates are found mostly at depths below 250 meters where high pressures cause them to remain frozen at temperatures above 0°C. Although they were only recognized because of extrusions from under the sediments in some locations, there are massive amounts of these materials just beneath the seafloor along the continental shelves. Methane hydrates are of interest to humans not only ecologically but also as a potential source of methane for energy production. Another interest

is the potential for the unwanted release of methane gases from these hydrates with warming temperatures. (Methane gas is a major greenhouse gas that, if released, could exaggerate global warming.) Following the large-scale release of methane from the 2010 *Deepwater Horizon* oil spill, however, aerobic methanotrophic bacterial communities grew rapidly and oxidized the methane before it could reach the atmosphere. John Kessler and colleagues propose that these bacteria could function to remove methane following other large releases in deep waters, keeping it from reaching the atmosphere.

Species found in association with gas hydrate habitats are similar to those at seeps and vents, including mussels, snails, shrimps, and tube worms (**Figure 8-14**). One polychaete worm species, *Hesiocaeca methanicola*, is uniquely adapted to the hydrates. It was discovered by Charles Fisher and colleagues living on the surface of mounds of gas hydrates about 500 meters under the surface. These so-called *ice worms* live in large colonies and sculpt the surface of the hydrates. They burrow into the hydrates for shelter and to feed off the hydrocarbons found there. These are the only organism—other than bacteria—that have been found living in direct association with the methane hydrates.

Further explorations of the deep sea have discovered that chemosynthetic habitats are more common than anticipated. One location of these habitats is associated with whale carcasses decomposing on the seafloor (**Figure 8-15**). As described above, after scavengers have removed soft tissues, bacteria continue to decompose the bones. The main source of energy for the microbes is the lipids, which can comprise much of the mass of the bones. As oxygen is depleted from the bones, anaerobic bacteria become abundant. During anaerobic metabolism sulfides are produced that provide an energy source for chemoautotrophic bacteria. These bacteria support an ecosystem similar to those found at hydrothermal vents. Grazers feed on the mats of

Figure 8-14 Polychaete ice worms *Hesiocaeca methanicola* living in association with methane hydrates.

Figure 8-15 Chemoautotrophic community supported by a whale-fall, six years after the whale's death. Note bacterial mats covering the skeleton.

bacteria and high densities of mussels and clams containing symbiotic chemosynthetic bacteria colonize the region. The resulting community can be very diverse. Over 400 species have been discovered on whale carcasses, 10% of which are endemic. It appears that an entire community has evolved that is dependent on decaying whale bones. Historically, sunken whale carcasses were apparently of such a density that adjacent carcasses were close enough for larval dispersal, though excessive whaling may have lowered that density substantially.

Even on the open seafloor away from seeps, vents, and animal carcasses, habitats have recently been discovered that support chemosynthetic faunas. Pogonophoran worms, related to vent tube worms, are buried in anoxic seafloor sediments rich in sulfides. The sulfides provide energy for chemosynthetic symbiotic bacteria within the worms. As another example, small mussels that house chemosynthetic symbionts colonize sunken wood in the deep sea. The distribution and commonness of such organisms and ecosystems are not well known due to the limited explorations of the deep sea.

8.4 Impacts of Deep-Sea Fisheries

Fish Harvest

People have been harvesting fishes at depths of 500 to 1,000 meters or more for centuries (including the Inuit off Greenland, Polynesians in the South Pacific, and Europeans in the Northeast Atlantic). These fisheries, using hook and line, probably had minimal impacts on fish populations due to their low efficiency and small scale. Eventually the development of longlines, baited with multiple hooks and

retrieved mechanically, made such fishing more efficient with a greater likelihood of excessive harvest. But it was only with the development of technology to pull trawl nets in deep waters and to handle and process fish at sea that fish populations began to be impacted on a large scale (see Chapter 11 for further discussion of the broad effects of improved fishing technology). The incentive for increased harvest at great depths over the continental shelf, slope, and seamounts was the continuing depletion of fish populations in shallower waters (see Chapter 11).

After the 1950s, fish harvest was expanded into deeper and deeper waters. A recent analysis of fisheries harvest data by Telmo Morato and colleagues found that the average depth at which marine bottom fish were caught by commercial fishers increased from about 100 meters in 1950 to nearly 150 meters in 2001. One reason for the shallower harvest depths in the early 1950s was the belief that the deep sea beyond about 500 meters depth held little life. Discovery of dense fish populations above seamounts gave incentive to move harvest further out to sea and into deeper waters. By the end of the 1950s, fish were being trawled as deep as 700 meters in the North Atlantic and North Pacific. By the 1960s drastic declines in numbers were occurring for some deep-water fishes. One of these was the pelagic armorhead found around seamounts in the North Pacific, which was harvested so heavily through the 1970s (50,000–200,000 metric tons per year) that it is unlikely that any substantial harvest has occurred since, although fishing still occurs on seamounts in this region. The armorhead is protected from harvest in U.S. waters around the Hawaiian-Emperor seamounts where stocks are considered to be depleted. It is likely that the only people who knew that the armorheads were being heavily harvested in the 1970s were the fishers, because the fishery was carried out in international waters, initially by Soviet trawlers, prior to international agreements to report and monitor the harvest.

By the 1980s, fish were commonly being harvested from 1,000 to 1,200 meters depth in regions of the North and South Atlantic and Pacific Oceans. Grenadiers (commonly called *rattails*) are in one of the most common deep-sea fish families (Macrouridae), especially prevalent on the upper continental slope (**Figure 8-16**). Large fisheries for several grenadier species developed in the Atlantic Ocean. Although attempts were made to limit the harvest, these were largely based on guesswork because the basic biology of the species, including great longevity and a slow maturation, were not known until 30 years after the fishery had begun. Alfonsinos are several species of berycid fishes harvested in deep waters of the Atlantic and Pacific, often marketed as red bream or Tasmanian snapper (see Figure 8-5b). Various species of deep-water sharks are also being harvested by commercial fishers.

Figure 8-16 A grenadier or rattail, one of the fish species harvested by deep-water trawls from the Atlantic Ocean deep sea floor.

In the Atlantic, lower numbers of seamounts and the more extensive deep areas over the continental shelf and slope provide about twice the area of habitat at 200 to 1,000 meters depth as the Pacific. With more extensive and continuous habitat, fishes in the North Atlantic do not tend to congregate as much as seamount-associated species. This, along with higher production by seasonal plankton blooms, results in lower vulnerability to rapid overfishing. Human ingenuity and a long history of fishing in the North Atlantic, however, have caused the declines of even the most resilient of fish species.

The longest history of bottom fish exploitation is for the Atlantic cod and other gadids. These species are mostly restricted to relatively shallow waters of the continental shelf, and their harvest and conservation are discussed in detail in Chapter 11. The first large-scale commercial fishing in the deep Atlantic was for Greenland halibut (*Reinhardtius hippoglossoides*) and roundnose grenadier (*Coryhaenoides rupestris*), at depths from 600 to 1,500 meters, by Soviet trawlers in the mid-1960s. By the 1970s other nations had joined in and the grenadier fishery peaked; the subsequent decline through the 1980s and 1990s led to an estimated reduction of the population to less than 1% of its original size by the early 2000s. The Greenland halibut has a life history that differs from many other deep-sea fishes in that it grows rapidly, typically does not live beyond 15 to 20 years, and does not tend to congregate. These aspects of its life history make it less likely to be overharvested; however, competition for harvest among European, former Soviet nations, and Canada led to a decline in the populations through the 1980s and 1990s.

Another major deep-water North Atlantic fishery is for the redfish, also called *ocean perch*. This fishery is actually made up of several species in the genus *Sebastes*. Like many deep-sea species, these are long-lived (up to about 75 years)

and reproduce late in life (at 8–12 years), making population recovery difficult. Additionally, most species tend to feed off the bottom, making them more vulnerable to capture by large trawl nets (the largest nets having a mouth opening of over 1 kilometer diameter). Poor monitoring by many nations of the identification of species harvested and exploited by both longline and trawling has led to difficulties in regulating harvest and conserving populations. Redfish have been harvested in the North Atlantic since the 1920s, but it was not until the mid-1990s that dramatic declines were observed in the size of the catch (by about 90%) as well as the size of the fish (by about 50%; **Figure 8-17**). Some populations appear to be recovering with increased protection; however, like many deep-sea fishes, the life history and ecology of the redfish could make long-term recovery in the presence of fishing questionable.

In the Pacific and Indian Oceans much of the deep-sea fishing is concentrated around seamounts because most

(a)

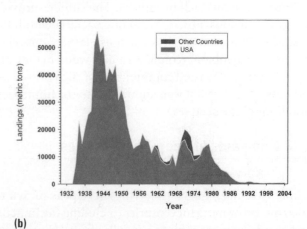

(b)

Figure 8-17 Fish associated with the flat plains of the sea floor, such as the Acadian redfish, *Sebastes fasciatus* **(a)**, are vulnerable to overharvest due to easy access with large trawl nets. This is indicated by the decline in commercial harvest landings off the Georges Bank region of the northeastern U.S. coast **(b)**.

of the seafloor in the open ocean is too deep to access for fishing. Harvest of fishes that congregate on seamounts can be especially problematic due to the life histories and ecology of these species, as described above. Once populations are reduced in one seamount group, the fishers move to another, with the same outcome. This type of serial depletion of stocks has been most well documented for the orange roughy, one of the most valuable and heavily exploited of seamount fisheries. The large exploitable populations were first discovered in the South Pacific off Australia and New Zealand in the 1970s; many of the populations were reduced to 10% to 20% of their original biomass in 10 to 15 years of fishing. Spawning aggregations on seamounts were especially vulnerable to overfishing. The pattern continued as fisheries expanded, initially around the South Pacific and eventually into other ocean regions, including the ridges and seamounts in the Atlantic and Indian Oceans. Due to the orange roughy's late maturity and great longevity and the nature of seamount ecosystems, it appears that recruitment of new individuals into exploited populations is very slow.

One of the most volatile and controversial fisheries in the deep sea is for the Patagonian toothfish (marketed under various names, including Chilean sea bass; see Chapter 2). This fishery provides a classic example of the difficulty in monitoring and managing deep-sea fisheries in international waters. This fish is larger than many other deep-sea fishes, reaching over 2 meters length and 130 kilograms. It is limited to cold deep waters, being in the cod icefish family Nototheniidae, with its range extending along the southern coast of South America into the Antarctic Sea. It lives around bottom relief in canyons, and on deep ridges, seamounts, and plateaus down to 2,000 meters depth. Trawl fishing for the toothfish began in the 1970s, and expanded to longlining in deeper waters in the 1980s. By the 1990s, annual harvest was being reported to be over 30,000 tons; however, due to illegal and unreported fishing the actual harvest level is unknown, possibly several times what has been reported. Problems with managing and limiting this fishery include difficulty in enforcement and the necessity for international agreements, issues inherent with international fisheries (see Chapter 12). Populations extend from national waters of Australia, New Zealand, southern South American, and African countries into international waters to the south. Although international agreements are in place to limit the harvest, these are difficult to enforce because pirate fishing is so lucrative and fishing vessels are often not monitored closely by the nations from which they originate. Ploys used to hide the landing of these fish, including changing the name under which is marketed, allow the toothfish to continue to be sold profitably around the world.

Crab Harvest

Numerous crab species are harvested in various regions of the world. Statistics accumulated by the Food and Agriculture Organization of the United Nations (FAO) report over 1 million metric tons of crab harvest each year since 2000. Many of these are caught in coastal and estuarine waters, the greatest harvest being for the swimming crabs in the family Portunidae. The largest of these fisheries, about 300,000 metric tons per year, is for the horse crab *Portunus trituberculatus* in western Pacific regions, primarily off China. The flower crab *Portunus pelagicus* also supports an important fishery, about 150 metric tons per year, in the Indo-Pacific. For each of these the level of harvest has increased dramatically over the past 50 years, but appears to be sustainable at current levels. In the western Atlantic about 100,000 metric tons of blue crab *Callinectes sapidus* are harvested annually, primarily from U.S. waters, where environmental and habitat issues have impacted the fishery, especially since 2000 (see Chapter 4).

Some of the most valuable crab fisheries are in deeper waters over the continental shelf. These include *Chionoecetes* (known by various names, including snow crab, queen crab, spider crab, and tanner crab) in cold waters of the North Atlantic and Pacific. About 100,000 tons per year have been harvested annually since 2000 down to depths of about 200 meters. Although overall harvest levels have remained relatively stable, overharvest has been an issue in some regions. For example, since 2000, overharvest has been a concern in for Canadian and Alaskan snow crab populations, resulting in stricter management, reductions in harvest, and limiting harvest only to males.

Some of the most valuable of the deep-water crab fisheries are for king crabs, the common name used for over 100 species in the family Lithodidae (**Figure 8-18**). Numerous species are harvested commercially from cold ocean waters either at high latitudes or at great depths. Some of the most

Figure 8-18 Trap harvest of highly valued king crabs must be closely monitored and regulated to avoid overharvest. Harvest of red king crab in Alaskan waters has stabilized through management after uncontrolled fishing in the 1970s.

valuable king crab fisheries are for *Paralithodes* species in the North Pacific. These include the red (*P. camtschaticus*) and blue (*P. platypus*) king crabs harvested in cold waters of the eastern and western North Pacific with large traps. Their value is in part due to their large size, up to 8 kilograms with a leg span of 2 meters. The Alaskan fishery in the Bering Sea has received much attention, in part due to the popularity of the documentary television series *Deadliest Catch*. Overharvest has resulted in the closure of fisheries for these species in some regions. Currently, the fishery in U.S. waters is considered to be stable, although populations are smaller than historic levels in the 1970s. Management methods are discussed further in Chapter 11.

In some ways, seafloor crab fisheries are more easily managed than harvest of deep-sea fish. Crabs are typically harvested by traps, thus there is no scouring of the seafloor as from trawl nets. And, crabs can be more easily selected for harvest by size or gender. In some instances traps can be developed to select by size, or the crabs can be sorted after capture. Crabs are not as likely to be killed or injured when brought to the surface as are fish, which are more sensitive to rapid pressure changes (see Chapter 11). Harvest in excess of what the population can sustain is an issue as it is with any fishery, however. A recurring problem with management is a rapid expansion of a valuable fishery prior to obtaining scientific knowledge about the species and its ecosystem. This often leads to overharvest and population declines after a short time before anything is done to limit harvest. For example, harvest of the red king crab off Alaska increased dramatically from about 4.5 million to 60 million kilograms from 1970 to 1980. Within a few years the king crab population had declined by 90%, forcing strict management of the fishery. Because little was known of the crab's ecology, it is difficult to attribute the decline to a single factor. Environmental and ecosystems changes have been linked to the decline, but excessive harvest is likely a strong contributor. Even with strict management measures populations have not returned to near their former levels.

For crabs in deeper ocean waters it is even more difficult to obtain biological information. Fisheries, therefore, can develop before anything is known about population size or life history. For example, the deep-sea red crab *Chaceon quinquedens* is long-lived, harvested at depths of 400 to 800 meters, but resides in waters as deep as 1,800 meters off the U.S. Atlantic coast. A fishery management plan was enacted only after the red crabs had been harvested commercially for over 20 years. Prior to this, the fishery fluctuated dramatically from year to year. Since implementation of the management plan in 2002, harvests have remained stable. The abundance of the males, targeted by the fishery because they are larger than females, is about the same as it was pre-harvest. Surveys by Richard Wahle and colleagues,

however, documented that the abundance of the largest males targeted by the fishery has declined significantly, changing the population from one dominated by crabs from 110–125 millimeters width to one now dominated by crabs 60–75 millimeters width. This has led to concerns about the future reproductive capacity of the population. This pattern of loss of the largest animals is common in fisheries for crabs and lobsters. Selection of the largest animals is easily accomplished due to segregation of individuals by size and gender, and the ease of size selection in the traps. The same phenomenon has been observed for many fish species (see Chapter 11). It is obvious from this discussion that we need to be more proactive and cautious before moving into new fishery ventures in the deep sea to replace markets for overharvested shallow-water or surface fishes.

Bottom Trawling

The difficulties inherent in the management and protection of deep-water species often leads to major declines in populations of commonly harvested species. This is due to many factors, including the slow growth, delayed reproduction, and longevity common in many deep-water fishes, along with the low biological production in deep waters, the difficulty of monitoring and studying these ecosystems, and the problems inherent in managing fisheries in international waters. Commercial harvesting can be more analogous to mining than fishing if proper limits are not established and populations are not allowed to recover for long periods of time before harvest resumes.

The effects of deep-sea fish harvest may even extend into depths beyond those where the fish are harvested. An analysis of deep-sea scientific trawl data from the 1970s to 2000s by David Bailey and colleagues found that total fish abundances declined significantly at all depths from 800 meters to 2,500 meters, even though the maximum fishing depth is about 1,600 meters. The greatest declines were for harvested species. No other obvious natural factors, such as environmental changes or changes in prey, were linked to these declines. Apparently, excessive fishing of these populations in shallower waters is translated throughout the depth of their distribution. Because little is known of the life history of many of these species, it is difficult to define reasons for these observations, and it is not known how prevalent this phenomenon is throughout the world's oceans. Possible explanations include movement of deep-sea fish between shallower and deeper waters. For example, some species move gradually into deeper waters as they age, so excessive harvest of shallow populations may be translated into deeper waters. Other species could move up and down the continental slope or ridges on a regular basis. If reproductive populations concentrate in shallower waters on ridges and seamounts, the reproductive success of these populations could be affected. Evidence provided by these studies lends further support to arguments for proceeding cautiously with the harvest of deep-sea species. It suggests that deep-sea protected areas are needed that include regions extending through a broad range of depths to thoroughly protect fish populations.

Effects on the targeted populations are not the only concerns with deep-sea fisheries. Animals captured by trawl nets from deep waters and hauled to the surface do not survive and are potentially impacted even if not targeted for harvest. There is also the possibility of impacts on deep-water habitats and ecosystems. It was long assumed that the seafloor was a vast smooth silt-covered surface with little relief and few attached organisms. It is now realized, however, that complex habitats and diverse ecosystems are being destroyed in regions of the seafloor where fishery harvest is commonplace before we even know what is there (**Box 8-1. Conservation Concern: Deep-Water Coral Reefs**).

Box 8-1   Conservation Concern: Deep-Water Coral Reefs

Until as recently as the late 1900s, it was generally accepted that the deep sea was mostly uniform and devoid of structure due to the long-term accumulation of organic matter and biogenic sediments from surface waters. As capabilities were developed to explore for canyons, seamounts, and ridges, and search for life there, new ecosystems were soon discovered. The first of these were the hydrothermal vent ecosystems in isolated geologically active areas of the deep sea. Eventually large regions covered with structure-building animals dominated by corals were discovered. The hermatypic corals associated with tropical reefs cannot survive in waters below the photic zone due to their dependence on photosynthesis, achieved by the symbiotic zooxanthellae (see Chapter 5); however, there are hundreds of known species of **ahermatypic corals** that are not restricted to shallow waters, but depend entirely on food captured by suspension feeding from the surrounding waters. These were largely unknown until recent advances in deep-sea exploration, but are now known to rival the hermatypic corals in species diversity. Deep-water corals represent the same major groups as shallow-water corals, including over 500 species of stony corals (Scleractinia), gorgonians (sea fans, Octocorals), hydrocorals, black corals (Antipatharia), zoanthids, and others (**Figure B8-1**). Associated with the corals are many of the other sessile organisms found in shallow water reef ecosystems. These include sponges, anemones, barnacles, tunicates, and crinoids (sea lilies). These attract other invertebrates, including echinoderms (e.g., brittle stars and starfish), mollusks (e.g., bivalves), and crustaceans (e.g., crabs and shrimps). Fishes often aggregate over these reefs, especially those on seamounts, sometimes to over 100 meters above the reef. These include the orange roughy, rockfishes, and the pelagic armorhead.

Many of the ahermatypic coral reefs are associated with seamounts, presumably due to the currents passing over these structures that transport food organisms to support suspension feeding animals. Other habitats exist along continental margins where strong currents prevent the accumulation of sediments and transport food organisms. Two such areas are in the Northeast Atlantic, along the coast of Norway where the North Atlantic current flows, and along the Atlantic coast of the southeast United States where the Gulf Stream produces currents down to over 500 meters depth. In these regions, reefs of stony corals have formed that are as thick as 30 meters or more and longer than 10 kilometers. Considering that these corals typically grow less than 2 centimeters per year, they are often thousands of years old. As with tropical reefs, the living portion of the coral is predominantly that which is exposed at the surface of the reef. The diversity of species associated with these reefs can rival that of tropical

(a)

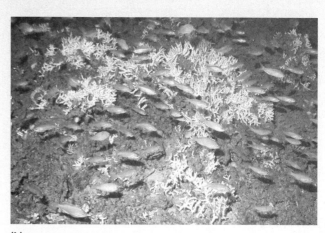

(b)

Figure B8-1   Discovery of diverse and vulnerable deep-water coral communities has led to increased limitations on bottom trawl fishing. **(a)** Gorgonian soft coral with rockfish resting in the branches and **(b)** *Oculina* stony corals with anthiid fishes.

coral reefs. For example, over 1,000 invertebrate species have been found on deep cold-water reefs in the northeast Atlantic. Often these cold-water reefs are characterized by a single dominant species of coral. For example, *Lophelia pertusa* reefs are found off Norway in the northeast Atlantic and North Carolina in the northwest Atlantic, growing predominantly at 200 to 1,000 meters depth. *Oculina* reefs are found off Florida at about 70 to 100 meters, where they support over 300 species of other invertebrates and serve as important fish habitat and spawning areas for fishes such as snappers and groupers.

Deep cold-water corals have been known for hundreds of years. Fishermen collected and used them as medicines and for making jewelry at least since the 1700s. Since the 1950s, black corals and gorgonians, so-called *precious corals*, have been harvested from seamounts and other deep waters for making jewelry and decorative objects, often resulting in local depletions and damage to the reefs during collections (**Figure B8-2**). The complexity of the ecosystems from which the deep-water corals originate has only been understood and appreciated since the 1990s, however, long after corals and fish were being harvested from these regions. The impact of modern fishing on deep-water coral ecosystems goes far beyond the removal of fish populations. The most convenient and profitable method of taking fish from the vicinity of these deep- and cold-water coral reefs is with trawl nets. As the nets are pulled across the bottom, they destroy or remove living organisms other than fish. The effect on the coral structure has been compared to a forest clear cut (**Figure B8-3**).

The degree of reef destruction caused by trawling on seamounts can be characterized by the orange roughy fishery off Australia and New Zealand. These seamounts contain reefs formed by dense colonies of stony corals, primarily *Solenosmilia* and *Goniocorella*. Trawl fishing on previously unfished seamounts could net over 1.5 metric tons of coral per hour. Over 1,700 metric tons of coral were removed in one year of fishing (this is about 40% of the weight of harvested fish). After two years of fishing only about 400 kilograms of coral were netted per hour, indicating that the reefs on the seamount were rapidly being destroyed. Seamounts that have been heavily fished for years, being trawled over hundreds or thousands of times, have been scraped clean of corals and other sessile organisms. The possibility of recovery is unlikely for structures that took hundreds or thousands of years to form. Knowledge of the extent of this damage and the uniqueness of these ecosystems was finally realized at the end of the 20th century. Australia and New Zealand then established marine protected areas around about 30 seamounts south of Tasmania and in New Zealand waters that had not been heavily fished.

In the northeastern Pacific off the Aleutian Islands, reefs containing concentrations of sponges and octocorals (sea fans, soft corals, and gorgonians) were heavily fished for rockfishes and mackerel. Over 2,000 metric tons of corals and sponges were removed by trawls between 1990 and 2002. After a protracted battle by conservation groups, much of the untrawled area was protected by U.S. law requiring protection of essential fish habitat. An area of over 1 million square kilometers is now protected from trawling. This was followed by protections of other untrawled areas off the Pacific coast of the U.S. mainland.

(a)

(b)

Figure B8-2   (a) Black coral, valued in making jewelry and figurines, is vulnerable to overexploitation because of its slow growth rate and, thus, is largely protected from harvest. **(b)** Black coral confiscated after being shipped from China to the U.S. Virgin Islands in violation of international trade laws.

Figure B8-3   Bottom fishing trawls can scour the bottom, destroying deep-sea reef ecosystems.

In the northeast Atlantic, almost 50% of the *Lophelia* reefs off Norway were destroyed by trawlers before their significance and extent were realized. Around 2000, Norway began providing protections, establishing six reef sites as marine protected areas. This was followed by efforts to protect other cold-water reefs off Ireland and Scotland. Although fishers resisted, protections were initiated by the European Union by 2003. In the northwest Atlantic, cold-water reefs found in canyons off Canada and Maine have been recently added to the list of areas protected from trawling. After most of the *Oculina* reefs off the southeast U.S. coast were destroyed by shrimp trawling, protective measures as essential fish habitat were finally instituted in 2003.

Although many areas of these diverse assemblages of ahermatypic corals have been destroyed, possibly beyond recovery, there is a push to protect much of the remaining untrawled areas of deep-water reefs through the creation of reserves. Over 10 million square kilometers are now under protection by coastal nations around the world within their national waters. Though many regions have been decimated, this could provide one of the best examples of proactive management of a critical marine habitat. Unfortunately, these protective measures only apply to deep-water reefs in national waters, mostly within 320 kilometers of a nation's coastline. In international waters, however, agreements for deep-water reef protection are rare. There are many regions, especially in the South Atlantic and Indian Oceans, where corals likely occur that are virtually unexplored and unprotected. In these waters, deep-sea trawling will continue to destroy these reefs until meaningful international agreements are reached that restrict such harvest. Species could be driven to extremely low levels or even to extinction before we have documented the biodiversity and ecology of these ecosystems. Areas that have been impacted for over 50 years by harvest of precious coral and fish, such as the Emperor Seamounts in the northwest Pacific, could have already experienced unknown extinctions. It remains to be seen whether deep-water coral reefs will become another example of human destruction of one of the Earth's major ecosystems, or one of the first examples of our ability to overcome our shortsightedness and greed to proactively protect a unique environment for the sake of conservation.

Because of the impact to habitats and ecosystems, and the unsustainability of most deep-sea fisheries, a general consensus has developed among scientists, as summarized in an analysis by Elliott Norse and others, that most commercial fishing in the deep sea should be stopped. They suggest that this could be achieved by removing subsidies (which are currently 25% of the value of the deep-sea catch—see Chapter 12) and supporting the rebuilding of more accessible and sustainable fisheries in waters closer to shore.

8.5 Habitat Impacts on the Deep Seafloor

Although we have made major advances in exploring the deep sea, it is still impractical and cost prohibitive to harvest or mine resources from the deep seafloor away from the continental margins and seamount regions that are within about 2,000 meters of the surface. This means that most regions of the deep sea are still protected from direct impacts of human activities. Because of the interconnectedness of the earth's ecosystems and the far-reaching effects of anthropogenic impacts, however, we now realize that no portion of the Earth's biomes is immune to the influences of human activities.

Offshore Oil and Gas Drilling

As the demand for petroleum resources has expanded and those that are accessible from land have become scarcer, we have expanded our search into areas beneath coastal waters. Now that much of the oil in shallow coastal waters is currently being exploited, drilling has moved out onto the continental shelf progressively further from shore. For example, in the Gulf of Mexico the number of offshore wells increased from six wells in 1992 to over 100 by 2006; most of these are drilled beneath waters from 500 to 1,500 meters in depth. Technology and expense still limits the amount of oil drilling beyond 1,500 meters depth, and the 2010 *Deepwater Horizon* oil spill in the Gulf of Mexico has set back deep-water drilling due to fears of the likelihood of similar incidents. Substantial oil reserves are known to occur beneath deeper waters out onto the continental slope; the deepest production oil well is in the Gulf of Mexico under waters over 3,000 meters depth (**Figure 8-19**).

Although new oil deposits are continually being produced naturally, oil cannot be considered a renewable resource because these deposits develop only on a geologic time scale. The major conservation concerns of oil drilling

Figure 8-19 Drilling platforms, such as British Petroleum's *Thunder Horse* floating oil platform in the Gulf of Mexico, can drill for oil thousands of meters beneath the sea surface.

are about the impacts the drilling and potential oil spills have on the environment and ecosystem in the waters, along the coast, and on the seafloor. Aesthetics values also come into play in that there is resistance to the presence of platforms and drilling rigs within sight of land. In U.S. coastal waters, a combination of these reasons has led to public opposition to drilling in coastal waters off California and northeastern states bordering the Atlantic. Political pressure for self-sufficiency in fuel production has led to a gradual expansion of access to oils beneath coastal waters. A debatable, but potential, benefit to the presence of oil rigs is their use as artificial reefs for attracting fish for harvest by recreational and commercial fishers. For example, over 4,000 drilling rig structures remain in place in the Gulf of Mexico, and *rigs-to-reef* programs allow toppling retired oil rigs as artificial reefs (see Chapter 6). The major negative concern about drilling and transporting oil within in coastal waters is still the potential impact of pollution from the oil following spills from ships or blowouts during the drilling process. The effects of such spills on coastal and nearshore ecosystems are addressed in Chapters 3 to 6.

Impacts of drilling under ocean waters are not limited to the effects on coastal ecosystems. Blowouts that occur during the drilling operation result in an uncontrolled release of oil until the oil well can tapped off or controlled, as was seen in the 2010 blowout in the Gulf of Mexico off the Mississippi River Delta. Not all of the oil stays confined as a defined slick at the surface. In deep waters, because of the high water pressure and density of the oil, a portion of the oil from a blowout stays in the water column or near the sea floor. As oil decomposes, its components become dispersed through the water column. Currents can disperse this oil far from the spill. Oil that becomes buried, especially in anoxic sediments, decomposes slowly and could have unknown effects on benthic organisms. The oil that stays in the water beneath the surface is transported away from the well by bottom water currents or as hydrocarbon plumes in the mid-water column. Although the dispersal by currents dilutes the oils effect on sea floor and open water ecosystems, the effects to sensitive species could be expanded due to an increase in the area of coverage. Little has been documented of the sensitivity of deep-water species to such contaminants. After the 2010 *Deepwater Horizon* spill, localized bottom areas of unknown extent were covered with oil at least a year after the spill, but the overall effect on bottom ecosystems is yet to be fully evaluated.

Other than the oil, other hydrocarbons are released during a deep water blowout. During the 2010 *Deepwater Horizon* spill (also see Box 3.3, Learning from History: *Torrey Canyon*, *Exxon Valdez*, and *Deepwater Horizon*), large amounts of methane were released along with the oil. At depths of this spill much of the methane would form hydrates, keeping it in place on the sea floor. Such hydrates serve as an energy source for bacteria and, along with the decomposition of the some of the oil, could have depleted oxygen from the bottom waters. Studies by John Kessler and colleagues following the spill concluded that the methane released in the spill was oxidized by bacteria to carbon dioxide within about four months after release, and that there was no measurable loss of methane gas to the atmosphere. Samantha Joye and colleagues, however, presented arguments (that were refuted by Kessler and colleagues) that the low-oxygen zones could result from possible explanations other than rapid bacterial decomposition of methane.

Chemical dispersants can be used to control dispersal of oil from a spill and keep it from washing ashore. Dispersal enhances the decomposition of the oil by increasing the contact area between the bacteria and oil, but does not remover the oil from the water (see Chapter 3). A controversial method applied for the first time during the 2010 Deepwater Horizon oil spill was the large-scale application of dispersants beneath the surface near the source of the spill. This served to disperse much of the oil into the ocean as small particles, which evidence suggests were more readily decomposed than mats of oil floating at the surface or resting on the sea floor. J. Vilcaez and colleagues estimated that 50% of the dispersed oil was degraded within one week following the spill, but the remaining 50% likely lasted for months. Major ecological effects were not noticeable as a direct result of the subsurface release of dispersant (over 1 million kilograms). A major component of the dispersant was detected by Elizabeth Kujawinski and colleagues months after its release, indicating that it had not undergone rapid decomposition as expected. Although concentrations of the residual chemicals were much lower than determined to be toxic to near-surface life, further studies are ongoing to determine its potential impact on deep-sea ecosystems.

The effect of deepwater blowouts on deep seafloor ecosystems is difficult to assess due to the time and expense of sending submersibles to survey extensive areas at great depths and limited information on ecosystem health before the blowout. 2011 surveys in the vicinity of the *Deepwater Horizon* blowout by Samantha Joye and colleagues found patchy oil deposits within an area 65 kilometers from the wellhead, some several centimeters thick, as well as evidence of suet from oil that was burned at the surface. The major source of the oil deposits appeared to be a slime-like material produced when bacteria colonized the oil at the surface, making it heavier causing it to sink to the sea floor. During the surveys, dead corals, crabs, tube worms, and sea stars were noted, and there was evidence of mortality of jellyfish from above the seafloor. The extent of seafloor ecosystem damage and long-term effects will be difficult to document and may not be known for years after the spill. Even with long-term monitoring, it will be difficult to distinguish a single cause and to distinguish effects of the spill from other environmental and anthropogenic impact on the deep-sea environment.

Avoiding the impacts of deep-water spills is best achieved by minimizing the chance of such spills. The *Deepwater Horizon* well blowout came as a surprise to many within the oil industry, because the probability of blowouts was believed to be miniscule after the development of blowout preventers. These devices are designed to be triggered during a blowout and clamp down to prevent oil and gas from being released out of control during such an emergency. Prior to the BP spill, the probability of an uncontrolled well blowout in offshore drilling operations was estimated at about 1 per 10,000 wells. Because the BP blowout resulted at least in part from a malfunctioning blowout preventer, the likelihood of such events in the future undoubtedly will be reconsidered. There are ongoing efforts to improve technology, tighten restrictions, and increase oversight to prevent the recurrence of such an accident.

Another potential source of oil leakage into seafloor habitats is undersea pipelines prone to material defects, corrosion, or impacts of ship anchors or bottom trawls. The probability of a leak developing in a major pipeline has been estimated at about 1 in 1,000. The effects of such leaks on marine ecosystems would depend on the severity of the leak and the sensitivity of nearby ecosystems. Modern technology of pipeline construction continues to reduce the likelihood of accidents but will never reduce that possibility to zero.

Damage from a major oil spill is not the only environmental concern of offshore drilling. The installation of the rig structure can damage the benthic environment; however, these effects are localized and can improve the habitat by providing shelter for organisms (whether the positives outweigh the negatives is debatable; see Chapter 6). A broader impact can be caused by the release of oil and wastes from the drilling process, including drilling muds and drill cuttings. Drilling muds are pumped down the oil well to lubricate and cool the drill bit. They circulate to the surface, carrying the drill cuttings. The long-term environmental effects of small releases of oil during the drilling operation are probably minimal (however, they may wash ashore polluting beaches and other coastal habitats). The drilling muds contain rock and clays that may be contaminated with zinc, copper, and lead, and they often contain oils to make the drilling process easier. Most of the drilling muds are recovered and re-used; however, the drill cuttings, leftover materials from the drilling process that may be contaminated with drilling muds and toxic metals, are often put back into the sea. The major harm from the drill cuttings that accumulate on the seafloor around drilling rigs is chemical contamination (e.g., by hydrocarbons, heavy metals, and sulfides), organic enrichment, and smothering. The materials that cover the sea floor can inhibit the natural settlement of invertebrates.

Some areas have already exhibited large environmental impacts from the deposition of drill cuttings. For example, 30 years of drilling in the North Sea have left over one million metric tons of drill cuttings around oil platforms. The low-oxygen and toxic sediments blanketing the surface can create a "dead zone" affecting bottom ecosystems for up to 5 km away from the oil platform. The duration of these environmental impacts is not known; however, it took several years for animals to return to some bottom habitats where platforms were removed. The extent of harm varies with the physical and biological nature of the bottom ecosystem. In regions where there are deep-sea corals, the local environmental harm would be at least as great as that of bottom trawling. In regions with greater bottom currents, the cuttings may be dispersed quickly, allowing for dilution and biodegradation of some of the contaminants. In areas of poor circulation, drill cuttings are more likely to accumulate and cause environmental impacts. Because deeper waters tend to have lower currents speeds and less resistance to physical impacts, the effects could be greater and of longer duration from deep-sea drilling. Organic enrichment could disrupt deep water with its already low natural productivity. It is also more likely that drill cuttings will be deposited around deep-sea rigs because the expense of transporting them to shore may be cost prohibitive.

Stricter regulations have been enacted around the world to limit the direct impact of deep-sea drilling on bottom ecosystems. In some regions the cuttings must be brought to shore for disposal (creating a potential land-pollution problem). In U.S. waters, the Environmental Protection Agency has established regulations limiting the materials used in drilling oils and monitoring the discharges around offshore oil rigs. Most industrialized nations now require an environmental impact assessment (EIS) prior to new drilling activities (although it was discovered after the 2010 *Deepwater Horizon* spill in the Gulf of Mexico that these were not always adequately enforced). Deep-water coral ecosystems would be especially sensitive to drilling procedures, requiring strict enforcement of EIS regulations. An increased appreciation of the biodiversity of deep-water soft-bottom ecosystems would go a long way in increasing their protection.

Impacts of Surface Fisheries on Deep-Sea Ecosystems

As discussed above, commercial fishing at the seafloor down to depths of over 1,000 meters below the surface has been documented to have substantial impacts on not only the fisheries species but also the ecosystems on which they depend; even beyond those depths, harvested species whose distribution extend into deeper waters may be impacted. But even though fishing is not currently feasible in the deepest regions of the ocean, the effects of fishing likely extends into these waters.

Technology is available to send nets, traps, and submersibles even into the deepest ocean regions and bring back animals to the surface; however, such expeditions are very few in number and are mostly limited to scientific studies. It is difficult, expensive, and time consuming to send and retrieve nets or traps to the deep seafloor from ships at the surface. Manned or remotely operated submersibles require expensive advanced technology (see Figure 8-1). Finally, most animals are too sparsely distributed in the deep sea to make it economically feasible to harvest them. This does not mean that the deep sea is unharmed by the effects of fishing, however.

The harvest of fishes from near-surface waters can influence deep-sea ecosystems by removing food resources, because much of the organic matter on which deep-sea animals depend originates in the ocean's surface waters. Deep-sea ecosystems are so poorly studied that little has been documented of these effects. Organisms likely affected include those that have evolved to depend on the fall of whales and other large animals from the surface (see Figure 8-15). Whaling and harvest of large predatory fishes may have dramatically reduced this source of food for deep-sea organisms. It is possible that the density of whale falls was reduced as much as sixfold over a period of about 50 years of heavy commercial whaling (see Chapter 10). As the distance between whale falls increased, larvae of organisms dependent on these fallen carcasses presumably would be unable to disperse and colonize new habitats. No data are available to document these effects; however, it is possible that some species have been driven to extinction already. Carcasses of large pelagic predators such as tunas and billfishes also would have been reduced from recent overfishing. Carcasses of beached whales have been experimentally hauled to sea and sunk in deep waters to follow the development of whale-fall ecosystems. On a larger scale, this practice could serve to enhance these ecosystems by providing stepping stones for species to disperse from other regions of the deep sea.

Seafloor Mining

As described above, the removal of hydrocarbons as oil and gas products has expanded into waters of over 3,000 meters depth. The development of hydrocarbons is primarily restricted to areas with large historic accumulation of organic matter from terrestrial sources, however, limiting them largely to continental shelf and slope waters. Less attention is given to other kinds of near-shore mining, but such mining affects benthic ecosystems. As discussed in Chapter 3, sands are mined for construction or beach replenishment in some regions, and there has been locally important mining of diamonds in water of the coast of Namibia.

Commercial mining of the deep seafloor below about 1,000 meters for minerals has not yet been worth pursuing due to the difficulty and expense of bringing up materials from the seafloor, and legal complications of mining beneath international waters in the open ocean (see Chapter 12). There are mineral resources in this region, however, that could warrant future mining if technologies are developed and the market value of minerals make mining profitable enough. These resources include manganese nodules, metal-containing sulfides, and methane hydrates. The potential value of deep-sea minerals, found mostly beneath international waters, has led to the development of international mining codes. The recently documented diversity of ecosystems in regions containing these minerals has accelerated controversies over mining these resources.

Potentially, the most valuable source of minerals from the deep sea is the boulder-like structures called **manganese nodules** or **polymetallic nodules** (**Figure 8-20**). These nodules were first taken from the seafloor in dredge collections made during the earliest deep-sea explorations in the late 1800s, but their potential commercial value and the mechanism of their formation wasn't realized until the 1960s. The nodules are formed as minerals dissolved in the seawater precipitate around some hard material, such as a shark's tooth or animal skeleton. The continued precipitation of minerals can eventually result in an irregular shaped nodule 10 centimeters or more in diameter. One of the most remarkable characteristics of these nodules is their extremely slow rate of growth. For example, in the deep Pacific Ocean, nodules form at a rate of about 1 to

Figure 8-20 Though currently not economically feasible to mine, manganese nodules are a potential source of metals.

2 centimeters per million years; thus, a large nodule can be tens of millions of years old. It is not known how nodules stay at the surface of the sediments as they accumulate for millions of years. Possibly, activity by animals in sediments surrounding the nodules keeps them at the surface, or currents may keep sediments washed away. Organisms found in the seafloor region around the nodules are mostly small benthic filter feeders and deposit feeders, including polychaete worms, isopods, shrimp-like tanaid crustaceans, and bivalves resting on or buried in the sediments. Larger grazers include echinoderms such as sea cucumbers and brittle stars.

The feasibility of mining manganese nodules depends not only on accessibility but also their mineral composition and distribution on the seafloor (**Figure 8-21**). Manganese is not the only commercially valuable mineral in the nodules; others include copper, nickel, and cobalt. Concentrations of these minerals vary regionally and their value can change rapidly and unpredictably. For example, the value of cobalt increased when it began to be used in the manufacture of strong metal alloys in jet aircraft engines, and high concentrations of cobalt in nodules in the south Pacific have made them potentially valuable for mining. Denser concentrations of nodules make them easier to mine. In the northeast Pacific, nodule concentrations are as high as 10 to

15 kilograms per square meter, and there are estimated to be billions of metric tons of commercially valuable minerals. The combination of these factors make these locations likely regions for future mining.

Potential environmental impacts of mining activities are numerous. For example, removal of the nodules takes away the only hard substrate available on the deep seafloor. Fauna directly associated with the nodules would likely either be killed outright or not survive the loss of habitat, with no hope of recovery of structures that took millions of years to develop. The collection of the nodules will remove about the top 5 centimeters of sediments from the seafloor, killing most of the sediment-dwelling polychaetes and crustaceans. The sediments will be distributed into the water column to settle back onto the seafloor. The effect of such disturbances will likely be a local devastation of the benthic fauna because deep-sea benthic animals are adapted to long-term environmental stability. Settling sediments could smother surface dwelling animals and disrupt the feeding and respiration of deposit feeders on the seafloor. The settling sediments would cover the thin layer of organic matter—the base of the benthic food web—with subsurface sediments depleted in organic matter.

Studies by Christian Borowski to assess the effects of small-scale mining-like disturbances of the deep seafloor

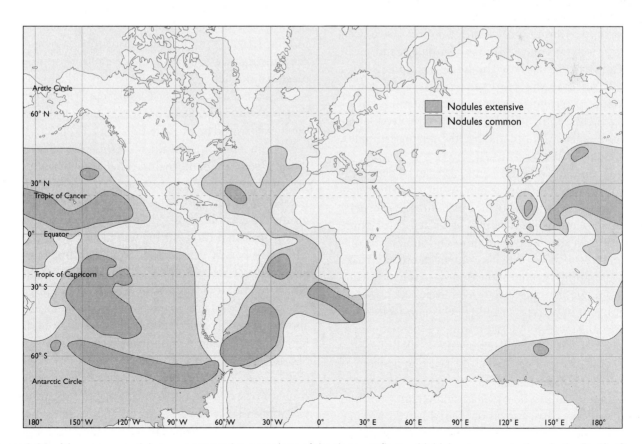

Figure 8-21 Manganese nodules are scattered over regions of the deep seafloor, with highest concentrations in the Pacific Ocean.

showed a 50% to 90% reduction of benthic fauna immediately after the disturbance. Three years after the disturbances the most prominent macroinvertebrates, especially polychaetes, had returned to their original abundance; however, the diversity remained low for at least seven years after the disturbance. The effect of massive disturbance that would result from commercial mining is unknown. Along with effects in the immediate area of mining, there also would be potential impacts from the release of tailings from the processing ships and suspension of sediments in the water column, especially on suspension- and particle-feeding organisms adapted to clear waters with naturally low turbidity. Because our knowledge of mechanisms that maintain the high species diversity in the deep sea is limited, it is unclear how such large-scale disturbances would affect that diversity. If large-scale manganese nodule mining commences in the near future the impacts will occur before we have the baseline knowledge to assess the effects.

The prospects of large-scale commercial mining of manganese nodules remain uncertain. The technology for mining manganese nodule from waters as deep as 3,000 meters is currently under development, some of which was adapted from methods developed for laying deep-ocean cables. The most current deep-sea mining methods consist of using collector vehicles that move across the seafloor on tank-like tracks, collecting the nodules by lifting them onto a mining platform at the sea surface by a hydraulic pump system. Hundreds of prospecting trips have been made to search for concentrations of nodules, and seven contractors have been licensed by the International Seabed Authority (see Chapter 12) to develop mining technologies. If international agreements over mining rights and controversies over environmental restrictions are resolved, large-scale commercial mining could begin, by some estimates, as soon as the 2020s.

Manganese and other minerals also can be deposited away from manganese nodule regions by similar mechanisms. These deposits are formed as **manganese crusts**, which are pavement-like coverings on undersea rock outcroppings associated with seamounts and ridges. The discovery of crusts rich in cobalt have piqued interest and have led to explorations by some industrialized countries. Mining technologies have not been well developed, however, and mining of these crusts would likely be more difficult than for manganese nodules. If mining of the manganese crust becomes a reality, it will remove the sessile organisms such as corals and sponges, and likely have major local impacts on these ecosystems.

Another location where metals can precipitate out on the deep seafloor is in regions associated with hydrothermal vents. These deposits develop when metal-rich waters beneath the seafloor are expelled by the vents and precipitate

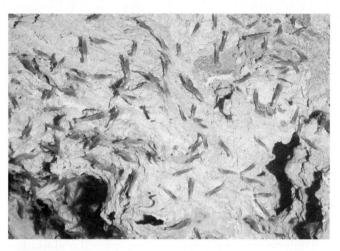

Figure 8-22 Sulfur deposits covered by shrimp in a volcanic region in deep waters of the western Pacific. Mining of such deposits could devastate ecosystems associated with deep sea vents.

out in **polymetallic sulfides**. Mining would likely target metals such as gold, silver, zinc, lead, and copper. Some of the largest sulfide deposits are in convergence zones between continental plates, where vent and volcanic activities are prevalent (**Figure 8-22**). Many of these are in waters controlled by coastal nations and therefore mining could commence without having to address the conflicts of international agreements. Plans have been made to initiate mining activities in inactive vent regions associated with the convergence zone off New Zealand. If mining takes place around active hydrothermal vents, the local effects on organisms would be severe because large amounts of material would be taken from which low concentration of the metals would be extracted.

Another potentially valuable resource to be taken from the deep sea is methane hydrates, believed to be the largest potential sources of combustible hydrocarbons on earth (**Figure 8-23**). Although large-scale mining is not yet economically feasible, if oil production declines at the currently projected rate, mining will likely become feasible within the next 50 years. Experimental drilling operations have been carried out in the western Pacific off the coast of Japan. Potential impacts of large-scale mining would include effects not only to methane hydrate fauna described above but also to other benthic ecosystems. A large-scale removal of methane hydrate deposits could cause slumping of the seafloor, possibly resulting in landslides and tsunamis. Large releases of methane during drilling operations could cause explosive harm during drilling operations and possibly enhance global warming (however, as discussed above, bacterial decomposition could limit the release into the atmosphere).

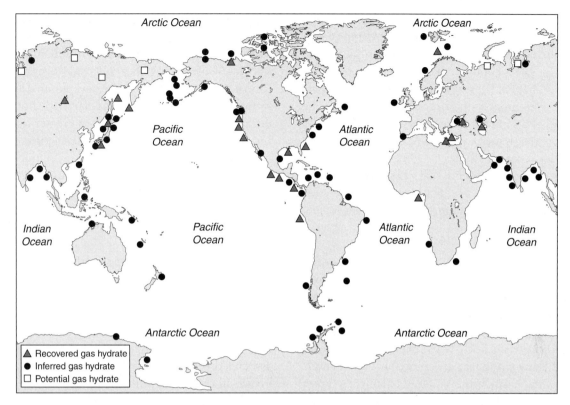

Figure 8-23 Locations of methane hydrates, considered to be a potential source of natural gas to be harvested.

Undoubtedly mining of minerals and hydrocarbons from the deep sea will continue to expand, with predicted and unanticipated impacts on deep-sea ecosystems. How rapidly that expansion will occur depends on many factors. One is economics; the pressure to mine will depend on value and availability of specific minerals. The implementation of international environmental regulations to minimize the impacts will require cooperation among many nations. Legal issues relative to conservation of deep-sea ecosystems are discussed in Chapter 12.

Ocean Pollution and Dumping

As long as humans have lived near the shore, the ocean has been considered a convenient place to dump our wastes. Even as recently as the 1960s the ocean was generally considered to be a virtually unlimited area into which we could dump almost anything without having to deal with fears of environmental impacts. Over 10 million tons of industrial wastes were being dumped into the oceans each year. Raw sewage was commonly pumped into coastal waters, and sewage sludge and dredge spoils were being dumped by barge-loads into deep-sea waters. The appropriate catch phrases became "out of sight, out of find," and "dilution is the solution to pollution." Beginning in the 1970s, with an increased global environmental awareness, many nations began limiting what they would put into the ocean, and

eventually international agreements were established that either prohibited or restricted what nations were dumping (see Chapter 12). Even with agreements in place, many tons of wastes are still put into the ocean intentionally and unintentionally each year. Trash items that are buoyant, such as plastics, often end up washing onto beaches or accumulating in open ocean circulation gyres, as discussed in previous chapters. Degrading plastics can reach the seafloor with unknown effects on deep-sea ecosystems. Other major sources of ocean pollution are lost and abandoned fishing nets and traps; these can cause the death of animals in the water column or on the seafloor when they become accidentally entangled or trapped. These impacts are discussed further in Chapter 11.

Shipwrecks

Many items disposed of at sea sink rapidly to the seafloor where effects are not readily observed and thus are often largely ignored. Shipwrecks have littered the seafloor for centuries; however, prior to the 1900s these were relatively few, and their components, mostly wood and other organic materials, probably did not have detrimental effects. In the 20th century, however, especially during wartime, large masses of metal, fuel, and toxic material reached the floor through shipwrecks. Their effects are probably local, but these have not been assessed globally. It is estimated that over 10,000 ships have sunk to the seafloor, weighing more than 40 million metric tons. Effects on shallow and coastal ecosystems include habitat destruction (e.g., tropical reefs; see Chapter 5), release of toxic chemicals (e.g., petroleum products; see Chapter 3), and potentially positive benefits of providing structure as artificial reefs (see Chapter 6). The intentional disposal of large structures such as ships in the deep sea (after toxic materials have been removed) has been considered; however, no studies have been carried out on the potential impact. The intentional sinking of drilling or mining structures in the deep sea will likely be considered as oil drilling and mining expand further into deeper waters.

Munitions and Nuclear Wastes

Possibly the most controversial type of dumping in the deep sea is of munitions and containers of radioactive and chemical wastes. For example, millions of tons of weapons, including chemical weapons, were dumped into waters off Europe from the end of World War II until the mid-1970s. Chemicals that decompose quickly when exposed to seawater (including nerve gas and tear gas) probably do not result in environmental harm; however, other chemical weapon materials, such as mustard gas, decompose slowly and may remain a hazard on the seafloor. Munitions can remain intact on the seafloor. The long-term environmental harm of chemicals and munitions is not known. Several incidents have occurred where European fishers have been exposed to mustard gas or munitions from trawling the seafloor.

As nuclear power generation developed following World War II, it became common practice to dispose of nuclear waste by encasing it in drums to be dropped to the deep seafloor. These drums were not made to contain the wastes for the tens-of-thousands of years they would remain radioactive. In fact, some of the drums were made to implode, releasing the wastes into the surrounding waters and seafloor sediments. Approximately 300,000 drums (over 100,000 metric tons) of low and intermediate-level radioactive waste were dumped into the deep north Atlantic and Pacific from the 1940s through the 1970s by the United States and European countries. Levels of dumping by the former Soviet Union are largely undocumented; however, Russian leaders admitted in the 1990s that they had dumped spent nuclear fuel, cooling water from reactors, and even retired reactor units containing high-level nuclear waste.

It is obvious that there was a lack of concern and/or awareness for conservation by the manner and location of dumping the nuclear wastes. For example, from 1946 to about 1960, the United States dumped almost 50,000 drums of low-level waste in waters of about 1 kilometer depth, just 50 kilometers off the California coast. This dump site off the Farallon Islands, an important habitat and breeding area for marine mammals and seabirds, was eventually declared a marine sanctuary. The waste was considered relatively low in reactivity, but the potential environmental hazard is not known. Bivalves and fishes collected from the region were shown by Thomas Suchanek to have elevated levels of some radionuclides. Removal of the barrels is not considered desirable, as it would likely pose a greater risk than leaving them undisturbed. The prevalence of such dumping around the world is unknown due to poor documentation.

By the 1980s, dumping of such wastes had been outlawed by nations with large nuclear-producing capabilities. These laws were not strictly adhered to, however, and as recently as 1993, a Russian vessel was observed dumping radioactive water into the Sea of Japan. It was not until 1993 that bans were fully enforced. Despite current bans on dumping, the potential harm of such wastes could remain for thousands of years due to the slow decomposition of nuclear materials. Most of the historic disposal sites have not been well studied; however, at U.S. dump sites, measurable levels of radioactivity have been found in organisms associated with benthic ecosystems, including anemones, brittle stars, sea cucumbers, and predatory fishes. This provides a mechanism for transfer throughout deep-sea ecosystems, and even to humans through deep-sea fisheries.

Whether the ban on deep-sea disposal of nuclear wastes will remain in place is unclear. As long as humans continue to accumulate these wastes (e.g., the United States accumulates over 2,000 metric tons each year), locations will be needed for disposal. It is argued—based on the lack of evidence that past disposal caused widespread contamination—that the deep sea may be the safest place for this disposal. One proposal is to bury the wastes in holes drilled into the seafloor in the mid-oceans, considered to be one of the most geologically stable regions on earth. Even though average level of exposure to marine organisms could be low, however, we have learned from other examples (e.g., metals and pesticides) how efficiently toxic materials can accumulate through marine food webs.

Sewage Sludge and Dredge Spoils

Most of the domestic wastes produced by humans are recycled (e.g., as fertilizer), placed in landfills, or released into fresh or coastal waters. As human populations have expanded, disposal of these wastes has increasingly created problems in these environments (see Chapters 3–6 for discussions of impacts of sewage discharge in coastal and nearshore ecosystems). One other option for disposal of these wastes is to create a **sewage sludge** (the solid remains after sewage treatment) that can be barged to sea and dumped onto the deep seafloor.

Sewage sludge is largely comprised of organic material; however, it can be contaminated with toxins, including toxic metals. The potential impacts of sewage sludge on the seafloor include toxic effects, smothering organisms, inhibiting feeding, increasing turbidity, and enhancing organic matter, resulting in eutrophication and hypoxia of seafloor organisms. Few of these impacts have been monitored for the millions of tons of sludge that have been dumped into the oceans. The most well-studied dump site is about 160 km off the coast of New York, where over 35 million metric tons of sewage sludge were dumped between 1986 and 1992 to settle onto the seafloor in 2,500 meter deep waters. Following dumping, macroinvertebrate abundance increased significantly, presumably in response to the increased organic matter. Silver concentrations were monitored as an indicator of metal contamination; levels were 20 times higher than at uncontaminated reference sites. The sludge material entered the food web through deposit feeders such as sea urchins and sea cucumbers. Environmental concerns resulted in a halt to dumping at this site in 1992. There are fears that dumping of sewage sludge could increase globally as waste production increases with human population growth. Currently it is not prohibited by international agreement.

Dredge spoils are the sediments scooped from the bottom of waterways and harbors to maintain water depths for boat traffic. These spoils are mostly sediments, but they may contain metals, hydrocarbons, pesticides, and other contaminants. The impacts of dumping dredge spoils at sea would be similar to those of sewage sludge, except that dredge spoils contain relatively little organic matter. The fine sediments in dredge spoils could increase turbidity substantially and be transported great distances by ocean currents (**Figure 8-24**). Currently there is little documented dumping of dredge spoils in the deep sea; however, millions of tons are dumped into coastal waters each year. In U.S. waters, states designate dumping sites and the EPA regulates the kind of dredge materials that can be dumped, eliminating those that contain pollutants.

Chemical Pollutants

By the time many of the pollutants that are introduced into the marine coastal or surface waters reach the deep sea, they have been decomposed or diluted sufficiently that their impact is likely minimal. Persistent chemical pollutants can accumulate to dangerous levels in deep-sea sediments or

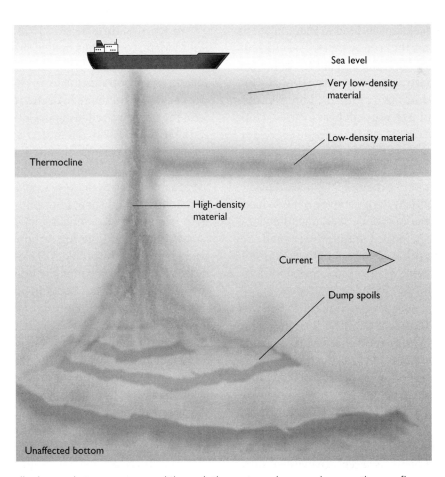

Figure 8-24 Dredge spoils dumped at sea can spread through the water column and across the sea floor.

food webs, however. For example, tributyl tin (TBT), an antifouling agent in marine paints problematic for its reproductive effects on mollusks in intertidal ecosystems (see Chapter 3), has been found in animals at depths of about 1,000 meters off Japan in concentrations similar to those found in shallow-water animals. This may be a reflection of concentrated ship traffic, because bans on the use of TBT in hull paints did not apply to larger ocean-going vessels greater than 25 meters length.

It is well documented that persistent organic pollutants (POPs) such as DDT and PCBs accumulate in food chains of nearshore coastal and terrestrial regions around the globe (see Chapter 7). They also can be rapidly transported into deep-sea waters. The mechanisms of transport include vertical circulation of water masses from the surface to deep sea, vertical migrations of organisms from the epipelagic to mesopelagic zones, and the sinking of animal wastes and remains of surface organisms to the deep sea. PCBs are biomagnified up the food chain and have been found at high concentrations in fecal pellets that sink to the seafloor. The transport of persistent pollutants to the deep sea is much more rapid than initially predicted by scientists. In the open ocean, these compounds remain in near-surface waters for less than a year on average. Because organic matter is rarely transported from the deep seafloor to the surface, persistent chemicals tend to accumulate in sediments and organisms that reside there. The efficient transport of POPs throughout the marine environment has resulted in substantial levels of exposure, especially for predators, in most oceanic environments. For example, during the 1970s when DDT was still heavily used, concentrations near the current U.S. FDA tolerance level for human consumption were found in mesopelagic lantern fish and benthic hakes in the eastern Pacific, and in cod in nearshore and deep waters (about 2,000 meters) in the northwest Atlantic. Potentially harmful levels of PCBs have recently been found in mesopelagic fishes, including lantern fish and hatchet fish, in the Gulf of Mexico and the North Atlantic. As with other deep-sea impacts, it is not known how these pollutants have affected these environments or ecosystems.

Mercury and other metals behave similarly to POPs in the ocean environment (see Chapter 7). Because they tend to bioaccumulate through the food chain, deep-sea animals that feed on the remains of surface organisms tend to have elevated concentrations. There is a trend of increasing mercury levels as depth increases, even in the planktivores. Mesopelagic fishes tend to have mercury concentrations several times higher than surface fishes that feed at the same trophic level. Greater longevity also results in increased mercury accumulation. Long-lived deep-ocean fishes, such as orange roughy and grenadiers, thus tend to have elevated

mercury levels. Large deep-water predators are particularly prone to mercury accumulation; for example, deep-water sharks in the Mediterranean exhibited mercury concentrations of about five parts per million, ten times levels considered acceptable for human consumption.

Carbon Sequestration

The release of carbon dioxide (CO_2) through human activities, primarily from the burning of fossil fuels, is considered to be a primary cause of global warming and climate change (see Chapter 1). One suggested mechanism for minimizing those impacts is to sequester the carbon dioxide beneath the land or the deep sea (**Figure 8-25**). Experimental studies of the feasibility of injecting CO_2 into subsurface oil and gas reservoirs, coal beds, and deep subsurface saline formations are being carried out on land in many regions. Some European countries have begun sequestering by injecting CO_2 into saline formations associated with gas reservoirs beneath the seafloor or beneath underwater rock formations. An investigation into the microbiology of the deep-ocean crust by Olivia Mason and colleagues discovered microbial activity, including carbon fixation, thousands of meters deep into rock on the deep seafloor. This provides evidence that carbon dioxide could be sequestered permanently by pumping it into deep rock layers beneath the seafloor.

At temperatures below 11°C, CO_2 can be injected in liquid form, conditions that exist below 500 meters water depth in most of the deep sea. At these temperatures and pressures, carbon dioxide forms into gas hydrates (forms similar to the methane hydrates mentioned previously), with the CO_2 encased in crystal lattice of water molecules. At the higher pressures at greater depths, below 3,700 meters, the liquid carbon dioxide is denser than water and forms "lakes" in seafloor depressions. If the carbon dioxide eventually dissolves in the waters of the deep sea, it would likely remain there for as long as several hundred years due to the slow vertical ocean circulation.

Although sequestering carbon dioxide in the deep sea could help solve the problem of elevated atmospheric CO_2, it could create other issues, including increasing acidity in the deep sea. There is an ongoing debate as to whether we should take a chance on impacting the deep sea in order to possibly slow carbon dioxide induced climate change. The choices may not be so simple, however, because other effects of increased CO_2 in the deep sea could be more direct. For example, the CO_2 could stress or kill deep-sea animals in the area of release. If dead animals attract scavengers to the area, the impacts would be expanded. Environmentalists argue that rather than taking the risk of unknown impacts of carbon dioxide sequestration in the deep sea, the better option is to move toward alternative forms of energy production that do not produce excessive CO_2.

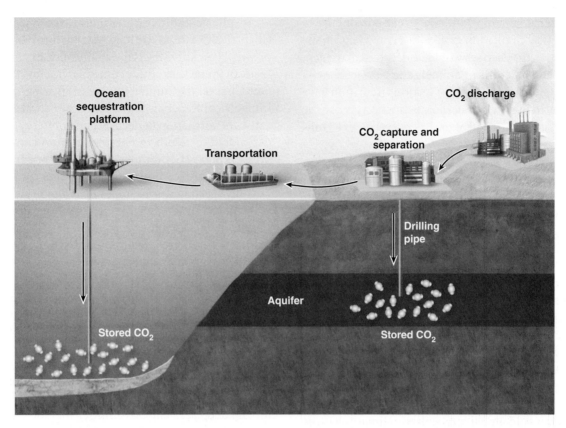

Figure 8-25 Various methods have been proposed for sequestering carbon beneath land or the deep seafloor to reduce levels of climate-changing carbon dioxide.

Another proposed method of transferring atmospheric carbon to the deep sea is through fertilization of surface waters (see Box 1-2, Conservation Controversy: Fertilizing the Ocean). The addition of iron has been shown to increase primary production in the photic zone; the sinking of dead plankton would in theory result in increased export of carbon to the deep sea. Even if iron fertilization were to be carried out on a scale large enough to successfully reduced atmospheric carbon and affect climate change, the amount and quality of carbon compounds sinking to the seafloor would be substantially altered. Quality could be changed because photosynthetic plankton (e.g., diatoms) have different nutrient concentrations than other sinking organic materials. The structure of the deep-sea ecosystem thus could be altered, favoring a different, and possibly less diverse, assemblage of organisms than is currently present. An increase in the quantity of organic matter sinking into the deep sea also could affect ecosystems by increasing the biomass of microbes, which could deplete the waters of oxygen and thereby produce dead zones similar to those found in coastal waters influenced by organic enrichment from river input (see Chapter 6). Debates over ocean fertilization reflect why we should proceed with caution when proposing simple solutions to the complex problems of global climate change.

Climate Change and the Deep Sea

Because the deep sea is physically further removed from atmospheric processes than any biome on Earth, one might predict that global climate change would have little effect on this apparently stable environment. Taking this view could be very misleading, however, because life in the deep sea, as discussed above, is largely dependent on physical and biological processes in the near-surface waters around the globe. There is strong evidence in the geologic record of the deep sea that past changes in the Earth's climate have resulted in dramatic changes in deep-sea life, including mass extinction events. Some of the potentially most influential physical effects of global climate change on the deep sea include changes in ocean temperature distributions and circulation patterns (see Chapter 1). Predicted effects of climate on primary production in epipelagic waters (see Chapter 7) would alter the transport of organic matter and nutrients upon which deep-sea life is largely dependent. Because of uncertainty concerning effects of climate change in shallower and near-surface ocean waters and how it may couple with ecosystems, any predictions are uncertain.

Potentially, one of the greatest impacts of climate change on the deep sea would be a disruption of thermohaline circulation (see Chapter 1, Box 1-1. Conservation

Concern: Global Warming and the Ocean). If sinking of waters in vertical circulation gyres is disrupted (e.g., in the North Atlantic), the deep sea would be deprived of oxygen, resulting in large die-offs of animals and a possible mass extinction event. Such changes in vertical circulation have been implicated in previous oceanic extinction events, the most recent being at the end of the Paleocene about 60 million years ago.

Change in production or abundance and distribution of organisms in the epipelagic could affect the function of deep-sea ecosystems. As discussed above, deep-sea organisms depend largely on the sinking of food material from the epipelagic zone in forms ranging from plankton and fecal matter to large pelagic predatory fishes and whales. Any of the biological changes discussed in Chapter 7 could be transferred to the deep sea because of these links. An increase in production in surface waters could result in an overall enhancement of biological activity and increase the transfer of carbon to the deep sea. The specific nature and distribution of effects would be impossible to accurately predict, however. Documentation of impacts as they occur would require more extensive studies of the deep sea to provide baseline data for comparison.

Not only would global atmospheric climate changes affect the deep sea, but deep-sea climate changes could affect the atmosphere. Frozen methane hydrates, as described above (see Figure 8-23), are dependent on cold temperatures as well as high pressures to maintain their stability in deep-sea sediments. An increase in temperature would change the depth at which these hydrates remain intact. For example, an increase in water temperature of 5°C would result in a change in the minimum depth of hydrate formation from the current 250 meters down to about 400 meters. Such a change would result in the destabilization of the hydrates between these depths, potentially releasing the methane gas to rise to the surface and into the atmosphere. Methane, considered one of the greenhouse gases, can trap about 20 times more heat in the atmosphere than carbon dioxide. A large influx of methane into the atmosphere thus would further enhance global warming, lowering the depth of methane hydrate stability even further and releasing more methane gas. The resulting domino effect could continue to increase atmospheric temperatures. The likelihood of this scenario developing with predicted changes in atmospheric and ocean temperatures is uncertain and hotly debated among scientists. Although many believe that the chances are remote, there is strong evidence that such events have occurred in the geologic past. For example, similar processes possibly contributed to mass extinction events at the end of the Paleozoic about 250 million year ago.

Increases in concentrations of CO_2 from anthropogenic or natural sources have effects beyond simply elevating atmospheric and ocean temperatures. As discussed in Chapter 2, CO_2 reacts with sea water to form acids, lowering ocean pH. Acidity affects the ability of stony corals to build reefs, an impact already documented in some tropical reefs (see Chapter 5). Similar, and possibly even more severe, impacts could occur with deep-water corals. This is because the formation of deep-water stony corals is strongly dependent on depth and ocean acidity. A combination of pressure and pH determines the solubility of aragonite (the form of calcium carbonate that comprises coral skeletons). Beyond a certain depth, aragonite is dissolved in sea water and thus stony corals will not form. An example of this effect can be seen under current ocean conditions. Deep-sea stony corals are largely absent from the North Pacific Ocean, presumably due to greater acidity resulting from lower circulation, oxygen depletion, and a buildup of CO_2. Extensive deep-water coral reefs are found in the North Atlantic in less acidic waters at comparable depths. Based on this evidence, it is predicted that a lowering of pH due to an anthropogenic increase in CO_2 in ocean waters would reduce the depth at which deep-sea corals can form.

The likelihood, degree, and impact of global climate change are extremely uncertain and this is especially true for the deep sea. The possibilities include warmer temperatures, slower circulation, declining oxygen levels, an accumulation of carbon dioxide and resultant elevation of pH, with potentially severe effects on deep-sea ecosystems. Despite the uncertainty, it would be risky to ignore the possibilities and assume everything will work out by itself. If humans take a precautionary approach and make efforts to control our carbon input into the atmosphere and oceans, we will not have to learn from experience which of the predictions are true.

8.6 Deep-Sea Conservation Measures

The deep sea, due to its large area and relative inaccessibility, is possibly the least impacted biome on Earth. For much the same reasons, however, less is known of the ecology of the deep sea than any other ecosystem. This lack of knowledge translates to a lack of awareness by the general public of the importance and potential harm of human actions to the deep-sea environment. Because much of the deep sea is in international waters long considered to be freely available for access and exploitation, it will take a cooperative effort of nations around the world to establish awareness and to develop protective measures. In many ways the deep sea is the final frontier on planet Earth. Hopefully, it will be treated with more respect than previous frontiers, which humans have typically exploited to near the breaking point before slowing down to look at what was lost.

Humans are at a turning point regarding conservation and the deep sea. We now have the capability to harvest fish below 1,500 meters and drill for oil or mine manganese nodules down to 3,000 meters. Because the average ocean depth is about 4,000 meters, humans can have a direct influence on much of the seafloor. Social, economic, and political factors will play a large role in determining how cautiously we proceed with potentially devastating actions. Even the deepest waters of the ocean are no longer immune to human influences. Manipulation of the atmosphere, coastal regions, and surface waters translates to largely unknown impacts in the deep sea. Further scientific studies are needed along with protection of the deepest regions of the ocean, so that we know what measures are needed before the deep sea begins to change for the worse.

The most successful method of protecting marine ecosystems in many regions has been the establishment of marine protected areas (MPAs). These mostly have been established in sensitive nearshore and coastal areas and tropical reef habitats. In recent years, however, there has been an increasing number of deep-sea areas afforded permanent protected status, primarily in regions with cold-water corals and hydrothermal vents. These protections typically include a prohibition of fishing, the activity currently most destructive to these ecosystems. Suggestions have been made that regions of seamounts and deep-water corals that already have been destroyed be designated as fishing areas, and that as much as possible of the remaining healthy habitat be focused on for protection. This is based on the premise that recovery of impacted habitats in the deep sea will be a long process due to the slow growth rate, late maturity, and variable recruitment of organisms residing there.

Little effort has been made to establish protections of regions of the deep sea dominated by abyssal plains; these regions are less impacted simply because they are mostly inaccessible. Legal protections of these ecosystems may not be as critical now; however, it is important to establish protections before incentives to harvest animals or remove minerals increase. Difficulties in achieving such protections include the need for international agreements throughout much of the range of these habitats, and the need to ensure that adequate areas are designated to protect deep-sea biodiversity. Questions concerning the needs of such reserves regarding size, numbers, and location are largely unanswered. Because individual organisms within populations are typically widely dispersed and often dependent on scattered resources, however, answering these questions is an important first step in designating deep-sea regions for protection.

As restrictions are established to protect deep-sea ecosystems and habitats, enforcement becomes a major issue. Enforcement is particularly problematic in the deep sea due to the large area involved and legal issues in international waters. These problems are characterized by valuable fisheries for species such as Patagonian toothfish, which commonly operate illegally with few repercussions, as discussed above. Some nations with deep-sea fisheries have begun to use observer systems, whereby fishers are required to report their positions via global positioning surveillance (GPS) on a regular basis. Landings are monitored upon return from the fishing trip to document the amount and types of fish harvested. Increasingly, the use of GPS and remote sensing are becoming practical ways to monitor deep-sea fishing activity. This does not eliminate illegal activities carried out covertly or with permission of fishing nations; however, strong international agreements would go a long way in increasing enforcement. Then, the question remains as to whether adequate international agreements can be established to achieve conservation goals. As will be discussed in subsequent chapters, there are reasons for both optimism and pessimism when it comes to international regulation and monitoring of marine resources.

STUDY GUIDE

▪ Topics for Review

1. Discuss the advantages and disadvantages of net sampling for collecting biological information on deep-sea organisms.
2. Why is the biomass in shelf and slope benthic habitats typically lower than in comparable habitats near shore?
3. How do Taylor cones enhance productivity and larval retention on seamounts?
4. Describe how vertically migrating plankton become trapped on seamounts.
5. What biological factors make seamount species more vulnerable to overharvest than other deep-sea species?
6. What feeding adaptations are typical of mesopelagic fishes for surviving in an environment with sparsely distributed food resources?
7. What factors make the density of life in the bathypelagic region more sparse than both the mesopelagic and the deep seafloor?
8. Describe how the source of energy differs among chemosynthetic ecosystems at hydrothermal vents, cold seeps, methane hydrates, and whale falls.
9. What survival mechanisms allow hydrothermal vent species to avoid extinction when the vents supporting their ecosystems expire?

10. Why are trawl net fisheries typically more harmful to deep-water ecosystems than hook-and-line or traps?

11. What factors led to the overharvest of fish species from seamount ecosystems before most government agencies were aware of the problem?

12. Explain why a great longevity and late age of reproduction makes conservation of deep-sea fishes more difficult.

13. Explain how excessive bottom fishing at depths to 1,500 meters may be affecting fish in deeper waters.

14. Explain how serial depletion has affected orange roughy populations on seamounts.

15. Why are crab fisheries less harmful to the seafloor environment than harvest of most bottom fishes?

16. Why are deep-water corals more prevalent in areas with relatively strong currents?

17. How would an elevation of ocean temperatures with global warming affect methane hydrate deposits in the deep seafloor?

18. What arguments are used in support of using the deep sea as a dumping ground for nuclear wastes?

19. Describe the mechanisms by which persistent organic pollutants reach deep-sea ecosystems.

Conservation Exercises

1. Describe how each of the following could impact seafloor or deep-sea ecosystems:
   a. disruption of thermohaline circulation
   b. mining of polymetallic sulfides from the seafloor
   c. mining methane hydrates
   d. pirate fishing in deep-sea waters
   e. deep-sea bottom trawling
   f. deepwater oil well blowouts
   g. release of oil drilling wastes on the seafloor
   h. excessive fishing in the epipelagic zone
   i. collecting manganese nodules
   j. dumping nuclear wastes
   k. dumping sewage sludge
   l. dumping dredge spoils
   m. sequestration of carbon in the deep sea
   n. adding iron to ocean surface waters to increase primary production
   o. reduction in pH of deep-sea waters

2. Describe conservation actions that have been taken to protect the following:
   a. deep-water coral reefs
   b. seafloor habitats around oil rigs
   c. deep-sea fisheries species
   d. seamount habitats
   e. whale fall ecosystems

FURTHER READING

Bailey, D. M., M. A. Collins, J. D. M. Gordon, A. F. Zuur, and I. G. Priede. 2009. Long-term changes in deep-water fish populations in the northeast Atlantic: a deeper reaching effect of fisheries? *Proceedings of the Royal Society B: Biological Sciences* 276:1965–1969.

Borowski, C. 2001. Physically disturbed deep-sea macrofauna in the Peru Basin, southeast Pacific, revisited 7 years after the experimental impact. *Deep-Sea Research II* 48:3809–3839.

Davies, A. J., J. M. Roberts, and J. Hall-Spencer. 2007. Preserving deep-sea natural heritage: Emerging issues in offshore conservation and management. *Biological Conservation* 138:299–312.

Etnoyer, P. J., J. Wood, and T. C. Shirley. 2010. How large is the seamount biome? *Oceanography* 23:206–209.

Fisher, C. R., I. R. MacDonald, R. Sassen, C. M. Young, S. A. Macko, S. Hourdez, R. S. Carney, S. Joye, and E. McMullin. 2000. Methane ice worms: *Hesiocaeca methanicola* colonizing fossil fuel reserves. *Naturwissenschaften* 87:184–187.

Genen, A., and J. F. Dower. 2007. Seamount plankton dynamics. Chapter 5. In T. J. Pitcher, T. Morato, P. J. B. Hart, M. R. Clark, N. Haggan, and R. S. Santos (eds.). *Seamounts: Ecology, Fisheries and Conservation.* Wiley-Blackwell Publishing, Ames, IA.

George, R.Y., and S.D. Cairns (eds.). 2007. *Conservation and Adaptive Management of Seamounts and Deep-Sea Coral Ecosystems.* Rosenstiel School of Marine and Atmospheric Science, University of Miami, Miami, FL.

Glover, A.G., and C.R. Smith. 2003. The deep-seafloor ecosystem: current status and prospects of anthropogenic change by the year 2025. *Environmental Conservation* 30:219–241.

Joye, S. B., I. Leifer, I. R. MacDonald, J. P. Chanton, C. D. Meile, A. P. Teske, J. E. Kostka, L. Chistoserdove,

R. Coffin, D. Hollander, M. Kastner, J. P. Montoya, G. Rehder, E. Soomon, T. Treude, and T. A. Villareal. 2011. Comment on "A persistent oxygen anomaly reveals the fate of spilled methane in the deep Gulf of Mexico." *Science* 332:1033.

Kessler, J. D., D. L. Valentine, M. C. Redmond, and M. Du. 2011. Response to comment on "A persistent oxygen anomaly reveals the fate of spilled methane in the deep Gulf of Mexico." *Science* 332:1033.

Kessler, J. D., D. L. Valentine, M. C. Redmond, M. Du, E. W. Chan, S. D. Mendes, E. W. Quiroz, C. J. Villanueva, S. S. Shusta, L. M. Werra, S. A. Yvon-Lewis, and T. C. Weber. 2011. A persistent oxygen anomaly reveals the fate of spilled methane in the deep Gulf of Mexico. *Science* 331:312–315.

Koslow, T. 2007. *The Silent Deep: The Discovery, Ecology and Conservation of the Deep Sea.* The University of Chicago Press, Chicago.

Kitchingman, A., S. Lai, T. Morato, and D. Pauly. 2007. How many seamounts are there and where are they located. Chapter 2. In T. J. Pitcher, T. Morato, P. J. B. Hart, M. R. Clark, N. Haggan, and R. S. Santos (eds.). *Seamounts: Ecology, Fisheries and Conservation.* Wiley-Blackwell Publishing, Ames, IA.

Kujawinski, E. B., M. C. K. Soule, D. L. Valentine, A. K. Boysen, K. Longnecker, and M. C. Redmond. 2011. *Environmental Science and Technology* 45:1298–1306.

Le Loc'h, F., C. Hilly, and J. Grall. 2008. Benthic community and food web structure on the continental shelf of the Bay of Biscay (North Eastern Atlantic) revealed by stable isotopes analysis. *Journal of Marine Systems* 72:17–34.

Mason, O. U., T. Nakagawa, M. Rosner, J. D. Van Nostrand, J. Zhou, A. Maruyama, M. R. Fisk, and S. J. Giovannoni. 2010. First investigation of the microbiology of the deepest layer of ocean crust. *PLoS ONE* 5(11):e15399. doi:10.1371/journal.one.0015399.

Morato, T., and M. R. Clark. 2007. Seamount fishes: ecology and life histories. In T. J. Pitcher, T. Morato, P. J. B. Hart, M. R. Clark, N. Haggan, and R. S. Santos (eds.). *Seamounts: Ecology, Fisheries and Conservation.* Wiley-Blackwell Publishing, Ames, IA.

Morato, T., R. Watson, T. J. Pitcher, and D. Pauly (editors). 2006. Fishing down the deep. *Fish and Fisheries* 7:23–33.

Morgan, L. E., C.-F. Tsao, and J. M. Guinotte. 2007. Ecosystem-based management as a tool for protecting deep-sea corals in the USA. *Bulletin of Marine Science* 81:39–48.

Mullineaux, L. S., and S. W. Mills 1997. A test of the larval retention hypothesis in seamount-generated flows. *Deep-Sea Research* 44:745–770.

Norse, E. A., S. Brooke, W. W. L. Cheung, M. R. Clark, I. Ekeland, R. Froese, K. M. Gjerde, R. L. Haedrich, S. S. Heppel, T. Morato, L. E. Morgan, D. Pauly, R. Sumaila, and R. Watson. 2012. Sustainability of deep-sea fisheries. *Marine Policy* 36:307–320.

Pitcher, T. J., T. Morato, P. J. B. Hart, M. R. Clark, N. Haggan, and R. S. Santos (eds.). 2007. *Seamounts: Ecology, Fisheries and Conservation.* Blackwell Publishing, Ames, IA.

Probert, P. K. 1999. Seamounts, sanctuaries and sustainability: moving towards deep-sea conservation. *Aquatic Conservation: Marine and Freshwater Ecosystems* 9:601–605.

Rogers, A. D., A. Baco, H. Griffiths, T. Hart, and J. M. Hall-Spencer (eds.). 2007. *Seamounts: Ecology, Fisheries and Conservation.* Wiley-Blackwell Publishing, Ames, IA.

Smith, C. R., and A. R. Baco. 2003. Ecology of whale falls at the deep-sea floor. *Oceanography and Marine Biology: an Annual Review* 41:311–354.

Suchanek, T. H., M. C. Lugunas-Solar, O. G. Raabe, R. C. Helm, F. Gielow, N. Peek, and O. Carvacho. 1996. Radionuclides in fishes and mussels from the Farallon Islands Nuclear Waste Dump Site, California. *Health Physics* 71:167–178.

Smith, C. R., F. C. De Leo, A. F. Bernardino, A. K. Sweetman, and P. M. Arbizu. 2008. Abyssal food limitation, ecosystem structure and climate change. *Trends in Ecology and Evolution* 23:518–528.

Vilcaez, J., L. Li, S. Hubbard, and T. Hazen. 2010. *Biodegradation of deep-sea oil spill at the Gulf of Mexico: an estimate of half life time.* Abstract; presented at *American Geophysical Union,* Fall Meeting; December 13–17, 2010: San Francisco.

Wahle, R. A., C. E. Bergeron, A. S. Chute, L. D. Jacobson, and Y. Chen. 2008. The Northwest Atlantic deep-sea red crab (*Chaceon quinquedens*) population before and after the onset of harvesting. *ICES Journal of Marine Science* 65:862–872.

White, M., I. Bashmachnikov, J. Aristegui, and A. Martins. 2007. In T. J. Pitcher, T. Morato, P. J. B. Hart, M. R. Clark, N. Haggan, and R. S. Santos (eds.). *Seamounts: Ecology, Fisheries and Conservation.* Wiley-Blackwell Publishing, Ames, IA.

Marine Endangered Species: Conservation, Protection, and Recovery

The term *endangered* is given to species or populations considered in some way imperiled with extinction. Although extinctions have occurred on Earth as long as there has been life, the current rate of extinctions is much higher than that which would occur because of natural processes (that is, without the influence of human activities). Extinctions have been relatively rare for marine species compared to terrestrial and freshwater organisms in recent times. One reason for the rarity of documented extinctions is a lack of monitoring in the marine environment. Also, human impacts have had less direct influence on the marine environment than terrestrial and freshwater environments because of lower accessibility. Finally, many marine species have broader distributions than terrestrial species; therefore, even if a species is driven to extinction locally it is more likely that populations of the same species survive elsewhere. If factors causing the local extinction event are corrected, then recolonization could reasonably be expected. This is not to minimize the endangerment of marine species. With increased human access and influence on the oceans the number of marine species considered endangered has increased rapidly.

The marine species historically most vulnerable to the risk of extinction are often those that use shore habitats during some portion of their life, because this is where habitat loss and human exploitation are most likely. These include marine mammals, such as seals and sea otters, that come ashore for reproduction; seabirds, all of which must at least come to shore to roost and lay eggs; and sea turtles, which must come onto beaches to lay eggs and rear young (**Figure 9-1**). Even though protection of coastal habitats and endangered species has increased globally, we must now deal with a new set of impacts that are even more difficult to address, including climate change, pollution, and other activities that cause unintended harm or death of species vulnerable to endangerment.

9.1 Legal Protection

Though many countries grant legal protection to species considered at risk of extinction, protections are not uniform around the world. Where legal mechanisms are in place, many factors are weighed to decide whether a species is sufficiently endangered or worth the costs required to provide adequate protection. These include not only biological factors but also economic, political, and social considerations. For example, there may be financial losses for a commercially harvested species or economic costs of habitat protection that are considered too great to tolerate. Once a species is classified as endangered, the amount of legal protection afforded varies widely among nations. International agreements help to provide some common standards; however, enforcement can be lax or nonexistent, and not all nations agree to participate in agreements.

Figure 9-1 An endangered Kemp's ridley sea turtle returns to the Gulf of Mexico after nesting on Padre Island, Texas.

Variability in conservation attitudes and ethical rights of animals around the world has resulted in disagreements over how much protection should be given, especially to marine mammal species that are not currently biologically endangered.

■ National Endangered Species Legislation

An increased awareness of environmental issues and species loss in the United States in the 1960s and 1970s led to pressure from citizens and environmental groups, resulting in passage of the United States Endangered Species Act (U.S. ESA) by Congress in 1973. This Act, and subsequent amendments, establishes laws to protect imperiled species from extinction. This could include any species of plant or animal residing in the United States, its territories, or waters. There also are mechanisms for including marine species outside of U.S. waters to allow protections from trade and harmful actions approved by U.S. government agencies or carried out by U.S. citizens.

Any U.S. citizen or group can recommend, with supporting evidence, to the United States Fish and Wildlife Service (USFWS) or the National Marine Fisheries Service (NMFS) that a species be *listed* or considered for endangered species status. The USFWS deals with terrestrial or freshwater organisms, including mammals typically classified as marine, such as walruses, polar bears, seals, sea otters, and manatees. The NMFS handles most other marine and anadromous organisms. These agencies are charged with reviewing the evidence and making a decision as to the status of the species. The criteria considered in making that determination include loss of critical habitat, over-utilization (e.g., excessive harvest), declines from disease or predation, inadequate regulations for protection, and any other factor affecting its continued existence. Species considered for listing are classified as *candidate* species. The

ESA gives a broad definition to the term *species,* allowing the inclusion of not only a true taxonomic species, but also subspecies, and "distinct population segments."

A petition to consider a species for listing must provide extensive evidence for the USFWS or NMFS to evaluate. The listing process is extensive, providing time for scientific evaluation and public review and comments. A final decision is not made until at least a year, often much longer, after a species is proposed for listing. If a species is eventually listed, it is assigned to one of several categories: an *endangered* species classification is given if the species is "in danger of extinction throughout all or a significant portion of its range." If a species is considered "likely to become an endangered species within the foreseeable future throughout all or a significant portion of its range" it is classified as *threatened.* A *species of concern* is one for which there are concerns about its status, but there is insufficient information to indicate a need for further listing.

The goal of the Endangered Species Act is to provide protection and means to conserve endangered and threatened species and their ecosystems. Habitat critical to the survival of threatened and endangered species must be defined, and federal agencies cannot authorize or carry out activities that destroy or adversely modify this habitat. A recovery plan is to be created by the USFWS or NMFS and implemented for all endangered species; however, because of funding and backlogs, it takes on average about six years to get a recovery plan in place for a listed species (though sometimes species are given priority to have the listing process accelerated). For species classified as endangered, the recovery plan defines the criteria that must be met in order for the species to be *downlisted* to threatened. For threatened species, criteria are specified for *delisting* or removal from the List of Endangered and Threatened Wildlife.

The U.S. Endangered Species Act establishes laws specifically prohibiting *taking* of species classified as endangered. The legal definition of taking is broad, considered as "to harass, harm, pursue, hunt, shoot, wound, kill, trap, capture, or collect, or to attempt to engage in any such conduct." Because much taking of marine species is unintentional, stipulations are given for the incidental take of endangered species. For example, permits are to be obtained by commercial fishers if they expect to incidentally take endangered species while fishing.

Marine species protected under the ESA include all six U.S. species of sea turtles. Three of these are endangered throughout their range, but all have populations classified as endangered. About 20 marine mammals are listed. Birds listed include eight endangered albatross populations and two gulls. Although there are no penguins in U.S. territory, the Galapagos penguin is listed as endangered and several are being considered for listing. Some anadromous fishes, including eight populations of salmon and sturgeons, have been listed. There are only two entirely marine fish species in U.S. waters listed as endangered, although fish species in other regions can be listed. The only marine invertebrates listed as threatened by the U.S. ESA are elkhorn and staghorn corals (see Chapter 5). The only marine plant listed is Johnson's seagrass *Halophila johnsonii*, found along the east coast of Florida.

Although many marine mammals are included on national and international endangered species lists, the United States provides additional considerations for protection of marine mammals, regardless of their endangered status, through the Marine Mammal Protection Act (MMPA). This Act was passed in 1972 in response to the outcry of citizen groups opposed to the killing or harming of marine mammals. The stated need for special protection included the depletions and endangered status of some populations as a result of human activities, the lack of adequate knowledge of the ecology and population dynamics, and the observation that marine mammals have "proven themselves to be resources of great international significance, aesthetic and recreational as well as economic." Although other endangered species might satisfy the first two of these criteria, because of the perceived high intelligence and emotional appeal, marine mammals are singled out by the United States to be worthy of the most strict protection of any animal group, even when populations are healthy. A depleted classification is given for 29 marine mammals, indicating that the species is "below its optimum sustainable population." Representatives of three mammal orders include marine species (though the term *marine mammal* is an ecological rather than phylogenetic grouping). This chapter considers the sirenians (e.g., manatees and dugongs) and marine carnivorans (e.g., pinnipeds and sea otters). Chapter 2 discusses the conservation issue for polar bears. The cetaceans, including dolphins and whales, receive special protections through the MMPA and international laws, and are addressed in more detail in Chapter 10.

Other nations have developed endangered species protections through national legislation, often modeled after the U.S. ESA. In Canada, the major government legislation protecting endangered species is the Species at Risk Act, passed in 2002. Legal protections are given and recovery plans are developed similar to those in the U.S. ESA. The listing process differs in that it is a two-step process. An independent scientific advisory committee assesses the status of species at risk; then the federal government decides whether to accept the assessment and add the species to the list of protected species. For species denied legal listing, no specific legal protections are given and no recovery actions are taken. Environmental and socioeconomic tradeoffs are considered in developing an action plan for recovery of species considered at risk. Based on an analysis of Canadian endangered species listing decisions, Arne Mooers and colleagues proposed that government agencies made discretionary decisions based too heavily on economics rather than on science, and that endangered protections are needed for more marine species such as the Atlantic cod, affected by overfishing (see Chapter 11), and beluga whales, affected by coastal pollution (see Chapter 10). Special considerations are given to marine mammals under the Marine Mammal Regulations of the Fisheries Act of Canada; for example, pinnipeds cannot be hunted or disturbed except for what is considered subsistence harvest.

The Australian Endangered Species Protection Act (ESPA) was implemented in 1992. Although it only includes protection of species on federally managed lands and waters and activities by or permitted by federal agencies, this would apply to most marine habitats. The ESPA allows for consideration of species protection and for protecting communities or eliminating practices that are considered threatening. Nominations for listing can be made by the public, to be assessed by a scientific committee, with the ultimate decision for listing made by the Environment Minister. Social or economic considerations are not supposed to affect the decision for listing, but may be considered in developing recovery plans. The inclusion of threatened communities for consideration provides a mechanism not available through the U.S. ESA that could, in theory, provide specific considerations of marine communities such as coral reefs, seagrasses, or mangroves; however, community listings have been rarely approved. Some harmful practices have been eliminated for seabird conservation through the ESPA. Of the 20 albatross species in Australian waters, 17 are listed as threatened, mostly because of incidental catch on longlines (see below). After longline fishing was listed as a threatening process, better fishing methods were instituted to limit the incidental catch. Broader legislation was passed in 1999 as the Environment Protection and Biodiversity Conservation Act (EPBC Act) that encompassed protection of the biodiversity of Australia as well as assessment and protection of the environment. Amendments to the EPBC Act included the establishment of the Great Barrier Reef Marine Park (see Chapter 5).

Several conservation agreements have been developed to address conservation and biodiversity issues in the European Union (EU), though there is no agreement limited exclusively to endangered species. Protection of biodiversity is one goal of the European Union's 6th Environment Action Programme. The EU Biodiversity Action Plan established a goal in 2002 to stop the loss of biodiversity within the EU by 2010. A Habitats Directive was implemented for environmental protection. Much of the protection was to be achieved in part by the establishment of the Natura 2000 network to protect threatened species and habitats, including as a marine network of Special Protection Areas. A second directive of legislation in the Habitats Directive was to establish strict protection of animal species in defined communities. The Directive requires that EU nations provide strict protection of all cetaceans in European waters.

International Agreements

Many nations work with international organizations to establish endangered species protections. Endangered species lists of the United States and other nations include some marine species not found in waters under their jurisdiction to assist with management and protection of those species, for example, to regulate citizens fishing in international waters or to restrict trade laws. There is no international endangered species act, however, nor is there any international government entity in existence that would have enforcement powers for protecting endangered species. It is up to individual countries to protect species within their marine waters and work within established international agreements.

Many nations, particularly smaller developing countries, do not have strong endangered species laws or the ability to effectively enforce laws that are in place. International agreements can provide the incentive and some support for protection of marine species both in national and international waters. An unfortunate aspect of international agreements is the general lack of enforcement capabilities. Often protection is forced by threats or the imposition of sanctions. Nations, however, cannot be forced to join international organizations or sign international environmental agreements, and there may be no mechanism to impose protective measures for at-risk species. Therefore, where local laws are not in place or enforced, many species have been greatly affected; however, the desire to participate in the global economy has provided an incentive for nations to increase species protection measures to gain cooperation of larger industrialized nations with a stronger conservation ethic.

Convention for International Trade in Endangered Species (CITES)

Despite the imperfections of international environmental agreements, several have had a strong positive influence on protecting biodiversity. One of the most successful in dealing specifically with species at risk of extinction is the Convention for International Trade in Endangered Species (CITES). This international agreement among governments was established in 1974 to ensure that plants and animals are not endangered with extinction because of trade in specimens. This trade can be live animals or plants, but it also can include food products, tourist curios, furs, animal skins or other parts. Over 170 nations are Parties (members) to CITES and agree to restrict or limit trade of designated species. This can be a very effective conservation measure. Because member nations, even those without strong conservation measures, cannot profit from international trade in species considered endangered by CITES, it limits their harvest, whether carried out through legal or illegal means. Even non-member nations may find it difficult to locate an international market for listed species. As with any international agreement, membership is voluntary; however, agreements are legally binding among member nations. The major difficulty with implementing laws established by CITES agreements is often with enforcement. The most effective enforcement can be the threat of a trade embargo by one member country on another if CITES restrictions are not followed. (However, this can conflict with international trade laws; see Chapter 12.)

CITES indicates the degree of protection needed by grouping species into Appendices. Appendix I includes species considered threatened with extinction. For these, trade is prohibited except in exceptional circumstances. Appendix II species are not necessarily threatened with extinction, but trade is restricted to avoid uses that would affect their survival. In order to add species to the list, recommendations are voted on by all member nations at meetings every 2½ years. Appendix III contains species protected by a member nation, so that other members can assist with restricting trade in that species.

Over 5,000 animal species and 28,000 plant species are protected from international trade by CITES; about 600 of these are in Appendix I. The majority of the listed species are terrestrial and freshwater species; however, numerous marine species are protected. For some, an entire animal group is protected, including cetaceans, sea turtles, and corals. As a result of the coral group listing under Appendix II, the cnidarians are the largest protected group, with over 2,000 species listed. The listing of the large whales has limited harvest by whaling nations, because it takes away the international market for whale products (see Chapter 10). CITES protection for sea turtles has substantially reduced trade in turtle shell products. Other listed groups include whale sharks, white sharks, basking sharks, and seahorses. Multiple efforts have been made, through pressure by environmental groups, to list marine species on Appendix I that are affected by overfishing and other environmental impacts. For example, in 2010, proposals were unsuccessful to list 6 shark species, over 30 types of coral, and the Atlantic bluefin tuna (see Chapter 11).

International Union for Conservation of Nature (IUCN)

The International Union for Conservation of Nature (IUCN; also known as The World Conservation Union) is an international organization formed in the 1940s, comprised of hundreds of nations, government agencies, and non-governmental organizations (NGOs). Its mission is to "influence, encourage, and assist societies throughout the world to conserve the integrity and diversity of nature." Their central mission is stated as conserving biodiversity. The IUCN Red List of Threatened Species is considered a standard reference for species at risk of extinction, providing a catalog of the conservation status of species. Species are classified based on data provided by government, NGOs, and other scientific studies. The IUCN works closely with other organizations. For example, Birdlife International is designated as the authority for seabirds and provides categorizations and documentations for the Red List. Categories include *critically endangered, endangered,* and *vulnerable.* Species for which adequate data are not currently available to make a confident assessment are given a *data deficient* designation. The list also includes information on major habitats, threats, and conservation needs for species. The IUCN supports projects dealing with sustaining biodiversity, and can provide guidance in developing national species protection, conservation, and management plans. An evaluation by Ana Rodrigues and colleagues indicates that the Red List, along with the supporting data that have been compiled, has become a powerful tool for international conservation. There are some concerns, however, with the mechanism for assigning marine species to the Red List, because broad distributions can lead to widely varying conservation status among regions. This presents a dilemma when classifying species such as sea turtles that have a near-global distribution.

Although endangered species laws continue to function as some of the most powerful conservation laws available, with the new focus on ecosystem-based conservation, many nations and international organizations are focusing conservation efforts more heavily on protecting the habitats that support biodiversity. Rather than using the species as a mechanism to protect the ecosystem, the philosophy is to protect the ecosystem in order to save the species. More ecosystem-based organizations are discussed along with other legal efforts in conservation in Chapter 12. Regardless of changes in the philosophy of conservation, there always will be a need for species-based conservation, and species will continue to serve a useful purpose as an indicator of ecosystem health. Specific case histories are provided in the following sections to help in understanding where there have been failures and successes in conservation of marine species prone to endangerment. The examples are chosen to illustrate methods applied to recover species that have become endangered and protect species sensitive to endangerment.

9.2 Endangered Species Case Histories

Marine Invertebrates

Relative to their diversity, few marine invertebrates are classified as endangered. This is in part because of high reproductive rates, broad distributions, and high levels of tolerance for many invertebrate species. It is also likely a result of poor availability of data and focus of endangered species protections on larger, so called "charismatic megafauna" (e.g., marine mammals and sea turtles). One marine group that is relatively vulnerable to endangerment is the mollusks, including bivalves and limpets, which are relatively accessible for harvest in coastal and intertidal areas (see Chapter 3). The black abalone *Haliotis cracherodii* and white abalone *H. sorenseni* are large mollusks found in rocky intertidal areas of southwest California and the Baja peninsula, endangered from disease and overharvest in the 1970s and 1980s (**Figure 9-2**). Their recovery is being inhibited by increasing distance between male and females for spawning because of low population densities, illegal harvest, loss of habitat, and natural predation by sea otters, crabs, and urchins. Protection measures include the establishment of Marine Protected Areas in California and prohibiting harvest. Success of protected areas is indicated by larger abalone sizes in areas where human access is limited.

The invertebrate groups most widely classified as imperiled are the tropical reef-building corals. The numerous anthropogenic and environmental factors affecting these species makes it difficult to define cause-and-effect relationships for declines (see Chapter 5). Of about 850 coral species assessed by IUCN, about 25% are considered by IUCN as threatened; another 20% are listed as near threatened and 17% are data deficient. Two historically common corals in the Caribbean, the elkhorn coral *Acropora palmata* and staghorn coral *A. cervicornis* (**Figure 9-3**), are IUCN critically endangered; these are the only corals listed as threatened by the U.S. ESA. Recovery of these

Figure 9-2   Diseased (left) and healthy (right) black abalone.

(a)

(b)

Figure 9-3 Historically common branching corals, **(a)** elkhorn coral and **(b)** staghorn coral, have undergone precipitous declines in the Caribbean and are listed as critically endangered by the IUCN.

populations will be difficult because multiple factors will need to be addressed, including coral bleaching, pollution, disease, fish predation, physical damage from fishing gear, and factors related to climate change (see Chapter 5).

Seabirds

The IUCN estimates that over 100 of 328 recognized species of seabirds are threatened or endangered with extinction. As discussed in previous chapters, threats include effects of bioaccumulation of persistent pollutants and loss of nesting and roosting habitats on shorelines, in mangroves, and on dunes. Where habitat protections have been established, seabird populations sometimes show signs of recovery but can continue to be affected by other impacts.

One of the main advantages of nesting on small islands is the avoidance of predators. Because of their isolation, smaller islands tend to be colonized by fewer mammalian and reptilian predators that could prey on eggs, chicks, or adults. The intentional and unintentional introduction of exotic predators to islands, however, can cause devastating

harm to species that are not adapted to avoid predation. Seabirds that have evolved with predators, such as native mammals or terrestrial crabs, are less likely to be affected by exotic introductions. Many seabirds nest on the ground or in burrows where they are readily accessible to predators. Invasive species are considered to be one of the largest terrestrial threats to breeding colonies of seabirds, and the most influential group of invasive species is rats.

Rats have likely been unintentionally transported to islands for as long as there have been seafaring human populations. Earlier explorers accidentally caused the extermination of many populations of island animals when rats from their ships came ashore and successfully colonized the island, and ships continue to transport rats around the world. Recent studies have suggested that about 25% of all seabird species are preyed upon by invasive rodents, typically on the eggs and young in nesting colonies; and there are at least 10 documented extirpations of seabird populations following the introductions of rats. An analysis by Holly Jones and colleagues found that the most common rodent invaders are the three prevalent *Rattus* species native to Europe and Asia, now established on 90% of the world's large islands and island chains. The species of seabirds most heavily harmed are those that reach relatively small sizes and nest in burrows or crevices. Two of the most affected are the auks (Alcidae) and storm-petrels (Hydrobatidae); in some regions storm petrels are found only on rat-free islands. Larger birds that nest on the ground or in trees, such as gulls, frigatebirds, and albatrosses, are less likely to be affected.

The only long-term solution to endangerment of vulnerable seabird populations by rat predation is total eradication, typically involving a broad application of biodegradable rat poisons. In New Zealand, with no native rodents, rats were first introduced by human explorers over a millennium ago. By the mid-1980s they were present on over 140 New Zealand islands. Eradication programs were initiated in the 1960s and have expanded to include more than 100 islands. At least some of these programs were successful in increasing numbers of fledgling in small burrow-nesting seabirds. More recently, eradication programs have been carried out on other small islands, including the Channel Islands off California. In the Aleutian Islands of Alaska, government and NGOs have been cooperating to remove rats. One of first islands to be targeted was the 2,700 hectare Rat Island (**Figure 9-4**), rat infested since 1780 after a Japanese shipwreck. Poison was dropped from helicopters over a week in 2007. By 2009 the island was rat-free, with no apparent long-term harm to native species, and there is evidence that nesting bird populations are recovering. These intensive efforts can be controversial, especially with animal rights groups or when harmful effects of rat invasions have not been thoroughly documented. For the Rat Island

Figure 9-4   Rat Island, in the Alaskan Aleutians. Norway rats, introduced by a shipwreck in the 1780s, plagued the island until 2009 when they were removed by a rat eradication program as part of efforts to restore seabird habitat to the island.

eradication, reports of the deaths of over 400 birds as a result of the release of excessive poison has been controversial. Arguably, however, the long-term benefit to seabird populations outweighs the risks of intensive application of poisons.

Another source of endangerment to seabird populations is from incidental catch, or **bycatch**, by fishing gears. The most harmful to seabirds is longline fishing, where a heavy fishing line is drifted at the sea surface with a series of short lines and baited hooks attached. These are typically fished in open-ocean waters with thousands of hooks attached to a single line. Longlines are typically used to target large migratory predators, such as tuna, billfish, and sharks (see Chapter 11) but can incidentally entangle seabirds. Longline fishing is considered a serious global threat to some seabird populations, especially albatrosses and large petrels (**Figure 9-5**). The IUCN identifies about 60 species of seabirds affected by longline fisheries; almost half are considered threatened with extinction. Seabirds are attracted to longline bait, which is typically some type of small fish. As

the baited hook is placed into the water from the boat, it is accessible to seabirds that get hooked or entangled as they dive for the bait. As the weighted longline sinks below the surface, the bird is carried down with it and drowns. Seabirds flying at surface also can be caught as baited hooks are brought back on board the vessel. Hundreds of thousands of seabirds are killed by longline fisheries each year, tens of thousands of which are albatrosses.

Numerous methods have been developed to minimize the bycatch of seabirds with longlines. These include avoiding times and areas when seabirds forage (e.g., by fishing at night), dying the bait blue to make it more difficult to see against the water, or shielding boat lights to make the bait less visible. Access to the baited hooks can be limited by setting the gear through chutes or along the side of the boat (rather than over the stern), thawing the bait to make it less buoyant, or adding additional weight to sink the line quicker. Bird-scaring devices include noise-making devices and streamer lines (**Figure 9-6**). Improved bird handling techniques and

Figure 9-5   Seabirds are vulnerable to bycatch on longlines. Laysan albatross *Diomedea immutabilis,* classified as vulnerable by the IUCN, caught on a baited fishing hook.

Figure 9-6   Red streamers on a fishing longline to scare birds and minimize seabird bycatch.

laws requiring the release of live birds can reduce the mortality of hooked birds. One or more of these methods may be appropriate for different fisheries because of differences in fishing methods and bird behaviors; thorough studies are needed to develop the best method for a given fishery. However, the ease of application and low expense of these methods should make them an appealing method for protecting seabirds.

Some nations, including Japan, the United States, Canada, Australia, and New Zealand, have implemented regulations to restrict seabird bycatch, especially in cases where it involves imperiled species. But because longlines are often fished in international waters, multinational agreements are needed to solve the seabird bycatch problem. Several agreements are in place, each with its own limitations. The Food and Agriculture Organization (FAO) of the United Nations developed an *International Plan of Action for Reducing the Incidental Catch of Seabirds in Longline Fisheries* in 1999, which requires nations who sign on to assess longline fisheries and develop a National Plan of Action to limit seabird bycatch. The U.S. Endangered Species Act provides specific protections to seabirds classified as endangered or threatened. The Migratory Bird Treaty Act (MBTA) implements treaties between the United States, Great Britain, Japan, and the former USSR to provide protection to any migratory bird, which would include all seabirds found in the open oceans; however, it is debated whether this agreement extends into international waters. The USFWS has not been willing to enforce the MBTA in cases of bycatch mortality of seabirds in open-ocean fisheries; however, bycatch reduction programs in federal fisheries management legislation establish legal methods by which bycatch could be reduced.

Although protected areas are rarely established solely for protection of seabirds, some include restrictions to minimize seabird bycatch, including limitations on longline fishing. For example, most of the world's populations of Laysan albatross *Phoebastria immutabilis* and black-footed albatross *P. nigripes* nest in regions of the northwest Hawaiian Islands protected as a marine sanctuary. Establishing protected areas in international waters of the open ocean may not be currently feasible because of the difficulty in defining protected areas and enforcing protections. International agreements would have to be reached, requiring a consensus among fishing nations—something that has been difficult to achieve (see Chapter 12). Despite the difficulties of designating protected areas for seabirds, incidental catch can result in fisheries restrictions for seabird protection. For example, even low levels of bycatch of the U.S. endangered short-tailed albatross *Phoebastria albatrus* in the North Pacific can result in fisheries closures. In the Alaskan trawl fishery, if two individuals are killed during a five-year period, a review process would be initiated that could result in shutting down the fishery.

Diving seabirds can be caught and killed incidentally in gill nets and drift nets, long stationary net panels fished beneath the surface (see Chapter 11). This bycatch is mostly in coastal waters, a result of outlawing of most open-ocean drift net fishing, and has not been well documented. It likely does not affect endangered seabird species but may kill thousands of birds each year in U.S. waters alone, with unknown numbers of seabirds killed worldwide. The reduced use of gill nets in marine waters around the world in response to bycatch issues for sea turtles, marine mammals, and fishes has likely reduced seabird bycatch.

Implementing measures globally to totally eliminate bycatch of seabirds in fisheries is unlikely because of economic, political, and legal issues. If effective, economically viable, and commercially practical methods are developed for reducing bycatch in the most harmful fisheries, however, encouraging or requiring their use may be realistic. Placing observers on vessels is the most efficient way to monitor success and enforce regulations. Adequate funding to thoroughly monitor these fisheries, however, is unlikely in the near future and is not always legally possible. For example, the U.S. National Marine Fisheries Service currently does not have the authority to put observers on vessels for the sole purpose of monitoring seabird bycatch other than species listed under the Endangered Species Act. Few observer programs are now in place around the world for monitoring effects of longlining on seabirds. Progress clearly has been made to protect seabirds from incidental death because of human activities. Scientific knowledge has been gained to define measures capable of almost entirely eliminating seabird bycatch. Pressure needs to be applied to encourage further implementation and enforcement of adequate national and international laws and agreements. Encouraging voluntary compliance with regulations by providing economic incentives could result in reductions in bycatch. Establishing eco-labeling programs have been successful in other fisheries to reduce bycatch, such as dolphin bycatch in the tuna fishery (**Box 9-1. Assessing Impacts on Endangered Species: Counting Fishery Bycatch**), and could be successful to reduce seabird bycatch.

■ Penguins

Penguins are concentrated in the Antarctic and other cold-water regions of the Southern Oceans where historically they have been less impacted by issues of habitat loss and human exploitation than other marine birds and mammals. They are vulnerable to other anthropogenic effects, however, including global warming, ecosystem disruptions, and loss of food resources. Other threats include habitat destruction, disturbance at breeding colonies, oil spills,

Box 9-1   Assessing Impacts on Endangered Species: Counting Fishery Bycatch

One of the most controversial seabird conservation issues in U.S. waters has been albatross bycatch in the north Pacific in longline fisheries for tuna and swordfish. In the Hawaii-based U.S. fishery, over 800 Laysan albatrosses, IUCN vulnerable (see Figure 9-5), and about 2,000 black-footed albatross (**Figure B9-1**), IUCN endangered, were caught and killed per year between 1991 and 2000. These data are considered relatively reliable because they are based on NMFS data from observer programs and logs required of participants in the fishery; however, this fishery accounts for less than 3% of the longline fishing effort in the North Pacific. As much of the fishery is in international waters, other nations, primarily Japan and Taiwan, also have a longline fishing fleet; these account for a much larger proportion of the fishing effort, and presumably a larger seabird bycatch. How to estimate the total effect on albatross populations presents a dilemma because the only data on fishing effort (number of boats and time spent fishing) are available for the international fishery.

Assessment methods were developed by Rebecca Lewison and Larry Crowder to work around these limitations. They used NMFS observer and fishery log data for the U.S. longline fishery based in Hawaii and extended that analysis to include harvest from Japanese and Taiwanese fisheries, assuming albatross bycatch would be similar for all longline vessels. Although admittedly the estimates obtained are not precise, they provide a way of assessing the potential effect of longline fishing on vulnerable seabird populations, thus establishing a model for what could be applied to other fisheries in international waters elsewhere. Considering all fleets operating in the North Pacific, the estimated black-footed albatross mortality was from 5,000 to 10,000 individuals per year. At these rates, substantial population declines are predicted over the next three generations (about 60 years). There is strong evidence from other programs, however, that bycatch reduction measures could effectively reduce longline mortality. After the mandated use of bycatch reduction methods—including streamer lines, weighted lines, thawed and blue-dyed bait, and side-setting of baits—an analysis by Eric Gilman and colleagues found that bycatch was reduced by almost 70% in the Hawaii longline fishery, to less than 300 per year for all albatross species. One NMFS study showed that, in Alaskan waters, seabird bycatch declined by about 60% after streamer lines were used (**Figure B9-2**). In Australian and New Zealand waters, mandatory bycatch reduction methods were successful in similar longline fisheries. To assume that international agreements cannot be reached in

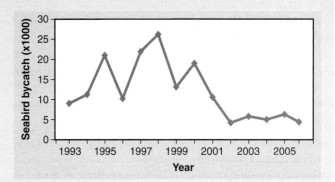

Figure B9-2   Bycatch of seabirds from 1993 to 2006 in Alaskan longline groundfish fisheries. The use of avoidance measures such as streamer lines were implemented voluntarily in 2002 and mandated in 2004. Note the subsequent dramatic reduction in bycatch.

regulating seabird bycatch would be a mistake. Radical restrictions have been made that have resulted in large reductions of seabird bycatch in the past, including outlawing drift net use in international waters in the 1990s (see Chapter 11), and eliminating most egg and feather harvest in the early 1900s.

Sea turtles are another endangered group of animals for which records of incidental harvest are difficult to obtain. A compilation of global fishery records by Bryan Wallace and colleagues estimated over 85,000 sea turtles were captured by gill nets, longlines, and trawls from 1990 to 2008. They argue that this is probably an underestimation, however, because the data are based on on-board observer programs that cover less than 5% of the fishing effort and often do not include small-scale fishing activities that could have a great cumulative bycatch. In some regions there is no reporting of bycatch information (including the western Indian Ocean and West Africa). In the U.S. Gulf of Mexico, shrimp fishers are not required to report sea turtle catch, and captured turtles are released overboard dead or alive (see below). One way that changes in catch are documented is through autopsies of dead turtles washed ashore. Wallace and colleagues surmise that the true number of sea turtles caught globally with commercial fishing gears is over 100 times the numbers reported. That estimate would equate to almost 10 million sea turtles caught in commercial fishing gear in the past 20 years.

A message from these studies is that reliable data on international endangered species are difficult to obtain. Because protected species are not harvested legally for commercial purposes, much of the harvest is unintentional and often unreported. Fishers have incentives to underreport incidental harvest because actions that affect endangered species may limit their ability to continue fishing. Even when solutions are readily available, they may be difficult to implement in international waters due to a lack of enforcement and a lack of understanding by fishers of their importance. Strong international laws limiting bycatch along with funding for monitoring, enforcement, and education are needed to protect endangered species whose range extends throughout much of the oceans.

Figure B9-1   IUCN endangered black-footed albatross *Phoebastria nigripes*.

marine pollution, and fishing bycatch. Warming temperatures are considered the major concern for the long-term health of Antarctic populations because of loss of sea ice and declines in food resources.

There are 18 species of penguins, ten of which are being reviewed for inclusion on the U.S. Endangered Species List. The IUCN classifies five species as least concern, seven species vulnerable, two near-threatened, and four endangered: the northern rockhopper *Eudyptes moseleyi*, the erect-crested penguin *Eudyptes sclateri*, the yellow-eyed penguin *Megadyptes antipodes*, and the Galapagos penguin *Spheniscus mendiculus*. The IUCN has indicated 10 of the 18 species as decreasing in numbers. Following are summaries of the status, threats, and protections of one of the most well known penguins, and four of the most endangered.

The emperor penguin *Aptenodytes forsteri* is the largest penguin species, and was the species featured in the popular 2005 *March of the Penguins* documentary. Because much of the year is spent on sea ice, for mating, rearing chicks, and molting, the emperor penguins are vulnerable to loss of sea ice because of global warming. Regions of the Antarctic Peninsula that they inhabit are considered to be in one of the most rapidly warming regions of the world and some loss of ice has already occurred. Warming of waters of the Southern Ocean could result in as much as 25% loss of ice cover, and early ice break up in summer could affect platforms used by growing chicks and molting adults. Predicted temperature increases and sea ice declines are expected to reduce krill populations in the Southern Ocean. Emperor penguins feed directly on krill and on fish and squid that depend on krill. Because they currently face relatively few other threats to survival, the emperor penguin is classified as least concern by IUCN.

The distribution of the rockhopper penguins is limited to the volcanic islands of the French Southern Territories in the southern Indian Ocean and St. Helen in the South Atlantic. The IUCN endangered northern rockhopper penguin *Eudyptes moseleyi* (**Figure 9-7**) was recently designated as a separate species from the vulnerable southern rockhopper penguin *E. chrysocome*. Populations of *E. moseleyi* have recently undergone rapid declines for unknown reasons. Total population size is unsure, but is estimated to have decreased by 90% since the 1950s. It is speculated that ocean warming, overfishing of prey such as squids and octopuses, competition from fisheries bycatch, and predation by introduced mice on eggs and young could all play a role in their decline.

The erect-crested penguin breeds on Bounty and Antipodes Islands off New Zealand. Populations appear to have declined by over 50% from the 1970s to the 1990s. Currently there are about 75,000 breeding pairs. As with many endangered species, precise reasons for declines are

Figure 9-7 IUCN endangered rockhopper penguins *Eudyptes chrysocome* on Macquarie Island, Australia.

not known but could include many of the factors affecting other penguins: ocean warming, change in oceanographic conditions, and excessive harvest of food resources by humans. The breeding islands have been designated as nature reserves and conservation efforts include removal of cattle and rat eradication.

The breeding habitat of the yellow-eyed penguin is on the southeast coast of New Zealand's South Island and smaller islands off the coast. Degradation of the forest/scrub habitat where it breeds is considered a major cause of recent population declines. Current populations are about 2,000 breeding pairs. Recovery efforts include protection and restoration of habitat, fencing of nesting areas from trampling by livestock, and establishment of a Yellow-Eyed Penguin Trust to assist with conservation efforts. Harms continue, however, with predation by introduced ferrets and cats, bycatch in fishing nets, and food shortages because of changes in sea temperature.

The Galapagos penguin is endemic to the Galapagos Islands. Populations have gone through dramatic changes in the past 30 years in association with El Niño events. Although El Niños are natural occurrences, changes in frequency and intensity of such events could endanger species such as the Galapagos penguin with low population sizes (currently about 1,000 individuals) that are also threatened by increased human populations, tourism, introduced mammalian predators (cats, rats, and dogs), fishing, and pollution. Conservation actions include the establishment the Galapagos National Park and Marine Reserve, predator controls, discouraging net fishing within its foraging area, and limiting human disturbance in breeding areas.

Sea Turtles

Sea turtle populations around the world are considered to some degree threatened or endangered with extinction; CITES lists all species on Appendix I, with no trade allowed between member nations. The IUCN categorizes six of the seven sea turtle species as vulnerable, endangered, or critically endangered globally (**Figure 9-8**). Important conservation measure for sea turtles are protecting nesting habitats on beaches (discussed in Chapter 3) and feeding habitats in seagrass beds (see Chapter 6) and coral reefs (see Chapter 5). Taking sea turtles from the open ocean is also a major conservation issue, however. Some populations have begun to show signs of recovery, but others have continued long declines. Populations of leatherbacks *Dermochelys coriacea* in the Pacific have declined over 95% and nesting female loggerheads *Caretta caretta* have declined over 80%. There may have been as many as 90 million adult green sea turtles

Chelonia mydas in the tropical western Atlantic before the arrival of European explorers to this region; currently there are only about half a million. It is unlikely that these turtles will ever return to their former abundance because of a lack of habitat, protected nesting sites, and sea grasses as food resources.

In recent times, a major reason for sea turtle harvest was to supply the commercial markets for food or for their shells. Turtle shells and tortoiseshell jewelry and decorations have been made from hawksbill turtles *Eretmochelys imbricata*, popular in China and Japan (**Figure 9-9**). International pressures and recent CITES agreements have made the import of such shells (primarily from Indonesia) illegal, removing the market for these products; however, a black market for some of tortoiseshell products still exists.

Sea turtles began to disappear around the world as harvest and environmental impacts became commonplace. Sea turtles are extremely vulnerable to population reductions because of their long life and slow rate of reproduction; female sea turtles often do not mature and come ashore to nest until they are about 30 years old. Currently sea turtles are protected from intentional harvest in most countries. Illegal harvest and trade could be in the tens of thousands in some regions, however, including island nations in the West Pacific and Mexico. There is also some legal harvest by indigenous groups in some countries, including Australia, Central American countries, and Caribbean island nations. Others have recently banned sea turtle harvest, including Cuba and Mexico.

As controls over targeted harvest of turtle eggs and adults increased, unintentional catch of sea turtles became a

(a)

(b)

Figure 9-8 IUCN critically endangered **(a)** leatherback and **(b)** hawksbill sea turtles.

Figure 9-9 Shells of green sea turtles being sold in Japan in the 1980s. Current laws and agreements make this practice illegal.

Figure 9-10 Green sea turtle incidentally hooked on a fishing line. This turtle was released alive, but effects of this kind of stress are not known.

major reason for population declines. Fishing nets and lines often catch sea turtles incidentally as bycatch (**Figure 9-10**). Sea turtles are hooked or entangled by longlines, trawls, or gill nets, and often held under water until drowned (**Figure 9-11**). Gill nets (called drift nets in the open ocean; see Chapter 11) have been outlawed in international waters, although they are still used in coastal waters in many regions. In regions where it is prevalent, such as the Indian Ocean and southwest Atlantic, gill netting kills large numbers of sea turtles. Gill nets have been largely replaced by longlines in the open ocean and thus likely have a greater effect on sea turtle populations, especially larger species such as leatherbacks. In nearshore waters of the U.S. Atlantic and Gulf of Mexico, the most serious bycatch problem with sea turtles is with the trawl fisheries. How many sea turtles are killed by these fishing gears is unknown but could be greater than 500,000 per year (see Box 9-1, Assessing Impacts on Endangered Species: Counting Fishery Bycatch).

Shrimp trawls are bag-shaped nets pulled through the water and over the bottom in shallow coastal waters (see

Chapter 11). When sea turtles are caught in a trawl and held under water for the duration of the trawl set, they often drown. Prior to enactment of measures to limit turtle by-catch in the U.S. Gulf of Mexico, over 10,000 sea turtles died annually because of incidental capture in shrimp trawls (**Box 9-2. Turtles, Trawls, and TEDs**). Sea turtles also can be attracted to the bait or accidentally become hooked and entangled on longlines designed for harvesting large migratory fishes in the open ocean. Sea turtles can die before they are released, especially if they are held under water unable to get to the surface for a breath of air. Although sea turtles can often remain underwater for more than two hours without breathing, an active or stressed turtle must surface more frequently to avoid drowning. As with seabirds, the sea turtle mortality caused by long line fishing is difficult to monitor.

U.S. law requires that the National Marine Fisheries Service establish a protection plan for all listed species under the Endangered Species Act. When fisheries affect sea turtles, some action thus is required to stop or minimize the impact. This can lead to controversy with the fishers and others dependent on the fisheries because taking action may affect their livelihood negatively. Such controversies are exemplified by longline fishers.

Swordfish fishers off Hawaii use longlines extending up to 100 km with up to 8,000 hooks in the water. Between 1994 and 1999 over 100 leatherback and 400 loggerhead turtles were caught by these fishers. This was of particular concern because as few as 3,000 reproductive female leatherbacks may remain in populations worldwide. When the National Marine Fisheries Service was slow in acting to stop this bycatch of an endangered species, in 2001 a U.S. federal judge ruled that the fishery must be closed. In 2004 the fishery was reopened with restrictions on fishing methods. Rules required only mackerel fish bait and a circle hook be used, replacing the J-hooks and squid that were more likely to hook sea turtles. Circle hooks are larger, more difficult to swallow, and the points curve inward, which means that turtles are not as easily snagged or hooked in the gut; and the turtles can bite off the mackerel in pieces instead of swallowing it whole as they do for squid. In addition, a limit was placed on the number of fishing days, and fishers had to be monitored by observers. Once 16 leatherbacks or 17 loggerhead turtles were hooked, the entire fishery would be closed for the year. In 2008 this was increased to 19 leatherbacks and 46 loggerheads after it was documented that the hooked sea turtles were rarely being killed; at the same time the limit on fishing days was removed. Lawsuits were filed, unsuccessfully, by conservation groups to stop the opening of the fishery. An analysis by Andrew Read documented that circle hooks also successfully reduced bycatch in other areas of the North Atlantic, the Gulf of Mexico, and the eastern

Figure 9-11 Green turtle killed from entanglement in a fishing net.

Box 9-2 Turtles, Trawls, and TEDs

When studies began showing that mortalities from shrimp trawls were harming sea turtle populations, it became obvious that something would have to be done to avoid shutting down the fishery to abide by the U.S. Endangered Species Act. The effect was most serious in the Gulf of Mexico, where shrimp are one of the most valuable fisheries and the sea turtle that was affected was one of the most endangered, the Kemp's ridley *Lepidochelys kempii* (**Figure 9B-3**).

Ridley turtles are distinctive in that they form massive synchronous nesting aggregations, called *arribadas*. Kemp's ridley turtles migrate from as far away as European Atlantic waters to coastal Mexico. Most nest on Mexico beaches, with the majority nesting on three beaches in the state of Tamaulipas. About a day before the arribada they move within a few meters of the beach. Then on some unknown cue they begin moving onshore in massive numbers. The arribada can last for several days or weeks, with turtles coming onto beaches, often in broad daylight, in tens of thousands. One of the best pieces of evidence of historic spawning events is a 1947 video taken on a Mexican beach from which it was estimated that 40,000 Kemp's ridleys nested in a single day. Presumably, concentrated nesting evolved as a predator-swamping strategy; however, one predator that cannot easily be "swamped" because of their great intelligence and large population size in coastal regions is humans. Before protections were enacted, it was common for 80% to 100% of the Kemp's ridley eggs to be taken from the nests by humans. As the arribadas declined over the years, egg harvest was so intense that on some beaches confrontation would ensue among people demanding to take eggs from a single nesting turtle. Numbers of Kemp's ridleys declined dramatically, most likely because of egg harvest but perhaps also some commercial harvest in the northern Gulf of Mexico of unknown amounts (they were sold along with other species as canned turtle). It became obvious that the Kemp's ridley would soon be driven to extinction if actions were not taken. From 1963 through 1974, laws were enacted in the United States and Mexico prohibiting harvest of Kemp's ridleys in their waters. Recovery was very slow, however, and from 1978 to 1991 only an average of 200 Kemp's ridleys nested each year.

In 1996 the Mexican government outlawed taking eggs from beaches. Although some poaching of eggs for local use has continued and unknown numbers of adults are taken for illegal trade in skins and shells, it is believed that this exploitation is not enough to keep the turtle populations from recovering. But it became obvious that nesting site protection alone would not solve the turtle's decline. Other sources of mortality were apparently keeping the adults from returning to nest.

Eventually it was accepted that the largest probable source of mortality for Kemp's ridley turtles since the ban on directed harvest of eggs and adults was the incidental capture and killing of turtles during trawl fishing operations. The increased mechanization of the shrimp fishery in U.S. waters in the 1970s corresponded directly to a decline in nesting Kemp's ridleys at Rancho Nuevo, Mexico, the only major nesting population remaining. Another piece of evidence was an increase in turtle strandings on Gulf of Mexico beaches, attributed to bycatch in shrimp trawls, because most strandings were in areas with the greatest shrimping efforts (in the western Gulf off Louisiana and Texas). Methods to eliminate this bycatch were searched for in earnest. Intense conflicts developed among shrimpers, conservationists, and the U.S. federal government. Finally, in 1991, the USFWS and NMFS implemented an endangered species recovery plan that included limiting the incidental catch in the shrimp trawl fishery. The approved method of reducing the bycatch was a technology that had been developed by shrimp trawlers on the U.S. east coast to eliminate larger jellyfish from their shrimp catch, which also efficiently eliminated sea turtles from the trawl. It was labeled as a **turtle excluder device** (commonly called **TED; Figure 9B-4**). Studies showed that the TED effectively excluded sea turtles and other large items from the trawl net by shunting them through a circular grate in the middle of the net. Smaller organisms, such as shrimp, pass through the grate to be captured at the end of the net.

Although studies carried out by the NMFS showed that shrimp loss from using TEDs was minimal, there continued to be a resistance to their use by the shrimp fishers. Some

Figure B9-3 Endangered Kemp's ridley turtle nesting on the beach at Padre Island, Texas.

Figure B9-4 A loggerhead turtle escaping a shrimp trawl net through a turtle excluder device (TED). Many coastal nations, including the United States and Mexico, now require TEDs in shrimp trawl nets.

were convinced that the TEDs reduced their shrimp catch; other argued that they caught few or no sea turtles and thus there was no problem (failing to take into account that thousands of shrimpers, even though each caught few turtles, could harm sea turtle populations). Due to Kemp's ridley mortality, NMFS was faced with the prospect of closing the shrimp fishery in order to protect the sea turtles as required by U.S. Endangered Species laws.

U.S. law now requires that U.S. shrimpers fishing in waters where they are likely to encounter sea turtles use TEDs in their trawl nets. Although there is evidence that some shrimpers get around the restrictions by not using TEDs or tying the opening of the TED closed during trawling operations, the requirements appear to be helping with Kemp's ridley recovery. Still, U.S. beach strandings continue, as bycatch in trawls and other fishing gears continues to be the greatest threat to Kemp's ridley populations. (Some have attributed an increase in strandings following the 2010 *Deepwater Horizon* oil spill to be the early opening of shrimping seasons or a movement of shrimpers into waters frequented by the turtles.) In the 1990s, Mexico banned trade in sea turtle products, began to require the use of TEDs, and closed the shrimping season during nesting periods.

As various conservation measures have been put in place, and despite difficulties with enforcement, the Kemp's ridley populations have begun to rebound. In 2009, nesting counts were over 20,000 in Mexico and almost 200 on the Texas coast (**Figure B9-5**). In the United States, a captive rearing program was initiated to help establish a nesting population on Padre Island, Texas. Several camps have been set up on Mexico beaches where intensive monitoring and protection efforts are being used to expand the nesting population.

U.S. shrimpers continue to argue that using the TEDs affects their ability to catch shrimp, in part because of the expense and inconvenience. This led to the U.S. attempting to use trade restriction to encourage or force other countries (e.g., Thailand) to implement the use of turtle excluder devices in their shrimp fishers. This was of limited success because of conflicts between environmental laws and trade agreements through the World Trade Organization as to whether the United States can impose its environmental laws on other nations. Eventually, international agreements were reached requiring nations in the Americas, Southeast Asia, and the Indian Ocean to implement sea turtle conservation measures. Enforcement and monitoring of such conservation measures can be problematic. For example, the U.S. repealed Mexico's certification to export wild-harvested shrimp in March 2010 after a NMFS study found that TEDs used by Mexico did not conform to U.S. standards; however, the certification was reinstated seven months later.

Whether Kemp's ridley conservation efforts can be considered a success story is still uncertain. Increases in nesting appear to be continuing, but it is likely that a combination of environmental, habitat, and fishery effects will keep the arribada from ever returning to their former numbers. Specific factors that will likely continue to plague the Kemp's ridley turtle include a loss of nesting beaches, poaching, and a loss of food sources and feeding habitat, for example, from pollution effects such as coastal dead zones and oil pollution.

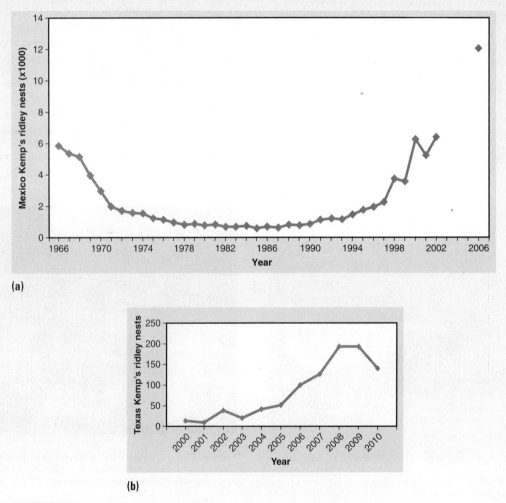

(a)

(b)

Figure B9-5 Kemp's ridley turtle nests found on **(a)** Mexican beaches and **(b)** Texas beaches, indicating recovering populations in response to protections to the turtles and nesting beaches.

Pacific; however, in some fisheries the catch rate for target fish declined along with the turtle bycatch.

The various methods applied to reduce sea turtle bycatch from fisheries have undoubtedly led to an overall reduction in sea turtle mortality. For example, an analysis of NMFS data collected from fisheries in all U.S. waters by Elena Finkbeiner and colleagues estimated a reduction in sea turtle bycatch mortality of 90% from 1990 to 2007 (from 70,000 to less than 5,000). The greatest number of turtle deaths (about 98%) throughout this period were from shrimp trawls in waters of the southeastern U.S. coast and Gulf of Mexico (Box 9-2, Turtles, Trawls, and TEDs), for which mortalities are likely underestimated due to the low number of on-board observers. Although bycatch mortality has not been eliminated, these efforts provide an excellent example of how fishers are able to protect non-target species and minimize conflicts without being forced to halt fishing. Hopefully, such efforts will provide incentives for fishers around the world to become more proactive in limiting bycatch in order to preserve their livelihoods while protecting biodiversity and saving lives of endangered species.

Endangered Fishes

Despite the great impact of harvest on fish populations, relatively few marine fishes are considered to be in eminent danger of biological extinction. Several factors lead to this situation. These include the life history characteristics common of marine fishes, such as high fecundity (thousands to tens-of-thousands of eggs per year) and high longevity; broad distribution ranges and high mobility are common with few barriers to isolate populations in the open ocean. Also, direct anthropogenic damage to habitats is still relatively low compared to freshwater and terrestrial ecosystems. Fishes most vulnerable to extinction are those with a limited distribution, dependence on a sensitive habitats, small population sizes, low egg production, and/or high rates of exploitation. The IUCN Red List includes over 300 fish species, a low percentage of the roughly 20,000 total marine fish species. One reason for the low number of listings is deficiency in the data; for example, little information on population health is available for marine fishes that are not harvested. There are likely more endangered marine species than national and international listings suggest. Another factor leading to the low number of fish listings is the resistance to list species for which commercial fishing has been profitable (for example, bluefin tuna and Atlantic cod; see Chapter 11), and the difficulty in obtaining precise population data. The number of marine fish species considered endangered is likely to increase with more thorough biological surveys and if conservation issues regarding habitat loss and overfishing are not adequately resolved.

Two groups of fish were given special consideration in IUCN assessments of species endangered with extinction. Elasmobranchs (mostly sharks and rays) are considered vulnerable because of recent increases in bycatch and directed harvest, in part to support the shark fin market (see Chapter 11). Of the species evaluated, 17% were considered threatened and an additional 13% near threatened; however, almost half of species considered were viewed to have inadequate data to make a judgment on status. A thorough assessment also has been attempted for the groupers. These and other marine fish species listed by IUCN are those associated with tropical habitats, including coral reefs (see Chapter 5). About 12% of groupers were considered threatened, another 14% near threatened, and about 30% were considered of unknown status because of inadequate data. Groupers are threatened by overfishing, especially of spawning aggregations and for the luxury and live-fish markets. About 25% of seahorse species assessed by IUCN are considered vulnerable or endangered (the remainder are categorized as data deficient), mostly by habitat loss and collections for traditional medicines and curios (see Chapter 5). All seahorses are listed on Appendix II of CITES, prohibiting international trade. The humphead wrasse *Cheilinus undulatus* (**Figure 9-12**), discussed in Chapter 5, is considered endangered by IUCN and on Appendix II of CITES.

All six *Pristis* sawfish are classified as critically endangered on the IUCN Red List and on Appendix II of CITES. Sawfish have toothed rostrum (the "saw") extending from their upper jaw, used to injure or impale fish in feeding. This rostrum was often cut off to remove sawfish from fishing nets even if they were not targeted for commercial harvest. Often the intent was to kill the sawfish, because it

Figure 9-12 The humphead wrasse *Cheilinus undulates,* classified as endangered by the IUCN, largely a result of its harvest for aquarium and live-fish food markets.

was considered a nuisance for entangling and damaging the nets. Anglers would often cut off the saw as a souvenir also. Rarely would these amputated fish survive to be caught later, and it is not known whether they could feed successfully. Directed fisheries for sawfish occur in some regions of the world. Commercial uses include: fins sold on the shark-fin market, rostrums as curios or traditional medicine, meat for consumption, or skins for leather. Another factor affecting sawfishes was habitat loss. Most species live in shallow coastal waters and estuaries, and many of the habitat issues discussed in Chapters 3-6 affected their survival, especially of the young. Finally, as with many elasmobranchs, late age at maturity (around 10 years) and low reproductive rate (about 20 offspring per year) slow down population recovery.

The smalltooth sawfish *Pristis pectinata* (**Figure 9-13**) is one of only two marine species native to U.S. waters that are classified through the ESA as endangered. This species has a broad distribution range around the world, living over sandy or muddy bottoms near the shore. It is severely threatened throughout much of its distribution range. In U.S. waters, populations are mostly limited to southwest Florida waters, in the region of the Everglades and Florida Keys, where it used to be plentiful. An analysis by Pablo del Monte-Luna and colleagues illustrated that the lack of sawfish capture in other regions where they were historically abundant should not be used as evidence for extinction and an excuse to discontinue habitat recovery efforts.

In the Indian River region, where there are historical accounts of a fisherman catching 300 smalltooth sawfish in a season, it was probably extirpated by the 1980s. Throughout U.S. waters, NMFS estimates the populations have been reduced by 95%. Recovery efforts for the smalltooth sawfish include many actions taken for conservation of southwest Florida ecosystems. Gill nets have been banned in state waters and taking of sawfish is prohibited through the Endangered Species Act. But sawfish populations continue to be harmed by anthropogenic impacts including not only

pollution and habitat loss but also entanglement in fishing lines and ropes. The amount of illegal taking of sawfish or removal of rostrums as curios is not known.

One final fish example demonstrates the difficulty of identifying causes of species endangerment. The totoaba *Totoaba macdonaldi* is the largest fish in the drum family (Sciaenidae), reaching sizes of 135 kilograms, with its range restricted to Mexico's Gulf of California. Populations of the totoaba began a systematic decline from the 1940s through the mid-1970s until the fishery was closed. Mexico, the United States, and IUCN declared the totoaba as endangered. In the 1990s, reserves were established to further enhance its conservation; however, numbers of totoaba have remained low.

Because the decline in totoaba populations corresponded to increasing fishery activity, it was long assumed that the population collapse was solely a result of excessive harvest. But concurrent with fishing increases, there was also a loss of spawning habitat in the Colorado River Delta. A decrease in river flow into the Colorado River because of diversions within the U.S. raised salinity in the estuary beyond the optimum for larvae that use it as a nursery. Even after fishing was banned, poaching has continued and young totoaba are incidentally caught in the shrimp trawl fishery. These considerations reflect a long-standing tendency to only consider anthropogenic factors in determining declines of fish populations (see Chapter 11). Diego Lercari and Ernesto Chàvez, however, considered that oceanographic processes also could play a role in the population collapse. They found strong correlations between fish catch and natural climate fluctuations. Such fluctuations have been found to have far reaching effects on factors such as rainfall and river input, water temperatures, and abundances of prey species, and thus could play a role in the decline of the totoaba. It appears that there are multiple human and natural causes explaining the decline of the totoaba. Even if a combination of factors affected populations, it is still likely that without the anthropogenic effects the populations would have experienced a more moderate decline and recovered more rapidly. Even when cause-and-effect relationships cannot be documented, it is important that a precautionary approach is taken to protect and enhance recovery of endangered species. Using endangered species as indicators of ecosystem impacts provides a legal rationale for protecting critical marine ecosystems around the world.

The bocaccio *Sebastes paucispinis* is a large rockfish (Sebastidae) in the eastern Pacific from Baja California to the Gulf of Alaska. This is the first fish with high fecundity (tens-of-thousands to millions of eggs produced) residing in moderately deep marine waters (about 50 to 500 meters) to be classified as endangered under the U.S. ESA. It appears

Figure 9-13 U.S. ESA Endangered smalltooth sawfish *Pristis pectinata*, vulnerable to capture in nets and entanglement in fishing lines.

to have declined to less than 10% of its historic abundance in Washington State's Puget Sound population, the only population considered endangered. Impacts appear to be from a combination of heavy fishing, bycatch in salmon fisheries, and possibly adverse environmental conditions. Recovery is inhibited because of its longevity and late maturity. Recovery efforts include tighter restrictions on harvest.

Sirenians: Sea Cows and Manatees

Steller's Sea Cow

The Steller's sea cow *Hydrodamalis gigas* is one of many mammals that evolved to large enough size (probably about 9 meters long and 10 metric tons weight) that there was little need to fear natural predators. Other characteristics that evolved along with the large size include slow reproduction, great longevity, high visibility, slow movements, and likely a limited ability to dive below the surface. It probably spent much of its time grazing on kelp fronds near the surface (see Chapter 6). These characteristics and behaviors made it highly vulnerable to an intelligent predator capable of using weapons (i.e., humans). Although undocumented, the Steller's sea cow was possibly driven to extinction throughout much of its original range because of hunting by aboriginal peoples. By the time Europeans arrived in the North Pacific the sea cow's range was limited to regions near islands uninhabited by humans. The remaining population of possibly 1,500 sea cows, limited to the Commander Islands (in the Bering Sea about 175 kilometers off the Russian Kamchatka Peninsula), was discovered by Georg Wilhelm Steller, a naturalist on a Danish expedition. Hunting began unabated by sailors, seal hunters, and fur traders traveling past the islands toward Alaska. The skins were used to make boats, the meat was eaten, and fat was used in lamps and as a butter substitute. By 1768, 27 years after Steller's discovery, the sea cow was extinct. This extinction demonstrates the sensitivity of large mammals to human exploitation even without modern technologies. Such species need thorough management and monitoring of any exploitation, even when carried out for subsistence needs by indigenous groups.

Manatees

Without special protections, the manatees and dugongs could easily have been driven to extinction along with the sea cow, because of their large size, slow swimming speed, and shallow habitats where most time is spent eating and resting. Loss of habitat is probably the greatest cause of recent declines for manatees and dugongs (see Chapter 6), but poaching for meat and hide in Central and South America also could be affecting population recovery. In Florida waters, collisions with boats are a major source of

manatee injuries and mortalities. Although some manatees survive the collisions, as evidenced by propeller scars on their body or flukes (**Figure 9-14a**), others are killed; about 300 deaths were documented from boat collisions in 2006. There are probably only about 3,000 West Indian Manatees remaining. Population recovery is slow because manatees do not mature until five to nine years for females and males, respectively; and only one calf is born every two to five years.

The Florida subspecies of the West Indian manatee (*Trichechus manatus latirostris*) is protected in the United States under the Endangered Species Act, the Marine Mammal Protection Act, and the Florida Manatee Sanctuary Act. In 2007, the IUCN classified it as endangered based on low population size and the potential for decline due to warming waters resulting from climate change and increases in boat traffic. The USFWS estimates the Florida manatee's populations currently as stable at about 3,800 individuals. To

(a)

(b)

Figure 9-14 **(a)** Manatees are vulnerable to injury by boat propellers due to their slow movements and habit of floating near the surface. **(b)** Protective measures include education programs, such as this warning for boaters on the southwest Florida coast.

minimize collisions, boat speed zones have been established in the regions manatees frequent.

As with other endangered marine species, the USFWS or NMFS (for the manatee it is the USFWS) establish criteria by which the species (or in this case subspecies) can be removed from the Endangered Species List or downlisted from endangered to threatened. For the Florida manatee, requirements for downlisting include reducing threats by: identifying minimum spring flows and warm-water refuges (being tropical species, manatees do not tolerate cold temperatures and springs provide relatively warm waters in winter); identifying important manatee areas and foraging habitat; and reducing human *take* (which would include any injury). Population benchmarks for downlisting include a 10-year time period with greater than 90% adult survival, at least 40% of adult females accompanied by calves, and a positive population growth. Once the Florida manatee is downlisted to threatened it can be considered for delisting. Requirements for delisting include reducing or removing the threats mentioned above by protecting the habitats, reducing or eliminating human take, and achieving the population benchmarks for an additional ten years.

The Florida manatee situation provides an example of how endangered species protection plans can lead to controversies among the public. Protection has a monetary cost, and some do not view the estimated $10 million dollar expenditure identified by the recovery plan as a good use of tax payer dollars, or believe that protections could be achieved more economically. Some think that recreational use of Florida waters should take precedence and are unwilling to abide by restrictions of boating practices specified for manatee protection. Others may be uneducated about rules for manatee protection, despite intensive education campaigns around south Florida boating areas (Figure 9-14b). Restricting coastal activities and development may have an economic cost to some businesses, resulting in resistance.

Controversies are not limited to public conflicts. Agencies also can disagree on the status and protection of imperiled species such as the manatee. In 2007, the IUCN changed the status of the Florida manatee from vulnerable to endangered. This was in conflict with proposals by the Florida Fish and Wildlife Conservation Commission to downlist the manatee from endangered to threatened through its own Endangered Species Act. Accusations were made that the state agency was being overly influenced by developers who would benefit from looser restriction on building marinas and docks along Florida's waterways.

This is but one example of conflicts that are almost inevitable with species protected by multiple endangered species and marine mammal protection laws and various government and international agencies and commissions.

Adequate protection of these species involves science, education, politics, economics, and social considerations.

Pinnipeds

The pinnipeds comprise three families, including the walrus (Obobenidae), the "eared seals," including sea lions and fur seals (Otariidae), and the "true seals" (Phocidae; see Chapter 2). Conservation issues for the various species are similar because of their biology and behaviors. Pinnipeds spend most of their time at sea where they feed, but most species come to shore into rookeries for reproduction and rearing of young. Pinnipeds have been valued for fur and sometimes meat and blubber, and almost all species have been targeted by sailors, fishers, and hunters (sealers) at some time in recent history, usually causing large population reductions. Pinnipeds are easily harvested on land, where they have little ability to resist hunters, or at sea, where they concentrate and regularly surface for air. Most pinnipeds that historically supported commercial harvest are in the groups commonly referred to as seals.

Indigenous groups, such as Native Americans and Europeans, have been harvesting seals along the northern oceans for thousands of years, and this has been an important source of food and part of their culture. A small amount of this subsistence harvest is still allowed in many regions of the world, but is not generally considered to be large enough to affect populations. Commercial sealing began and intensified as people gained the ability to sail at sea for extended periods and reach isolated coastlines and small islands where seals had historically gained refuge from predators. The first records of commercial sealing are in the early 1500s by Europeans. By the late 1700s, whalers began to target herds of fur seals (Otariidae, Arctocephalinae; **Figure 9-15**) in the Southern hemisphere for their furs, and harvests increased. Before the turn of the century, tens

Figure 9-15 Fur seals have long been valued for their fur, resulting in heavy exploitation for hundreds of years. Here are Antarctic seals with an elephant seal lying in the background.

of thousands of seals were being harvested annually in the Southern Ocean by over 100 sealing vessels from Europe and North America. Sealing boomed in the extreme South Pacific and Arctic in the early 1800s, with vessels sailing out of Australia and New Zealand. By 1820 unregulated harvest had reduced seal populations so that sealing in this region was no longer profitable; by 1830 few sealers remained. Eventually many nations outlawed sealing, but not until some species had been reduced to remnant populations. The mechanisms to establish international conservation agreements did not exist until the early 1900s and then only in a few isolated cases; conservation attitudes were uncommon until the second half of the 20th century. This is one of many examples demonstrating a period of extreme risk for marine mammal populations. By the mid to late 1800s, humans had developed the capabilities of harvesting virtually every animal in a population; however, legal means of establishing checks and balances had not yet been established that would force protection of sensitive populations. These would only come with the development of a global conservation ethic in the next century. For example, the northern fur seal *Callorhinus ursinus* barely survived overexploitation several times before protections to populations were established (**Box 9-3. Learning from History: Northern Fur Seals**).

Much of the historical problems with seal population declines have been because of hunting onshore but also at sea for some species. Because of population declines and endangered species and marine mammals protections, relatively few nations continue to harvest seals; most are harvested from the Canadian North Atlantic, but Greenland Norway, Denmark, and Russia have subsistence or commercial harvest of some seals in the North Atlantic and Arctic. Namibia harvests fur seals along a portion of the South Atlantic coast of Africa. Russia and native indigenous groups harvest small numbers of seals in the North Pacific. With protection, many seal populations have rebounded; however, for some these rebounds have not been long-lived. Many populations have recently declined for unknown reasons. Possible reasons for declines include overharvest of food resources, habitat loss, increased predation, pollution, and climate change; specific examples are discussed in Box 9-3, Learning from History: Northern Fur Seals.

Monk Seals

The monk seals of the genus *Monachus* include two of the most endangered seal species and one that was recently driven to extinction. The Caribbean monk seal *Monachus tropicalis* (sometimes called the West Indian seal) is one of only two marine mammal species known to be driven to total extinction within the past two centuries (the other

being the Japanese sea lion *Zalophus japonicus,* extinct by the 1970s due to excessive harvest). The Caribbean monk seal was the only seal native to the Caribbean Sea and Gulf of Mexico. The last sighting was in 1952, and in 2008 the U.S. NMFS declared the species as officially extinct. Documented exploitation of the Caribbean monk seal began with the harvest of eight seals during Columbus' 1494 voyage. Although populations were possibly not abundant even prior to this date, commercial harvest led to it extremely low abundances by the late 1800s. Without international protections, opportunistic killings continued and habitat loss contributed to its eventual extinction.

The Hawaiian monk seal *Monachus schauinslandi* (**Figure 9-16**) is one of the most endangered seal species. About 1,100 individuals remain in the population, which has declined about 60% since the 1950s (**Figure 9-17**). There are six main reproductive populations in the northwest Hawaiian Islands, with small numbers on the main Hawaiian Islands. The species is classified as endangered by the IUCN and U.S. Endangered Species Act, depleted under the MMPA, and on Appendix I of CITES. As with other seals, the original population decline was because of excessive harvest by sealers and whalers in the 1800s. In the early 1900s, the northwest Hawaiian Islands were classified as a U.S. National Wildlife Refuge, giving the seals protection from directed harvest. Population surveys beginning in the 1950s showed rapid declines in beach counts through the mid-1970s of about 50%. Additional protection measures were then established through the MMPA and the ESA, including habitat protections to the beaches and coastal waters. The establishment of the Northwest Hawaiian Islands Coral Reef Ecosystem Reserve in 2000 (see Chapter 5) further restricted activities within the seals habitat. Despite all of these protections, population declines continued at about 5% per year from 1985 to 1993. Populations remained stable through 2000, and then continued declining to the 2010 population of about 1,100 individuals. On a positive note, monk seals have recently begun returning to the main Hawaiian Islands, some hauling out on popular beaches, and from 2006 to 2008 a total of 43 pups were observed there.

Why Hawaiian monk seals have not recovered despite intensive protection measures is not known but many hypotheses have been presented. Juvenile survival is a vital component of population recovery, and the transition from weaning to feeding independently is a critical stage. It appears that food limitation may be reducing population growth. This is supported by observations of juveniles in poor condition with low juvenile survival at many rookeries. Possible reasons include oceanographic changes that result in low productivity in coastal waters. Monk seals feed primarily on bottom invertebrates, and harvest of lobsters (now illegal in the Northwest Hawaiian Islands) could

Box 9-3 Learning from History: Northern Fur Seals

The northern fur seal *Callorhinus ursinus* is the only fur seal species found in the North Pacific Ocean. Breeding occurs on several islands, three in Russian waters (Commander, Kuril, and Robben Islands) and three in U.S. waters (2 off Alaska, Pribilof Islands in the Bering Sea and Bogoslof Island in the Aleutians; 1 off southern California, San Miguel Island; **Figure B9-6**). Northern fur seals are distinctive for their thick fur (desirable by sealers) and that adults spend about 80% of their time foraging at sea. The tendency for fur seals to congregate over ocean features, such as seamounts, canyons, and the continental shelf break, to take advantage of elevated productivity made them more vulnerable to harvest at sea.

Historically, the major decline in northern fur seals was because of unregulated harvest, primarily for the fur industry. Harvest began after Russian fur traders discovered the primary breeding grounds of the seals in the Pribilof Islands in the 1780s. The slaughter of the seals in rookeries led to its near extermination by the end of the century. The Pribilof Islands became a preserve of the Russian-American company that took control of regulating harvest in 1799. In the 1860s the northern Pacific fur seals numbered in the millions. The United States took control of the islands in 1867 with purchase of Alaska and continued regulating harvest. The harvest of island populations began to be poorly managed and herds were declining. In the late 1800s the United States attempted to sustainably manage the harvest of fur seals on the Pribilof Islands by leasing the harvest rights to a single commercial company, and outlaw harvest of fur seals from the open waters of the eastern Bering Sea by any nation. This plan backfired when

other nations excessively harvested seals in offshore open waters, where no nation had power to regulate the harvest.

American schooners were the first to develop techniques (actually adapted from those used by coastal indigenous peoples) in the early 1880s to efficiently harvest seals from the open ocean (**Figure B9-7**). Even with only sailing ships, the fur seals were easy targets because they spend extended periods migrating along predictable routes in large slow-moving aggregations. Hunters would leave the mother ship in small boats, called dories, to enter the swimming herds of seals where they would shoot the seals or harpoon them with long-handled spears. A crew of three on a dory could take up to 50 seals in a day. The seals would be skinned and the pelts stored on board to be auctioned onshore at the end of the hunting season. These pelts would eventually end up in Europe, where the luscious furs were popular for clothing both women and men. Two full-time sealing fleets developed, one from Canada and one from the United States. The vessels were privately owned and there were no government regulations on their operations. The main restrictions on harvest were the Pacific Ocean weather (resulting in the loss of 16 vessels in 14 years) and competition among U.S., Canadian, and Russian seal hunters. The preferred targets were large pregnant females (evidence that hunters did not have long-term sustainability of harvest on their minds). This seal fishery is memorialized in the Jack London novel *Sea Wolf*, which provides a historically authentic account of the operations on a sealing schooner.

The pelagic harvest of the seals started depleting the herds at an alarming rate; however, no nation had the power to stop it because the open ocean was considered common

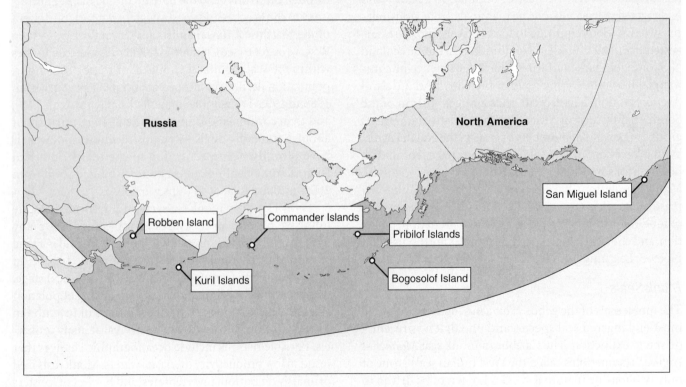

Figure B9-6 Locations of northern fur seal breeding colonies and extent of their winter range (dark gray).

Figure B9-7   A pelagic sealing schooner, the *Anni E. Point*, 1895.

property with free access to all. The United States eventually tried to stop Canadian vessels in the Bering Sea, but this led to conflicts and threats of military confrontation. The conflict was resolved by establishing an international agreement for the Bering Sea, including a closed season with no harvest within a 100-km radius of the Pribilofs and prohibition of the use of firearms. This forced the sealers further to sea, where they simply followed the migrating herds from rookery to feeding grounds.

Finally, in 1911, the North Pacific Fur Seal Treaty, between Russia, Great Britain, Japan, and the United States, outlawed all open-water sealing and allowed the United States to regulate harvest of onshore herds in U.S. territory. This treaty is historically significant in that it was the first international wildlife conservation treaty. In 1912 the United States took advantage of this conservation opportunity and an Act of Congress prohibited commercial harvest of northern fur seal for five years to allow populations to recover. Only Pribilof Island Aleut tribes were allowed to continue harvest, and then only for subsistence needs.

Protections resulted in a rapid recovery of the fur seal populations. By the 1950s they were believed to be at preharvest levels, and limited harvest was allowed. In 1956 an experimental harvest of females was begun in an attempt to increase the herd's productivity. After a decade it became apparent that the experiment was not working and it was discontinued; however, unexpectedly, the seals continued to decline. Pup production eventually was cut in half. As the adult population continued its decline, the northern fur seals were designated as depleted under the Marine Mammal Protection Act, and all commercial harvest was stopped on the Pribilofs by 1984. Only subsistence harvest of about 2,000 seals annually is now allowed by Alaskan natives. Despite these efforts, as indicated from estimates by Rod Towell and colleagues, populations still have not recovered for uncertain reasons. Possible

factors include changes in prey because of commercial fishing in the region, and possible increases in predation by growing killer whale populations, after declines in their preferred prey (see Box 10-2, Are the Great Whales Important Prey for Killer Whales?). Incidental take of fur seals in fishing gear and interactions with marine debris could be affecting populations also. Climate change cannot be ruled out, as studies have shown that climate variability, especially related to storm activity, may alter dispersal and survival of migrating pups. To add further confusion to the issue, a recently established population on Bogoslof Island has been increasing at the same time as the Pribilof populations decline. The northern fur seal is listed as vulnerable under the U.S. ESA and by the IUCN, and threatened by Canada.

Several lessons can be drawn from the history of fur seal exploitation and protection. One is the need for protection of vulnerable marine species from unregulated exploitation in areas where the congregate for breeding, rearing, or feeding. Even with protections of vulnerable life stages, however, an open harvest of a valuable resource will almost inevitably lead to overexploitation. Hunters or fishers will rarely operate for the conservation of the resource when competing with others for what they consider is their share. For large, long-lived mammals, total protection may be required for many years for populations to recover. Even then, if the ecosystem on which they depend is modified by human actions, populations will often behave in unpredictable ways. Unfortunately, the ecosystem on which the northern fur seals depend is a large fraction of the Northern Pacific Ocean. Human modifications of this ecosystem through fishing, pollution, and climate change could be more damaging to the long-term survival of this species than even the extreme levels of exploitation they experienced over a century ago. This adds to the growing evidence that we have entered a new era of complexity in marine conservation.

Figure 9-16 The endangered Hawaiian monk seal.

have affected growth and survival. Ocean conditions can increase erosion rates also, resulting in the loss of available haul-out beaches on some islands.

Injury or bycatch with fisheries gears could be a problem for monk seals swimming and feeding away from shore as well. Monk seals have been caught on longline hooks, but these are now prohibited within 80 km of the northwest islands. Recreational shore fishers and gill netters fishing from the main Hawaiian Islands could harm monk seals, but these are not allowed in the northwest islands. Entanglement in marine trash could be a significant factor in Hawaiian monk seal declines as they are more commonly entangled than any other seal species in proportion to their population size. Up to 25 seal entanglements per year have been reported and there is an unknown number of unreported entanglements. But even low numbers of deaths of adult seals could affect populations. Debris

removal programs have been in place since 1982 to remove beach trash and potential entangling debris. NMFS biologists have worked to disentangle over 200 monk seals from marine debris (**Figure 9-18**).

Natural ecosystem and population interactions could affect population growth and inhibit recovery for species such as the Hawaiian monk seal with extremely low numbers. For example, increased predation by sharks has been observed in some areas, especially on young pups. Up to 30% of pups can be killed by sharks, but this appears to occur only in isolated incidences. Changes in population makeup also could affect pup survival; when there are more males than females in a rookery, male aggression ensues in competition for mates. This can result in the death of female and immature seals. Documentation of the potential role that each of these factors could play in Hawaiian monk seal survival has led to additional measures of protection. In regions where predation was high, predatory sharks (Galapagos sharks *Carcharhinus galapagensis*) have been removed using hook-and-line and harpoons, and some pups were relocated to other areas; increased pup survival resulted. Restoration of beaches lost by erosion is being considered. In populations where pup survival was low because of food limitation, undersized weaned pups were captured, brought into captivity and fed for 8 to 10 months, and returned to the wild. About 60% of 104 captured pups were healthy enough to be returned. Ideally such interferences would not be recommended for species protection, and they can be controversial; however, critically imperiled species may need extreme measures to boost populations to levels at which they can become naturally stable.

As with other endangered pinnipeds, populations of Hawaiian monk seals appear to be affected by a combination

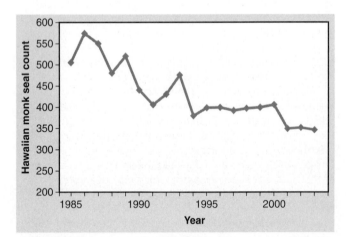

Figure 9-17 Population trends based on beach counts for the Hawaiian monk seal. Reasons for recent declines have been difficult to identify.

Figure 9-18 Entanglement in fishing nets is one factor in recent monk seal declines. Here a diver attempts to rescue an entangled seal.

of anthropogenic and natural occurrences. In a healthy, stable population natural changes and variability in ecosystem interactions would likely be tolerated in the long run. Conservation of critically endangered mammals, however, with total populations declining to approach 1,000 or less, may require intensive management efforts to enhance recovery. Human actions put the species in its current condition and human interference in the "natural order" may be necessary to assist the recovery of the species; whether the cost in terms of money and other resources warrants such actions is continually debated. Unfortunately, because of the irreversibility of extinction, we don't get a second chance.

The Mediterranean monk seal *Monachus monachus,* with individuals numbering less than 500, is even more critically endangered than the Hawaiian monk seal. It lives in the Mediterranean Sea and a few locations in the eastern Atlantic, though its original distribution was throughout the Mediterranean and Black Sea and the northwest coast of Africa. Original declines were because of excessive harvest with clubs, spears, and nets, peaking in the Middle Ages. Most of the reduction in the species' range, however, was in the 19th and 20th centuries with habitat loss and pollution resulting from the industrial revolution, two world wars, an explosion in tourism and development, and large-scale fishing. Fishers often would kill the monk seals for fear they were eating their catch and damaging fishing nets. Had it not been for the Mediterranean monk seal developing a behavior of seeking refuge and bearing young in remote caves accessible only beneath the water, they might have been driven to total extinction.

Current threats to the Mediterranean monk seal are habitat loss from coastal development and tourism, accidental entanglement in fishing gears, loss of food (fish, squid, and octopus) due to overfishing, and unintentional harassment by scuba divers and tourists. There are still reports of fishers killing seals that get into their nets. Unpredictable environmental events and disease outbreaks are of particular concern because of the low population size and isolation of the species in two remaining populations (one in the northeastern Mediterranean and one along the Atlantic coast of northwest Africa). About two-thirds of the Atlantic population died in 1997 from an epidemic possibly caused by a toxic algae bloom; the current population size is about 150, one-half of what it was in 1997. Conservation efforts for the Mediterranean monk seal include establishing protected areas and no-fishing zones, and rescuing orphaned and wounded seals, all in coordination with education and public awareness campaigns. Funding of such efforts has been difficult to establish, possibly because of the seals' low visibility. Protected areas are considered inadequate because they consist of only four scattered reserves. An extended network of reserves is considered necessary for the species' long-term survival. Capture, captive breeding, and translocation have been considered but are controversial. Conservation organizations have resisted the idea, believing that no colony is large enough to afford the loss of animals to establish a breeding program, and that monk seals are so sensitive to human disturbance that this would be an additional threat to the species. They also point out that there have been no successful efforts to breed the Mediterranean monk seal in captivity.

It is questionable whether there is enough public support for the Mediterranean monk seal to provide the cost and effort required to save it from extinction. This species is attempting to survive adjacent to one of the most heavily developed and populated areas of the world, and adequate protections would undoubtedly result in substantial economic costs and inconveniences to some. Unfortunately, this may be a species too many believe we can do without. Conservationists and scientists are not ready to give up, however, and hopefully this seal species can remain viable at the reduced population levels that currently exist in the short term, with expectations for population expansion in the long term. Several groups continue to work toward achieving conservation goals by concentrating efforts toward conservation and awareness campaigns for the Mediterranean monk seal.

Elephant Seals

Other than the monk seals, most of the phocid seals have populations that are currently in relatively good health. The southern and northern elephant seals (*Mirounga leonina* and *M. angustirostris;* **Figure 9-19**) were both heavily exploited in the 16th and 17th centuries, primarily for

Figure 9-19 Driven almost to extinction by hunting in the 1800s, the southern elephant seal recovered rapidly with protection from harvest.

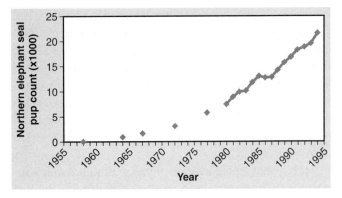

Figure 9-20 Population trends for northern elephant seals in the Channel Islands based on NMFS counts from aerial photographs. Because of rapid population growth they are no longer classified as endangered, but are still protected under the Marine Mammal Protection Act.

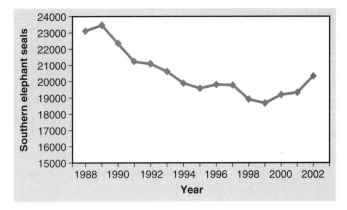

Figure 9-21 Population trends for southern elephant seals on Macquarie Island, Australia based on counts of breeding females. Following declines through the 1990s, for unknown reasons, the population appears to have stabilized.

blubber and due to their large size. Males are over 2,000 kilograms for northern elephant seals and up to 4,000 kilograms for southern elephant seals; females are about a quarter of that size. Fewer than 1,000 northern elephant seals likely survived, on a single rookery on Guadalupe Island of Mexico, by the late 1800s. Beginning in the early 1900s, both Mexico and the United States began protecting populations. Since protections were enacted, populations have recovered dramatically, with new rookeries established along the coast of Mexico and California (**Figure 9-20**). The northern elephant seal feeds in the North Pacific Ocean, at depths from 200 to 800 meters, on squids, octopuses, and various bottom fishes. Current populations are over 100,000. In California, populations continue to grow but will eventually be limited by available beaches for haul-out. Northern elephant seals are listed by the IUCN as least concern, and were removed from CITES Appendix II because of healthy, growing populations. Potential concerns about the long-term health of populations remain, however, because of their limited range and low genetic variability resulting from low population size. This **genetic bottleneck effect** could make them susceptible to disease or pollution. There are also concerns about the potential impacts of climate change; the El Niño event of 1997–1998 resulted in mortality of pups likely as high as 80%.

Southern elephant seals, distributed around the Antarctic and extreme southern oceans, were also hunted to near extinction in the 19th century. With protections, populations recovered rapidly until the 1950s. Since then, however many populations declined for unknown reasons, until they appeared to stabilize by the end of the 20th century (**Figure 9-21**), and currently there are about 600,000 individuals. This decline was possibly because the populations rebounded above carrying capacity and depleted resources before populations could stabilize. (Southern elephant seals

feed primarily on fishes and squids from 400 to 1,000 meters depth.) Climate changes or possible interactions with competing species could provide other explanations. Breeding sites are currently protected by international agreements through the United Nations. The IUCN lists its conservation status as least concern.

Conservation efforts for both species of elephant seals can be considered successful. Once protections were established populations rebounded rapidly. Why they recovered more rapidly than some other seal populations is not known. Several factors could come into play. Restrictions on harvest were successful in part because the marketability of seal products declined when the value of blubber decreased dramatically in the early 1900s, as replacement products were found or developed (see discussion of whale harvest in Chapter 10). Isolated populations were no longer worth pursuing. It is also less likely that food resources were overharvested as they have been in epipelagic and shallower waters; the elephant seals gain an advantage by diving into deep waters and feeding on animals less likely to be targeted by fisheries. Many of the southern elephant seal haul-out sites were in areas with little development or human presence in or near Antarctic waters. Still, these species could easily have gone the way of the Caribbean monk seals and Steller's sea cow and reached total extinction. Credit must be given to national and international efforts to protect populations during a time when conservation ethics were less prevalent.

Antarctic Seals

The lobodontine seals include four species found in Antarctica and the extreme Southern Ocean, the crabeater seal *Lobodon carcinophagus,* the Ross seal *Ommatophoca rossi,* the leopard seal *Hydrurga leptonyx,* and the Weddell seal *Leptonychotes weddellii* (**Figure 9-22**). Populations of each of

Figure 9-22 Populations of the Weddell seals and other lobodontine Antarctic seals have stabilized with protection through the international Convention for the Conservation of Antarctic Seals.

these species are considered to be stable. The crabeater seal is the world's most abundant seal with about 30 million individuals, feeding on krill in the extremely productive Southern Ocean. The Ross seal is relative low in abundance, with probably 100,000 to 200,000 individuals. This is probably a natural situation; these seals are more solitary and do not form large rookeries. The leopard seal is the largest of the lobodontine seals, with males sometimes weighing over 400 kilograms. It feeds on a large variety of animals supported by the Antarctic productivity, including krill, squid, fish, penguins, and smaller seals. Population size is around 200,000 to 400,000 individuals.

The Antarctic lobodontine seals are one of the few groups of mammals whose populations are likely the largest they have ever been. Probably their numbers have increased in response to the dramatic reduction in populations of baleen whales from commercial whaling (see Chapter 10), as one of the primary food items for these whales is the Antarctic krill. If whale populations rebound with the increased protection, numbers of Antarctic seals are likely to decline because of competition for food. The Antarctic seals never went through the dramatic population declines of many of the other seal species from human harvest; they were largely protected from hunting because of the inaccessibility and expense of operating in Antarctic regions. There has been some recent interest in taking leopard seals, and in the 1980s Soviet sealers harvested over 600. This type of harvest is strongly opposed today by conservation groups. The seals are protected by international agreements protecting the Antarctic region, including the Antarctic Treaty and the Convention for the Conservation of Antarctic Seals.

Northern Seals

The phocine seals (Phocidae, subfamily Phocinae), commonly called the *northern seals,* include 10 species found in northern hemisphere waters (**Figure 9-23**). All are marine species, except the Baikal seal in Lake Baikal, Siberia (the only freshwater pinniped). The northern seals comprise by far the largest seal populations, including the most abundant pinniped in the world, the harp seal *Pagophilus groenlandicus,* with about 8 million individuals. Most species have populations between 100,000 and one million. The only species classified as endangered by IUCN is the Caspian seal *Pusa caspica,* with about 100,000 individuals. Other species are classified as least concern or have insufficient data for listing, except for the hooded seal *Cystophora cristata,* classified as vulnerable. All of the northern seals support some hunting, commercial and/or subsistence. Nations that hunt northern seals are those that encompass cold-water regions, the most common habitat of northern seals; the primary seal-hunting nations are Canada, Greenland, Iceland, Norway, Finland, and Russia. Some of this hunting is carried out by indigenous groups, including the native tribes of Alaska and Canada. Another source

(a)

(b)

Figure 9-23 Most of the northern seals, **(a)** such as this hooded seal, are not considered endangered and many are heavily harvested. **(b)** Although whitecoat pups are generally protected, older harp seals are commercially harvested along some shores of the northern Atlantic and Arctic.

of human-caused mortality of northern seals is culling of seals with the intent of protecting harvested fish populations from seal predation or to limit damage to fishing nets. Incidental entanglement of seals in fishing nets is common in some regions. Environmental effects on populations include prey loss because of fishing, entanglement in debris, coastal development, pollution, oil and gas exploration, global warming and loss of sea ice.

The most endangered of the northern seals, and the species with the largest recent population declines, is the Caspian seal. Intensive harvest has continued since the early 1800s, with 50,000 to 300,000 individuals taken in most years through the 1980s. In the 1930s populations were still over one million, before beginning a decline that continues today. Reproducing females have declined by more than 70% since 1955, over about three generations. Changes affecting the Caspian seal include reduction in prey fish by overfishing, loss of breeding habitat from ice reductions because of climate change, and possibly oil drilling operations. Hunting stopped temporarily in 1996, but has continued since 2004 at 3,000 to 4,000 annually. Environmental groups protested in 2007 when a hunting quota of 18,000 seals was set, more than the estimated annual pup production. Several thousand may also be killed per year by fishery bycatch, especially in gill nets. From 10% to 40% of pups may die from natural predation by wolves and sea eagles (*Haliaeetus*). Since the 1970s there have been mass mortality events for Caspian seals because of infection by a canine distemper virus. Seals must overcome numerous other potential threats, including introduction from ballast water of an exotic jellyfish that eats zooplankton and causes fish declines; pollution by heavy metals and pesticides, including DDT, that have decreased reproductive rates; low ice cover that has limited breeding grounds; and development of one of the largest oil fields in the world in its ecosystem.

Measures to protect the Caspian seal have been implemented, including harvest quotas and areas protected from sealing. One of the most significant measures was prohibiting female harvest on breeding grounds, implemented in 1966; however, this caused a switch to harvesting pups. In 1970, quotas were set specifically for harvest of pups, which continues today at levels that the IUCN views as unsustainable, even as populations continue to decline. Conservation plans have been enacted but do not include binding laws to reduce harvest. In order to save the Caspian seal, a total ban on harvest may be needed, along with protected areas in the sea, on ice, and on shore. This species is attempting to survive near one of the most heavily exploited and industrialized regions of the world. Economic hardships and continued environmental impacts in this region likely will hinder conservation measures for years to come.

Although none of the species of northern seals in the northern Atlantic are classified as endangered, seal harvest in this region is one of the most volatile issues in marine mammal protection (**Box 9-4. Animal Welfare, Hunter Rights, and Conservation: Clashes Over the Harp Seal**). The greatest controversy has been conflicts over seal harvest in Canada primarily off Newfoundland. Attention was brought to this issue when animal welfare groups took this cause on in 1969 after discovering that young "whitecoat" pups were being clubbed to death to provide furs for the luxury fur market. This high profile campaign was able to recruit celebrities who could garner media attention. Photos of the bloody harvest that made it into the news media eventually affected the market for furs, thus reducing the commercial value of the seals. In 1983 Canada responded to the pressure by banning the killing of the whitecoat seal pups. Animal welfare groups continued to protest, however, because most of the harvested seals are still younger than a year old (the white coat is molted in about 3 to 4 weeks). Most of the harvest is of harp seals, but commercial harvest is also allowed for gray seals and hooded seals (Figure 9-23).

Sea Lions

Sea lions include six extant species of otariid pinnipeds, with each species forming rookeries in one specific geographic region. These include: the Steller sea lion *Eumetopias jubatus* in the North Pacific, the Australian sea lion *Neophoca cinerea*, the South America sea lion *Otaria flavescens*, the New Zealand sea lion *Phocarctos hookeri*, the California sea lion *Zalophus californianus*, and the Galapagos sea lion *Neophoca wollebaeki*. Three species (Stellar, Australian, and Galapagos) are classified by the IUCN as endangered, one is vulnerable (New Zealand), and two (South American and California) are of least concern. As with other pinnipeds, sea lions historically have been harvested by aboriginals and commercial sealers. For the Galapagos, New Zealand, and Australian sea lions, the endangered status reflects their current limited geographic range; populations range from 10,000 to 40,000 individuals. Current threats to Australian and New Zealand sea lions are primarily from bycatch in fisheries. Both are protected by national laws that have resulted in restrictions on harmful fishing operations. New Zealand requires Sea Lion Escape Devices (SLEDs) to release sea lions incidentally caught in the squid fishery. The Galapagos sea lion is vulnerable to large fluctuations during El Niño events when prey species decline.

Reasons for declines in populations of the Steller sea lion (**Figure 9-24**) are not clear. They are distributed across the Northern Pacific Ocean, distinguished as an eastern population extending up the North American coast, and a western population extending across the Northern Pacific from Alaska across to northern Japan. Eastern populations

Box 9-4   Animal Welfare, Hunter Rights, and Conservation: Clashes Over the Harp Seal

Harp seals have been harvested for thousands of years by native peoples of the Atlantic Arctic region. Around the 1500s Europeans began harvest for oil rendered from the blubber. In the 1700s French-Canadian explorers and English settlers in Newfoundland were taking about 15,000 to 20,000 thousands seals annually. In the 1800s the introduction of schooners increased harvest to over 100,000 per year, and by the mid-1800s about 500,000 were harvested annually. During the last 40 years of the 19th century over 12 million seals were harvested, mostly for oil. As substitutes for oil from blubber increased in the 1900s, harp seal became more valued for the fur. Catches declined, but then increased in the 1950s to over 300,000 per year (**Figure 9B-8**). An increasing harvest of adult seals led the Canadian government to limit the length of the hunting season and protect adult females from harvest in the 1960s. In the 1970s harvest quotas allowed harvest of about 100,000 to 200,000 harp seals per year. By the 1980s marketing of furs became more difficult due to pressure applied by animal welfare groups. Europe banned import of whitecoat furs in 1983, and by 1987 Canada had banned killing of whitecoats altogether. Canadian harvest switched to post-weaning pups called *beaters*. The quota for harp seal harvest was curtailed for several years following the controversy over whitecoat harvest. Then in 1996 harvest increased and the quota was raised to about 300,000 per year by the early 2000s. Commercial harvest by Greenland and fishery bycatch increased the mortality by about 100,000. There is also a subsistence harvest in Arctic regions. The IUCN estimates that the allowable level of harvest is not sustainable.

In order to make rational decisions about the allowable level of harp seal harvest that would allow the population to be maintained, Mike Hammill and Garry Stenson argue that a precautionary management approach be taken. In resource management, **precautionary approach** refers to defining in advance rules for making management decisions based on caution, especially when information is less certain (see Chapter 11). For Northwest Atlantic harp seals this means defining population sizes at which certain harvest restrictions will be taken. For example, when populations are over 70% of their

maximum (about 4 million), management can be based on ecosystem and socioeconomic considerations. If the population drops below 70%, the management strategy should be to restrict harvest in order to return the population to 70%. If the population drops below 50% of maximum (about 3 million), then substantial conservation measures are required, including a drastic reduction in harvest. If the population goes below 30% (about 1.7 million seals), then all removals should be stopped until populations recover.

Several factors complicate managing populations through harvest restrictions. For example, an unknown number of seals are killed but not counted in the harvest numbers. Some of these are fisheries bycatch throughout coastal water of the north Atlantic. Tens-of-thousands are caught in fishing nets during some years, but precise numbers are difficult to obtain because fishers do not always report seals killed in their fishing operations. Subsistence hunting by aboriginal peoples is also difficult to monitor as much of this hunting does not require permits. Other unreported seal mortalities occur during the sealing operations. Animals considered "struck and lost" are those that dive or sink beneath the water after being struck by the club. An analysis by Becky Sjare indicated that these rates can be as high as 50% when seals are taken from the water, but less than 5% when taken from the ice.

A major complicating factor in conservation of harp seals is protection of food resources. Major prey species of the harp seal include fish such as capelin and herring, which are heavily harvested in the North Atlantic. In years when fish populations are low the seals tend to move to regions with healthy fish populations. Continued declines of fish in the North Atlantic could result in rapid seal declines. Although seals have been blamed for the loss of cod in the North Atlantic, cod declines were likely primarily a result of excessive commercial fishing (see Chapter 11).

Potential environmental impacts on the harp seal are similar to those for other seal species, including global climate change and loss of sea ice on which they depend for pupping, molting, and resting. Oil presents a danger, especially in areas with high tanker traffic; in 1969 a ruptured tank coated over 10,000 seals with oil, resulting in the death of many pups. Bioaccumulation of heavy metals and persistent organic pollutants is a danger, as it is with all predatory mammals; high levels of DDT and PCBs have been documented although they appear to be on the decline.

Management of the harp seals will remain a highly contentious issue regardless of the health of the populations as long as hunting is allowed. Because of ethical concerns, debates are not just about protecting ecosystems and defining how many animals can be sustainably taken from the population. Sealers and government managers must deal with resistance to harvest for ethical reasons even if they disagree on such issues. Seals are killed either by firearm or by striking with a club called a *hakapik*, a heavy wooden club with a hammer head. Clubbing is the preferred method of sealers to kill young seals. Graphic visual images have been used by animal welfare groups to make an emotional appeal against seal harvest. The clubbing appears brutal; however, a veterinary group commissioned to evaluate killing methods reported that if done

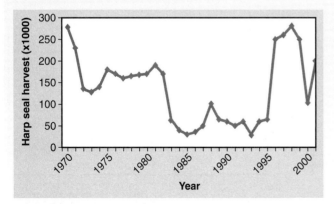

Figure B9-8   Commercial harvest of harp and hooded seals in Canada and Canadian waters has fluctuated dramatically, largely as a result of the volatile fur market, animal welfare conflicts, and management controversies.

properly, clubbing is humane because loss of consciousness is rapid. Animal welfare groups argue that the methods used by the sealers do not always kill the seals quickly and thus the seals often times suffer unduly. It is not feasible to monitor all sealing and it is unknown how often killing methods are applied properly and death of the seal is ensured prior to skinning them. Beyond the cruelty arguments, there is the debate as to whether marine mammals should be killed to support a market for luxury products such as furs.

It is doubtful that sealers and animal welfare groups will ever come to an agreement on the ethical issues regarding seal harvest; however, sealers must realize that public opposition can be a strong force. Even if government agencies continue to allow seal harvest, the sealing industry is dependent on international trade in furs and public perceptions can have a powerful effect on that trade. This can be seen by the ban on import of whitecoat products in Europe in the 1980s and the eventual prohibition of whitecoat harvest in most commercial harvests. The primary

argument by sealers in opposition to ethical arguments is that the seals represent an economic resource and source of jobs for Canadians in a region that has had a depressed economy. The value of the seal pelts in 2005 was estimated at over $16 million. Animal welfare groups argue that the Canadian government has subsidized the hunt with government programs that assist the seal hunters, including upgrading processing plants, promoting the hunt, and establishing markets.

As with many marine mammal conservation issues, these debates cannot be solved solely through a scientific evaluation. From a purely conservation perspective, most agree that the seals can withstand some harvest and sealing can provide an economic resource for the sealers from the fur, meat, and oil (considered a healthy food supplement). From the animal welfare perspective: the seals do suffer during the kill, young immature animals are being harvested to provide a luxury item, and the commercial harvest is not a subsistence hunt necessary for the survival of the hunters.

have remained healthy in recent years and continued growing as a response to increased protection. Western populations, considered healthy in the early 1960s, began declining for unknown reasons, however, until by 1989 they had been reduced by over 60% to about 110,000 individuals (**Figure 9-25**). An analysis of anthropogenic causes of declines by Shannon Atkinson and colleagues found several direct threats that could have contributed to declines, including commercial hunting (legal in Alaska from 1959–1972, when about 45,000 pups were taken), bycatch in fishing nets (about 1,000/year in the 1960s–1980s, but about 30/year since the mid-1990s), and shooting by fishers to eliminate a perceived competitor for fish resources (legal prior to passage of the MMPA but currently illegal); however, none of these are likely substantial threats to population recovery. Subsistence harvest by native Alaskans is not considered a

significant cause of decline or threat to recovery because of low numbers (150–200 sea lions taken annually since 1996). More likely, anthropogenic threats to recovery were identified as a decline in food sources because of overharvest of fish such as pollock, cod, and herring (see Chapter 11) as well as environmental pollution (PCB and DDT levels have been found in some sea lions at levels that could cause physiological problems; see Chapter 7). Other possible factors causing the decline and inhibiting the recovery include recent environmental variability, including oceanographic regime shifts in the North Pacific that affect biological production and distribution of food organisms. Environmental factors that affect the nutrition and health of seals could result in decreased survival and lower pup production, which has been observed over a significant portion of the western sea lion's range. Another proposed effect is

Figure 9-24 Steller sea lions, here hauled out on Middleton Island, Alaska in 1978, appear to be vulnerable to overharvest of their food resources and impacts of global pollution.

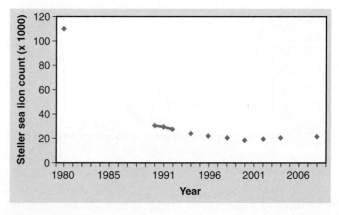

Figure 9-25 Stellar sea lion population trends based on NOAA/NMFS counts of non-pups at haul-out sites. Although protected from harvest, efforts to recover populations have been largely unsuccessful.

increased predation by killer whales resulting from ecosystem shifts caused by environmental change and overfishing. Declines in seal populations—a preferred prey for killer whales—in the Northern Pacific could result in greater dependence on other prey, including sea lions.

What can and should be done to assist with the recovery of the Steller sea lion populations is a difficult dilemma because it is not simply a matter of stopping direct impacts by humans. Although U.S. Endangered Species laws require that action be taken to recover endangered populations, and the federal government cannot permit actions that jeopardize the continued survival of an endangered species, without a strong definition of the causes of the sea lion decline it will be difficult to implement dramatic changes in fisheries harvest solely to protect sea lion populations. As discussed in Chapter 11, however, there are further reasons supporting the re-evaluation of fisheries harvest in this region. Reduction of environmental pollutants is even more problematic because this is a global issue without a simple solution. Until further studies are completed, it can best be said that anthropogenic threats are likely acting in concert with other natural changes affecting Steller sea lion populations.

South American sea lion populations are estimated at over 250,000 individuals. In most regions they are considered relatively stable. Sealing, however, has reduced populations in the southern Patagonia region by about 85% since the 1940s. The Peruvian population is sensitive to El Niño events and underwent an 80% decline after the 1997 El Niño, but eventually recovered. An increase in the frequency of these events could affect the survival of these populations.

Populations of the California sea lion recovered so well from harvest in the 19th and 20th centuries that current populations are about 350,000 (**Figure 9-26**). A recent decline of about 20% in 15 years could be a result of overshooting the ecosystem's carrying capacity. The major issue with the California sea lion is dealing with conflicts. Large groups of sea lions can damage docks and boats, and individuals sometimes steal fish from nets. Government agencies have limited ability to control populations due to protections by the MMPA. Conflicts between the MMPA and U.S. ESA are presented when sea lions prey on endangered salmon. Relocations of sea lions are preferable, but translocated individuals often return. A special ruling was passed by NMFS in 2008 to allow the limited lethal removal of sea lions (up to 85 per year) when harming Columbia River salmon runs, as a measure to balance the conservations of an endangered fish (as required by the ESA) with the protection of all sea lions (as stipulated by the MMPA). This policy was overturned in federal court in 2011, however, based on a ruling that the NMFS had not given adequate explanation of why there was a need to kill sea lions, but

(a)

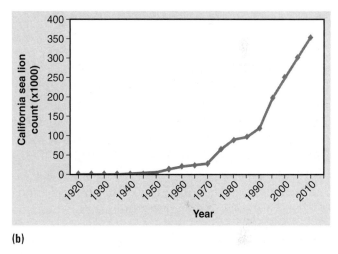

(b)

Figure 9-26 **(a)** California sea lions, the performing "seals" of circuses and marine parks. **(b)** Populations have increased dramatically with protection in U.S. and Mexican waters, to the point that it may be inhibiting the recovery of endangered salmon populations.

had not considered fisheries as a significant negative impact on the salmon; further, NMFS did not explain why current levels of predation by sea lions would damage salmon populations.

The fact that the California sea lion is a highly visible marine mammal (it is the performing "seal" of circuses, zoos, and marine parks) has engendered an emotional attachment, adding public support for its continued total protection. Fishery interests are strong as well, and perceptions of competition with humans for fishery resources could continue to lead to conflicts such as these when abundant populations of pinnipeds (and other marine mammals) are given total protection in ecosystems that have historically been used by humans for fishing and recreation.

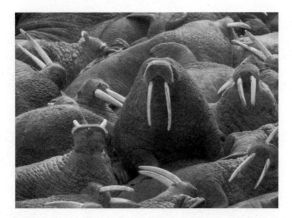

Figure 9-27 Walruses, here protected in the Togiak National Wildlife Refuge, Alaska, have stable populations in the Arctic but could be vulnerable due to the loss of sea ice with global climate change.

The Walrus

The walrus *Odobenus rosmarus* is the only extant species in the pinniped family Odobenidae (**Figure 9-27**). It has a circumpolar distribution in Arctic and sub-Arctic regions, separated into three subspecies. Only the elephant seals are larger pinnipeds. Walruses spend much of their time around the sea ice where they hunt for benthic bivalves, their preferred food. Walruses have been an important resource for indigenous Arctic peoples, providing meat, fat, skin, bone, and tusks. During the 19th and 20th centuries walruses were heavily harvested commercially for blubber and ivory from tusks. Pacific walrus populations have recovered well, to about 200,000 individuals. Atlantic populations barely survived commercial harvest and have only recovered to about 20,000 individuals; there are fears this population could be declining. Populations used to extend down the length of the Canadian Atlantic coastline, but they are now considered extinct in Atlantic Canada. The only legal harvest of walruses currently is a subsistence harvest by indigenous peoples. About 5,000 Pacific walruses are harvested each year in Alaska and Russia. The greatest long-term concerns about walrus survival, however, are effects of global climate change. Walruses rely on the pack ice for aggregation during reproduction and bearing young, and for resting near feeding grounds. The walrus is classified as data deficient by IUCN due to difficulties in estimating population sizes, and is likely to be classified as either endangered or threatened following a review for listing under the U.S. Endangered Species Act. CITES restricts trade in walrus ivory to protect populations from commercial harvest.

Sea Otters

The sea otter *Enhydra lutris* (Carnivora, Mustelidae; **Figure 9-28a**) is an important keystone component of coastal regions in the North Pacific Ocean (see Chapter 2). The sea otter experienced the same historical patterns of population

(a)

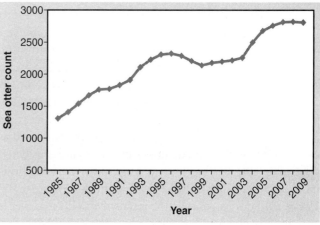

(b)

Figure 9-28 **(a)** California sea otters, although still classified as endangered internationally, have recovered in several stable populations around the Pacific. **(b)** Increasing population trends in U.S. and Mexican waters have resulted in conflicts over competition for clams between the otters and fishers.

declines as pinnipeds and other mammals that were commonly harvested for furs. Indigenous groups had sustainably harvested sea otters from North Pacific coastal waters for thousands of years. Then, beginning in the 1700s, Russian trappers systematically removed most of the sea otters from the northwestern and northern Pacific, and American trappers moved up the North American coast in the northeast Pacific. By the mid-1800s about one million sea otters had been harvested and population numbers were so low it was no longer profitable to hunt in most regions. By the early 1800s the Russian-American Company, which controlled harvest of seals and sea otters in the northern Pacific, began protecting the sea otters from harvest in some regions. The Alaska population recovered to over 100,000 until the U.S. bought Alaska in 1867 and then quickly resumed hunting. As the otters became rarer the price of furs increased, keeping the fur trade profitable, until only 1,000 to 2,000 individuals remained. The 1911 Treaty for Preservation and Protection of Fur Seals included a moratorium on sea otter harvest. Populations began to recover rapidly in most of the sea otter's range.

Off California, protection of a remnant sea otter population, discovered in 1938, resulted in a population expansion of about 5% per year through the mid-1970s, followed by a 30% increase through the mid 1980s. Even with a strong recovery, it was feared that the population was too isolated and vulnerable to something catastrophic such as an oil spill, so a recovery plan was implemented that included translocating some otters to other areas to establish separate breeding populations. It was eventually realized that most of the otters were not remaining at the translocation site and the program was abandoned. Not everyone was glad to see the sea otters recovering off California. Harvesters of clams and abalone were concerned that the otters would compete and reduce their harvest. To address this concern, an "otter-free" management zone was established where any otters that left the parent range and moved into the otter-free zone were returned. The discovery that many of the otters returned to the parent range were dying led to the abandonment of this program. Eventually, the U.S. Fish and Wildlife Service determined that the relocation program was violating the ESA, under which the otters were protected. The otter population grew to over 2,000 animals by the mid-1990s and by 2007 the California population was up to about 3,000 (**Figure 9-28b**) (compared to historic populations of 16,000). As the sea otters recovered, the controversy continued and lawsuits were filed by fishermen against the USFWS contending that the abandonment of the otter-free zone would harm the shellfish fishery. On the other hand, conservation groups complained that the USFWS was acting too slowly to implement a valid plan that allowed for the expansion of the sea otter populations, and filed papers to keep any more otters from being translocated. A situation had developed where fishers had become accustomed to the conditions without sea otters. When more natural conditions returned that included sea otters, they perceived that there was unfair competition with the otters for seafood resources. The battle is presented as conservation versus economic interests. In this case, the Endangered Species Act is powerful enough that the otters may win, even if it takes legal battles by conservation groups.

In the Aleutian Islands, sea otter populations had recovered to over 60,000 by the 1980s and were considered stable. By 2000, however, their numbers had declined to about 6,000 for unknown reasons. Hunting was no longer allowed, except for a small harvest by native Alaskans, not considered a substantial mortality source. It is possible the sea otters are experiencing impacts similar to those of the sea lions described above. An increase in killer whale predation is suspected as a major cause of the decline; although the sea otter is not the preferred prey of killer whales, declines of seals and sea lions may have caused a prey switch. Since 2005, populations have increased but still remain below what is considered the carrying capacity for the population. Other Alaskan sea otter populations have been more stable, except for Prince William's Sound, where the *Exxon Valdez* oil spill likely killed thousands of individuals (see Chapter 3).

Russian sea otter populations appear to have stabilized, and populations are slowly returning to Mexico and Japan. Along the Canadian and U.S. coast south of Alaska there are several stable populations. Recent efforts have been made to translocate sea otters to regions within this range to eventually establish a more continuous distribution. Current world population numbers for the sea otter are about 107,000; it remains as endangered on the IUCN Red List. Oil spills continue to be a major threat. Lacking a layer of blubber, sea otters are very sensitive to oil pollution when the fur loses its insulation ability. Fisheries bycatch causes unknown numbers of mortalities, and El Niño events can reduce food availability.

9.3 Future Prospects for Endangered Species Conservation

Although the number of endangered marine invertebrates and fishes is accelerating, the species for which the most protection efforts are given are marine mammals, sea turtles, and seabirds. Commonalities that make species in these groups susceptible to endangerment include life history traits that result in low reproductive rates. Behaviors that make them vulnerable to harvest and habitat loss are equally important, however, including the need to come ashore for bearing young and to come to the surface for respiration. Many endangered species are dealing with similar conservation issues. Through the early 1900s, excessive harvest of marine mammals and sea turtles for meat, blubber, skins, or shells was prevalent. Most, but not all, species survived this onslaught—some only barely—with greatly reduced numbers or isolated populations that had escaped excessive harvest. Many species recovered in the 1900s under protections provided through national and international agreements, with the assistance of reduced markets for marine animal products and pressures from conservation organizations and animal welfare groups.

Although targeted hunting continues for some species, this is not currently the major cause of population declines for imperiled species. Unintentional catch in fishing gear kills hundreds of thousands of marine mammals, birds, and sea turtles each year. By the beginning of the 21st century, many exploited populations had returned to dramatic sizes, but others continued to struggle, either never rebounding from excessive harvests or beginning a new cycle of decline after a strong rebound. We have moved beyond the days of simple solutions to population declines of marine species. Today's problems are complex and more difficult to define and solve. They are represented by a suite of issues that hinge on habitat and ecosystem protection, including climate change, ocean pollution, and habitat loss as well as many that are related to fisheries harvest, including incidental catch, overharvest of prey species, and illegal harvest (discussed further in other chapters). In order to maintain diversity of marine species it is imperative that strong environmental, conservation, and endangered species laws remain in place and are enforced. We must resist the temptation to give up on any species for economic or political reasons, and realize the importance of species survival as indicators of the health of the global ecosystem on which all life depends.

STUDY GUIDE

Topics for Review

1. Describe the shift in the primary focus of conservation laws and agreements since the implementation of endangered species protections in the 1970s.
2. What biological factors likely explain lower rates of endangerment of marine species than freshwater aquatic species?
3. Explain why the proportion of marine mammals, seabirds, and sea turtles that are considered endangered is much higher than for other marine vertebrates.
4. How does the definition of *species* used in the U.S. ESA differ from the common biological definition?
5. How does the definition of *taking* used in the U.S. ESA and MMPA differ from the common dictionary definition?
6. How does the mechanism for classifying endangered species differ between Canada and the United States?
7. How do Australian endangered species protection laws enhance biodiversity protections over those provided by the U.S. ESA?
8. How do CITES restrictions differ for Appendix I and Appendix III species?
9. What types of information are provided in the IUCN Red List?
10. What group of marine invertebrates is considered most greatly imperiled by the IUCN and why?
11. What factors make sirenians (sea cows, manatees, and dugongs) particularly vulnerable to overharvest?
12. What factors have likely led to recent declines in some unharvested seal species after a recovery from overharvest?
13. What factors led to the rapid overharvest of northern fur seals in the Pacific Ocean in the late 1800s and what legal mechanisms were used to reduce the harvest?
14. How have animal welfare groups influenced management of northern seals in the Atlantic?
15. Define *precautionary approach* as it applies to management of northern seals.
16. Compare the perspective of sealers and animal welfare groups regarding the harvest of harp seals.
17. Describe conflicts that have developed as a result of recent recovery of California sea lion and southern sea otter populations.
18. Describe how predator-prey interactions are hypothesized to have affected Aleutian Island sea otter populations.
19. Compare the perspective of sealers and animal welfare groups regarding the harvest of harp seals.

Conservation Exercises

1. Describe the significance of each of the following to imperiled species and describe measures being taken to reduce impacts:
   a. exotic rodents
   b. longline fishing bycatch
   c. shrimp trawl bycatch
   d. loss of sea ice
2. What do the following acronyms represent? Describe the significance of each to imperiled species.
   a. ESA
   b. MMPA
   c. CITES
   d. IUCN
   e. MBTA
   f. TED
   g. WTO
3. What factors are likely important contributors to recent declines of the following species?
   a. northern rockhopper penguin
   b. erect-crested penguin
   c. hawksbill turtles
   d. Kemp's ridley turtles
   e. smalltooth sawfish
   f. totoaba
   g. bocaccio

h. Steller's sea cow
i. Caspian seal
j. Steller's sea lion

4. Describe conservation efforts for recovery of these imperiled species and their success or failure.
   a. black abalone
   b. storm petrels

c. yellow-eyed penguin
d. Galapagos penguin
e. leatherback turtles
f. Kemp's ridley turtles
g. West Indian manatee
h. Caspian seal

FURTHER READING

Antonelis, G. A., J. D. Baker, T. C. Johanos, R. C. Braun, and A. L. Harting. 2006. Hawaiian monk seal (*Monachus schauinslandi*): status and conservation issues. *Atoll Research Bulletin* 543:75–101.

Atkinson, S., D. P. Demaster, and D. G. Calkins. 2008. Anthropogenic causes of the western Steller sea lion *Eumetopias jubatus* population decline and their threat to recovery. *Mammal Review* 38:1–18.

Baisre, J. A. 2010. Setting a baseline for Caribbean fisheries. *Journal of Island & Coastal Archaeology* 5:120–147.

BirdLife International. 2010. *Eudyptes sclateri. In: IUCN 2011. IUCN Red List of Threatened Species.* Version 2011.1 [online]. Retrieved September 27, 2011, from http://www.iucnredlist.org

BirdLife International. 2010. *Spheniscus mendiculus. In: IUCN 2011. IUCN Red List of Threatened Species.* Version 2011.1 [online]. Retrieved September 27, 2011, from http://www.iucnredlist.org

Del Monte-Luna, P., J. L. Castro-Aguire, B. W. Brook, J. Cruz-Aquero, and V. H. Cruz-Escalona. 2009. Putative extinction of two sawfish species in Mexico and the United States. *Neotropical Ichthyology* 7:509–512.

Doroff, A., and A. Burdin. 2010. *Enhydra lutris. In: IUCN 2011. IUCN Red List of Threatened Species.* Version 2011.1 [online]. Retrieved September 27, 2011, from http://www.iucnredlist.org

Finkbeiner, E. M., B. P. Wallace, J. E. Moore, R. L. Lewison, L. B. Crowder, A. J. Read. 2011. Cumulative estimates of sea turtle bycatch and mortality in USA fisheries between 1990 and 2007. *Biological Conservation* 144:2719–2727.

Fish, M. R., I. M. Cote, J. A. Gill, A. P. Jones, S. Renshoff, and A. R. Watkinson. 2005. Predicting the impact of sea-level rise on Caribbean sea turtle nesting habitat. *Conservation Biology* 19:482–491.

Fitzgerald, S., M. Perez, and K. Rivera. *Summary of seabird bycatch in Alaskan groundfish fisheries, 1993–2006* [online]. National Marine Fisheries Service. Retrieved August 29, 2011, from http://access.afsc.noaa.gov/reem/ecoweb/html/ecocontribution.cfm?id=52

Frazier, J., R. Arauz, J. Chevalier, A. Formia, J. Fretey, M. H. Godfrey, R. Márquez-M., B. Pandav, K. Shanker. 2007. Human–turtle interactions at sea. *In:* Plotkin, P. T. (ed.). *Biology and Conservation of Ridley Sea Turtles.* John Hopkins University Press, Baltimore.

Gales, N. 2008. *Phocarctos hookeri.* In IUCN 2010. IUCN *Red List of Threatened Species.* Version 2010.1 [online]. Retrieved August 21, 2011, from http://www.iucnredlist.org

Gelat, T., and L. Lowry. 2008. *Eumetopias jubatus.* In *IUCN 2011. IUCN Red List of Threatened Species.* Version 2010.1 [online]. Retrieved September 27, 2011, from http://www.iucnredlist.org

Gilman, E., N. Brothers, and D. R. Kobayashi. 2005. Principles and approaches to abate seabird by-catch in longline fisheries. *Fish and Fisheries* 6:35–49.

Gilman, E., D. Kobayashi, and M. Chaloupka. 2008. Reducing seabird bycatch in the Hawaii longline tuna fishery. *Endangered Species Research* 5:309–323.

Godfrey, M. H., and B. J. Godley. 2008. Seeing past the red: flawed IUCN global listing for sea turtles. *Endangered Species* 6:155–159.

Goldsworthy, S., and N. Gales. 2008. *Neophoca cinerea. In: IUCN 2010. IUCN Red List of Threatened Species.* Version 2011.1 [online]. Retrieved September 27, 2011, from http://www.iucnredlist.org

Hammill, M. O., and G. B. Stenson. 2007. Application of the precautionary approach and conservation reference points to management of Atlantic Seals. *ICES Journal of Marine Science* 64:702–706.

Harkonen, T. 2008. *Pusa caspica. In: IUCN 2011. IUCN Red List of Threatened Species.* Version 2011.1 [online]. Retrieved September 28, 2011, from http://www.iucnredlist.org

Heppel, S. S., P. M. Burchfield, and L. J. Pena. 2007. Kemp's ridley recovery: How far have we come, and where are we headed? *In:* Plotkin, P. T. (editor). 2007. *Biology and Conservation of Ridley Sea Turtles.* John Hopkins University Press, Baltimore.

Jones, H. P., B. R. Tershy, E. S. Zavaleta, D. A. Croll, B. S. Keitt, M. E. Finkelstein, and G. R. Howald. 2008. Severity of the effects of invasive rats on seabirds: a global review. *Conservation Biology* 22:16–26.

Klein, D. R. 2004. Management and Conservation of Wildlife in a Changing Arctic Environment. Chapter 11. In: *Arctic Climate Impact Assessment.* Cambridge University Press. Retrieved August 21, 2011, from http://www.acia.uaf.edu/PDFs/ACIA_Science_Chapters_Final/ACI_Ch11_Final.pdf

Kovacs, K. 2008. *Pagophilus groenlandicus. In: IUCN Red List of Threatened Species.* Version 2011.1 [online]. Retrieved September 28, 2011, from http://www.iucnredlist.org

Lea, M.-A., D. Johnson, R. Ream, J. Sterling, S. Melin, and T. Gelatt. 2009. Extreme weather event influence dispersal of naïve northern fur seals. *Biology Letters* 5:252–257.

Lercari, D., and E. A. Chavez. 2007. Possible causes related to historic stock depletion of the totoaba, *Totoaba macdonaldi* (Perciformes: Sciaenidae), endemic to the Gulf of California. *Fisheries Research* 86:136–142.

Lewison, R. L., and L. B. Crowder. 2003. Estimating fishery bycatch and effects on a vulnerable seabird population. *Ecological Applications* 13:743–753.

Lewison, R. L., and L. B. Crowder. 2006. Putting longline bycatch of sea turtles into perspective. *Conservation Biology* 21:79–86.

Mangel, M. 2010. Scientific inference and experiment in Ecosystem Based Fishery Management, with application to Steller sea lions in the Bering Sea and Western Gulf of Alaska. *Marine Policy* 34:836–843.

Miller, S. L., M. Chiappone, L. M. Rutten, and D. W. Swanson. 2008. *Proceedings of the 11th International Coral Reef Symposium.* Ft. Lauderdale, FL, 7–11 July 2008.

Mooers, A. O., L. R. Prugh, M. Festa-Bianchet, and J. A. Hutchings. 2007. Biases in legal listing under Canadian endangered species legislation. *Conservation Biology* 21:572–575.

Moore, J. E., B. P. Wallace, R. L. Lewison, R. Zydelis, T. M. Cox, and L. B. Crowder. 2009. A review of marine mammal, sea turtle and seabird bycatch in USA fisheries and the role of policy in shaping management. *Marine Policy* 33:435–451.

Norfleet, B. 2011. Legal killing of sea lions halted. *The SandBar* 10 (2):8–10.

Peckham, S. H., D. M. Diaz, A. Walli, G. Ruiz, L. B. Crowder, and W. J. Nichols. 2007. Small-scale fisheries bycatch jeopardizes endangered Pacific loggerhead turtles. *Plos One* 2(10): e1041. Doi:10.1371/journal.pone.0001041

Plotkin, P. T. (ed.). 2007. *Biology and Conservation of Ridley Sea Turtles.* John Hopkins University Press, Baltimore.

Read, A. J. 2007. Do circle hooks reduce the mortality of sea turtles in pelagic longlines? A review of recent experiments. *Biological Conservation* 135:155–169.

Read, A. J., P. Drinker, and S. Northridge. 2006. Bycatch of marine mammals in U.S. and global fisheries. *Conservation Biology* 20:163–169.

Rodrigues, A. S. L., J. D. Pilgrim, J. F. Lamoreux, M. Hoffman, and T. M. Brooks. 2005. The value of the IUCN Red List for conservation. *Trends in Ecology and Evolution* 21:71–76.

Seitz, J. C., and G. R. Poulakis. 2006. Anthropogenic effects on the smalltooth sawfish (*Pristis pectinata*) in the United States. *Marine Pollution Bulletin* 52:1533–1540.

Sjare, B., and G. B. Stenson. 2006. Estimating struck and loss rates for harp seals (*Pagophilus groenlandicus*) in the northwest Atlantic. *Marine Mammal Science* 18:710–720.

Sutherland, J. 1992. Jack London and the *Sophia Sutherland.* Appendix I. In London, J. 1992. *Sea Wolf.* Oxford University Press. New York.

Towell, R. G., R. R. Ream, and A. E. York. 2006. Decline in northern fur seal (*Callorhinus ursinus*) pup production on the Pribilof Islands. *Marine Mammal Science* 22:486–491.

Towns, D. R., I. A. E. Atkinson, and C. H. Daugherty. 2006. Have the harmful effects of introduced rats on islands been exaggerated? *Biological Invasions* 8:863–891.

U.S. Fish and Wildlife Service. 2001. Florida manatee recovery plan (*Trichechus manatus latirostris*), third revision. U.S. Fish and Wildlife Service. Atlanta, Georgia.

Wallace, B. P., R. L. Lewison, S. L. McDonald, R. K. McDonald, C. Y. Kot, S. Kelez, R. K. Bjorkland, E. M. Finkbeiner, S. Helmbrecht, and L. B. Crowder. 2010. Global patterns of marine turtle bycatch. *Conservation Letters* 3:131–142.

Woinarski, J. C. Z., and A. Fisher. 1999. The Australian Endangered Species Protection Act 1992. *Conservation Biology* 13:959–962.

Zador, S. G., A. E. Punt, and J. K. Parrish. 2008. Population impacts of endangered short-tailed albatross bycatch in the Alaskan trawl fishery. *Biological Conservation* 141:872–882.

Conservation of Cetaceans

The order Cetacea is made up of about 90 marine mammal species (**Figure 10-1**), including the whales, dolphins, and porpoises (see Chapter 2). Many populations of smaller cetaceans are in healthy condition, though some are endangered by pollution, harvest, loss of prey, and other environmental impacts. The so-called *great whales*, the baleen whales and some large toothed whales, were greatly reduced by large-scale commercial whaling, and most have still not fully recovered. As with other marine mammals, excessive commercial harvest has been replaced in recent years by a variety of environmental factors including habitat deterioration, loss of prey, ship strikes, and noise disturbances. Although commercial whaling has largely been eliminated, it is estimated that hundreds of thousands of cetaceans are unintentionally caught in fishing gear. Overall, the IUCN considers about 25% of cetaceans to be threatened and about 10% endangered or critically endangered; however, more than half of cetacean species are considered data deficient. Endangered species include some of the largest baleen whales, such as the blue whale *Balaenoptera musculus* (**Figure 10-2a**), fin whale *Balaenoptera physalus*, sei whale *Balaenoptera borealis*, North Atlantic right whale *Eubalaena glacialis* (Figure 10-2b), and western gray whale *Eschrichtius robustus*. Species considered to be recovering include the humpback whale *Megaptera novaeangliae* (Figure 10-2c) and the southern right whale *Eubalaena australis*, moved into the least concern category by IUCN in 2008. Several smaller dolphins

and porpoises are threatened in coastal waters mostly by incidental catch in fishing gears; the vaquita *Phocoena sinus* is possibly the most likely to be driven to extinction, from entangling in gill nets in the Gulf of California, Mexico. CITES lists all species on either Appendix I (no trade allowed) or Appendix II (restricted trade). All whales historically harvested commercially are listed on Appendix I to ensure consistency in collaborations with International Whaling Commission (IWC) restrictions on commercial harvest, discussed in detail below.

10.1 Legal Protections of Cetaceans

The effects of 19th and 20th century cetacean harvest as a large poorly-controlled industry continue to be felt today. Legal protections were too slow to be implemented and enforced to protect many populations from dramatic declines. Luckily, few, if any, species were driven to total extinction. Though human-caused death is still problematic, large-scale commercial harvest has been largely stopped. Many of the greatest conservation concerns today are related to unintended effects by humans through pollution, habitat impact, unintended harvest, and other activities that are more difficult to control through legal means than is targeted harvest. Ethical concerns bring further attention to cetacean conservation issues, making broad legal protections more likely than for most other groups of marine animals.

Figure 10-1 A pod of killer whales.

The Marine Mammal Protection Act (MMPA) was largely initiated to address issues related to cetacean conservation, setting guidelines to keep populations at **optimum sustainable levels**, levels at which they can maintain themselves through reproduction, clarified in the MMPA as "usually close to maximum" (see Chapter 9). Measures are to be taken to recover any population that is below its optimum sustainable level so that it continues to function as a "significant element in the ecosystem." Fifteen species or populations of cetaceans are designated as endangered under the U.S. Endangered Species Act (ESA); these and four others are MMPA depleted.

The MMPA provides maximum protection to cetaceans by establishing a moratorium on the "taking" of marine mammals in U.S. waters or by U.S. citizens on the high seas. The definition of *taking* is similar to that used in the Endangered Species Act (see Chapter 9), including "to harass, hunt, capture, or kill," or attempt to "harass, hunt, capture, or kill" any marine mammal. Further clarification was provided in the 1994 amendments to the MMPA by defining *harassment* as "any act of pursuit, torment, or annoyance which has the potential to injure . . ." or "disturb a marine mammal or marine mammal stock in the wild." The harassment does not have to be intentional. For example, operating a boat in waters containing cetaceans or other marine mammals can be viewed as harassment because it is likely to change behavior.

Exceptions to full protection by the MMPA are given to designated groups. For example, the moratorium on taking

(a)

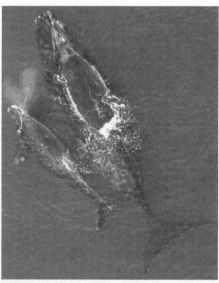

(b)

(c)

Figure 10-2 IUCN endangered **(a)** blue whales and **(b)** North Atlantic right whales (mother and calf). **(c)** The humpback whale is classified as least concern based on recent population increases.

does not apply to harvest by indigenous groups who take marine mammals such as whales or seals for subsistence harvest. Scientists or photographers may apply for permits to research or photograph marine mammals using methods that might be considered low-level harassment. A permit can be requested for taking marine mammals from the wild for display (e.g., by marine parks and aquariums), but this is very tightly regulated and monitored. 1994 amendments to the MMPA specifically addressed the increasingly popular whale-watching industry; for example, it prohibited approach to humpback whales in Hawaii to no closer than 100 yards (91 meters). The 1994 amendments also allows for an unspecified amount of taking incidental to commercial fishing, with a goal of "reducing incidental serious injury and mortality of marine mammals to insignificant levels approaching a zero mortality and serious injury rate," while closely monitoring marine mammal mortality.

There are few nations with laws as strong as the Marine Mammal Protection Act of the United States. New Zealand modeled their Marine Mammals Protection Act of 1978 after the U.S. MMPA, including similar restrictions on *taking* marine mammals. Most other nations with strong conservation laws incorporate marine mammal protection within those laws. For example, in Australia, the Environment Protection and Biodiversity Conservation Act and the 2010 Australian Government Marine Mammal Conservation Initiative provided funding for "non-lethal whale research and other marine mammal conservation initiatives." There is no international law specifically designated for marine mammal protection, though marine mammals are considered by IUCN and CITES. The most important international agreement protecting large cetaceans is through the International Whaling Commission (IWC). Although the IWC was originally developed to protect the interest of whalers, it has recently become a more conservation-focused organization. Both CITES and IWC have been important players in whaling controversies.

10.2 Cetacean Harvest

Uncontrolled harvest brought many of the largest whales dangerously close to extinction. Although most commercial whaling has stopped, few marine issues stir up as much international controversy. The issues are often presented as a dichotomy of total protection of endangered species versus wanton slaughter of large, intelligent marine mammals. The issues, however, are much more complex than this.

Historic Use of Whales

The first humans to use whales for food, clothing, tools, and other resources likely took advantage of individuals that floated dead onto shore or swam ashore and got stranded on the beach. In more recent centuries, most coastal nations have some history of cetacean harvest. In some cultures, including indigenous Japanese and North American, whale meat was valued as food; however, European whalers prized whales primarily for oil obtained from blubber, and typically discarded the meat. Baleen (commonly called *whalebone*) was used as a flexible construction material in many products eventually replaced by plastics, including umbrella ribs, skirt hoops, corset stays, buggy whips, and even construction materials. Sperm whales were prized primarily for the oil contained within the spermaceti organ in the large head, ideal for candle making and as a lubricant. A compaction sometimes found in the sperm whale intestine, **ambergris**, is associated with the beaks of squid and cuttlefish, common food items of sperm whales. Ambergris was highly valued as a fixative for perfumes and component in medicines. Although sailors sometimes made artistic carvings on the large teeth of sperm whales (**Figure 10-3**), the teeth were not generally considered commercially valuable.

Targeted whale harvest began by at least 1000 AD, when European Basques and Canadian Inuit were taking whales in coastal waters. By the Middle Ages the most intensive harvest was likely off Europe, including intensive harvest from the English Channel. With increased global exploration came an expansion of whaling. By the late 1600s, North American sperm whaling was dominating global catch.

Early whalers attacked whales from rowing skiffs with hand-harpoons. Baleen whales typically did not resist the attack and floated to the surface when killed (this was especially true of the right whale, and thus it was the "right" whale to target). Early nearshore whaling with small boats and harpoons may have reduced local coastal populations, and may have resulted in the only extinction of a whale species—the Atlantic gray whale—by European whalers. However, early coastal whaling probably caused little long-term harm to whale populations with global

Figure 10-3 Scrimshaw, artistic carvings made on sperm whale teeth by whalers.

distributions. Still, many of the large baleen whales could be easily targeted during their predictable migrations along coastlines, and some effects are evident today. For example, right whales have never returned in large number to feeding grounds in the western North Atlantic near Newfoundland and Labrador after being harvested in the thousands during the 16th and 17th centuries. For most whales there are few data on early harvest or pristine population sizes, as harvest numbers prior to the 1700s were rarely documented.

The Great Whales and Commercial Whaling

Historic Whaling

Early commercial whaling was limited mostly to coastal regions because the whales had to be brought to shore for processing and rendering of the blubber into oil by heating in *try pots*. Eventually, stations were established in isolated locations for processing the whales, and these contributed greatly to ocean exploration (**Figure 10-4**). The development of methods to render blubber onboard and store the oil on the ship opened the entire ocean to whaling in the 1800s (**Figure 10-5**). Sailing vessels would go to sea for years in search of whales, with only occasional stops in port. The introduction of steam-powered vessels and cannon-fired harpoons in the mid-1800s greatly increased the whalers' capabilities, allowing them to target even the largest whale species (such as blue whales and fin whales), which were too fast and evasive for sailing ships and skiffs.

During commercial whaling, the large baleen whales were valued mostly for the oils used as fuel for heating and other oil-based products. The replacement of whale oil by easily obtained petroleum products around the late 1800s resulted in a decline in the value of whale products. Lower profitability and a sparse distribution due to excessive

Figure 10-5   Model of a 19th century whaling vessel. Note the try-works in the middle of the deck, where the blubber was heated to liquefy the oil and separate it from the flesh.

harvest contributed to the subsequent decline in whaling. The reduced interest in whaling was short lived, however. As modern diesel-powered factory ships were developed in the early 1900s, whaling became much more efficient and once again was profitable.

As whaling industries expanded throughout the world, undoubtedly the whales were undergoing declines, at least in regions where they were commonly taken. Making scientific surveys of whale populations, however, was neither feasible nor considered necessary until well into the 20th century. Data that would eventually substantiate the declines of whale populations were harvest reports by the whaling industry for commercial reasons. Although these data are far from perfect due to undocumented whaling, incomplete reporting, and unreported deaths of whales that were not taken (e.g., those lost at sea), they do provide an index of population size. For example catch-per-unit-effort (CPUE), measured as the number of whales harvested per days at sea, can provide an index of population declines (by assuming that the harder the whales are to find the fewer there are to be found).

By the end of the 1800s the bowhead whale was near extinction, and the gray whale and right whales were being harvested at levels that were not sustainable, causing populations to decline considerably. North Atlantic right whales, once likely supporting the largest cetacean harvest, was no longer a substantial component of the commercial whalers' catch. It became the first whale to receive international protection, in 1935, and remains one of the most endangered of the great whales. The sperm whale became one of the most valued whales and by 1850 American whalers were harvesting about 10,000 annually. Although petroleum products drove down

Figure 10-4   Whalers at an Arctic whaling station in the early 1800s attempting repairs to a whaling vessel damaged by Arctic ice. Note harpoons and oil casks in foreground and whale oil rendering operations in background.

the price of whale oil, with increased efficiency of harvest it retained its marketability as a lubricant and for other specialized uses. Whale meat was still used for consumption in some cultures, particularly in western Pacific nations, and eventually began to be used in domesticated animal feeds.

The Modern Whaling Era

Modern whaling ships enabled whalers to switch from harvesting smaller depleted species to targeting the largest whales (**Figure 10-6**). The turning point was around 1910, when whalers took 176 blue whales worldwide; subsequently, with improving technology and the development of profitable markets, harvest exploded, increasing by several thousand each year. In 1931 about 30,000 blue whales were taken from a world population that numbered probably fewer than 200,000. Although even among the whalers there was concern that eventually populations would be decimated, no good mechanism was in place to control harvest in international waters where most of the blue whales were being taken. Much of the harvest shifted to the Antarctic on populations that had not been known or accessible to early whalers. These whales were considered common property resources in international waters; no entity had jurisdiction over whale harvest in this region, and whaling nations were not willing to force regulations on their whalers. There also was no international body in place through which whaling countries could make multilateral agreements to limit harvest. Whaling was truly a free-for-all for any nation that had the desire and could afford the ships to harvest the whales.

Harvest of blue whales continued at the rate of about 15,000 per year through the 1930s. With increased concern over depletion of populations, agreements were made among whaling nations to limit harvest. For example, whalers agreed not to kill females with calves, and some minimum size limits for take of blue whales and other species were enacted. In the 1940s, some whale species whose populations were critically low could not be legally harvested; however, if not reported, there was no way to enforce these agreed restrictions.

As the blue whales began to disappear and become more difficult to find, the whalers switched to harvesting fin whales. Fin whales are smaller than blue whales, but populations were greater (probably around 500,000 pre-exploitation). By the late 1930s fin whale harvest had surpassed that of the blue whales and averaged over 20,000 per year. Something extreme would be needed to slow the harvest of the great whales. World War II was able to do that briefly, as many whaling ships were shifted to other duties, which gave the whale populations a reprieve. As soon as the war ended, however, commercial harvest resumed.

Concern among whaling nations that they would soon drive themselves out of business finally led to international communications regarding how to control wanton harvest of whales. Talks led to the formation of the International Whaling Commission (IWC) in 1946. The IWC has been much criticized for being slow in acting to stop commercial whaling; however, this was the first international agency developed to cooperate in establishing a sustainable harvest of an international resource. The IWC was created to protect the interests of the whalers by encouraging the orderly development of the whaling industry, not to stop commercial whaling. The data collected were provided by

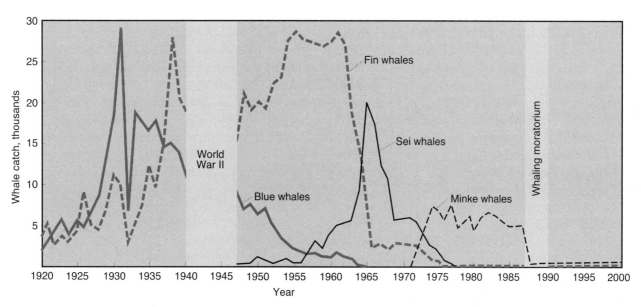

Figure 10-6 Catch history for baleen whales in the Antarctic from 1920–2000. Note that as blue whales declined, harvest switched to fin whales; as fin whales declined, smaller species were harvested until eventually the whaling moratorium stopped most harvest.

the whalers themselves, and agreements were developed by whaling nations.

Early in its existence, the IWC came to agreements on some restrictions to limit whale harvest; however, the resolutions were not binding or adequately enforced. Whaling nations that did not belong to IWC had no restrictions, and nations commonly threatened to leave IWC if they did not think their needs were being met. The eventual outcome was that whale harvest continued to expand and was soon proceeding at record levels. By the time the IWC was formed, blue whale numbers had already declined substantially. Catch per-unit-effort (catch per boat per day) declined every year after the mid-1930s as whalers continued to kill several thousand per year through the 1950s. By 1963, 80% of blue whales harvested were sexually immature. In 1966 only 70 blue whales were killed, not due to tight harvest restrictions, but because the whales were too difficult to locate. Finally, IWC nations agreed to totally protect the blue whale from harvest. By then blue whales were in such low numbers they were no longer profitable for commercial whalers to pursue. Still, some non-IWC nations continued killing blue whales through the early 1970s.

As the blue whales were undergoing their precipitous decline, other species of large baleen whales were targeted to take up the slack. Over 20,000 fin whales were harvested each year for 15 years after World War II. Over 300,000 fin whales—probably over half the total population—were killed in just a decade. As the fin whales began to disappear, whalers transferred much of their efforts to harvest of sei whales. Overexploitation came quickly; the 25,000 sei whales harvested in 1965 were probably about one-third of the remaining population.

Sperm whales harvest also continued after World War II. On average, about 15,000 were taken annually through the 1950s, increasing to over 20,000 through the 1960s, followed by a rapid declined through the 1970s. Though spermaceti oil was no longer used as heating oil, it retained its value for other purposes. For example, millions of pounds of spermaceti oil were used in the automatic transmissions of automobiles in the 1950s and 1960s because of its ability to withstand high temperatures and pressures, and it became valuable to the United States for use in lubrication of some machinery, and in the leather tanning, cosmetic, and garment industries.

With populations of all the largest baleen whales depleted, only the smaller minke whale had abundant populations. Through the 1980s it was the only baleen whale being harvested in substantial numbers. Now that many of the most powerful and influential nations, including the United States and most European countries, were no longer interested in whaling, and strong feelings about conservation had developed internationally, the IWC was pressured to put a halt to commercial whaling. Arguments in support of whaling were becoming harder to sell, not only because of conservation issues, but also because demand declined as substitutes were found for almost all whale products. Japan offered the most resistance to a whaling moratorium because there was still a market for whale meat in Japan and in pet food elsewhere. In addition, Japan argued that minke whale populations were still healthy.

The IWC charter allowed non-whaling nations to have an equal voice as whaling nations, and eventually non-whaling nations came to dominate voting power within the IWC. This led to the passage of a moratorium on commercial whaling, implemented in 1986, to take full effect in 1988. The moratorium was not intended to completely outlaw commercial whaling, but to halt it for some indefinite period of time until a valid management plan could be implemented and whale populations were healthy enough to withstand a sustainable harvest. Although the IWC maintained no direct enforcement powers, there was great political pressure for all nations to stop the commercial harvest of whales. Japan was the final country to go along with the moratorium, when the United States threatened economic sanctions and a prohibition of fishing in U.S. waters. Further protection to whales was given in 1994 when the IWC adopted the Southern Ocean as a sanctuary where no whaling would be allowed.

Looking back at pre-moratorium commercial whaling and the actions of the IWC, we can attempt to evaluate why management of whaling was such an apparent failure. One obvious reason is that restrictions that were passed were not easily implemented without an agency established to enforce them. The country from which the whalers originated was responsible for controlling their whaling industry, but when this was achieved it was often only after great political pressure from other nations.

Another incentive for nations to continued whale harvest was based on the economics of whaling. Although whaling eventually became unprofitable for most nations, for those that had invested in modern whaling vessels there was little economic incentive to stop whaling. Using Japan as an example, by the time the moratorium was put into place, they had a considerable investment in whaling vessels. To simply retire those vessels would mean the loss of that investment. For a country to make new whaling ships might not be sensible, but to continue operating the ones already in service maintained jobs and gave a return on the initial investment.

The international nature of whaling presented an additional difficulty in controlling whaling. Even if restrictions are followed by most countries, the international accessibility of whales made it difficult to monitor harvest and to come to agreements to which all countries would abide.

As with other fisheries, technology became so efficient that without a sincere desire to limit whale harvest, whalers had the capability of easily decimating populations, especially those of the largest, most visible, and most vulnerable species. Foresight and restraint are needed to manage a resource that is so easily overexploited.

Post-Moratorium Whaling

In order to achieve success in getting the IWC commercial whaling moratorium implemented, some concessions had to be made. These exceptions, regarded by many as loopholes, gave certain groups or nations the rights to continue a limited harvest of whales. The first of these exceptions is that whaling can be carried out for the purpose of scientific research. Whaling nations argued that, in order to adequately study the whale populations and collect scientific data on age, growth, and feeding, lethal sampling was necessary. To pay for the cost of the research, the whalers would be allowed to sell the whales caught under the scientific research exemption. Another exception was for indigenous cultures that practiced subsistence whaling using traditional methods; for example, native groups in regions including Alaska, Canada, and Pacific island nations are allowed to kill a small number of whales each year for food, culture, and trade. As a further stipulation, a country could continue small-scale commercial whaling of abundant species by filing an official objection to the moratorium on harvest. Under this exception, Norway resumed commercial whaling for the minke whale in 1993. Other countries have threatened to resume commercial whaling by dropping out of the IWC. Iceland quit the IWC in 1992, but was exposed to great political pressure not to resume commercial whaling. By 2002, Iceland had rejoined the IWC with the stipulation that they be exempt from the commercial whaling moratorium. They began, under international pressure and national controversy, to plan future commercial whaling.

Since the moratorium was put into place, the bulk of the whale harvest has been the take of minke whales based on the exceptions described above. The largest harvests have been by Japan and Norway, about 1,000 per year combined. Beginning in 2000, Japan extended whaling under its scientific permit to also harvest about 10 sperm whales and 50 each of Bryde's whales and sei whales. Although these harvests are within the bounds of the exemptions to the moratorium, evidence has been presented that Japan whalers have taken whales not reported within their allowable quotas (**Box 10-1. Whaling: The Ongoing Debate**). In addition, the Antarctic sanctuary established by IWC has not always been honored by countries still harvesting whales.

The commercial whaling moratorium remains in place; however, it could be repealed by a two-thirds vote of the IWC membership. (IWC membership is open to any recognized nation in the world, and each member nation has one voting representative at IWC meetings.) Such a vote would initiate a policy of establishing what are deemed sustainable quotas on harvest of some of the great whales. In order to influence the vote, both whaling and non-whaling nations have been accused of paying for memberships and applying political pressure to smaller nations that would otherwise have little interest in joining IWC. Recent history is that around half of the IWC nations have been voting in favor of resuming commercial whaling. The debate will likely continue as long as there are divergent views on the conservation and rights of whales (see Box 10-1, Whaling: the Ongoing Debate).

Although most nations with a whaling history come to the conclusion that large-scale commercial whaling should

Box 10-1 Whaling: The Ongoing Debate

Now that some (but definitely not all) whales appear to have rebounded from the excessive harvest of the 19th and 20th centuries, the prospect of whether commercial whaling should ever be resumed has returned to the IWC agenda. Pro-whaling nations argue that recovering populations provide a resource that humans should have the right to harvest. Many non-whaling nations and organizations present biological, ethical, and historic arguments against resumption of large-scale commercial whaling even if populations recover to healthy levels. Viewing the issues based solely on scientific considerations does not reflect the reality of conservation decisions. Regardless of one's viewpoint on such issues, it is important to understand divergent perspectives that govern decision making on such contentious issues as whaling.

From an economic perspective, conservationists argue that there is currently no whale product for which there is not a cheaper commercial substitute. Oils are available from plants or petroleum products, baleen products are replaceable by plastics, and whale meat is not needed for survival, except arguably in a few small subsistence cultures. A substitute was recently discovered for even the mostly highly valued of whale oils; spermaceti oil from sperm whales can be replaced by oil of a near-identical chemical makeup from the jojoba shrub *Simmondsia chinensis* (native to deserts of the southwestern U.S. and Mexico, now being grown commercially for use in the cosmetics industry).

Pro-whaling nations (Japan being the most influential) argue, from a legal perspective, that the IWC moratorium on whale harvest was never intended to be permanent; and, now that some populations have recovered, commercial harvest should resume and trade should no longer be restricted. Economic arguments are countered by pointing out that, even though whale products are no longer a unique commodity and whaling has an almost negligible economic impact on Japan's economy, without whaling, jobs and people's livelihoods would be lost. Furthermore, even if whales are considered an unneeded "luxury" product, prohibiting whaling presents a double standard because there are many luxury fishery products that humans could just as easily do without that are more endangered than some whales. For example, swordfish and many tuna are high-end products that could be replaced by other less-endangered fishes.

Another factor in the economic debate is the sustainability of markets for whale products. Currently the only large market for whale meat is Japan, and some analyses indicate that whaling could not be supported economically without government subsidies. Countries where whale meat is not a traditional food, such as Norway, cannot profit from increased whaling without being able to trade with Japan. Concerns about the safety of consuming some whales due to increases in bioaccumulation of toxins (see Chapter 7) could reduce the current market.

Even if whaling is economically viable, whales may be worth more economically alive than dead, through whale-watching (**Figure B10-1**). In many former whaling nations whale watching has become a profitable tourist industry. Care must be taken, however, to avoid an uncontrolled expansion of whale watching, as it is not necessarily a harmless activity due to its potential to disturb whales and disrupt normal behavior. For example, excessive vessel traffic could disturb feeding, nursing, socializing, or resting. Many countries where whale watching is popular have begun to pass laws regulating such activities; some guidelines are voluntary, but mandatory restrictions on how close vessels can get to whales are now in place in many regions. In U.S. waters, the Marine Mammal Protection Act defines any disturbance of the whales as "takings," and, thus, illegal. It is difficult to define, however, what constitutes a disturbance, to determine how close an approach can be made, or to determine the best response when the whale makes a move to approach the vessel. Laws sometimes lag behind the rapid expansion of a profitable whale-watching business. For example, there was a rapid increase in whale- and dolphin-watching tours off Argentina and Chile recently, before laws could be implemented limiting such activities.

From the ethical perspective, debates over whaling are often centered on the argument that whales are social animals of higher intelligence and thus deserve special protection. Animal "intelligence" is difficult to define because it is often based on human standards, and this can make the logic of ethical arguments difficult to defend. For the United States and many other industrialized nations, however, concerns over the welfare of the whales are primary arguments making the idea of whaling unpopular. In these nations the vast majority of the general public tends to be opposed to whaling on ethical grounds. The lack of economic incentives combined with conservation values and ethical concerns result in virtually no public or political support for future commercial whaling in these nations.

The ethics of whaling are viewed differently in countries with a stronger recent whaling tradition. For example, surveys of the public in Japan have shown that many do not view whales as deserving of special rights beyond those afforded to other mammals that humans use for food. Keiko Hirata argues that the Japanese public's perception of the whale historically has been as more of a "fish" than a mammal, and thus many do not have a special affinity for whales. These people are more likely to consider conservation-based arguments than arguments for protecting the "welfare" of whales (i.e., to end suffering irrespective of conservation status). To address cruelty arguments, whalers argue that modern whaling methods minimize suffering and the explosive devices used cause rapid death. This argument is refuted by animal welfare groups, who present evidence to the contrary from observers on whaling vessels.

From the perspective of whaling management, history has provides evidence that humans are often unable to control or adequately regulate harvest of vulnerable international resources such as whales. Arguments for the whaling moratorium and CITES trade restriction for recovering whales thus is supported by our inability to limit harvest when there was an open commercial fishery. Arguments that opening harvest and trade could lead to abuses are based on recent documentation of illegal harvest and international trade in whale products. For example, genetic studies of whale meat sold in restaurants and markets in Japan and South Korea have shown that some of the species in markets and restaurants were those for which harvest and/or trade is illegal through IWC and CITES. There is also an unknown amount of unauthorized trade of legally harvested whale products. Documentation was provided through genetic analysis by Scott Baker and colleagues of products sold in sushi restaurants in Seoul, South Korea and Los Angeles, where meat from fin, sei, and Antarctic minke whales was being sold that had been illegally imported from Japan. Pro-whalers counter arguments that whaling is unmanageable by saying that lessons have been learned from failures of the past, and that sustainable management could be carried out if adequate monitoring provisions are set up under the oversight of the IWC.

Figure B10-1 After the whaling moratorium came the realization that the great whales, such as these feeding humpback whales, could be of more value for watching than for harvesting.

From a biological perspective, the life history of whales makes them poor candidates for the establishment of a sustainable fishery. Whales have a very low reproductive rate and thus the populations do not grow rapidly to replace those that have been harvested. Most of the great whales mature late in life compared to fishes, reproduce only every two to four years, and produce few offspring at a time. An individual whale can live 50 to 100 years or more. Some whales have not recovered as expected following over 40 years of protection from harvest. Even if whales are not driven to extinction, the disruption of ocean food webs has unpredictable effects on marine ecosystems.

To counter biological arguments, pro-whalers point out that the whale species they wish to harvest can withstand harvest and still maintain healthy populations. They argue that the minke whale populations are large enough that there is no biological reason that harvest should be prohibited. CITES trade agreements were designed to protect endangered species, and the minke whale is currently not endangered based solely on biological criteria. They argue that the more rapidly reproducing, smaller whales such as the minke have moved into the feeding niche of the larger whales and now inhibit their recovery. Thus, harvesting minke whales could enhance the recovery of the great whales. As pointed out in the text, however, this argument has been countered by recent estimates of historic minke whale population sizes based on genetic studies.

The one area of the whaling debate for which there is a broader agreement among nations is the indigenous exemption of the whaling moratorium. Contentious issues arise regularly, however, regarding what groups or nations should be granted this exception. For example, in the 1980s the United States argued that the Makah tribe of the U.S. Pacific Northwest had reserved their right to hunt whales by treaties signed in the 1800s (**Figure B10-2**). Although gray whales were considered endangered by U.S. law, the Makah were given permission to hunt up to 5 per year beginning in 1998, even though they had hunted no whales since the 1920s. It was agreed that primitive methods would be used, including hand-hewn cedar canoes and harpoons (however, high powered rifles were to be used to kill the whales to minimize suffering). Although no whale was taken, whaling nations and NGOs objected, and environmentalists made attempts to physically hinder the hunts. In a media-hyped incident in 2000, anti-whalers harassed, and in one case swamped the boats of, the Makah whalers. Public pressure resulted in court actions against the U.S. NMFS for approving the Makah harvest, ruling that an adequate environmental assessment had not been provided before the Makah hunt was approved. The Makah were ordered to stop whaling unless they could win an exception to the Marine Mammal Protection Act. Because this would require prohibitive cost and effort to carry out environmental impact studies, for all practical purposes it outlaws Makah whaling.

These cases demonstrate the problems with defining *subsistence* and *indigenous*. Subsistence harvest is limited to animals taken for food as a necessity for survival (see Chapter 11). As a working definition, the United Nations considers indigenous peoples to be "those which, having a historical continuity with pre-invasion and pre-colonial societies that developed on their territories, consider themselves distinct from other sectors of the societies now prevailing on those territories, or parts of them." This would obviously include groups such as Native American tribes or Australian Aboriginals; however, Norway and Japan have not been able to sell argument that they should have rights to harvest whales under this ruling because whaling has been a part of their culture for thousands of years. (Even many Japanese do not realize that eating whale meat only became common in Japan following World War II). In 2002, Japan unsuccessfully attempted to have the IWC disallow requests for subsistence harvest by any group, arguing that it represents a double standard. The IWC continues to make a distinction, not considering Japan or Norway's whale harvest as subsistence, because the market for whale meat is a commercial high-end luxury market, and modern methods could harm population health.

Although it is doubtful that the days of uncontrolled commercial whaling will ever return, it is also unlikely that total protection of all great whales from targeted harvest will be implemented in the near future. The balance between the two extremes will likely continue to change through time. As a conservation success story, however, the post-moratorium harvest of biologically endangered species has been minimal and it is doubtful that harvest continues to contribute to the endangerment of the great whales. The conservation focus has now largely shifted to effects of environmental changes and harm that is incidental to other human activities.

Figure B10-2 Makah Indians whaling off the coast of Washington state in the late 1800s. An attempt by the Makah to return to their whaling traditions in the 1980s was controversial and ultimately unsuccessful.

not be resumed, they typically agree to allow harvest by indigenous groups that use traditional means to take low numbers of whales. Those that are given rights to harvest a limited number of whales under indigenous exceptions to the IWC moratorium include indigenous Alaskans of various tribes, who harvest up to 50 bowhead whales, and occasionally low numbers of beluga, grey, and minke whales. Indigenous tribes in Russia harvest beluga whales, and about 100 gray and 5 bowhead whales per year. Indigenous communities in Canada harvest about 6 bowhead whales, 800 belugas and 350 narwhal per year. Pilot whales are still harvested off Canada and the Faroe Islands of the North Atlantic. In the Faroe Islands the whales are herded into bays and onto the beach, or taken by harpoons from small boats; practices regularly protested by animal welfare groups. From 2 to 60 sperm whales are harvested annually by Indonesia, a non-IWC nation that does not necessarily abide by IWC regulations. CITES restrictions on trade in whale products, however, would keep them from trading to other IWC nations.

◼ Whale Recovery

Even with protections from commercial whaling, many populations of great whales obviously have not recovered to historical levels; however, the degree to which they have recovered is difficult to assess. In order to precisely assess whale recovery, an estimate of the population size prior to the beginning of large-scale whaling is needed as a baseline for comparison. This is a difficult task because reliable population estimates were impossible until recently, and even now are debated. The most common method of *back-calculating* whale population size prior to whaling has been analysis of whaling records. Using estimates of the number of whales harvested, whaling effort, and knowledge of whale life history parameters (e.g., growth and reproduction rates), rough estimates of historic population sizes can be obtained. Unfortunately, many variables can bias population estimates. For example, whaling records can be incomplete or inaccurate due to unrecorded harvest. An unknown number of whales could have been injured and lost but not recorded as catch (loss could have been as high as 50% in some whale fisheries). In addition, some whalers killed whale calves to lure the mother closer to the ship, adding an additional source of unreported mortality.

Despite the uncertainty over historic population numbers, statistical analysis of harvest records provide adequate information to reliably conclude that declines of the great whales were dramatic after the implementation of commercial whaling. Line Bang Christensen developed a reconstruction of global historic abundances of cetaceans and other marine mammals (**Figure 10-7**). The estimated decline in cetacean biomass was over 80% since the beginning

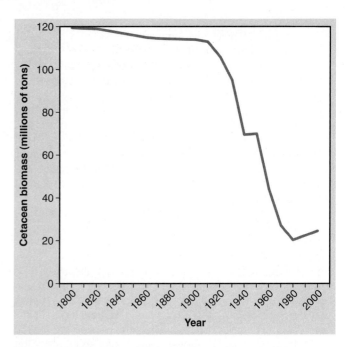

Figure 10-7 Estimated global decline in biomass of cetaceans from 1800 to 2000, based on an analysis of historical data. This dramatic decline and poor recovery reflects the fact that populations of the largest great whales were the most heavily exploited and have been the slowest to recover.

of commercial whaling, with most of that decline due to harvest of the great whales from the early 1900s until the initiation of the commercial whaling moratorium in the 1980s. The most dramatic declines were for the blue whale (almost 99%) and right whales (over 90%). The estimated decline in total numbers of the great whales was about 65%, largely due to an estimated harvest of over 3 million individuals since 1800.

Though whale population estimates are difficult using even current methods, more precise data are available on the rates of population recovery following the moratorium on commercial whaling. These can provide population trends as an assessment of recovery rates and the success of conservation efforts. For example, humpback whales, protected from commercial harvest since the 1960s, have recovered from a population size of a few thousand to about 60,000 currently, and are likely increasing.

Fin whale populations declined from 400,000 to a low of 50,000 in the 1970s, and although commercial harvest ended in 1976, populations have recovered only to about 90,000. Current direct causes of fin whale mortality from human actions include a small harvest by Japan (10–50 whales per year) and Iceland (about 10 per year). Fin whales are one of the most commonly reported whales in vessel collisions: about 5 were recorded off the U.S. coast from 2000 to 2004. It is unlikely, however, that collision-caused deaths inhibited population recovery. Although fin whales have naturally low population growth rates, it is uncertain why populations have not recovered more rapidly.

The North Atlantic right whale, first protected from commercial harvest in 1935, is possibly extinct in the eastern Atlantic and is at about 300 individuals in the western Atlantic. At current population sizes, recent mortalities of about three per year from fishing gear entanglement and one per year from vessel collisions could result in decreased populations (see below). Southern right whale populations, classified by IUCN as least concern, are much healthier at about 3,000 individuals, up from about 1,500 in 1997.

The blue whale has slowly recovered from commercial whaling to populations of 10,000 to 25,000 individuals. Prior to modern whaling with deck-mounted harpoons, population sizes were likely at least 10 times this number. Although commercial harvest of blue whales was outlawed by IWC in 1966, some continued to be taken through the 1970s.

The sperm whale, with pre-whaling populations of 1 million or more, were reduced to about 700,000 by 1880 and to 350,000 by the 1990s. Numbers appear to have remained at about that level since, although some populations are still affected by mortality from entanglement in fishing nets. Populations are currently largely excluded from human interactions in much of their range, but natural population growth rates of about 1% per year mean recovery will continue to be slow.

Minke whales survived commercial whaling better than the other baleen whales. Because of their smaller size, they were not targeted by commercial whalers until the 1950s when larger whales became harder to find. Although populations were reduced by 1982 when the commercial moratorium was put in place, they appear to have recovered well. They are currently the most heavily harvested of the baleen whales, through a combination of subsistence harvest, "scientific research" harvest by Japan, "experimental" whaling by Iceland, and through an exemption under protest of the IWC moratorium by Norway. Total reported harvest is about 1,000 minke whales per year. Due to unreported harvest, however, those numbers are likely higher. Evidence of unreported harvest was obtained in DNA profiling surveys by Scott Baker and colleagues in analysis of minke whale products in South Korea. They found that there were about 80% more whales in the market than were officially reported to have been harvested. And, in Japanese markets the number of minke whales taken from protected coastal populations was shown to be higher than reported, indicating either higher rates of harvest for scientific whaling or greater amounts of incidental catch in fishing nets than was reported.

Population estimates and recovery rates for minke whales is a controversial topic. Based on a concept called the Krill Surplus Hypothesis, minke whales may be affecting the recovery of other great whale populations. This hypothesis is based on the premise that killing of the great whales in the Antarctic would have resulted in higher availability of krill for minke whales. Greater food resources could increase minke populations, which in turn might inhibit the recovery of other great whales through competition. If true, this hypothesis provides an argument supporting pro-whalers' proposal for increased harvest of minke whales to release food resources and assist in recovery of other krill-dependent whale species. Recent studies, however, have not supported this hypothesis. Population estimates based on genetic analysis by Kristen Ruegg and colleagues have shown that pre-whaling abundance of minke whales was similar to current population levels at around 700,000 individuals. This suggests that it is not simply food availability that controls minke whale populations. Other environmental and ecosystem effects are likely to exert an influence as well. These data suggest that removing more minke whales is not a solution to recovery of the great whales (see Box 10-1, Whaling: the Ongoing Debate).

The recent development of genetic analysis methods has provided a new way to estimate historic population size of whales. DNA sequence variation takes into account data on genetic mutation rates to obtain a population history of a species. Through the application of DNA analysis, Stephen Palumbi and Joe Roman estimated pre-whaling population sizes for the great whales that were about an order of magnitude greater than traditional population estimates; for example, humpback whales numbers of about 1.5 million rather than 115,000. If whale populations were historically higher, recovering whale populations are further from their original population sizes than currently assumed and recovery rates may need to be reassessed for making conservation decisions.

Although there has been much concern over the impact of commercial whaling on whale population levels, until recently there has been relatively little discussion of the ecosystem effects of removing most of the largest animals in the ocean from the communities that they often likely controlled. In the North Pacific, whale biomass was reduced 60%, from an estimated 10 million tons to about 4 million tons, within about 100 years. In the Antarctic, biomass was reduced by about 85%. Changes in biomass and whale community composition likely altered energy flow and trophic interactions in ocean food webs. The whales' trophic role may have been replaced by other predators, such as penguins and seals, that feed on krill in the Antarctic. The large population growth of fur seals after the cessation of whaling has coincided with declines in penguins that are also krill predators. In the North Pacific the loss of whales could have resulted in an increase in a major fishery species, pollock *Theragra chalcogramma*, due to reduced competition for food as well as initiating a series of changes in pinniped and sea otter populations (**Box 10-2. Are the Great Whales Important Prey for Killer Whales?**)

The loss of sperm whales could have affected open ocean systems through a **trophic cascade**: fewer sperm whales result in increased numbers of squid, causing declines in prey items of squid. Effects on seafloor ecosystems also are possible due to dependence on sunken whales as a food source, as discussed in Chapter 8. These effects of whale declines are mostly speculative. Defining the role that whale loss has had on ecosystem changes is difficult due other interacting factors, including environmental effects, fish harvest, and changes in climate and oceanographic conditions. In addition, very little is known of the composition of ocean ecosystems prior to whaling to establish a baseline for comparison.

Takings of Dolphins and Porpoises

The reasons for harvesting smaller cetaceans, such as dolphins and porpoises, includes not only commercial uses as food, oil, leather, or bait, but also the removal of animals due to their perceived competition with fishers for prey species. Although this harvest is not monitored as closely by IWC or CITES, evidence of its potential effect is available through surveys carried out in specific regions. For example, in Japan the *drive and net* fishery (in which dolphins are driven in to nets and killed, to supply the Japanese market for small cetacean meat) resulted in a dramatic reduction in populations of the striped dolphin *Stenella coeruleoalba* by the early 1980s. Following this decline, fishers switched to killer whales and bottlenose and other dolphin species. Beluga whales have been depleted in some areas by hunting classified as subsistence based on their use for local consumption. Also, in Madagascar the harvest of over 3,000 individuals of several dolphin species over five years, for consumption and sale of meat, apparently depleted local populations. Although severe overharvest has been reduced in most regions, conservation issues persist. One of these is the lack of international regulations or agreements to control the take of small cetaceans. In localized coastal areas, unregulated and undocumented hunting can affect cetacean populations. There is no international regulation regime for the harvest of small cetaceans in international waters, although some nations, including the United States, protect cetaceans from take by prohibiting landings in their ports through marine mammal protection regulations.

Currently, the most common human-caused mortality of cetaceans is likely through incidental catch (bycatch) in

---

Box 10-2   Are the Great Whales Important Prey for Killer Whales?

In Chapter 9, the hypothesis was presented that killer whale predation has affected pinnipeds and sea otter populations in North Pacific ecosystems. According to this hypothesis, recent changes in killer whale feeding resulted from the loss of the great whales as a food resource after commercial whaling. This would force an expansion of predation by killer whales to include pinnipeds (including the Steller sea lion) and sea otters. To test this hypothesis, it important to know how much killer whales depended on large whales as a food source prior to whaling, or whether they have always depended on pinnipeds. Otherwise, the decline in pinnipeds and sea otters could be due to reasons other than increased killer whale predation. One argument against this hypothesis is that attacks by killer whales on larger whales are rarely observed (**Figure B10-3**); thus, they could not have been an important event even when whale numbers were larger. An analysis of whale numbers and reported predation events by Daniel Doak and colleagues led them to conclude that, even if killer whale predation on great whales is a relatively common occurrence, the chance of humans observing such events would be extremely low. These authors also point out that when great whales were abundant they could have provided an important source of food for killer whales, with little effect on populations of great whales, even when only juveniles or portions of adults are consumed (the tongue appears to be the preferred food choice). Doak and colleagues argue that there is no solid proof, but it is plausible that killer whales are important in controlling pinniped and sea otter populations, and that no other hypothesis explains the changes in marine mammal populations in the North Pacific as well as this one. Scientific proof of the importance of the great whales in these and other community interactions may never be available unless a recovery of great whale populations can be achieved.

Figure B10-3   A young gray whale being attacked by a killer whale in Alaskan waters. Such sightings have become more frequent as biologists and whale watchers spend more time on the water.

fishing nets. This is more difficult to monitor than directed harvest, because the cetaceans killed incidentally by drowning in fishing nets are not typically harvested or reported. Some of the better documented large-scale cetacean bycatch include: the tuna fishery in the eastern tropical Pacific (see below); the kill of Dall's porpoises *Phocoenoides dalli* in the Japanese salmon fishery in the North Pacific; pelagic drift net fisheries in the northeast Atlantic, South Pacific, and Mediterranean; and coastal gill net fisheries off the Americas and Europe. In most cases this bycatch is considered a nuisance to fishers; however, in the some regions (e.g., coastal areas of Peru, Sri Lanka, and the Philippines) markets developed for incidentally caught cetaceans led to directed fishing. Warning devices have been developed that limit some cetacean bycatch; however, this can be an unaffordable cost to subsistence fishers. Often there is resistance to bycatch reduction programs because an individual fisher may experience little bycatch, leading to the perception that it is not a problem. They do not understand the enormous impact of many thousands of small fishers, each taking a few cetations. Fishing also affects cetacean populations through the removal of prey species (see Chapter 11).

Possibly the most controversial source of cetacean mortality is culling to remove animals thought to be competing for harvestable fish species. The best documented of these have been for belugas in the St. Lawrence River of Canada, killer whales in Iceland and Greenland, and dolphins in Japan. Fear of damage to fishing gear by entanglement is also a reason for killing cetaceans. Although some fishers use acoustic devices to keep dolphins away, this is not always recommended because it can exclude dolphins from the best feeding areas. Whether or how much cetaceans affect the availability of fish to fishers is difficult to determine; however, the cetaceans are a convenient scapegoat due to their high visibility to fishers, even when predatory fish consume more fish than cetaceans.

10.3 Unintended Anthropogenic Harm to Cetaceans

Pollution

Global pollution has the potential of having a greater effect on cetaceans than even extensive whale harvest. Pollution impacts are much harder to define and control than harvest, bycatch, or collisions. Any of the sources of ocean pollution discussed in previous chapters potentially could affect whales and dolphins. Because they are feeding at the top trophic level of food chains, bioaccumulation is especially problematic for the Odontocetes (toothed cetaceans) including dolphins, porpoises, sperm whales, and killer whales. High concentrations of pesticides, heavy metals, PCBs, and other persistent organic compounds have been

measured in the tissues of many cetaceans, as discussed in more detail in Chapter 7.

Although the global impact of pollutants on cetacean populations is difficult to know, in some coastal whale populations there is strong evidence of the pollution effects. For example, over one-fourth of the documented recent deaths of beluga whales in the Saint Lawrence Estuary were attributed to cancer, likely a result of the high levels of industrial pollution. High levels of mercury and dioxin found in some whale meat sold for consumption in Japan has possibly hindered the trade and marketing of these products. It would be ironic if the toxins we are expelling into the environment globally end up indirectly saving the whales from harvest. It would be tragic, however, if those toxins end up poisoning the whales that have otherwise been saved. High levels of persistent pollutants such as PCBs have been found in dolphins and whales far from pollutions sources, and are sometimes associated with reproductive abnormalities and disease. Biomagnification is especially severe for killer whales that feed at the top of ocean food chains. Oil pollution can have toxic effects on cetaceans from breathing contaminated air or eating contaminated prey, and cause mechanical damage to baleen and thus impair baleen whales' ability to feed. Following the 2010 *Deepwater Horizon* spill, about 100 cetacean carcasses were recovered for which deaths were attributed to the oil spill; however, Rob Williams and colleagues argue that actual deaths could have been as much as 50 times higher since all carcasses would not likely be recovered.

Mass die-offs of cetaceans, though rare and in restricted areas, have been documented in recent years in various locations around the world including the Mediterranean Sea, the Gulf of California, the Black Sea, and the Persian Gulf. Many times a source of pollutants causing the death eventually becomes evident. In some cases, the die-offs have been associated with toxins from harmful algae blooms by dinoflagellates (see Chapter 7). In other cases they were the result of the release of toxic substances. Die-offs become apparent when dead cetaceans float ashore or apparently healthy cetaceans beach themselves before eventually dying from exposure; the cause of these strandings is often difficult to identify (**Box 10-3. Noise Pollution and the Mysteries of Cetacean Strandings**).

Another potential impact to cetaceans is from a source not typically considered a pollutant: anthropogenic noise that functions as noise pollution. Noise transmitted into ocean waters from human activities has been attributed to strandings, increased stress, avoidance and other changes in behavior, and masking of sounds used in communication. Specific effects of noise pollution have been difficult to prove because no physical substance is left behind as evidence; however, physical damage to the internal organs,

Box 10-3   Noise Pollution and the Mysteries of Cetacean Strandings

Cetaceans have been stranding themselves on beaches for unknown reasons at least as long as humans have lived near the coast; thus, it is presumed to be a natural phenomenon. In some situations it is a single whale or dolphin that is injured or in poor health, or washed ashore already dead. Many mass cetacean strandings, however, have been documented where tens or hundreds of apparently healthy individuals of the same species swim onto shore and remain alive but unable to return to the water, eventually succumbing to the stresses of exposure. In some cases, animals returned to the water swim back to shore and beach themselves again, possibly due to the instinctive desire to remain with rest of the pod even if it is stranded onshore. The most commonly documented mass stranding around the world are of toothed whales, such as pilot, sperm, or beluga whales.

Hypothesized natural causative factors of strandings include variations in climate, current patterns, bottom features, earthquakes, and errors in navigation using magnetic configurations (whales may use crystals of magnetite in their brains as a compass for navigation). A confused or ill member of a pod (infection of the inner ear can cause disorientation) could lead others to be stranded due to high group cohesion. Individuals could come ashore to seek safety from predators such as killer whales and then become stranded. It is doubtful that any one of these could explain all strandings, and it is likely that there are different causes for stranding and the simultaneous occurrence of multiple events may be necessary to cause a mass stranding.

Through scientific studies, progress has been made in defining the cause of some strandings, linking them to specific events. Care must be taken, however, not to extrapolate those causes to other events without evidence. Looking for patterns or common factors associated with stranding is one step toward establishing cause-and-effect relationships. For example, Karen Evans and colleagues analyzed strandings in the region around Tasmania, where they are relatively common (records since the 1920s show years with more than 20 mass strandings, some involving over 300 individuals). They found clear 12 to 14 year cycles in the numbers of strandings that could be linked to climate patterns; for example, more stranding occurred when nutrient-rich waters were closer to the coast. Although these studies do not define a cause, it suggests that strandings may be related to the distribution of food sources as affected by oceanographic conditions. Other studies provide evidence of links between the location of strandings and magnetic variations, current patterns, and seafloor features.

The fraction of individuals that are lost to stranding is unknown due to the difficulties of making precise estimates of cetacean population sizes. If most strandings are the result of natural events, then strandings may not be of conservation concern because it is likely that populations have tolerated the loss of some individuals to strandings throughout their existence. If, however, anthropogenic factors have increased the frequency of strandings, conservation measures should be considered. Evidence of human effects includes an apparent increase in the frequency of strandings in recent years and the close link of some stranding to specific human actions. Defin-

ing the historical frequency is difficult because the increased number of reported strandings would likely be affected by increases in coastal populations, boating activities, awareness, and reporting ability. The best way to define whether human activities have increased the frequency of cetacean strandings is to look for associations with specific activities and stranding incidences. The most likely candidate and most commonly hypothesized anthropogenic cause of stranding is noise pollution.

Although not traditionally considered as pollution, excessive noise can affect the behavior and survival of cetaceans. Whales and dolphins have sensitive hearing, allowing them to communicate, find prey, or navigate through the use of sound. It is not surprising that such adaptations would develop in cetaceans, because sounds travel much farther and faster in the ocean than through the air, and sight is often of limited use underwater (especially in turbid waters or at great depths). Attention has been drawn to noise pollution and cetaceans in recent years after a link between the military application of sonar and whale strandings was documented and widely reported in the news media. Scientists and environmentalists began to wonder if the mystery of whale strandings had been at least partially solved.

The strongest association between whale strandings and noise pollution is for beaked whales (Ziphiidae) and the use of mid-frequency active sonar (MFAS) in naval ships (intended for submarine detection). Behavioral responses could lead to strandings when whales swim away from the noise into shallow water. But the proposed effect on beaked whales is hemorrhaging around the brain and inner ear or gas bubble formation caused by decompression sickness from surfacing too rapidly (as with scuba divers rapidly surfacing from depth). Beaked whales might be prone to such conditions because they feed at greater depths than most other cetaceans. Much attention has been drawn to this issue through the news media, and the issue has been hotly debated, leading to court cases attempting to limit the U.S. Navy's use of this technology. In 2008, the U.S. Supreme Court ruled that the U.S. Navy could continue training with MFAS sonar but should develop methods less likely to be harmful to whales.

Angela D'Amico and colleagues analyzed historical data on strandings and use of MFAS in an attempt to clear up the issue through a scientific analysis. They found that 93% (126) of the 136 reports of beaked whale massed strandings since 1874 occurred from 1950 to 2004, since the development of MFAS; about 73% of these were after 1980 (**Figure B10-4**). (About one-half of these *mass strandings* were of two individuals each; 82% contained four or fewer). This does not necessarily indicate a cause–and-effect relationship but could simply reflect an increase in reporting; however 10% of the 126 strandings coincided directly with naval activity that likely included the use of MFAS. Further evidence of the effect of MFAS is provided in the few cases where there is evidence of trauma that could have led to whale strandings and deaths. A total of 27 other strandings were near ships or naval bases with no evidence of sonar use (the associations with naval personnel were possibly because they were the only inhabitants in the area of the stranding). The conclusions drawn from this study were that

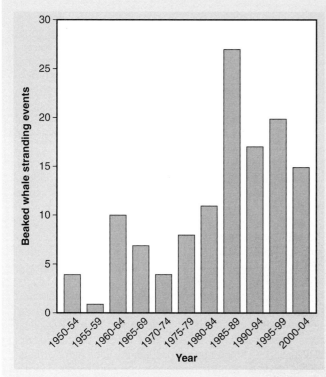

(a)

(b)

Figure B10-4   **(a)** Reported stranding events for beaked whales from 1950 to 2004. There was an increase in the use of naval sonar, implicated in some strandings, during this time period; however, increased awareness and monitoring effort also would elevate observed events. **(b)** A beaked whale, showing tusk-like teeth, a secondary sexual characteristic in males. Death and stranding may result from stress when surfacing too rapidly from deep waters if startled by military sonar.

some of the post-1950 beaked whale strandings were likely the result of naval sonar activities, but that most sonar activities take place without strandings. A co-occurrence of several factors, therefore, is likely necessary for beaked whale strandings.

Studies by Peter Tyack and colleagues provided direct observations of the response of beaked whales to Navy sonar. Whales tagged to monitor movements and sound production (as an indication of foraging activities) were observed to stop

feeding and move out of foraging areas near sonar transmission, but gradually returned when transmissions were stopped. When the whales were exposed to the playback of sounds produced by the sonar (or sounds of killer whales) they stopped foraging for food and made a long, slow ascent to the surface away from the sound. Tyack and colleagues concluded that Navy sonar can disrupt the behavior of beaked whales, even at relatively low sound levels; however, the risk of stranding can be reduced through monitoring and management.

Whether strandings caused by sonar activity have a long-term effect on the whale populations remains unanswered. Based on documented stranding, the total numbers of deaths are unlikely to have a substantial effect, except possibly on small isolated populations. There is no reliable way of estimating the number of undetected strandings around the world, however, and effects of noise on cetaceans are not always evidenced by strandings. It also must be considered that naval sonar could cause injury or death without strandings, and mortalities at sea would be almost impossible to identify; therefore, it is unknown how common such incidences could be.

Other sources of noise pollution, though at lower intensities, are ubiquitous in the ocean. Boat and industrial noises have created a new sound environment to which cetaceans must adjust; it is most intense in populated coastal regions but also heard throughout the open ocean. At close distances there could be physiological effects. Temporary hearing loss can occur in dolphins at sound levels comparable to those induced by boat noises. Explosions used in coastal construction or during oil exploration produce shock waves that could cause organ damage or rupture ears or lungs. At lower intensities, sounds could mask those produced by cetaceans for communication. Noise pollution could affect communication in large baleen whales, such as blue whales and fin whales, which produce mating calls and other communications transmitted over hundreds of kilometers distance. Noises from boats, military sonar, and seismic surveys have been associated with changes in feeding and diving behavior and sound production in various cetacean species.

Sounds can affect cetaceans by displacing them from a preferred habitat. For example, gray whales have been displaced from breeding lagoons for years by industrial sounds, dredging, and shipping activities. Seismic surveys have displaced and changed behaviors of cetaceans including gray whales, killer whales, harbor porpoises, and sperm whales. Boat noises from whale-watching activities has affected the behavior of humpback whales, resulting in increased restrictions in some regions on how close whales can be approached. Indirect harm to cetaceans is also possible where noise pollution affects prey species. For example, noise can affect the behavior of fishes by scaring them from an area, decreasing their activity, or causing physical damage, all of which could affect the success of feeding by cetaceans and other predators.

Other than where injuries or mortalities are documented, it is difficult to determine how much noise pollution affects the health of individual cetaceans or the stability of populations. This is a situation where the precautionary approach to conservation is needed to minimize the potential harm by taking measures to reduce ocean noise pollution. In an effort to reduce noise pollution, the U.S. NMFS has established minimum levels for industrial noise in areas where cetaceans are present.

Figure B10-5 U.S. and Canadian vessels participating in sonar surveys of the Arctic sea floor. Observers are stationed so that activities can be modified if marine mammals are spotted in the vicinity.

Safety zones are commonly established in which visual observers scan for cetaceans during seismic survey activities. Sound production is stopped if cetaceans approach within a certain distance (**Figure B10-5**). The effectiveness of such policies is questioned, however, because some cetaceans cannot be observed at the surface, especially deep-diving whales like the beaked whales.

In a review of noise effects on cetaceans, Linda Weilgart recommends reducing the effects of noise pollution by limiting overall noise levels in the ocean environment. These include construction of quieter ships, using quieter alternatives to the air guns used for seismic surveys, and altering naval sonar to reduce risk to cetaceans, all of which are technologically feasible. Other actions already being taken include distancing noise pollution from areas important to cetaceans for breeding, feeding, and migration. For example, the use of sonar by the Spanish military is not allowed in whale-inhabited regions of the Canary Islands. In Australian waters, a Marine Mammal Protection Zone was established where oil and gas exploration is not allowed and boat traffic is seasonally restricted. In humpback whale breeding areas off Brazil, seismic surveys are prohibited. In the United States, lawsuits have been filed by environmental groups in an attempt to restrict the use of military sonar in regions frequented by whales through endangered species and MMPA legislation.

With increased attention to noise pollution, additional measures have been recommended for protecting cetaceans. These include additional Marine Protected Areas, with a surrounding noise buffer zone. Cargo ship traffic could be minimized by filling ships to capacity, and seismic surveys could be minimized by avoiding redundancy through data sharing. Monitoring can be expanded to define the effects of anthropogenic noises on cetaceans and determine the sources of noise pollution. Overall, Weilgart recommends a precautionary approach by reducing overall noise levels in the oceans and keeping noises from biologically important areas. Such efforts will require national management efforts and international cooperation.

especially the inner ear, can provide ample evidence. Some cetaceans appear to be more sensitive than others, apparently due to their preferred habitats and feeding behaviors (Box 10-3, Noise Pollution and the Mysteries of Cetacean Strandings).

Climate Change

Changes in the community composition of marine ecosystems can potentially affect populations of marine species, including cetaceans, through changes in the abundance and availability of food resources. This could be true even for populations of the great whales, which are capable of migrating great distances but dependent on seasonal availability of food resources in specific ocean regions. The whales most vulnerable to the effects of climate change would be those currently most endangered by other anthropogenic factors. For example, the North Atlantic right whale, at population numbers that make it one of the most highly endangered of the great whales, is fighting to maintain current population levels while being harmed by ship strikes and net entanglements. During breeding the right whales depend on abundant plankton food resources, including large blooms of *Calanus* copepods. Analyses by Charles Greene and Andrew Pershing discovered that in years when these resources are reduced due to oceanographic and climatic factors, reproductive rates for right whales are reduced, in part due to females skipping reproduction in years when there has been inadequate food to support their three-year reproductive cycle. Whether climate change will enhance or reduce production of food that affects right whale reproduction is unclear. A warming trend could increase copepod production and a cooling trend likely would reduce production; however, an increase in climate variability could have an overall effect of reducing reproduction rates for right whales. For species with large healthy population levels, such vascillations probably would not have long-term effects on population size. But with highly endangered species, including but not limited to the North Atlantic right whales, there is fear that even proportionately small changes in production rates would turn around their already slow recovery and push the species closer to extinction.

An unanticipated effect of increased anthropogenic input of carbon dioxide into the environment, commonly associated with global warming and lowering ocean pH (see Chapter 2), is its impact on ocean sound transmission. Keith Hester and colleagues found that lower ocean pH significantly decreases absorption of lower-frequency sounds, creating a noisier ocean. The greater transmission of anthropogenic noises could affect cetaceans by elevating levels of noise pollution and interfering with communication.

Ship Collisions

Although whale strandings garner much attention because of their high visibility, ship collisions (commonly called *ship strikes*) likely are a more common cause of deaths of the great whales since the moratorium on commercial whaling (**Figure 10-8**). Some of the increase in whale stranding could be attributed to collisions between ships and whales. Although lethal ship strikes to cetaceans were documented even before the days of motorized vessels (a collision with a sperm whale and sinking of the whaleship *Essex* in 1820 was an inspiration for Herman Melville's novel, *Moby Dick*), the rapid increase in the number and speed of modern vessels, along with the increased vulnerability of endangered populations makes ship strikes a more serious conservation issue. Most ship collisions with whales occur in shallow coastal waters where whales spend more time near the surface while feeding, mating, or nursing. Collisions with ferries, cargo ships, and cruise ships are documented annually in waters of the United States, Europe, Japan, and elsewhere. Ship strikes also have killed fin and sperm whales in the Mediterranean Sea, right whales off Argentina, and sperm whales off the Canary Islands. Mortalities from vessel collisions are considered to be a factor in declines of populations of small dolphins and porpoises in some regions; the common occurrence of wounds and scars attests to the frequency of collisions.

Right whales appear to be particularly sensitive to injury or death from collision with vessels (**Box 10-4. Right Whales in the Wrong Place**). They show little response

Figure 10-8 Humpback whale struck by a cruise ship off Alaska. Whale–ship collisions have become more frequent with increases in shipping traffic. Whales often do not respond to avoid ships.

to the noise or visual presence of the boat, appearing to have become habituated to such activities. This might seem remarkable for an intelligent mammal; however, the great whales evolved in a world with few predators, none of which approached them from the surface. Thus, they would not have evolved an instinctive avoidance (this might also explain the ease with which right whales and others could be approached by whaling vessels to be harpooned). There is little deterrence of right whales, humpback whales, or minke whales with acoustic alarms. In fact, it has been known to make them approach more closely, and causes right whales to surface quickly where they are exposed to a greater risk of collision.

Incidental Effects of Fisheries on Cetaceans

Currently, incidental harvest causes far more deaths of whales and dolphins than does commercial whaling. The numbers that are killed are unknown, but U.S. scientists have estimated that over 60,000 whales, dolphins, and porpoises die annually from being caught in fishing gears. There is also incidental catch and killing of larger whale species in fishing nets. For example, humpback whales are vulnerable to capture in fishing nets during their movements through coastal waters; most humpback whales off the east coast of North America have scars from encounters with fishing nets. Marine mammals are particularly vulnerable to gill nets. Many countries have outlawed their use in coastal waters; however, there are many exceptions. For example, in the North Sea it is estimated that as many as 10,000 harbor porpoises *Phocoena phocoena* may be caught per year in nets used to catch cod and other species. Fishers have a vested interest in limiting this type of incidental catch because a large cetacean can cause the loss of a fishing net.

In some regions incidental harvest is limited by reducing or stopping fishing during certain seasons. In the Gulf of Maine, fish bans, observers on fishing boats, and the use of *pingers* have reduced porpoise bycatch. Pingers are acoustic alarms that can reduce dolphin mortality by over 90% in ocean fisheries where gill nets are still legal. One of the problems, however, is monitoring and enforcing the use of pingers. Because coastal gill netters are often small-scale fishers, the manpower is not available to monitor each fisher. The moratorium on the use of open ocean gill nets (called *drift nets*) has also been difficult to enforce (see Chapter 11). Fishing vessels from several countries have been caught using drift nets in the Pacific Ocean and Mediterranean Sea since the agreement was made.

One of the highest profile of recent marine mammal conservation issues has been controversies over incidental harvest of dolphins in tuna purse seines. Purse seines are large surrounding nets common for capturing large

Box 10-4   Right Whales in the Wrong Place

The North Atlantic right whale is the most endangered of the great whales, with populations of about 300 individuals. Loss of a small number of individuals could have an effect on population survival. If current rates of population declines continue, the North Atlantic right whale could be extinct by 2200. The elimination of a few female deaths per year, however, could reduce mortality rates enough to result in a slow population recovery. The way to accomplish this is by reducing deaths resulting from human activities, mostly attributed to two sources: vessel strikes and entanglement in fishing gear.

Ship strike mortalities are especially problematic near coastlines and bays where shipping traffic is heavy. The most practical way to decrease the likelihood of deaths from vessel strikes is to route vessels around whale habitats and reduce vessel speeds. Although slower speeds do not eliminate strikes, they do substantially reduce the likelihood of lethal injuries. One of the regions where the chance of collisions is most likely is in feeding areas in the Western Scotian Shelf/Gulf of Maine region at the border between the United States and Canada, the primary feeding grounds for these whales. The Canadian Right Whale Conservation Area has been established in a portion of this region. Within this area there is a program set up to warn vessels of the likely presence of right whales; however, no restrictions are in place for speed reductions, though they are recommended in the conservation area. Traffic schemes have reduced the risk of collisions by about 10% to 30% for baleen whales using the conservation area. A study by Angela Vanderlaan and colleagues, however, found no clear evidence of vessels slowing when passing through the conservation area. Many vessels travel at near to or greater than 25 kilometers per hour, speeds at which a collision has a greater than 50% probability of resulting in death of the whale. They argue that the greatest reduction in collisions between vessels and whales, with minimal disruption in operation of vessels, would be achieved through speed restrictions. In both the United States and Canada recommendations for reduced speeds are in place; however, it appears that legal restrictions will be necessary to substantially reduce the number of lethal collisions between the endangered North Atlantic right whale and marine vessels.

The other primary source of North Atlantic right whale mortalities is entanglement in fishing gears, including ropes associated with fixed-gear fisheries (gill nets or traps placed stationary in the water; see Chapter 11; **Figure B10-6**). Many different ropes are used with these gears, but two types are of concern: buoy lines, attached from the fishing gear to the surface, and groundlines, attaching traps together in a series. The fishery that is of particular interest due to its large size, mainly found along the coast of the extreme northeast United States and southeastern Canada, is the American lobster trap fishery. Over 4.5 million lobster traps are set in these fisheries by about 13,000 licensed fishers. Even though the rate of encounter and mortality of whales is relatively low, the large number of lines in the water led to concerns that resulted in the restriction of the lobster fishery in U.S. waters. These restrictions require that traps in U.S. waters replace floating groundlines with sinking groundlines to reduce encounters with right whales, which spend most their time near the surface. These laws have been controversial because the lobster fishers had to replace their floating lines with more expensive sinking lines in an effort that many believed was not necessary. An analysis by Sean Brillant and Edward Trippel in the Bay of Fundy lobster fishery (where sinking groundlines are not required), found that few groundlines rose far enough above the bottom that they would likely be a threat to entangling whales. They suggested further studies are needed on the diving behavior of the right whales that lead to encounter with the lines. Buoylines, attached to floats at the surface, may be a greater risk to whale entanglement, and future studies should focus on defining regulations that would reduce this risk.

Figure B10-6 Biologists work to disentangle a right whale from ropes. Such efforts are dangerous, often unsuccessful, and frequently too late. An unknown number of cetaceans have died as a result of gear entanglement.

schooling fish species such as some tunas (**Figure 10-9**). The purse seine fishery for the yellowfin tuna *Thunnus albacares* in the eastern Pacific Ocean is one of the primary suppliers of tuna for the canned tuna market. Dolphins often swim with tunas in this region, and presumably the two species assist each other in locating prey fishes. Hundreds of thousands of dolphins were killed in the tuna purse seine fishery in the 1960s and 1970s (**Figure 10-10**). When news of this reached the public through NGOs and the news media, there was a large public outcry. For example, school children flooded the canned tuna companies with letter writing campaigns deploring the killing of the dolphins. This protest was one of the initiatives that led to the Marine Mammal Protection Act. The MMPA, along with other laws, has required U.S. fishers to reduce the total number of dolphins killed in the yellowfin tuna fishery each year. Unfortunately, after stringent U.S. laws were passed, many U.S. tuna fishers got out of the tuna fishery altogether, selling their boats to fishers in Mexico where regulations were not as strict. As dolphin mortality in the United States declined, it began to increase in the foreign fleet.

Eventually, U.S. purse seine fisheries developed techniques to limit dolphin mortality and continue to fish for the yellowfin tuna in the eastern Pacific. One of these techniques is called *backdown*, where dolphins are allowed to escape from the net by forcing one edge of the float line below the water by pulling away from the net with the boat. Often the fishing crew monitors the net from a small boat, assisting struggling dolphins out of the net. U.S. fishers are also required to use a small mesh panel in the part of the net that the dolphins swim over so that the dolphins do not entangle their snouts in the net and drown.

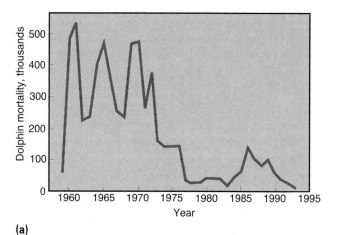

(a)

(b)

Figure 10-10 **(a)** Annual mortality of dolphins in the tuna purse seine fishery in the eastern tropical Pacific Ocean, 1960–1993. Increased regulation and public awareness through dolphin-safe tuna campaigns forced fishers to limit their dolphin bycatch. **(b)** Spotted dolphins trapped in a purse seine net. Even if released alive, these dolphins have been severely stressed and may not survive or be able to avoid predators.

Figure 10-9 Tuna purse seine boat. The smaller boat on the stern is released to pull the net and encircle the school of tuna. Dolphins that associate with the captured tuna can be stressed or killed during the fishing operations.

In an attempt to force the Mexican fishery to limit dolphin harvest, the U.S. government proclaimed that Mexico exports to the United States would not be allowed unless dolphin bycatch was reduced. The World Trade Organization (WTO), however, ruled that the United States could not use trade sanctions to force its environmental regulations on other countries by restricting trade (this ruling has remained controversial; see Chapter 12). The United States got around this problem by creating an ecolabelling program; in order for tuna to be marketed as *dolphin-safe* (as most U.S. canned tuna is currently), dolphin mortality must be eliminated. The United States would not be required to import unmarketable tuna to be sold on the U.S. market. This indirectly forced other nations to change fishing practices. In the United States, the purse seine vessel must have a qualified observer on board to document that no dolphins were killed during the tuna fishing operation. This has kept U.S. fishers from intentionally setting their nets on dolphins. There is still concern that dolphin mortalities continue, however, due to the stress of being pursued, circled, and released by the purse seine fishers, especially in countries where monitoring is poorly enforced.

Part of the debate concerning the dolphin-tuna issue is whether the death and harassment of the dolphins affect population health. The high mortalities of dolphins prior to restrictions on the purse seine fishery resulted in declines of two dolphin populations (the eastern spinner dolphin *Stenella longirostris orientalis* and the northeastern pantropical spotted dolphin *Stenella attenuata attenuata*) to about 25% of their pre-fishery levels. This resulted in a depleted classification under the U.S. MMPA. After the protective legislation passed in the 1990s, reported bycatch mortality was less than 0.1% and a rapid population recovery was anticipated. Population recovery through 2010 has been much slower than expected, however. Possible factors affecting the recovery include stress and mortality as a result of harassment from chasing and encircling the dolphins, especially affecting females that were nursing young. A study by Katie Cramer and colleagues found support for this argument. They found that purse seine fishing was associated with a decline in the proportion of females with calves, which could result from separation of mothers from calves, increased predation, or induced abortions. Increased mortalities also could be from unreported bycatch, effects of climate change, or El Niño events.

10.4 Captive Cetaceans and Animal Rights

For most people, the only chance for direct exposure to whales and dolphins is in sea parks or aquariums; this visibility assists conservation efforts by enhancing awareness among the general public. But keeping cetaceans in captivity is viewed by some as unethical and unsupportive of conservation. Live capture of small cetaceans is most common in small island nations or less-developed countries where regulations are lax or poorly enforced. In recent years, bottlenose dolphins and other cetaceans have been captured off Cuba, Mexico, Bahamas, Cambodia, Myanmar, and other nations with little disclosure of numbers or evaluations of population status. In the United States, special permits are required to keep any marine mammal in captivity. Sea parks are not capable of maintaining facilities to keep large baleen whales. Dolphins and killer whales are readily kept in captivity; however, some people believe that conditions are too unnatural and stressful even in the best of facilities. In the United States, those wishing to keep captive cetaceans must be granted a special permit by the National Marine Fisheries Service as an exemption to the Marine Mammal Protection Act, and must comply with standards for handling, care, and transport under the Animal Welfare Act. Although performance in dolphin shows

Figure 10-11 Wild dolphin feeding and swim-with-dolphin programs, though still present in some regions, are illegal in the United States under the Marine Mammal Protection Act.

was historically the major reason for keeping captive dolphins, in recent years swim-with-dolphin programs have become popular in many locations around the world. For example, such programs are common in some areas of the Caribbean, with many of the dolphins supplied from Cuba. Some environmental and animal rights groups have actively petitioned against captive-dolphin shows and swim-with-dolphin programs. They argue that capture of the dolphins in the wild can be cruel and sometimes cause death of dolphins, and that it is unethical to keep dolphins confined, sometimes in stressful conditions. Furthermore the feeding and swim-with-dolphin programs can be dangerous to humans, due to dolphin aggression, and unhealthy for dolphins (**Figure 10-11**), and the dolphins are sometimes abused by humans. Laws and regulations become difficult to implement when the business ventures that rely on such activities are put in place before government regulations can be enacted. Once a strong economic incentive is established to continue such programs, the resistance to change becomes stronger. Prohibition or strict limits have been place on captive-dolphin facilities in many countries such as Brazil, South Africa, Australia, and Israel. According to U.S. federal law, it is still legal to take dolphins from the wild with a permit; however, no permits have been granted in the United States since 1989 because injured animals rescued from strandings and captive births have been adequate to supply dolphins for the captive dolphin industry (**Figure 10-12**).

Whether dolphins or whales should be kept in captivity presents a controversial ethical argument. Some animal welfare groups argue that keeping mammals of such intelligence in captivity is analogous to imprisonment. Marine park representatives counter by saying that the animals are treated well and have much social interaction with others,

Figure 10-12 Trained bottlenose dolphins that had been washed from their aquarium in Gulfport, Mississippi by storm surge from Hurricane Katrina in 2005. Some of the dolphins were in poor health, and all were captured and returned to captivity for fear they would not have skills to survive on their own in the wild.

including humans. In some parks with open access to the ocean, dolphins will not leave even if given a chance, suggesting this is a desirable arrangement. This is countered by arguments that it is the dolphins' desire to stay near a reliable source of food and it may be the only life to which they are accustomed. Once removed from the wild, cetaceans may not be able to successfully socialize or feed if released (**Box 10-5. Freeing Willy**).

Concerns over cetaceans in captivity largely hinge on whether they should be afforded special treatment not afforded to other marine animals (for example, there is little controversy over keeping fishes in captivity). This begs the questions: is the purported intelligence of whales and dolphins a valid reason for humans to treat them differently than other animals? Are they able to feel emotions, such as love and compassion, similar to those that humans feel? If so, is this a reason for special considerations; and should they be accorded "inalienable" rights that are comparable to those we give to other humans? These questions cannot be answered by science; they will always be influenced strongly by perceptions about animal ethics and cultural points of views.

In North America, surveys indicate that the majority of the general public agrees that dolphins and whales should be accorded special protections and considerations. The U.S. Marine Mammal Protection Act thus receives broad support from political leaders and the general public, and the overwhelming majority supports the moratorium on commercial whaling. Still, decisions regarding special rights are rarely simple. In the peak of the whaling era, it is likely that whalers with morals similar to those of Americans

living today did not think twice about killing whales for their livelihood. No doubt some realized the intelligence and special nature of whales. Even today, a whale that becomes entangled and causes the loss of a fishers' net, or dolphins that are perceived to be competing with fishers for a profitable fishery product, are not viewed in the same way as a whale watcher perceives these animals. Groups involved in harvesting whales or using whale products may view them differently than they do fishes, but often argue that they are no more special than other mammals that are killed and eaten by the majority of humans, even in countries strongly opposed to whaling (see Box 10-1, Whaling: The Ongoing Debate). Regardless of what is thought as the best treatment for marine mammal populations from a conservation perspective, some people will always be at odds when dealing with the emotional arguments over marine mammal use or protection.

10.5 Cetacean Conservation Solutions

There is no one simple solution to cetacean conservation dilemmas that will ensure the protection of sensitive populations, recovery of declining populations, and protection of habitat and food resources. The critical issues of cetacean conservation differ from issues important for many other large marine animals in that directed harvest of cetaceans is rarely the most critical concern and ethical concerns are important in driving conservation decisions. Cetaceans populations are often more difficult to recover due to their k-selected life history (see Chapter 2), including factors such as low reproduction rates due to low offspring production and high parental care, large body size, great longevity, and need for large amounts of food resources.

Randall Reeves and colleagues, through the IUCN, make specific recommendations to solve cetacean conservation problems. These include the sustainability of cetacean catch. Although there is strong resistance from many of the industrialized western countries, substantial targeted catch continues in many regions, including the Arctic, Japan, Norway, the Faroe Islands, Peru, and the West Indies. Although whaling is closely regulated, illegal harvest continues. Further regulations are needed for take of smaller cetaceans, along with scientific monitoring to assess population impacts.

A second recommendation is to reduce incidental mortality in fishing nets by modification of gear and using deterrent devices. Because deterrents do not always work and may have negative effects on cetacean survival, scientific studies are needed to assess their desirability. Developing methods to reduce bycatch requires research, and will not be used by small-scale fishers if too expensive or difficult to use (see Chapter 11). Education is needed along with regulations.

Box 10-5 Freeing Willy

Possibly the most well-publicized attempt to return a whale from captivity to the wild is presented by the "Free Willy" saga. The killer whale, given the name Keiko by its trainers, was captured in Iceland in 1979 and transported to a marine parks in Ontario and then Mexico before starring in the 1993 movie, *Free Willy*. Following the production and release of the movie, Keiko was discovered in poor conditions in a Mexico City aquarium. The Free Willy Keiko Foundation was established through fund raising efforts and Keiko was airlifted to an Oregon aquarium for rehabilitation, with the goal of returning him to the wild. This was not expected to be a simple process because Keiko had lived almost his entire life in captivity and it was feared he may be unable to survive in the wild. Whale experts predicted that a naïve whale released into the ocean would not have adequate survival or social skills to intermingle with wild killer whales. So, Keiko was airlifted once more (**Figure B10-7**) to Iceland, where his handlers attempted to teach him to catch live fish and to integrate him into a pod of wild killer whales. After escaping

from a training session, Keiko swam over 1,300 kilometers to the Norway coast. He followed a fishing boat into a bay where he interacted with people swimming there. Keiko's trainers arrived, finding him emaciated, and coaxed him away from people, planning, unsuccessfully, that he would be adopted by a passing pod. The *Free Willy* saga ended when Keiko beached himself and died, apparently ill from pneumonia. Willie was buried on land, and a memorial was set up in Halsa, Norway nearby.

Looking back, it is debated whether the rehabilitation and release of Keiko was a good idea. After such a long period in captivity, apparently the killer whale was unable to learn to feed, socialize with other whales, and adapt to living in the wild. The efforts cost more than $20 million, money that some felt would have been better spent on protecting wild whale populations. Groups that worked with the rehabilitation and recovery argue that the attempt served a benefit by teaching us more about killer whales, and assisted conservation efforts by bringing attention to the plight of all whales.

(b)

(a)

Figure 10B-7 **(a)** Keiko, the killer whale of the *Free Willy* movies, was weighed and loaded into a transport tank at the Oregon State Aquarium in 1998. **(b)** The U.S. Air Force was hired by the Free Willy Keiko Foundation to transport Keiko in a cargo loader to Iceland for return to the ocean.

Another way to reduce mortality is through rescue and release efforts for cetaceans trapped in fishing nets or stranded on beaches. These are particularly useful to protect populations of endangered whales. In the United States and Canada there is government support for such efforts, and programs have been established in the Mediterranean Sea. Even when populations are not endangered, assisting stranded or entangled cetaceans is beneficial for increasing public awareness and increasing scientific knowledge.

Cetacean-oriented tourism, such as whale- and dolphin-watching programs, need to be managed to avoid harm or stress. Although more guidelines are being established, they need to be more broadly implemented by government agencies. Further scientific studies are needed to make decisions about establishing guidelines.

More and larger marine protected areas are needed. Many of those that are in place need further restrictions to exclude activities such as fishing and high-speed vessel traffic that can be harmful to cetaceans. Education campaigns are needed to enforce restrictions, convincing locals of opportunities in nature tourism and the role protected areas play as breeding areas or nurseries for fisheries species. In some cases, closure of an area for a specific time when cetaceans are present is adequate to protect feeding, migrating, or breeding populations.

Environmental pollution needs to be reduced. Of course, this is important not only to cetaceans but to all marine environments and organisms. Further studies are needed, however, to define the effects of pollution on cetaceans. Government and NGOs could take advantage of people's special affinity for cetaceans to enhance the public's support of pollution reduction measures.

Noise pollution needs to be minimized. Although it is unrealistic to expect a return to natural sound conditions in the oceans, there are many practical ways to reduce sound by improved machinery and boat motors. Governments need to treat noise as a pollutant and incorporate noise effects in environmental impact studies. Examples can be followed where noise restrictions have been implemented; for example, during oil exploration and drilling activities in Alaska, seismic operations had to be suspended when endangered bowhead whales were in the vicinity.

Cetacean conservation is an international issue, and will require continued and enhanced efforts and cooperation at local, national, regional, and international levels. Although the IWC and CITES have established an international structure for protection of the great whales, they do not address management issues for smaller cetaceans. There are regional organizations that address cetacean conservation issues as a portion of their charge; for example, the Convention for the Conservation of Antarctic Marine Living Resources (CCAMLR), the North Atlantic Marine Mammal Commission (NAMMCO), and the Convention on Migratory Species (CMS). A global international commission dealing specifically with dolphins, porpoises, and small whales could be modeled on other international organizations focused on marine animal protection. Although aggressive measures taken by activist organizations are considered extreme by some (see Chapter 12), numerous NGOs have taken an active role in education and awareness campaigns concerning cetacean conservation. Compromises will continue to be necessary in cetacean conservation since extreme views sometimes serve to turn some away from conservation entirely. The most dire need, as it is with other marine mammals and endangered species, is to address the sources of incidental harm to cetaceans through fishing, coastal development, environmental pollution, ocean transportation, and industrial and military activities.

STUDY GUIDE

Topics for Review

1. What factors are considered in determining if a species is at optimum sustainable levels according to the U.S. Marine Mammal Protection Act?

2. Give examples of actions that would be considered *harassment* according to the U.S. Marine Mammal Protection Act.

3. What were the primary commercial products that attracted whalers to the harvest of sperm whales?

4. What characteristics of whale behavior made them vulnerable to excessive harvest in nearshore waters prior to the development of modern whaling technology?

5. For what product were large baleen whales primarily targeted in the 1800s?

6. What was the primary goal of the originators of the International Whaling Commission (IWC); how have the goals changed?

7. What factors led to limited success of early restrictions on whaling implemented by the IWC?

8. Under what premises are exemptions given under the IWC moratorium to allow harvest of whales by native Alaskans, Japanese whalers, and Norwegian whalers?

9. How do CITES agreements help to limit the harvest of whales?

10. How have genetic studies been used to reveal illegal harvest of whales?

11. What biological factors make the great whales unable to support a large sustainable harvest?

12. Compare the use of whaling records and genetic analysis as means of estimating historic population sizes for whales.

13. Compare the success of recovery since large-scale commercial whaling of the blue, fin, right, and sperm whales.

14. What environmental factors have been proposed by scientists as influencing whale beach strandings?

15. How can cetacean-oriented tourism be harmful to whales and dolphins?

16. Describe how education may be more useful than legal enforcement in cetacean conservation.

17. How can changes in oceanographic conditions in feeding areas potentially impact endangered whale populations?

■ Conservation Exercises

1. Give arguments from the perspective of each of the following groups to defend the given actions. Present counterarguments from a conservation perspective.
  a. whalers in the early 1900s harvesting blue whales
  b. native Alaskans harvesting bowhead whales
  c. Japanese whalers harvesting minke whales
  d. Makah Indians harvesting gray whales
  e. commercial fishers killing dolphins to avoid losses of fish to dolphin predation
  f. ship captains travelling through whale-inhabited waters at high speeds
  g. tuna purse-seine fishers operating in regions where dolphins associate with tunas
  h. marine parks running swim-with-dolphin programs
  i. conservation groups working to return the killer whale Keiko to the wild

2. Discuss the scientific debate concerning each of the following and how it relates to conservation.
  a. historic population sizes of the great whales
  b. current population sizes of minke whales
  c. the importance of great whales and pinnipeds as prey for killer whales
  d. the association of noise pollution and whale beach strandings
  e. the effect of chasing dolphins in the tuna purse seine fishery

FURTHER READING

Baker, C. S., J. G. Cooke, S. Lavery, M. L. Dalebout, Y.-U. Ma, N. Funahashi, C. Carraher, and R. L. Brownell. 2007. Estimating the number of whales entering trade using DNA profiling and capture-recapture analysis of market products. *Molecular Ecology* 16:2617–2676.

Baker, C. S., D. Steel, Y. Choi, H. Lee, K. S. Kim, S. K. Choi, H.-U. Ma, C. Hambleton, L. Psihoyos, R. L. Brownell, and N. Funahashi. 2010. Genetic evidence of illegal trade in protected whales links Japan with the US and South Korea. *Royal Society Journal Biology Letters* 6:647–650.

Bearzi, G., C. M. Fortuna, and R. R. Reeves. 2008. Ecology and conservation of common bottlenose dolphins *Tursiops truncates* in the Mediterranean Sea. *Mammal Review* 39:92–123.

Bradshaw, C. J. A., K. Evans, and M. A. Hindell. 2006. Mass cetacean strandings—a plea for empiricism. *Conservation Biology* 20:584–586.

Brillant, S. W., and E. A. Trippel. 2010. Elevations of lobster fishery groundlines in relation to their potential to entangle endangered North Atlantic right whales in the Bay of Fundy, Canada. *ICES Journal of Marine Science* 67:355–364.

Cerchio, S., N. Andrianarivelo, Y. Razafindrakoto, M. Mendez, and H. C. Rosenbaum. 2009. Coastal dolphin hunting in the southwest of Madagascar: status of populations, human impacts and conservation actions. Presented at the First International Conference on Marine Mammal Protected Areas; March 30–April 3, 2009: Maui, HI.

Cramer, K. L., W. L. Perryman, and T. Gerrodette. 2008. Declines in reproductive output in two dolphin populations depleted by the yellowfin tuna purse-seine fishery. *Marine Ecology Progress Series* 369:273–285.

Christensen, L. B. 2006. Marine mammal populations: reconstructing historical abundances at the global scale. *Fisheries Centre Research Reports* 14(9), The Fisheries Centre, University of British Columbia.

Croll, D. A., R. Kudela, and B. R. Tershy. 2006. Ecosystem impact of the decline of large whales in the North Pacific. Pages 202–214 in, Estes, J. A., D. P. Demaster, D. F. Doak, T. M. Williams, and R. L. Brownell (eds.). *Whales, Whaling and Ocean Ecosystems*. University of California Press, Los Angeles.

D'Amico, A., R. C. Gisiner, D. R. Ketten, J. A. Hammock, C. Johnson, P. L. Tyack, and J. Mead. 2009. Beaked whale strandings and naval exercises. *Aquatic Mammals* 35:452–472.

Doak, D. F., T. M. Williams, and J. A. Estes. 2006. Great whales as prey: using demography and bioenergetics to infer interactions in marine mammal communities. Pages 231–244 in, Estes, J. A., D. P. Demaster, D. F. Doak, T. M. Williams, and R. L. Brownell (eds.). *Whales, Whaling and Ocean Ecosystems.* University of California Press, Los Angeles.

Estes, J. A., D. P. Demaster, D. F. Doak, T. M. Williams, and R. L. Brownell (eds.). 2006. *Whales, Whaling and Ocean Ecosystems.* University of California Press, Los Angeles.

Evans, K., R. Thresher, R. M. Warneke, C. J. A. Bradshaw, M. Pook, D. Thiele, and M. A. Hindell. 2005. Periodic variability in cetacean strandings—links to large-scale climate events. *Biology Letters* 1:147–150.

Greene, C. H., and A. J. Pershing. 2004. Climate and the conservation biology of North Atlantic right whales: the right whale at the wrong time? *Frontiers in Ecology and the Environment* 2:29–34.

Hester, K. C., E. T. Peltzer, W. J. Kirkwood, and P. G. Brewer. 2008. Unanticipated consequences of ocean acidification: a noisier ocean at lower pH. *Geophysical Research Letters* 35, L19601, doi:10.1029/2008GL034913.

Hirata, K. 2005. Why Japan supports whaling. *Journal of International Wildlife Law and Policy* 8:129–149.

Lukoschek, V., N. Funahashi, S. Lavery, M. L. Dalebout, F. Cipriano, and C. S. Baker. 2009. High proportion of protected minke whales sold on Japanese markets due to illegal, unreported or unregulated exploitation. *Animal Conservation* 12:385–395.

Palumbi, S. R., and J. Roman. 2006. The history of whales read from DNA. Pages 102–115 in, Estes, J. A., D. P. Demaster, D. F. Doak, T. M. Williams, and R. L. Brownell (eds.), *Whales, Whaling and Ocean Ecosystems.* University of California Press, Los Angeles.

Read, A. J., P. Drinker, and S. Northridge. 2006. Bycatch of marine mammals in U.S. and global fisheries. *Conservation Biology* 20:163–169.

Reeves, R. R., B. D. Smith, E. A. Crespo, and G. Notarbartolo di Sciara (compilers). 2003. *Dolphins, Whales and Porpoises: 2002–2010 Conservation Action Plan for the World's Cetaceans.* IUCN/SSC Cetacean Specialist Group. IUCN, Gland, Switzerland and Cambridge, UK. xi + 139 pages.

Ruegg, K. C., E. C. Anderson, C. S. Baker, M. Vant, J. Jackson, and S. R. Palumbi. 2010. Are Antarctic minke whales unusually abundant because of 20th century whaling? *Molecular Ecology* 19:281–291.

Tyack, P. L., W. M. X. Zimmer, D. Moretti, B. L. Southall, D. E. Claridge, J. W. Durban, C. W. Clark, A. D'Amico, N. DiMarzio, S. Jarvis, E. McCarthy, R. Morrissey, J. Ward, and I. L. Boyd. 2011. Beaked whales respond to simulated and actual Navy sonar. *PLoS ONE* 6(3): e17009. doi:10.1371/journal.pone.0017009.

Vanderlaan, A. S. M., C. T. Taggart, A. R. Serdynska, R. D. Kenney, and M. W. Brown. 2008. Reducing the risk of lethal encounters: vessels and right whales in the Bay of Fundy and on the Scotian Shelf. *Endangered Species Research* 4:283–297.

Weilgart, L. S. 2007. The impacts of anthropogenic ocean noise on cetaceans and implications for management. *Canadian Journal of Zoology* 85:1091–1116.

Williams, R., S. Gero, L. Bejder, J. Calambokidis, S. D. Kraus, D. Lusseau, A. J. Read, and J. Robbins. 2011. Underestimating the damage: interpreting cetacean carcass recoveries in the context of the Deepwater Horizon/BP incident. *Conservation Letters* 4: 228–233.

Marine Fisheries: Overharvest and Conservation

Despite all of the environmental issue affecting the world's marine ecosystems, often the most substantial impact to marine populations is harvest for human food. The effects of excessive harvest on coastal marine populations have been discussed in the previous chapters. This chapter focuses primarily on the impacts of excessive harvest on fish populations in the open ocean and ways that these effects can be minimized (**Figure 11-1**). The mechanisms of management are complex and involve numerous legal, social, and economic considerations that are addressed primarily in Chapter 12.

In order to understand how to adequately deal with issues related to excessive fish harvest, it is necessary to have an understanding of the parties involved and how the problems have developed through time. The first portion of this chapter, therefore, characterizes the people involved in fish harvest, and continues with a description of fisheries harvest and development of management methods.

11.1 Fisheries Characterization

The term **fishery** commonly refers to aspects of harvesting and managing aquatic organisms. It can refer specifically to a species being harvested (e.g., the orange roughy fishery), the methods of harvesting (e.g., a gill net fishery), or the ecosystem from which the animals are harvested (e.g., a coral reef fishery). Fisheries are not limited to animals classified as fish but can include marine mammals (e.g., whale fisheries), crustaceans (e.g., lobster fisheries), mollusks (e.g., clam fisheries), or other invertebrates (e.g., sea urchin fisheries), and even algae (e.g., seaweed fisheries). When related species are fished in the same region by similar methods, they are often grouped for monitoring and regulation purposes (e.g., the Gulf of Mexico snapper fishery). Broader groupings are used to distinguish true fishes ("finfish") from mollusks and crustaceans ("shellfish"). Marine fisheries are usually managed as **stocks**, groups considered to be distinct units, typically distinct populations. Historically stocks were often designated by political boundaries (for example, the Canadian and U.S. Atlantic cod fisheries), even though they may have comprised the same population. Most management organizations now strive to manage at the population level, however, establishing international agreements when stocks overlap politically designated boundaries. For example, North Atlantic bluefin tuna are managed as a single unit through international agreement among fishing nations around the world.

Fisheries management refers to the regulation and monitoring of a fishery, including setting harvest limits, developing management plans, and protecting fish habitat. In recent years, fisheries management has been applied broadly by management agencies to include scientific, social, political, and economic considerations. Fisheries

Figure 11-1 Fish trawl catch taken from North Pacific waters off Alaska.

agencies are not limited to managing harvested species but are charged with scientific research, habitat protection and monitoring, and the protection of endangered species.

A common measure used to monitor fisheries is the amount of fish caught, measured as the biomass or number of individuals caught and brought to shore; these are called **fisheries landings**. This is a convenient point of reference from an economic and management perspective because the harvest is typically sold and weighed or counted at this point. Marine fisheries landings are typically reported to government management agencies, for example, the National Marine Fisheries Service (NMFS) in the United States or the Department of Fisheries and Oceans (DFO) in Canada. By international agreement, most nations with substantial marine fisheries also report landings to the Food and Agricultural Organization (FAO) of the United Nations for data compilation.

The completeness and reliability of landings data vary among species and geographic regions. For example, it is difficult to document all harvest when there are numerous small-scale fisheries selling their harvest in local markets. For large-scale fisheries, on-board observers are used to ensure compliance with regulations and a complete reporting of harvested animals, including bycatch that may be discarded overboard. For example, the U.S. NMFS deploys observers in over 40 different fisheries. These observer programs are likely to become more frequent as technological advancements reduce their costs. As one example, the United States, Canada, and Australia have implemented video-based electronic monitoring programs in some fisheries, using video cameras fitted with global positioning system (GPS) units that are activated when the fishing gear is deployed. Despite these efforts, a large fraction of the harvest globally is likely undocumented (such illegal,

unreported, and unregulated fishing harvest will be discussed later in this chapter).

A person who catches fishery organisms is termed a **fisher**, replacing the traditional term *fisherman* for increased inclusiveness. Fishers are typically distinguished as **commercial** (fishing to sell) and **recreational** (fishing for sport). **Subsistence fishers** harvest fish for consumption by them and their families or local community (**Figure 11-2a**). They are often distinguished by the use of customary or traditional fishing practices. If the fish are sold it is to provide basic needs. The majority of subsistence fishers are members of indigenous groups or inhabitants of developing countries or small island nations. Harvest by these groups may be given special protection and consideration because of past agreements or treaties and to avoid competition with fishers using more efficient modern technologies.

Artisanal is another category used widely by international organizations (e.g., the FAO of the UN) to classify fishers. Although precise definitions vary, artisanal fishers typically use simpler technology, fishing in a limited geographic range and often for subsistence needs (Figure 11-2b). These artisanal fisheries, however, are not strictly limited to subsistence harvest. They also may include small-scale commercial fishing using simpler methods (e.g., handmade traps, handlines, cast nets, or small gill nets; see below). What is considered artisanal varies among regions around the world; it is used most often in reference to developing countries using traditional methods.

The bulk of fisheries harvest from the oceans would be classified as commercial fisheries, those harvested to be sold for profit on the commercial market. There is a large variation in the kinds of fishers that might be classified as commercial. **Small-scale fishers** use simple methods (such as small traps or nets from small motorized boats) and sell to a local market (Figure 11-2c). Fishing methods may be relatively simple; however, as fish become more difficult to find and harvest, even small-scale fishers are forced to use more advanced technology, including high-powered vessels with mechanically operated equipment, monofilament gill nets, modern trawling gear, and technology such as GPS and acoustic fish finders. Many of these small-scale commercial fishers fish for a living because of a family or community tradition, and many are family operations. Commercial fishing is attractive to many coastal residents because it gives them the lifestyle they desire, a life of independence and freedom on the sea. Because of these intangible attractions, fishers may continue to fish even if they make only a moderate income or overharvest the resource. As families grow with each generation, the number of people desiring this livelihood grows; they may claim that family tradition as their "license" to harvest fish. This can lead to excessive harvest even when large industrial

(a)

(b)

(c)

(d)

(e)

Figure 11-2 Fishers are classified according to their goals and capabilities: **(a)** a subsistence fisher fishing with a gill net, Jamaica; **(b)** artisanal fishing traps and nets, Guadeloupe, French West Indies; **(c)** small-scale fishers, Japan; **(d)** mid-scale fishing vessels, shrimp trawlers, northern Gulf of Mexico of Mississippi; **(e)** large-scale factory trawler.

fishing vessels are kept away from a region to protect artisanal and small-scale fisheries.

As fishers expand to enable the harvest of more fish, their goals often include moving up to the next category, the **mid-sized fishers**. Although there is no well-defined boundary between small-scale and mid-sized fisheries, mid-sized fishing vessels can travel greater distances, catch more fish before returning to port to offload, and typically must hire a larger crew. For these vessels, modern technology is important to enable locating and capturing the fish, and there is typically a larger refrigerated hold, where the fish are kept, and living quarters on the boat. Because these vessels are a significant investment, mid-sized fishers feel an increased economic pressure to maintain a stable level of harvest. If fishing allowances decline too much, the fishers may be faced with unaffordable boat payments. There is less flexibility in switching fishing gears, because the boat is usually designed for a specific type of fishing.

The largest of the commercial fishing operations are the large-scale **industrial fishers**. These fisheries are dominated by **factory trawlers**, the largest of fishing vessels, able to travel the worlds' oceans in search of large quantities of fish. These vessels pull large trawl nets (see below), catching tons of fish in a single haul of the net. The most modern electronic technology and most efficient of fishing methods are used. The fish are processed, packaged, and frozen on board, ready for market when the boat arrives at port. A large crew is needed to operate the boat and fishing gear, and even more to handle and process the fish once they are brought on-board. Factory trawler vessels are floating factories, typically remaining at sea for weeks at a time. They have living quarters, recreational facilities, and a staff to take care of cooking, housekeeping, and boat maintenance. The boats and gear represent a large financial investment and the fishers rely on a steady harvest to maintain their operations. These fishers are the most susceptible to a fishery closure or major reduction in fishing quotas; therefore, there can be great pressure on fisheries managers to keep quotas at levels that can support these large fishing operations (see Box 11-3, Learning from History: Boom and Bust of Atlantic Cod Fisheries later in this chapter).

Recreational fishers or **sport fishers**, by definition, are fishing for the purpose of recreation (**Figure 11-3**). Many people fishing under a sport license or permit, however, are catching fish primarily for personal consumption. Fishes harvested under a recreational license may not be legally sold. Sport fishers may be fishing in part or totally for aesthetic reasons: the enjoyment, challenge, and social appeal of fishing. Recreational harvest limits and methods are usually restrictive enough that it is not practical to rely on fishes caught recreationally as a reliable source of food. One of the means of limiting and partitioning harvest of recreational fisheries is to restrict it to inefficient methods, such as hook-and-line fishing (**angling**) with a specified type of hook, lure, or bait. Sport fishing's popularity is relatively recent, and is largely limited to regions where the economy can support the purchase of angling equipment and people can afford the leisure time. The popularity of recreational angling is centered in North America, where the number of people participating increased substantially through the last half of the 20th century. In 2009 almost 12 million people participated in recreational marine fishing in the United States, catching hundreds of millions of fish. Recreational fishing has recently increased in popularity in other regions of the world, largely as a sector of the tourist industry. Even if the locals cannot afford recreational fishing, it can provide a source of income through sale of fishing boats, angling tackle, or payment to fishing guides and other ancillary businesses. Recreational fishers are a diverse group. Many fish from the shore or piers using

(a) (b)

Figure 11-3 Sport fishers fish for recreation or personal consumption: **(a)** Kagoshima Bay, Japan; **(b)** Florida charter boat fishers.

simple fishing tackle; others have small boats enabling them to access nearshore or estuary habitats; and others have seaworthy vessels capable of targeting large oceanic predators, (e.g., tuna, sailfish, and marlins). The popularity of **catch-and-release** fishing has increased as conservation and animal welfare issues have become more culturally ingrained. In some sport fisheries over 50% of the catch is released alive voluntarily (though concerns remain over the mortality because of stress, even if the fish is released alive; see below).

For other fisheries, various user groups are interested in harvesting the same species or group of species. A potential source of conflict is the allocation of the resource among these groups. For example, recreational and commercial fishers may each argue that they should have priority or sole access to a fishery (e.g., the snappers or striped bass in U.S. waters). Arguments made by commercial fishers for preference include the need to harvest fish for their livelihood and that they provide fish for the general non-fishing public to eat. Recreational fishers counter that commercial fishing allocates the resource to a few people who are allowed to catch large numbers of what is considered a public resource. They argue that, because individual harvest limits are low, there is less chance of overharvest by recreational anglers. Although this is often true, one must consider that anglers may greatly outnumber commercial fishers. In some U.S. coastal fisheries the recreational harvest often outweighs commercial harvest, for example, for striped bass, red drum, and yellowfin tuna. Finally, recreational fishers point out that angling methods are not as destructive to the habit and ecosystem as commercial gears such as trawls and gill nets. From an economic perspective, the money spent on angling, including boats, fishing tackle and gear, and other costs associated with the fishing trip may outweigh the economic gain from commercial fishing. The result of such conflicts is usually a compromise, where both recreational and commercial fishers receive an allocation of the resource. Unfortunately, if decisions are made as a response to political pressures, each group may be allocated more than is desirable for conservation purposes.

For many fisheries there is no need for recreational allocations, such as for non-predatory species (e.g., herrings and anchovies), those inaccessible to angler harvest (e.g., deep-water fisheries), or when the fishery is needed as a source of food (e.g., in developing countries). But allocation among various segments of the commercial fisheries can lead to controversy. Commercial fisheries often are managed by setting quotas to define the total allowable catch. These quotas can be regulated on a *first-come-first-served* basis, where the harvest is not restricted among individuals, but the fishery is monitored and then closed when the overall quota is attained. Such overall quota regulations

favor the larger, more efficient vessels and thus can lead to conflicts among user groups, especially when subsistent, artisanal, or small-scale fishers are protecting their rights to harvest because of tradition, the need for food, or income as a livelihood. Conflict resolutions may lead to special allocations for specific user groups or the elimination of industrial fisheries from a traditional fishery. For example, Native Americans in the U.S. Pacific Northwest have treaty rights that allow them exclusive commercial harvest in certain rivers, and Native American tribes in Washington state were recently given rights to half the harvest of shellfishes in Puget Sound. In India, artisanal fishers were successful in having large industrial fishing fleets removed from their coastal waters in the 1990s after the Indian government attempted to use allocations to foreign vessels as a means of relieving international debt.

11.2 Fisheries Harvest Methods

To understand how fishing has affected marine organisms and ecosystems, it is important to be familiar with harvest methods. Although the basic principles of harvesting animals from the sea have been in place for hundreds or thousands of years, advances in marine vessel and electronic technology have made the deployment of gears much easier and have allowed for a dramatic expansion in the size of gears and the locations and depths where they can be used. An expression of the long history of fishing can be seen in the almost endless variety of nets, traps, and other gears that have been used to catch marine organisms. Despite this variety, most fishery gears can be placed into one of several broad categories, as detailed below.

The expansion of fishing gears, along with an increase in the number of people fishing, has led to worldwide problems of excessive harvest of commercial and sport species. This is not, however, the only conservation concern relative to fishing. Few, if any, fishery gears are selective enough to harvest only the targeted species and sizes of organisms with no damage to the environment. Most fisheries, therefore, have to deal with habitat damage and unintentional injury or harvest of non-targeted organisms (**incidental harvest** or **bycatch**). The FAO estimates that over 25 million tons of marine organisms are caught and discarded as bycatch each year globally: about one-third of the total global catch. (This number is a rough estimate, however, because much of the bycatch is not reported and most is discarded at sea.) A portion of this wasted catch reflects a lack of concern by fishers or management agencies, but eliminating or limiting bycatch is often legally and practically difficult. One of the major areas of research in fisheries is the development of methods to limit this bycatch. Efforts are most often successful when imperiled species are involved. When harm to non-targeted species cannot be addressed by gear restrictions even lucrative fisheries may be closed. Many commercial fisheries have seasonal bycatch caps, which, when reached, cause an automatic closure of the fishing season, even if the harvest limits of targeted species have not been met. (Chapters 9 and 10 thoroughly discuss bycatch issues involving endangered species and marine mammals).

Although many of the greatest problems related to fisheries harvest have arisen with advancements in technology in the past several decades, similar issues likely have been of concern, at least locally, throughout the history of fishing. Humans have been taking animals from the ocean for as long as they have been living near the coastal waters. *Homo sapiens* were eating fish and mollusks 70,000 years ago along the southern coast of Africa, and as populations expanded along the coasts of the world, humans continued to develop new fishing methods. Although early methods of catching fish and "shellfish" were primitive compared to modern commercial methods, most currently used harvest gears are modifications and expansions of methods used hundreds or thousands of years ago. Many traditional methods remain in use today, especially for subsistence harvest in developing countries. However, as coastal fish populations declined, large scale commercial fisheries developed to expand access to valuable resources. Traditional methods, such as coastal weirs and traps, were no longer efficient enough for commercial fishers to make a living, forcing the development of methods to allow fishers to move progressively further from shore and into deeper waters. The continued increase in coastal human populations led to further competition for fisheries resources, an additional pressure on fishers to develop more efficient methods of catching larger numbers of fish.

As boat technology was enhanced, fish harvest methods improved dramatically. Motor-driven vessels and modern technology allowed handling of nets over 100 meters diameter or lines containing thousands of hooks; the ability to process and freeze tons of fish on board allowed vessels to remain at sea longer. A large percentage of the ocean became accessible to harvest of its living resources, leading to widespread overharvest. Limitations on harvest became a necessity. There are a number of ways limits can be applied, including restrictions on the number of fishers or how much each fisher can catch. Such limits, however, can be difficult to implement because of social concerns (a perceived freedom to fish) and difficulties of monitoring and enforcement (see Chapter 12). Therefore, much of the initial efforts to limit harvest were through restrictions on the kinds and sizes of fishing gear used. This is management by **regulated inefficiency**, defined as limiting efficiency such that each fisher is unable to harvest excessive amounts of fish. Gear limitations also serve the purpose of minimizing

environmental damage that can result from fishing activities because the most efficient harvest methods often are the most destructive of habitat and ecosystems. Examples of methods and conservation issues will be presented below for various harvest gears.

Dredging

There is strong evidence from shell middens that many prehistoric coastal populations depended greatly on abundant resources that could be scraped or excavated from the substrate in shallow marine areas (see Chapters 3 and 4). Bivalves such as clams and oysters are easily taken by excavating with simple tools in shallow or intertidal areas. Middens evidence hundreds or thousands of years of consumption of bivalves and limpets in some regions. Simple digging and scraping tools were eventually modified to **rakes** or **tongs** handled manually from boats to scrape bivalves from soft bottoms (**Figure 11-4a**). Eventually, technological advances led to mechanization of these harvest operations.

Motorized vessels and industrial technologies led to the development of **dredges** not only to excavate sediments for maintaining channels or mining sand (see Chapter 3) but also to target bivalves residing within the substrate (Figure 11-4 b, c). Dredges designed specifically for shellfish harvest have rigid frames pulled along the bottom to scrape and retrieve various kinds of mollusks. The dredge design varies with the species being harvested. A **scallop dredge** has a cable attached to a metal frame, to which are attached a scraping bar and a bag constructed of metal rings. As the dredge is pulled along the bottom, the scallops are scraped by the bar from the bottom into the bag. An **oyster dredge** has rake-like teeth on the scraping bar to dislodge the attached oysters. **Hydraulic dredges**, such as those used to harvest surf clams along the western North Atlantic coast, utilize a hydraulic device that washes the clams out of the sediment and into a collecting basket or conveyor belt to the boat deck.

Conservation concerns over the use of dredges include effects on populations of bottom organisms and potential harm to habitat. When clams or oysters are dredged from bottom sediments other organisms are taken and the substrate is disturbed. Larval oysters require a hard substrate to settle onto, and in soft bottom ecosystems the shells of dead oysters may be the only substrate available. Dredging that removes the dead as well as the living oyster shells slows the recovery of that oyster bed. Returning empty oyster shells to the bottom enhances settlement of the next generation of oysters (see Chapter 4).

Spears and Harpoons

The first harvest of fish by humans was likely with stone- or bone-pointed spears designed for taking various small

(a)

(b)

(c)

Figure 11-4 Dredging devices for harvesting bivalves: **(a)** hand-tonging for oysters in Chesapeake Bay in the 1960s, **(b)** an oyster dredge in operation, **(c)** recovering a clam dredge.

animals. These tools evolved over time and, as seaworthy boats were developed, surface dwelling fish and marine mammals became vulnerable to harpoon fishing. Harpoon fishing for large predatory fishes such as swordfish was common well into the 20th century, but eventually died off when the number of larger fish declined.

Whalers took whales and other marine mammals with hand-held harpoons from small skiffs for centuries (**Figure 11-5a**). These could be dangerous operations, especially for sperm whalers. As the industry expanded into modern industrial whaling for the largest and fastest whales, cannon fired harpoons were developed. These eventually evolved into the modern harpoon fired from the deck of a whaling ship, with an explosive tip designed to accelerate the death of the whale (see Chapter 10 for discussion of whaling and conservation issues).

Spears continue to be used for harvest, primarily by divers using spearguns in subsistence and sport fisheries (Figures 11-5b and c). The development of scuba greatly enhanced the ability of selecting individuals for harvest, especially in clear waters around shallow reef habitats, and for species with little instinctive avoidance response to humans

in the water. Spear fishing causes relatively little harm to the environment or non-targeted animals compared to net, trap, or hook harvesting; however, spears can be used to target the largest individuals or rare species, with devastative effects on populations (for example, removal of the largest groupers from coral reef ecosystems; see Chapter 5).

Hook-and-Line Gears

Hooks made from wood, shell, bone, horn, or stone attached to lines made from various materials, including plant fibers and animal parts, have been used by humans for fishing for thousands of years. As technology developed toward the current metal hooks and monofilament line, angling for fish became more and more efficient. Hook-and-line fishing is the predominant catch method for sport fishers but is also used in commercial fisheries for predatory species.

A baited metal hook attached to a line is a simple and efficient device for catching predatory fish that causes relatively little harm to the physical environment. Major variations in hook-and-line fishing include the method of deployment, the number of hooks, and the kind of bait

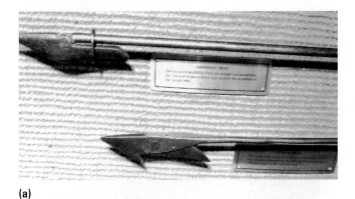

(a)

(b)

(c)

Figure 11-5 Fishers using fishing harpoons and spears do not take large numbers of individuals, but can target the largest or most valuable fish: **(a)** 19th century whaling harpoons; **(b)** spear fishers, Maui, Hawaii; **(c)** subsistence spear fisher, Jamaica

used. **Handlines** are deployed and retrieved by hand, typically from a boat. Although handlining was used extensively in coastal fisheries such as Atlantic cod fishery, it is now considered too labor intensive and inefficient for most commercial fisheries. Subsistence fishers around the globe, however, continue to use hand lines in coastal fisheries. Sport anglers fish with monofilament nylon lines wound onto hand-turned reels. Commercial fishers often use reels of similar design, but power operated; multiple hooks can be attached to a line lowered vertically to the bottom as a **downrigger**. When fish inhabit waters near the surface and are too fast or too widely dispersed to be captured efficiently by nets pulled through the water, hook-and-line capture may be the most efficient harvest method. **Pole-and-line fishing**, using a single hook attached to a line on a simple pole, is not widely used in commercial fisheries today. Prior to the development of purse-seining gears in the mid-1900s, however, tuna were commonly harvested commercially by pole-and-line fishing (**Figure 11-6a**). The tuna were attracted to the boat by **chumming**, throwing live bait such as anchovies or sardines overboard. As the tuna congregated to feed, they were caught on barbless hooks attached to artificial lures by fishers standing at the edge of the boat. The tunas were rapidly thrown onto the boat one after another, a labor intensive, dangerous operation. Another hook-and-line fishing method used in some commercial fisheries is **trolling**, where an artificial lure is pulled through the water from a moving boat to attract predatory fish such as mackerels, salmon, or tuna. Multiple lines are set from a single boat and the hooked fish are brought aboard the boat using a hand- or power-operated reel. The most common method of recreational fishing is by angling, using a pole and line or **casting rod** and reel, and natural or artificial baits attached to a hook. This method is less efficient than most commercial methods and, as a form of regulated inefficiency, assists in the partitioning of recreational harvest among many anglers.

The most efficient way to fish near the surface in open water is using a **longline**, multiple hooks attached to a single line with floats attached to keep it at the surface (Figure 11-6b). Each hook is attached to the long line by a leader; leaders vary in length from 25 to 100 centimeters or more and are spaced at intervals from 1 to 10 meters apart depending on the species targeted. A single longline can stretch for several kilometers and contain thousands of hooks. Longlines are also used to capture some bottom species (e.g., cod, halibut); the line is sunk to the bottom with an anchor, with floats attached to the ends at the surface. Longlines are often used at night and may be allowed to remain in the water fishing overnight before being checked the next day.

(a)

(b)

Figure 11-6 Commercial hook-and-line fishing methods: **(a)** pole-and-line fishing for tuna in the 1950s; **(b)** a rack rigged with longline gear ready to be baited.

Although hook-and-line fishing causes minimal harm to the environment compared to other fishing methods, it can harm populations of fish and other species that are caught incidentally or without legal restrictions. One advantage to hook-and-line fishing is that it is not necessarily a lethal method. If a fish is being caught solely for sport or is an untargeted size or species, there is a chance it can be released to survive. This **catch-and-release** fishing has gained in popularity among sport anglers; however, not all fish survive the stress of being hooked, fought, landed, and handled before being released. For virtually all catch-and-release fisheries that have been monitored closely there is evidence of some mortality. Mortality estimates range

from 10% to 50%, depending on a number of factors that include the degree of sensitivity of the species, what type of hook is used, how the fish is hooked, how long the fish is fought before it is caught, and how the fish is handled before it is released.

To minimize mortality of released fish, fishers are encouraged to limit how long they fight the fish, to use hooks that are not as likely to harm the fish (e.g., barbless hooks or circle hooks that the fish are less likely to swallow), to release the fish immediately, and to avoid using live bait that the fish is likely to swallow along with the hook. Many organizations that run fishing tournaments for large pelagic species (e.g., marlins and sailfish) consider a fish caught if the leader between the line and hook can be touched while the fish is hooked, enabling release of the fish without taking it from the water.

For hook-and-line fisheries of bottom species (e.g., groupers and snappers) in waters of moderate depths, a special handling procedure is recommended. Because of the reduction in pressure as the fish is brought to the surface, the fish's air bladder expands, sometimes releasing gases in the body cavity and forcing the stomach through the mouth. A released fish has difficulties swimming and will float at the surface vulnerable to predators until the gases are expelled (most bony fishes have no connection from the swim bladder to the gut and must expel gases from the air bladder slowly through the circulatory system). Anglers are often advised to vent the gas from the fish's body cavity using a modified hypodermic syringe or other venting tool.

For longline fishers, with thousands of lines in the water for long periods of time, often overnight, it is more difficult to monitor what is being caught and to release it before it is injured or dead. Incidental catch of non-targeted fishes on longlines is problematic; for example, billfish and sharks are commonly hooked in tuna and swordfish longline fisheries. Some fishes are better able to survive catch and release; for example, 60% of blue sharks are released alive in some longline fisheries. In many longline fisheries the bycatch fish are removed by either knocking the fish against the side of the vessel or running the hooked fish into rollers of the power block (a hydraulic powered pulley system) to pop the fish off of the hook, potentially causing considerable damage to the fish. Longline bycatch of endangered species includes sea turtles and seabirds, which is of particular concern. Catch can often be controlled through variations in hook size, bait, and the manner in which the line is deployed from the boat (see Chapter 9).

Eliminating predatory fishes that are similar in size to target species from longline catch is more problematic;

however, methods have been developed to take advantage of differences in the biology of predatory species. For example, Rodrigo Vega and Roberto Licandeo found that using monofilament nylon leaders rather than multifilament leaders containing wires in south Pacific swordfish fisheries reduced shark bycatch. This was attributed to the shark's ability to bite through the monofilament line; however, there also may be an effect of sharks avoiding multifilament lines because of strong visual acuity. In some instances shark bycatch can be reduced by regulating the depth at which the hook is set and avoiding sets during diurnal periods when sharks are active.

Size selection also can be problematic with hook-and-line harvest. The largest fish can be selected in sport trophy fisheries or by hook size selection or fishing location in commercial fisheries. Removing the largest fish from the population can decrease the reproductive potential and drive the fish to smaller average sizes through artificial selection (e.g., swordfish and cod; see discussion below). Despite these disadvantages, hook-and-line fishing is still generally less harmful to the environment than most other harvest methods. Few fishing methods provide as great a chance of releasing fish with a high change of survival.

Traps and Weirs

Ancient cultures developed numerous types of portable fish traps from grasses, sticks, and other organic materials. Some of the earliest known fish traps were woven basket-like structures. As techniques to work with natural materials improved, traps became more diverse. With the development of technologies to work with metals and plastics and the use powered motors to aid in lifting, traps have become much larger and more efficient. Modern traps include hundreds of designs, ranging from simple handmade traps fished in shallow estuaries to large metal crab traps weighing tons fished at hundreds of meters depth in the open sea.

Modern fish traps are usually made of wire or a metal grillwork covering a relatively rigid frame made of wood, metal, or synthetic materials. The trap is designed so that the fish enter by way of a funnel-shaped opening through which they cannot easily escape. Although some fish enter a trap to gain shelter, the traps are usually baited to attract the fish. Often a line is attached to a float at the surface to identify its location and allow for retrieval. Fish traps are often used around structures, such as reefs, to avoid potential damage to the habitat from net fishing (see Chapter 5).

Rigid, portable traps designed to harvest crustaceans are referred to as **pot traps**. These work on a similar principle to fish traps but vary greatly in size and shape depending on the fishery. They are typically made of a rigid frame covered with a grid, mesh, or netting. Pots can be fished individually or in a series attached together by a groundline, identified by a float at the surface. Traditional pot trap fisheries include the U.S. Atlantic lobster fishery (**Figure 11-7a**), and blue crab fisheries in U.S. Atlantic and Gulf of Mexico (see Chapter 4). With increased boat capacity and mechanization, fisheries using large traps up to several meters in diameter and weighing hundreds of kilograms have developed for crabs, including king crabs from Alaska (Figure 11-7b) and similar species in coastal regions around the world (see Chapter 8 for a discussion of these fisheries).

A trap net is a trap made of netting forming a series of connected funnels, with the final funnel closed at the end. Fishes that swim into the net are channeled through the funnels until they are trapped at the end of the net. Often these trap nets have long walls of netting (called wings) attached at the opening to channel fish into the trap. Along the northeast U.S. coast this type of trap is called a **pound net**. **Fishing weirs** operate in a similar manner to trap nets; walls made of sticks, stones, or other rigid structures that guide the fish to the trap.

The use of fishing weirs goes back thousands of years in many coastal regions around the world. Similar structures are still used in subsistence fisheries, such as in regions of Africa and eastern Asia. Trap nets and weirs were used extensively in U.S. coastal areas from the 1800s through the mid-1900s to harvest fish that travelled on coastal migrations, such as Atlantic herring *Clupea harengus* and Atlantic mackerel *Scomber scombrus* (Figure 11-7c). Fish that travel through narrow channels, such as anadromous salmon, are especially vulnerable to harvest with weirs and traps. As coastal fish stocks have been depleted, the use of trap nets and weirs has been prohibited in many fisheries.

When coastal fish populations were larger, trap nets were used extensively by local fishers. Traps, however, can overharvest fishes that migrate along the coast to reach feeding or reproductive grounds. Problems with trap nets along the U.S. coast were dealt with much earlier than for most fisheries. The U.S. Fish Commission recommended closing the New England trap net fishery as early as the 1870s, and Pacific salmon trap nets were prohibited in the 1950s. As coastal fish populations continued to decline, trap nets became highly restricted in many fisheries in the United States and around the world.

One conservation advantage to trap fishing is that non-target animals can often be released alive, especially if the traps are checked frequently. This allows for releasing fish that are restricted from harvest or unmarketable for other reasons. Harvested crustaceans (e.g., crabs and lobsters) can survive in pot traps for long periods of time, and have an excellent chance of survival after release. Release

(a)

(b)

(c)

Figure 11-7 Traps are not as destructive to habitat but can harvest large numbers of fish without requiring large powerful vessels: **(a)** setting lobster traps; **(b)** king crab pots; **(c)** a 19th-century herring fishing weir in coastal Maine.

of undersized lobsters, larger adults, or those carrying eggs has been successfully used in management of U.S. Atlantic fisheries. For example, in Maine waters of the United States, only lobsters from 86 to 127 millimeters in carapace length can be kept; in New Jersey the range is 86 to 133 millimeters, and other states have similar restrictions.

An advantage to traps or pots is that most designs do not harm the environment if they are deployed properly. However, traps especially those with small mesh sizes, set around structures such as reefs are not very selective in their harvest and the take of juveniles and smaller species can be excessive (see Chapter 5). Escape gaps have been designed that allow for the escape of smaller fishes; however, they do not permit the escape of non-targeted fishes of similar sizes as targeted species. If traps are not checked regularly, mortality of fish can be high. In deeper waters, bringing traps to the surface can cause harm to the fish because of pressure changes similar to those experience by hooked fish brought to the surface. An additional conservation issue with trap fisheries is the impact of lost or abandoned traps. The rate of trap loss varies widely among fisheries and the most typical reason for loss is adverse weather or water conditions. Lost traps may continue fishing (called **ghost fishing**) as long as the trap remains intact and catch rates can be high. For example, in dungeness crab fisheries of North America, the ghost catch can be as high as 7% of the commercial catch. In tropical reef habitats the death of a diversity of species from ghost harvest by traps is of particular concern, and in some fisheries traps must have a degradable panel to enhance escape of fish from ghost traps.

Net Fishing

Soon after coastal cultures developed the capability of weaving plant fibers into nets it is likely they began using them for harvesting fish. Nets were placed in channels, attached to poles to dip fish from the water, thrown over fish in shallows, or pulled through the water. These techniques continue to be used and are applied with modern throw nets, cast nets, seines, trawls, and gill nets.

Gill and Drift Nets

A **gill net** is designed to block a migration corridor or movement pathway of fish by entangling fish in the mesh of the netting (**Figure 11-8**). They consist of a long wall of netting set stationary, floating at the surface, or resting on the sea bottom. This wall of netting passively catches fish as they swim into the net, entrapping them with mesh openings large enough for the fish to put its head through, but small enough that the fish cannot pass through entirely. When the fish tries to back away from the net, it becomes

(a)

(b)

Figure 11-8 Gill nets can efficiently trap large numbers of fish or other animals, including those not targeted for harvest: **(a)** gill net being deployed; **(b)** removing salmon from gill net.

entangled at the gill cover just behind its head or by its fins or spines. Other animals also can be entangled by appendages or wrapped up in these nets and become incidental mortalities (e.g., marine mammals, sea turtles, and seabirds; see Chapters 9 and 10).

In coastal waters, gill nets are typically anchored to the bottom or attached to the shore. The net can cover the entire width of a narrow channel. Even in broader bays or inlets, many fish migrate along the edges in shallow waters where they are vulnerable to gill nets set from the banks. Being stationary and light when made of monofilament line, the gill net does not require a powerful vessel to deploy and can be fished efficiently to harvest mid-sized or large commercially valuable species that inhabit shallow coastal waters. Although gill nets are designed to be selective for certain sizes of fish, larger individuals can become entangled in the nets. It is typical for fish to intermingle in multispecies schools in coastal areas; therefore, the nets catch more than just the targeted species. And, the struggles of animals entangled in the gill net can attract predators, which also may become entangled. Because of these issues, gill nets have been outlawed or tightly regulated in many regions of the world. For example, by the 1990s most U.S. coastal states had banned gill nets in inlets, estuaries, and often in coastal waters, with the exception of some restricted fisheries.

When gill nets are set floating near the surface in the open ocean they are called **drift nets**. Drift nets are typically much larger, and can extend for kilometers, providing an all too efficient means of catching fish and other animals swimming near the ocean's surface. These nets are lightweight, easy to handle and transport, and virtually invisible to fish, especially when fished at night. The ability to easily manufacture nets with synthetic filaments and small mesh size led to a rapid expansion of fisheries in international waters from the 1960s through the 1980s, with few restrictions on size or use of the nets. The lack of selectivity in catching not only large fish, such as sharks and billfish, but also marine mammals, birds, and sea turtles, led to widespread international opposition from environmental organizations, condemning the drift nets as *walls of death*. Some of the most influential observations were the documentation of large incidental catches of cetaceans in drift nets in the Pacific and Mediterranean (see Chapters 9 and 10).

In the late 1980s, a number of coastal nations began enacting legislation to outlaw or restrict the use of drift nets. International fishery management agencies followed suit to begin restricting drift net use in international waters of the open ocean. In 1991 the United Nations called for "a moratorium on large-scale pelagic driftnet fishing." One of the last major fishing regions where drift-net fishing was banned was in European Community (EC) waters. Although the European Union placed a ban on drift net fishing in 2002, enforcement has been difficult because of a lack of acceptance of the rules, especially in the Mediterranean, and legal challenges stemming from the lack of a clear definition of what constitutes a drift net.

Where gill and drift nets are still legally used, efforts have been made to minimize the death of marine mammals and other endangered species. Although monitoring nets to release incidental catch may be feasible, it can be cost prohibitive to small-scale fishers and impractical for large drift nets. Acoustic alarms have successfully reduced dolphin mortalities (see Chapter 10). The lack of requirements, poor enforcement, and expense to small-scale fishers has limited the use of these pingers.

When gill nets or other net gears are lost or abandoned (e.g., during a storm or when entangled on some structure), the gear can continue to catch and kill animals. Gill nets are of most concern because they continue catching fish when lost (trawls, are designed to catch fish only when pulled through the water). The synthetic fibers comprising most gill nets are non-biodegradable, so the net may continue fishing as a *ghost net* for months or years after it has been lost (**Figure 11-9**). Although fishers may desire to recover the net, it can be cost prohibitive or impractical if it sinks to the bottom in deep waters or cannot be located, though the recovery of nets has increased with the use of GPS. Even though the rate of loss is low and gill nets have been eliminated from many fisheries, there is still widescale use of fishing nets around the world. An example is in European fisheries where, even though less than 1% of nets are lost, the total length of lost netting is over 200 kilometers per year. Net loss rates can be much higher when fished below the surface in deep waters. Because of the high loss rate (over 20,000 nets in a year) and difficulty of recovery, fishing with gill nets below 200 meters depth was banned in the northeast Atlantic by the European Community in 2006. The number of animals caught in abandoned gill nets varies widely depending on habitat, conditions, and the situation of the net (e.g., spread out or balled up). The ghost catch of fishes and invertebrates is estimated to range from less than 1% to about 4% of the commercial harvest in European fisheries, though often the greatest concern is entanglement of marine mammals, sea turtles, and seabirds in abandoned nets.

Finding and removing abandoned nets is difficult and expensive. Ongoing efforts have resulted in the removal of hundreds of nets that remain in relatively shallow waters or near the surface. In U.S. waters, NOAA uses satellite imagery and aerial flyovers to locate the nets and other large debris and a ship is sent to the area for removal. Ghost net retrieval programs have been enacted in European, South Korean, and Australian waters also. Based on an analysis of ghost fishing in European waters, James Brown and Graeme Macfadyen argue that the most effective way to minimize the effects of ghost fishing is not through net removal programs, which can be expensive and of limited success, but through improved management (e.g., requiring tagging of nets for identification), better communication (the majority of lost nets in European waters is a result of conflicts among fishers), and education (much of the harmful effect of ghost fishing can be eliminated if fishers retrieve gear rapidly). They also argue that the environmental impacts of trawl fishing (such as scouring the bottom habitat) are greater than that of lost gill nets. Removal programs are likely most effective for nets washed ashore or into shallows where they are more likely to entangle endangered sea turtles and marine mammals.

Seines

A **seine** is a net designed to encircle fish, entrapping them or enabling them to be forced to shore. Seines have small mesh openings that are designed not to entangle the fish, but to contain them until they are brought to shore or removed from the net. The bottom of the net is held down by a weighted *leadline*. The top of the net is held at the surface by a buoyant *floatline*. **Haul seines** are pulled through the water by hand or from a boat (**Figure 11-10a**). A **beach seine** is used in shallow waters next the shore and the fish are pulled up onto the shore. Because of water resistance of the fine-meshed net, it must be pulled slowly; thus, large fast-swimming fish are rarely caught. A net can be pulled somewhat faster and in deeper water when operated from a boat. Seines can be used to catch larger species if fish congregate in shallow bays where they might be trapped. For example, black drum up to 23 kilograms are harvested in coastal estuaries of the Gulf of Mexico using seines operated from boats.

Throw nets (or **cast nets**) are circular nets, typically 1 to 2 meters in diameter, thrown from the shore or from a small boat, that sink to the bottom, trapping the fish underneath (Figure 11-10b). Although the size and number of fish captured are limited and throwing a large cast

Figure 11-9 Ghost nets are nets that have been lost or abandoned, but continue to catch fish. Here, a sea turtle is entangled in a ghost net.

(a)

(b)

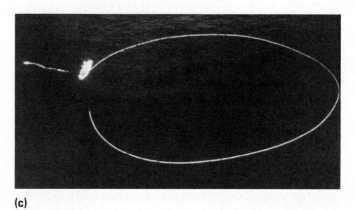

(c)

(d)

Figure 11-10 Seines and throw nets can be used to catch fish close to shore, now most commonly used to catch bait fish. **(a)** Haul-seine fishing for menhaden at Long Island in the 1800s; **(b)** a fisherman throwing a cast net. Large purse seines can efficiently encircle and capture large schooling fishes, **(c)** as with the purse seine encircling a school of fish (the skiff on the left is pulling the boat away from the net to avoid it becoming entangled), **(d)** and this purse seine catch of jack mackerel *Trachurus murphyi* off Peru.

net successfully takes practice and skill, coastal residents around the world still use these throw nets in subsistence fisheries or for bait harvest. In coastal waters of the southeastern United States, shrimp harvest by cast net is typically classified as a recreational fishery and the shrimp captured can be used for bait or personal consumption.

Purse seines are larger seine nets designed for harvesting large schooling fish in the open ocean. The net is deployed from a boat to encircle the school and is then pursed

at the bottom to be formed into a bowl shape containing the fish (Figure 11-10c). The purse seine gear consists of a long (up to about 2 kilometers), deep (about 100 to 250 meters) curtain of netting, with a floatline at the top, a weighted leadline at the bottom, and a cable forming a purse line running through a series of rings along the bottom. When a fish school is spotted, from the boat or airplane or helicopter, the fishing vessel is pulled along alongside the school and transfers one end of the net to a separate smaller boat. The two boats go in opposite directions, playing out the net into the water, making a circle until they meet each other. The fish that are enclosed follow the circular pattern formed by the net until the net is closed at the ends and pursed at the bottom using the purse line. The fish are drawn toward the boat by gradually pulling the net on board using a hydraulic-driven power block, making the circle of net increasingly smaller (Figure 11-10d). The trapped fish are taken on board using a large crane-operated dip net, for large fish like tuna, or a suction pump, for smaller pelagics such as herring and anchovies. Less energy is needed for harvest with purse seines than with trawl nets, because the fish are not physically pulled by the net but simply trapped until they can be removed. In many nearshore regions purse seine fisheries provide the largest harvests, including small clupeoids (e.g., anchovies, herrings, and menhadens), and larger scombrids (e.g., tunas and mackerels). The ability to target largely monospecific school of the targeted species results in much lower incidental catch than for most other large-scale fishery gears; however, predators may congregate with the targeted school. Shark bycatch is common, and incidental catch of dolphins in the tropical East Pacific purse-seine tuna fishery has led to tighter controls (see Chapter 10 for a discussion of the dolphin–tuna controversy).

Trawls

Trawls are cone-shaped nets pulled through the water, designed to filter out fish or crustaceans and accumulate them at the end of the net. Trawl sizes vary greatly. The smallest trawls can be handled by hand by one person from the back of a small boat. The largest trawls are those pulled by ocean-going **stern trawlers** that can have openings as large as 500 square meters (**Figure 11-11a**). Most trawl nets have similar basic components. Lines run from the boat to the **boards** or **doors** attached to each side of the net's mouth. The bottom of the net is held down by a weighted leadline, and the top of the net is held up by a floatline. As the net is pulled through the water, the water pressure pushes the otter boards outward, causing the net to remain open. For bottom **otter trawls**, the boards are designed to slide along the bottom, and rollers spaced along the net keep the net from tangling on rocks. From the mouth, the net tapers to the end (called the **cod end**), where the fish are trapped. The net is pulled through the water for durations of several minutes to hours. The net is then brought on board by a winch, the cod end is opened, and the fish are spilled onto the deck for sorting.

Trawls are typically not pulled fast enough to overtake large fast-swimming predators. Even in dark waters, a fish can sense the pressure wave pushed by the trawl as it moves through the water. For smaller or slow-moving species, however, trawls can be a very efficient method

(a)

(b)

(c)

Figure 11-11 **(a)** Trawl fishing nets can sieve large numbers of animal swimming over smooth bottoms; however, they sometimes kill large numbers of nontargeted animals, such as **(b)** this bycatch from shrimp trawl in the U.S. Atlantic. Devices have been designed to minimize bycatch. **(c)** This shrimp trawl contains a mesh panel by-catch reduction device (BRD) and a turtle excluder device (TED).

of commercial harvest. Many bottom-dwelling fishes are harvested by trawls, including cod, haddock, flounder, and rockfishes. Shrimp support one of the most valuable of the trawl fisheries in many coastal regions. Mid-water trawls are pulled away from the bottom to harvest animals residing in the mid water column (e.g., mackerels and squid).

The trawl is one of the most widely used—but also the most controversial—commercial harvest methods in coastal oceans around the world. The source of the controversies is that trawls catch not only the targeted species but also most other organisms too large to pass through the mesh openings and unable to avoid the net. Because trawls are pulled over broad areas (in some coastal regions there is hardly a square meter of bottom that is not trawled over at least once per year), there is a substantial amount of bycatch (Figure 11-11b). Much of this bycatch is juvenile fishes or crustaceans (such as shrimps and crabs), many of which support important commercial fisheries as adults (including red snapper *Lutjanus campechanus;* see Chapter 6 for controversies over management of adults populations of this species). Such small animals are typically not marketable so they are returned to the water after being caught; however, few survive.

The shrimp trawl fishery has one of the highest bycatch rates of any fishery. Hundreds-of-millions of pounds of fish bycatch can be discarded in the U.S. Gulf of Mexico shrimp fishery alone each year. Other bottom species that can be affected include echinoderms (such as sand dollars and brittle stars), soft corals, and other invertebrates living in the surface sediments. Sometimes the weight of organisms discarded is over 15 times greater than the weight of the shrimp harvest. Sea turtle bycatch has affected shrimp trawl fisheries more than any other, because of small population sizes and their endangered status (see Box 9-2, Turtles, Trawls, and TEDs).

In the United States and other industrialized countries, small fish bycatch is typically discarded dead overboard as unmarketable or of insignificant value compared to the targeted species. Thorough monitoring of the effects on the species comprising the bycatch mostly is undocumented, with the exception of a few that are commercially valued (e.g., red snapper in the Gulf of Mexico) and/or endangered (e.g., sea turtles; see Chapter 9). This is largely because of a general lack of reporting of the mass or species composition of the bycatch.

Trawl bycatch can be a portion of the commercial harvest in regions of the world where there is a market for these products or subsistence fisheries exist. For example, in China shrimp fisheries very little of the bycatch is discarded because it is used as fish food in the aquaculture industry. In Thailand some of the bycatch is used for *surimi* (fish paste) or locally consumed fish balls. In small-scale fisheries of

India most of the bycatch is dried for human consumption or poultry feeds. In Mozambique, local artisanal fishers collect bycatch from larger shrimp trawlers. Although a high bycatch rate in these fisheries still presents a problem, the amount of discard and waste is much lower and the use of the bycatch may take some pressure off other coastal fisheries. In the U.S. shrimp fishery, which has one of the highest bycatch and discard rates, there has been little success in establishing a profitable market for most of the bycatch.

The development of techniques to reduce the bycatch in trawl fisheries has been an area of active research in recent years. For animals larger than the targeted harvest, trap door escapes such as turtle excluder devices (TEDs) have been used successfully (see Chapter 9). Bycatch reduction devices (BRDs, pronounced *birds*) have been designed to operate on a similar principle to eliminate fish bycatch from shrimp trawls. For example, rigid grated BRDs are used in the trawl fishery for ocean shrimp (*Pandalus jordani*) off the U.S. northwest coast. This device allows the smaller fish to pass between bars that comprise a rigid panel in the net and into the net's cod end. Because most of the fish are much larger than the shrimp, they are guided by the grate out an escape exit at the end of the panel (Figure 11-11c). These BRDs have reduced bycatch in the ocean shrimp fishery by about 75% with very little loss of shrimp. State regulations now require the use of BRDs in the U.S. ocean shrimp fishery.

When much of the bycatch is of similar size as the shrimp or fish being caught, techniques that eliminate trawl bycatch by size will typically result in much loss of the targeted species. BRDs have been developed that separate the fish from the shrimp through a small opening at the top of the trawl net that allows small swimming fishes to escape. Another design has a mesh panel with larger opening in a portion of the net to allow fishes to escape. These have been of limited success in eliminating fish bycatch in the Gulf of Mexico shrimp fishery, where much of the fish bycatch is of similar size to the targeted shrimp. For example, the maximum reduction of bycatch of juvenile red snapper is about 25%. Trawled areas have been documented to have fewer larger snappers than untrawled areas. Only about 16% of fish species are excluded by approved BRDs in the northern Gulf of Mexico.

Habitat construction is another major conservation concern with bottom trawling. Because many of the trawled areas are over flat, soft-bottom low-relief substrates, it was long assumed that habitat impacts were minimal. It is now realized that disturbing the bottom substrate can cause substantial damage to the habitat of burrowing or attached animals. The discovery of widespread deep water reefs, many of which are on seamounts in international waters, that had been devastated by trawling led to increased concern and protection of bottom habitats from trawling (see Chapter 8 for a discussion of trawling, seamounts, and deep-water coral reefs). Even where there are no substantial living reef structures, loss of bottom relief used by fishes and benthic organisms has been substantial in intensively trawled areas. This can lead not only to a loss of species that created the structures (e.g., tube worms, algae, corals, and bryozoans), but also increased predation on fishes and invertebrates that depend on these structures.

Much scientific research has been carried out to document how trawling affects the biodiversity of low-relief bottom ecosystems. Some with an economic interest in trawl fisheries, however, continue to argue that trawling causes little ecosystem damage over smooth bottoms. They even suggest—without evidence—that it may improve things by "tilling the soil" like a gardener's plow. One difficulty in refuting the non-scientific arguments is in finding control sites that have yet to be exposed to trawling. John Gray and colleagues present evidence, based on numerous scientific studies, that trawling has a significant effect on bottom habitat through habitat destruction and homogenization. Even on bottoms dominated by sediments, variations are caused by patches of sand or stones, shells, animal tubes, holes, and animal remains or wastes. These provide important structures and refuges for macroinvertebrates and microscopic organisms. Trawls that scour the bottom homogenize this structure at and below the sediment surface. Gray and colleagues argue that the *Precautionary Principle* should be used more often in regulating trawl fisheries: if trawl fishing is to be allowed, then fishers and managers should be forced to demonstrate that it will not negatively affect the environment (as is required for chemical pollution and mining), rather than expecting ecologists to prove the harmful effect before trawling can be stopped.

Trawlers historically have avoided bottoms with hard or rocky substrates because of the fear of damaging their nets. Rocky bottoms, however, have been made more accessible to trawlers by the development of a technology called the **rockhopper**. Rockhoppers are large rollers attached to the bottom of the trawl net designed to hop over rocky bottoms. Now, areas that formerly were inaccessible are trawled, disturbing the bottom habitat. Rockhopper gear is now widely used in many groundfish (e.g., Atlantic cod) fisheries. An additional concern with rockhopper gear is that it can roll over and injure or kill large numbers of fish that are not caught in the net, resulting in a source of undocumented fish mortality.

Other than eliminating the use of bottom trawls in some regions, little effort has been made to develop methods that minimize the effect of commercial trawling on bottom habitats. Because the animals being targeted by bottom trawls are associated with the benthic habitat, the trawl must be pulled across the bottom to fish efficiently. In 2007 John Valdemarsen and colleagues, with the FAO, presented various methods by which the pressure on the bottom by components of trawl gears could be minimized. For example, lighter or redesigned trawl doors reduce the pressure applied to the bottom. Other ropes in contact with the bottom could be elevated by rolling discs. Acceptance of such methods would depend on the degree to which it affected trawl catch, and it is questionable whether they would have a substantial reduction on trawling impacts. Other alternatives to bottom trawling are to target fish with mid-water trawls when they are off the bottom or switching to stationary gears such as gill nets, longlines, or traps where feasible; however, as discussed above, these are no panaceas, as virtually all fishing gears have the potential for some type of damage to habitat or harm to ecosystems.

11.3 Marine Fisheries Overharvest

The ability to sustainably fish any marine population depends on the population's ability to produce a surplus of individuals that are available for harvest. A primary goal of regulations governing marine fisheries harvest is to achieve **sustainability**, defined as harvesting at levels that allow continued harvest indefinitely. This can be achieved if the population remaining after harvest is able to replenish itself by reproduction. In theory, all that is needed for sustainable management is a scientific determination of the surplus that can be taken without adversely affecting production. In reality, however, this is rarely ever as simple as it first appears. Scientifically, it is extremely difficult to define the surplus production, and even when possible that number is constantly changing because of biological and environmental variability. In addition, management decisions also must take into account political, social, economic, and technological issues. This section discusses variation in harvest by year, geographic region, ecosystem type, and biological grouping with a goal of describing trends and patterns that can be used to understand the problem of excessive fish harvest, and presents methods to be applied to achieve the goals of marine fishery sustainability.

Marine fisheries account for about 90% of the world fisheries catch (the remaining 10% is inland harvest from freshwaters). Marine fisheries landings have increased dramatically through the 1900s as human populations have expanded and fisheries harvest technology has advanced (**Figure 11-12**). Prior to the 1950s, overharvest problems were localized in a few regions, including the North Atlantic, the North Pacific, and the Mediterranean Sea. After World War II, however, as vessel and harvest technology improved dramatically, harvest exploded, increasing more than fourfold from less than 20 million tons in 1950 to over

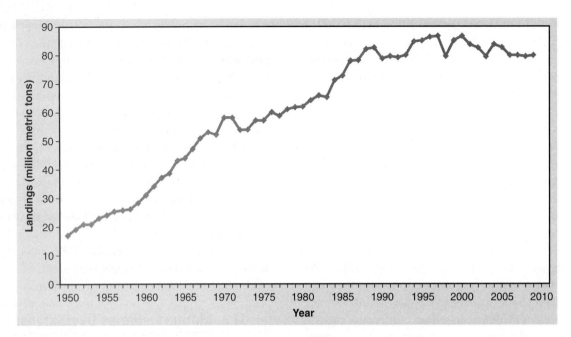

Figure 11-12 Reported global landings from capture fisheries, 1950–2009. Harvest increased, on average, 1.5 million tons per year from 1950 through 1990, followed be a leveling or decline in harvest, likely reflecting declining fishery populations.

80 million tons in 1990. A pattern developed that Daniel Pauly calls the *toxic triad:* catches were underreported, scientific advice was ignored, and the environment was blamed as a standard response to the collapse of marine fisheries. As fish populations became overexploited, fisheries moved on to other species and other geographic areas and expanded into deeper waters. Global landings leveled off after the late 1980s at about 80 to 85 million tons annually, and have remained below 80 million since 2005, despite continual increase in the amount of fishing effort and expansion into new fisheries. This pattern suggests that maximum harvest levels were being approached on a global scale. Because efforts continued to increase while catches remained fairly constant, the **catch-per-unit-effort** (**CPUE**) was declining (i.e., more effort was needed to catch the same amount of fish). CPUE is a common index used to monitor fish populations; a decline in CPUE generally indicates declining populations.

The FAO has kept a thorough global fish database since the 1950s based on landing data provided by fishing nations. Based on these data it is estimated that over 25% of the global harvested fish stocks are significantly overharvested or depleted and that about 50% are being harvested at a **maximum sustainable yield** (**MSY**), defined as the greatest harvest allowable while still maintaining healthy populations. FAO data also show that catch of the most heavily harvested fish populations has declined in weight by an average of 40% below mid-1980 levels.

Massive declines in total global fish harvest are being avoided by targeting more and different species that have yet to be overharvested.

Although a population's response to fishing varies widely, FAO data indicate that, on average, fishing using modern techniques can reduce a large fish population to about 20% of its original size in 15 years. This is collaborated by statistics showing that many of the large-scale fisheries implemented by the 1980s had caused a population **collapse** (a dramatic lowering in the total biomass) by the end of the 1990s. Currently, the world capacity for fish harvest (i.e., the amount that could be harvested with the current amount of fishing boats and gears) is estimated to be about twice as large as the resources can withstand. Harvest restrictions have kept the actual harvest below that capacity; however, some fisheries scientists argue that current fisheries harvest levels should be cut in half to avoid a collapse of global fisheries.

In order to understand why harvest regulations have not been successful in managing large-scale marine fisheries, it is necessary to understand the philosophical basis for many of those regulations. Historically, marine fisheries management was based on the assumption that optimal harvest for most fish population is achieved by keeping the population at about 40% to 60% of its pristine levels. This is based on basic population growth models that predict maximum population growth rate at 50% of the population's carrying capacity. This maximum sustainable yield approach to management has been widely criticized, as evidence by the publication of "An Epitaph for the Concept of Maximum Sustainable Yield" by P. A. Larkin, a Canadian fisheries biologist, in 1977.

> *Here lies the concept of MSY*
> *It advocated yields too high*
> *And didn't spell out how to slice the pie*
> *We bury it with the best of wishes*
> *Especially on behalf of fishes*
> *We don't know yet what will take its place*
> *But hope it's as good for the human race.*

Several shortcomings of the MSY philosophy were identified by Larkin. Fishing to the level of 50% of natural population levels is likely to select out the largest fish. These fish typically have the highest fecundity (egg production) and may produce more fit offspring. More recent studies have shown that excessive fishing of the largest fish, through artificial selection, drives the population genetically to smaller average sizes and reproductive rates. Fishing to MSY also reduces genetic variability in a population in other ways. The most easily caught fish may have desirable genes (e.g., for disease or temperature tolerance) that are removed from the population by fishing. Moreover, fishing one species to excess does not account for interactions among species in the ecosystem. For example, if a predator is overfished, its prey may become overabundant and then over-consume its prey; the result is a total restructuring of the ecosystem in unpredictable ways. As fishers move to harvesting smaller fishes at lower trophic levels, the food web is exposed to further disruption (**Box 11-1. Fishing down the Food Web**).

Possibly the greatest problem with allowing the MSY concept to drive fishery harvest restrictions is that the fish population is left vulnerable to unpredictable fluctuations. Many fishes, especially short-lived fast-growing species, can undergo dramatic population cycles even in the absence of fishing or other anthropogenic stresses (e.g., coastal herrings; **Box 11-2. Unpredictability in Pelagic Ecosystems**). Populations maintained at 50% or below of their original levels have less flexibility and are less likely to recover from a rapid collapse. Fishery data have shown that if populations reach 10% to 20% of their original levels, collapse is a serious concern.

Even if fish populations follow the predictive models, social and economic concerns can drive management decision-making. There can be a fine line between sustainable harvest and overharvest, and disagreements often lead to poor decision-making. When economic considerations outweigh science-based recommendations, a decision pattern can develop. This pattern, coined *Ludwig's ratchet* by T. Hennessey and M. Healey, results from an emphasis on short-term economic gains over long-term economic and ecological sustainability. A cycle begins when managers allow harvest of a population at an unsustainable level in order to reduce short-term economic and social pressures. When these harvest levels cause the population to decline and the fishery to crash, catch levels are reduced somewhat, but government financial assistance (**subsidies**) are implemented to allow overfishing to continue. If the population begins to recover, there is a rush to increase catches again, starting the cycle over again. This is a broad generalization and many species once overharvested recover slowly, if at all, especially if the habitat is destroyed during fishing (for example, with some bottom-trawl fishing), or the fishery is on long-lived species (for example, the orange roughy fishery discussed in Chapter 8). Hennessey and Healey, however, point out that the fishery for groundfish, such as cod, off the northeast U.S. coast has conformed well to this model (see Box 11-3, Learning from History: Boom and Bust of Atlantic Cod Fisheries later in this chapter). They argue that the solution to avoiding similar situations is to integrate science, management, and harvesting considerations by applying ecosystem-based management.

Another potential pitfall of management based on harvest statistics is inaccuracy of the data. Mathematical models used to estimate population size and allowable catch require estimates of adult **mortality** (natural and fishing), **survival** rates for young fish, **recruitment** (number of fish that reach maturity), **catchability** of the fish (what proportion are caught with a given effort), and an **age profile** for the fish in a population. Each of these variables is difficult to estimate accurately and highly variable with time, location, and species, not to mention variability inherent in the measurement process itself. When precise numbers are not available, assumptions are made to generate the "best available" data, and when data are lacking, this is simply an educated guess.

Misreporting of harvest amounts is another source of variability in fishery data. The reliability of reported catches varies considerably among fisheries; for example, those with many subsistence or small-scale vessels are more difficult to monitor than those with fewer but larger vessels. The accuracy of data also varies among nations and is dependent on the resources available for monitoring. When fisheries harvest is not recorded to the species level, it is difficult to monitor population status. For example, harvest in the Indian and central Pacific Ocean is often reported in categories such as herrings and snappers, rather than by species.

Fisheries landings are often under-reported because of harvest that can be categorized as **illegal, unreported, or unregulated (IUU)**. Unreported discards are largely comprised of fishes caught incidentally in trawl fisheries, and it is common for this to comprise as much as 40% of the

Box 11-1 Fishing Down the Food Web

Since the development of modern technologies that allow fishers to target the largest fish, which are often the most profitable to harvest, there has been a dramatic decline in many populations of apex predators in ocean ecosystems. These species cannot withstand a large sustained harvest because of their slow growth rates and late age at maturity, and ecosystems cannot produce a large biomass of large apex predators because of the inefficiency of energy transfer up the food web (see Chapter 2). As the large top trophic level predators disappear, fishers feel pressure to move down the food web to fish for the prey at the second trophic level. This harvest not only affects the prey species but also inhibits the recovery of the predators. As the second trophic level species decline, the pattern continues. The fishery moves on down the food chain, taking smaller and smaller fish (**Figure B11-1**). The filter feeding clupeoids (herrings, anchovies, etc.) at the bottom of the fish food chain are not as valued for harvest as top predators; however, large global markets have been developed, including the production of fish meal used in feeding livestock and farmed fish (see below). This pattern keeps the ecosystem from recovering and forces it into simple food webs with low species diversity.

This pattern of *fishing down the food web* was brought to the public's attention through publications by Daniel Pauly and colleagues in the late 1990s. They documented, using global fish harvest data reported to the FAO, that the mean trophic level (MTL) of fishes caught in global fisheries has been declining since the 1950s, moving from a dependence on large high trophic level predators to invertebrates and pelagic fishes at low trophic levels. There is a tendency to continue this practice, in part because the total harvest in an ecosystem initially increases because of the ability of lower trophic levels to support more biomass. Eventually, however, these catches begin to decline and the fishery becomes unsustainable if continued at high levels. Overfishing of the lower trophic level species inhibits the recovery of predatory species, even if fishing of the predators has been stopped. Additional ecosystem effects include the loss of forage for other animals such as seabirds or marine mammals.

There are numerous examples of ecosystems where this trend toward the harvest of smaller and smaller fish has been documented. For example, in many coral reef fisheries large predatory groupers and sharks have gradually disappeared from the fishery to be replaced by smaller grazing species, such as parrotfishes and damselfishes, with impacts on the entire ecosystem (see Chapter 5). After the collapse of the cod fisheries in the North Atlantic Ocean (see below), fishers began to target shrimp, a common food of the cod, more heavily. This means that, despite restrictions on catching cod, they may not recover because of a lack of shrimp to eat. Possibly the most practical way to reverse this trend and avoid widespread fishery collapses is to create large no-take marine reserves to protect all links within open ocean food webs.

Although well-documented examples of the fishing-down-the-food-web phenomenon exist, other studies suggest that this may not necessarily indicate a global trend. Trevor Branch and colleagues argue that the metric used to measure MTL is not reliable because it is based on reported commercial fisheries harvest, which is biased by factors such as harvest gear used and economic value of the fish. When less-biased scientific trawl survey and stock assessment data are used to estimate fish abundance, a different trend becomes apparent. Branch's analyses suggest that fisheries harvest is increasing at all trophic levels. Although higher-level predators are not being selectively eliminated, all trophic levels are at lower abundances than before. If this is true, ecosystem function may not have been impaired and recovery could be more likely if excessive fishing is controlled.

The controversies over how fishing is affecting marine ecosystems may not be resolved easily. This reflects the complexity of these ecosystems. Research, however, invariably points to the need for a scientific basis for management. Not only does the biomass of harvested populations need to be carefully monitored and regulated, but the health of every population in the entire food web, and the size and age structure of individuals that comprise each population must be considered. This level of ecosystem management may never be fully achieved for ecosystems supporting harvested species. But it is a goal for which fisheries managers can strive.

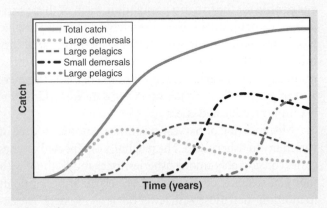

Figure B11-1 A hypothetical model showing how total catch can continue increasing as fishers continue *fishing down the food chain*. The largest predatory fish decline, leading to harvest of smaller prey species, keeping predators from returning even if protected from harvest.

total harvest. Because most bycatch is dead even if returned to the water, and thereby removed from the population, many fisheries now include bycatch in harvest statistics. The amount of harvest by illegal fishing is unknown and difficult to estimate. Illegal fisheries operate where fishing is not permitted, use banned technologies, or underreport landings. In some fisheries it is estimated that this harvest may be as much as three times the allowed fishing quotas. Even in legal fisheries, landings are often unreported. Reporting requirements are often nonexistent or unenforceable

Box 11-2 Unpredictability in Pelagic Ecosystems

There are still many unanswered questions concerning the effects of human activities on the open ocean. This is partly because a pelagic environment is naturally variable and unpredictable, influenced by changing weather and current patterns that can happen rapidly and without warning. Plankton rely on ocean currents for transportation to places desirable for survival. A simple change in the wind direction during the reproductive season, therefore, could alter where the eggs and larvae are transported. If they are transported into an area with inadequate food organisms or where they are vulnerable to predators, there could be a large impact on survival, affecting the next generation supporting fisheries. It is often not apparent that the population has been affected until years later when adult fishes are low in abundance.

Although fishing likely influences these population cycles, there is strong evidence that they also are governed by natural phenomena. A creative way to document historical population cycles in coastal clupeoids was developed by Tim Baumgartner and colleagues through the study of scale deposits in coastal sediments. The density of scales preserved in sediments from seafloor cores provided population indices for northern anchovies and Pacific sardines off California from the years 300 to 2000 AD (**Figure B11-2**). By dating these sediments, *hindcasts* of fish biomass were obtained throughout this time period. These data indicated extreme population fluctuations on an approximately 60-year cycle even before fisheries or other human effects. This suggests that pelagic ecosystems can be highly variable and unpredictable, making it difficult to discern whether human activities, such as environmental impacts or fishing, or natural fluctuations are causing ecosystem changes.

The sardine and anchovy cycles off California simply could reflect local changes. However, simultaneous booms in sardine and anchovy populations around the Pacific suggest that the populations are affected by global climatic events (**Figure B11-3**). Water temperature changes appear to be one

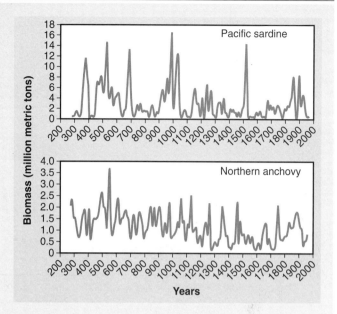

Figure B11-2 An index of Pacific sardine and northern anchovy population size off California from the 4th through 20th centuries, 300 to 1998, based on scale deposition rate in bottom sediments. These data document dramatic population fluctuations even when fish are not exposed to fishing pressure.

important factor controlling these cycles, and, in the Pacific, cooler waters tend to favor anchovies and warmer waters favor sardines. For example, in the past 100 years the Japanese sardines and anchovies have swapped places several times as the most important clupeoid in harvest weight as water temperatures vary (**Figure B11-4**). Although mechanisms controlling fish population cycles are still not completely understood, much progress has been made in recent years to determine important factors and the influence of fishing on population recovery.

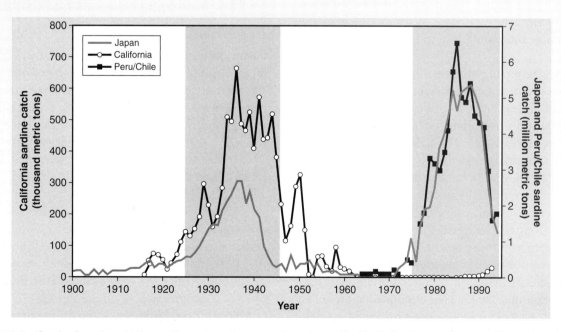

Figure B11-3 Catch of sardines in the northwest, northeast, and southeast Pacific, indicating synchronous fluctuations in populations, possibly as a reflection of global climate fluctuations.

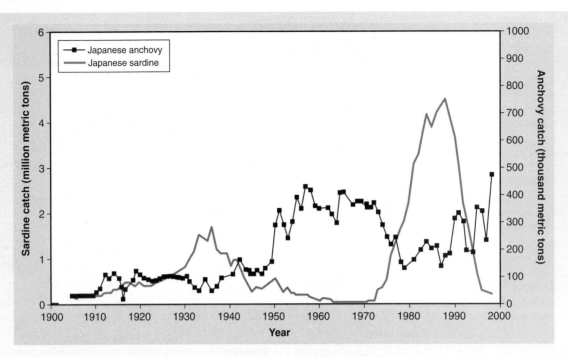

Figure B11-4 Comparisons of catches of herrings and anchovies by Japanese fleet indicating synchronous population changes but in the opposite directions.

in coastal subsistence and small-scale fisheries in developing nations. Even fisheries harvest by developed nations can go unreported. For example, studies by D. Zeller discovered that harvest by Arctic indigenous and subsistence fisheries from the United States, Canada, and Russia from 1950 to 2006 was 75 times greater than that reported to the FAO (actual harvest averaged 12,000 tons per year). There are concerns that such reports will give a false impression of underfished populations. Increased access to Arctic waters with loss of sea ice due to global climate change thus could lead to increased fishing and overharvest.

Even overreporting of harvest has resulted in a substantial data bias in some fisheries. For example, mid-level Chinese officials seeking promotions were recently found to be systematically overreporting landings to receive incentive awards (see Chapter 12). This contributed enough to reported global landings that the apparent moderate increases or leveling of global harvest through the 1990s may be based on a misrepresentation. This is significant because lower harvest data would indicate a reduced ability to catch fish (lower CPUE), evidence that fish populations were smaller than had been assumed previously.

There are a number of ongoing efforts to establish international agreements to help monitor and eliminate IUU fishing. For example, the FAO has attempted to broker an agreement that would require vessels fishing in international waters to report their fish harvest prior to landing, set

up regular inspections of fishing vessels, and have nations communicate information concerning IUU-associated vessels. In 2008, the European Union established measures to deter and eliminate IUU fishing by establishing a catch documentation plan and conditioning access to markets for fisheries products on the degree to which an importing country is reducing IUU fishing.

11.4 Managing Fisheries for Sustainability

Harvest Quotas

Historically, large commercial fisheries have been managed by setting total annual harvest quotas based on a maximum sustainable yield, as described above. The fishing season is opened at a defined time of the year, typically based on the biology and availability of the fishery species; for example, when they reach a certain size or are aggregated in schools. The close of the season may be at a predefined date, but in quota-based management is typically set to when the total allocated weight is harvested.

Size restrictions can be used to protect fish until they reach maturity and are able to reproduce. Gear restrictions, such as mesh sizes on gill nets or excluder devices for trawls, may be used to limit smaller individuals; however, as discussed above, often this is not feasible. Post-harvest sorting by size is only practical when the animals are not stressed or killed during fishing (e.g., trap fishing for crustaceans; see above).

Although **annual harvest quotas** are commonly used as a management tool, in recent years more thorough monitoring and flexibility have been incorporated. If a fishery continues to decline, adjustments are made to the quota. The population size is designated at which the entire fishery will be closed to harvest until recovery is achieved (see North Atlantic cod discussion below). Often these closures do not remain in place long enough for the population to recover to healthy levels, however.

Various modifications of the quota management system have been made to avoid overharvest while more efficiently allocating the fishery. **Individual fishery quotas (IFQs)** reduce fishing capacity by limiting the number of vessels in the fishery (**limited entry fisheries**). This partitions the overall harvest quotas to individuals within a fishery. They then have the flexibility of methods and times to harvest to maximize their profit from the fishery. Problems with individual allocations include deciding which fishers have a right to a quota and how to judiciously sell or transfer quotas to avoid the fishery being dominated by a few large corporations (see Chapter 12 for further discussion of limited entry and IFQ management).

In **community-based management**, quotas are allocated to local communities to make management decisions. Smaller communities tend to make better decisions to sustain fisheries because each person believes they have a vested long-term interest in the health of the resources. Community-based management has been implemented successfully in areas of Alaska and Canada, and indigenous groups around the globe often manage fishery resources successfully as a small community.

Even if quotas are set at excessive levels, once a population declines to a size at which fishing is no longer profitable (because fish become too difficult to catch), one would expect fisherman to stop fishing for economic reasons. In order to avoid job loss, however, governments often provide subsidies to allow fishing to continue. A **subsidy** can be defined as some grant or benefit given by the government to an enterprise considered beneficial to the public. Subsidies to fisheries allow for the continued harvest even if the fishery is not profitable. Examples of subsidies include price supports on fisheries products, tax breaks on fuel purchases, guaranteed loans on fishing vessel purchases, and construction of fishing ports. Subsidies are often applied to industries that are ancillary to the fisheries, including seafood wholesalers or retailers, processing plants, and boat manufacturers. Estimates of the global worth of fisheries subsidies varies widely depending on estimation methods; however, the value could be over 30 billion U.S. dollars annually worldwide. Eliminating subsidies would result in a reduction of the capacity of the world's fishery harvest; however, political and sociological factors weigh in on subsidy-related decisions. Fishers might be encouraged to go along with loss of subsidies if funds are used for retraining or re-employment in other areas.

An educated public can have a strong influence on fisheries management decisions in defining quotas and other regulation through consumer choices, electronic media, and political channels. Ecolabelling of seafood products can force the fishing and seafood industry to change harmful practices; for example, dolphin-safe labeling of canned tuna (see Chapter 10) and country-of-origin labeling for many seafood products. Because specific criteria for such classifications have not been established, however, there is no assurance that ecolabels are applied appropriately. Various non-governmental organizations (NGOs) have developed lists of seafood products that they judge are being sustainably produced with little environmental impact.

There is always some uncertainty about how fishery populations will respond to management actions or environmental variability. Scientists thus provide a range of recommendations for quotas and other harvest restrictions. This unavoidable uncertainty can be used as a rationale to keep high catch levels in place if managers succumb to pressure to choose the quota at the higher end of the range. Historically, such decisions often have been proven to be disastrous (see examples below). Scientists and environmentalists have begun to pressure management agencies and lawmakers to change this attitude by requiring that decisions be made using a **precautionary approach**. The precautionary approach is defined by the FAO as the *application of prudent foresight*. This includes recognition of the uncertainty and inaccuracy of our scientific knowledge and exercising caution to avoid outcomes that would affect the future health of marine ecosystems, for example, by choosing the lower end of a range of recommended harvest quotas.

A precautionary approach identifies potential problems before a crisis develops and emphasizes conservation principles. Conservationists argue that, under the precautionary approach, the users of ocean resources should bear the burden of proof that their level of exploitation will not harm the sustainability of the fisheries resources, rather than requiring this of fisheries managers as has been done historically. In order to implement a precautionary approach to management, specific regulations must be based on ecological and biological factors. But as long as there is a great demand for seafood, pressure to overharvest will continue. As long as socioeconomic and political considerations take precedence, the world will continue to deal with declines of fisheries resources.

National governments around the world have produced plans to deal with the dilemmas of mismanagement of fisheries. Implementation of these plans is typically slow, however, and most developing nations have been unable to take the measures needed to solve overharvest problems. Regional fishery management organizations, established through multinational agreements, serve an important role in trying to reduce marine fishery capacity in developing nations.

Although numerous broad international agreements have been established recently to combat unsustainable fishing, completely implementing and enforcing such agreements has been met with only limited success. For example, in 1999 an International Plan of Action for the Management of Fishing Capacity was adopted by the FAO. Through this plan nations agreed that global fishing capacity was too large. The agreement, however, did not establish specific criteria detailing exactly how to measure or reduce fishing capacity. It is clear that an overall change in the philosophy of how to manage living ocean resources is needed to avoid continuing the current patterns of overharvest. Hopefully, the changes that would make a difference are not too idealistic to be realized.

■ Ecosystem-Based Management

Ecosystem-based management has become a common catch-phrase for management to protect not only the fishery population but also the habitat and community upon which it depends for long-term survival. True ecosystem-based management is difficult to achieve, however, because it is virtually impossible to optimize populations of every species in a community, and managers are almost always dealing with incomplete information. Coastal and estuarine communities are the most thoroughly studied; however, because of the complexity of these communities, species interactions are still not well understood. Even less is known of open-ocean and deep-water ecosystems. Uncertainty over how ecosystems will react to environmental events (e.g., El Niño or global climate change) presents further challenges to managing marine fisheries for sustainability. Although it is well known that fish populations are sensitive to environmental changes, rarely have predictive capabilities been realized. Increased sensitivity of populations that have undergone excessive harvest adds another degree of uncertainty.

Several processes in marine systems have been identified that may help explain the unpredictable responses of marine ecosystems to fishing and other anthropogenic and natural effects. Some well-studied marine ecosystems undergo **regime shifts**, large transitions between alternative **stable states**, defined by the ecosystem composition during a given time period. Regime shifts can be driven by numerous natural factors, including natural population changes, invasions of alien species, climate change (e.g., for coastal pelagic ecosystems; see Box 11-2, Unpredictability

in Pelagic Ecosystems?), eutrophication, or overfishing. Exact causes of regime shifts are extremely difficult to pinpoint.

Mechanisms for regime shift because of overfishing include **trophic cascades**. A trophic cascade results from a **top-down** effect on the food chain. For example, overharvest of apex predators allows for an increase in their prey at the second trophic level. This then causes a decline in animals at the third trophic level, with the alternating cascade effect continuing down the food chain. This has been proposed as a reason for ecosystem changes in coral reef (see Chapter 5) and other coastal ecosystems. Essentially, fewer apex predators in the ecosystem causes an increase in intermediate predators, which then over-prey on herbivores or planktivores; thus, excessive harvest of apex predators indirectly results in an ecosystem dominated by algae or plankton. Marten Scheffer and colleagues propose that such a cascade affected the pelagic food web associated with fisheries in the North Atlantic. According to this hypothesis, overharvest of cod and other large predators caused an increase in the numbers of smaller fishes, crab, and shrimp due to lowered predation pressure; large-bodied zooplankton experienced a resultant decline and, finally, the phytoplankton increased in response to lowered zooplankton, reducing nutrient levels in the water. Of conservation concern is that, for unknown reasons, the ecosystem changes resulting from trophic cascade may not be easily reversed, even when excessive fish harvest is eliminated.

Whatever the mechanism, the loss of biodiversity in an ecosystem has implications that extend beyond management of fisheries. Analyses by Boris Worm and colleagues have shown that the loss of biodiversity in marine ecosystems often leads to increases in resource collapse and reductions in stability, water quality, and recovery potential. When biodiversity is restored, productivity and ecosystem stability increase. Ongoing loss in marine diversity is resulting in a reduction in the ocean's ability to provide food resources and maintain water quality. At the current rate of decline in biodiversity, an overall global collapse in fisheries would occur by the mid-21st century.

One of the simplest ways to avoid having to deal with the scientific uncertainty and uncontrollability pertaining to fishery harvest, and the resulting loss in biodiversity, is to eliminate harvest from at least a portion of the ecosystem. A total long-term closure of major fisheries is rarely achievable; however, protecting a portion of the fishery and its supporting ecosystem has been rapidly gaining acceptance among scientists and managers, and could be the greatest achievement towards controlling overexploitation of fisheries while maintaining functional marine ecosystems. No-fishing **marine protected areas** (MPAs) or **marine reserves** are being established around the world where removal of marine organisms is restricted. These provide habitat protection and refuges, allowing fishery populations to not only grow within the refuge, but also replenish overfished regions outside the refuge with offspring transported beyond the MPA. Additional advantages were presented by Irene Novaczek to include: protecting commercial fishery populations, limiting bycatch, protecting marine diversity, protecting critical life stages of fishery species, protecting and enhancing productivity, enhancing marine research, protecting artisanal fisheries, enhancing public education, and encouraging non-destructive use of the sea.

Although intuitively the premise of using no-fish reserves to protect and enhance fish populations is clear, documenting the success of such actions is an important step toward encouraging the expansion of reserves. One fear of fishers and managers is that reserves will reduce overall harvest for a species. Although models have predicted that reserves will enhance fishery yields for many species, direct evidence of success is limited by the number of reserves that have been in place for an extended period. For those for which data are available, the total number of fish has been observed to increase as much as 50% to 150%, and the number of species has increased on average by almost 25%. Many reserves are too new to judge the success of replenishing populations outside of the sanctuary; however, examples of success stories have increased in recent years. For example, a network of reserves off the Caribbean island of St. Lucia, established in 1995, has resulted in a 90% increase in the harvest in adjacent small-scale fisheries. As well, the first marine reserve in Scotland waters—fully protected in 2008—began to show fishery benefits within two years, as indicated by substantially greater increases in scallop populations within than outside of the reserve, associated with increased growth of kelp and other beneficial macroalgae.

Large MPAs have recently been developed off Australia and California. In Australia, a network of MPAs was established in the Great Barrier Reef region, with a goal of protecting at least one-third of each bioregion in the area. Off the coast of California, the Channel Island network of marine reserves is made up of 13 areas, selected to protect biodiversity and critical habitat for breeding fish. In U.S. waters off the northwest Hawaiian Islands, a large no-take marine reserve was established in 2006 (see Chapter 5). These reserves were established in an attempt to recover marine populations that have declined dramatically. Small no-take marine reserves are now being established around many Caribbean Islands that historically have been overfished, largely by subsistence and artisanal fishers. For example, in Jamaica a series of no-fishing reserves have been established scattered around the coast. Enforcement has been implemented slowly and the success of these reserves remains to be seen.

Aquaculture as a Solution to Overharvest

As global marine fisheries harvest leveled off or declined through the 1990s, the total production of seafood continued to increase. The source of this increased production was from **aquaculture**, raising aquatic organisms in captivity until harvested for market; about 50% of organisms are marine species, designated as **mariculture**. The greatest mariculture production is for salmon, shrimp, and bivalves (e.g., oysters, clams, and mussels). Farmed seafood production has grown dramatically from less than 10 to over 60 million metric tons per year from 1980 to 2008; it now comprises almost 50% of the total world fisheries production (**Figure 11-13**). Asia has been the source of most of this increase and now supports over 90% of world aquaculture production, and 60% percent is from China alone. Although Europe, North America, and Japan account for only 10% of production, they are the greatest importers of aquaculture products.

As aquaculture production begins to approach the harvest of wild organisms in volume, some are beginning to tout aquaculture as the panacea for global overharvest problems. These statistics can be misleading, however, because fisheries products produced by aquaculture can cause ancillary problems to marine ecosystems, reducing the availability of wild marine fisheries species. For example, shrimp aquaculture in coastal regions often results in the removal of natural coastal habitats, such as mangroves, which are crucial for the survival of wild organisms (see Chapter 4). Salmon, currently the top aquaculture species, are commonly farmed in floating pens in bays and coastal habitats (primarily in Norway, Chile, the United Kingdom, and Canada; **Figure 11-14a**), potentially affecting marine ecosystems through organic pollution from fish wastes and introduction of antibiotics (though the use of antibiotics in salmon farming has been reduced substantially since the early 1990s) and other chemicals into the environment. The most commonly farmed marine species is the Atlantic salmon. When these are raised in coastal waters of the eastern Pacific, escaped farmed fish could potentially compete or interbreed with native salmon, and there are concerns over transfer of parasites such as sea lice (caligid copepods) to wild salmon populations.

Another complication with farmed marine fish is that most are carnivores, and therefore they are fed feeds that are high in animal proteins (freshwater farmed fishes such as tilapia and carp are natural herbivores or omnivores and feeds have a higher fraction of plant materials). The primary source of these proteins is from other wild-caught marine fishes, mostly clupeoids such as herrings, sardines, and anchovies, which are classified generically as **forage fish**. On average, it takes almost 2 kilograms of forage fish

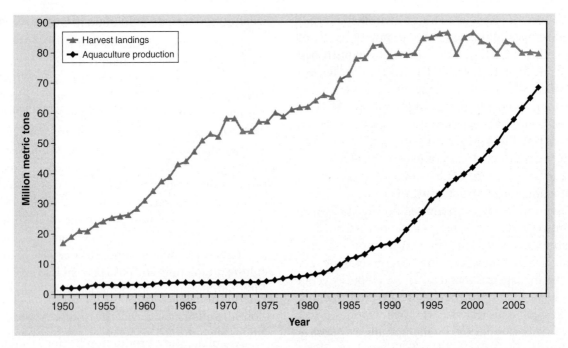

Figure 11-13 Global aquaculture production of all aquatic organisms compared to marine capture landings. Over the past two decades, harvest landings have stabilized but aquaculture production has increased at a rate of over 2.5 million tons per year.

to create 1 kilogram of farmed fish. When harvesting forage fish to feed carnivorous fish, aquaculture is not solving the overharvest problem because it is not contributing to net fish production but instead to excessive harvest while affecting marine food webs (see Box 11-1, Fishing Down the Food Web). Obviously, it would be more efficient for humans to eat the forage fish rather than feed them to farmed fish. Although herrings and anchovies are edible, however, they are not nearly as desirable as the farmed salmon and currently could not be marketed profitably on a global scale as human food. (However, direct consumption of forage fish has increased in recent years, to about 10%–20% of the harvest, especially in low-income groups or where animal protein is in short supply and farm raised fish are unaffordable). One way to reduce the use of harvested forage fish to feed farmed fish is to increase the amount of soy protein in the fish feeds. Some progress has been made in this direction; however, the quality of soy protein for raising fish is inferior to fish meal, and thus more likely to reduce their growth and disease resistance. Around the turn of the 21st century, over half of the fish meal and fish oils being produced were used in fish foods for aquaculture. Increases in food conversion efficiency are predicted to decrease the use fish meal and oils in feeds; however, global expansion in aquaculture is likely to increase the demand for these products. The availability of fish meal could limit future increases in farming of carnivorous fish such as salmon.

Another effect of aquaculture expansion is harm to estuaries and coastal ecosystems (see Chapter 4). One way to minimize these effects and expand the possibilities for aquaculture production is through open-ocean aquaculture (Figure 11-14b). Destructive impacts would be less likely, but effects of wastes on bottom ecosystems are of concern. There are also many technological challenges to open-ocean aquaculture, including the production of cages that can withstand the storms and wave action. Maintenance expenses would be greater because food and other materials would have to be transported out to sea. In the United States, the NMFS proposed regulations on aquaculture in federal waters in 2010. They plan to permit from 5 to 20 operations by 2020 in the Gulf of Mexico to produce a estimated 30 million metric tons of fish per year. Farmed species will be limited to native species, and requirements will be established to minimize environmental impacts. Resistance from environmental groups and fishers over concerns that waste could affect ecosystems and harvest of wild fish is likely to lead to delays in implementing offshore aquaculture operations.

For some cultured marine species, there has been a recent trend to move inland for rearing in ponds and tanks. Shrimp aquaculture operations that are considered to be the least harmful to the environment are those away from the shore that do not require the destruction of coastal habitats (see Chapter 4). Some salmon farming is now also carried out in this manner. If environmental issues

Figure 11-14 The most efficient fish farming methods are extensive farming in ocean pens, such as **(a)** these floating net pens, most widely used for salmon rearing. **(b)** Methods are being developed for deepwater aquaculture, such as with this culture pen off Hawaii; note the feeding tube to surface.

can be resolved, aquaculture may serve as an environmentally friendly replacement for some seafood products. For some products aquaculture has already begun to replace wild harvest for economic reasons. For example, in the U.S. Gulf of Mexico, competition from less expensive imported cultured shrimp contributed to a 60% decline in the number of shrimpers over a decade time span. During that time period the price of imported shrimp declined over 30%, driving many U.S. shrimpers out of the market. U.S. shrimpers have been unsuccessful in their fight to stop imports of inexpensive shrimp due to requirements stipulated by U.S. trade laws.

Another controversy has developed in recent years over proposals to introduced genetic engineering techniques into aquaculture. Techniques to genetically modify

organisms through the insertion of novel genetic material has led to promises of more efficient aquaculture, along with fears of potential harm to natural ecosystems and populations. The test case in these issues is a proposal to allow transgenic growth enhancement of Atlantic salmon (by inserting genetic material from other fish species) for aquaculture. Despite assurances that the fish would be raised from sterilized eggs (though a small fraction of the fish produced could be reproductively viable) in a closed system with little fear of release into the environment, the threat of an accident resulting in the release of transgenic fish is of concern. If released transgenic fish or their offspring are present in large numbers, there is the chance of competition with wild fish; enhanced growth transgenics have a greater rate of feeding and take a larger share of food resources. Migratory behavior also could be altered, with more rapidly growing fish migrating from streams earlier, which could alter timing and use of resources with unknown ecosystem effects. The risk of these and other unknown effects is impossible to determine precisely.

To address some of these concerns, proposals have been made recently to spawn the fish in Canada and rear them in Panama. Alison van Eenannaam and William Muir argue that this would virtually guarantee safety because all fish would be female and 99.7% would be sterile, the rearing facility would be land-based and screened, waters in Panama are tropical and too warm for salmon survival were some to escape, and the nearest mates and spawning streams are several thousand kilometers away. Still, the U.S. FDA has hesitated to approve the use of these transgenic fish as food in the United States because of environmental and human health concerns (there are reports of increased levels of allergens compared to wild salmon).

In making decisions concerning the rearing of genetically modified marine organisms, extreme precautions must be taken to ensure containment of the genetic material. Some fear that potential economic incentives could eventually override environmental concerns in gaining broad approval for the use of transgenic fish in aquaculture. Environmentalists have complained that the approval process through the U.S. FDA does not take into account potential environmental problems but only considers the safety of transgenics for consumption by humans.

11.5 Conservation of Fishery Populations

Although ecosystem-based management is a desirable goal, single-species management has been the most common approach for large international fisheries. So many biological, environmental, and human factors come into play that fisheries have to be managed on a case-by-case basis, and the best management strategy is not always obvious. Specific examples will be used to demonstrate how complicated management and conservation of fisheries populations can be.

Coastal Pelagic Fisheries

Coastal pelagics are fishes that school in open waters, away from structures or the bottom but near the shore, so that they can take advantage of productive coastal waters. These are primarily forage fish in the families Clupeidae (herrings, sardines, and menhadens) and Engraulidae (anchovies), grouped as clupeoids. There are over 300 species, with maximum sizes ranging from about 2 to 75 centimeters in length, mostly distributed in temperate and tropical coastal regions around the world. The life history characteristics make clupeoids ideal species for harvest. Most are small and mature early in life, often when they are one year old. Longevity is short, ranging from one to several years. Because of rapid maturity and great egg-production (tens- to hundreds-of-thousands of eggs per female), enough fish usually remain to reproduce even when they are heavily harvested. Larval and juvenile life stages are not as susceptible to harvest since they are small and often live in sheltered coastal areas away from adults.

Clupeoids have historically been harvested with trawls; however, the most efficient harvest method is the purse seine, to which they are vulnerable when aggregated near the surface feeding on plankton blooms or spawning. The relatively small size makes clupeoids easy to take on board the fishing vessel from the seine net using large dip nets or pumps. Historic markets include canned, dried, pickled, or salted fish; however, because they are considered small, bony, and oily, there has never been a large fresh-fish market. Some herrings are harvested primarily for the female gonads, marketed as **sac roe**. For example, the Alaska herring fishery switched almost entirely to harvest for the roe beginning in the 1970s as a salted herring roe market developed in Japan. Most current clupeoid harvest supports a **reduction fishery**; the fish are ground, cooked, and processed into a press cake of fishmeal, fish solubles, and oil (**Figure 11-15**). This fish meal is used in animal and human food. The largest user in the United States is the chicken industry, but the meal is also used in other livestock and pet foods, and about half of the global production is now used in aquaculture (see the previous section Aquaculture as a Solution to Overfishing). The oils are used for human consumption in cooking oils and margarines and in processed foods such as cookies and cakes (most commonly in Europe and South America). In the United States, historically the oils have been largely used for marine lubricants, paints, and in the rubber industry. Clupeoids are rich in omega-3 fatty acids, considered beneficial to human health,

Figure 11-15 Much of the coastal pelagic fish harvest is processed into oils and meals. Anchovy fishmeal at a Peruvian processing plant.

and low in accumulated toxins (e.g., mercury), leading to an increasing market as fish-oil supplements. With recent increases in fish farming, by the early 2000s over 80% of fish oil production was used in fish foods for aquaculture.

The clupeoid fisheries are typically regulated by coastal countries because most species are taken within national waters and processed on shore. The products extracted in the reduction fishery are undifferentiated by species and easily shipped to support the global market. About 27% of global fish landings are clupeoids used for reduction. Coastal processing facilities have sprung up and disappeared around the world over the past century as new populations were discovered and exploited.

The harvest of clupeoids exhibited a dramatic global increase beginning in the 1950s after harvest and processing technologies were developed that allowed large catches (**Figure 11-16**). These advancements included diesel and gasoline engines, hydraulic power blocks for manipulating the nets, lighter and faster aluminum boats, spotter planes for sighting the fish schools, more durable nylon seines, chilled holding tanks, and large fish pumps to transfer the fish from the net to the hold. World catch increased from about 13 to 20 million metric tons per year from the 1970s to 1985. Harvest fluctuated around 22 million tons before eventually increasing to the current annual catch of about 31 million tons, over 35% of global marine fisheries landings. The pattern for harvest of individual species is not nearly as stable, and is characterized by wild fluctuations, resulting from a short life span and dependence on plankton commonly affected by environmental fluctuations (see Box 11-2, Unpredictability in Pelagic Ecosystems).

Analysis of historic landing data by species exhibits a common pattern of harvest: a rapid increase in landings,

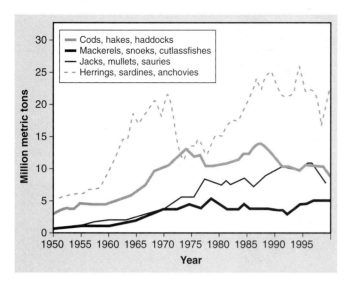

Figure 11-16 World catch of various fish groups. Note the large annual fluctuation, and dominance of clupeoids (herrings, sardines, and anchovies) in most years.

inadequate monitoring and controls, and an eventual collapse of the fishery. Following the collapse the fishing industry moves on to other species, with a repetition of the same cycle. Even when the best scientific data are used in an attempt to control harvest, our limited ability to predict the reaction of fish populations to exploitation and environmental variability leads to management dilemmas and overharvest problems. These issues are demonstrated below through examples of some of the most historically valuable fisheries.

U.S. Sardine and Menhaden Fisheries

The Pacific sardine *Sardinops sagax* was one of the first clupeoids to be harvested intensively in U.S. waters (the same species also is harvested by Japan in the northwest Pacific and Peru and Chile in the southeast Pacific). This fishery began in the 1800s off California to supply the canned sardine market, with the wastes used as fish meal and oils. Fishery landings began to increase rapidly in the 1920s and reached levels fluctuating around 225,000 metric tons per year by 1930 (see Figure B11-3). By the 1930s, concerns that the fishery might be overharvested resulted in laws passed to restrict harvest, including limiting the building of new processing plants. The fishing industry skirted these restrictions by creating offshore floating processing plants, which were technically not under the state's jurisdiction. Because this fishery was an important source of income in coastal California during the depressed U.S. economy of the 1930s (it was a focal point of John Steinbeck's 1945 novel, *Cannery Row*), there was great political and social pressure to allow increased harvest as a means of job creation. By the mid-1930s catches had exploded to 450,000 metric tons per year,

and were stable for about 10 years. Although many were concerned that the fishery was being overharvested, no one predicted the collapse that followed. By 1950 catches had been cut in half, by the 1960s they had declined to pre-1920 levels, and then virtually disappeared. Even with minimal harvest the populations were slow to recover. Recovery by the mid-1980s allowed harvests to increase but to only a fraction of the former harvest.

Reasons for the dramatic decline and slow recovery of the Pacific sardine populations have been much debated. Although natural population cycles probably contributed to the decline (see Box 11-2, Unpredictability in Pelagic Ecosystems), fish harvest likely played a major role in the population reduction and delayed the recovery. This example demonstrates how socioeconomic factors can limit the capabilities of scientists and managers to institute management policies that would be best for fish population health.

As Pacific sardine populations declined, U.S. fishers looked for other sources of clupeoids to supply the market for fish meal and oils; they found it in the Atlantic menhaden *Brevoortia tyrannus*. This fishery already had a long history on the U.S. Atlantic coast. The commercial fishery started in the early 1800s, primarily located in the northeastern states, to provide an alternative to whale oil in the production of lubricants, lamp fuel, soap, and paints. The volume of oil produced from menhaden was much greater than that taken from whales. The purse seine was developed soon after the U.S. Civil War, which allowed this fishery to expand, and steam-powered vessels resulted in further expansion in the late 1800s. By the 1900s, menhaden were being used in fertilizer, animal feed, and in manufacturing cosmetic products such as perfume and fingernail polish. Technological advancements resulted in further harvest increases through the 1900s.

As the technology improved and markets were established, the U.S. Atlantic menhaden fishery expanded from the New England region down the coast (**Figure 11-17a**). Then, as the Pacific sardine fishery declined in the 1940s, menhaden landings increased from 180,000 to 320,000 metric tons (Figure 11-17b). The Chesapeake Bay region became a center of Atlantic menhaden harvest. The total U.S. harvest peaked in the 1950s at 500,000 to 700,000 tons per year; afterward the harvest began its plummet as the number of fish in the population declined. By the late 1990s landings had declined by 60% and the Atlantic menhaden no longer dominated the U.S. Atlantic fisheries. Since then, harvest levels have never returned to their peak levels but have fluctuated around 225,000 tons, and estimates of population size have remained relatively low. It is now apparent that menhaden harvest in the 1950s was at levels beyond the maximum sustainable yield.

(a)

(b)

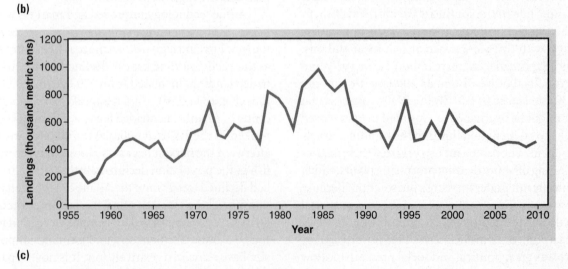

(c)

Figure 11-17 Menhadens have been the most productive fishery in U.S. waters. **(a)** Here menhaden are pumped from the purse seine into the hold. **(b)** After collapsing in the 1960s, harvest of the Atlantic menhaden has never fully recovered. **(c)** Gulf menhaden have taken up the slack, with harvest levels remaining relatively stable.

Despite the declines in harvest, menhaden still supports the largest single-species fisheries on the U.S. east coast. Because menhaden are an integral component of the Atlantic coast ecosystems, overharvest could affect not only the menhaden but the other organisms within their food webs. Because of their abundance, many predatory fish depend on menhaden. Menhaden serve an important function of filtering plankton from the waters, improving water clarity, and transferring nutrients up the food chain. The decline of menhaden already may be affecting other fish species throughout this ecosystem.

As the Atlantic menhaden fishery declined, the fishery for the Gulf menhaden *Brevoortia patronus*, a nearly identical species in the Gulf of Mexico, picked up the slack (Figure 11-17c). Although Gulf menhaden catch began increasing around 1950, harvest levels were only one-third of those for the Atlantic menhaden at this time. However, collapse of the Atlantic menhaden fishery in the 1960s was concurrent with a rapid increase in harvest of the Gulf menhaden, and by 1970 Gulf menhaden harvest had reached over 700,000 tons, greater than the Atlantic menhaden or Pacific sardine harvest had ever been. In 1984 harvest peaked at almost 1 million metric tons before declining and stabilizing around 400,000 to 600,000 tons, where it has remained through the first decade of the 2000s.

The Gulf menhaden remained the single largest fishery in U.S. waters, and harvest made up around 20% of the total U.S. fish harvest after the 1960s. Some consider the Gulf menhaden fishery an ideal fishery based on several factors. Bycatch levels are low (1%–3% by weight), there is no waste because all parts of the fish are used (for fish meal, oils, etc.), there is no competition between recreational and commercial fishers, the number of boats is low enough to easily monitor (because of the expense of the initial investments in a vessel, spotter plain, and processing plant), and (at least so far) the fishery has been sustainable. Although overharvest and population collapse of the Gulf menhaden has been feared by some, this has not materialized, and fisheries managers believe that the current yield may be sustainable. Caution must be taken, however, because anomalous environmental events can affect fish populations in unpredictable ways.

Some biologists argue that even if the menhaden fishery is sustainable, harvest could negatively affect coastal ecosystems. Fewer menhadens to remove algae from the water could result in increased nutrients in bottom waters, increased hypoxia, and an enhanced *dead zone* (see Chapter 6). If fishing is reducing populations substantially, it could allow for increases in plankton populations, including those that cause harmful algae blooms. Presented as possible evidence of such effects is that, as Gulf menhaden harvest has increased, harmful algal blooms (HAB) and the coastal dead zone have increased. Numerous other factors (discussed in Chapter 6); however, could explain these observations.

Peruvian Anchovy Fishery

During the same time period that menhaden fisheries were expanding off the U.S. coast, harvest of another clupeoid was accelerating at an even more rapid rate to supply the global market for fish meal and oils. The Peruvian anchovy (*Engraulis ringens*) fishery, supported by the great coastal upwelling off the Pacific coast of South America (see Chapter 1), soon became the largest fishery in the world. It currently makes up almost 10% of global fish harvest and about 25% of the global reduction fisheries.

The history of this fishery follows a similar boom and bust pattern as the clupeoids off the U.S. coast, and provides an additional example of the problems and difficulties with managing and conserving coastal marine fisheries. In the 1940s, catches of the Peruvian anchovy totaled about 30,000 metric tons per year. After the introduction of the purse seine into the fishery in the mid-1950s, however, catches doubled each year until they reached 3.5 million metric tons in 1960 (**Figure 11-18**). Limits were not set and catches continued to increase as more and more boats entered the fishery. Catches reached 9 million tons in 1964, making Peru the largest fish harvesting nation in the world. At this point scientists estimated that the fishery had reached a maximum sustainable yield. After a decline in harvest in 1965, however, harvest continued to expand, reaching over 13 million tons in 1970, almost 25% of the global fisheries harvest for that year.

Despite this world-record harvest, people employed in the anchovy fishery began to have economic difficulties because the fishing industry had overcapitalized and there were now too many boats and processing plants to make a profit. Although managers began to set quotas and limit harvest during the spawning season, the socioeconomic pressure to continue excessive levels of harvest was overwhelming. Conservation measures appeared to be too little and too late and the fishery began its decline. In 1972 a combination of factors worked to spell doom for the fishery. This was a strong El Niño year, causing a direct effect on coastal regions in the eastern Pacific that support the anchovy fishery (see Chapter 1). High water temperatures from the El Niño forced the larger fish to concentrate in cool pockets where they were easy to catch. Therefore, the individuals with the greatest reproductive potential were being hit the hardest. At the same time the number of young smaller fish was too low to replenish the overharvested population.

In 1972 catches dropped below 5 million tons and finally the fishery was closed, but for less than one year. By 1973 harvest was just over 2 million tons, resulting in severe

(a)

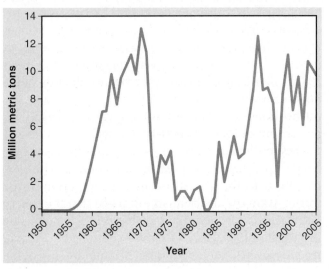

(b)

Figure 11-18 **(a)** Purse seining boat with hold full of Peruvian anchovies for use in the reduction fishery. Note the purse seine net to the left and another fishing boat in the background. **(b)** Peruvian anchovy landings from 1950 to 2005. Dramatic fluctuations in harvest have resulted from a combination of factors, including overfishing, increased regulation, and natural environmental events.

economic hardships for the region. Through the mid-1980s the populations never recovered and the Peruvian government tried to support fishers switching to harvest of other species, including sardines. After harvest levels reached a low of under 500,000 tons, in 1984 the fishery finally began to recover and harvest levels were over 9 million tons by 1994. For the next decade harvest fluctuated around 6 million tons. In the early 2000s harvest fluctuated around 8 to 10 million tons.

The Peruvian and Chilean governments have recently made efforts to avoid future population collapses of the Peruvian anchovy and the resulting economic hardships. For example, in Peruvian waters the fishery is monitored

closely; offshore fishing boats are equipped with satellite tracking systems to monitor the fishery. Closed seasons have been implemented from August to October (to protect the fish during reproduction) and in February to March (to allow the young anchovies time to grow). Entry of new fishing vessels into the fishery has been closed. The fish populations are monitored closely. If sizes of the fish harvested are too small, the fishery is closed, and hydroacoustic gear is used to monitor the size of the anchovy populations. In addition, scientists are working toward establishing stations to monitor the ocean environment, looking for patterns to predict future natural population increases and decreases.

The effects of anchovy population declines are not limited to fish communities. Far-reaching effects were documented throughout the marine coastal ecosystems of the tropical eastern Pacific. For example, seabirds (including cormorants and boobies) declined from about 20 million to 5 million, primarily because of the loss of anchovies as a source of food. Islands off Peru that had abundant nesting bird populations became devoid of seabirds. This affected the mining of guano from these islands for fertilizer as after the anchovy collapse guano deposition declined considerably. Effects on other parts of marine ecosystems are not well understood, but scientists are encouraging an ecosystem-based approach to management. We are a long way from routinely assessing impacts of this and other clupeoid fisheries on entire ecosystems, and debates continue as to the relative impacts of fishing and the environment in population collapses (see Box 11-2, Unpredictability in Pelagic Ecosystems).

Lessons from Clupeoid Fisheries

Clupeoid fisheries provide important lessons concerning interactions of the environment, conservation, and fisheries. One lesson is that fisheries management must deal with unpredictability. Clupeoid populations are greatly influenced by climatic events, some of which have been identified. We do not, however, have a predictive understanding of how environmental events interact with fish harvest to influence the collapse and recovery of these populations. It is clear that a continued harvest of overfished populations inhibits their recovery. Monitoring the fishery and environmental conditions is important if we will ever be able to predict changes in the population. If better predictive capabilities are developed, managers will need to react quickly and avoid management decisions based primarily on socioeconomic concerns. An additional lesson is that fish populations do not function in a vacuum. To avoid long-term effects, an ecosystem-based approach to management must be instituted, considering not only the fish populations but the biological community and the physical environment.

Groundfish Fisheries

Fishes that live on or near the bottom in relatively shallow areas of the ocean are categorized by fisheries managers as **groundfish**. There are hundreds of species of groundfish globally, with fisheries providing an important source of foodfish in coastal regions. Groundfish have been harvested from coastal areas for hundreds of years, and the development of fisheries was important in the exploration and settlement of new lands, in particular those bordering the North Atlantic. Fisheries there and in the North Pacific have recently supported large international markets for frozen fish products.

Commonly harvested groundfish include cods and haddocks (Gadidae), flatfishes (Pleuronectiformes), and drums (Sciaenidae). Gadids that support large groundfish fisheries live in waters shallow enough to be accessed by trawl nets, generally less than 100 meters depth. Other than during reproduction, when they may school near the surface, the fish remain at the bottom feeding on invertebrates and other smaller fishes.

Methods of harvest for groundfish vary. Traditionally, many species were harvested by hook-and-line, often handlines from small boats in coastal areas. Although hook-and-line fishing lasted well into the 20th century as the preferred fishing method, as the fish became harder to find and fishing methods became more efficient, there was a switch to trawls as the preferred harvest method. After large factory trawlers were developed in the 1950s, harvests began a dramatic increase. Through the 1960s factory ships from various countries, including the Soviet Union, Japan, Germany, and Spain, were harvesting groundfish in coastal regions around the world. In the 1970s, as countries began extending their claim to national waters within which they could exclude international harvest (see Chapter 12), smaller and mid-sized trawling became more important.

Groundfish are usually sold for direct human consumption, making their value greater than for clupeoids but less than for large pelagic or reef species. Before modern refrigeration technology was developed, the fish were preserved by salting and drying, sometimes on islands or coasts far away from the homeport, before being transported to market. Today most groundfish are marketed fresh or frozen, often simply called whitefish. One of the largest markets is for battered and frozen fish fillets, sold as frozen fish sticks and patties, or breaded and fried fish in fast-food restaurants. These same fish are also commonly processed into *surimi*, a gel-like paste made from minced, processed fish muscle proteins. In the United States about 68 million kilograms of surimi products are eaten each year, largely as imitation crab meat. In Japan, where surimi originated as a means of using less desirable fish, it is marketed in many forms, historically as *kamaboko* in the form of small loafs (In Japan, about 7 kilograms of surimi products are eaten per person each year).

Cumulative global landings of groundfish can rival those of the clupeoids; in 1975 global landings were over 13 million metric tons. But as global landings of clupeoids increased to almost double that amount, landings of groundfish stabilized or decreased, fluctuating around 11 to 12 million tons through the 1990s. In part because of the decline of the cod fishery (**Box 11-3. Learning from History: Boom and Bust of Atlantic Cod Fisheries**), the

---

Box 11-3   Learning from History: Boom and Bust of Atlantic Cod Fisheries

The Atlantic cod *Gadus morhua* lives near the bottom in waters over the continental shelf of the North Atlantic. It prefers cold water and is therefore most heavily concentrated along the northeast coast of the United States, through Canadian waters, along Greenland and Iceland, and through the European Atlantic. This species has supported one of the most important food fisheries in nations bordering the North Atlantic Ocean throughout recorded history. During the reproductive season cods migrate into shallow coastal waters, making them an easy catch by hook-and-line or trawl fishing. During the remainder of the year they move into deeper waters, but typically less than 40 meters depth, where they are still vulnerable to bottom trawls.

In the northwest Atlantic, cod concentrate along shallow areas called *banks,* which are large shoals where nutrient-laden Gulf Stream waters mix with North Atlantic waters, resulting in high productivity (see Chapter 1). These banks are distributed from Georges Bank off Massachusetts north to the Grand Banks off Newfoundland. To the first European settlers of North America, cod were so concentrated along these banks that the populations appeared to be virtually unlimited. These dense populations with a high reproductive potential appeared to be impossible to overfish. Fishers and scientists noted that a single 1-meter-long female cod can produce 3 million eggs in an annual spawning season; larger fish can produce as many as 10 million eggs. This gave the impression that no matter how many cod were caught, there would always be more than enough fish from the next generation to replace them. In 1873, Alexander Dumas reflected this opinion by his statement in *Le Grande Dictionnaire de Cuisine:*

> *"It has been calculated that if no accident prevented the hatching of the eggs and each egg reached maturity, it would take only three years to fill the sea so that you could walk across the Atlantic dryshod on the back of cod."*

Although this extrapolation may be technically true, it is obviously makes the unrealistic assumption that each egg produced grows to reach maturity. The mortality rate of eggs and larvae of cod (and most other marine fish) is higher than 99% and very few that survive the larval stage survive to adulthood to reproduce. The impression of near-infinite reproductive potential strongly influenced management of cod for many years, however.

Cod are relatively easy to catch even with simple fishing methods. They are not strong swimmers, and typically meander over the bottom in search for food, opportunistically lunging at prey they encounter. Because they are not selective feeders, they will take nearly any kind of bait or lure and because they are not strong swimmers they do not put up much of a fight when caught. Even the earliest of the commercial cod fishermen, the European Basque, thus could easily catch the abundant cod by hook-and-line fishing from sailing vessels or small rowboats. Hook-and-line fishing remained the preferred method for catching cod well into the 1900s. Once trawl gears were developed, fishers came to realize that, after being pursued for about 10 minutes, a cod would tire enough that the trawl net would overtake it and even large fish were easily captured. The white and tender muscles of weak-swimming fishes such as cod have long been considered most desirable for general human consumption.

The cod is arguably the most influential fish in history. Coastal cultures have developed around the cod, it resulted in the discovery of a continent, it has caused conflicts (and nearly a shooting war) between nations, and it has controlled economic booms and busts in coastal regions. Cod have been harvested for centuries by European countries in their coastal waters. As early as the ninth century the Norse were traveling by boat to Iceland and Norway to catch and dry cod and trade it with Europe. In the 10th century the Vikings traveled from Norway to Iceland to Greenland to Canada, surviving by eating dried cod along the way. The Basques of the 10th century (located in what is now northern Spain) traveled to North American waters in search of cod (and whales). The Basques were able to preserve the cod by salting for transport back to Europe; and by the year 1000 they had developed an international trade in cod caught throughout the North Atlantic.

There was little documentation of the Basque fisheries off North America. In fact, other Europeans of the time could not figure out where the large numbers of cod the Basques were selling were coming from, because the Basques were not seen in the major fishing grounds off Europe. There is now good archaeological evidence that the Vikings and Basque "discovered" North America 500 years before Columbus but simply were not interested in sharing this discovery because they did not want to share the cod.

Portugal was another great seafaring, fishing nation and their fishers were catching and salting cod from European waters by the 1300s. Portuguese explorers made it to the coast of Newfoundland in 1452, reporting large schools of cod off the Grand Banks. There is good evidence that British fishers had discovered the cod of Newfoundland and were drying fish on the rocky shores by sometime in the 1480s, but these fishers also were not willing to share their secret fishing grounds. When British explorer John Cabot reached Newfoundland in 1497 and claimed it for England, he noted the large schools of cod. In the 1520s Jacques Cartier "discovered" the mouth of the St. Lawrence River and claimed it for France. He actually remarked on the presence of hundreds of Basque fishing boats. Although they had been fishing this region for 500 years, they were more interested in keeping their fishery secret than laying claim to new lands.

Cod have always been more abundant in the Western Atlantic off North America than in the Eastern Atlantic off Europe. As settlers began arriving at the *New World* through the 1500s the cod fishery expanded. The British settlers of North America in the early 1600s named their point of landfall Cape Cod because of the abundant cod populations found there. The first successful settlements in New England developed a winter fishery for cod because these fish come into the shallow waters to spawn in winter months. The cod provided not only food but also an important trade commodity. The trade in fish from North America grew throughout the century. For example, in 1700 Boston exported 5 million kilograms of winter-cured cod; about 300 ships were leaving Boston each year with cod for the West Indies.

By the end of the 1600s the French had 100 ships being loaded with cod caught off Newfoundland. The British continued fishing for cod off North America until the United States began to establish its independence. Throughout the 1700s, France, England, and New England territories argued over who had rights to the productive fishing grounds of the Grand Banks; but regardless of who was fishing there, harvest continued to expand.

The British still had 300 fishing vessels off Newfoundland at the end of the 1700s, but eventually were unable to compete with North American fishers who did not have to transport their fish across the Atlantic. The British fleet was reduced to 15 vessels by the 1820s.

Development of the schooner, a much faster fishing boat, in the early 1700s helped U.S. fishermen expand the cod fishing industry (**Figure B11-5**). Through the 1800s the cod fishery continued to expand and dominated the commerce of New England and Newfoundland. When the lucrative market in the West Indies was lost after their slave-based economy disappeared, consumers in the United States took up the slack.

The 1800s were the glory days of the cod fishery. Fishing was mostly by hook and line. U.S. ships would carry 7-meter deckless skiffs called *dories* out to the banks and drop them off with two-man crews (a storm or large wave could easily disrupt the dory, throwing the fishermen into the cold waters; even as late as the 1970s, over 100 Gloucester cod fishermen were lost per year). Although in some regions nets and longlines were used to catch cod, they were outlawed in the United States, being viewed as unfair competition. Although pre-1900 catch data are not totally reliable, it is estimated that landings of the Northern cod increased from about 100,000 to 300,000 tons per year through the 1800s. As effort continued to increase, harvest increased. It appeared that the fishery was inexhaustible. Thomas Huxley, one of the 19th century's most eminent scientists, commented in 1883, "I still believe the cod fishery . . . and probably all the great fisheries are inexhaustible; that is to say that nothing we do seriously affects the number of fish." In the late 1800s, however, catch per effort declined (the rate of fishing was increasing more rapidly than harvest rates), likely a result of lower productivity because of climate changes (see below).

In the early 1900s cod populations remained healthy and were even able to increase as fishing continued. New, more efficient harvest technologies were now being introduced into the fishery, however. After the otter trawl was introduced

Figure B11-5 In the days of sailing ships fishing was dangerous, inefficient, and overharvest was unlikely. Here, fishing schooners transported dories to handline for cod on the Grand Bank.

into the New England cod fishery in 1905, landings increased to almost 350,000 tons, and in less than a decade scientists began to fear the impact of trawl fishing on the cod populations. A U.S. congressional committee was formed to investigate the damage the otter trawl was having on the cod fishery. World War I provided a reprieve for the cod; however, the cod fishery was back in full swing by 1920 and continued increasing for about 10 more years to almost 400,000 tons by 1930. Landings continued increasing as more efficient technology was developed and the fishermen exploited new fishing grounds. Despite the development of modern fishing methods, sailing schooners were used in the North American cod fisheries up until World War II, though they had been phased out long before then by European fishers.

Continued improvements in technology allowed for a greater expansion of the cod fishery. The steam engine, the diesel engine, on-board freezing of fish, filleting machines, and improvement in nets and net-handling technology eventually all played a role. The cod received another reprieve with World War II, and landings declined to around 200,000 tons. After the war and as technology and seafaring ability improved, however, competition for the cod became volatile.

As competition for cod harvest increased, countries began to claim rights to restrict fishing in waters off their coasts. For example, since 1919 Iceland had set its fishing limit to 3 miles (5.6 kilometers) from the coastline. Beyond this, any nation could fish, considering the fish to be a common resource (see Chapter 12). Iceland increased its coastal limit to 4 miles (7.4 kilometers) in 1952 and 12 miles (22 kilometers) in 1958; other countries followed suit. When Iceland decided to extend its limit to 50 miles (93 kilometers), it was not to the liking of the British who had fished these waters for centuries, so the restrictions were ignored. Iceland's coast guard sent out vessels with trawl-wire cutters to disconnect the nets from the British vessels, which responded by attempting to ram Iceland's vessels. British navy frigates were sent to protect their trawling vessels. Meetings between British and Icelandic prime ministers resolved the issue and Britain agreed not to send their fishing vessels into Icelandic waters.

As battles over fishing rights developed, competition among North Atlantic countries and others with international fishing capabilities expanded. Some nations gained what many considered an unfair advantage by developing large factory trawlers. In the 1950s the British introduced a trawling vessel over 75 meters in length into the fishery, much larger than any other fishing vessels. Fish could be cleaned, filleted, and frozen onboard the boat. Other countries followed suit and factory trawlers grew to over 135 meters with a capacity of 4,000 tons. Eventually the Soviets built ships with 8,000-ton capacities. The large fishing nations were now traveling throughout the North Atlantic to catch cod and other groundfish.

In the 1950s harvest began to accelerate and by 1968 the harvest of cod had reached over 3 million tons in the North Atlantic (**Figure B11-6**). Overharvest was apparent as catches plummeted. By 1975 catches had declined dramatically, especially in the northwest cod stocks. Although quotas were set, the populations did not recover.

Confrontations between Iceland and Britain were renewed when Iceland again expanded its fishing rights to

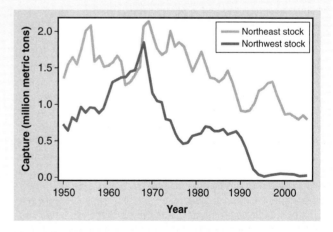

Figure B11-6 Atlantic cod harvest in the northwest Atlantic, largely by North American fishers, and the northeast Atlantic, largely by Icelandic and European fishers, from 1950 to 2005. Signs of excessive harvest were largely ignored, especially in the northwest Atlantic, thus delaying setting harvest restrictions until a population collapse was imminent.

200 nautical miles (370 kilometers) from the coast. The British ignored the restrictions and sent fishing vessels into Iceland's proclaimed waters. Iceland deployed coast guard vessels, helicopters, and planes while British frigates accompanied their fishing vessels. Iceland broke off relations with Britain until 1976, when the issue was once again resolved. This series of conflicts is often referred to as the *Cod Wars* (even though no shots were ever fired). This series of events led up to the eventual establishment of a 200-mile **Fishery Conservation Zone (FCZ)** by other nations. The proclamations evolved into international agreements through the Law of the Sea to allow all nations to establish a 200-mile **Exclusive Economic Zone (EEZ)** within which they had exclusive rights to most resources (see Chapter 12).

Both the United States and Canada began a rapid expansion of fishing fleets to take advantage of their newly declared 200-mile FCZ. They were able to claim sole rights to the cod fisheries in the Grand Banks. Canada provided huge subsidies and the United States provided guaranteed loans to fishers to invest in new fishing vessels. Scientific data indicating that the cod fisheries were overexploited were either ignored or discounted, and the available stock was consistently overestimated. Canadian catches remained unregulated until the early 1970s. A moderate increase in harvest at the end of the 1980s provided a rationalization that populations were still healthy; however, this was likely a result of new fishing grounds being discovered, technology making fishing more efficient, and increases in the number of boats fishing (for example, the U.S. fishing fleet doubled from 1977 to 1983).

With the increase in fishing effort the cod fishery began a rapid decline through the 1980s and early 1990s (Figure B11-6). The Canadian government eventually cut the cod quota, but only by 10% so that jobs would be protected. Things continued to get worse and in 1992 the Newfoundland cod fishery was completely closed, putting tens of thousands of Newfoundlanders out of work. The United States soon followed suit and shut down the George's Bank cod fishery in 1993. By 1995 the United States had paid $60 million in subsidies to New England fishers; Canada had paid $600 million to Newfoundlanders.

Following the closure of the fishery, management agencies began to look optimistically for a population recovery. The Department of Fisheries and Oceans of Canada (DFO) estimated a two-year increase in the cod population and projected a partial recovery of the fishery by as early as 1994. Many doubted DFOs prediction of such a rapid recovery, in part because cod do not reproduce until they are five to seven years old. In actuality the Canadian cod populations showed only a slow, moderate recovery after the 1992 closure. By the early 2000s it was still estimated that Canadian cod populations were less than 1% of what they were in the early 1960s, and a fishing moratorium remained in many regions. By 2010 there was still little evidence of recovery and fisheries have remained closed or severely restricted. It is feared that even small quotas in limited areas could stress populations and inhibit their recovery.

Not long after the northwest Atlantic cod collapse, harvest in the northeast Atlantic began to plummet. Off Iceland and Britain cod catches fell by over 60% from the 1980s to the 1990s. Scientists recommended cutting cod catch by 40%. Spawning stocks in the North Sea declined by as much as 90% from the 1970s to 2002. When cod catches continued to decline, European scientists convinced the European Union (EU) to ban most cod fishing in the North Sea and Baltic Sea to see if populations would recover. When increased recruitment of young cod in 2005 gave indications of a possible recovery in the North Sea, fishing quotas were once again increased by the EU, but not without controversy. Declines in the northeastern Atlantic were not as serious as in the northwest and limited harvest continued; however, closures of some regions did not lead to a rapid recovery, and additional restrictions were implemented by the EU in the North Atlantic in the early 2000s. Management controversies will likely always be present in these fisheries. Political pressure will be great to return to fishing upon any sign of population recovery, while environmentalists and ecologists will counter with arguments to proceed with caution.

Although it is largely accepted that overfishing was the primary cause of the demise of the Atlantic cod, it is still questioned whether other factors played an important role. The role of environmental factors in the decline has been considered based on several possible mechanisms. For example, global climate change is associated with a decrease in the atmospheric ozone layer, resulting in an increase in the amount of ultraviolet (UV) radiation reaching the Earth. While this UV radiation does not penetrate to the depths where adult cod reside, their larvae float near the surface. This increased radiation has a potentially deadly effect on the larvae and has been shown to cause damage to their DNA. An analysis by Stephen Simpson and colleagues provides support to theories that temperature change could be affecting fish communities in the European North Atlantic. Along with a temperature increase of about 1.3°C, cod and pollack, cold-water species, have declined by 50% while warmer water species, such as hakes, have doubled.

Biological factors could play a role in declining cod populations also. One consideration is whether an increase in harp seal populations (see Chapter 10) could cause a decline in cod survival in Canadian waters. Harp seals do sometimes eat cod but they show a preference for other fish species. Harp seals are not dependent on cod, as is evidenced by their population increasing after the cod had declined. It is unknown whether the harp seals could play a role in inhibiting cod recovery. Overfishing may have changed the North Atlantic ecosystem to a degree that is difficult to determine which factors are most important to the recovery of cod populations.

Debates continue today over the likely reasons for cod population collapses and recoveries. George Rose developed models based on historical commercial harvest, and inferred that three separate declines in Atlantic cod populations resulted from interactions between overfishing and lower productivity as a result of climate change (that is, lower water temperatures lowering productivity). The first decline through the 1800s resulted from climate change during the Little Ice Age (1800–1880). The more recent population collapse in the 1960s was attributed to overfishing, and the collapse in the 1980s likely resulted from a combination of overfishing and climate change. Biological factors such as a decline in cod prey species, especially capelin (*Mallotus villosus*), and increased predation by harp seals also could have played a role. Several scientific assessments have suggested that rebuilding of cod populations is currently being hindered by low population growth as a result of low cod population size or low individual growth rates caused by low temperature or less food availability. Current directed fishing and bycatch likely continue to delay population recovery.

As the harvest of Atlantic cod declined, the large international market for surimi and battered fish products used in grocery and fast-food markets needed a new source of fish. Another gadid fish, *Theragra chalcogramma,* the Alaska pollock (also known as walleye pollock or Pacific pollock; **Figure B11-7**) has been the primary source of these fish products over about the past 40 years. In 1960, catches were less than 200,000 tons; but increased to over 1 million tons by 1970, and over 1.5 million tons by the mid-1980s (**Figure B11-8**). Since the 1970s the Alaska pollock has alternated with the Peruvian anchovy as the most heavily harvested fish species in the world. In U.S. waters, as this fishery expanded controversies between the factory trawlers and smaller trawlers were common. A total allowable catch quota was established, causing the fishery to develop into a *derby fishery.* When the season opened, all boats rushed to catch as many fish as possible as fast as possible until quotas were met, often within two months. In the 1990s the United States initiated an individual quota system, partitioning quotas among the fishers. The Russian fishery for Alaskan pollock has not been as well managed, with much of the quota being sold to other countries and the fishery not as tightly monitored. Although pollock populations have been exposed to similar exploitation pressure as other

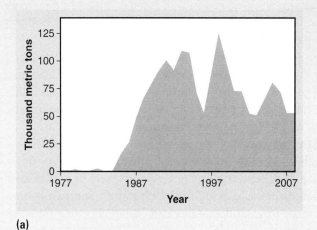

(a)

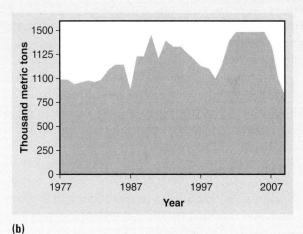

(b)

Figure B11-8 Alaska pollock landings by U.S. fishers in **(a)** the Gulf of Alaska, and **(b)** the east Bering Sea. There is cautious optimism that lessons learned from the collapse of other fisheries are resulting in more sustainable management.

large groundfish populations, they have not undergone the same crash as exhibited by the Atlantic cod. Although current annual quotas of about 1.5 million tons appear to be sustainable, a precautionary approach requires the ability to rapidly adjust harvest levels if signs of population declines are discovered.

The Alaska pollock fishery has largely replaced other groundfish as the major supplier of "whitefish" products for the international market. Pollock also provide more fish for the surimi market than any other species; on average, about 10% of the Alaska pollock harvest is processed into surimi each year. The ecosystem impacts of the heavy fishing pressure on the pollock are unclear and still debated. In the early 1990s there was concern that pollock fishing had caused the decline of seabirds and the endangered Steller sea lion by removing pollock as their major source of food (see Chapter 9). Because data were not available to show that the pollock fishery was the primary reason for the decline of the sea lion, the fishery has continued. If the fishery was proved to be a major cause of the sea lion decline, it would present a major management dilemma over the economic effects of shutting down a major fishery to support protections of an endangered species.

Figure B11-7 Pollock trawl fisheries provided a replacement for overharvested Atlantic cod.

stability of global groundfish catch is misleading. At a regional level a pattern of overharvest and decline is evident, in particular in the North Atlantic.

In the northwest Atlantic, harvest of groundfish peaked in the 1960s at over 2.5 million tons per year. Catches declined to under 2 million tons by mid-1970s, around 1 million tons by the 1980s, and under one-half million tons by the mid-1990s. Population declines caused the closure of many fisheries. The most heavily fished groundfish species, including cod, haddocks, and flatfishes, declined by more than 80% below natural population levels. The demise of these fisheries is associated with rapidly increasing harvest after the United States and Canada claimed rights to fisheries within 200 nautical miles (370 kilometers) of the coastline. The economic impacts of this fishery collapse were severe and had obvious effects on fishing communities all along the East coast of North America. In Canada 50,000 fishers had to quit fishing and depend largely on government support. In the U.S. New England region about 14,000 fishers lost their jobs each year through the 1990s. Without a large government bailout, the remaining U.S. fishers ended up in fierce competition for the shrinking harvest quota.

As harvest of groundfish in the Northwest Atlantic was plummeting in the late 1960s and 1970s it was booming in the North Pacific. Landings rose from less than 2 to over 5 million tons from 1960 to the mid-1970s, and fluctuated around 7 million tons in the mid-1980s. Annual harvest in the Northwest Pacific declined substantially in the 1990s, but the fishery there has not collapsed. Although the Alaskan fishery in the Eastern Pacific is held up as a model fishery, some fear its overharvest. As the profitability of the fishery increased, fierce competition developed. Rather than limiting the number of fishers and establishing an individual quota system, the seasons were made progressively shorter for the most profitable fisheries. The Alaskan pollock fishery became as short as two months; the Pacific halibut season, normally six months, was shortened to two days, with several thousand fishers vying to catch as much as they could during this so-called **derby fishery**. Eventually an individual quota system (see Chapter 12) was established in some of the U.S. groundfish fisheries.

The decline in the North Pacific groundfish fishery was accompanied by an increase in groundfish harvest in the Indian Ocean. Although this fishery is still much smaller, catches more than quadrupled, from around 200,000 tons in the 1950s to almost 1 million tons in the 1990s. A pattern similar to that of clupeoid fisheries was developing; as one fishery declines another increases to give the appearance of global stability. If currently harvested populations are not managed sustainability and into the future, and there are no new populations to target, there will be noticeable effect on global seafood availability.

Highly Migratory Predators

Some of the largest and most highly prized fishes in the ocean are large apex predators that move great distances for reproduction and in search of food. These highly migratory predators include tunas (Scombridae), billfishes (sailfish and marlins, Istiophoridae; and swordfish, Xiphiidae), and sharks. These fishes swim in and out of the waters within the EEZ of coastal nations and may spend much of their life in international waters. This behavior complicates management and conservation and often requires the formation of international commissions to develop binding agreements (see Chapter 12). Because of the great value of these large fish to commercial fishers and the general lack of enforced regulations, populations have declined dramatically. Peter Ward and Ransom Myers compared longline catch from a 1950s scientific survey with commercial longline catch in the early 2000s and found that large predators, including billfishes, sharks, and tuna had declined in abundance by about 20% on average. The large predators declined not only in number, but also in average size. Populations were reduced by about 90% from 1950 levels. The most likely reason for such declines was overfishing.

A study by Bruce Collette and an international group of scientists using methods of IUCN (see Chapter 9) found that 7 of 61 tuna and billfish species analyzed (11%) would be classified as threatened, including 3 tuna and 2 marlin species. The authors recommend that harvest of southern and Atlantic bluefin tuna (see below), the most endangered of the species studied, be stopped until populations can rebuild; however, it is questionable whether this will be achieved without a ban on international trade through CITES.

Tunas

Some tuna species are primarily harvested by purse seine, including the controversial fisheries that have to deal with the incidental harvest of dolphins (see Chapter 10). The greatest landings are for the skipjack tuna *Katsuwonus pelamis* and yellowfin tuna *Thunnus albacares* (**Figure 11-19**); however, the most highly valued tunas are not as vulnerable to purse seining and are primarily targeted by hook-and-line gears. These large tunas are some of the fastest, most efficient long distance swimmers in the ocean. Physiological differences between the tunas and gadids (e.g., cods and pollocks) result in very different muscle types. Tuna muscles are highly desired as a luxury food based on their color, texture, and high fat content. This is especially true for those marketed to be eaten raw as sashimi. These tunas demand some of the highest prices of any fish. Although

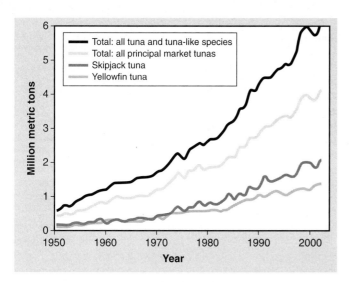

Figure 11-19 World landings of tuna separated as total and the principal species marketed as canned or fresh tuna. Skipjack and yellowfin tuna support the largest harvest by weight, primarily harvested by purse seine for the canned tuna market. Rapid increases have led to concerns of eventual population collapses.

their popularity originated in Japan it has spread around the world.

The most highly prized pelagic predator is the bluefin tuna (*Thunnus thynnus* in the Atlantic and *T. orientalis* in the Pacific and Indian Oceans; **Box 11-4. Conservation Controversy: The World's Most Valued Fish**). They spend much of their lives in the pursuit of fish in the near-surface waters of the open ocean and can reach lengths up to 4 meters and weights over 600 kilograms in the Atlantic. Tunas are most vulnerable to fishing when migrating or feeding in schools; during migrations they may move predictably through coastal areas and during feeding they drive prey fish to the surface, which attracts fish-eating birds that pinpoint their location.

Commercial fishers can afford to travel around the world in search of a large bluefin tuna because of its great value. Excessive fishing pressure has resulted in overfishing. In the western Atlantic it is estimated that bluefin tuna populations are only 5% of their 1980 size. Although the fisheries have been regulated by international agreements

Box 11-4   Conservation Controversy: The World's Most Valued Fish

The bluefin tuna has been prized as a gamefish since the early 1900s, and until the 1960s the fishery was still limited to subsistence fishing, sport fishing, and small-scale commercial operations. Commercial fishers initially caught the tuna with harpoons but eventually switched to longlines. By the late 1970s the Japanese sashimi market had caused the bluefin tuna to become the most highly prized seafood product in the world (**Figure B11-9**). In several instances, single tunas sold for over U.S. $100,000 at a Tokyo fish auction.

The Atlantic bluefin tuna fishery is managed in separate stocks for the eastern and western Atlantic. One of the most famous of the tuna fisheries was in the northwest Atlantic off the banks of the North American East coast (the same area targeted by cod fishers). These populations were showing signs of overfishing by the late 1960s. Measures by ICCAT were inadequate to keep the fishery from declining through the 1970s and 1980s. By 1992 it was estimated that this western Atlantic bluefin tuna population was only 10% of its 1975 level. Fishers did not want to accept these numbers, however, and continually fought for higher quotas based on their perception that the fishery had begun to recover. ICCAT too often compromised to give in to their demands, ignoring the advice of their scientific committee.

Historically, bluefin tuna harvest has been greater in the eastern than western Atlantic. The massive schools of large tunas that enter the Mediterranean have been fished for thousands of years. A traditional method of catching tuna called *la tonnara* has occurred in Sicily since the about the 10th century. An elaborate trap system is set up nearshore, spanning over a kilometer, where migrating tuna are known to pass. The tunas are corralled into a system of chambers where they swim in circles until escorted into the harvesting chamber; there they

are lifted to the surface manually in a large net for *la mattanza* (the killing) with harpoons. Hundreds of fish weighing up to 200 kilograms were trapped in the nets. Thousands could be killed in a single season. As the bluefin tuna became valued as an international resource, large scale purse seining began in the northeast Atlantic, and gillnetters and longliners started competing with trap fishing in the Mediterranean. The trap catch declined drastically in the 1960s. Still, until the mid-1900s there were 20 or more *tonnare* throughout Sicily, before declining to zero in the early 2000s.

Cumulative harvest values for the Atlantic bluefin tuna showed dramatic increases to what were then record levels of over 35,000 metric tons in the mid-1950s (**Figure B11-10**). Harvest fluctuated dramatically, with most of the catch supporting European markets for canned and fresh fish. Catches declined to less than half of that number by 1970. Large increases in fishing pressure caused an acceleration of catches in the mid-1970s to levels fluctuating around 20,000 to 25,000 tons. The increase in demand to supply the valuable sashimi market resulted in a rapid acceleration in bluefin catches in the 1990s to over 50,000 metric tons. As value of the tuna amplified, longlining in the Mediterranean accelerated and there was an unknown amount of illegal fishing. Further restrictions were initiated as populations declined and ICCAT initiated a management scheme.

Similar management issues have plagued the southern bluefin tuna fishery. Japan and Australia, the principal harvesting nations, have worked together on agreements to limit harvest, with limited success. Harvests increased moderately to 10,000 to 20,000 tons in the 1950s to supply the canned tuna market. When Japan developed super-cold freezers, fish could be rapidly frozen and kept fresh for the sashimi market.

(a)

(b)

(c)

Figure B11-9 The bluefin tuna **(a)** has become one of the world's most profitable fishery species. **(b)** Here frozen bluefin tunas are auctioned at the Tokyo Tsukuji fish market, the largest wholesale bluefin tuna market in the world. **(c)** Bluefin steaks are valued for their texture, taste, and appearance as maguro sashimi.

As value increased, harvest accelerated dramatically to over 50,000 tons per year through much of the 1960s, making up for the decline of Atlantic bluefin tuna and keeping the global market supply relatively stable. Catches began declining as the tunas became more difficult to find. Australia, Japan, and New Zealand organized agreements and established quotas at about 15,000 tons in hopes of allowing populations to recover. Catches increased in the late 1990s, however, as Taiwan, Indonesia, and others not a part of the quota agreement began to participate in the fishery. Currently it is estimated that tuna populations are only 9% of their 1960 levels.

With declines in allowable harvest because of increased regulations and reduced populations, a new trend has developed in tuna fisheries. Despite strict quotas on harvest, there are few regulations on **tuna ranching**, farming tuna in floating pens. Efforts to raise tuna in pens were initiated in the 1970s in Japan. It took several years to successfully develop methods, and they require that young fish (up to about 0.5 kilograms) be captured in the wild. Only in 2002 were researchers able to raise young successfully from eggs produced by tuna in captivity. As yet, this is a drawn out and inefficient operation and tunas do not reach maturity until several years in age. It is more efficient to net fish in the wild and then fatten them in pens in preparation for marketing. All commercial tuna farming currently is based on such methods, adding additional harvest pressure to wild tuna populations.

The first commercial farming efforts started in the Mediterranean when large spent tunas caught in traps were transferred to pens for a short time to be fattened before being sold to the Japanese market. Then in the mid-1990s fishers developed methods to harvest small to mid-sized tunas by purse seine, transfer them to floating cages, and feed them until they were at marketable sizes. The methods were refined and adopted in Australia. Purse seine fishers target tuna 15 to 25 kilograms. After capture, a cage is attached to the seine net and the fish are driven in by scuba divers. This floating cage is then towed slowly toward the coast, sometimes taking several weeks. There the tuna are fed wild-caught squids, herrings, or anchovies for 3 to 10 months. The fish are taken from the nets gently and killed quickly to avoid damage to the muscle tissue that would decrease their value for sashimi. The tunas are either put on ice or flash frozen at low temperatures and rapidly transported to Japan or other markets. Agreements with Japan allowed expansion of the tuna ranching operations using these methods and Australia soon dominated the tuna ranching industry, producing over 3,000 tons per year by 2001 (Figure B11-10b). Mexico and Mediterranean nations rapidly adopted the methods used in Australian waters and tuna ranching exploded; by 2004 Mexico, Australia, and about five Mediterranean nations shared most of the 13,000 to 14,000 tons of annual bluefin tuna farming production. Although rules were adopted by ICCAT in the early 2000s requiring the catch of juvenile bluefin tuna to be reduced by 60%, reporting irregularities have made this nearly impossible to monitor and enforce. For example, studies by ICCAT have shown that unreported catches are prevalent, and several nations that are not members of ICCAT are ranching tuna but have no obligation to report their harvests. Another bias in reporting is the practice of re-export through another country to avoid reporting a portion of the harvest. The poor reporting of harvest suggests that is currently impossible to get accurate data on bluefin tuna harvest in the Mediterranean. Even if extreme competition makes tuna ranching unprofitable for some nations, it is likely to continue as a result of subsidies. Conservation organizations complain that

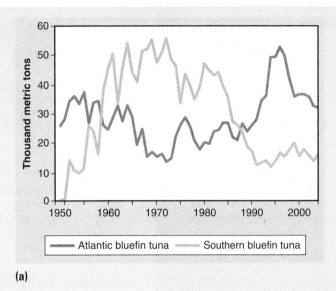

(a)

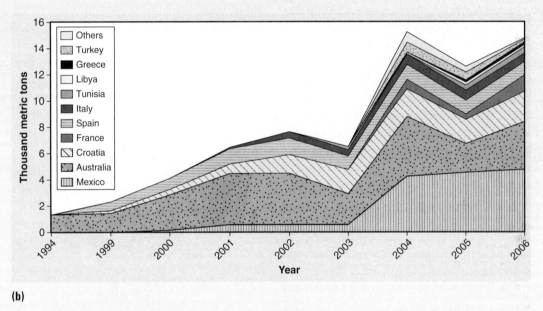

(b)

Figure B11-10 **(a)** Harvest of the Atlantic and southern (Pacific) bluefin tuna has been difficult to regulate due to the value and need for international agreements to establish regulations. **(b)** Aquaculture production has been controversial due to its reliance on wild tuna as stock and wild fish as food. Data obtained from Japanese import statistics indicate that farmed production of bluefin tuna now rivals wild capture. Although Mexico and Australia dominate, Mediterranean nations have increased production.

unknown amounts of government subsidies are given to tuna ranching ventures as aquaculture operations (more appropriate descriptions are *on-growing* and *fattening*), even though the fish are caught from the wild and only kept in pens for six to seven months before being harvested.

One way to improve monitoring of bluefin tuna harvest is to track imports of tuna into Japan, the nation to which almost all production is exported. Such analysis provided evidence of harvest of farmed tuna that was not reported in data collected by ICCAT. Although reported bluefin tuna catches declined by 50% from 1997 to 2004, imports of tuna caught and fattened in farms increased dramatically during the same period. With

tuna ranching operations expanding to other regions of the world (including Mexico), management dilemmas are likely to remain problematic.

In addition to tuna conservation issues, the increase in tuna farming has led to concerns over potential ecosystem and environmental impacts. Although aquaculture has potential for relieving pressure from the wild-caught fishes, in the case of farmed tuna the source is wild populations. Further, the food for the tuna is wild-caught animals, including sardines, anchovies, mackerel, and squid. For fattening, the tuna are overfed so that the conversion ratio can be as high as 15-to-20 to 1; that is, it takes up to 20 kilograms of food fish to produce one kilogram of tuna. Tuna ranching methods thus encourage not only excessive harvest of important apex predators but also prey species low on the food chain.

Further environmental impacts occur in direct association with the farming operations. Unconsumed feed and fish feces sink to the bottom beneath the cages, resulting in an artificial enrichment of the benthic environment. By studying the biological community in bottom sediments near the cages, Ruben Vita and Arnaldo Marin found that the benthic ecosystems were highly impacted to a radius 5 meters out from the cages; however, moderate effects extended out to a 220-meter radius. The benthic communities in this region gradually recovered after tunas were removed from the cages, except for the area beneath the cages that did not fully recover within a six-month time period.

The bluefin tuna dilemma provides an example that highlights the difficulties presented when economics drive the expansion of an overexploited international resource. Numerous examples have been presented in this and previous chapters, including whales, fur seals, cod, reef fishes, and orange roughy, that demonstrate a similar pattern. National laws and international agreements are hardly able to keep up with the ingenuity of fishers and fish farmers when large profits are to be made. Political leaders and government agencies are invariably slow in acting when making decisions that are going to result in economic harm to their constituents. We must hope that the reactions are not too little and too late to save the world's most valued fish from biological endangerment; and that a true precautionary approach to management and conservation of international resources can eventually be achieved.

throughout this period, it has been very difficult to develop a consensus on harvest limits, in part because coming up with precise population estimates is difficult. Some ship owners registered their fishing vessels under nations that are not members of international management conventions (referred to as operating under *flags of convenience*), and thus avoided abiding by regulations.

Unfortunately, management generally has not been implemented using a precautionary approach. For example, the International Convention for the Conservation of Atlantic Tunas (ICAAT) (also see Chapter 12) has representatives from up to about 30 nations interested in the fishery. ICCAT is responsible for implementing management plans, setting annual quotas in the eastern and western sectors of the North Atlantic that are divided among the member nations, with European nations harvesting primarily from eastern stocks and North Americans from western stocks. Getting this many nations with different goals and conservation ethics to come to an agreement is difficult. The ICCAT, thus, has been criticized for dragging its feet on establishing agreements that are likely to stop the decline of the large tuna. There have been several unsuccessful efforts in the past 20 years to force better management of the valued bluefin tuna by proposing that it be listed on CITES Appendix I (see Chapter 9), which would eliminate international trade among member nations, including all major tuna fishing and importing nations. This would shut down most legal fishing and tuna ranching because it would remove the lucrative international marker.

Swordfish

The swordfish *Xiphias gladius* is the only species in the family Istiophoridae. The swordfish fishery has changed dramatically since the 1800s when they were harvested in the North Atlantic with harpoons from the front deck of boats

(**Figure 11-20**). This fishery was well-publicized because of the simple methods used and the potential size of harvested fish, with individuals sometimes over 500 kilograms. Through the early 1900s several thousand large swordfish could be taken each year in coastal New England waters.

As the number of large fish accessible by harpoon fishing declined and fishing technology advanced in the mid-1900s, fishermen began switching to pelagic longlines

Figure 11-20 Spearfishing for swordfish, as portrayed here in the late 1800s, was sustainable because fish were difficult to pursue and harvest.

as their primary fishing gear. The fishery was expanded throughout much of the North Atlantic and into the Caribbean Sea and Gulf of Mexico, where young fish were more common. Large swordfish became increasingly rare in the catches even in the North Atlantic. To avoid excessive harvest of smaller fish, size restrictions were enacted; however, many of the smaller fish still died before being removed from the longline hook. Although the average size of harvested swordfish declined after the switch to longlining, the total weight of swordfish landed began to increase rapidly. When the U.S. FDA found in 1969 that some swordfish filets had mercury levels of more than 0.5 parts per million, the market for swordfish plummeted.

By the mid-1970s, health authorities concluded that levels of mercury in recently-caught swordfish were not high enough for concern and the market rapidly recovered. Landings increased rapidly and the average weight of individual swordfish plummeted from 122 to 52 kilograms from the early 1960s to 1980 (**Figure 11-21**). Commercial landings continued to increase to a peak of 20,000 metric tons in 1987. Through the 1990s, swordfish catches gradually declined, reaching 10,000 tons by 2001. Increasingly smaller fish were being caught; the average fish weighed under 40 kilograms. About 60% of those caught were under three years old and immature (a female swordfish is at least 5 years old and almost 2 meters long when it is first capable of reproducing). ICCAT (which oversees swordfish as well as tuna harvest in the north Atlantic) was unable to institute strict management of swordfish harvest. NGOs began to make efforts to apply pressure for stricter management of swordfish harvest. The "Give Swordfish a Break" campaign, initiated in 1998, encouraged many east coast restaurants and cruise lines to remove swordfish from their menus. This and other pressures encouraged ICCAT to initiate a 10-year recovery plan for swordfish in 1999. This plan

included quotas on swordfish harvest, and swordfish nursery areas were closed to fishers in U.S. waters. Swordfish populations began to show a recovery in the 2000s. Within four years, a stock assessment survey estimated that the population had returned to 94% of its healthy stock size; however, this is based on fish biomass. The average fish size is still low. Conservation groups are pressuring fishers to use larger hook sizes to limit catch of smaller fish to avoid future dramatic population fluctuations.

Shark Harvest and Conservation

The mention of sharks often conjures up images of large vicious predators likely to attack anything they come upon, including humans. Through the mid-1900s a primary reason for shark harvest was fear—largely irrational—of danger to humans. Since then, its popularity as a food item has increased, resulting in large harvest of many shark populations. Because of their slow population growth rates and k-selected life history, sharks recovery slowly from overharvest (see Chapter 2) and a large sustained harvest is not possible. The most current concern with sharks thus is not shark attacks on humans but an increase in human attacks on sharks through fishing.

Shark Attacks

Although humans have a rational fear of any large predator, including sharks, this fear has been enhanced in the past 40 years with the release of dramatic movies such as *Jaws*, and focus by the news media on infrequent shark attacks. Attacks by sharks on humans are actually extremely rare occurrences. The odds of an unprovoked attack by a shark on a person swimming in coastal ocean waters are extremely low relative to other dangers. Recent increases in reported shark attacks are likely because of increases in coastal tourism and expansion of communication technologies that allow greater ease of reporting. As a result the United States typically reports the most shark attacks, over 60% of the approximately 60 to 80 reported per year from 2000 to 2007. The true number of attacks worldwide is unknown because of the lack of a convenient reporting method; however, estimates of probabilities of experiencing an attack are extremely low, thousands of times less than the chance of drowning.

Most shark attacks are likely a case of mistaken identity. When sharks attack large prey they typically make an initial attack to wound the prey and then circle around before returning. In most attacks on humans the shark does not return after the initial lunge. Although the attack can cause a serious wound or even result in death, most attacks on humans are not fatal (reported fatalities range from about 1 to 10 per year). This suggests that the attacking shark is looking for a large fish or marine mammal, such as otter or

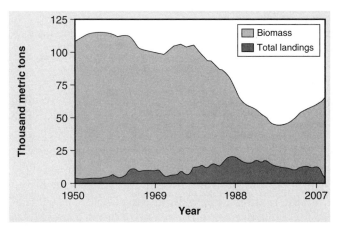

Figure 11-21 North Atlantic swordfish landings and estimated population biomass. The lower biomass in the 1980s and 1990s reflects reduced populations and smaller fish sizes.

seal, and mistakenly attacks the silhouette of a human. Only in isolated cases is there a need for concern of unprovoked attacks when swimming in coastal waters.

Shark Fisheries

Fishing for sharks began to gain popularity in the 1960s among anglers looking for a trophy fish and shark-fishing tournaments became popular in coastal areas of the United States; however, few were eating the sharks that were caught and it is unlikely that sport fishing had a large long-term effect on shark populations. One possible exception is the species most recognized as a predator, popularized in the movie *Jaws:* the great white shark *Carcharodon carcharias.* Following the *Jaws* movies, the reputation of sharks as dangerous to humans increased and it became popular to pursue great white sharks as a trophy for display; a large set of jaws could sell for thousands of dollars. Even a low level of harvest can affect populations of large apex predators like the great white shark. Concerns over species endangerment have led to listing by CITES on Appendix II, requiring strict regulation of trade in products from the great white shark. Protections from unreasonable harvest or trade have now been implemented off South Africa, Australia, and the United States. Despite restrictions, an unknown amount of illegal harvest still occurs. This was documented by Mahmood Shivji and colleagues using genetic analysis of illegally marketed shark fins confiscated from a dealer on the U.S. east coast. In the catch were not only fins from large great white sharks (likely to be sold as trophies) but also from very young white sharks (likely to be sold as food).

Until the later decades of the 20th century, shark meat was largely considered undesirable for consumption and there was little value for sharks in the commercial seafood market. Improvements in handling methods for harvested sharks and an increase demand for seafood products began to improve the ability to market shark products. Shark fisheries began to increase in the 1960s and accelerated through the 1990s (**Figure 11-22a**). The largest harvests have been in Pacific and Indian Ocean waters by Southeast Asian nations (e.g., Indonesia and India). The largest market for shark meat has been in Europe, where it was used to replace overharvested cod and other groundfish in processed fish products.

Another large market supporting the international shark fishery has expanded dramatically in the last few decades. Shark fin soup has been a part of traditional Chinese cuisine for centuries, valued as a status symbol because of its expense and difficulty to obtain. Demand for shark fins also results from beliefs that they are good for health and impart strength to those who consume them. There is now a widely held belief in China that shark fin soup has value as a general cure all. Hong Kong has been the primary center for shark trade in recent history; however, the opening of

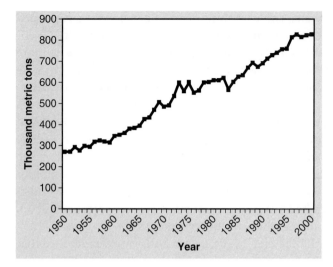

(a)

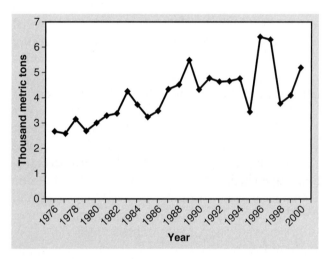

(b)

Figure 11-22 **(a)** Global harvest of chondrichthyans (sharks, skates, and rays). Explosive growth in harvest internationally has led to concerns that many populations are being overharvested. The most wasteful fishing practice is shark finning, in which the fins are removed and the body often discarded. World export of shark fins **(b)** indicates a continued increase despite international pressure to restrict finning practices.

the mainland China economy in the late 1980s has resulted in a rapid increase in markets and trade in shark fins, much still being routed through Hong Kong. By 1990 the reported export of shark fins was over 4,000 metric tons per year (Figure 11-22b). Most of these fins are from sharks caught by fishers, not just from waters in the western Pacific but from international waters around the world. The value of the fins could be over 100 times that of the meat and thus it was more profitable for fishers to maximize the number of fins they could obtain without having to handle and store the carcass. Profits were maximized by cutting off the dorsal, pectoral, and lower caudal fins from sharks to be sold as a *set* (a set of fins from the most desired species, oceanic whitetip *Carcharhinus longimanus*, blue, and mako sharks have varied in value from about U.S. $50 to $80 in the Hong Kong market). It is not common practice to kill the sharks before removing the fins, and death of the finned shark returned to the water is an imminent, but prolonged, process as the sharks sink to the bottom unable to swim. Asian countries, primarily China, remain the largest harvesters and purchasers of shark fins, though fishers from many nations are involved in shark fishing.

As the public became aware of the wasteful and cruel practices of the shark finning fishery, an international discussion began over how to limit the excessive harvest it promoted. The most popular solution was to recommend that the entire carcass would be retained for any harvested sharks, even when the primary market was for the fins. The United States took the lead on these restrictions; however, it was at least two decades until the U.S. Shark Fin Prohibition Act was passed in 2002 (**Figure 11-23**). This Act required that U.S. vessels fishing in United States or international waters, and foreign vessels in U.S. waters, could not possess shark fins without having the remainder of the shark carcass onboard the boat. Several nations have followed suit with similar laws, including South Africa, Brazil, Costa Rica, and the European Union. The number of sharks harvested, however, remains largely unregulated in the waters of most nations and in international waters. Somewhere between 25 and 75 million sharks are harvested or killed each year, as estimated based on the shark fin market. Harvest could be as much as four times larger than official catch statistics because of high levels of illegal, unregulated, and unreported (IUU) harvest. Analyses by Shelley Clarke and colleagues suggest that harvest levels are close to or exceed the maximum sustainable yield for many shark species. There are

Figure 11-23 A NOAA agent counting confiscated illegally harvested shark fins.

hopes that recent reports of toxins in shark fin products (e.g., high levels of mercury) and support of shark conservation among younger consumers will pressure agencies to apply further restrictions to shark harvest and finning practices.

Although the market for shark fins likely has the greatest impact, there are other profitable markets for shark products. For example, the dietary supplement market, developed based on reports that sharks do not develop cancer, has expanded despite a refuting of these claims by scientific studies. An additional reason for health claims are scientific studies that a protein contained in shark cartilage inhibits the development of new blood cells needed for the growth of cancer tumors. The protein cannot be incorporated into the body through the ingestion of shark cartilage, however, and there is no evidence of ingested shark cartilage pills having significant health benefits in humans. This effect was evident only when shark cartilage extracts were injected into rabbits and mice, and have led to additional research studies on the development of potential anticancer drugs. Conservation organizations are discouraging the targeting of sharks for the shark cartilage market based on the unsubstantiated claims.

From 1950 to 1996 reported global shark harvest increased from under 300,000 to about 800,000 metric tons per year. Many larger pelagic sharks that are targeted for meat or for shark fin soup have exhibited population declines and are considered overfished. The World Conservation Union (IUCN) has listed over 50 species as vulnerable or near threatened and 10 species as endangered, most because of overfishing and a lack of management. Precise population numbers are difficult to obtain, and IUCN lists over 100 species as data deficient.

Incidental bycatch of sharks provides yet another source of mortality for sharks in the open ocean. Because sharks commonly prey on the same items as other pelagic predators, bycatch is especially high in pelagic longline fisheries for swordfish and tunas, with sharks comprising up to 25% of the total catch. Most of these sharks are discarded either because they are prohibited from harvest or lack significant commercial value. As catches of sharks are often not reported in the harvest data for other pelagics (e.g., billfish and tunas in longline fisheries), the overall effect on global populations of sharks is not known. Not all sharks that are hooked are killed; for example, about 70% of longline-caught blue sharks can be released alive. Few regulations limit the bycatch of sharks, and there has been limited success in developing methods to reduce numbers killed from incidental harvest. Cutting the line to release sharks hooked on longlines can be of some success. Experimental studies have shown positive results in deterring sharks from taking baited hooks by placing a powerful magnet above the shaft of the hook, taking advantage of shark's sensitive electromagnetic perception. Although the number of sharks incidentally caught with longlines has declined, it is unclear how much of this effect is because of longline bycatch. One of the most impacted of sharks is the blue shark *Prionace glauca*, because of excessive targeted harvest and incidental harvest in longline and gill net fisheries. Though the decline in the Atlantic has been estimated at 60%, this statistic is uncertain. Blue shark abundance in the tropical Pacific Ocean in the 1990s was estimated at only 13% of what it was in the 1950s and individuals have declined in average size from 52 to 22 kg.

Although many of the sharks targeted for harvest in the shark fin market are larger pelagic species, there also has been an increasing market for smaller coastal bottom species. One of the largest such fisheries is for the spiny dogfish *Squalus acanthias*. These are small sharks that congregate in large schools in cool coastal waters around the world. Historically, the dogfish was considered a nuisance fish throughout much of its range, as it is caught in the nets of commercial trawlers. But as other fisheries started declining from overharvest, the dogfish became a readily available replacement. A variety of profitable markets exist for the spiny dogfish. They have been used as pet food, fishmeal, liver oil, and fertilizer. In Europe they are sold in fish and chip restaurants under a variety of names; for example, in England they are marketed as rock salmon. It is also the most commonly used shark for dissection by students. In U.S. waters many fishers operating in the depleted groundfish fisheries switched to catching dogfish through the 1990s. It was sold as *cape shark* to increase its marketability. The fishery was declared overfished in 1998, but it was not until 2002 that a management plan was implemented. Some are concerned that the quotas may be too liberal and unsustainable in the long run.

11.6 Climate Change and Fisheries

It has now become apparent that fishery populations have been dealing with climate change throughout their existence, long before humans were harvesting a substantial portion of their production. As discussed above, the evidence of dramatic changes in populations associated with natural environmental variability is strong for some of the most prominent fishery species including various sardines and anchovies, and groundfish such as the Atlantic cod. With the predicted increase in variability in climate and oceanographic conditions from human-induced climate change (see Chapter 1), fish population fluctuations are likely to be even more extreme and more frequent. For example, an increase in El Niño conditions would likely affect populations of Peruvian anchovies in the southeastern Pacific.

Unfortunately, predicting precisely when, where, and how much climate variability will affect fishery populations is still an unachievable goal. Managers, therefore, must strive to develop flexible and responsive management plans. Thorough monitoring is needed to enable scientists to catch population changes quickly, such that excessive harvest does not add additional stress that eliminates the population's ability to rebound. We must continue to move away from the philosophy that a maximum sustainable yield can be defined for fish populations over the long term when large natural population fluctuations are inevitable.

Beyond the variability, fisheries must be ready to deal with long-term shifts in distributions of fishery species, their food resources, and overall productivity. Regional temperature changes could be sudden and dramatic, moving conditions outside the tolerance range of some species. A rapid adjustment in the quotas and locations of harvest may be necessary. Less dramatic changes in climate could have effects on behavior, growth, development, or reproduction that take years to be realized in the adult populations. The additional effects of harvest must be anticipated if populations are to adjust to such changes.

Even when population changes are not in evidence, harvest must be carried out such that fishery populations are maintained in healthy condition. There is a need to minimize some of the major effects of excessive fishing on a marine population, such as modification of the age and size structure, depletion of local segments of the population, altering behavior or life history patterns, and reducing the complexity of the ecosystem. We should not, however, make the mistake of focusing too much on climate change and oceanographic variability such that they become the scapegoat for all the problems in marine fisheries management. Scientists, managers, and fishers must continue to make efforts to reduce overharvest driven by economics, politics, and greed. Habitat and ecosystem health must remain a primary goal if global fisheries are to be sustained. It is not just an idealistic goal to presume that a precautionary approach to fishery management decisions can become the norm around the world.

STUDY GUIDE

Topics for Review

1. What is the difference in a fishery stock and a fish population?
2. What role does the FAO play regarding international fisheries?
3. Discuss similarities and differences in subsistence, artisanal, and small-scale fishers.
4. Describe different economic roles served by recreational fishing.
5. Describe arguments given by sport and commercial fishing in support of being given preference in fishing allocations.
6. How does the allocation of a single seasonal quota favor large-scale fishers?
7. Discuss net removal versus education as a means of dealing with impacts of ghost nets.
8. Why are bycatch reduction devices less successful in removing fish than sea turtles from shrimp trawl bycatch?
9. Describe and explain the pattern of global fishery landings from 1950 to 2000.
10. Describe how *fishing down the food web* affects marine ecosystems.
11. How do fish population cycles affect fisheries management decisions?
12. How are no-fishing marine protected areas used to enhance fisheries?
13. Describe harmful effects of net-pen aquaculture on marine ecosystems.
14. How does farming of salmon and tuna affect wild forage-fish populations?
15. What are the environmental concerns of farming genetically modified fishes?
16. What factors likely contributed to the decline in Pacific herring fisheries through the mid 1900s?
17. Describe potential ecosystem effects of Atlantic menhaden overharvest.
18. What population factors make the Gulf menhaden an *ideal* species for fishery harvest?
19. How did environmental and economic factors combine to cause the collapse of the Peruvian anchovy fishery?
20. How did changes in fishery harvest technology and climate change likely combine to effect the collapse of the Atlantic cod fishery?
21. Describe how conflicts over cod fishing rights led to an expansion of coastal fishing rights.
22. What biological factors make overfishing more likely for bluefin tuna than Alaskan pollock or Gulf menhaden?
23. How does bluefin tuna ranching impact wild tuna populations?
24. What factors have led to difficulties in monitoring bluefin tuna farming?
25. How has fishing affected the size structure of swordfish populations?
26. Describe non-traditional markets that have led to excessive harvest of sharks.

Conservation Exercises

1. Describe the importance of the following with respect to conservation of fisheries populations:
   a. regulated inefficiency
   b. maximum sustainable yield
   c. Ludwig's ratchet
   d. subsidies
   e. IUU
   f. fishing down the food web
   g. IFQs
   h. community-based management
   i. precautionary approach
   j. ecosystem-based management
   k. trophic cascades
   l. derby fisheries
   m. FCZs

2. Compare the following fishing methods regarding impacts on population structure of targeted fishery populations:
   a. spearguns
   b. longlines
   c. pot traps
   d. gill nets
   e. purse seines

3. Compare the following regarding fishery bycatch and environmental effects:
   a. longlines
   b. pot traps
   c. drift nets
   d. purse seines
   e. bottom trawls
   f. hydraulic dredges

FURTHER READING

Alder, J., B. Campbell, V. Karpouzi, K. Kaschner, and D. Pauly. 2008. Forage fish: from ecosystems to markets. *Annual Reviews in Environment and Resources* 33:153–166.

Anderson, C. N. K., C. Hsieh, S. A. Sandin, R. Hewitt, A. Hallowed, J. Beddington, R. M. May, and G. Sugihara. 2008. Why fishing magnifies fluctuations in fish abundance. *Nature* 452:825–838.

Baumgartner, T. R., A. Soutar, and V. Ferreira-Bartrina. 1992. Reconstruction of the history of Pacific sardine and northern anchovy populations over the past two millennia from sediment of the Santa Barbara Basin. *California Cooperative Fisheries Investigations Reports* 33:24–40.

Beamish, R. J., G. A. McFarlane, and A. Benson. 2006. Longevity overfishing. *Progress in Oceanography* 68:289–302.

Branch, T. A., R. Watson, E. A. Fulton, S. Jennings, C. R. McGilliard, G. T. Pablico, D. Ricard, and S. R. Tracey. 2010. The trophic fingerprint of marine fisheries. *Nature* 468:431–435.

Brander, K. 2010. Climate change and fisheries management. *In* Grafton, R. Q., R. Hilborn, D. Squires, M. Tait, and M. J. Williams (eds.). *Handbook of Marine Fisheries Conservation and Management.* Oxford University Press, New York.

Brander, K. M. 2007. The role of growth changes in the decline and recovery of North Atlantic cod stocks since 1970. *ICES Journal of Marine Science* 64:211–217.

Breen, P. A. 1987. Mortality of dungeness crabs caused by lost traps in the Fraser River Estuary, British Columbia. *North American Journal of Fisheries Management* 7:429–435.

Brown, J., and G. Macfadyen. 2007. Ghost fishing in European waters: impacts and management responses. *Marine Policy* 31:488–504.

Caddell, R. 2010. Caught in the net: Driftnet fishing restrictions and the European Court of Justice. *Journal of Environmental Law* 22:301–314.

Catarci, C. 2004. World markets and industry of selected commercially-exploited aquatic species with an international conservation profile. *FAO Fisheries Circular* C990.

Chavez, F. P., J. Ryan, S. E. Lluch-Cota, and M. Niquen. 2003. From anchovies to sardines and back: multidecadal changes in the Pacific Ocean. *Science* 299:217–221.

Clarke, S., E. J. Milner-Gulland, and T. B. Cemare. 2007. Social, economic, and regulatory drivers of the shark fin trade. *Marine Resource Economics* 22:305–327.

Collette, B. B., K. E. Carpenter, B. A. Polidoro, M. J. Juan-Jorda, A. Boustany, D. J. Die, C. Elfes, W. Fox, J. Graves, L. R. Harrison, R. McManus, C. V. Minte-Vera, R. Nelson, V. Restrepo, J. Schratwieser, C.-L. Sun, A. Amorim, M. Brick Peres, C. Canales, G. Cardenas, S.-K. Chang, W.-C. Chiang, N. de Oliveira Leite, H. Harwell, R. Lessa, F. L. Fredou, H. A. Oxenford, R. Serra, K.-T. Shao, R. Sumaila, S.-P. Wang, R. Watson, and E. Yanez. 2011. High value and long life—double jeopardy for tunas and billfishes. *Science* 333:291–292.

Cowan, J. H., C. B. Grimes, W. F. Patterson, C. J. Walters, A. C. Jones, W. J. Lindberg, D. J. Sheehy, W. E. Pine, J. E. Powers, M. D. Campbell, K. C. Lindeman, S. L. Diamond, R. Hilborn, H. T. Gibson, and K. A. Rose. 2010. Red snapper management in the Gulf of Mexico: science- or faith-based. *Reviews in Fish Biology and Fisheries.* 21:187–204.

Daskalov, G. M., A. N. Grishin, S. Rodionov, and V. Mihneva. 2007. Trophic cascades triggered by overfishing reveal possible mechanisms of ecosystem regime shifts. *Proceedings of the National Academy of Sciences* 104:10518–10523.

DeStefano, V., and P. G. M. Vanderheijden. 2007. Bluefin tuna fishing and ranching: a difficult management problem. *Mediterranean Journal of Economics, Agriculture and Environment* 2:59–64.

Etheridge, L. 2010. Gulf of Mexico aquaculture plan. *Water Log* 30(3):15–17.

Franklin, H. B. 2007. *The Most Important Fish in the Sea: Menhaden and America.* Island Press, Washington.

Gallaway, B. J., and J. G. Cole. 1999. Reduction of juvenile red snapper bycatch in the U.S. Gulf of Mexico shrimp trawl fishery. *North American Journal of Fisheries Management* 19:342–355.

Gaylord, B., S. D. Gaines, D. A. Siegel, and M. H. Carr. 2005. Marine reserves exploit population structure and life history in potentially improving fisheries yields. *Ecological Applications* 15:2180–2191.

Gray, J. S., P. Dayton, S. Thrush, and M. J. Kaiser. 2006. On effects of trawling, benthos and sampling design. *Marine Pollution Bulletin* 52:840–843.

Hannah, R. W., and S. A. Jones. 2007. Effectiveness of bycatch reduction devises (BRDs) in the ocean shrimp (*Pandalus jordani*) trawl fishery. *Fisheries Research* 85:217–225.

Heales, D. S., D. T. Brewer, P. M. Kuhnert, and P. N. Jones. 2007. Detecting declines in catch rates of diverse trawl bycatch species, and implications for monitoring. *Fisheries Research* 84:153–161.

Hennessey, T., and M. Healey. 2000. Ludwig's ratchet and the collapse of New England groundfish stocks. *Coastal Management* 28:187–213.

Howarth, L. M., H. L. Wood, A. P. Turner, and B. D. Beukers-Stewart. 2011. Complex habitat boosts scallop recruitment in a fully protected marine reserve. *Marine Biology* 158:1767–1780.

Ingolfsson, O. A., and T. Jorgensen. 2006. Escapement of gadoid fish beneath a commercial bottom trawl: relevance to the overall trawl selectivity. *Fisheries Research* 79:303–312.

Kurlansky, M. 1997. *Cod: A Biography of the Fish that Changed the World.* Penguin Putnam, New York.

Larkin, P. A. 1977. An epitaph for the concept of maximum sustained yield. *Transactions of the American Fisheries Society* 106:1–11.

Lauck, T., C. W. Clark, M. Mangel, and G. R. Munro. 1998. Implementing the precautionary principle in fisheries management through marine reserves. *Ecological Applications* 8:S72–S78.

Lee, A., and R. Langer. 1983. Shark cartilage contains inhibitors of tumor angiogenesis. *Science* 221:1185–1187.

Longo, S. B. 2010. Mediterranean rift: socioecological transformations in the Sicilian bluefin tuna fishery. *Critical Sociology* 1-20. DOI:10.1177/0896920510382930.

Ludwig, D., R. Hilborn, and C. Walters. 1993. Uncertainty, resource exploitation and conservation: lessons from history. *Science* 260:36–37.

Mandelman, J. W., P. W. Cooper, T. B. Werner, and K. M. Lagueux. 2008. Shark bycatch and depredation in the U.S. Atlantic pelagic longline fishery. *Reviews in Fish Biology and Fisheries* 18:427–442.

Miyake, P. M. 2005. Summary report on international marketing of bluefin tuna. Presented at GFCM/ICCAT Working Group on Sustainable Tuna Farming/Fattening Practices in the Mediterranean; March 16–18, 2005: Rome.

Novaczek, I. 1995. Possible roles for marine protected areas in establishing sustainable fisheries in Canada. Pages 31–36 *in* N. L. Shackell and J. H. M. Willision, (eds.). *Marine Protected Areas and Sustainable Fisheries.* Centre for Wildlife and Conservation Biology, Acadia University, Wolfville, Nova Scotia, Canada.

Pauly, D. 2009. Beyond duplicity and ignorance in global fisheries. *Scientia Marina* 73:214–224.

Pauly, D., V. Christensen, J. Dalsgaard, R. Froese, and F. Torres. 1998. Fishing down marine food webs. *Science* 279:860–863.

Planque, B., J.-M. Fromentin, P. Cury, K. F. Drinkwater, S. Jennings, R. I. Perry, and S. Kifani. 2010. How does fishing alter marine populations and ecosystems sensitivity to climate? *Journal of Marine Sciences* 79:403–417.

Rose, G. A. 2004. Reconciling overfishing and climate change with stock dynamics of Atlantic cod (*Gadus morhua*) over 500 years. *Canadian Journal of Fisheries and Aquatic Sciences* 61:1553–1557.

Scheffer, M., S. Carpenter, and B. de Young. 2006. Cascading effects of overfishing marine systems. *Trends in Ecology and Evolution* 20:579–581.

Schlag, A. K. 2010. Aquaculture: an emerging issue for public concern. *Journal of Risk Research* 13:829–844.

Schroeder, D. M., A. W. Schultz, and J. J. Dindo. 1988. Innershelf hardbottom areas, northeastern Gulf of Mexico. *Gulf Coast Association of Geological Societies Transactions* 38:535–541.

Shelton, P. A., A. F. Sinclair, G. A. Chouinard, R. Mohn, and D. E. Duplisea. 2006. Fishing under low productivity conditions is further delaying recovery of Northwest Atlantic cod (*Gadus morhua*). *Canadian Journal of Fisheries and Aquatic Sciences* 63:235–238.

Sheppard, C. 2006. Trawling the sea bed. *Marine Pollution Bulletin* 52:831–835.

Shivji, M. S., D. D. Chapman, E. K. Pikitch, and P. W. Raymond. 2005. Genetic profiling reveals illegal international trade in fins of the great white shark, *Carcharodon carcharias.* *Conservation Genetics* 6:1035–1039.

Simpson, S. D., S. Jennings, M. P. Johnson, J. L. Blanchard, P.-J. Schon, D. W. Sims, and M. J. Genner. 2011. Continental shelf-wide response of a fish assemblage to rapid warming of the sea. *Current Biology* 21:1565–1570.

Sundstrom, L. F., M. Lohmus, and R. H. Devlin. 2010. Migration and growth potential of coho salmon smolts: implications for ecological impacts from growth–enhanced fish. *Ecological Applications* 20:1372–1383.

Tsanebtum M., M. A. Palma, B. Milligan, and K. Mfodwo. 2010. The European Council Regulation on illegal, unreported and unregulated fishing: an international fisheries law perspective. *The International Journal of Marine and Coastal Law* 25:5–31.

Valdemarsen, J. W., T. Jorgensen, and A. Engas. 2007. Options to mitigate bottom habitat impact of dragged gears. *FAO Fisheries Technical Paper 506.* Food and Agriculture Organization of the UN, Rome.

Van Eenennaam, A., and W. M. Muir. 2011. Transgenic salmon: a final leap to the grocery shelf? *Nature Biotechnology* 29:706–710.

Vega, R., and R. Licandeo. 2009. The effect of American and Spanish longline systems on target and non-target species in the eastern South Pacific swordfish fishery. *Fisheries Research* 98:22–32.

Vita, R., and A. Marin. 2007. Environmental impact of capture-based bluefin tuna aquaculture on benthic communities in the western Mediterranean. *Aquaculture Research* 38:331–339.

Ward, P., and R. A. Myers. 2005. Shifts in open-ocean fish communities coinciding with the commencement of commercial fishing. *Ecology* 86:835–847.

Wells, R. J. D., J. H. Cowan, and W. F. Patterson. 2008. Habitat use and the effect of shrimp trawling on fish and invertebrate communities over the northern Gulf of Mexico continental shelf. *ICES Journal of Marine Science* 65:1610–1619.

Worm, B., E. B. Barbier, N. Beaumont, J. E. Duffy, C. Folke, B. S. Halpern, J. B. C. Jackson, H. K. Lotze, F. Micheli, S. R. Palumbi, E. Sala, K. A. Selkoe, J. J. Stachowicz, and R. Watson. 2010. Impacts of bio-diversity loss on ocean ecosystem services. *Science* 314:787–790.

Zeller, D., S. Booth, E. Pakhomov, W. Swartz, and D. Pauly. 2011. Arctic fisheries catches in Russia, USA and Canada: baselines for neglected ecosystems. *Polar Biology* DOI:10.1007/s00300-010-0952-3.

Ocean Conservation Laws, Agreements, and Organizations

CHAPTER

12

Solutions to many of the environmental and conservation issues discussed in previous chapters hinge on limiting human use or access to the physical environment or resources occupying that environment (**Figure 12-1**). Many of the failures to achieve such goals stem from the historical use of the ocean as a common resource with few restrictions on human actions. The realization that rules are needed to control exploitation of the ocean's resources has gradually developed as human populations have grown exponentially and technological developments have exploded. It is now well documented that humans often do not work for the common good of conservation of marine resources when their livelihood or survival is in jeopardy. Even when substantial profits are being gained from resource exploitation, greed and competition result in actions taken for short-term gain at the cost of long-term sustainability. Moderation can be encouraged through public pressure, especially when it affects the market for seafood and other marine products; however, enforceable rules, laws, and agreements are needed at local, national, international, and global levels if marine conservation goals are going to be realized in the long run. This chapter discusses the cultural evolution of philosophies concerning the use of common resources and how this has led to successes and failures in the use of modern science, public involvement, and legal mechanisms to encourage and force increased conservation of marine resources.

12.1 Tragedy of the Commons and Marine Resources

A large proportion of marine conservation issues involve disputes over shared use of common resources, especially with regards to regulation of fisheries and protection of consumable resources. The term "**commons**," in this context, refers to some region or thing that is equally available to all members of a defined group (e.g., citizens of a nation or the human population). A specific item or area is called **common property**, and items considered of some value for the taking are called **common property resources** (e.g., fishes harvested from international waters or minerals mined from the deep sea floor). International common property resources are not considered the exclusive province of any one person, group, or nation, but are available for the use of all people; a national common property resource would be equally available to all citizens of a nation (in waters controlled by that nation).

When human populations were low enough that overutilization was improbable, sharing common resources was achievable with few conflicts. But when a designated resource is limited, equitable sharing as common property becomes problematic. For example, is there a presumption of equal sharing, or does the resource go to those better at exploiting it (e.g., the fishers with a bigger net or faster boat)? For valuable, limited resources, the latter philosophy

Figure 12-1 A U.S. Coast Guard vessel.

leads to a race for everyone to grab their *fair share* before others can—and everyone who gets a *fair share* gets it at the expense of others. If the human population is capable of overexploiting the resource, conservation will be impossible. On the other hand, although the philosophy of equal sharing may appear to be a worthy goal, history has shown that few will work to harvest a resource and encourage conservation if they cannot profit because the resource must be shared with everyone. Either one of these extremes is impractical and unmanageable. There must be some mechanism to establish rules for sharing and encourage or force responsible, wise use and protection of common resources for long-term sustainability.

Global commons are regions that, through international agreements, are not claimed by any individual, nation, or group of nations. These commons include the atmosphere. Though countries may claim the air space over their territories, they do not claim the rights to the air, clouds, and weather that pass over their countries. Most of the atmosphere is far above the air space of individual countries. Although we have experimented with changing weather patterns (e.g., through cloud seeding), no serious international conflicts have resulted. Air pollutants that modify the atmosphere do not remain where they are released and commonly cross national borders and often end up in the oceans. Atmospheric pollutants also are likely the primary cause of current global climate change (see Chapter 1). Regulating these pollutants requires powerful agreements that have been difficult to implement. Deep space, though still virtually unexplored except for the few missions to the Moon and planets, is considered a commons. Satellites circling in low-Earth orbits have the potential to overutilize this region, and decisions eventually will need to be made on how to allocate the most popular geosynchronous orbits.

International agreements have been made to limit the space to peaceful exploration and to keep nuclear weapons from being deployed; however, there is still potential for conflicts over the demilitarization of space.

The "**high seas**"—the open ocean away from the territorial waters of coastal nations—historically were considered as commons in part because they were only accessible to the few with the ability and the courage to navigate them. Any nation proclaiming exclusive rights found it impossible to defend those rights. Until recently the resources of the open ocean and sea bottom were largely inaccessible. As these resources have become progressively more accessible, however, nations and international commissions have struggled to establish agreements that avoid conflicts and partition limited resources equitably. Even within coastal waters, resources are often designated as common resources to citizens of the nation with economic and fishing rights to those waters. Today the problems often boil down to questions over how to partition a limited amount of resources among an increasingly expanding human population. Conservation based on the assumption that individuals will operate in the interest of the *common good* often leads to failure.

In 1968, Garret Hardin formulated the concept of the **tragedy of the commons**, presenting arguments to address the problems of human population growth and the overuse of the Earth's resources. He argued that decisions made by individuals are not necessarily the best for human society as a whole, and that we need to examine which of our freedoms are defensible. His argument is presented by way of an example.

Hardin asks us to imagine a pasture that is without rules or customs for sharing. It is open to anyone who has the desire and ability to raise livestock. There are no restrictions to limit the number of livestock a person might put on the common pasture. The inevitable outcome is that herdsmen will continue to add livestock as long as they can achieve some gain, and they may need to maximize their livestock just to make a living. It is unlikely that any individual herdsman will seriously consider the impact of his actions on the land (due to overgrazing) and on the other herdsmen competing for use of the land. As Hardin eloquently states,

> *"Therein is the tragedy. Each man is locked into a system that compels him to increase his herd without limit—in a world that is limited. Ruin is the destination toward which all men rush, each pursuing his own best interest in a society that believes in freedom of the commons."*

This concept can be expanded beyond this pastoral example to any resource that is considered a commons. We can restate Hardin's example in terms of ocean resources. For example: Imagine an ocean that is without rules or customs for sharing. It is open to anyone who has the desire and ability to harvest fish. There are no limits on the number of fish a person might take from the common ocean and, therefore, fishers will continue to take fish as long as they can achieve some gain—and may need to maximize their harvest in order to make a living. But each fisher does not consider the impact of his actions on the ocean's resources (due to overfishing), and on the other fishers competing for use of the ocean.

Several options can be used to overcome this tragedy (meaning *dilemma* in this context) and make common resources sustainable. In terrestrial systems, private ownership of the land provides an incentive for protection and sustainable use of resources. (Although this may operate in the interest of economic sustainability, it might not be in the interest of preserving natural ecosystems; thus, at least a portion of the terrestrial commons is protected for use or is regulated for ecosystem sustainability.) However, the ocean's dynamic nature and its lack of defined boundaries and accessibility historically made private ownership implausible. There are exceptions in coastal areas, where land is managed by leasing, for example, for harvesting oysters and other mollusks or farming fish. In most ocean regions beyond the coastline, the concept of *freedom of the seas* became established and engrained in most coastal cultures. In common areas, rules are needed for sharing limited resources; however, the variance in personal opinions and cultures makes the establishment of common rules one of the most difficult aspects of governing the commons. Even when a consensus is reached on rulemaking, enforcement of the rules can be problematic, especially for international agreements where there is no defined enforcement body.

In fisheries, rules for sharing can be enhanced by encouraging restraint through gear restrictions. A **forced inefficiency** restricts the types of fishing gears that are allowed so that a few individuals do not have the capability to harvest a large fraction of the sustainable catch of a species. (Forced inefficiency also can be used to avoid damage to the environment or untargeted species; see Chapter 11). Seasonal **quotas** that establish a **total allowable catch** (**TAC**) provide specific rules for sharing by limiting how many fish can be taken by an individual or group. A third way to make common resources sustainable is through the establishment of **cooperative institutions**. This gives access rights to a group rather than an individual, and the group legislates its own restraints. This is based on the premise that individuals in a small group are more likely to agree to work together sustainability for the long-term benefit of everyone in the group. Such management has been applied by giving coastal communities responsibility for managing coastal fisheries in

their region. This **community-based management** is most commonly applied for small indigenous groups; however, one creative application of the principle has been applied to some fisheries where environmental non-governmental organizations (NGOs) pay for quotas or easements in a region to implement environmentally sustainable fishing practices. For example, this is being attempted in regions of California through the NGOs The Nature Conservancy and Environmental Defense Fund.

Numerous examples have been given in previous chapters of the application of these methods for successfully managing and conserving ocean resources. The struggle to apply methods for sharing continues, however, and the tragedy of the commons remains for many ocean resources. This is especially true in international waters of the open ocean, where historic views of an inherent right of unlimited access and use continue to prevail.

12.2 Marine Conservation Laws and Agreements

Until the late 19th century, low accessibility of the open oceans limited conflicts over resource use. The number of conflicts increased as countries developed capabilities to navigate the world's oceans, to harvest the living resources of the ocean to feed exponentially growing populations, to extract oil and minerals from the sea floor to feed our ever-growing demands for sources of energy (**Figure 12-2**), and to use the ocean as a dump for ever increasing anthropogenic wastes. Gradually, mechanisms were developed, in the form of international and national laws and agreements, to resolve or avoid these conflicts. Doctrines addressing freedom of the seas have been established since the 17th century, long before the public awareness of conservation issues. At the same time, nations have developed

Figure 12-2 A major impetus for the United States to expand its coastal rights in the 1940s was for access to oil and gas resources.

policies to determine how ocean resources are used within ocean waters they control.

Treating ocean resources as common property worked fine when human populations were low and the technology to harvest tons of fish per day from the ocean had not been developed. A relatively few people willing to take the risk to go to sea and throw a hook or harpoon at their prey could not catch a large enough fraction of the animals they targeted to impact a population. These resources, thus, sustained themselves and were functionally inexhaustible. Even as populations expanded around the globe, ocean fishing remained a difficult and dangerous occupation, and large open ocean fish populations remained relatively unimpacted (**Figure 12-3**). Eventually fishers developed capabilities to harvest large fractions of fish populations (see Chapter 11). Modern technology now enables the tracking of fish with sonar and satellite imagery. Factory ships can travel around the globe processing their catch as they go. Coastal fishing fleets expand until they are able to harvest large fractions of a population in a single fishing season. And human populations continue to grow exponentially, with much of that growth in coastal regions where there is a near insatiable need for fish products.

Some of the most severe abuses of the freedom of the seas were attitudes that led to early exploitation of vulnerable marine mammals and reptiles. For example, unregulated international harvest of seals, whales, and sea turtles drove species to near extinction before restrictions could be enacted (see Chapters 9 and 10). Once modern fishing technology was developed, large-scale systematic depletion of ocean fish populations began (see Chapter 11). Although rarely have fish populations been driven to biological extinction, we have witnessed the decline of one fish population after another in international waters, in part due to the inability of management agencies and commissions to regulate fish harvest in the commons.

The use of the global commons also applies to seafloor resources, for mining minerals and extracting hydrocarbons as an energy source. Although current technology and value does not make it profitable to mine the deep sea floor, we continue to move into progressively deeper waters to extract resources (see Chapter 8). Although mined resources are not renewable within a practical time frame, the debate continues over how to share the resources now and in the future, while minimizing environmental impacts.

Marine Species Protections

Many of the immediate conservation concerns in regions considered to be common international waters are related to fisheries and marine mammal harvest and protection, which have been largely addressed in previous chapters. The United Nations and other international commissions help to bring countries together and encourage agreements; however, neither the U.N. nor any other group has enforcement capabilities in these international waters. Often enforcement is carried out through threats of trade sanctions by one or more nations against the offending nations. The numerous commissions and other groups involved in establishing international agreements operate under different charges and are mostly comprised of representatives of nations with a vested interest in the species or resource of interest.

Species whose survival is considered to be imperiled to some degree are protected internationally through several organizations. Although these organizations may not have legal powers within national waters, the agreements are typically designed to protect endangered species throughout their range. For example, the Convention on International Trade in Endangered Species (CITES) enforces endangered species protections by restricting trade in products originating from listed species (see Chapter 9). The

Figure 12-3 Fishers using low technology, like these U.S. cod fishers in the 1800s, could fish sustainably in the ocean commons with limited impacts on large ocean fish populations.

International Whaling Commission (IWC) restricts whaling through international agreements that include whaling and non-whaling nations. The moratorium on commercial whaling includes restrictions applied throughout the ranges of all great whale species, with only well-defined exceptions for certain nations or groups (see Chapter 10).

Species that spend much of their lives in international waters or move through waters of multiple nations are largely managed by international commissions. These commissions are comprised of representatives from nations that are interested in the fishery. Most, but not all, nations participate in these agreements; however, enforcement of restrictions can be problematic (see Chapter 11). Most of the fisheries governed by these commissions are targeted towards large migratory predators. For example, the International Commission for the Conservation of Atlantic Tunas establishes agreements not only for tunas but for a total of 30 species, including swordfish, marlins, sailfish, and mackerels—and some progress has been made to include sharks.

■ Ocean Commons: International Agreements

For broader and more complex global conservation issues, a simple harvest restriction is inadequate to provide reasonable protections. Through the 20th century an international awareness developed of the need for broad agreements to deal with partitioning the resources of the global ocean commons. Although the power of such agreements is limited by the lack of a strong international body to enforce rules and regulations, there has been some success, primarily working within the United Nations. More specific agreements have been established through numerous international commissions, and NGOs have worked to educate and encourage compliance through various means. The following sections address the establishment of broad international agreements that address conservation of marine environments and resources.

Freedom of the Seas Doctrines

Ever since the development of vessels capable of crossing the oceans to discover "new worlds," humans have been wrestling with the question of whom—if anyone—controls access to the sea. Early attempts to gain control of the open oceans include a proclamation in 1494 by Pope Alexander VI that divided the Atlantic Ocean between Spain and Portugal. It was not long before it was realized that claiming an ocean was much different than claiming lands, because defending these claimed resources was impossible. Ocean-going nations began to legally recognize the freedom of the seas through doctrines enacted at least as early as the 17th century. But even with the acceptance of this philosophy, limits needed to be established to define the borders between national waters and the high seas. The earliest proclaimed borders to national waters were established based on limits to the ability of nations to monitor and protect their coastlines. In the 17th century, European countries widely accepted the *cannon-shot rule*. Countries could claim dominion over territorial seas as far out from the coast as could be reached by a projectile fired from a cannon, which at the time was approximately three nautical miles (5.5 kilometers). This became the basis for what became a traditional three-mile territorial sea limit in Freedom of the Seas doctrines. The remainder of the ocean, the **high seas**, was considered "free to all, belonging to none," and, thus, open to use and exploitation by anyone.

Although there was variability in proclaimed rights of access in coastal waters, Freedom of the Seas doctrines largely remained in place regarding access to fisheries and other ocean resources for as long as accessibility to these resources was limited. Development of conflicts over open access to the oceans largely followed technological advances through the early- to mid-1900s as reflected in improvements in seafloor drilling (see Chapter 8) and vessel and fishing technology (see Chapter 11). Conflicts over harvesting ocean resources became imminent.

The first serious challenge to Freedom of the Seas doctrines was the made by the United States in 1945, in response to interests by U.S. companies to exploit oil resources on the continental shelf. The proclamation of access rights included any oil, gas, or minerals that could be extracted from this region. Argentina soon followed suit, claiming rights over resources on and over its continental shelf. Discovering, exploiting, and transporting offshore oil was now becoming a lucrative business. For European nations, conflicts became commonplace because several nations bordered the same waters and had used those waters for centuries. Nations could not come to an agreement on how to carve up the shelf for oil exploration in the North Sea.

Access to fishing rights drove further conflicts over exclusive right to coastal waters. By the 1960s, fishing fleets were roaming the oceans, processing their catch at sea, and depleting coastal fish stocks. Fishing became a free-for-all in the world's richest fishing grounds. As soon as a coastal country would set a limit, it would be contested by countries profiting from fishing in those waters. Conflicts developed, including the Cod Wars in the North Atlantic (see Chapter 11) that led to the establishment of a 200-mile (370-kilometer) Fishery Conservation Zone by several nations to keep foreign fishing fleets away from their coasts.

Eventually access rights became an issue well beyond coastal waters. Not only were valuable fisheries for migratory predators being discovered, but technologies were being developed to mine the deep sea floor (see Chapter 8).

The discovery of huge expanses of the sea floor covered with mineral-rich manganese nodules led to debates over who had the right to mine these once technology made it profitable. Access to the open ocean moved beyond debates over the extraction of resources. During the Cold War period, nuclear submarines traversed the deep sea and plans were made for missile systems to be placed on the sea floor. The need for an international agreement regarding the use and exploitation of the world's oceans was imperative.

Conflicts led to an international awareness of ocean access and protection issues. In his 1967 speech, Arvid Pardo, Malta's Ambassador to the United Nations, asked that people look around at what we were doing to our ocean resources. He called for "an effective international regime over the seabed and ocean floor" and warned of "the escalating tension that will be inevitable if the present situation is allowed to continue." This speech set in motion a global effort to conserve and regulate the use of ocean resources. Although the initiatives started as a desire to regulate the seabed, they eventually led to efforts to regulate the use of the sea and all of its resources. In the 1970s the United Nations set about to update the Freedom of the Seas doctrines, including a treaty to ban nuclear weapons from the sea floor. The U.N. General Assembly adopted declarations stating that seabed resources are the "common heritage of mankind." Although the proclamations were vague and did not define how the ocean recourses would be shared, they initiated a realization that there must be dialogue, negotiations, and agreement to consistently manage the oceans.

United Nations Convention on the Law of the Sea

The Third United Nations Conference on the Law of the Sea convened in New York in 1973 to put together a comprehensive agreement on the use and conservation of the world's ocean resources. These negotiations involved 160 nation-states and continued for the next nine years. In 1982 the **United Nations Convention on the Law of the Sea** (commonly called **UNCLOS** or simply the **Law of the Sea**) was adopted. This was an unprecedented effort by the international community to regulate all resources and uses of the oceans, and resulted in a substantial reduction in the freedom of the seas. By signing the Convention, states agree not to take any action that would defeat its objectives and purposes. Ratification of the Convention is an agreement to be bound by its provisions, without exception. When 60 states had ratified the Convention in 1994, it came into force. To date there is still not 100% ratification by signatories, and non-ratifying nations include the United States, Thailand, several North African nations, and Columbia.

The most important features of the Law of the Sea Treaty deal with navigation rights in coastal waters and the high seas, territorial limits for coastal states, economic

jurisdiction, laws concerning how seabed resources are to be divided, rights of ship passage in a state's territorial waters, conservation and management of fisheries and other living marine resources, environmental protections from pollution and other impacts, regimes for ocean research, and procedures for settling disputes through an International Tribunal for the Law of the Sea.

It was understood that The Convention on the Law of the Sea could not regulate the use of ocean resources until the disputes over limits of coastal nations could be settled. Although international laws had established that coastal states had jurisdiction to waters in a band along its coastline, countries claimed limits ranging from 3 to 200 nautical miles. Those states that possessed large ocean-going ships wanted narrow limits to protect their access to coastal waters around the world. Smaller states wanted a broad limit to protect resources in their coastal waters. Countries with a broad continental shelf wanted broad limits to give them exclusive access to oil resources off their coast. Eventually a compromise was reached and different zones were established within which coastal states would have different rights and responsibilities.

Territorial seas were declared out to a 12-mile (22-kilometer) limit from shore. Within this region, states can enforce any law, regulate its use, and exploit any resource found there. Other countries are allowed the right of "**innocent passage**," for example, if shipping routes to a region pass through another state's territorial seas. A contiguous zone is set out to a 24-mile (44-kilometer) limit from coast, within which states can prevent certain violations and enforce police powers. For example, the U.S. Coast Guard can pursue vessels for illegal activities (such as drug smuggling) in this area.

As limits were being set on use of the seas, a major debate ensued concerning passage through narrow straits between countries, because these would fall within the territorial seas of one or more countries. Larger countries wanted the same freedom of navigation as if on the high seas. Coastal nations wanted only innocent passage rights to be declared. The compromise was to allow *transit passage* (unimpeded navigation) but otherwise following international regulations, and no threat or use of force is to be used (**Figure 12-4**).

The most important limit set concerning the use and conservation of ocean resources was the 200-mile limit declared as the **Exclusive Economic Zone** (**EEZ; Figure 12-5**). Within the EEZ the coastal state has jurisdiction rights to exploit, develop, manage, and conserve all resources in the waters or on the sea floor. This is a very significant declaration, as both the majority of fisheries harvest and the most hydrocarbon reserves under the sea floor are within 200 miles of the coast. Regarding fisheries, the Law of the

Figure 12-4   The international Law of the Sea establishes rights of *transit passage* through straits such as the Strait of Gibraltar between Africa and Europe.

Sea states that coastal nations have "sovereign rights for the purpose of exploring and exploiting, conserving and managing...the fishery resources contained therein." This resulted in the expulsion of foreign fishing vessels from the waters of many countries, unless some compensation agreement could be made. Not as widely recognized is the stipulation that there is also a responsibility for setting limits on harvest to establish an optimum use without overfishing and depletion of the fishery. (Unfortunately, this policy is virtually unenforceable but does establish a

moral obligation.) Coastal states are obliged to allow other states to harvest any surplus of the allowable catch that goes unharvested. Another responsibility that comes with these rights is the obligation to prevent pollution (another stipulation not easily enforced) and to facilitate scientific research in the EEZ.

A final compromise was needed to satisfy countries with mining and drilling interests on broad continental shelves extending beyond the 200-mile limit. These countries did not want to lose access to valuable hydrocarbon

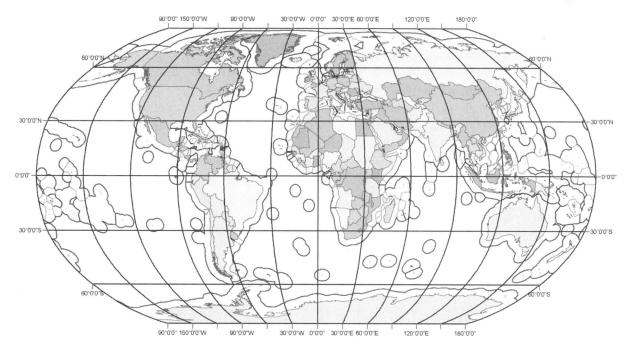

Figure 12-5   The Law of the Sea established global Exclusive Economic Zones (EEZ) extending to 200 nautical miles from shore, within which nations have rights to use and obligations to manage resource use.

and mineral resources in what they considered an extension of the continental land mass. The compromise allows countries the rights to resources on or under the seafloor out to 350 miles (650 kilometers) from the shore, or further if the continental shelf extends that far; but they must share revenue from mineral resources exploited beyond 200 miles.

The next major issue to be addressed by the Law of the Sea was the sharing of the deep-sea commons. Agreements were largely driven by the future potential for deep-sea mining of manganese nodules (see Chapter 8). Deep-sea resources beyond the EEZ of coastal nations are declared by the United Nations Law of the Sea to be "the common heritage of mankind." This was one of the most divisive issues, and countries had a difficult time coming to common ground on how to assure conservation but still gain future profits from these resources. An International Seabed Authority was created to develop a system and administer the rules. Sides were drawn between developed and developing countries. Developed countries argued that private mining companies, as a consortium, should be granted license by an international authority to exploit seabed resources. Developing countries argued that because these resources are the common heritage of all mankind, a public enterprise should be established to mine the international seabed. By compromise, a complex system has been set up to allow public and private enterprises to participate in future mining.

Another issue addressed by the Convention on the Law of the Sea was ocean pollution. The six main sources of ocean pollution identified by the Convention are: land-based and coastal activities; continental shelf drilling; seabed mining; ocean dumping; pollution originating from vessels; and atmospheric pollution affecting the ocean. The Convention states that coastal states are obligated to protect and preserve the marine environment and urges them to cooperate in setting rules and standards. Coastal states also are empowered to enforce their standards to limit pollution from sources such as dumping, land-based sources, or seabed activities. For pollution originating from ships on the high seas, the state under which the ship is registered and whose flag it flies must enforce international rules regardless of the ship's location. The International Seabed Authority is given the power to assess potential impacts of any deep-sea operation. States are to be held responsible for pollution caused by anyone working under their jurisdiction.

U.N. Conference on Environment and Development (UNCED)

The Conference on the Law of the Sea did not establish specific regulations for managing living ocean resources in international waters. It, therefore, has been largely left up to international commissions to establish agreements to limit fishery harvest and adhere to the general principles established in the Law of the Sea. Realization of this shortfall led the United Nations to expand its involvement in resolving ocean resource problems. In 1992, the United Nations Conference on Environment and Development (UNCED) placed a large emphasis on protecting the ocean's environment while developing its living resources. The program of action adopted by the Conference was one of "sustainable development," a concept that has driven and continues to drive marine fisheries management policy in international and national waters. In order to implement the program of action, several areas have been recognized where work is needed. One of these is the need to stop the depletion of fish stocks due to harvest levels that are too high and fishing fleets that are too large, both in national and international waters. Another is the unintentional bycatch and discard of organisms by fishers. Unfortunately, although efforts are being made in this direction, much remains to be accomplished.

An area that was not adequately addressed in the original Law of the Sea was management of fish stocks that move among waters of more than one nation and into international waters of the high seas; for example, tunas that congregate in U.S. national waters to reproduce but spend the remainder of their lives feeding in international waters (see Chapter 11), or swordfish that regularly move back and forth from the EEZ to international waters. Cooperation is expected. International fishers on the high seas still have rights to fish for these resources but are expected to take into consideration the interests of the coastal nations. These vague expectations were not always realized. For example, conflicts between Russia and the United States over harvest of Alaska pollock resulted in excessive harvest in the region between the nations' EEZs (a region commonly called the *donut hole;* see Figure 12-5). UNCED adopted agreements on these **straddling** and **highly migratory** fish stocks. The agreement obliged states to adopt a precautionary approach to fisheries exploitation by not taking the maximum the population can provide, but being conservative and cautious in setting harvest limits. This agreement also expanded the powers of coastal states to enforce the proper management of fisheries resources. The power of these agreements is limited by a lack of specific requirements and the lack of a specified definition for vague terms such as *precautionary* and *proper* that may be interpreted differently by different countries.

A Fish Stocks Conference was held from 1993 to 1995 to deal specifically with issues of straddling stocks. The resulting U.N. Fish Stocks Agreement (UNFSA) went into effect in 2001, requiring that fish stocks be managed by regional fisheries management organizations (RMFOs) comprised of members of coastal nations as well as distant-water fishing fleets. This resulted in the establishment of the North East Atlantic Fisheries Commission (NEAFC) and other such organizations to manage these fisheries. Although the situation has improved in some fisheries since then, these commissions typically have not worked in the best interest of sustainability and conservation (see Chapter 11).

Coastal Ocean Commons Laws: Case Studies

An estimated one billion people throughout the world rely on fish as their primary source of protein, and 200 million people are involved in fishing or fisheries-related industries. The great majority of these fish are harvested from waters within the 200-mile EEZ of nations around the world. Each nation is obligated by the Law of the Sea to monitor and regulate fisheries within their waters. Some have done a better job than others. The harvest of fisheries products is currently dominated by eight nations that account for more than half of the total fisheries harvest. These countries are, in order of fisheries production, China, Peru, Indonesia, the United States, Japan, India, Chile, and Russia (**Figure 12-6**). These nations have productive coastal waters and/or are capable of developing large fishing fleets that operate in international waters. Though much of the catch is by industrialized countries, about 25 of the 30 nations most dependent on fish as a source of protein are considered developing countries.

The proclamation of a 200-mile EEZ internationally rapidly changed the nature of international fisheries. Now any nation wishing to fish within another nation's EEZ would have to come to some compensation or trade agreement. Some nations moved to protect the rights of

their citizens by designating harvest within the EEZ to be the exclusive right of the nation's citizens. The removal of foreign fishers from coastal waters was largely viewed as an opportunity to properly manage the resource to avoid overfishing; however, this opportunity was rarely realized because fishers, with the support of their governments, began expanding their fishing capabilities rapidly to take advantage of the new opportunity.

Although the United States and Canada are not as dependent on fisheries products as a source of food as are many other nations, the popularity of seafood products has continued to increase since the 1970s after the full potential of coastal fisheries and the health benefits of seafood consumption were realized. No foreign vessel is allowed to take fish that citizens are capable of fishing for and have a desire to harvest. Expectations of expanded profits from fisheries and government supports of fishing vessels resulted in too many fishers vying to harvest the same fish. Fishers received loans and subsidies from the government in a rush to exploit coastal fisheries almost as if they were inexhaustible resources. Despite warnings from conservation groups and fisheries scientists, harvest continued to expand. In the United States alone, landings of 3 to 4 million metric tons per year through the mid-1970s expanded to over 7 million metric tons by 1989. Many of the coastal fisheries became unsustainable. Catch regulations were established, mostly annual species quotas based on concepts of maximum sustainable yield (see Chapter 11). Entry into the fishery was unlimited, which meant that any citizen had the right to enter the fishery and catch fish. Fisheries often became free-for-alls until quotas were reached. Wasteful fishing practices were often applied; for example, sorting fish at sea by throwing those that were not of the optimum size overboard in order to bring the fish that brought the best price to port. As quotas were reached increasingly earlier in the fishing season and each fisher's share of the catch was increasingly smaller, there was increased pressure to elevate the total allowable catch. In some fisheries, lengths of seasons were decreased continually in order to avoid increasing allowable catch. It became a race to catch fish before the season ended. An extreme example was the season for the Pacific Halibut fishery, which decreased to 24-hours with over 3,000 vessels in the fishery. In other fisheries, because of politics and the influence of people with a vested interest in the fisheries (e.g., boat and restaurant owners and seafood processors), quotas were set above what would be considered a sustainable limit.

To limit the free-for-all of the open access fisheries, some fisheries are now regulated by **limited entry**, restricting the number of fisheries allowed to harvest a particular species. Sometimes the fisheries are assigned **individual fishery quotas** (**IFQ** or **IQ**), splitting the total allowable

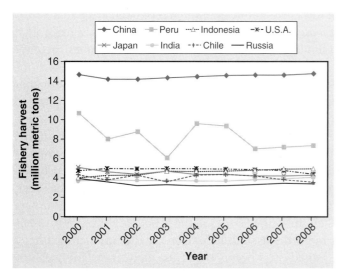

Figure 12-6 Fisheries harvest for the top fishing nations, comprising over half of the global harvest. Harvest by China accounts for over one-third of the landings of these nations. Peru's harvest reflects landings of Peruvian anchovies.

catch among those in the fishery. This simplifies management and enforcement by limiting the number of fishers. Implementing this kind of regulations has been controversial for many fisheries, because many believe it violates the rights of access to the common property resources of the oceans. The ability of any new fisher to enter the fishery becomes problematic if quotas can be transferred (or sold) when a person leaves the fishery. This **individual transferable quota (ITQ)** system also can lead to dilemmas over how to avoid a few large companies monopolizing the fishery by accumulating individual quotas.

U.S. Fisheries Management

As the United Nations was developing the framework for the Law of the Sea, the United States was working on its own national fisheries management policy to regulate fisheries in its coastal waters. There was a feeling among fishers that overharvest by foreign vessels in international waters off the coast was largely responsible for the U.S. overharvest and that current international laws did not adequately regulate that fishing. There was also the realization that coastal fish stocks were not being managed adequately and habitats critical to their survival were being lost. Many coastal communities were having economic hardships and were declining because of the loss of fisheries resources. Laws were needed to claim and protect what were considered national marine resources.

After extensive debate and discussion, in 1976 the U.S. Congress passed the Fishery Conservation and Management Act (FCMA; often called the Magnuson Act for the sponsor of the bill). The goal of this Act was "to provide for the conservation and management of the fisheries." Jurisdiction was declared over all resources to 200 nautical miles from the U.S. coast and a system was set up for establishing fishery regulations at the federal level through federal agencies (e.g., NOAA) and citizen-based fishery commissions; state jurisdiction was given over resources within three miles of the coast. Unfortunately, economic and political interests and efforts to protect common access rights led to overharvest of many fisheries. After several re-authorizations of the FCMA, management is now more science and ecosystem focused; however, a cumbersome system resulting in slow reactions to overharvest problems continues to plague management (**Box 12-1. Case History: Establishing U.S. Marine Fisheries Management Principles**).

Canada Fisheries Laws

Prior to 1960, Canada's fisheries were considered *underdeveloped* (that is, fish populations could sustain a larger harvest). As fisheries expanded during the 1970s and 1980s, new regulations were established, primarily based on seasonal quotas. Given exclusive rights to resources in the

EEZ by the Law of the Sea, Canada initially benefitted with increased harvest. Expansion of the fisheries eventually led to overharvest of the largest fisheries, however, the most notable being the Atlantic cod (see Chapter 11). Along the Canadian Atlantic coast, a slow reaction to the increasing problems of excessive harvest resulted in the collapse of the fisheries that resulted in the 1990s fishing moratoriums. But increases in shellfish populations resulted in increased fisheries for shrimp, snow crabs, and lobster that partially replaced the groundfish fisheries (although landing weights are much lower, the value of these fisheries is much higher than groundfish). On the west coast of Canada the collapse of salmon fisheries followed a similar pattern, and fishing became dominated by groundfish (e.g., cods, pollocks, and flatfishes).

In Canada, the federal government has had exclusive authority over fishery regulations since 1867. The Minister of Fisheries and Oceans was formed to exercise that authority under the Fisheries Act, establishing accountability for the protection and sustainable use of Canadian fisheries resources. The Oceans Act of 1997 defined and expanded the powers given through the Fisheries Act. Charges include placing limits on recreational, commercial, and aboriginal fishing, including setting catch, seasons, and license requirements. The Fisheries Act deals with managing fisheries, and also authorizes scientific research and sets up provisions to protect fish and aquatic habitats from pollution. Fish habitat is protected throughout the fish's life cycle from "harmful alteration, disruption or destruction." One of the complaints of the Fisheries Act is that its focus is reactive. It imposes penalties if there is damage to fish or habitats, rather than requiring planning or regulating activities that would damage fish populations. Also, limited resources result in law enforcement that is often inadequate.

Regulation of Canadian marine fisheries went through a similar cycle as in the United States, with harvests accelerating beyond sustainability due to an emphasis on technological improvements and fishery expansion. The goal of management following the declaration of the EEZ was to control fishing and allow overfished populations to recover; however, a focus on economic and social concerns (e.g., avoiding job loss and economic hardships) over conservation and rigid adherence to fisheries models contributed to the eventual collapse of the fisheries. Canada has been much more accepting of limited entry licensing of fishers than the United States. License buyback programs sometimes were used to help decrease the number of fishers. Limited entry licenses or ITQs can be transferred when a fisher leaves the fishery, and thus may obtain a great value. Although limited entry has made fishing more profitable to licensed fishers, it has resulted in numerous conflicts over rights to harvest. Because ITQs do not necessarily result in

Box 12-1   Case History: Establishing U.S. Marine Fisheries Management Principles

The passage of the U.S. Fishery Conservation and Management Act (FCMA) established rights to resources within a 200-mile limit from the coastline, called a **Fishery Conservation Zone (FCZ)**, several years before passage of the Law of the Sea established similar rights within the 200-mile EEZ for all nations. Upon enacting the FCMA, rather than immediately eliminate foreign vessels from declared FCZ waters, foreign fishing was phased out over several years. The intended outcome of conserving fisheries by protecting them from overharvest was not realized. As U.S. fishers rapidly moved in to replace foreign fishers within the EEZ, harvest of fisheries continued to accelerate.

The FCMA gave U.S. states the responsibility of fisheries management within three miles of their coastline, with the federal government establishing regulations in the remainder of the EEZ. States are required to work with the federal government to avoid contradictions in state and federal management. Various arrangements are made between bordering states and between federal and state agencies for fisheries that span two regions. For some fisheries the states are given primary authority over management of the fish stocks (e.g., salmon); for others the federal government has authority (e.g., most groundfish fisheries).

In federal waters, management was set up through regional fishery management councils. These councils were charged with preparing, monitoring, and revising management plans. The council setup allows representatives from the coastal states, the fishing industry, consumer and environmental organizations, and other interested persons to be involved directly in management. The goal was to give stakeholders in the fisheries substantial control over management. Eight regional councils were established, three along the Atlantic coast, one in the Caribbean (Puerto Rico and U.S. Virgin Islands), one in the Gulf of Mexico, two on the Pacific coast (including Alaska), and one in the western Pacific (Hawaii and U.S. island territories). The number of members on each council ranged from 7 (Caribbean) to 21 (mid-Atlantic; **Figure B12-1**). Voting members of each council include the head of each state's marine management agency, a representative from the National Marine Fisheries Service, and members appointed by the Secretary of Commerce from a list submitted by the Governors of each State in the region. This list is to include people who are "knowledgeable regarding the conservation and management, or the commercial or recreational harvest, of the fishery resources of the geographical area concerned." On the one hand, this system maximized local participation in managing the resource, but on the other, it

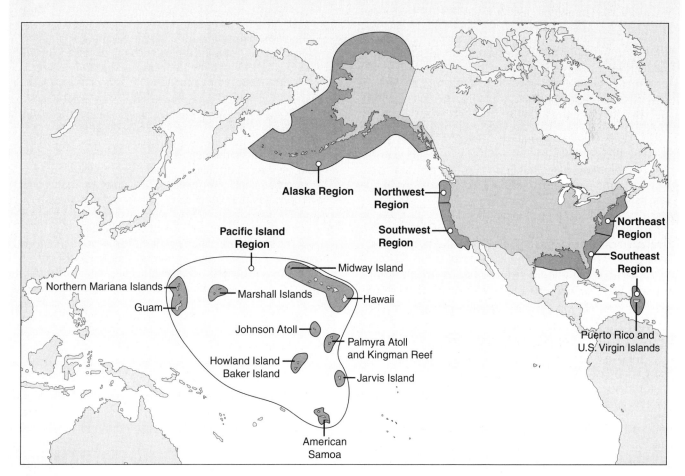

Figure B12-1   Fisheries management regions for the U.S. National Oceanic and Atmospheric Administration (NOAA).

allowed people that would benefit from a large fishery harvest to manage the fishery, which sometimes led to accusations of conflict of interest. With revisions of the Act, there have been attempts to keep people with a vested interest in a particular fishery from making decisions on that fishery.

The power to implement management plans is shared between the regional councils and a government agency. Ultimate control over management is through the U.S. Department of Commerce, working through the National Marine Fisheries Service, a branch of the National Oceanic and Atmospheric Administration (NOAA). Although the councils write and revise fishery management plans, NMFS provides scientific advice and makes sure the plans meet the legal obligations of the FCMA. It is not unusual for NMFS to return a plan to the councils for revision. The Department of Commerce is charged with the final approval of the plan. The plan is then implemented by NMFS and the U.S. Coast Guard. The U.S. Congress oversees the process by reviewing and reauthorizing the Act on a regular basis. The Congress also must approve funding of the regional councils, NMFS, and the Coast Guard.

The Fishery Conservation and Management Act was passed with the realization that specific management principles would be needed in order to allow harvest of fisheries in the EEZ. First, harvest would need to be sustainable in the long term, so that harvest could continue indefinitely with the resource renewing itself. The principle of Maximum Sustainable Yield (MSY) was developed based on the philosophy that catch should be as large as possible for economic reasons, while still allowing fish populations to sustain themselves. It was further stated that management would be regulated to achieve an optimum sustainable yield (OSY), the yield that would bring the most benefit (economically) to the fishers, while still preventing overfishing (for example, setting fish size restrictions so that fishers could catch fish at the most profitable sizes). Management was to be carried out to maximize the benefits of the fishery to U.S. citizens, but to harvest at such a level that the populations would continually renew themselves. It is further stated that the "best scientific information available" would be used to manage the fisheries.

These are worthy goals in principle, but, as discussed in previous chapters, they often have not worked in reality. The management strategy was based on the best available models, which we now realize was flawed in theory and unrealistic in application (see Chapter 11). Some of the underlying problems include: the best available scientific information is not always capable of predicting changes in fish populations; precisely monitoring and controlling the harvest is not always possible; and economic and political interests can influence management decisions so that harvest level is not optimized.

Although the FCMA governs management within the 200 miles of the coast for most species, there are some exceptions. Management of highly migratory predators (e.g., tuna

and swordfish) is to be handled by international agreements such as ICCAT (see below). Other countries may fish in the U.S. EEZ for these species if allowed through international agreements. Another exception is anadromous species, mostly salmon, that spawn in U.S. fresh waters and then move out to sea to feed. The United States claims rights to harvest these fish wherever they may swim, even if it is into waters of another nation (such as Canada) or into international waters. International councils are established to decide how to coordinate claims with other countries over straddling stocks.

Since the passage of the FCMA in 1976 there have been regular reviews and revisions. The initial priorities of the Act were to develop national fisheries and maximize **fishing capacity** (by increasing the number and/or size of vessels). By the early 1990s, however, many fisheries were being overexploited and fishery capacity was excessive. Major amendments were made in 1996 as the Sustainable Fisheries Act, which establish a priority of controlling catch and limiting fishing capacity. This required better conservation of living marine resources. First, it stipulates that the biology of the fish species will be considered in determining fishing rates. For example, a fish that has a slow rate of reproduction may not be able to withstand as large a harvest, or a fish that matures late in life must be protected longer before harvest begins. A second requirement is that overfished populations are to be allowed to rebuild. This means that fish harvest must be stopped or reduced on populations that are deemed to be overfished. Bycatch rates must be determined and minimized on all ocean wildlife. Essential fish habitat is to be defined and protected throughout the fish's life cycle. One of the problems with previous management plans was that they focused only on the adult stage of fishes and did not consider the ecosystem where they resided. As discussed earlier, fishes have widely varying habitat requirements at different stages of their life cycles. Fishery management is now to include ecosystem and habitat issues. One way this partially can be achieved is through no-fish Marine Protected Areas. Additional standards include minimizing adverse economic impacts on fishing communities. Individual Fishing Quotas have been established for some fisheries, but no new IFQs were established from 1996 to 2006 out of fear that local fishing communities would be economically affected.

A 2007 reauthorization of the FCMA required tougher harvest restrictions, including setting annual harvest quotas below those needed to rebuild overfished stocks; IFQs were made more difficult to implement. Research programs were set up to study methods for integrating ecosystem consideration into management. Although ecosystem-based management is not required, it is a movement away from problems of management based on old principles. The U.S. system controlling fisheries remains slow, however, and too often is driven by reaction to crises and short-term biological or economic concerns.

lower total catches, their success in stopping overharvest has been limited.

In 1996, Canada instituted the Oceans Act, considered the first such national act in the world to deal with comprehensive ocean management. It operates on the principles of a precautionary approach, integrating ecosystem-based management to support sustainable fisheries. This requires setting fishery quotas below what is considered sustainable based on traditional single-species models, but there is no clear definition of how it will be achieved. Although implementation of some policies has been slow, plans are moving forward to develop an extensive network of marine sanctuaries and ecosystem-based principles are being applied to some fisheries. Because a relatively large percentage of Canadians depend on fisheries in some way economically, however, there continues to be a great social pressure to expand employment within fishery sectors.

European Fisheries Policies

Many European nations have a long history of fishing as well as a long history of independence. Borders of European nations extend from the Baltic and North Seas through the eastern Atlantic Ocean and into the Mediterranean and Black Seas. Of the 27 member states of the current European Union, 22 have access to the sea (see Figure 12-5). Cumulative fisheries harvest by these nations is second only to China. Although each nation has worked to manage its own fisheries, because adjacent countries often vie for the same fishery resources, cooperative management measures are necessary. The E.U. Common Fisheries Policy (CFP, discussed below) originated with the 1957 Treaty of Rome as part of the agricultural policy agreed upon by six nations. The Policy protected fishing rights in nearshore coastal waters for local fishermen, but agreed that otherwise fishers would have access to waters of other European nations. As a result of conflicts such as those resulting in the Cod Wars (see Chapter 11), the North East Atlantic Fisheries Commission (NEAFC) was established by 14 nations in the 1960s. Their conventions and eventually the Law of the Sea came to guide E.U. fisheries management. Management objectives were to consider conservation measures, science-based management, marine reserves, and limiting harvest. As with North American fisheries, however, management was largely driven by the desire to increase harvest for economic and food production goals, with little concern for sustainability or ecosystem health.

After the passage of the Law of the Sea, Europe followed suit with other countries to claim jurisdiction over fisheries out to 200 miles. Negotiations resulted in the Common Fisheries Policy (CFP), established in 1983, as the European Union's instrument for management of fisheries and aquaculture. The E.U. was given jurisdiction over all fisheries matters, while also considering biological, economic, and social dimensions in managing the fisheries. The charges of the E.U. CFP are divided into four main categories. The first of these is the conservation of fish stocks. The CFP is to set quotas on fish harvest that allow a sustainable yield and divide the quotas among the member states. Each state has flexibility in terms of how they distribute the catch of their quota. Scientific fish stock assessments were carried out in the northeast Atlantic to guide these decisions, but these typically were not available in the Mediterranean.

The bycatch problem also is recognized and there are to be restrictions set on fishing gears to avoid bycatch; for example, restricting net mesh size or outlawed fishing methods are enforced throughout the E.U. The CFP is charged with regulating fishing vessels, ports, and processing plants. Marketing of fisheries products is incorporated also and the CFP is charged with organization of the E.U. common fisheries market. Finally, the CFP handles fisheries agreements with other countries, including negotiations through international fisheries organizations, for example, for highly migratory species. A major issue for the E.U. since the passage of this policy has been fishing overcapacity, as too many vessels have been harvesting the resources, making it difficult to limit catches and still ensure profitability for the fishers. In response, guidelines were set for fishing fleet limits for each E.U. member state in the 1980s.

In the early 21st century, efforts were made to initiate an ecosystem-based approach to management. For example, quotas were established for multispecies fisheries based on species catch combinations rather than single species independently. Long-term projections have been made for fishery stocks to establish sustainable fisheries. Fishing subsidies have been reduced, especially those given in the form of grants to modernize or construct new fishing vessels. It appears likely that the necessary methods for sustainable management of fisheries are now in place in the E.U. if measures can be taken to return the fish stocks to a healthy starting point for future management.

China Fisheries Policies

Many of the dominant fishing nations in the world are Asian nations that border on the western Pacific Ocean. The 21 nations comprising the Asia-Pacific Economic Cooperation (APEC) have almost 75% of the world's fishery harvest capacity and participate in the majority of the global fisheries trade. These nations also have the highest per capita fish consumption, reflecting the importance of fishery products as a source of protein. Because of this heavy fishing pressure, commercially harvested species in this region are estimated to be only 10% to 30% as abundant as they were in the 1960s. The most industrialized of these countries (e.g.,

Japan, China, and Russia) do not limit their harvest to the eastern Pacific and have global fishing fleets.

Fisheries landings by China have increased more rapidly than for any other country in the past 20 years. China's reported fisheries landings more than doubled through the 1990s. By 2002, reported annual landings were over 45 million metric tons, the largest production in the world. Historically, China also has depended heavily on aquaculture, which increased more than fivefold from 1990 to 2004; however, much of this production is for freshwater species such as tilapia and carp. Almost 300,000 fishing vessels are licensed to fish in China's waters (**Figure 12-7**) and over 1,600 fish on the high seas in the Pacific, Atlantic, and Indian Oceans.

National laws obligate China to implement management regulations adopted by international regional management commissions. China has ratified the Law of the Sea Treaty and instituted corresponding recommendations within its national fisheries policies. Management of China's fisheries are governed by the Fisheries Law of the People's Republic of China, adopted in 1986 and managed by the Bureau of Fisheries. The Fisheries Law was formulated to enhance protection and develop reasonable utilization of fishery resources as well as develop aquaculture. It includes developing fisheries on the high seas and managing fishing capacity in national waters, in part by limiting the number of licensed fishers. Restrictions on fishing gears and methods are established, and protected areas can be designated. Fees from those profiting from the fishery can be applied to increase fishery resources. One controversial article of the China policy is the distribution of "material awards to units and individuals who make outstanding

contributions to the increase . . . of fishery resources." This not only encourages excessive overharvest but also has purportedly led to exaggerated landings reports. Overreporting of harvest led to overestimates of global fishery landings when the data were compiled by FAO in international landing statistics (see Chapter 11). Revisions of the Fisheries Law in 2000 specified that catch limits would be based on Law of the Sea agreements and used total allowable catch (TAC) quotas to manage fisheries. Additional requirements were made for permitting and logging catches for vessels operating on the high seas.

China has exhibited a similar trend in overfishing as other nations with the increased development of fishing technology. Stock assessment capabilities are poor and overfishing is typically identified by a sharp reduction in catch-per-unit-effort (CPUE). The response to overfishing in the 1990s was to implement seasonal closures of fishing in specified regions. The South China Sea is China's most heavily fished coastal region, and is a region with numerous potential conflicts due to competing claims among neighboring nations (**Figure 12-8**). A complete closure of fisheries annually for two- to three-month periods was instituted in the hope of allowing population recoveries. Beginning in 1998 China established a *zero-growth* policy for its fisheries, and in 2002 set a *minus growth* policy, removing tens-of-thousands of vessels and hundreds-of-thousands of fishers through vessel buyback and job transfer programs. Other measures included restrictions on net mesh size and protections of spawning stocks and juveniles. Fines were increased to discourage illegal fishing activities believed to be one of the greatest impediments to sustainable management. Efforts also are being made to comply with U.N. agreements

Figure 12-7   A fishing port on an island in the South China Sea near Hong Kong, 1983. Hundreds of thousands of vessels are currently licensed to fish in China waters.

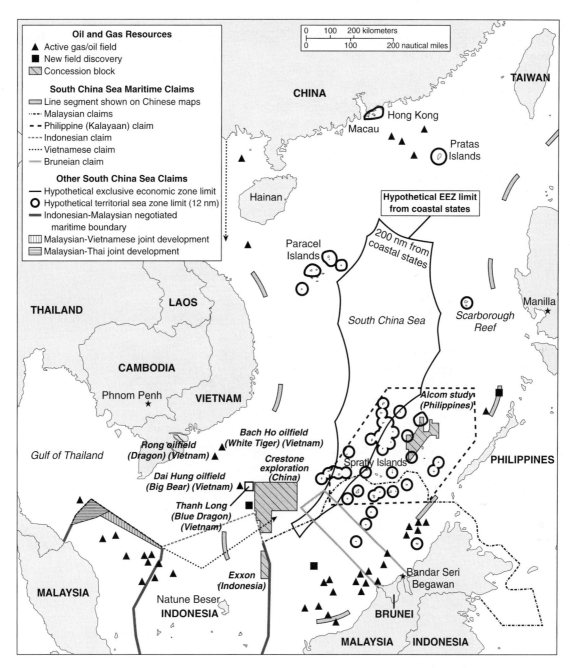

Figure 12-8 The South China Sea, indicating competing claims to the commons.

by reducing shark and seabird bycatch and eliminating high seas drift net fishing.

Japan Coastal Fisheries Regulation

With limited land available for agriculture production, throughout its history Japan has relied heavily on fisheries products. In 2002 the use of fisheries products in Japan was over 11 million metric tons, with a consumption of over 37 kilograms per capita, one of the highest levels in the world. Economic and industrial advancements in the second half of the 20th century allowed Japan to develop one of the most modern fishery fleets in the world as coastal fisheries expanded (**Figure 12-9**). Overfishing in Japan's coastal waters resulted in a large increase in dependence on imported seafood products; in 2010 imports provided almost 50% compared to 30% in 1992. Fisheries in international waters have continued to keep Japan's harvest levels relatively high, but the declaration of a global EEZ removed Japan's vessels from many coastal fisheries of other nations.

(a)

(b)

Figure 12-9 Japan has been a dominant participant in artisanal fishing in national waters **(a)** and large-scale fisheries in international waters **(b)** since the mid-1900s.

The Fishery Law of 1949 established the general principles of fisheries management in coastal Japan and is still in force. It gives the local fishers much of the power in establishing policies in coastal waters using a system of fishing rights and licenses. All uses of marine coastal resources in a region must be coordinated through various agencies, from the federal to the prefectural to the local level. This system has resulted in some innovative fishery management organizations where decisions about fishing rights and allocations are made by a group of fishers from local fisheries cooperatives. Although the decisions are based on science provided by the government, and the government must approve the decisions, they are implemented from a bottom-up approach (originating locally), rather than a top-down approach (originating from the federal government).

Unfortunately, the traditional form of local fisheries management in Japan is not well coordinated with ecosystem-based management, and thus may not be sustainable in the long run. The defined goal of management aims, formed during hard times soon after World War II, is developing fisheries productive to enhance the food supply and assist fishers economically. The Basic Fisheries Act, passed in 2001, established new rules to address some of the key problems in Japanese fisheries, including ecosystem conservation. Management decisions are typically not inclusive of interests of other users of the ecosystem, such as environmentalists and recreational users, however, and there is little monitoring of the ecosystem other than fisheries species.

An innovative exception to traditional fisheries management in Japan was recently implemented in northern

Japan in a region designated as the Shiretoko World Natural Heritage Site that could serve as a model for marine resource management. This is a diverse ecosystem including fishes, sea lions and seals, and sea birds including sea eagles. A committee has been established to coordinate interests of fishers, environmentalists, scientists, land and river managers, education, tourism, and other interests. Fishers monitor the ecosystem and provide data for analysis through government agencies. Based on scientific advice, fishers incorporated their recommendations into a Marine Management Plan. The community-based format of this strategy has resulted in voluntary adjustments in fishing gears and catches to enable sustainability and environmental protections, including the establishment of no-fishing zones.

Concerns about Japan's fisheries have been expressed following the March 11, 2011, earthquake and tsunami. A large portion of marine fishing vessels and ports were lost in a region that accounts for about 20% of Japan's marine fisheries production. This could reduce the fishing pressure locally on marine fishery populations. How much or for how long this will affect Japan's volume of fish harvest or how rapidly affected fisheries will recover is unknown. Although long-term effects on fish populations are not expected as a result of radiation contamination at levels released, contamination in some coastal fishes following the incident resulted in a temporary closure of fishing for human health concerns and fishes are being monitored due to concerns over potential bioaccumulation.

Southeast Asia

The Southeast Asian region comprises 11 separate nations, including 6 of the top 10 fishing nations in the world: Indonesia, Philippines, Myanmar, Thailand, Vietnam, and Bangladesh (**Figure 12-10**). The large area of coastal seas in this region has resulted in a great dependence on fisheries for thousands of years. Marine fishery populations currently support about 10 million fishers, and possibly 100 million people depend directly on the fisheries. Combined harvest comprises about 17% of world fisheries landings (almost 14 million metric tons). Management mechanisms in this region were established based on small-scale inefficient fisheries. Therefore, as technological developments increased fishing capacity, virtually all fisheries in the region became fully exploited or overharvested. *Fishing down the food web* (see Chapter 11) has become a major issue as large fish disappear and smaller species or juveniles support many fisheries, a pattern typical of tropical reef-based ecosystems (see Chapter 5). Some successful community-based management plans are operational on a small scale, but most have eroded over time as the region has become more dominated by larger-scale fisheries.

Each nation has its own management system; however, Southeast Asian nations are largely underdeveloped and resources are rarely available for strong regulations or protection of national boundaries, corruption is common, and the availability of scientific data is limited. Functionally, many fisheries in this region are open-access for larger commercial fishing boats that regularly

(a)

(b)

Figure 12-10 Coastal nations of Southeast Asia have been dependent on local fishery resources for thousands of years, and subsistence and artisanal fisheries, **(a)** as evidenced by these fishing boats in Indonesia, continue to provide an important source of food and employment. **(b)** Local fish markets such as this one in Bangkok, Thailand are prevalent.

avoid restrictions on catch limits or gear use. Fishing has become largely unprofitable, especially for poor coastal communities. The establishment of closed seasons and delineated fishing zones rarely has been successful due to poor enforcement.

In the early 21st century, movements have been made by some nations to develop a co-management scheme, whereby national governments cooperate with each other in managing fishery resources. Many Southeast Asian nations have ratified the Law of the Sea Treaty and are members of multinational organizations. The most influential is probably the Asia-Pacific Fisheries Commission (APFIC), sponsored by FAO. Most agreements are still in the development stage, however, and it remains to be seen if they can overcome historic management problems in this region. Community-based management is being encouraged, but competition with large-scale fishers targeting the same species remains problematic. Efforts to deal with illegal, unreported, and unregulated fishing (IUU) are being considered internationally, but adequate mechanisms have yet to be established. One of the most promising developments has been the incorporation of an ecosystem management approach in some regions through the development of Marine Protected Areas and fish sanctuaries, supported by APFIC. In order to establish ecosystem recovery, other environmental issues will need to be addressed regarding coral reef, mangrove, and sea grass ecosystems (see Chapters 4–6).

India Fisheries Development

India's fisheries have always provided a critical source of protein for its citizens, and many of these fisheries are still for artisanal or subsistence harvest (**Figure 12-11**). As with Southeast Asian nations, the transition from traditional

Figure 12-11   In India, many subsistence fishers have been forced to move away from traditional fishing methods and modernize to motorized vessels to compete with large trawl fisheries and aquaculture.

fishing methods to motorized and modernized vessels from the 1960s through the early 2000s has been devastating to fisheries that are governed by open access rights with little formalized management. Most fishing crafts continue to be unmotorized and depend on haul seines, gill nets, and hook-and-line; these have a difficult time competing with motorized vessels pulling trawl nets. In order to become more competitive, traditional fishers have begun to adopt new fishing methods that encouraged an expansion of coastal fisheries. Marketing of bycatch by trawlers have flooded the market for small fish traditionally sold by local subsistence fishers; large protests resulted in some protections of local community fishers. This has become a true tragedy of the commons.

Management of fisheries in Indian waters is divided between national and state governments. Management in the EEZ is divided among several national agencies and states are responsible for waters out to 12 miles. Beginning in the 1950s a series of five-year plans encouraged excessive harvest. This resulted from an emphasis on economic and food production concerns over conservation of sustainable fisheries, including government subsidies put toward enhancing ports and modernizing fishing vessels. Traditional Indian fishers formed the National Fishworkers Forum to protect their interests in competing with trawl fishers, but have had limited success in influencing government legislation. As overharvest became prevalent at the beginning of the 21st century, plans began to consider conservation measures. In order to protect the rights of subsistence fishers but remove them from overfished coastal waters, support was given for upgrading of vessels to participate in offshore fishing. Local fishers, however, are now concerned about recent trends to support development of aquaculture at the expense of wild fisheries harvest.

Australia and New Zealand

Although landings are not as great for Australia and New Zealand as for the more populated Asian countries in productive tropical waters, marine fisheries harvest is an important component of their economy. Coastal fisheries are supported in part by the Great Barrier Reef system; however, these reefs are not as open to unrestricted harvest as other western Pacific tropical reefs (see Chapter 5). Harvest of deep-water species, such as orange roughy, has increased dramatically since the early 1990s (see Chapter 8).

Management of fisheries in Australia was reorganized under the Australia Fisheries Management Authority (AFMA) in the late 1980s due to problems with open-access fisheries. The new management model mandated ecologically sustainable development of fisheries and a focus on the use of ITQs for most fisheries. By 2004, this model was questioned when assessments indicated that the number of overfished stocks was increasing. Many of these were trawl-based fisheries, including deep-water orange roughy. An assessment of the fishery showed that economic decisions had been driving management by avoiding precautionary decisions. Ecologically risky decisions, such as excessive quotas, resulted in overfished populations.

A new Harvest Strategy Policy (HSP) was established in 2007 to encourage conservation by applying risk assessment and decision rules to make management decisions, something relatively new in fisheries. Fisheries capacity was reduced by a government buyback program. Although overall landings have declined, the boats remaining in the fishery have exhibited higher catch rates and better fish prices as a result of reduced competition. It is too early, however, to judge the success of this policy until enough time has passed for populations to rebuild.

In New Zealand, recent policy changes have included a heavy emphasis on ITQ management since the 1980s. Fisheries are primarily deep-water fisheries dominated by large-scale fishing vessels. Fish stocks are relatively healthy, though overfishing is an issue. Conflicts have resulted due to a management system dominated by the national government working through the Ministry of Fisheries. Current efforts include allowing more involvement of fishers and the fishing industry in decision making as well as incorporating a more ecosystem-based approach to management.

West Africa

The marine coastal regions of West Africa house some of the most productive waters of the world. This is largely due to oceanographic conditions that support large upwelling, similar to the Pacific coast of South America (see Chapter 1). The fisheries are productive, the biodiversity rich, and there are rich oil and gas reserves and precious minerals beneath the sea floor. Fishing is carried out in this region by 22 coastal nations as well as about 50 foreign nations. More than 4.5 million people in the region depend on fisheries in some way for their livelihood, and fish provide about 30% of the protein supply in this region. Numerous subsistence fishers comprise some of the poorest communities. This region is plagued by large population growth, resulting in increased needs for fish products and increases in local fisheries. Overharvest, however, is largely to support increasing international trade of the most valuable fishery products.

Local, small-scale fishers harvest primarily in a region within five miles of the coast using traditional labor-intensive methods (**Figure 12-12**); however, technological advances (e.g., motor-driven boats and GPS) have resulted in expansions of this sector of the fishery. The industrial fishery employs mid-sized trawlers and purse seiners in offshore waters; however, it is common for these fishers to move into coastal waters reserved for local fishers. Many nations have signed agreements with European nations, allowing them to fish in this region on productive stocks of migratory predators such as tunas and swordfish, pelagics such as sardines and anchovies, and groundfish. Since 1970, approximately 25% to 50% of the harvest has been by foreign fishing fleets. This is in contrast to the strategy used by other nations to reserve fisheries in the EEZ for their own citizens, and can lead to excessive harvest to the benefit of the government rather than the fishing public (though there are efforts in some nations to channel some

Figure 12-12 Expanding populations and competition among local artisanal fishers, such as these in a crowded port in Ghana, and international industrial fishers has led to overexploitation of many fishers.

of the profits back into the local fisheries). Over 70% of the fisheries stocks are not considered fully exploited or over-exploited. Problems with habitat destruction and pollution are also important concerns.

Management of fisheries in West African nations is largely controlled by national governments in a top-down approach, with little input from local fishers; resources for stock assessments are largely unavailable. Management regulations established through national fisheries laws, therefore, are rarely science-based, and there are few species-specific management plans. Small-scale fisheries are poorly regulated and often operate as open-access fisheries. There are intense social pressures to expand fishing as a solution to poverty, resulting in many fisheries having twice the fishing capacity than needed to support a sustainable fishery. Management is enhanced in some communities through application of their own restrictions, often based on tradition or superstition, that serve to limit harvest in some regions, at certain times of the year, with some gear types, and on some size classes or species. Although tuna harvest is regulated through ICCAT, undersized fish are commonly harvested in coastal fisheries. Restrictions on fishing techniques and gear types are often poorly enforced, and IUU fishing is prevalent. Some promising recent developments include the incorporation of Marine Protected Areas into management, and there are now at least 15 separate MPAs spread among waters of 7 nations. Some nations also have begun accepting assistance from the European Union to develop management plans. Efforts by international organizations such as the FAO and NGOs such as the World Wildlife Fund are establishing some agreements to address issues of poverty and mismanagement in the region that lead to overfishing and environmental impacts.

Western South America

Another coastal region with a large contribution to global fish harvest is the west coast of South America, largely dominated by the Peruvian anchovy fisheries of Peru and Chile, as discussed in Chapter 11. In separate coastal artisanal fisheries of Chile, some of the largest inshore fisheries are for invertebrates, the most important being loco, *Concholepas concholepas* (a muricid snail similar to abalone). Problems of overharvest in this poorly controlled open-access fishery led to the establishment of a system based on individual permits and catch quotas. When this system failed to control overharvest, a new management regime for coastal artisanal fishers was instituted based on a type of community-based management, whereby a limited number of fishers are given permits through a local organization to work in a specific region to fish using a specific method. These local organizations are required to complete regular population surveys, coordinated by a consultant, upon which their management plans are based. Although regulations are established by the national government, the local organization imposes its own rules on access, penalties for violations, and distribution of benefits among members. This type of community-based management is called **territorial use rights in fisheries** (TURF). Although there has been resistance to change in some regions and survey and report quality is sometimes questionable, TURF-managed regions have generally shown less overharvest than neighboring areas. Further refinement of this management system has been recommended, including better coordination among adjacent management regions; however, it has been successful in bringing local fisheries management in coastal Chile under control. This provides a possible model for application in other fisheries; however, care must be taken in transferring this system to other regions without the social, economic, ecosystem structure to make it successful.

Other Regions

Other coastal regions of the world with lower biological productivity cannot sustain as large a harvest as the southeastern Pacific, eastern Atlantic, or tropical western Pacific regions, which support fisheries in the above examples. These regions, thus, are less likely to be dependent on coastal fisheries. Similar local problems with sustainable management of fisheries and conservation of ecosystems are common globally, however, as described in previous chapters. Management of tropical fisheries associated with reef habitats has been hampered in most regions by ecosystem impacts in conjunction with inadequate harvest restrictions, most notably in the western Pacific and Caribbean (see Chapter 5). Efforts are being made throughout these areas to protect habitats and incorporate sustainable fisheries in these regions through gear limitations, stricter harvest regulations, and establishment of community-based management (**Figure 12-13**). Sustainable fisheries in intertidal areas and estuaries has been hampered by loss of seagrass, mangrove, salt marsh, and kelp ecosystems in conjunction with poor management and protection of fisheries (discussed in Chapters 3–6).

The case studies presented here show a common pattern of inadequate legal restrictions and poor enforcement of those that exist. Too often economic, social, and political considerations have resulted in the allowance or encouragement of excessive fisheries harvest in conjunction with poor environmental protection. There has been a gradual realization that unlimited free access to the common resources of coastal marine environments is no longer a valid option for sustainability. Increases in coastal populations have resulted in expansion of fisheries to support

Figure 12-13 In Jamaica and other Caribbean nations, community involvement and the establishment of no-fishing sanctuaries are becoming more prevalent in an attempt to recover overfished coastal fisheries.

jobs and coastal food production at levels above or near the point of causing ecosystem collapse.

Some changes made in fisheries governance in localized coastal regions are promising. A well-monitored system of local community-based management can be successful. This success is likely because smaller groups are better able to work together for the common good of the long-term stability of "their" ecosystem and fishery. Ecosystem-based management, however, appears to need an outside influence from those with conservation interests outside the fishery, and requires a government that is willing to relinquish the power and profits to be gained immediately from excessive harvest of coastal resources for the greater goal of future sustainability. In conjunction with a community/ecosystem-based approach, marine sanctuaries can enhance biodiversity and possibly feed neighboring ecosystems with excess production. The greater acceptance of sanctuaries in coastal regions encourages locals to provide policing for enforcement of harvest restrictions. Sanctuary protection may be successful because a total prohibition of fishing activity in a relatively small, protected area is a black-and-white law that can be more readily monitored and enforced than vague harvest restrictions. As one moves from the coastal commons into the international commons of the high seas, the scale changes dramatically. Enforcement becomes problematic and the tragedy of the commons is prominent. International agreements are needed on a multinational or even global scale.

Global Environmental Agreements

Because of the interconnectedness of all ecosystems, environmental impacts on the ocean environment cannot be separated easily from those affecting the Earth as a whole. National laws dealing with water pollution in freshwater environments and habitat impacts in nearshore environments are critical for conservation of coastal ocean ecosystems (see Chapters 3–6). Even international laws not intended to address environmental issues often have unintended effects on the ocean environment (**Box 12-2. Case Study: Conflicts Between Trade Laws and Conservation**). In the open ocean, atmospheric pollutants are particularly influential, especially regarding persistent organic pollutants (POPs), metals such as mercury, and carbon dioxide and other *greenhouse gases* (see Chapters 1 and 7). Several attempts have been made to establish global agreements to minimize global environmental impacts and limit atmospheric pollutants, with mixed results.

One of the first of global attempts to establish meaningful agreement for environmental protection at the global level was the United Nations Conference on Environment and Development, commonly called the Earth Summit, held in Rio de Janeiro, Brazil in 1992. Its primary goal was to reconcile worldwide economic development with environmental protection. This was the largest gathering of countries in history, with representatives from 172 countries (including 108 heads of state). The major topics of discussion were biodiversity, global climate change, and sustainable development. Major achievements were a Climate Change Convention, which led to the Kyoto Protocol, and a Convention on Biological Diversity. Follow-up summit meetings were held in New York in 1997 and in Johannesburg, South Africa in 2002, broadening the focus on issues such as biodiversity, water, energy, health, and agriculture. Unfortunately, participation declined with each successive meeting and the agreements established were considered weak. Although some agreements were made to address the sustainable use and conservation of ocean resources and habitats, specific commitments were lacking and conservation measures weak. The greatest conflicts hindering agreements were between larger industrialized countries and poorer developing countries, which were reluctant to accept environmental restrictions without increased economic aid.

The **Kyoto Protocol** was adopted in Kyoto, Japan in 1997, and was an agreement primarily addressing reducing greenhouse gas emissions that cause air pollution and lead to global climate change. The Protocol has been signed and

Box 12-2   Case Study: Conflicts Between Trade Laws and Conservation

International agreements made for economic or social reasons sometimes can become controversial because of their indirect impacts on environmental protection or conservation. In some cases these impacts result in a direct conflict with international environmental agreements. Recent global trade agreements have resulted in some conflicts with marine conservation agreements. The Global Agreement on Trades and Tariffs (GATT), established in 1943 as an international agreement on trade among nations, was replaced in the 1990s by agreements made within the World Trade Organization (WTO). The WTO is the only global international organization dealing with rules of trade between nations, and most of the world's trading nations have signed and ratified WTO agreements. The WTO does not have specific agreements dealing with environmental issues, but a number of agreements have provisions that deal with environmental concerns, including ocean resource issues. One complaint by environmentalists has been that WTO trade agreements do not consider conservation of the environment and sustainable development. WTO has emphasized that it is not an environmental agency and that agencies that specialize in environmental issues should deal with such conflicts through multilateral environmental agreements already in place.

A series of conflicts between the United States and the WTO led up to trade-and-the-environment controversies. The first of these was the tuna–dolphin dispute (see Chapter 9). The United States wanted to ban the import of all tuna from Mexico unless they met the dolphin protection standards set in U.S. law for tuna purse seine fishers. In 1991, Mexico filed a complaint through GATT. GATT ruled that the United States could not embargo imports of tuna products because Mexican regulations on tuna fisheries did not meet U.S. regulations, arguing that one country could not take trade action in an attempt to enforce its own laws in another country. Conservation groups protested the ruling. But GATT officials argued that this violated free trade laws and—if allowed—it would open the door to a flood of issues in which one country wants to force another to follow its environmental laws. The disagreement would be better settled through direct negotiations between the countries on that specific issue. The United States was allowed to use *dolphin-safe* labeling to let the consumers choose whether to purchase imported tuna caught using methods considered harmful to dolphins. This functionally stopped the importation of canned tuna products to the United States that could not be certified as dolphin-safe, due to the lack of a market for those products.

The second controversial issue involving trade and ocean resources was the shrimp–turtle dispute (see Chapter 9). In this case the United States attempted to prohibit the import of shrimp from Thailand, India, Malaysia, and Pakistan unless they could certify that shrimp were being harvested with turtle-safe methods, including the use of turtle excluder devices (TEDs;

see Chapter 9), as was required of U.S. shrimpers under the Endangered Species Act. These four Asian countries brought a complaint against the United States and won their case. In this case, the WTO stated that under WTO rules a country can take trade action to protect the environment or endangered species and refuse trade on a shipment-by-shipment basis. The United States lost the case because they were discriminating among WTO members by providing other countries, mainly those in the Caribbean, with assistance and a longer transition period to start using TEDs. The shrimp trade controversy has continued as U.S. imports of inexpensive shrimp produced by aquaculture in Asian and South American countries accelerated and began to outcompete wild harvested U.S. shrimp (see Chapter 11). As prices declined dramatically, imports reached greater than 90% of the U.S. shrimp supply by 2008. The U.S. Department of Commerce ruled that shrimp were being *dumped* at prices below fair market value and that a duty could be imposed on shrimp exported from those countries (but this did not reduce the demand for shrimp imports). The controversy between shrimp imports and TEDs was reopened in 2010 when the United States banned, temporarily, imports from Mexico because they were not compliant with U.S. and Mexico turtle protection laws regarding the use of TEDs.

At first glance these decisions send a mixed message about how the WTO plans to deal with trade disputes related to environmental protection. The WTO has been hesitant to incorporate its own environmental standards on trade, stating that it will not impede actions based on other environmental agreements. If the countries have not agreed on environmental restrictions, however, WTO members cannot impose trade restrictions on a product purely because of the way it has been produced (including how fish are caught); and a country cannot impose its own standards on another country. Pressure on WTO, however, eventually resulted in the creation of a Trade and Environment Committee to study the relationship between trade and the environment, and make recommendations about any changes that might be needed in trade agreements. Efforts are being made by WTO to negotiate with nations to reduce fisheries subsidies that lead to harmful fishing practices (see Chapter 11). There are still no specific WTO agreements on trade and the environment; however, the WTO states that members can take trade-related measures to protect the environment if they prove they are not misusing measures to achieve protectionist goals.

The simplest way to address trade issues may be through ecolabeling (see Box 12-3, Conservation Concern: What Seafood Should I Eat?). If imported seafood products produced in an environmentally safe manner can be identified through ecolabeling and the public is educated on the issues, pressure will be applied to fishers and fish farmers internationally to work toward improvements to gain certification that will increase the marketability and value of their products.

ratified by 187 nations, with the notable exception of the United States, the second largest producer of the world's greenhouse gases (see Chapter 1). A major disagreement concerns negotiating a consensus between developed and poorer developing countries over how much of the burden of change each should carry. It has been difficult to balance two extreme philosophies. Developing nations tend to argue that the largest polluters should bear the brunt of the responsibility for change, while industrialized nations are likely to argue that all nations should cut back on emissions by a similar percentage.

Possibly the most successful of the international agreements on global climate issues was the Stockholm Convention on Persistent Organic Pollutants, established in 2001 (see Chapter 7). Although the largest industrialized nation—the United States—has not ratified the Treaty that was established through this convention, national laws have established similar restrictions. This agreement placed a global ban or restrictions on the use of some of the persistent pollutants that are most harmful to marine ecosystems, with the flexibility of adding new chemicals to the list in the future.

12.3 Marine Environmental and Conservation Organizations

As has been demonstrated in previous chapters, addressing complex marine conservation issues is not easy. Due to the complicated legal, political, and social concerns, there are numerous national, international, and non-governmental organizations working to monitor, enforce, and encourage the management and conservation of marine environments and resources. Multiple groups have divergent interests in taking, using, viewing, or protecting these same resources because they are considered of scientific, conservation, commercial, recreational, or aesthetic value. Many conservation and management issues are addressed through **government organizations**, which are those legally affiliated with some political entity, most typically a national agency. **International governmental organizations** (IGOs; also called intergovernmental organizations) comprise multiple nations that work together to establish agreements and provide education. The most influential of these, with representation by most sovereign states of the world (a total of 192 member states), is the United Nations (U.N.). The U.N. works to facilitate international cooperation on legal, social, economic, and security issues. The U.N. includes several subsidiary organizations that specifically address environmental and marine issues. **International commissions** are bodies established by **international conventions** or agreements, set up and run by representatives of member nations to manage a specific resource (for example,

whales, tunas, or salmon). **Non-governmental organizations** (NGOs) are legally formed organizations that have no government status. These can operate at local, national, or international levels. Many NGOs work to promote some environmental or marine conservation cause. They may work in conjunction with government agencies but typically pride themselves in avoiding political influences.

Government Organizations

Most coastal nations have some level of organization that serves to establish and enforce rules and regulations and facilitate research and education regarding conservation, and management of marine resources and the environment. In fact, several organizations may be charged with establishing laws, enforcing management, or educating their citizens concerning fisheries or other ocean resources. The operation and power of agencies that serve these functions vary widely with the political and legal systems of nations. The United States' structure is discussed as an example, and the case studies below describe actions by agencies in other countries.

In the United States, various agencies have marine-related charges at the federal level and each coastal state has their own government organizations that function primarily within the three-mile limit designated by the Fishery Conservation and Management Act. Some of these agencies have been mentioned in previous chapters in reference to estuarine, coastal, and nearshore issues (see Chapters 3–6). At the federal level, the primary agency involved in marine issues is the National Oceanic and Atmospheric Administration (NOAA), established in 1970 under the Department of Commerce to "conduct research and gather data about the global oceans, atmosphere, space, and sun, and apply this knowledge to science and service that touch the lives of all Americans." NOAA is divided into five offices, two of which are directly involved in ocean resource research, management, and education.

The National Ocean Service of NOAA is charged with preserving and enhancing the coastal resources and ecosystems of the United States. Other priorities include: collecting data to measure the impact of climate change on the oceans; collecting and distributing water-related information, such as tide, water level, and environmental data; and collecting physical data, such as currents and meteorological data for enhancing marine commerce and transportation. The National Ocean Service has eight program offices that deal with marine issues, including marine sanctuaries and coral reef conservation.

The National Marine Fisheries Service (NMFS; commonly pronounced *nimfs*) or NOAA Fisheries provides "stewardship of living marine resources through science-based conservation and management, and the promotion

Figure 12-14 In the United States, the National Marine Fisheries Service (NMFS) under the National Oceanic and Atmospheric Organization (NOAA) not only establishes and enforces fisheries management plans, it also has an active fleet of research vessels to carry out scientific research and surveys.

of healthy ecosystems." NMFS does scientific research and works to establish and enforce management plans for fisheries harvest in U.S. waters (**Figure 12-14**). There has been a recent increase in focus on conservation of marine habitat and protection of non-harvested ocean species. NMFS is also charged with making decisions concerning listing of marine endangered species. Offices and scientific labs are distributed in coastal regions around the United States. Because NMFS represents the government in dealing with controversial marine fisheries issues, it frequently has to go head-to-head with fishers who disagree with management policies, and it has been accused of being too influenced by political pressures on some issues. NMFS works with Regional Fisheries Management Councils to establish fishing regulations (see above).

■ International Organizations

International governmental organizations (IGOs) operate through representatives of member nations. Membership of organizations at a regional level may be limited to a few nations within that region and function to resolve conflicts over shared resources that straddle borders or that are found in international waters. Organizations at the global level work to facilitate cooperation among nations regarding shared resources, and they may provide technological, scientific, and educational support to assist with utilization or conservation of environments and resources. IGOs facilitate sharing and conservation of fisheries and other resource use in the international commons of the open ocean. Because these are common property resources, membership is typically not limited to nations bordering the region (East Asian fishing nations have membership in

Atlantic fisheries organizations, for example). Membership can include nations with no interest in use of the resources under the charge of the organization (for example, the majority of the membership of the International Whaling Commission is non-whaling nations).

United Nations Organizations

The United Nations was established in 1945 "to maintain international peace and security; to develop friendly relations among nations; to cooperate in solving international economic, social, cultural and humanitarian problems and in promoting respect for human rights and fundamental freedoms; and to be a centre for harmonizing the actions of nations in attaining these ends." The U.N. does not set international law or serve as an international police, but it does have a system of organizations that coordinates efforts among countries to establish agreements on legal and conservation matters. Several organizations of the United Nations deal with ocean resource issues in whole or in part. These include: the Food and Agriculture Organization (FAO), the International Maritime Organization (IMO), the International Seabed Authority (ISA), the United Nations Environment Programme (UNEP), the World Trade Organization (WTO), and the United Nations Education, Scientific, and Cultural Organization (UNESCO).

UNEP sponsors several ocean programs. For example, the Regional Seas Programmes includes plans for 13 geographic regions designated to combat environmental problems through management of marine and coastal areas. The Caribbean Environment Programme works to assist and educate countries and to facilitate international cooperation in the Caribbean.

The Global Programme of Action for the Protection of the Marine Environment (GPA) functions to protect and preserve the marine environment from impacts of land-based activities. The UNEP Coral Reef Unit works with nations to increase support for reef conservation and sustainable use. Under UNESCO, the Intergovernmental Oceanographic Commission (IOC) coordinates the sharing of ocean-related knowledge, information, and technology among governments, and assists with cooperation among nations in the study of the oceans.

The Food and Agriculture Organization of the United Nations (FAO) has programs in Fisheries Resource, Policy, Industry, and Information. The FAO works to promote responsible fisheries management by nations, promotes increased contribution of fisheries to the world food supply, and assists with collecting and analyzing world fisheries data (see Chapter 11). The International Coral Reef Action Network (ICRAN) is the initiative of several organizations with the assistance of the United Nations Foundation, and functions as a global partnership of coral reef experts coordinating to monitor and conserve the health of the coral reefs around the world.

International Fisheries Commissions

Because any nation has access to resources in the global ocean commons, beyond national Exclusive Economic Zones, harvest can only be restricted through international agreements. International Commissions typically deal with a group of related species or a designated geographic area. There are many of these agencies, designated by an "alphabet soup" of acronyms. Only a selected few of the largest or most influential in conservation issues will be mentioned here.

The International Commission for Conservation of Atlantic Tunas (ICCAT) has been mentioned in previous chapters (see Chapter 11) as providing a mechanisms for countries to work together to develop management agreements to maintain populations of approximately 30 tuna and related species. ICCAT, established in 1969, is open to any nation that is a U.N. member; however, its 30 members are primarily nations bordering the Atlantic or Mediterranean or with a fishing fleet that fishes in the Atlantic (e.g., Japan and China). (Although open ocean fisheries are considered common resources, fishing nations are not obliged to share catch or profits with non-fishing nations.) Because of the multiple issues ICCAT must address, it has a complex structure that consists of numerous committees and working groups. As with any international organization, enforcement of regulations is often problematic. Threats or actions against an offending nation are often the most practical methods of enforcing agreements. For example, in the late 1970s the United States banned bluefin tuna imports from Panama, Honduras, and Belize after accusing these countries of not following ICCAT agreements on tuna harvest. ICCAT has recently been accused by conservation groups of being too lax in its agreements, allowing the harvest of too many fish, in particular the valuable species like the bluefin tuna (see Chapter 11).

The South Pacific Forum Fisheries Agency (FFA) was established in 1979 to deal with the management of highly migratory species (tunas, billfishes, etc.) in the Exclusive Economic Zone of countries in the South Pacific region (see Figure 12-5). Membership in the FFA includes Australia and New Zealand, but also many small island nations with limited capabilities to manage their marine fisheries resources or deal with other countries wishing to fish in their waters. A convention was signed by 12 member nations to establish the FFA so that a common front could be presented when dealing with marine fisheries issues. The FFA does not pass management regulations, but provides a means for countries to work together to manage fisheries in the South Pacific region. Some of the strategies

to accomplish this mission are collecting and distributing information, providing legal, economic, and technical advice, and assisting with developing fisheries management plans.

Membership in the North Pacific Anadromous Fish Commission (NPAFC), established in 1993, is comprised of the United States, Canada, Japan, and Russia. Each of these countries has anadromous fish stocks (primarily salmon) that migrate into the North Pacific Ocean. The primary goal of this Commission is to promote conservation of salmon and other species (including incidentally caught marine mammals and sea birds) that are harvested or affected by fishing in international waters of the North Pacific. Committees are set for administration, scientific research, and enforcement. Agreements give each country the authority to board, inspect, and detain vessels of any other member country that is observed violating agreements of the NPAFC. The NPAFC has recently agreed to prohibit fishing for salmonids in international waters because of their conservation status; however, salmon are harvested within the waters of each of these countries as governed by national laws and regulations.

International Conservation Commissions

Not all international marine-focused commissions are charged with regulating fisheries harvest or resource exploitation. Many recently formed commissions designate their mission with a focus on ecosystem or species protection. Organizations focused on endangered species and marine mammals were presented in previous chapters. These include the Convention for International Trade in Endangered Species (CITES; see Chapter 9), whose goal is to ensure that organisms are not endangered with extinction as a result of international trade. The International Whaling Commission (IWC), established in 1949, was one of the first international organizations to take an active role in management of resources in the ocean commons. Although it was originally established to develop and control the whaling industry, currently it is more focused on conservation of the whales through the commercial harvest moratorium (see Chapter 10). As controversies between whaling and non-whaling nations accelerated, IWC membership expanded to over 40 countries, many with no active interest in whale conservation or whaling, but recruited to support the agendas of larger nations (particularly the U.S. and Japan).

The Convention for the Conservation of Antarctic Marine Living Resources (CCAMLR) was enacted in 1982 to regulate harvest and research activities conducted in the Antarctic region, including conservation of "all species of living organisms" (**Figure 12-15**). The Commission includes over 20 countries interested in harvest or research in the

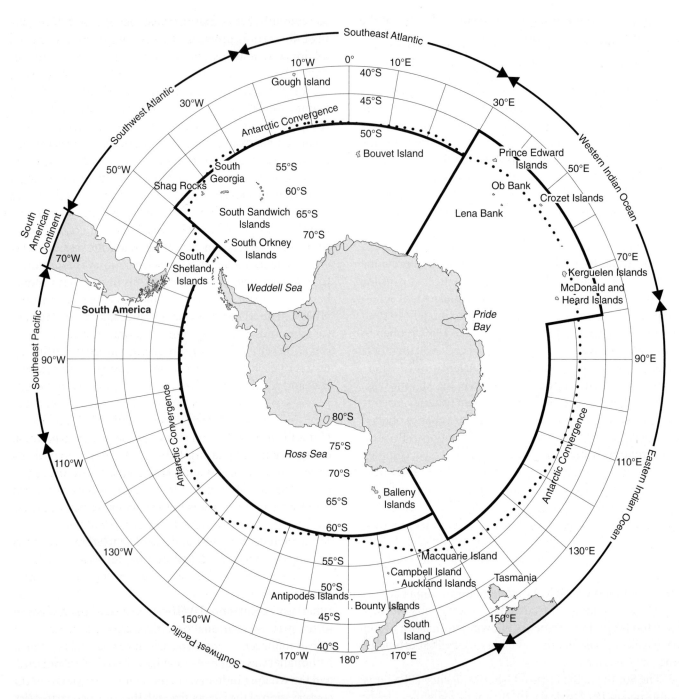

Figure 12-15 The Antarctic region, indicating areas regulated by the Convention for the Conservation of Antarctic Marine Living Resources (CCAMLR). Through this agreement, the primary focus of resource management in the Antarctic—arguably more than any other region globally—is on maintaining a viable healthy ecosystem.

Antarctic region. The CCAMLR is noteworthy among international marine resource organizations because it was established before the Antarctic ecosystem had been substantially impacted by humans, as well as for its emphasis on an *ecosystem approach*. Much of the effort by the CCAMLR is to maintain a viable ecosystem, including large krill populations that support fishes and marine mammals, including the great whales. A scientific committee collects data and makes recommendations to the Commission, which establishes harvest limits, seasons, and allowable fishing gears as well as designates protected species and areas.

Non-Governmental Organizations

Non-governmental organizations include a broad spectrum of organization types that operate at the local, national, or international level. Concerning ocean resource issues, most of the marine-focused NGOs are groups with a conservation or animal-protection interest. So that they can maintain their independence, NGOs typically do not accept funding from governments, corporations, or political entities, but depend primarily on private donations and membership fees to fund their causes. Although many of large environmental NGOs originated in North American or European countries, they have expanded internationally. Approaches to achieve goals of conservation or environmental protection range from legal to educational to scientific to confrontational.

Activist organizations use public actions to bring awareness of their concerns in order to initiate change. Greenpeace is one of the largest and most visible of the conservation-oriented NGOs worldwide (**Figure 12-16**). Their proclaimed mission is to use "non-violent, creative confrontation to expose global environmental problems, and force solutions for a green and peaceful future." Although Greenpeace states a commitment to non-violent methods, they have been known to use aggressive, high profile, and even illegal activities to proclaim their causes in order to gain attention from the news media. Some of Greenpeace's most prominent causes have been marine issues. For example, they have used their vessels to harass whalers, large-scale commercial fishers, and ships transporting controversial materials, including toxic or nuclear wastes. Greenpeace has contributed to successes in bringing action to ban drift net fishing on the high seas, place a moratorium on commercial whaling, ban the dumping of toxic wastes at sea,

Figure 12-16 Activist organizations such as Greenpeace play an influential, but often controversial, role in conservation of marine habitats and resources.

and establish better international management of marine fisheries. The Greenpeace Oceans Campaign recently has focused on overfishing, pirate fishing, commercial whaling, and coastal shrimp aquaculture. Most of their recent ocean campaigns serve primarily to present information and bring awareness to the public of actions that they believe negatively impact life in the oceans and other habitats.

The Sea Shepherd Conservation Society was formed by former Greenpeace members who wanted to take more direct action to stop what they considered were illegal activities. They proclaim to be "dedicated to protecting the marine environment with a particular focus on halting illegal fishing activities." The Sea Shepherds take an active role in forcing nations to comply with international law but have made some controversial decisions on their interpretation of these laws and the means they use to enforce them. Some of their more controversial actions include harassing and ramming whaling ships and fishing vessels, claiming responsibility for sinking some vessels. Recently, their efforts have been less violent, but they continue a confrontational approach to stop whaling, sealing, illegal fishing activities, shark finning, and habitat destruction.

Fishery NGOs work with local fishers to establish sustainable fisheries, while also working with the fishers on social and legal issues. For example, the International Collective in Support of Fishworkers (ICFS) focuses its efforts on assisting the small-scale, artisanal fishing sector around the world, primarily in less-developed regions with few resources for fisheries management. Recent activities have been carried out in Southeast Asia, Central and South America, India, and Africa. Recent agendas have included Marine Protected Areas, aquaculture, rights and duties of small-scale fishers, and economic, social, and cultural rights of fishworkers.

Many **environmental NGOs** are much less confrontational than Greenpeace or Sea Shepherds and take more of an educational or political role. The World Wide Fund for Nature (WWF), formerly called the World Wildlife Fund, and often referenced simply by its acronym, is a global NGO with a network of offices scattered around the world. Its goal is to "lead international efforts to protect endangered species and their habitats," and they do so through research and education, bringing a public awareness to issues related to conservation around the world. The focus of much of their efforts is on endangered species or ecosystems, including marine species such as the great whales and sea turtles.

Many marine-focused NGOs were established in conjunction with the development of the Internet and increased ease of communication. Many of these are **advocacy groups**, educating the public primarily from a conservation perspective. The proclaimed mission of the

Ocean Conservancy is to "protect ocean ecosystems and global abundance and diversity of marine wildlife." The Ocean Conservancy uses science-based advocacy, research, and education of the public to encourage people to speak out and protect the oceans. The focus of their education campaigns includes coral reef loss, fisheries overharvest, pollution of estuaries, and the endangerment of whales and sea turtles. Oceana, founded in 2001, is also dedicated to protecting the world's oceans, but it takes a more legal-focused approach. Oceana calls itself an "international advocacy organization dedicated to restoring and protecting the world's oceans through policy advocacy, science, law and public education." Oceana is based in Washington, D.C., and has offices around the United States, South America, and Europe, with members from 150 countries. Recent campaigns by Oceana include proposals to the European Union to stop waste oil release by ships, eliminating contamination in aquaculture salmon, limiting the impact by trawlers on European marine fisheries, and protecting sea turtles and highly migratory fish species. The Marine Stewardship Council is a "global organization working with fisheries, seafood companies, scientists, conservation groups and the public to promote the best environmental choice in seafood." They function independently of government agencies to certify fisheries as environmentally friendly and sustainable, and **ecolabel** seafood that is considered healthy to consume and from certified fisheries. Their label is one of the most widely recognized by consumers for seafood safety, and lists of certified seafood and fisheries provide guides for concerned seafood consumers (**Box 12-3. Conservation Concern: What Seafood Should I Eat?**).

12.4 The Future of the Ocean Commons

Can any portion of the ocean remain as a true commons, *free to all,* and still sustain a healthy environment and ecosystem? The answer may be a qualified *yes,* but only for now and only for some places. If history is any indicator, inevitably humans will have the capability to access, overuse, and dramatically alter all marine-associated communities, while removing or destroying most of the resources contained therein. The need for *rules for sharing* now has become obvious. Locally, nationally, and globally, current rules will have to be better enforced and new rules will need to be put in place.

Where coastal communities or states are given jurisdiction over marine resources, habitats, and ecosystems in estuaries, bays, intertidal, or coastal regions, systems of sharing must be put into place. Although small communities may self-regulate for the common good successfully, limits must be established. For valued resources, or where

access to other sources of food or income is limited, open access rarely will be sustainable. Limiting entry to fishing or other resource use can require some difficult decisions. But the alternative—complete open access—ultimately will lead to the loss of the resources to everyone.

Uncontrolled use of valuable resources in the commons will lead to overuse as soon as some combination of value, accessibility, and number of users reaches some difficult-to-define point. With management by a top-down system, handed down by a government or multinational commission, unambiguous rules must be established to limit access and exploitation. It may not matter from the conservation perspective if these limits are achieved by cutting "the pie" into increasingly smaller portions or continually reducing the number of users. What is most important is that the rules be enforceable and flexible. We cannot expect voluntary actions to be taken for the common good in the global community when competition for profits, livelihoods, and food resources are involved.

On the high seas, where no nation can establish rules that apply to all people, the world community must accelerate efforts to move away from the extreme application of the Freedom of the Seas philosophy. Numerous examples enforce the lesson that once a resource gains enough value, the development of innovations to exploit that resource is inevitable. Unlimited open access to the resources on and beneath the high seas is not compatible with conservation and sustainability. Multinational or global agreements will require continued compromises. Although education will play a prominent role, social and political pressures will continue to be necessary if a precautionary approach to ocean conservation is to be successful.

Uncontrolled *non-consumptive* use (e.g., for recreation) or pollution of the ocean environment must be considered as a form of *open access* requiring limitations. Rules to limit pollution or damage must be extended beyond the oceans into rivers and the atmosphere that transport nutrients and toxins into the coastal and open ocean environment. Although we cannot remove the human influence from the entire ocean, we must do our best to minimize that influence in large regions of the oceans through protected areas and sanctuaries. We must accept that the ocean environment is not static. It is continually changing due to both natural causes and human-induced changes. Of course, we must attempt to predict these changes but be ready to adjust and adapt quickly. Science must be valued as a means of achieving these and all conservation goals. Although scientific predictions are far from perfect, they provide the best methods that humans have been able to develop to guide our efforts to use and protect the marine and global environment and conserve its resources for future generations.

Box 12-3 Conservation Concern: What Seafood Should I Eat?

With increasing evidence of the health benefits of seafood and greater environmental awareness, even in nations with historically low seafood consumption, consumers are wondering how to select from an increasingly long list of seafood products. To assist with choices, the Marine Stewardship Council and other organizations provide certifications, buying guides, and lists of recommended seafoods through ecolabeling. Seafood ecolabeling provides a mark that give purchasers an indication that a product is better from an ecological and health perspective. For example, the Dolphin Safe logo (see Chapter 10) is an ecolabel (**Figure B12-2a**). Ecolabeling allows the consumer to voluntarily take action by eating only certified products; however, it is becoming more commonplace for resellers and governments to consider supplying only certified products. Although some ecolabels are established by governments (for example, Germany has had a government ecolabeling system since 1977), and others by the fishery or seafood industry, many consider those established by NGOs to be less prone to bias. The reliability of any ecolabeling program depends on the quality of the data on which it is based as well as opinions about what are considered to be *environmentally friendly* practices.

Of tunas, those commonly recommended for consumption include albacore (sold canned in the United States as *white tuna*) and skipjack tuna (sold canned as *light tuna;* Figure B12-2b) caught by trolling or pole fishing, due to low bycatch and less chance of overfishing or bioaccumulation of toxins due to its relatively short lifespan (see Chapters 7 and 11). Alaska wild salmon is recommended over farm-raised salmon due to concerns over environmental impacts and possible use of chemicals associated with farm-raised salmon (see Chapter 11). Tilapia (although a freshwater species, tilapia is typically sold as *seafood*) is an omnivorous fish that does not require high-protein feeds and pond farming does not affect marine habitats. Atlantic herrings are recommended for direct human consumption because they feed low on the food chain, their populations recover rapidly from harvest, and they are short lived and thus do not readily accumulate toxins (see Chapters 7

and 11). Wild Alaska pollock is recommended for consumption as a frozen or breaded seafood product due to the relative health of populations compared to other groundfish species (see Chapter 11).

Fishes that these lists often suggest to avoid include Patagonian toothfish (typically sold as Chilean seabass in the United States) because of problems with illegal fishing and marketing as well as excessive harvest (see Chapters 2 and 8). Orange roughy is not recommended due to environmental impacts of fishing on deep-water reefs and excessive harvest of a long-lived species (see Chapter 8). Farm-raised shrimp is included as a product to avoid because of the environmental impacts of shrimp aquaculture, particularly to mangrove habitats (see Chapter 4). Sharks and shark fins are not recommended due to overharvest and wasteful finning practices (see Chapter 11). Longline-harvested tunas, especially bluefin tuna, are considered a poor seafood choice because of excessive harvest, bioaccumulation of toxins such as mercury, and problems with bycatch (see Chapters 7 and 11). Grouper are to be avoided because of excessive harvest and effects on tropical reef ecosystems (see Chapter 5). Although it is often difficult to determine the origin of fish sold in markets or restaurants, some nations have improved requirements for seafood labeling. For example, the U.S. Department of Agriculture (USDA) established mandatory Country of Origin Labeling for seafood; however, processed seafood and wholesale markets are exempt from labeling requirements. The benefits of ecolabeling extend beyond those realized by the consumer. If there is adequate awareness and a push by the consumers, the fishery industry will be forced to use environmentally friendly practices to maintain the marketability of their products. A strong educational and ecolabeling program can force changes more rapidly than government regulations can be modified through legal process, such as happened as a result of dolphin-safe tuna protests. Boycotts such as the 1998 "Give Swordfish a Break" campaign (see Chapter 11), have also been successful in forcing management changes.

(a)

(b)

Figure B12-2 Ecolabeling programs, **(a)** the dolphin-safe tuna logo being one of the first and most successful, give consumers the ability to choose seafood products produced or caught in an environmentally sustainable manner. Trade name restrictions give additional information to the educated consumer. **(b)** As an example, *white tuna* is a trade name for albacore tuna, a fish considered an *environmentally friendly* seafood by ecolabeling organizations.

STUDY GUIDE

Conservation Exercises

1. Describe how the concept of the *tragedy of the commons* affects the following:
   a. harvest of fisheries resources on the high seas
   b. harvest of fisheries resources in national waters
   b. scuba diving on a popular reef dive site
   c. mining of the deep sea
   d. pen farming fish in a coastal bay

2. Describe how the following can be used to address the *tragedy of the commons*:
   a. forced inefficiency
   b. TAC quotas
   c. cooperative institutions

3. How do the following international commissions assist in the conservation of marine species in the ocean commons?
   a. CITES
   b. IWC
   c. ICCAT
   d. NPAFC
   e. CCAMLR

4. Compare the following mechanisms for conservation and limiting excessive fisheries harvest in coastal waters.
   a. community-based management
   b. marine protected areas
   c. limited entry
   d. territorial use rights

5. Describe how the Law of the Sea and UNCED address rights for the following:
   a. innocent passage of ships into a nations waters
   b. national fishing rights in coastal waters
   c. national interests in mining the continental shelf
   d. deep-sea mining
   e. limiting ocean pollution
   f. strandline fish stocks

6. Compare the following legislation with regards to the delegation of fisheries rights in coastal waters.
   a. the U.S. FCMA
   b. Canada Oceans Act

   c. E.U. Common Fisheries Policy
   d. China Fisheries Policy
   e. Japan Fisheries Law
   f. Asia-Pacific Fisheries Commission
   g. India government legislation
   h. Australia Harvest Strategy Policy
   i. West African national management laws

7. Describe how the following have promoted marine conservation efforts.
   a. U.S. Sustainable Fisheries Act
   b. Shireteko, Japan Marine Management Plan
   c. establishment of marine sanctuaries in Southeast Asia
   d. Australia Harvest Strategy Policy

8. Compare strategies used by the following NGOs to promote marine conservation issues.
   a. Greenpeace
   b. Sea Shepherds
   c. WWF
   d. Ocean Conservancy
   e. Oceana
   f. Marine Stewardship Council

9. Of the following pairs, which would you consider the most *environmentally friendly* to consume and why?
   a. bluefin tuna sashimi or canned light tuna
   b. tilapia or grouper filets
   c. broiled Chilean seabass or canned white albacore tuna
   d. fish sticks made from Alaska pollock or Atlantic cod
   e. filets from wild Alaskan salmon or Atlantic salmon from Chile
   f. frozen orange roughy or mahi mahi
   g. shark steaks or dried herring

10. How have the trade disputes affected the following?
    a. purse seine tuna harvest in the Eastern Tropical Pacific
    b. shrimp trawling in Thailand and India
    c. shrimp trawling in Mexico

FURTHER READING

Adams, C., W. J. Keithley, and S. Versaggi. 2005. The shrimp import controversy. Chapter 15. In Schmitz, A., C. B. Moss, T. G. Schmitz, and W. W. Koo (eds.). *International agricultural trade disputes: Case studies in North America*. Michigan State University Press, East Lansing, Michigan.

Borgese, E. M. 1983. The law of the sea. *Scientific American* 162:1243–1248.

Burk, W. T. 1995. Implications for fisheries management of U.S. acceptance of the 1992 Convention on the Law of the Sea. *American Journal of International Law* 1995:792–806.

Cleveland, H. 1993. The global commons. *The Futurist* 27:9–13.

Connor, R. and B. Shallard. 2010. Evolving governance in New Zealand fisheries. Pages 347–359. In Grafton, R. Q., R. Hilborn, D. Squires, M. Tait, and M. Williams (eds.). *Handbook of marine fisheries conservation and management.* Oxford University Press, New York.

Daw, T., and T. Gray. 2005. Fisheries science and sustainability in international policy: a study of failure in the European Union's Common Fisheries Policy. *Marine Policy* 29:189–197.

Frost, H. 2010. European Union fisheries management. Pages 471–484. In Grafton, R. Q., R. Hilborn, D. Squires, M. Tait, and M. Williams. *Handbook of marine fisheries conservation and management.* Oxford University Press, New York.

Galdorisi, G. 1995. The United States and the Law of the Sea: A window of opportunity for maritime leadership. *Ocean Development and International Law* 26:75–83.

Grafton, R. Q., R. Hilborn, D. Squires, M. Tait, and M. Williams. 2010. *Handbook of marine fisheries conservation and management.* Oxford University Press, New York.

Hardin, G. 1968. The tragedy of the commons. *Science* 162:1243–1248.

Holland, D. S. 2010. Governance of fisheries in the United States. Pages 382–392. In Grafton, R. Q., R. Hilborn, D. Squires, M. Tait, and M. Williams (eds.). *Handbook of marine fisheries conservation and management.* Oxford University Press, New York.

Hotta, M. 2004. *Review of fisheries management in the Pacific—Japan. Review of the state of world marine capture fisheries management: Pacific Ocean.* FAO Fisheries Technical Paper. No. 488/1.

Johnson, R. 2011. Japan's 2011 earthquake and tsunami: food and agriculture implications. *U.S. Congressional Research Report* 7-570 www.crs.gov.

Jones, K., D. J. Harvery, W. Hahn, and A. Muhammad. 2008. U.S. demand for source-differentiated shrimp: A differential approach. *Journal of Agricultural and Applied Economics* 40:609–621.

Liu, X. 2003. *Review of fisheries management in the Pacific—China. Review of the state of world marine capture fisheries management: Pacific Ocean.* FAO Fisheries Technical Paper. No. 488/1.

MacDonald, J. M. 1995. Appreciating the precautionary principle as an ethical evolution in ocean management. *Ocean Development and International Law* 36:255–286.

Makino, M. 2010. Japanese coastal fisheries. Pages 287–298. In Grafton, R. Q., R. Hilborn, D. Squires, M. Tait, and M. Williams (eds.). *Handbook of marine fisheries conservation and management.* Oxford University Press, New York.

Marffy-Mantuano, A. 1995. The procedural framework of the agreement implementing the 1982 United Nations convention on the Law of the Sea. *American Journal of International Law* 89:814–824.

Martin, G. S., A. M. Parma, and J. M. Orensanz. 2010. The Chilean experience with territorial use rights in fisheries. Pages 324–337. In Grafton, R. Q., R. Hilborn, D. Squires, M. Tait, and M. Williams (eds.). *Handbook of marine fisheries conservation and management.* Oxford University Press, New York.

McGinn, A. P. 1998. Rocking the boat: conserving fisheries and protecting jobs. *Worldwatch Paper 142,* 92 pages. Worldwatch Institute, Washington, DC.

McGinn, A. P. 1999. Safeguarding the health of oceans. *Worldwatch Paper 145,* 87 pages. Worldwatch Institute. Washington, DC.

McLoughlin, R. and N. Rayns. 2010. Pages 338–346. In Grafton, R. Q., R. Hilborn, D. Squires, M. Tait, and M. Williams (eds.). *Handbook of marine fisheries conservation and management.* Oxford University Press, New York.

Mensah, T. A. 1998. The International Tribunal and the protection and preservation of the marine environment. *Environmental Policy and Law* 28:216.

Miller, C. J., and J. L. Croston. 1999. WTO scrutiny v. Environmental objectives—assessment of the International Dolphin Conservation Program Act. *American Business Law Journal* 37:73–125.

Munro, G. R. 2010. The 1982 U.N. Convention on the Law of the Sea and beyond: the next 25 years. Pages 646–658. In Grafton, R. Q., R. Hilborn, D. Squires, M. Tait, and M. Williams (eds.). *Handbook of marine fisheries conservation and management.* Oxford University Press, New York.

Nandakumar, D., and N. Nayak. 2010. Coastal fisheries in India: current scenario, contradictions and community responses. In Grafton, R. Q., R. Hilborn, D. Squires, M. Tait, and M. Williams (eds.). *Handbook of marine fisheries conservation and management.* Oxford University Press, New York.

Parsons, L. S. 2005. Ecosystem considerations in fisheries management: theory and practice. *International Journal of Marine and Coastal Law* 20:318–422.

Parsons, L. S. 2010. Canadian marine fisheries management: a case study. Pages 393–414. In Grafton, R. Q., R. Hilborn, D. Squires, M. Tait, and M. Williams (eds.). *Handbook of marine fisheries conservation and management*. Oxford University Press, New York.

Safina, C. 1995. The world's imperiled fish. *Scientific American* 273(5):46–53.

Satia, B. P., and A. M. Jallow. 2010. West African coastal capture fisheries. Pages 258–273. In Grafton, R. Q., R. Hilborn, D. Squires, M. Tait, and M. Williams (eds.). *Handbook of marine fisheries conservation and management*. Oxford University Press, New York.

United Nations. 1982. *United Nations Convention on the Law of the Sea*. U.N. Document A/Conf.62/122. United Nations, Geneva.

United Nations Food and Agriculture Organization. 2007. Fishery and Aquaculture Statistics. In *FAO Yearbook*. ISSN 2070-6057. Retrieved September 14, 2011, from ftp://ftp.fao.org/docrep/fao/012/i1013t/i1013t.pdf

Ward, T., and B. Phillips. 2010. Seafood ecolabelling. Pages 608–617. In Grafton, R. Q., R. Hilborn, D. Squires, M. Tait, and M. Williams (eds.). *Handbook of marine fisheries conservation and management*. Oxford University Press, New York.

Weber, P. 1993. Abandoned seas: Reversing the decline of the oceans. Worldwatch Paper 116, 66 pages. Worldwatch Institute, Washington, DC.

Weber, P. 1994. Fish, jobs, and the marine environment. Worldwatch Paper 120. Worldwatch Institute, Washington, DC.

Williams, M. J. and D. Staples. 2010. Southeast Asian fisheries. Pages 243–257. In Grafton, R. Q., R. Hilborn, D. Squires, M. Tait, and M. Williams (eds.). *Handbook of marine fisheries conservation and management*. Oxford University Press, New York.

Index

Credits

Unless otherwise indicated, all photographs and illustrations are under copyright of Jones & Bartlett Learning.

Chapter 1

1-2 Courtesy of NDGC/NOAA; **1-3** Courtesy of NOAA; **1-6** Courtesy of OAR/National Undersea Research Program (NURP)/NOAA; **1-7** Courtesy of NOAA; **1-8** Courtesy of Liam Gumley, Space Science and Engineering Center, University of Wisconsin-Madison; **1-10** Courtesy of Liam Gumley, Space Science and Engineering Center, University of Wisconsin-Madison; **1-13** Courtesy of Jacques Descloitres, MODIS Land Rapid Response Team at NASA GSFC; **1-12a** and **b** Courtesy of NASA's Earth Observatory; **1-16** Courtesy of NOAA; **1-18** Adapted from Couper, A., ed. *The Times Atlas of the Oceans*. Van Nostrand Reinhold, 1983; **1-20** Adapted from Sverdrup, H.U., et al. *The Oceans*. Prentice-Hall, 1942; **1-21** Reproduced from NOAA/PMEL/TOA; **1-22** Adapted from McCartney, M.S., *Oceanus* 37 (1994): 5–8; **B1-1** Reproduced from James Hansen/GISS/NASA; **B1-2** Reproduced from NOAA; **1-23** Adapted from Emery, K.O., *Sci Am.* 221 (1969): 106–122; **1-24** Courtesy of Estuary to Abyss 2004/NOAA Office of Ocean Exploration/NOAA; **1-25** Reproduced from NOAA; **1-27** Courtesy of NASA/Goddard Space Flight Center, The SeaWiFS Project and GeoEye, Scientific Visualization Studio; **B1-3** Reproduced from NASA Earth's Observatory.

Chapter 2

2-2a Courtesy of NOAA/DOC; **2-2b** Courtesy of NASA Kennedy Space Center; **B2-1a** Photo by Jean DeMarignac. Courtesy of the Monterey Bay National Marine Sanctuary/NOAA; **B2-1b** Photo by Jean DeMarignac, Monterey Bay National Marine Sanctuary. Courtesy of NOAA; **2-4a** and **b** Courtesy of Randolph Femmer/life.nbii.gov; **2-8c** Courtesy of NOAA; **2-9** Courtesy of Dr. Dwayne Meadows, NOAA/NMFS/OPR; **2-11a** Courtesy of NOAA; **2-13a** Courtesy of Matt Wilson/Jay Clark, NOAA NMFS AFSC/NOAA/DOC; **2-13b** © Paul Yates/ShutterStock, Inc.; **2-13c** Courtesy of Jamie Hall/NOAA/DOC; **2-16a** Courtesy of Operation Deep Scope 2005 Expedition/NOAA/DOC; **2-17a** Courtesy of NOAA; **2-21a** © susan flashman/ShutterStock, Inc.; **2-21b** © Aqua Image/age fotostock; **2-22a** © Andre Seale/age fotostock; **2-22b** Courtesy of Southwest Fisheries Science Center, NOAA Fisheries Service; **2-22c** © Eric Prine/age fotostock; **2-22d** © neelsky/ShutterStock, Inc.; **2-22e** © Emily Veinglory/ShutterStock, Inc.; **2-24** Adapted from Bond, C.E. *Biology of Fishes*. Saunders, 1979; **2-25** Modified from: http://www.epa.gov/glnpo/atlas/glat-ch4.html, accessed 1-26-2011; **2-26** Adapted from A C Hardy, *Fisheries Investigations* 7 (1924): 1–53.

Chapter 3

3-2 Modified from Martinez et al. 2004. *Coastal Dunes*; **3-4** © John Pagliuca/ShutterStock, Inc.; **3-8b** Courtesy of NOAA Restoration Center, Erik Zobrist/NOAA/DOC; **B3-1** © Dima_Rogozhin/ShutterStock, Inc.; **3-10** Photo by Steve Lonhart. Courtesy of Monterey Bay National Marine Sanctuary/NOAA; **3-13b** Photo by Dr. Dwayne Meadows, NMFS/OPR. Courtesy of NOAA/DOC; **3-14** © Loren Rodgers/ShutterStock, Inc.; **3-15** Courtesy of the U.S. Fish & Wildlife Service; **3-19** Courtesy of the U.S. Army Corps of Engineers; **3-20a** © iStockphoto/Thinkstock; **3-22** Courtesy of NOAA/DOC; **3-23a** and **b** Courtesy of USGS; **3-25a** © Design Pics/Valueline/Thinkstock; **3-25b** © Robert DeGoursey/Visuals Unlimited, Inc.; **3-30** Photo by Nancy Sefton. Courtesy of NOAA/DOC.; **B3-3a** © PA Photos/Landov; **B3-3b** Courtesy of the U.S. Coast Guard; **B3-4a** and **b** Courtesy of NOAA; **B3-4c** Courtesy of U.S. Navy; **3-34** © Martin Fowler/ShutterStock, Inc.; **3-35** © Luke Schmidt/ShutterStock, Inc.; **3-36** Courtesy of NASA/JPL.

Chapter 4

4-2a Courtesy of NASA/Goddard Space Flight Center Scientific Visualization Studio; **4-2b** and **c** Courtesy of NASA; **4-2d** Courtesy of NASA's Earth Observatory; **4-5a** © iStockphoto/Thinkstock; **4-5b** Courtesy of USGS; **4-6** Modified from U.S. EPA, *National Estuary Program Coastal Condition Report*; **B4-1** ©2009 Melisa Beveridge; **4-9** Data from NOAA; **B4-2** Modified from FAO and Maryland Department of Natural Resources, Virginia Marine Resource Commission; **4-11** Courtesy of Jacques Descloitres, MODIS Rapid Response Team, NASA/GSFC; **4-12a** and **b** Adapted from Edwards, J.M. and R.W. Frey, Senckenbergiana Maritima 9 (1977): 215–259; **4-13a** and **b** Courtesy of NOAA/DOC; **4-14** Courtesy of the U.S. Fish & Wildlife Service; **4-17** Courtesy of NOAA/DOC; **4-18**

Courtesy of NASA/JPL/NGA; **4-19** Courtesy of USGS; **4-22** Courtesy of NASA/GSFC/JPL, MISR Team; **4-31** Courtesy of USGS; **4-32** Courtesy of NASA; **4-34** Courtesy of NASA's Earth Observatory; **4-37** © 2011 CourtneyPlatt.com; **4-38** Courtesy of NASA's Earth Observatory.

Chapter 5

5-2c Courtesy of NOAA/DOC; **5-3** Adapted from Lerman, M. *Marine Biology: Environments, Diversity, and Ecology.* Benjamin-Cummings, 1986; **5-6b** Courtesy of NOAA/DOC; **5-7c** Courtesy of NOAA CCMA Biogeography Team/NOAA/DOC; **B5-2** Courtesy of USGS; **5-9** Courtesy of NOAA; **5-10** Photo by B. Williams, Ross et al. Courtesy of NOAA/HBOI; **5-11** © WaterFrame/Alamy; **5-15a** Courtesy of NOAA/DOC; **5-15b** Photo by William Folsom, NOAA, NMFS. Courtesy of NOAA/DOC; **5-19** Courtesy of Berkley White/NOAA; **5-22a** Courtesy of the Florida Keys National Marine Sanctuary/NOAA/DOC; **5-22b** Courtesy of NOAA; **5-24** Courtesy of NASA/GSFC/LaRC/JPL, MISR Team; **B5-4** Courtesy of NOAA/DOC; **B5-5** Modified from NOAA National Marine Sanctuaries and Friedlander and DeMartini 2002; **5-25a** Courtesy of NOAA/DOC; **5-25b** © Matthew Oldfield/Photo Researchers, Inc.

Chapter 6

6-3 Modifed from U.S. EPA; **6-4** Courtesy of the NERRS/NOAA/DOC; **6-5** Courtesy of NASA; **6-9** © Rich Carey/ShutterStock, Inc.; **6-10** © Douglas Faulkner/Photo Researchers, Inc.; **6-11** © Peter Scoones/Photo Researchers, Inc.; **6-14** Courtesy of Alexandre Meinesz, Université de Nice Sophia-Antipolis; **B6-1a** Courtesy of NASA/Goddard Space Flight Center Scientific Visualization Studio; **B6-1b** Courtesy of Jeff Schmaltz/NASA; **6-15** Courtesy Jeff Schmaltz, MODIS Land Rapid Response Team/GSFC/NASA; **6-16** Courtesy of NOAA; **6-18a** Courtesy of SWFSC/NOAA/DOC; **6-18b** Courtesy of Monterey Bay Aquarium Research Institute /NOAA/DOC; **B6-2a** Courtesy of NOAA/DOC; **6-19** Courtesy of NOAA/DOC; **6-21** © iStockphoto/Thinkstock; **6-23** Reproduced from *PNAS*, December 4, 2007, vol 104, no 49, 19163–19164; **6-24a** Photo by Shane Anderson. Courtesy of NOAA/DOC; **6-24b** and **c** Courtesy of NOAA/DOC; **6-25** © SuperStock/age fotostock; **B6-3** Modified from Steneck et al. 2002. Fig. 3; **B6-4** Modified from Steneck et al. 2002. Fig. 3; **6-26** Courtesy of the Monterey Bay National Marine Sanctuary/NOAA/DOC; **6-27** Courtesy of NOAA Restoration Center.

Chapter 7

7-2a Reproduced from NOAA; **7-3** Courtesy Jacques Descloitres, MODIS Land Rapid Response Team/GSFC/NASA; **7-4** Courtesy of NASA/Goddard Space Flight Center, The SeaWiFS Project and GeoEye, Scientific Visualization Studio; **7-6** Courtesy of Norman Kuring, Ocean Color Group/GSFC/NASA; **B7-1a** Courtesy of NOAA/DOC; **B7-1b** Photo by Ross, et al. Courtesy of NOAA/DOC; **7-8** Courtesy of NOAA National Marine Fisheries Service, Pacific Islands Fisheries Science Center; **7-9a** Photo by Emily Stone. Courtesy of the

National Science Foundation; **7-9b** Photo by Donald LeRoi, SWFSC/NOAA. Courtesy of the National Science Foundation; **7-9c** Courtesy of NOAA/DOC; **7-10** Modified from U.S. EPA 2007 *Air Trends Report: Lead*; **B7-2a** Data from Mercury Levels in Commercial Fish and Shellfish (1990–2010), USFDA and from Storelli MM and Marcotrigiano GO. Total mercury levels in muscle tissue of swordfish (*Xiphias gladius*) and bluefin tuna (*Thunnus thynnus*) from the Mediterranean Sea (Italy). *Journal of Food Protection* 2001;64(7):1058–1061; **B7-2b** Data from Endo T., et al. Mercury contamination in the red meat of whales and dolphins marketed for human consumption in Japan. *Environmental Science & Technology* 2003;37(12)2681–2685; **7-11** Adapted from Epel D. and Lee W.L. Persistent chemicals in the marine ecosystem. *The American Biology Teacher* 1970; 32(4)207–211; **7-12** Modified from Global Education Projects; **B7-3** Courtesy of NOAA/DOC; **7-13a** Courtesy of the SeaWiFS Project/GFSC/NASA/ORBIMAGE; **7-13b** Courtesy of NASA Earth Observatory; **7-14a** Reproduced from *PNAS* April 11, 2000, vol 97, no 8, 4303–4308; **7-14b** Courtesy of USGS; **B7-4** Courtesy of NOAA/DOC; **7-15** Photo by Claire Fackler, NOAA National Marine Sanctuaries. Courtesy of NOAA; **7-16** Courtesy of NOAA; **7-17a** and **b** Courtesy of Janet Nye, NEFSC/NOAA; **7-18** Courtesy of NOAA/DOC.

Chapter 8

8-1 Courtesy of NOAA/DOC; **8-2** Courtesy of NOAA/DOC; **8-3** Reproduced from Mullineau L.S. and Mills S.W. A test of the larval retention hypothesis in seamount-generated flows. *Deep Sea Research Part I: Oceanographic Research Papers* 1997;44(5):745–770 with permission from Elsevier; **8-4a** and **b** Courtesy of the Monterey Bay Aquarium Research Institute/NOAA/DOC; **8-5a** Courtesy of NURP/NOAA/DOC; **8-5b** Courtesy of Life on the Edge 2004 Expedition: NOAA Office of Ocean Exploration/NOAA/DOC; **8-6a** Photo by Personnel of NOAA Ship DELAWARE II. Courtesy of NOAA/DOC; **8-6b** Courtesy of NOAA/DOC; **8-7a** Courtesy of NOAA Okeanos Explorer Program, INDEX-SATAL 2010; **8-7b** Courtesy of New Zealand-American Submarine Ring of Fire 2005 Exploration/NOAA Vents Program/NOAA/DOC; **8-8** Adapted from Couper, A., ed. *The Times Atlas of the Oceans.* Van Nostrand Reinhold, 1983; **8-9a** Courtesy of NURC/UNCW/FGBNMS and NOAA/DOC; **8-9b** Photo by Dr. Les Watling, University of Maine. Courtesy of NOAA/DOC; **8-10a** Photo by Monika Bright, University of Vienna. Courtesy of NOAA/DOC. **8-10b** Photo by Dr. Bob Embley, NOAA PMEL. Courtesy of NOAA/DOC; **8-11** Courtesy of New Zealand American Submarine Ring of Fire 2007 Exploration/NOAA/DOC; **8-13** Courtesy of NOAA/DOC; **8-14** Courtesy of Gulf of Mexico 2002, NOAA/OER/DOC; **8-15** Photo by Craig Smith, University of Hawaii. Courtesy of NOAA/DOC; **8-16** Courtesy of Monterey Bay Aquarium Research Institute/NOAA/DOC; **8-17a** and **b** Courtesy of NOAA/DOC; **8-18** © Cyril Hou/ShutterStock, Inc.; **B8-1a** Photo by Ed Bowlby, NOAA/Olympic Coast NMS. Courtesy of NOAA/DOC; **B8-1b** Courtesy of NOAA/Harbor Branch; **B8-2a** Courtesy of NURC/UNCW and FGBNMS/NOAA/DOC; **B8-2b** Courtesy of the U.S. Fish & Wildlife Service; **B8-3** Courtesy of

NEW ZEEPS 2006 Expedition, NOAA/NIWA; **8-19** Courtesy of the Bureau of Ocean Energy Management, Regulation and Enforcement; **8-20** © Institute of Oceanographic Sciences/NERC/Photo Researchers, Inc.; **8-21** Adapted from Cronan, D.S. *Marine Manganese Deposits.* Elsevier, 1977; **8-22** Courtesy of Submarine Ring of Fire 2006 Exploration/NOAA/DOC; **8-23** Modified from USGS; **8-24** Adapted from Pequegnat, W.E., et al., *Procedural Guide for Designation Surveys of Ocean Dredged Material Disposal Sites*, EL 81-1. U.S. Government Printing Office, 1981; **8-25** Modified from Oilgae.

Chapter 9
9-1 Courtesy of the National Park Service; **9-2** Courtesy of the Bureau of Ocean Energy Management, Regulation and Enforcement; **9-3a** Courtesy of NOAA/DOC; **9-3b** Courtesy of NOAA/DOC; **9-4** Courtesy of the U.S. Fish & Wildlife Service; **9-5** Courtesy of NOAA/DOC; **9-6** Courtesy of NOAA/DOC; **B9-1** Courtesy of NOAA/DOC; **B9-2** Reproduced from Fitzgerald S., et al. Summary of Seabird Bycatch in Alaskan Groundfish Fisheries, 1993 through 2006. Alaska Fisheries Science Center 2008. Courtesy of NMFS/NOAA; **9-7** Courtesy of Dr. Robert Ricker, NOAA/NOS/ORR and the DOC; **9-8a** Courtesy of NOAA/DOC; **9-8b** © Brian Lasenby/ShutterStock, Inc.; **9-11** Courtesy of NOAA/DOC; **B9-3** Courtesy of the National Park Service.; **B9-4** Courtesy of NOAA/DOC; **B9-5a** Data from NOAA and Heppell S.S., et al. A population model to estimate recovery time, population size, and management impacts on Kemp's ridley sea turtles. *Chelonian Conservation and Biology* 2005;4(4):767–773.; **B9-5b** Data from: NPS/Padre Island National Seashore; **9-12** Photo by David Burdick. Courtesy of NOAA/DOC; **9-13** © Ariel Bravy/ShutterStock, Inc.; **9-14a** © Reinhard Dirscherl/Visuals Unlimited, Inc.; **9-15** Photo by Pat Mcmillan. Courtesy of the National Science Foundation; **B9-6** Reproduced from NOAA Fisheries; **B9-7** Courtesy of the Gulf of Maine Cod Project/National Archives/NOAA/DOC; **9-16** © iStockphoto/Thinkstock; **9-17** Data from NOAA/NMFS Pacific Islands Fisheries Science Center; **9-18** Courtesy of NOAA/DOC; **9-19** Photo by Rebecca Shoop. Courtesy of the National Science Foundation; **9-20** Modified from NOAA/NMFS Alaska Fisheries Science Center; **9-21** Modified from Australia Department of the Environment and Heritage; **9-22** Photo by Laura Hamilton. Courtesy of the National Science Foundation; **9-23a** © Fred Bruemmer/Peter Arnold, Inc.; **9-23b** © iStockphoto/Thinkstock; **B9-8** Data from Arctic Climate Impact Assessment, Arctic Council and the International Arctic Science Committee, 2004; **9-24** Photo by Captain Budd Christman. Courtesy of NOAA/DOC; **9-25** Courtesy of NOAA; **9-26a** Courtesy of NOAA/DOC; **9-26b** Courtesy of Monterey Bay National Marine Sanctuary/NOAA; **9-27** Photo by Bill Hickey. Courtesy of the U.S. Fish & Wildlife Service; **9-28a** Photo by Tania Larson. Courtesy of USGS; **9-28b** Courtesy of USGS.

Chapter 10
10-1 Photo by Donald LeRoi. Courtesy of the Southwest Fisheries Science Center and the National Science Foundation; **10-2a** Photo by Dan Shapiro. Courtesy of NOAA/DOC; **10-2b** Courtesy of NOAA/DOC; **10-2c** Courtesy of NOAA; **10-4** Courtesy of NOAA/DOC; **B10-1** © Vicki Beaver/Alamy; **B10-2** Courtesy of NOAA/DOC; **10-7** Modified from Cristensen (2006) Figure 99; **B10-3** Photo by John Durban. Courtesy of SWFSC/NOAA/DOC; **B10-4a** Modified from D'Amico et al. (2009), Figure 3; **B10-4b** Photo by Dr. Brandon Southall, NMFS/OPR. Courtesy of NOAA/DOC; **B10-5** Courtesy of USGS; **10-8** Courtesy of NOAA; **B10-6** Courtesy of the Florida Fish & Wildlife Conservation Commission/NOAA/DOC; **10-9** Photo by Jose Cort. Courtesy of NOAA/DOC; **10-10b** Courtesy of NOAA/DOC; **10-11** Photo by Commander James W. O'Clock. Courtesy of NOAA/DOC; **10-12** Courtesy of NOAA/DOC; **B10-7a** and **b** Courtesy of the U.S. Department of Defense.

Chapter 11
11-1 Courtesy of the Alaska Fisheries Science Center/Marine Observer Program/NOAA/DOC; **11-2e** © Flavio Massari/ShutterStock, Inc.; **11-4a** Courtesy of NOAA/DOC; **11-4b** Photo by Bob Williams. Courtesy of NOAA/DOC; **11-4c** Courtesy of NOAA/DOC; **11-6a** Courtesy of NOAA/DOC; **11-6b** Photo by David Csepp, NOAA/NMFS/AKFSC/Auke Bay Lab. Courtesy of NOAA/DOC; **11-7a** Photo by G.W. Coffin. Courtesy of NOAA/DOC; **11-7b** Photo by David Csepp, NOAA/NMFS/AKFSC/ABL. Courtesy of NOAA/DOC; **11-7c** Courtesy of NOAA/DOC; **11-8a** and **b** Photo by Karen Ducey, NMFS. Courtesy of NOAA/DOC; **11-9** Courtesy of Doug Helton/NOAA/NOS/ORR/ERD; **11-10a** Courtesy of NOAA/DOC; **11-10b** Photo by William B. Folsom, NMFS. Courtesy of NOAA/DOC; **11-10c** Courtesy of South Pacific Commission/NOAA/DOC; **11-10d** Photo by C. Ortiz Rojas. Courtesy of NOAA/DOC; **11-11a** Photo by Allen M. Shimada, NMFS. Courtesy of NOAA/DOC; **11-11b** Courtesy of NOAA/DOC; **11-11c** Courtesy of NMFS SEFSC Pascagoula Laboratory/NOAA/DOC; **11-12** Data from FAO Fisheries and Aquaculture Department; **B11-1** Modified from FAO; **B11-2** Adapted from Baumgartner TR, et al. Reconstruction of the history of Pacific sardine and northern anchovy populations over the past two millennia from sediments of the Santa Barbara basin, California. *CalCOFI* 1992; Vol. 33; **B11-3** Modified from NOAA, modified from Kawasaki 1992, data source: NMFS, FAO; **B11-4** Modified from FAO; **11-13** Data from FAO Fisheries and Aquaculture Department; **11-14a** and **b** Courtesy of NOAA/DOC; **11-15** Photo by Jose Cort. Courtesy of NOAA/DOC; **11-16** Date from FAO; **11-17a** Courtesy of NOAA/DOC; **11-17b** and **c** Modified from FAO; **11-18a** Photo by Jose Cort. Courtesy of NOAA/DOC; **11-18b** Modified from FAO; **B11-5** Courtesy of NMFS/NOAA/DOC; **B11-6** Modified from FAO; **B11-7** Courtesy of NOAA/DOC; **B11-8a** and **b** Modified from NOAA/NMFS; **11-19** Modified from FAO; **B11-9a** Courtesy of NOAA; **B11-10a** Modified from Destefano 2007—FAO data; **B11-10b** Modified from Matsuda, Y., Integrated coastal zone management with sustainable aquaculture, Food & Fertilizer Technology Center for the Asian and Pacific Region, FAO data; **11-20** Courtesy of NMFS/NOAA/DOC; **11-21** Modified from NOAA/NMFS; **11-22a** and **b** Modified from FAO Fisheries and Aquaculture Department; **11-23** Courtesy of NOAA/DOC.